An Introduction to LINEAR ANALYSIS

D1716310

DONALD L. KREIDER, Dartmouth College
ROBERT G. KULLER, Wayne State University

ADDISON-WESLEY (CANADA) LIMITED, DON MILLS, ONTARIO

DONALD R. OSTBERG, Indiana University
FRED W. PERKINS, Dartmouth College

An Introduction to LINEAR ANALYSIS

ADDISON-WESLEY PUBLISHING COMPANY, INC., READING, MASSACHUSETTS, U. S. A.

QA
402
K7

This book is in the
ADDISON-WESLEY SERIES IN MATHEMATICS

LYNN H. LOOMIS, *Consulting Editor*

Copyright © 1966
Philippines Copyright 1966

ADDISON-WESLEY PUBLISHING COMPANY, INC.

Printed in the United States of America

All rights reserved. This book, or parts thereof,
may not be reproduced in any form without
written permission of the publisher.

Library of Congress Catalog Card No. 65–23656

ADDISON-WESLEY PUBLISHING COMPANY, INC.
READING, MASSACHUSETTS · Palo Alto · London
NEW YORK · DALLAS · ATLANTA · BARRINGTON, ILLINOIS

ADDISON-WESLEY (CANADA) LIMITED
DON MILLS, ONTARIO

1-21-67

preface

For the student. Tradition dictates that textbooks should open with a few remarks by the author in which he explains what his particular book is all about. This obligation confronts the technical writer with something of a dilemma, since it is safe to assume that the student is unfamiliar with the subject at hand; otherwise he would hardly be reading it in the first place. Thus any serious attempt to describe the content of a mathematics text is sure to be lost on the beginner until after he has read the book, by which time, hopefully, he has discovered it for himself.

Still, there are a few remarks which can be addressed to the student before he begins the task of learning a mathematical dicipline. Above all, he should be told what is expected of him in the way of prior knowledge, and just what special demands, if any, are going to be made of him as he proceeds. In the present case the first of these points is easily settled: We assume only a knowledge of elementary calculus and analytic geometry such as is usually gained in a standard three-semester course on these subjects. In particular, the reader should have encountered the notion of an infinite series, and know how to take a partial derivative and evaluate a double integral. Actually almost two-thirds of this book can be read without a knowledge of these last two items, while the first is covered quickly, but (for our purposes) adequately, in Appendix I. In short, we have kept formal prerequisites to a minimum. At the same time, however, we demand that the reader possess a certain amount of that elusive quality called mathematical maturity, by which we mean the patience to follow mathematical thought whither it may lead, and a willingness to postpone concrete applications until enough mathematics has been done to treat them properly.

This last demand is a reflection of the fact that initially much of our work may seem rather abstract, especially to the student coming directly from calculus. Thus, even though we have made every effort to motivate our arguments by referring to familiar situations and have illustrated them with numerous examples, it may not be out of place to reassure those students not interested in mathematics *per se* that every one of the topics we discuss is of fundamental importance in applied mathematics, physics, and engineering. Indeed, without falsifying fact, this book could have been entitled "An Introduction to Applied Mathematics" or "Advanced Engineering Mathematics," and might have been save that the material covered is of real value for the student of "pure" mathematics too. Nevertheless, most of the ideas which we have treated grew out of problems encountered in classical

B+T (math.)
141564

physics and the mathematical analysis of physical systems. As such these ideas lie at the foundations of modern physics and, to a lesser extent, modern mathematics as well.

But even more important, the subject which we have chosen to call "Linear Analysis" is, when viewed as an entity, one of the most profound creations of the human mind, and numbers among its contributors a clear majority of the great mathematicians and physicists of the past three centuries. For this reason alone it is worthy of study, and as our discussion proceeds we can only hope that the student will come to appreciate the beauty and power of the mathematical ideas it exploits and the remarkable creativity of those who invented them. If he does, his efforts and ours will have been well rewarded.

For the instructor. As its title suggests, this book is an introduction to those branches of mathematics based on the notion of linearity. The subject matter in these fields is vast, ranging all the way from differential and integral equations and the theory of Hilbert spaces to the mathematics encountered in constructing Green's functions and solving boundary-value problems in physics. Needless to say, no single book can do justice to such a variety of topics, particularly when, as in the present case, it attempts to start at the beginning of things. Nevertheless, it is the firm conviction of the authors that the notion of linearity which underlies these topics and ultimately enables them to be classified as branches of a single mathematical discipline can be developed in such a way that the student will gain a real understanding of the issues at stake.

Since we have assumed nothing more than a knowledge of elementary calculus and analytic geometry, the first two chapters of the book are devoted to an exposition of the rudiments of linear algebra, which we develop to the point where differential equations can be studied systematically. Anyone with a background in linear algebra should be able to begin at once with Chapter 3, using the first two chapters for reference.

Chapters 3 through 6 constitute an introduction to the theory of ordinary linear differential equations comparable to that taught in most first courses on the subject. Following the usual preliminaries, we introduce the notion of an initial-value problem and state the fundamental existence and uniqueness theorems for such problems. With these theorems as our real starting point, we proceed to show that the solution space of a normal nth-order homogeneous linear differential equation *must* be n-dimensional, and use this fact to obtain a complete treatment of the Wronskian. Then come equations with constant coefficients, solved by factoring the operators involved, after which we turn our attention to the method of variation of parameters. Here the algebraic point of view begins to pay real dividends, since we are in a position to see this method as a technique for inverting linear differential operators. This leads naturally to the notion of Green's functions and their associated integral operators, which we then treat in detail for initial-value problems. These ideas arise again in Chapter 5 when we study the Laplace transform, and our approach is such that we are able to give an integrated treatment of what all too often strike the student as unrelated techniques for solving differential equations. The sixth, and last, in this sequence of chapters extends the survey of linear differential equations beyond the customary beginning course by proving the Sturm separation and comparison theorems and the Sonin-Polya theorem. At this point we anticipate our later needs by using these results to study the behavior of solutions of Bessel's equation long before these solutions have been exhibited in series form. Finally, continuing in the same spirit, we introduce the method of undetermined coefficients for generating power series expansions of functions defined by equations with analytic coefficients.

In Chapters 7 and 8 the setting changes to Euclidean spaces, and metric concepts are introduced for the first time. We begin by proving the standard results for finite dimensional spaces, and then proceed to discuss convergence in finite and infinite dimensional spaces. Here we introduce the notion of an orthogonal basis in an infinite dimensional Euclidean space together with the concept of orthogonal series expansions in function spaces. Our point of view is that these ideas are straightforward generalizations of concepts familiar from Euclidean n-space, and we have made every effort to present them as such. In Chapters 9 and 11, we illustrate this theory by introducing several of the classical (Fourier) series of analysis, first relative to the trigonometric functions, and then, in succession, relative to the Legendre, Hermite, and Laguerre polynomials. (Chapter 10 is in the nature of a digression and is devoted to the study of convergence of Fourier series.)

In Chapter 12, we pull the various threads of our story together by introducing two-point boundary-value problems. We define eigenvalues and eigenvectors and discuss the eigenvalue method for solving operator equations. As usual we begin with the finite dimensional case, which is reduced to a problem in elementary algebra via the characteristic equation, and then generalize to symmetric operators on function spaces. The question of the existence of eigenfunction bases is treated in a theorem (left unproved) of sufficient generality to cover the boundary-value problems considered in the chapters which follow. We conclude this discussion by returning to the subject of Green's functions to establish their existence and uniqueness for problems with unmixed boundary conditions.

The last three chapters of the book use these results to solve boundary-value problems involving the wave, heat, and Laplace equations. The physical significance of these equations is discussed and the method of separation of variables is applied to reduce the problems considered to Sturm-Liouville systems which fall under our earlier analysis. The question of the validity of the solutions obtained is settled for the wave equation by appeal to earlier results on the convergence of Fourier series. Various forms of Laplace's equation are then considered, and the elegant theory of harmonic polynomials is introduced. Finally, cylindrical regions make their appearance, and the Frobenius method is developed to the point where Bessel's equation can be solved and orthogonal series involving Bessel functions constructed.

The book ends with four appendices containing material which would have been unduly disruptive in the body of the text. There we provide a discussion of pointwise and uniform convergence which is sufficient for our needs, a brief treatment of determinants, and a development of vector field theory to the point where uniqueness theorems for boundary-value problems can be proved.

Having outlined what is *in* the book, a few words may be in order concerning what is *not*. First, this is *not* a text in linear algebra. Thus, even though we do present much of the material usually taught in a first course in linear algebra, a few familiar topics have been omitted as unnecessary for the analysis we had in view. Second, we have said nothing whatever about numerical approximations, finite difference equations, and the like. Here our decision was guided by the belief that this material properly belongs in a course on numerical analysis, and any attempt to introduce it here would have resulted in an unwieldy book far too large to appear decently in public. Finally, for similar reasons we have avoided all topics which require a genuinely sophisticated use of operator theory, such as integral equations and the Fourier transform and integral. Logically such material ought to appear in a course following one based on a text such as this.

Given the modest level of preparation which we have assumed, we have made every effort, particularly in the earlier chapters, to motivate what we do by slow and careful explanations. Indeed, throughout the book we have been guided by the feeling that it is better to err on the side of too much explanation than too little. We have also tried to keep the discussion sharply in focus at all times by giving formal definitions of new terminology and precise statements of results being proved or used. For the most part, theorems stated in the text are proved on the spot. Those which are not comprise results whose proofs were felt to be either too difficult for a book at this level or unenlightening in view of our objectives. Such statements are usually accompanied by a reference to a proof in the literature.

In its present form this book is sufficiently flexible to be used in one of several courses. For instance, Chapters 1 through 6 plus parts of 7 and 15 provide material for a combined course on (ordinary) differential equations and linear algebra at the introductory level. On the other hand, Chapters 7 through 11 are logically independent of everything which precedes them, save Chapter 1, and can be used to give a course on series expansions and convergence in Euclidean spaces. By omitting Chapter 10 and adding Chapter 12, the first few sections of Chapter 2, and portions of Chapters 13 through 15, one obtains ample material for a one-semester course in boundary-value problems suitable for students who are able to solve elementary differential equations. Further there is more than enough material (though not exactly of the traditional sort) for several of those courses which go under the name of "engineering mathematics." In fact, this book was written primarily for such courses, and was motivated by the belief (or hope) that engineers ultimately profit from mathematics courses only to the extent that these courses present an honest treatment of the ideas involved.

For everyone. The internal reference system used in the text works as follows: Items in a particular chapter are numbered consecutively as, for example, (3–1) to (3–100). The first numeral refers to the chapter in question, the second to the numbered item within that chapter.

Throughout the book we have followed the popular device of indicating the end of a formal proof by the mark ▌ in the belief that students derive a certain comfort from a clearly visible sign telling them how far they must go before they can relax.

As usual, sections marked with an asterisk may be omitted without courting disaster. Everything so marked is either a digression which the authors had not the strength to resist, or material of greater difficulty than that in the immediate vicinity.

As a gesture toward scholarly respectability, we have included a short bibliography comprising those books which the authors personally found especially useful, and for the convenience of those inveterate browsers of books we have prepared an index of special symbols used in the text (see pp. xvi-xvii). Finally, a diagram showing the logical interdependence of the various chapters appears after the table of contents.

Debts and acknowledgements. Collectively and individually the authors are indebted to a large number of people who at long last can be publicly thanked:

First, the numerous students who have used portions of this material more or less willingly at Dartmouth College and Indiana University over the past several years, and whose comments have been far more valuable than they ever imagined.

Second, the surprisingly large number of professional colleagues whose advice has been sought, sometimes unknowingly, and who have been more than generous in answering questions and furnishing criticism. In particular, special thanks are due Professors H. Mirkil of Dartmouth College, G. C. Rota of Massachusetts Institute of Technology, and M. Thompson of Indiana University, and also Mr. L. Zalcman, presently at M. I. T.

And third, Mrs. Helen Hanchett of Hanover, New Hampshire, and Mrs. Darlene Martin of Bloomington, Indiana, for their patience, good nature, and unfailing accuracy in preparing typewritten versions of the manuscript too numerous to count.

Thanks are also due, and hereby given, Dartmouth College for assistance rendered in preparing a preliminary version of the manuscript and the Addison-Wesley staff for seeing the book through press.

Lastly, thanks of a very special sort to our several wives for their constant support and encouragement as well as their equally constant insistence that we get on with things and finish the job.

Conclusion. It seems to be one of the unfortunate facts of life that no mathematics book can be published free of errors. Since the present book is undoubtedly no exception, each of the authors would like to apologize in advance for any that still remain and take this opportunity to state publicly that they are the fault of the other three.

January, 1966

D. L. K.
R. G. K.
D. R. O.
F. W. P.

logical interdependence of chapters

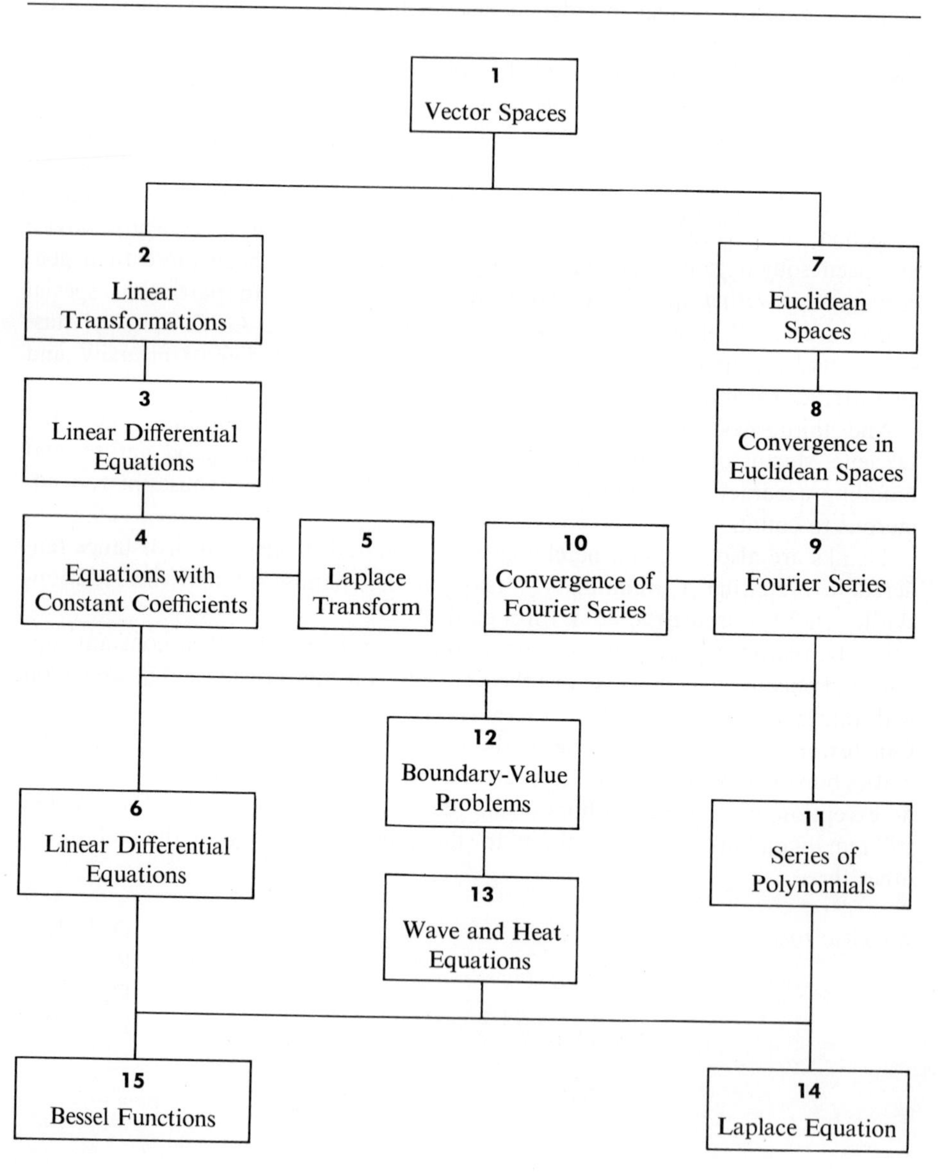

contents

4 EQUATIONS WITH CONSTANT COEFFICIENTS

5 THE LAPLACE TRANSFORM

6 FURTHER TOPICS IN THE THEORY OF LINEAR DIFFERENTIAL EQUATIONS

7 EUCLIDEAN SPACES

8 CONVERGENCE IN EUCLIDEAN SPACES

index of symbols used

SYMBOL	PAGE INTRODUCED	MEANING
$\mathcal{R}^2$	1	Real two-space
$\mathcal{C}[a, b]$	3, 258	Space of continuous functions on $[a, b]$
$\mathbf{0}$	6	Zero vector
$\mathcal{R}^n$	7, 257	Real n-space
$\mathcal{P}$	8	Space of polynomials with real coefficients
$\mathcal{R}^\infty$	9	Space of sequences of real numbers
$\mathcal{C}^n[a, b]$	13	Space of n times continuously differentiable functions on $[a, b]$
$\mathcal{P}_n$	13	Space of polynomials of degree $<n$
$\mathcal{W}_1 \cap \mathcal{W}_2$	14	Intersection of $\mathcal{W}_1$ and $\mathcal{W}_2$
$\mathcal{S}(\mathfrak{X})$	14	Subspace spanned by $\mathfrak{X}$
$\dim \mathcal{V}$	23	Dimension of the vector space $\mathcal{V}$
$\mathcal{G}^2$	33	Space of two-dimensional geometric vectors
$\sim$	37	Equivalent
$A: \mathcal{V}_1 \to \mathcal{V}_2$	42	Linear transformation from $\mathcal{V}_1$ to $\mathcal{V}_2$
$\mathcal{I}(A)$	42	Image of the linear transformation A
D	44	Differentiation operator
$\mathcal{C}^\infty[a, b]$	51	Space of infinitely differentiable functions on $[a, b]$
$\mathfrak{N}(A)$	55	Null space of the linear transformation A
$\mathfrak{T}_n$	269	Space of trigonometric polynomials of "degree" $\leq 2n + 1$
$P_n(x)$	280, 410	Legendre polynomial of degree n
$\mathfrak{X}^\perp$	289	Subspace perpendicular to $\mathfrak{X}$
$\mathcal{PC}[a, b]$	307, 332	Space of piecewise continuous functions on $[a, b]$
$\sim$	314	Formal series expansion
ℓ_2	323	Space of all "square summable" sequences of real numbers
$\overline{\mathcal{W}}$	324	Closure of $\mathcal{W}$
$\overline{\mathcal{S}(\mathfrak{X})}$	325	Closed subspace generated by $\mathfrak{X}$
$f(x_0^+), f(x_0^-)$	330	right- (left-) hand limit of f at x_0
f_E, f_O	336	Even (odd) part of f
E_f	350	Even extension of f
O_f	351	Odd extension of f
$\mathcal{G}_2^w(-\infty, \infty)$	437	Space of square integrable functions on $(-\infty, \infty)$ with respect to the weight function $e^{-x^2/2}$
$H_n(x)$	438	Hermite polynomial of degree n
$\mathcal{G}_2^w[0, \infty)$	444	Space of square integrable functions on $[0, \infty)$ with respect to the weight function e^{-x}

Symbol	Page	Meaning
$\mathcal{O}$	55	Trivial subspace
(α_{ij})	64	$m \times n$ matrix
$[A: \mathcal{B}_1, \mathcal{B}_2]$	64	Matrix of A with respect to bases $\mathcal{B}_1$ and $\mathcal{B}_2$
$\mathcal{L}(\mathcal{V}_1, \mathcal{V}_2)$	69	Space of linear transformations from $\mathcal{V}_1$ to $\mathcal{V}_2$
$\mathfrak{M}_{mn}$	69	Space of $m \times n$ matrices
$\mathcal{C}^n(I)$	86	Space of n times continuously differentiable functions on the interval I
L	86	Linear differential operator
$W[y_1, \ldots, y_n]$	111	Wronskian of $y_1, \ldots, y_n$
$\mathcal{L}[f]$	179	Laplace transform of f
$\mathcal{E}$	183	Space of piecewise continuous functions of exponential order
$\mathfrak{F}$	183	Space of real-valued functions defined on intervals of the form (s_0, ∞) or $[s_0, \infty)$
$u_a(t)$	194	Unit step function
$f * g$	206	Convolution of f and g
$\Gamma(x)$	211, 599	Gamma function
$\delta(t)$	223	Dirac delta function
$\mathbf{x} \cdot \mathbf{y}$	256	Inner product of $\mathbf{x}$ and $\mathbf{y}$
$\lVert\mathbf{x}\rVert$	261	Norm or length of $\mathbf{x}$
$d(\mathbf{x}, \mathbf{y})$	264	Distance between $\mathbf{x}$ and $\mathbf{y}$
δ_{ij}	269	Kronecker delta
$L_n(x)$	444	Laguerre polynomial of degree n
$L_n^{(\alpha)}(x)$	447	Laguerre polynomials
$\mathcal{C}_\varphi(S)$	560	Space of continuous functions on the unit sphere dependent only on colatitude
$P_n(\cos\varphi)$	562	nth surface zonal harmonic
$\mathcal{C}(S)$	567	Space of continuous functions on the unit sphere
$u_{mn}(\varphi, \theta), v_{mn}(\varphi, \theta)$	567	Spherical harmonics
$\mathcal{H}_m$	575	Space of homogeneous harmonic polynomials of degree m
$I(\nu)$	592	Indicial polynomial
$\mathrm{Re}(\nu)$	592	Real part of ν
$J_p(x)$	599, 600	Bessel function of order p of the first kind
$Y_0(x)$	603	Bessel function of order zero of the second kind
$Y_n(x)$	604	Bessel function of order n of the second kind
$I_n(x)$	616, 630	Modified Bessel function
$\lvert a_{ij}\rvert$	685	Determinant of order n
$\sigma(p)$	683	Signature of the permutation p
$\mathbf{a}_1 \times \mathbf{a}_2$	702	Vector product of $\mathbf{a}_1$ and $\mathbf{a}_2$
$\mathrm{div}\ \mathbf{F}$	712	Divergence of $\mathbf{F}$
$\boldsymbol{\nabla}$	718	Del
∇^2	718	Laplacian

1
real vector spaces

1–1 INTRODUCTION

The Cartesian plane of analytic geometry, denoted by $\mathcal{R}^2$, is one of the most familiar examples of what is known in mathematics as a *real vector space*. Each of its points, or *vectors*, is an ordered pair (x_1, x_2) of real numbers whose individual entries, x_1 and x_2, are called the *components* of that vector. Geometrically, the vector $\mathbf{x} = (x_1, x_2)$ may be represented by means of an arrow drawn from the origin of coordinates to the point (x_1, x_2) as shown in Fig. 1–1.*

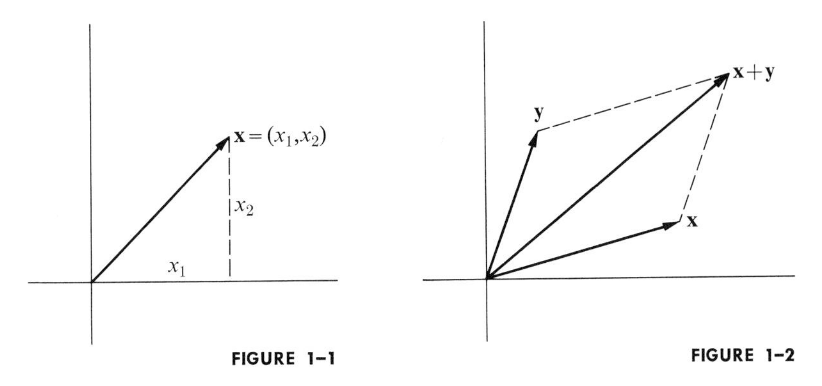

FIGURE 1–1

FIGURE 1–2

If $\mathbf{x} = (x_1, x_2)$ and $\mathbf{y} = (y_1, y_2)$ are any two vectors in $\mathcal{R}^2$, then, by definition, their *sum* is the vector

$$\mathbf{x} + \mathbf{y} = (x_1 + y_1, x_2 + y_2) \tag{1–1}$$

obtained by adding the corresponding components of $\mathbf{x}$ and $\mathbf{y}$. The graphical interpretation of this addition is the familiar "parallelogram law," which states that the vector $\mathbf{x} + \mathbf{y}$ is the diagonal of the parallelogram formed from $\mathbf{x}$ and $\mathbf{y}$ (see Fig. 1–2). It follows at once from (1–1) that vector addition is both associative

* Throughout this book we shall use boldface type (i.e., **x**, **y**, . . .) to denote vectors.

and commutative, namely,

$$\mathbf{x} + (\mathbf{y} + \mathbf{z}) = (\mathbf{x} + \mathbf{y}) + \mathbf{z}, \tag{1–2}$$

$$\mathbf{x} + \mathbf{y} = \mathbf{y} + \mathbf{x}. \tag{1–3}$$

Moreover, if we let $\mathbf{0}$ denote the vector $(0, 0)$, and $-\mathbf{x}$ the vector $(-x_1, -x_2)$ obtained by reflecting $\mathbf{x} = (x_1, x_2)$ across the origin, then

$$\mathbf{x} + \mathbf{0} = \mathbf{x}, \tag{1–4}$$

and

$$\mathbf{x} + (-\mathbf{x}) = \mathbf{0} \tag{1–5}$$

for every $\mathbf{x}$. Taken together, Eqs. (1–2) through (1–5) imply that vector addition behaves very much like the ordinary addition of arithmetic.

As well as being able to add vectors in $\mathcal{R}^2$, we can also form the *product* of a real number α and a vector $\mathbf{x}$. The result, denoted $\alpha\mathbf{x}$, is the *vector* each of whose components is α times the corresponding component of $\mathbf{x}$. Thus, if $\mathbf{x} = (x_1, x_2)$, then

$$\alpha\mathbf{x} = (\alpha x_1, \alpha x_2). \tag{1–6}$$

Geometrically, this vector can be viewed as a "magnification" of $\mathbf{x}$ by the factor α, as illustrated in Fig. 1–3.

The principal algebraic properties of this multiplication are the following:

$$\alpha(\mathbf{x} + \mathbf{y}) = \alpha\mathbf{x} + \alpha\mathbf{y}, \tag{1–7}$$

$$(\alpha + \beta)\mathbf{x} = \alpha\mathbf{x} + \beta\mathbf{x}, \tag{1–8}$$

$$(\alpha\beta)\mathbf{x} = \alpha(\beta\mathbf{x}), \tag{1–9}$$

$$1\mathbf{x} = \mathbf{x}. \tag{1–10}$$

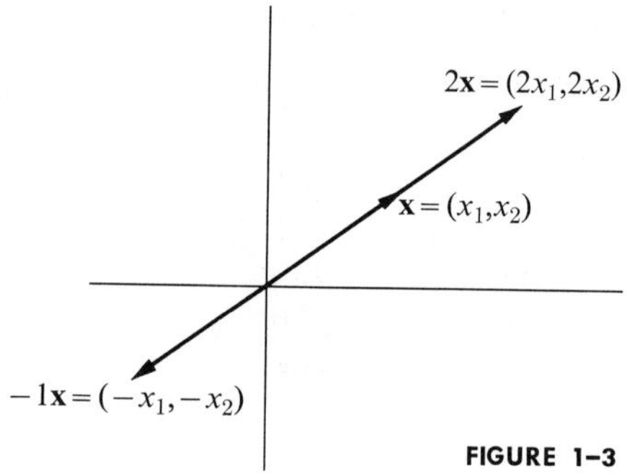

$2\mathbf{x} = (2x_1, 2x_2)$ $\mathbf{x} = (x_1, x_2)$ $-1\mathbf{x} = (-x_1, -x_2)$

FIGURE 1–3

The validity of each of these equations can be deduced easily from the definition of the operations involved, and save for (1–9), which we prove by way of illustration, the equations are left for the student to verify. To establish (1–9), let $\mathbf{x} = (x_1, x_2)$ be an arbitrary vector in $\mathcal{R}^2$, and let α and β be real numbers. Then by repeated use of (1–6), we have

$$\begin{aligned}(\alpha\beta)\mathbf{x} &= \big((\alpha\beta)x_1, (\alpha\beta)x_2\big) \\ &= \big(\alpha(\beta x_1), \alpha(\beta x_2)\big) \\ &= \alpha(\beta x_1, \beta x_2) \\ &= \alpha\big(\beta(x_1, x_2)\big) \\ &= \alpha(\beta\mathbf{x}),\end{aligned}$$

which is what we wished to show.

The reason for calling attention to Eqs. (1–7) through (1–10) is that they, together with properties (1–2) through (1–5) for vector addition, are precisely what make $\mathcal{R}^2$ a real vector space. Indeed, these equations are none other than the axioms in the general definition of such a space, and once this definition has been given, the above discussion constitutes a verification of the fact that $\mathcal{R}^2$ is a real vector space. But before giving this definition, we pause to look at another example.

This time we consider the set $\mathcal{C}[a, b]$ consisting of all real valued, continuous functions defined on a closed interval $[a, b]$ of the real line.* For reasons which will shortly become clear we shall call any such function a vector, and, following our general convention, write it in boldface type. Thus $\mathbf{f}$ is a vector in $\mathcal{C}[a, b]$ if and only if $\mathbf{f}$ is a real valued, continuous function on the interval $[a, b]$.

At first sight it may seem that $\mathcal{C}[a, b]$ and $\mathcal{R}^2$ have nothing in common but the name "real vector space." However, this is one of those instances in which first impressions are misleading, for as we shall see, these spaces are remarkably similar. This similarity arises from the fact that an addition and multiplication by real numbers can also be defined in $\mathcal{C}[a, b]$ and that these operations enjoy the same properties as the corresponding operations in $\mathcal{R}^2$.

Turning first to addition, let $\mathbf{f}$ and $\mathbf{g}$ be any two vectors in $\mathcal{C}[a, b]$. Then their sum, $\mathbf{f} + \mathbf{g}$, is defined to be the function (i.e., vector) whose value at any point x in $[a, b]$ is the sum of the values of $\mathbf{f}$ and $\mathbf{g}$ at x (see Fig. 1–4). In other words,

$$(\mathbf{f} + \mathbf{g})(x) = \mathbf{f}(x) + \mathbf{g}(x). \quad (1\text{–}11)$$

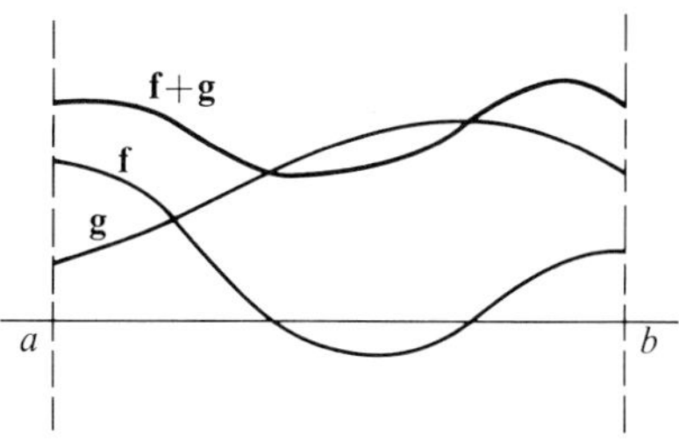

FIGURE 1–4

At this point it is important to observe that since the sum of two continuous functions is continuous, this definition is meaningful in the sense that $\mathbf{f} + \mathbf{g}$ is again a vector in $\mathcal{C}[a, b]$.†

It is now easy to verify that apart from notation and interpretation Eqs. (1–2) through (1–5) remain valid in $\mathcal{C}[a, b]$. In fact,

$$\mathbf{f} + (\mathbf{g} + \mathbf{h}) = (\mathbf{f} + \mathbf{g}) + \mathbf{h} \quad (1\text{–}12)$$

and

$$\mathbf{f} + \mathbf{g} = \mathbf{g} + \mathbf{f} \quad (1\text{–}13)$$

follow immediately from (1–11), while if $\mathbf{0}$ denotes the function whose value is

---

* The *closed interval* $[a, b]$ is the set of all real numbers x such that $a \leq x \leq b$; i.e., $[a, b]$ is the interval from a to b, end points included. By contrast, if the end points are not included in the interval, we speak of the *open interval* from a to b, and write (a, b).

† For a proof of this fact see, for example, C. B. Morrey, Jr., *University Calculus with Analytic Geometry*, Addison-Wesley, 1962, p. 89.

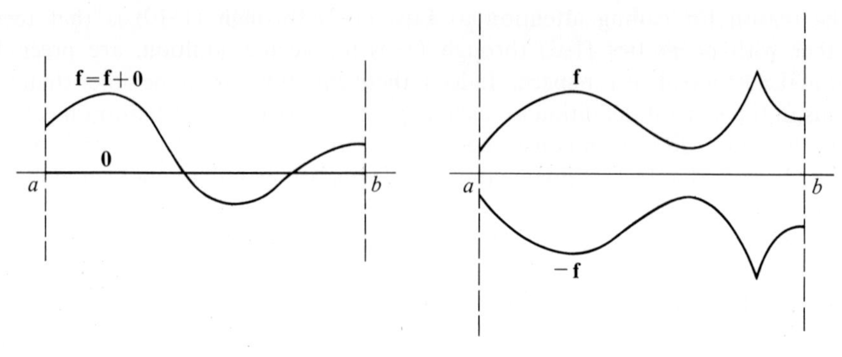

FIGURE 1–5 FIGURE 1–6

zero at each point of $[a, b]$, then

$$\mathbf{f} + \mathbf{0} = \mathbf{f} \tag{1–14}$$

for every $\mathbf{f}$ in $\mathcal{C}[a, b]$ (Fig. 1–5). Finally, if $-\mathbf{f}$ is the function whose value at x is $-\mathbf{f}(x)$ (i.e., $-\mathbf{f}$ is the reflection of $\mathbf{f}$ across $\mathbf{0}$, as shown in Fig. 1–6), then $\mathbf{f} + (-\mathbf{f})$ has the value zero at each point of $[a, b]$, and we have

$$\mathbf{f} + (-\mathbf{f}) = \mathbf{0}. \tag{1–15}$$

We have seen that the sum of two vectors in $\mathcal{R}^2$ is found by adding their corresponding components (Eq. 1–1). A similar interpretation of vector addition is possible in the present example and may be achieved as follows. If $\mathbf{f}$ is any vector in $\mathcal{C}[a, b]$, we agree to say that the "component" of $\mathbf{f}$ at the point x is its functional value at x. Of course, every vector in $\mathcal{C}[a, b]$ then has infinitely many components, one for each x in the interval $[a, b]$, but once this fact has been accepted it becomes clear that Eq. (1–11) simply states that the sum of two vectors in $\mathcal{C}[a, b]$ is obtained by adding corresponding "components," just as in $\mathcal{R}^2$.

Next, if $\mathbf{f}$ is any vector in $\mathcal{C}[a, b]$ and α is an arbitrary real number, we define the product $\alpha\mathbf{f}$ to be the vector in $\mathcal{C}[a, b]$ whose defining equation is

$$(\alpha\mathbf{f})(x) = \alpha\mathbf{f}(x). \tag{1–16}$$

In other words, $\alpha\mathbf{f}$ is the function whose value at x is the product of the real numbers α and $\mathbf{f}(x)$. The similarity between this multiplication and the corresponding operation in $\mathcal{R}^2$ is clear, since $\alpha\mathbf{f}$ is also formed by multiplying each "component" of $\mathbf{f}$ by α. (Figure 1–7 illustrates this multiplication when $\alpha = 2$.)

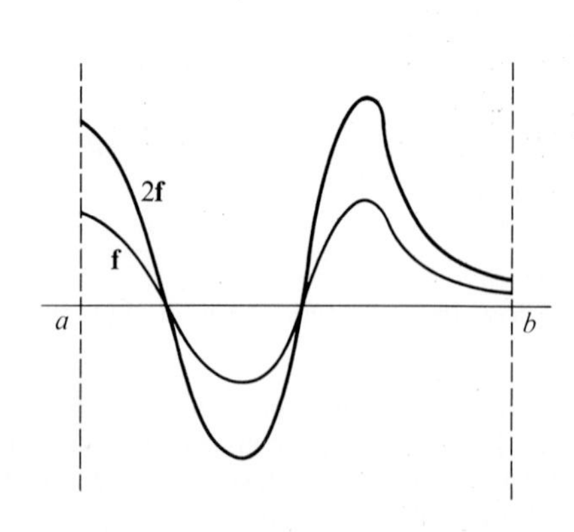

FIGURE 1–7

The analogy with $\mathcal{R}^2$ is now complete, for Eqs. (1–7) through (1–10) are also valid in $\mathcal{C}[a, b]$. We restate them here as

$$\alpha(\mathbf{f} + \mathbf{g}) = \alpha\mathbf{f} + \alpha\mathbf{g}, \tag{1–17}$$

$$(\alpha + \beta)\mathbf{f} = \alpha\mathbf{f} + \beta\mathbf{f}, \tag{1–18}$$

$$(\alpha\beta)\mathbf{f} = \alpha(\beta\mathbf{f}), \tag{1–19}$$

$$1\mathbf{f} = \mathbf{f}, \tag{1–20}$$

and leave their proofs as an exercise.

This is perhaps an appropriate point at which to warn the unsuspecting reader that the space $\mathcal{C}[a, b]$ is much more than an idle example. Indeed, a great deal of our later work will be devoted to a study of this and similar vector spaces, and the student will do well to understand it before reading further.

EXERCISES

In Exercises 1 through 5 compute the value of $\mathbf{x} + \mathbf{y}$ and $\alpha(\mathbf{x} + \mathbf{y})$ for the given vectors $\mathbf{x}$ and $\mathbf{y}$ in $\mathcal{R}^2$ and real number α. Illustrate each of your computations with an appropriate diagram.

1. $\mathbf{x} = (0, 2)$, $\mathbf{y} = (-1, 1)$, $\alpha = 3$
2. $\mathbf{x} = (\frac{1}{2}, 1)$, $\mathbf{y} = (1, -2)$, $\alpha = -2$
3. $\mathbf{x} = (-\frac{1}{2}, \frac{1}{3})$, $\mathbf{y} = (-2, -1)$, $\alpha = -1$
4. $\mathbf{x} = (5, -2)$, $\mathbf{y} = (-3, 2)$, $\alpha = \frac{1}{2}$
5. $\mathbf{x} = (-5, -2)$, $\mathbf{y} = (-1, -1)$, $\alpha = -3$

In Exercises 6 through 10 compute the value of $\mathbf{f} + \mathbf{g}$ and $\alpha(\mathbf{f} + \mathbf{g})$ for the given vectors $\mathbf{f}$ and $\mathbf{g}$ in $\mathcal{C}[-1, 1]$ and real number α. Illustrate each of your computations with an appropriate diagram.

6. $\mathbf{f}(x) = 2x$, $\mathbf{g}(x) = x^2 - x + 1$, $\alpha = 2$
7. $\mathbf{f}(x) = \tan^2 x$, $\mathbf{g}(x) = 1$, $\alpha = -1$
8. $\mathbf{f}(x) = e^x$, $\mathbf{g}(x) = e^{-x}$, $\alpha = \frac{1}{2}$
9. $\mathbf{f}(x) = \dfrac{x+3}{x-2}$, $\mathbf{g}(x) = -\dfrac{x-2}{x+3}$, $\alpha = \dfrac{1}{5}$
10. $\mathbf{f}(x) = \cos^2 x$, $\mathbf{g}(x) = \sin^2 x$, $\alpha = -3$

In Exercises 11 through 20 determine whether or not the indicated function belongs to $\mathcal{C}[-1, 1]$; if it does not, state why not.

11. $\sin \dfrac{x+1}{x-1}$
12. $|x| = \begin{cases} x & \text{if } x \geq 0 \\ -x & \text{if } x < 0 \end{cases}$
13. $\tan x$
14. $\tan(2x + 1)$
15. $\ln |x|$
16. $\dfrac{2x(x+1)}{x^2 + 3x + 2}$

17. $\mathbf{f}(x) = \begin{cases} 1 & \text{if } x \geq 0 \\ -1 & \text{if } x < 0 \end{cases}$

18. $\mathbf{f}(x) = \begin{cases} x - \frac{1}{2} & \text{if } x \geq 0 \\ -x - \frac{1}{2} & \text{if } x < 0 \end{cases}$

19. $\mathbf{f}(x) = \begin{cases} x - 1 & \text{if } x \geq 0 \\ x + 1 & \text{if } x < 0 \end{cases}$

20. e^{-x}

Solve each of the following vector equations for $\mathbf{x}$.

21. $2\mathbf{x} + 3(2, 1) = (0, 0)$
22. $4\mathbf{x} + 3\mathbf{x} = 2(-1, 3)$
23. $(\frac{1}{2} + \frac{2}{3})\mathbf{x} + 5(\frac{1}{6}(3, 4)) = \frac{1}{5}(2, -1)$
24. $\mathbf{x} - 3(2, 1) = 4\mathbf{x}$

Solve each of the following vector equations in $\mathcal{C}[-1, 1]$ for $\mathbf{f}$.

25. $2\mathbf{f} - 1 = \sec^2 x$
26. $\ln \mathbf{f} = 1$
27. $e^{\mathbf{f}} = x + 2$
28. $2\mathbf{f} + e^{-x} = e^x$
29. Prove Eqs. (1–2) through (1–5) and (1–7) through (1–10) of the text, and illustrate each with a diagram.
30. Prove Eqs. (1–12) through (1–15) and (1–17) through (1–20) of the text.
31. Let $\mathbf{x}$ be a vector in $\mathcal{R}^2$, and let α be a real number. Prove that $\alpha\mathbf{x} = \mathbf{0}$ if and only if $\alpha = 0$ or $\mathbf{x} = \mathbf{0}$.
32. Let $\mathbf{x} \neq \mathbf{0}$ be a vector in $\mathcal{R}^2$, and suppose $\alpha\mathbf{x} = \mathbf{x}$. Prove that $\alpha = 1$.

1–2 REAL VECTOR SPACES

With the examples of the preceding section in mind we now give the definition of a real vector space.

Definition 1–1. A *real vector space* $\mathcal{V}$ is a collection of objects called *vectors*, together with operations of addition and multiplication by real numbers which satisfy the following axioms.

Axioms for addition. Given any pair of vectors $\mathbf{x}$ and $\mathbf{y}$ in $\mathcal{V}$ there exists a (unique) vector $\mathbf{x} + \mathbf{y}$ in $\mathcal{V}$ called the *sum* of $\mathbf{x}$ and $\mathbf{y}$. It is required that

(i) addition be *associative*,

$$\mathbf{x} + (\mathbf{y} + \mathbf{z}) = (\mathbf{x} + \mathbf{y}) + \mathbf{z},$$

(ii) addition be *commutative*,

$$\mathbf{x} + \mathbf{y} = \mathbf{y} + \mathbf{x},$$

(iii) there exist a vector $\mathbf{0}$ in $\mathcal{V}$ (called the *zero vector*) such that

$$\mathbf{x} + \mathbf{0} = \mathbf{x}$$

for all $\mathbf{x}$ in $\mathcal{V}$, and

(iv) for each $\mathbf{x}$ in $\mathcal{V}$ there exist a vector $-\mathbf{x}$ in $\mathcal{V}$ such that

$$\mathbf{x} + (-\mathbf{x}) = \mathbf{0}.$$

Axioms for scalar multiplication. Given any vector $\mathbf{x}$ in $\mathcal{V}$ and any real number α there exists a (unique) vector $\alpha\mathbf{x}$ in $\mathcal{V}$ called the *product*, or *scalar product*, of α and $\mathbf{x}$. It is required that

$$\text{(v)} \qquad \alpha(\mathbf{x} + \mathbf{y}) = \alpha\mathbf{x} + \alpha\mathbf{y},$$

$$\text{(vi)} \qquad (\alpha + \beta)\mathbf{x} = \alpha\mathbf{x} + \beta\mathbf{x},$$

$$\text{(vii)} \qquad (\alpha\beta)\mathbf{x} = \alpha(\beta\mathbf{x}),$$

$$\text{(viii)} \qquad 1\mathbf{x} = \mathbf{x}.$$

The student should not be discouraged by the formality of this definition; it looks much worse than it really is. The issue here is simply that in order to deserve the name, a real vector space must have a number of elementary and eminently reasonable properties in common with $\mathcal{R}^2$. We have already seen this happen in the case of the function space $\mathcal{C}[a, b]$, and before embarking on the general study of this subject we give several additional examples of such variety as to convince the reader that real vector spaces are very common objects indeed in mathematics. Still others will be found in the exercises at the end of this section. In each case we leave the verification of Axioms (i) through (viii) as an exercise to aid the beginner in assimilating the various requirements of the definition.

EXAMPLE 1. The real numbers, with the ordinary definitions of addition and multiplication, form a real vector space. In this case Axioms (i) through (viii) are all familiar statements from arithmetic.

EXAMPLE 2. Let n be a fixed positive integer, and let $\mathcal{R}^n$ denote the totality of ordered n-tuples $(x_1, \ldots, x_n)$ of real numbers. If $\mathbf{x} = (x_1, \ldots, x_n)$ and $\mathbf{y} = (y_1, \ldots, y_n)$ are two such n-tuples, and α is a real number, set

$$\mathbf{x} + \mathbf{y} = (x_1 + y_1, \ldots, x_n + y_n) \tag{1–21}$$

and

$$\alpha\mathbf{x} = (\alpha x_1, \ldots, \alpha x_n). \tag{1–22}$$

Then $\mathcal{R}^n$ becomes a real vector space. It is clear that $\mathcal{R}^1$ is the vector space of Example 1 above, and that $\mathcal{R}^2$ and $\mathcal{R}^3$ are the vector spaces studied in analytic geometry.

Note that there is nothing in the least mysterious in the fact that n is allowed to assume values greater than three in this example. It is true, of course, that pictures of the usual sort cannot then be drawn, but this is no serious shortcoming. In fact, geometric intuition from 3-space, if used circumspectly, is still reasonably accurate in $\mathcal{R}^n$ when $n > 3$, despite the lack of a visual representation for these spaces.

EXAMPLE 3. Let $\mathcal{P}$ denote the set of all polynomials in x with real coefficients, and let polynomial addition and multiplication by real numbers be defined as in high school algebra. Then $\mathcal{P}$ is a real vector space.

This completes our list of basic examples. As we shall see, each of the real vector spaces $\mathcal{C}[a, b]$, $\mathcal{R}^n$, and $\mathcal{P}$ has distinctive features not shared by either of the others, and for this reason these three spaces will occupy a special place in our later work. And since we have already mentioned the future importance of $\mathcal{C}[a, b]$, it is only fair to add that $\mathcal{R}^n$ and $\mathcal{P}$ will in turn receive a great deal of attention.

Before we continue, a few general remarks concerning the definition of a real vector space are in order. First, since vector addition is associative, any finite sum of the form

$$\mathbf{x}_1 + \mathbf{x}_2 + \cdots + \mathbf{x}_n$$

is well defined as it stands and need not be festooned with parentheses to avoid ambiguity. Moreover, the commutativity of vector addition implies that the value of such a sum does not depend upon the order of the summands.

Secondly, in the interest of simplicity we shall write $\mathbf{x} - \mathbf{y}$ from now on in place of the unwieldy expression $\mathbf{x} + (-\mathbf{y})$. This, of course, involves the usual change in terminology, for we then say that $\mathbf{x} - \mathbf{y}$ is obtained by *subtracting* $\mathbf{y}$ from $\mathbf{x}$.

And finally, we call attention to the reasonably obvious fact that our insistence upon using real numbers in the definition of a vector space is quite unnecessary. Complex numbers, for instance, would do just as well, in which case we would have what is unsurprisingly known as a complex vector space. Even further generalizations are possible. However, these more general vector spaces will not appear in this book, and so we agree that the term "vector space" will always mean real vector space, as defined above. Furthermore, the term *scalar* will be used now and again as a synonym for the words "real number" in consonance with the standard terminology of the subject.

EXERCISES

1. With addition and scalar multiplication as defined in the text, prove that $\mathcal{R}^n$ is a real vector space.
2. Prove that $\mathcal{P}$ is a real vector space. (Recall that the vectors in $\mathcal{P}$ may be written in the form

$$a_0 + a_1 x + \cdots + a_n x^n,$$

where $a_0, \ldots, a_n$ are real numbers.)
3. Find the value of $\alpha_1\mathbf{x}_1 + \alpha_2\mathbf{x}_2 + \alpha_3\mathbf{x}_3$ in $\mathcal{R}^2$ when

(a) $\mathbf{x}_1 = (2, 1)$, $\mathbf{x}_2 = (-1, 2)$, $\mathbf{x}_3 = (1, 0)$, $\alpha_1 = 1$, $\alpha_2 = 2$, $\alpha_3 = -1$;

(b) $\mathbf{x}_1 = (\frac{1}{2}, 3)$, $\mathbf{x}_2 = (-\frac{1}{3}, \frac{1}{2})$, $\mathbf{x}_3 = (-1, 2)$, $\alpha_1 = 2$, $\alpha_2 = 9$, $\alpha_3 = -2$;

(c) $\mathbf{x}_1 = (-2, 2)$, $\mathbf{x}_2 = (-\frac{1}{2}, -3)$, $\mathbf{x}_3 = (\frac{2}{3}, -\frac{1}{2})$, $\alpha_1 = \frac{1}{2}$, $\alpha_2 = \frac{2}{3}$, $\alpha_3 = 2$.

4. Find the value of $\alpha_1\mathbf{x}_1 + \alpha_2\mathbf{x}_2 + \alpha_3\mathbf{x}_3$ in $\mathcal{R}^4$ when

   (a) $\mathbf{x}_1 = (2, 0, -1, 3)$, $\mathbf{x}_2 = (1, -2, 2, 0)$, $\mathbf{x}_3 = (2, -1, 3, 1)$, $\alpha_1 = -1$, $\alpha_2 = 0$, $\alpha_3 = 1$;

   (b) $\mathbf{x}_1 = (1, 2, 3, 4)$, $\mathbf{x}_2 = (0, 2, 1, -2)$, $\mathbf{x}_3 = (2, 6, 7, 6)$, $\alpha_1 = 2$, $\alpha_2 = 1$, $\alpha_3 = -1$;

   (c) $\mathbf{x}_1 = (-\frac{1}{2}, 2, 1, -3)$, $\mathbf{x}_2 = (1, -2, \frac{1}{2}, 5)$, $\mathbf{x}_3 = (\frac{2}{3}, 0, 0, \frac{1}{2})$, $\alpha_1 = -2$, $\alpha_2 = \frac{1}{2}$, $\alpha_3 = -1$.

5. Find the value of $\alpha_1\mathbf{p}_1 + \alpha_2\mathbf{p}_2 + \alpha_3\mathbf{p}_3$ in $\mathcal{P}$ when

   (a) $\mathbf{p}_1(x) = x^2 - x + 1$, $\alpha_1 = 2$,
   $\mathbf{p}_2(x) = 3x^3 + 2x - 1$, $\alpha_2 = -1$,
   $\mathbf{p}_3(x) = -x^3 + 2x$, $\alpha_3 = -2$;

   (b) $\mathbf{p}_1(x) = 2x^4 - 4x^2 + 1$, $\alpha_1 = \frac{1}{2}$,
   $\mathbf{p}_2(x) = -x^4 + 2x^3 + x^2 - x + 2$, $\alpha_2 = 2$,
   $\mathbf{p}_3(x) = -x^4 + 4x^3 - 2x - \frac{5}{2}$, $\alpha_3 = -1$;

   (c) $\mathbf{p}_1(x) = \frac{1}{2}x^3 - 2x^2 + \frac{1}{5}$, $\alpha_1 = 3$,
   $\mathbf{p}_2(x) = 2x - 1$, $\alpha_2 = -\frac{1}{4}$,
   $\mathbf{p}_3(x) = x^2$, $\alpha_3 = 2$.

6. Does the set of all polynomials in x with *integral* coefficients form a real vector space with the usual definitions of addition and multiplication by real numbers? Why?

7. Let $\mathcal{C}$ denote the set of all complex numbers, i.e., all numbers of the form $a + bi$, a and b real, and $i = \sqrt{-1}$. With addition of complex numbers defined by

$$(a + bi) + (c + di) = (a + c) + (b + d)i,$$

   and scalar multiplication defined by

$$\alpha(a + bi) = \alpha a + \alpha bi,$$

   prove that $\mathcal{C}$ is a real vector space. Compare this space with $\mathcal{R}^2$.

8. Let $\mathcal{R}^\infty$ be the set consisting of all infinite sequences

$$\mathbf{x} = (x_1, x_2, \ldots)$$

   of real numbers. If $\mathbf{y} = (y_1, y_2, \ldots)$ is another such sequence, define $\mathbf{x} + \mathbf{y}$ by

$$\mathbf{x} + \mathbf{y} = (x_1 + y_1, x_2 + y_2, \ldots);$$

   while if α is a real number, define $\alpha\mathbf{x}$ by

$$\alpha\mathbf{x} = (\alpha x_1, \alpha x_2, \ldots).$$

   Prove that $\mathcal{R}^\infty$ is a real vector space.

9. With addition and scalar multiplication defined as in $\mathcal{C}[a, b]$, determine which of the following sets of functions is a real vector space:

   (a) all functions which are continuous everywhere on $[a, b]$ except at a finite number of points;

   (b) all functions which are zero everywhere on $[a, b]$ except at a finite number of points;

(c) all functions which are different from zero at all but a finite number of points of $[a, b]$;

(d) all continuous functions such that $\mathbf{f}(a) = \mathbf{f}(b)$;

(e) all continuous functions such that $\mathbf{f}(a + x) = \mathbf{f}(b - x)$;

(f) all continuous functions which are zero on some closed subinterval of $[a, b]$;

(g) all continuous functions which are zero on a *fixed* subinterval of $[a, b]$.

10. Let $\mathcal{R}^+$ be the set consisting of all *positive* real numbers, and define "addition" and "scalar multiplication" in $\mathcal{R}^+$ as follows: If x and y belong to $\mathcal{R}^+$, let

$$x + y = xy,$$

where the product appearing on the right-hand side of this equality is the ordinary product of the real numbers x and y; if α is an arbitrary real number, let

$$\alpha x = x^\alpha.$$

Prove that $\mathcal{R}^+$ is then a real vector space.

11. Let $\mathcal{V}$ be the set of all ordered pairs of real numbers. If $\mathbf{x} = (x_1, x_2)$ and $\mathbf{y} = (y_1, y_2)$ are any two elements in $\mathcal{V}$, define

$$\mathbf{x} + \mathbf{y} = (x_1 + y_1, 0),$$

and let scalar multiplication be defined as in $\mathcal{R}^2$. Is $\mathcal{V}$ a real vector space? Why?

12. Repeat Exercise 11, this time with addition defined as in $\mathcal{R}^2$ and scalar multiplication changed to

$$\alpha\mathbf{x} = (x_1, 0).$$

13. Use the associativity of vector addition to prove that

$$(\mathbf{w} + \mathbf{x} + \mathbf{y}) + \mathbf{z} = \mathbf{w} + (\mathbf{x} + \mathbf{y} + \mathbf{z}).$$

14. Is the operation of subtraction in a vector space associative or commutative?

1–3 ELEMENTARY OBSERVATIONS

In this section we note a number of immediate consequences of Definition 1–1, all of which are so elementary that they will be used without explicit mention in the future. The first of these concerns the zero vector and asserts that this vector behaves very much as one might expect. Specifically,

$$0\mathbf{x} = \mathbf{0} \quad \text{for every } \mathbf{x}, \tag{1–23}$$

and

$$\alpha\mathbf{0} = \mathbf{0} \quad \text{for every } \alpha. \tag{1–24}$$

To prove the first of these assertions set $\alpha = \beta = 0$ in $\alpha\mathbf{x} + \beta\mathbf{x} = (\alpha + \beta)\mathbf{x}$. This gives

$$0\mathbf{x} + 0\mathbf{x} = (0 + 0)\mathbf{x} = 0\mathbf{x}.$$

Now subtract $0\mathbf{x}$ from both sides of this equation, and then use the fact that

$0\mathbf{x} - 0\mathbf{x} = \mathbf{0}$ to obtain

$$0\mathbf{x} + (0\mathbf{x} - 0\mathbf{x}) = 0\mathbf{x} - 0\mathbf{x}$$

and

$$0\mathbf{x} + \mathbf{0} = \mathbf{0}.$$

Hence $0\mathbf{x} = \mathbf{0}$.

The proof of (1–24) is similar; this time set $\mathbf{x} = \mathbf{y} = \mathbf{0}$ in $\alpha(\mathbf{x} + \mathbf{y}) = \alpha\mathbf{x} + \alpha\mathbf{y}$.

Next, an equally elementary proof establishes the fact that the vector $-\mathbf{x}$ and $(-1)\mathbf{x}$ are one and the same. Indeed, since $1\mathbf{x} = \mathbf{x}$ and $0\mathbf{x} = \mathbf{0}$, we have

$$\mathbf{x} + (-1)\mathbf{x} = 1\mathbf{x} + (-1)\mathbf{x} = (1 - 1)\mathbf{x} = 0\mathbf{x} = \mathbf{0}.$$

Now subtract $\mathbf{x}$ from both sides of this equation to obtain $(-1)\mathbf{x} = -\mathbf{x}$, as asserted.

Finally, it may be of interest to prove that in any vector space $\mathcal{V}$ the vector $\mathbf{0}$ is unique and $-\mathbf{x}$ is uniquely determined by $\mathbf{x}$, in the sense that they are the only vectors in $\mathcal{V}$ possessing their particular properties. This is the burden of

Lemma 1–1. *If $\mathbf{0}'$ is a vector in $\mathcal{V}$ such that $\mathbf{x} + \mathbf{0}' = \mathbf{x}$ for every $\mathbf{x}$ in $\mathcal{V}$, then $\mathbf{0}' = \mathbf{0}$. Similarly, if $\mathbf{x}'$ is any vector in $\mathcal{V}$ such that $\mathbf{x} + \mathbf{x}' = \mathbf{0}$, then $\mathbf{x}' = -\mathbf{x}$.*

Proof. If $\mathbf{x} + \mathbf{0}' = \mathbf{x}$ for every $\mathbf{x}$ in $\mathcal{V}$, we have, in particular,

$$\mathbf{0} + \mathbf{0}' = \mathbf{0}.$$

On the other hand, the zero vector has the property that $\mathbf{0} + \mathbf{x} = \mathbf{x}$ for every $\mathbf{x}$. Hence

$$\mathbf{0} + \mathbf{0}' = \mathbf{0}',$$

and it follows that

$$\mathbf{0} = \mathbf{0}'.$$

The second statement of the lemma follows from the sequence of equalities

$$\mathbf{x}' = \mathbf{0} + \mathbf{x}' = (-\mathbf{x} + \mathbf{x}) + \mathbf{x}' = -\mathbf{x} + (\mathbf{x} + \mathbf{x}') = -\mathbf{x} + \mathbf{0} = -\mathbf{x}. \blacksquare$$

EXERCISES*

1. Prove that $\alpha\mathbf{x} = \mathbf{0}$ implies that $\alpha = 0$ or $\mathbf{x} = \mathbf{0}$. [This is the converse of (1–23) and (1–24).]
2. Prove that $\alpha\mathbf{x} = \beta\mathbf{x}$ if and only if either $\mathbf{x} = \mathbf{0}$ or $\alpha = \beta$.
3. Establish the equality $(-\alpha)\mathbf{x} = -(\alpha\mathbf{x})$ for any α and any $\mathbf{x}$.

---

* Exercises marked with an asterisk are intended to be somewhat more challenging than the rest.

*4. Prove that the commutativity of addition in a vector space is a consequence of the remaining axioms, as follows:

(a) Use Axioms (i), (iii), and (iv) to prove that $\mathbf{x} + \mathbf{z} = \mathbf{y} + \mathbf{z}$ implies $\mathbf{x} = \mathbf{y}$. (This result is sometimes known as the *Cancellation law* of vector addition.)

(b) Justify each step in the following sequence of equalities:

$$(\mathbf{0} + \mathbf{x}) + (-\mathbf{x}) = \mathbf{0} + [\mathbf{x} + (-\mathbf{x})] = \mathbf{0} + \mathbf{0} = \mathbf{x} + (-\mathbf{x}).$$

Now apply (a) to show that $\mathbf{0} + \mathbf{x} = \mathbf{x}$ for all $\mathbf{x}$ in $\mathcal{V}$.

(c) With the aid of the result just proved, justify each step in the following sequence of equalities:

$$[(-\mathbf{x}) + \mathbf{x}] + (-\mathbf{x}) = (-\mathbf{x}) + [\mathbf{x} + (-\mathbf{x})] = -\mathbf{x} + \mathbf{0} = -\mathbf{x} = \mathbf{0} + (-\mathbf{x}).$$

Now apply (a) to show that $-\mathbf{x} + \mathbf{x} = \mathbf{0}$ for all $\mathbf{x}$ in $\mathcal{V}$.

(d) Use (b) and (c) to prove that $\mathbf{z} + \mathbf{x} = \mathbf{z} + \mathbf{y}$ implies $\mathbf{x} = \mathbf{y}$.

(e) Use Axioms (v) and (vi) of Definition 1–1 to expand $(1 + 1)(\mathbf{x} + \mathbf{y})$ in two ways, and then apply (a) and (d) to deduce that $\mathbf{x} + \mathbf{y} = \mathbf{y} + \mathbf{x}$ for all $\mathbf{x}$ and $\mathbf{y}$.

1–4 SUBSPACES

Now that we are well armed with examples, we begin the systematic study of real vector spaces. To do so, we introduce the important notion of a subspace of a vector space.

Definition 1–2. A subset $\mathcal{W}$ of a vector space $\mathcal{V}$ is said to be a *subspace* of $\mathcal{V}$ if $\mathcal{W}$ itself is a vector space under the operations of addition and scalar multiplication defined in $\mathcal{V}$.†

Before giving examples, we consider the problem of determining when a given collection of vectors actually is a subspace. One way of doing this, of course, is to verify that the set of vectors in question satisfies all of the requirements of Definition 1–1. However, the step by step verification of the axioms in this definition is both time consuming and tedious, and we now show how this procedure can be substantially shortened. Specifically, we establish the following:

Subspace Criterion. *If every vector of the form*

$$\alpha_1 \mathbf{x}_1 + \alpha_2 \mathbf{x}_2 \tag{1–25}$$

belongs to $\mathcal{W}$ *whenever* $\mathbf{x}_1$ *and* $\mathbf{x}_2$ *belong to* $\mathcal{W}$, *and* α_1 *and* α_2 *are arbitrary scalars, then* $\mathcal{W}$ *is a subspace of* $\mathcal{V}$.

---

† The term "subset" as used here and in similar contexts in the future means that every vector in $\mathcal{W}$ also belongs to $\mathcal{V}$. Moreover, we always assume that there is at least one vector in $\mathcal{W}$—a fact which is sometimes expressed by saying that $\mathcal{W}$ is *nonempty*.

To prove this assertion, we must show that $\mathcal{W}$ satisfies Definition 1–1. But by first setting $\alpha_1 = \alpha_2 = 1$ in (1–25), and then setting $\alpha_2 = 0$, we deduce in turn that (a) the sum of any two vectors in $\mathcal{W}$ again belongs to $\mathcal{W}$, and (b) $\alpha\mathbf{x}$ belongs to $\mathcal{W}$ for every real number α and every $\mathbf{x}$ in $\mathcal{W}$.*

From (b) it follows in particular that $-\mathbf{x}$ belongs to $\mathcal{W}$ whenever $\mathbf{x}$ does, and that $\mathcal{W}$ also contains the zero vector. Thus $\mathcal{W}$ satisfies Axioms (iii) and (iv) of Definition 1–1. Finally, we observe that the remaining axioms certainly hold in $\mathcal{W}$, since they are valid everywhere in $\mathcal{V}$. Hence $\mathcal{W}$ is a subspace of $\mathcal{V}$. ▮

EXAMPLE 1. Every vector space has two subspaces: (a) the whole space, and (b) the subspace consisting of the zero vector by itself, called the *trivial subspace*. A subspace of $\mathcal{V}$ which is distinct from $\mathcal{V}$ is called a *proper subspace*.

EXAMPLE 2. If $\mathcal{W}$ is the subset of $\mathcal{R}^3$ consisting of all those vectors whose third component is zero, then the above criterion implies at once that $\mathcal{W}$ is a subspace of $\mathcal{R}^3$. When the components of each vector in $\mathcal{R}^3$ are viewed as its ordinary x, y, z-components, then $\mathcal{W}$ is just the (x, y)-plane in 3-space.

EXAMPLE 3. Let $\mathcal{C}^1[a, b]$ denote the set of all functions which possess a continuous derivative at every point of the interval $[a, b]$; i.e., the so-called *continuously differentiable functions* on $[a, b]$. Since a differentiable function is continuous, each function in $\mathcal{C}^1[a, b]$ also belongs to $\mathcal{C}[a, b]$. But both the scalar multiple of a continuously differentiable function and the sum of two such functions are continuously differentiable. Hence $\mathcal{C}^1[a, b]$ is closed under addition and scalar multiplication and thus is a subspace of $\mathcal{C}[a, b]$. More generally, if $\mathcal{C}^n[a, b]$ denotes the set of all *n times continuously differentiable functions* on $[a, b]$, then $\mathcal{C}^m[a, b]$ is a subspace of $\mathcal{C}^n[a, b]$ whenever $m \geq n$.

EXAMPLE 4. Let $\mathcal{P}_n$ be the set consisting of all polynomials of the form

$$a_0 + a_1x + \cdots + a_{n-1}x^{n-1},$$

where $a_0, \ldots, a_{n-1}$ are arbitrary real numbers, and n is a *fixed* positive integer; i.e., $\mathcal{P}_n$ consists of all polynomials with real coefficients of degree $<n$, together with the zero polynomial. Then $\mathcal{P}_n$ is a subspace of $\mathcal{P}$ and also of $\mathcal{P}_m$ whenever $m \geq n$.

Now that we know what a subspace is, it is natural to ask how one might go about finding all subspaces of a given vector space. In general, this is a hard problem, but for certain spaces the answer can readily be given. For instance, it is not difficult to show that the only *nontrivial, proper* subspaces of $\mathcal{R}^3$ are lines and planes through the origin (see Exercise 4, below). And once this observation

---

* Mathematicians summarize these two facts by saying that $\mathcal{W}$ is *closed* under vector addition and scalar multiplication. In this language the above criterion becomes the statement that *a nonempty subset of a vector space is a subspace if and only if it is closed under vector addition and scalar multiplication.*

has been made one is struck by the fact that the intersection of any two subspaces of $\mathcal{R}^3$ is again a subspace of $\mathcal{R}^3$. Actually, this is true in general, as is shown in

Lemma 1–2. *If $\mathcal{W}_1$ and $\mathcal{W}_2$ are subspaces of $\mathcal{V}$, then the set consisting of all vectors belonging to both $\mathcal{W}_1$ and $\mathcal{W}_2$ is a subspace of $\mathcal{V}$.*

Proof. Let $\mathcal{W}$ be the set in question, and note that $\mathcal{W}$ contains the zero vector since this vector belongs to both $\mathcal{W}_1$ and $\mathcal{W}_2$. Now let $\mathbf{x}_1$ and $\mathbf{x}_2$ be any two vectors in $\mathcal{W}$. Then $\mathbf{x}_1$ and $\mathbf{x}_2$ belong to $\mathcal{W}_1$ and to $\mathcal{W}_2$, and hence so does $\alpha_1\mathbf{x}_1 + \alpha_2\mathbf{x}_2$ for any pair of real numbers α_1 and α_2. This implies that $\alpha_1\mathbf{x}_1 + \alpha_2\mathbf{x}_2$ belongs to $\mathcal{W}$, and the assertion that $\mathcal{W}$ is a subspace of $\mathcal{V}$ now follows from the subspace criterion. ▮

The subspace $\mathcal{W}$ of this lemma is known as the *intersection* of $\mathcal{W}_1$ and $\mathcal{W}_2$, and is denoted $\mathcal{W}_1 \cap \mathcal{W}_2$ (read "$\mathcal{W}_1$ intersect $\mathcal{W}_2$").

We now return to the problem of finding all subspaces of an arbitrary vector space $\mathcal{V}$. Rather than attempt a frontal assault on this problem, it turns out to be much more profitable to proceed as follows: Let $\mathcal{X}$ be *any* (nonempty) subset of $\mathcal{V}$. Then, as was noted above, there is at least one *subspace* of $\mathcal{V}$ containing $\mathcal{X}$, namely $\mathcal{V}$ itself. This being so, we attempt to find the "smallest" subspace of $\mathcal{V}$ containing $\mathcal{X}$, where by this we mean *that* subspace of $\mathcal{V}$ which contains $\mathcal{X}$, and which in turn is contained in every subspace of $\mathcal{V}$ containing $\mathcal{X}$. To show that such a subspace actually exists, consider the totality of *all* subspaces of $\mathcal{V}$ which contain $\mathcal{X}$, and let $\mathcal{S}(\mathcal{X})$ denote the set of vectors belonging to *every one* of these subspaces; i.e., $\mathcal{S}(\mathcal{X})$ is the intersection of these subspaces. Reasoning as in the proof of Lemma 1–2, we see that $\mathcal{S}(\mathcal{X})$ is a subspace of $\mathcal{V}$, and from its very definition it is clear that there is no subspace of $\mathcal{V}$ which contains $\mathcal{X}$ and is *properly* contained in $\mathcal{S}(\mathcal{X})$. Thus $\mathcal{S}(\mathcal{X})$ is the desired subspace. It is called the subspace of $\mathcal{V}$ *spanned* by $\mathcal{X}$ and, as we shall see, is uniquely determined by the set $\mathcal{X}$.

All this is well and good, but unless we can discover an easy method for finding $\mathcal{S}(\mathcal{X})$ in terms of the vectors belonging to $\mathcal{X}$, we will have made little progress on the problem of surveying the subspaces of $\mathcal{V}$. Fortunately (and this is the reason for introducing $\mathcal{S}(\mathcal{X})$ in the first place) such a method is easy to derive. To do so, we introduce the following definition.

Definition 1–3. An expression of the form

$$\alpha_1\mathbf{x}_1 + \cdots + \alpha_n\mathbf{x}_n, \tag{1–26}$$

where $\alpha_1, \ldots, \alpha_n$ are real numbers, is called a *linear combination* of the vectors $\mathbf{x}_1, \ldots, \mathbf{x}_n$.

And now we can describe $\mathcal{S}(\mathcal{X})$: *it is the set of all linear combinations of the vectors in* $\mathcal{X}$. Thus once $\mathcal{X}$ is known, so is $\mathcal{S}(\mathcal{X})$, and (1–26) gives the form of each of its vectors.

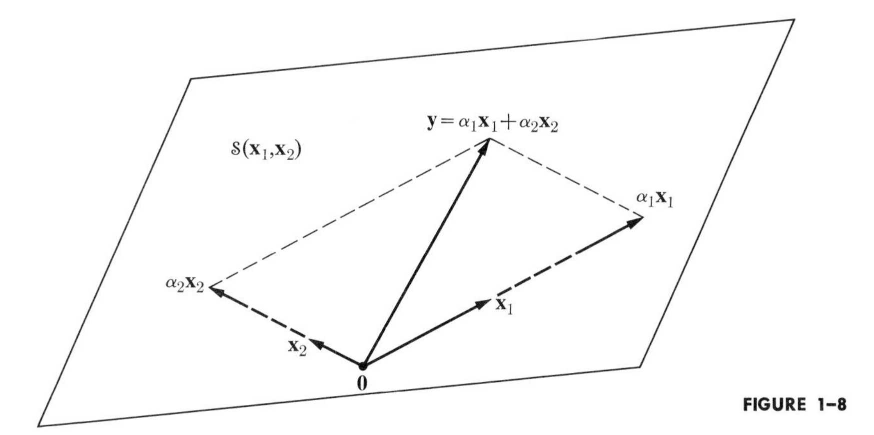

FIGURE 1–8

Before proving this assertion, let us look at some examples in $\mathcal{R}^3$. First let $\mathcal{X}$ consist of a single nonzero vector $\mathbf{x}$. Then $\mathcal{S}(\mathcal{X})$ is the line through the origin determined by $\mathbf{x}$. But the points on this line are simply all scalar multiples $\alpha\mathbf{x}$ of $\mathbf{x}$, and our assertion holds in this case. Next, let $\mathcal{X}$ consist of two nonzero, noncollinear vectors, $\mathbf{x}_1$ and $\mathbf{x}_2$. In this case, $\mathcal{S}(\mathcal{X})$ is the plane through the origin determined by $\mathbf{x}_1$ and $\mathbf{x}_2$. (Why?) But every linear combination of the form $\alpha_1\mathbf{x}_1 + \alpha_2\mathbf{x}_2$ certainly lies in this plane, and, conversely, we can "reach" any vector $\mathbf{y}$ in this plane by a linear combination of $\mathbf{x}_1$ and $\mathbf{x}_2$, as indicated in Fig. 1–8. Thus $\mathcal{S}(\mathcal{X})$ is again the set of all linear combinations of the vectors in $\mathcal{X}$.

We now prove the general result.

Theorem 1–1. *Let $\mathcal{X}$ be a (nonempty) subset of a vector space $\mathcal{V}$. Then the subspace of $\mathcal{V}$ spanned by $\mathcal{X}$ consists of all linear combinations of the vectors in $\mathcal{X}$.*

Proof. In the first place, the set of all linear combinations of vectors in $\mathcal{X}$ is closed under addition and scalar multiplication, and hence is a subspace $\mathcal{W}$ of $\mathcal{V}$. Moreover, the equation $\mathbf{x} = 1\mathbf{x}$ shows that each $\mathbf{x}$ in $\mathcal{X}$ is a linear combination of vectors in $\mathcal{X}$, thus proving that $\mathcal{X}$ is contained in $\mathcal{W}$. Finally, every subspace of $\mathcal{V}$ which contains $\mathcal{X}$ must contain all vectors of the form (1–26) by virtue of the fact that a subspace is closed under addition and scalar multiplication. In other words, $\mathcal{W}$ is contained in every subspace of $\mathcal{V}$ containing $\mathcal{X}$, and it follows that $\mathcal{W} = \mathcal{S}(\mathcal{X})$. ▮

EXERCISES

1. Determine which of the following subsets are subspaces of the indicated vector space. Give reasons for your answers.

   (a) The set of all vectors in $\mathcal{R}^2$ of the form $\mathbf{x} = (1, x_2)$.

   (b) The zero vector together with all vectors $\mathbf{x} = (x_1, x_2)$ in $\mathcal{R}^2$ for which x_2/x_1 has a constant value.

(c) The set of all vectors $\mathbf{x} = (x_1, x_2, x_3)$ in $\mathcal{R}^3$ for which $x_1 + x_2 + x_3 = 0$.

(d) The set of all vectors in $\mathcal{R}^3$ of the form $(x_1, x_2, x_1 + x_2)$.

(e) The set of all vectors $\mathbf{x} = (x_1, x_2, x_3)$ in $\mathcal{R}^3$ for which $x_1^2 + x_2^2 + x_3^2 = 1$.

2. Repeat Exercise 1 for the following subsets.

(a) The subset of $\mathcal{P}_n$ consisting of the zero polynomial and all polynomials of degree $n - 1$.

(b) The subset of $\mathcal{P}$ consisting of the zero polynomial and all polynomials of even degree.

(c) The subset of $\mathcal{P}$ consisting of the zero polynomial and all polynomials of degree 0.

(d) The subset of $\mathcal{P}_n$, $n > 1$, consisting of all polynomials which have x as a factor.

(e) The subset of $\mathcal{P}$ consisting of all polynomials which have $x - 1$ as a factor.

3. Repeat Exercise 1 for the following subsets of $\mathcal{C}[a, b]$.

(a) The set of all functions in $\mathcal{C}[a, b]$ which vanish at the point x_0 in $[a, b]$.

(b) The set of all nondecreasing functions in $\mathcal{C}[a, b]$.

(c) The set of all constant functions in $\mathcal{C}[a, b]$.

(d) The set of all functions $\mathbf{f}$ in $\mathcal{C}[a, b]$ such that $\mathbf{f}(a) = 1$.

(e) The set of all functions $\mathbf{f}$ in $\mathcal{C}[a, b]$ such that $\int_a^b \mathbf{f}(x)\,dx = 0$.

4. Prove that the only proper subspaces of $\mathcal{R}^3$ different from the trivial subspace are lines and planes through the origin.

5. Prove that $\mathcal{P}_n$ is a subspace of $\mathcal{P}$.

6. Consider the set of all infinite sequences $\{x_1, \ldots, x_n, \ldots\}$ of real numbers which have only a finite number of nonzero entries. Prove that this set is a subspace of the vector space defined in Exercise 8 of Section 1–2.

7. Determine which of the following vectors belong to the subspace of $\mathcal{R}^3$ spanned by (1, 2, 1) and (2, 3, 4).

(a) (4, 7, 6) (c) $(\frac{1}{2}, 1, 1)$ (e) (2, 9, 4)

(b) $(-\frac{1}{2}, -\frac{1}{2}, -\frac{3}{2})$ (d) (2, 9, 5) (f) $(0, \frac{1}{3}, -\frac{2}{3})$

8. Determine which of the following vectors belong to the subspace of $\mathcal{R}^3$ spanned by (1, −3, 2) and (0, 4, 1).

(a) (3, −1, 8) (c) $(\frac{1}{2}, 1, \frac{3}{2})$ (e) $(\frac{1}{3}, -1, \frac{2}{3})$

(b) (2, −2, 1) (d) $(2, -\frac{9}{2}, 0)$ (f) $(\frac{2}{3}, 3, -\frac{4}{3})$

9. Determine which of the following polynomials belongs to the subspace of $\mathcal{P}$ spanned by $x^3 + 2x^2 + 1$, $x^2 - 2$, $x^3 + x$.

(a) $x^2 - x + 3$ (c) $4x^3 - 3x + 5$ (e) $-\frac{1}{2}x^3 + \frac{5}{2}x^2 - x - 1$

(b) $x^2 - 2x + 1$ (d) $x^4 + 1$ (f) $x - 5$

10. Let $\mathbf{f}$ and $\mathbf{g}$ be the functions in $\mathcal{C}[0, 1]$ defined by

$$\mathbf{f}(x) = \begin{cases} 0 & \text{if } 0 \le x \le \frac{1}{2}, \\ x - \frac{1}{2} & \text{if } \frac{1}{2} \le x \le 1, \end{cases} \qquad \mathbf{g}(x) = \begin{cases} -x + \frac{1}{2} & \text{if } 0 \le x \le \frac{1}{2}, \\ 0 & \text{if } \frac{1}{2} \le x \le 1. \end{cases}$$

Find $\mathcal{S}(\mathbf{f})$, $\mathcal{S}(\mathbf{g})$, and $\mathcal{S}(\mathbf{f}, \mathbf{g})$.

11. Let $\mathbf{f}$ be the function whose value is 1 at every point of the interval $[a, b]$. Find the subspace of $\mathcal{C}[a, b]$ spanned by $\mathbf{f}$.

12. Find the subspace of $\mathcal{R}^3$ spanned by each of the following sets of vectors.
    (a) $(2, 1, 3)$, $(-1, 2, 1)$
    (b) $(1, 0, 2)$, $(2, 1, -2)$
    (c) $(-1, 1, 2)$, $(0, 1, 0)$, $(2, 4, 1)$

13. Find the subspace of $\mathcal{P}$ spanned by each of the following sets of vectors.
    (a) x^2, $x(x + 1)$
    (b) $x + 1$, $x^2 - 1$
    (c) 1, $x - 2$, $(x - 2)^2$

14. Prove that the intersection of any collection of subspaces of a vector space $\mathcal{V}$ is a subspace of $\mathcal{V}$. (By the *intersection* of a collection of subspaces of $\mathcal{V}$ we mean the totality of vectors common to all of the subspaces in question.)

15. (a) Prove that $\mathcal{S}(\mathcal{S}(\mathfrak{X})) = \mathcal{S}(\mathfrak{X})$ for any $\mathfrak{X}$.
    (b) Prove that $\mathcal{S}(\mathfrak{X}) = \mathfrak{X}$ if and only if $\mathfrak{X}$ is a subspace of $\mathcal{V}$.

16. Let $\mathcal{W}_1$, $\mathcal{W}_2$, and $\mathcal{W}_3$ be subspaces of a vector space, and suppose that $\mathcal{W}_1$ is a subspace of $\mathcal{W}_2$, and that $\mathcal{W}_2$ is a subspace of $\mathcal{W}_3$. Prove that $\mathcal{W}_1$ is a subspace of $\mathcal{W}_3$.

17. Let $\mathfrak{X}_1$ and $\mathfrak{X}_2$ be two sets of vectors in $\mathcal{V}$, and let $\mathfrak{X}_1 \cap \mathfrak{X}_2$ be the set of vectors belonging to $\mathfrak{X}_1$ *and* $\mathfrak{X}_2$. Furthermore, let us agree that if $\mathfrak{X}_1$ and $\mathfrak{X}_2$ have no vectors in common, $\mathcal{S}(\mathfrak{X}_1 \cap \mathfrak{X}_2)$ is the trivial subspace of $\mathcal{V}$. Show that $\mathcal{S}(\mathfrak{X}_1 \cap \mathfrak{X}_2)$ is a subspace of $\mathcal{S}(\mathfrak{X}_1) \cap \mathcal{S}(\mathfrak{X}_2)$. Give an example in $\mathcal{R}^3$ where these two subspaces are distinct, and one where they are identical.

18. Prove that the following two subsets span the same subspace of $\mathcal{R}^3$.
    (a) $(1, -1, 2)$, $(3, 0, 1)$
    (b) $(-1, -2, 3)$, $(3, 3, -4)$

19. Prove that the functions $\sin^2 x$, $\cos^2 x$, $\sin x \cos x$ span the same subspace of $\mathcal{C}[a, b]$ as 1 $\sin 2x$ and $\cos 2x$.

20. Let $\mathcal{W}_1$ and $\mathcal{W}_2$ be subspaces of $\mathcal{V}$, and let $\mathcal{W}$ denote the set of all vectors which belong to $\mathcal{W}_1$ or $\mathcal{W}_2$, or both. Prove that $\mathcal{W}$ is a subspace of $\mathcal{V}$ if and only if one of the $\mathcal{W}_i$ $(i = 1, 2)$ is contained in the other.

21. (a) Let $\mathcal{W}_1$ be the subset of $\mathcal{C}[-a, a]$ consisting of all functions $\mathbf{f}$ such that $\mathbf{f}(x) = \mathbf{f}(-x)$, and let $\mathcal{W}_2$ be the subset consisting of all $\mathbf{f}$ such that $\mathbf{f}(x) = -\mathbf{f}(-x)$. Prove that $\mathcal{W}_1$ and $\mathcal{W}_2$ are subspaces of $\mathcal{C}[-a, a]$.
    (b) Give a graphical description of the functions belonging to each of these subspaces.
    (c) What is $\mathcal{W}_1 \cap \mathcal{W}_2$?
    (d) Prove that every vector $\mathbf{f}$ in $\mathcal{C}[-a, a]$ can be written in one and only one way as a sum of a vector in $\mathcal{W}_1$ and a vector in $\mathcal{W}_2$. [*Hint:* Consider the functions $(\mathbf{f}(x) + \mathbf{f}(-x))/2$ and $(\mathbf{f}(x) - \mathbf{f}(-x))/2$.]

22. Let $\mathcal{W}_1$ and $\mathcal{W}_2$ be subspaces of $\mathcal{V}$, and let $\mathcal{W}_1 + \mathcal{W}_2$ be the set of all vectors in $\mathcal{V}$ of the form $\mathbf{x}_1 + \mathbf{x}_2$, where $\mathbf{x}_1$ belongs to $\mathcal{W}_1$ and $\mathbf{x}_2$ to $\mathcal{W}_2$. Show that $\mathcal{W}_1 + \mathcal{W}_2$ is a subspace of $\mathcal{V}$.

23. Let $\mathcal{W}_1$ be the subset of $\mathcal{P}_n$ consisting of all polynomials which have zero as a root, and let $\mathcal{W}_2$ be the subset of $\mathcal{P}_n$ consisting of the zero polynomial and all polynomials of degree zero (i.e., constant polynomials). Prove that $\mathcal{P}_n = \mathcal{W}_1 + \mathcal{W}_2$ (see Exercise 22), and show that each vector in $\mathcal{P}_n$ can be written in one and only one way as the sum of a vector in $\mathcal{W}_1$ and a vector in $\mathcal{W}_2$.

24. If $\mathfrak{X}_1$ and $\mathfrak{X}_2$ are two sets of vectors in $\mathcal{V}$, then $\mathfrak{X}_1 \cup \mathfrak{X}_2$ (read "$\mathfrak{X}_1$ *union* $\mathfrak{X}_2$") is the set of vectors belonging to $\mathfrak{X}_1$ or to $\mathfrak{X}_2$ (or both). Prove that

$$\mathcal{S}(\mathfrak{X}_1 \cup \mathfrak{X}_2) = \mathcal{S}(\mathfrak{X}_1) + \mathcal{S}(\mathfrak{X}_2),$$

and illustrate this result by examples chosen from $\mathcal{R}^3$ (see Exercise 22).

25. Let α_1 and α_2 be real numbers and consider the linear equation

$$\alpha_1 x_1 + \alpha_2 x_2 = 0$$

in the unknowns x_1 and x_2. A vector (c_1, c_2) of $\mathcal{R}^2$ is said to be a *solution* of this equation if the substitution of c_1 and c_2 for x_1 and x_2 respectively reduces this equation to an identity. Show that the set of solutions of the given equation is a subspace of $\mathcal{R}^2$. Describe this subspace graphically.

26. Show that the set of all simultaneous solutions of the pair of linear equations

$$\alpha_1 x_1 + \alpha_2 x_2 = 0 \quad \text{and} \quad \beta_1 x_1 + \beta_2 x_2 = 0$$

is a subspace of $\mathcal{R}^2$, and give a geometric description of this subspace.

27. (a) Suppose that the vector $\mathbf{x} = (c_1, c_2)$ is a solution of the pair of linear equations

$$\begin{aligned} \alpha_1 x_1 + \alpha_2 x_2 &= \gamma_1, \\ \beta_1 x_1 + \beta_2 x_2 &= \gamma_2. \end{aligned} \tag{I}$$

Prove that every solution of this pair of equations is of the form $\mathbf{x} + \mathbf{y}$, where $\mathbf{y}$ is a solution of

$$\begin{aligned} \alpha_1 x_1 + \alpha_2 x_2 &= 0, \\ \beta_1 x_1 + \beta_2 x_2 &= 0. \end{aligned} \tag{II}$$

(b) Conversely, with $\mathbf{x}$ and $\mathbf{y}$ as in (a), prove that every vector of the form $\mathbf{x} + \mathbf{y}$ is a solution of (I).

(c) Give a geometric description of the solutions of (II), and then use it and the above results to obtain a description of the solutions of (I).

*28. Let $\mathfrak{X}$ be an arbitrary subset of a vector space $\mathcal{V}$, and let $\mathbf{x}$ and $\mathbf{y}$ be vectors in $\mathcal{V}$. Suppose that $\mathbf{x}$ belongs to the subspace $\mathcal{S}(\mathfrak{X}, \mathbf{y})$ but not to $\mathcal{S}(\mathfrak{X})$. Prove that $\mathbf{y}$ then belongs to $\mathcal{S}(\mathfrak{X}, \mathbf{x})$. (This result is sometimes known as the *exchange principle*.)

1–5 LINEAR DEPENDENCE AND INDEPENDENCE; BASES

Consider the subspace $\mathcal{S}(\mathbf{x}_1, \mathbf{x}_2, \mathbf{x}_3)$ spanned by three nonzero, coplanar vectors in $\mathcal{R}^3$, no two of which are collinear (Fig. 1–9). In this case it is perfectly clear that the given set contains more vectors than are needed to span the plane $\mathcal{S}(\mathbf{x}_1, \mathbf{x}_2, \mathbf{x}_3)$, since any two of them suffice in this respect. But at least two vectors are always necessary to span a plane in $\mathcal{R}^3$, and hence we obtain a "minimal" subset of $\mathbf{x}_1, \mathbf{x}_2, \mathbf{x}_3$ which spans $\mathcal{S}(\mathbf{x}_1, \mathbf{x}_2, \mathbf{x}_3)$ by discarding any one of the given vectors. This example suggests that it may be possible to reduce any finite set of vectors $\mathbf{x}_1, \ldots, \mathbf{x}_n$ to a minimal subset which continues to span $\mathcal{S}(\mathbf{x}_1, \ldots, \mathbf{x}_n)$. This is in fact the case, as we show in Theorem 1–2, but before doing so we introduce some useful terminology.

Definition 1–4. A vector $\mathbf{x}$ is said to be *linearly dependent* on $\mathbf{x}_1, \ldots, \mathbf{x}_n$ if $\mathbf{x}$ can be written in the form

$$\mathbf{x} = \alpha_1\mathbf{x}_1 + \cdots + \alpha_n\mathbf{x}_n,$$

where the α_i are scalars. If, on the other hand, no such relation exists, $\mathbf{x}$ is said to be *linearly independent* of $\mathbf{x}_1, \ldots, \mathbf{x}_n$.

Thus $\mathbf{x}$ is linearly dependent on $\mathbf{x}_1, \ldots, \mathbf{x}_n$ if and only if $\mathbf{x}$ is a linear combination of $\mathbf{x}_1, \ldots, \mathbf{x}_n$ (Definition 1–3), and hence if and only if $\mathbf{x}$ belongs to the subspace spanned by them (Theorem 1–1). In particular, each of the vectors $\mathbf{x}_i$, $1 \leq i \leq n$, is linearly dependent on $\mathbf{x}_1, \ldots, \mathbf{x}_n$, since it belongs to the subspace spanned by these vectors. We also note that the equation $\mathbf{0} = 0\mathbf{x}$ implies that *the zero vector is linearly dependent on every vector.*

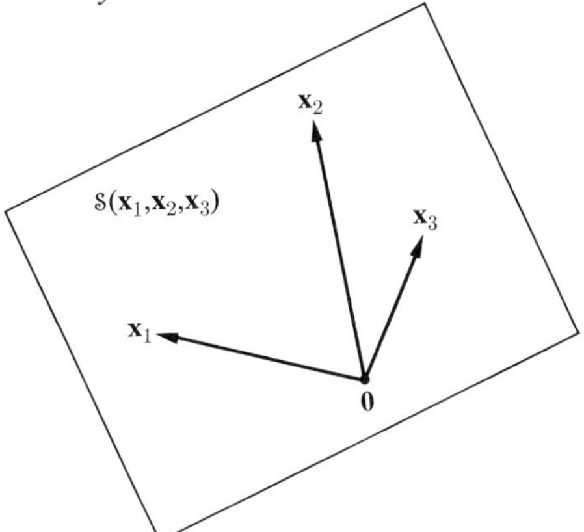

FIGURE 1–9

It is convenient to extend the terminology of Definition 1–4 to include finite *sets* of vectors by saying that such a set is *linearly independent* if no one of its vectors is linearly dependent on the remaining ones. If this is not the case, we say that the set is *linearly dependent*. Finally, when referring to a linearly independent (or dependent) set $\mathbf{x}_1, \ldots, \mathbf{x}_n$, we shall often relax our terminology and say that the vectors themselves are linearly independent (or dependent).

How does one determine in practice whether a set of vectors is linearly dependent or independent? The easiest method is given by the following test for linear independence.

Test for linear independence. *The vectors* $\mathbf{x}_1, \ldots, \mathbf{x}_n$ *are linearly independent if and only if the equation*

$$\alpha_1\mathbf{x}_1 + \cdots + \alpha_n\mathbf{x}_n = \mathbf{0}$$

implies that $\alpha_1 = \cdots = \alpha_n = 0$.

For instance, $\mathbf{x}_1 = (1, 3, -1, 2)$, $\mathbf{x}_2 = (2, 0, 1, 3)$, $\mathbf{x}_3 = (-1, 1, 0, 0)$ are linearly independent in $\mathcal{R}^4$ since the equation

$$\alpha_1\mathbf{x}_1 + \alpha_2\mathbf{x}_2 + \alpha_3\mathbf{x}_3 = \mathbf{0}$$

implies that

$$\begin{aligned} \alpha_1 + 2\alpha_2 - \alpha_3 &= 0, \\ 3\alpha_1 \qquad\quad + \alpha_3 &= 0, \\ -\alpha_1 + \alpha_2 \qquad\quad &= 0, \\ 2\alpha_1 + 3\alpha_2 \qquad\quad &= 0, \end{aligned}$$

from which it easily follows that $\alpha_1 = \alpha_2 = \alpha_3 = 0$.

We leave the proof of the above test as an exercise (see Exercise 9 below), with the strong recommendation that it be done.

And now we are ready to show how one can weed the extraneous vectors from any finite set $\mathbf{x}_1, \ldots, \mathbf{x}_n$ without disturbing $\mathcal{S}(\mathbf{x}_1, \ldots, \mathbf{x}_n)$. The basic idea is obvious; just get rid of as many linearly dependent vectors as possible.

To accomplish this we begin with the vector $\mathbf{x}_n$. If $\mathbf{x}_n$ is linearly dependent on $\mathbf{x}_1, \ldots, \mathbf{x}_{n-1}$, then

$$\mathbf{x}_n = \alpha_1\mathbf{x}_1 + \cdots + \alpha_{n-1}\mathbf{x}_{n-1},$$

and we can rewrite the expression

$$\mathbf{x} = \beta_1\mathbf{x}_1 + \cdots + \beta_n\mathbf{x}_n$$

for an arbitrary vector in $\mathcal{S}(\mathbf{x}_1, \ldots, \mathbf{x}_n)$ in the form

$$\mathbf{x} = (\beta_1 + \alpha_1\beta_n)\mathbf{x}_1 + \cdots + (\beta_{n-1} + \alpha_{n-1}\beta_n)\mathbf{x}_{n-1}.$$

This proves that $\mathbf{x}$ is already a linear combination of $\mathbf{x}_1, \ldots, \mathbf{x}_{n-1}$, and hence that $\mathcal{S}(\mathbf{x}_1, \ldots, \mathbf{x}_{n-1}) = \mathcal{S}(\mathbf{x}_1, \ldots, \mathbf{x}_n)$. In this case we drop the vector $\mathbf{x}_n$ from the set $\mathbf{x}_1, \ldots, \mathbf{x}_n$. If, on the other hand, $\mathbf{x}_n$ is *not* linearly dependent on $\mathbf{x}_1, \ldots, \mathbf{x}_{n-1}$, we keep it.

If we repeat this procedure with each of the $\mathbf{x}_i$ in turn, dropping $\mathbf{x}_i$ if it is linearly dependent on the remaining vectors in the (possibly modified) set, keeping it otherwise, it is clear that we obtain a linearly independent subset of $\mathbf{x}_1, \ldots, \mathbf{x}_n$ which spans the subspace $\mathcal{S}(\mathbf{x}_1, \ldots, \mathbf{x}_n)$. This, of course, is what we started out to show, and we have proved

Theorem 1–2. *Every finite set of vectors* $\mathfrak{X}$ *contains a linearly independent subset which spans the subspace* $\mathcal{S}(\mathfrak{X})$.

(Note, however, that in general $\mathcal{X}$ contains many such subsets. This was the case for instance in the example given at the beginning of this section. Other examples will be found in the exercises below.)

Linearly independent sets enjoy a special status in the study of vector spaces, and among such sets those which span the entire space are particularly important. Such sets are named in

Definition 1–5. A finite linearly independent subset $\mathcal{B}$ of a vector space $\mathcal{V}$ is said to be a *basis* for $\mathcal{V}$ if $\mathcal{S}(\mathcal{B}) = \mathcal{V}$.

As an example, we cite the vectors $\mathbf{i} = (1, 0, 0)$, $\mathbf{j} = (0, 1, 0)$, $\mathbf{k} = (0, 0, 1)$, which form a basis for $\mathcal{R}^3$. We shall prove this assertion in Example 1 below, and now merely wish to observe that every vector $\mathbf{x} = (x_1, x_2, x_3)$ in $\mathcal{R}^3$ can be written in *one and only one* way as a linear combination of these basis vectors, namely, $\mathbf{x} = x_1\mathbf{i} + x_2\mathbf{j} + x_3\mathbf{k}$. This last property actually serves to characterize a basis in a vector space, as we now show.

Theorem 1–3. *A set of vectors $\mathbf{e}_1, \ldots, \mathbf{e}_n$ is a basis for a vector space $\mathcal{V}$ if and only if every vector in $\mathcal{V}$ can be written uniquely as a linear combination of $\mathbf{e}_1, \ldots, \mathbf{e}_n$.*

Proof. First suppose that $\mathbf{e}_1, \ldots, \mathbf{e}_n$ is a basis for $\mathcal{V}$. Then the $\mathbf{e}_i$ span $\mathcal{V}$, and hence every vector in $\mathcal{V}$ can be written in *at least one* way as

$$\mathbf{x} = \alpha_1\mathbf{e}_1 + \cdots + \alpha_n\mathbf{e}_n. \tag{1–27}$$

To show that this is the only such expression possible, let

$$\mathbf{x} = \beta_1\mathbf{e}_1 + \cdots + \beta_n\mathbf{e}_n \tag{1–28}$$

be another. Then, subtracting (1–28) from (1–27), we obtain

$$\mathbf{0} = (\alpha_1 - \beta_1)\mathbf{e}_1 + \cdots + (\alpha_n - \beta_n)\mathbf{e}_n. \tag{1–29}$$

But since $\mathbf{e}_1, \ldots, \mathbf{e}_n$ is a basis for $\mathcal{V}$, these vectors are linearly independent. Hence, by our test for linear independence, each of the coefficients in (1–29) is zero, and it follows that $\alpha_1 = \beta_1, \ldots, \alpha_n = \beta_n$, as desired.

Conversely, suppose every vector in $\mathcal{V}$ can be written *uniquely* as a linear combination of $\mathbf{e}_1, \ldots, \mathbf{e}_n$. Then these vectors certainly span $\mathcal{V}$, and we need only prove their linear independence in order to show that they are a basis for $\mathcal{V}$. To accomplish this, we observe that $\mathbf{0} = 0\mathbf{e}_1 + \cdots + 0\mathbf{e}_n$ and that our assumption concerning the uniqueness of such expressions implies that this is the *only* representation of $\mathbf{0}$ as a linear combination of $\mathbf{e}_1, \ldots, \mathbf{e}_n$. Thus if $\alpha_1\mathbf{e}_1 + \cdots + \alpha_n\mathbf{e}_n = \mathbf{0}$, we must have $\alpha_1 = \cdots = \alpha_n = 0$, and the test for linear independence now applies. ▮

EXAMPLE 1. The vectors

$$\begin{aligned}\mathbf{e}_1 &= (1, 0, \ldots, 0),\\ \mathbf{e}_2 &= (0, 1, \ldots, 0),\\ &\vdots\\ \mathbf{e}_n &= (0, 0, \ldots, 1)\end{aligned}$$

are a basis for $\mathcal{R}^n$, since $\mathbf{x} = x_1\mathbf{e}_1 + \cdots + x_n\mathbf{e}_n$ is the only way of expressing the vector $\mathbf{x} = (x_1, \ldots, x_n)$ as a linear combination of $\mathbf{e}_1, \ldots, \mathbf{e}_n$. This particular basis is called the *standard basis* for $\mathcal{R}^n$.

EXAMPLE 2. Again in $\mathcal{R}^n$, let

$$\begin{aligned}\mathbf{e}'_1 &= (1, 0, \ldots, 0),\\ \mathbf{e}'_2 &= (1, 1, \ldots, 0),\\ &\vdots\\ \mathbf{e}'_n &= (1, 1, \ldots, 1),\end{aligned}$$

where, in general, $\mathbf{e}'_i$ is the n-tuple having 1's in the first i places and 0's thereafter. Then $\mathbf{e}'_1, \ldots, \mathbf{e}'_n$ is a basis for $\mathcal{R}^n$. To prove this let $\mathbf{x} = (x_1, \ldots, x_n)$ be given, and let us attempt to find real numbers $\alpha_1, \ldots, \alpha_n$ such that $\mathbf{x} = \alpha_1\mathbf{e}'_1 + \cdots + \alpha_n\mathbf{e}'_n$. In order that such an equality hold we must have

$$\begin{aligned}(x_1, \ldots, x_n) &= \alpha_1(1, 0, \ldots, 0) + \alpha_2(1, 1, \ldots, 0) + \cdots + \alpha_n(1, 1, \ldots, 1)\\ &= (\alpha_1, 0, \ldots, 0) + (\alpha_2, \alpha_2, \ldots, 0) + \cdots + (\alpha_n, \alpha_n, \ldots, \alpha_n)\\ &= (\alpha_1 + \alpha_2 + \cdots + \alpha_n, \alpha_2 + \cdots + \alpha_n, \ \ldots \ , \alpha_n),\end{aligned}$$

which leads to the system of equations

$$\begin{aligned}\alpha_1 + \alpha_2 + \cdots + \alpha_n &= x_1,\\ \alpha_2 + \cdots + \alpha_n &= x_2,\\ &\vdots\\ \alpha_n &= x_n.\end{aligned}$$

Hence

$$\begin{aligned}\alpha_1 &= x_1 - x_2,\\ \alpha_2 &= x_2 - x_3,\\ &\vdots\\ \alpha_{n-1} &= x_{n-1} - x_n,\\ \alpha_n &= x_n,\end{aligned}$$

which simultaneously shows that $\mathbf{x}$ can be written as a linear combination of $\mathbf{e}'_1, \ldots, \mathbf{e}'_n$, and that the coefficients of this relation are uniquely determined. Thus the $\mathbf{e}'_i$ are a basis for $\mathcal{R}^n$, as asserted.

EXAMPLE 3. The polynomials $1, x, x^2, \ldots, x^{n-1}$ form a basis for the vector space $\mathcal{P}_n$ since each polynomial in this space can be written in one and only one way in the form $a_0 + a_1x + \cdots + a_{n-1}x^{n-1}$.

EXAMPLE 4. Let $\mathbf{p}_1(x), \ldots, \mathbf{p}_n(x)$ be any finite set of polynomials in $\mathcal{P}$, and let d be the maximum of the degrees of the $\mathbf{p}_i(x)$. Then no linear combination of these polynomials is of degree greater than d, from which it follows that $\mathbf{p}_1(x), \ldots, \mathbf{p}_n(x)$ is *not* a basis for $\mathcal{P}$, since $\mathcal{P}$ contains polynomials of arbitrarily high degree. *Thus* $\mathcal{P}$ *does not possess a basis* in the sense of Definition 1-5.

In Examples 1 and 2 we exhibited two distinct bases for $\mathcal{R}^n$, and in each case found that the total number of vectors involved was the same. This was no coincidence, for it can be shown that any two bases in a vector space $\mathcal{V}$ always have the same number of elements. In other words, *the number of vectors in a basis for* $\mathcal{V}$ (provided $\mathcal{V}$ has a basis) *is an intrinsic property of* $\mathcal{V}$ *itself.* We shall prove this important result in Section 7, along with certain other facts about bases in vector spaces, and mention it here only in order to justify the following definition.

Definition 1-6. A vector space is said to be of *dimension n* if it has a basis consisting of n vectors, and is said to be *infinite dimensional* otherwise.* We denote the fact that $\mathcal{V}$ is n-dimensional by writing $\dim \mathcal{V} = n$.

On the strength of the above examples we can assert that both $\mathcal{R}^n$ and $\mathcal{P}_n$ are n-dimensional, and that $\mathcal{P}$ is infinite dimensional.

EXERCISES

1. Find all linearly independent subsets of the following sets of vectors in $\mathcal{R}^3$.
   (a) $(1, 0, 0)$, $(0, 1, 0)$, $(0, 0, 1)$, $(2, 3, 5)$
   (b) $(1, 1, 1)$, $(0, 1, 1)$, $(0, 0, 1)$, $(6, 4, 7)$
   (c) $(1, 1, 1)$, $(2, 2, 2)$, $(1, 0, 0)$, $(0, 0, 1)$
   (d) $(0, 0, 0)$, $(1, 2, 1)$, $(1, 3, 3)$, $(1, 4, 6)$
2. Find all linearly independent subsets of the following sets of vectors in $\mathcal{P}_4$.
   (a) $1, x - 1, x^2 + 2x + 1, x^2$
   (b) $x(x - 1), x^3, 2x^3 - x^2, x$
   (c) $2x, x^2 + 1, x + 1, x^2 - 1$
   (d) $1, x, x^2, x^3, x^2 + x + 1$

---

* By convention, the vector space consisting of just the zero vector is assigned dimension 0.

3. Are the vectors $(0, 2, -1)$, $(0, \frac{1}{2}, -\frac{1}{2})$, $(0, \frac{2}{3}, -\frac{1}{3})$ linearly independent in $\mathcal{R}^3$? If not, find a linearly independent subset which has a maximum number of elements.
4. Prove that each of the following sets of vectors is a basis for $\mathcal{R}^2$.
   (a) $(1, 0)$, $(0, -1)$
   (b) $(\cos\theta, \sin\theta)$, $(-\sin\theta, \cos\theta)$, $0 \leq \theta \leq 2\pi$
   (c) $(\alpha, 0)$, $(0, \beta)$, α, β nonzero real numbers
   (d) $(1, 1)$, $(0, 1)$
5. Express each of the following vectors in $\mathcal{R}^2$ as a linear combination of the vectors in the various bases of Exercise 4:
$$\mathbf{i},\ \ \mathbf{j},\ \ \mathbf{i} + \mathbf{j},\ \ \alpha'\mathbf{i} + \beta'\mathbf{j}.$$
   [Recall that $\mathbf{i} = (1, 0)$ and $\mathbf{j} = (0, 1)$.]
6. Prove that each of the following sets of vectors is a basis for $\mathcal{R}^4$.
   (a) $(1, 0, 0, 0)$, $(0, 1, 0, 0)$, $(0, 0, 1, 0)$, $(1, 1, 1, 1)$
   (b) $(1, 1, 0, 0)$, $(0, 0, 1, 1)$, $(-1, 0, 1, 1)$, $(0, -1, 0, 1)$
   (c) $(2, -1, 0, 1)$, $(1, 3, 2, 0)$, $(0, -1, -1, 0)$, $(-2, 1, 2, 1)$
   (d) $(1, -1, 2, 0)$, $(1, 1, 2, 0)$, $(3, 0, 0, 1)$, $(2, 1, -1, 0)$
7. Express $(2, -2, 1, 3)$ as a linear combination of the vectors in the various bases of Exercise 6.
8. (a) Show that the functions 1, $\sin^2 x$, $\cos^2 x$ are linearly dependent in $\mathcal{C}[-\pi, \pi]$.
   (b) Show that the functions 1, $\cos x$, $\cos 2x$ are linearly independent in $\mathcal{C}[-\pi, \pi]$.
9. Prove that the vectors $\mathbf{x}_1, \ldots, \mathbf{x}_n$ are linearly independent if and only if the equation
$$\alpha_1\mathbf{x}_1 + \cdots + \alpha_n\mathbf{x}_n = \mathbf{0}$$
   implies that $\alpha_1 = \cdots = \alpha_n = 0$.
10. Prove that the vectors (a, b) and (c, d) are linearly independent in $\mathcal{R}^2$ if and only if $ad - bc \neq 0$.

*11. Prove that the vectors (x_1, x_2, x_3), (y_1, y_2, y_3), (z_1, z_2, z_3) are linearly independent in $\mathcal{R}^3$ if and only if
$$\begin{vmatrix} x_1 & y_1 & z_1 \\ x_2 & y_2 & z_2 \\ x_3 & y_3 & z_3 \end{vmatrix} \neq 0.$$

*12. Show that the functions $\sin x, \sin 2x, \ldots, \sin nx$ are linearly independent in $\mathcal{C}[-\pi, \pi]$ for any positive integer n. [*Hint:* Let $\sum_{k=1}^{n} \alpha_k \sin kx = 0$; multiply by $\sin jx$, where $j = 1, \ldots, n$, and integrate from $-\pi$ to π.]

*13. Show that the functions $1, \sin x, \cos x, \sin 2x, \cos 2x, \ldots, \sin nx, \cos nx$ are linearly independent in $\mathcal{C}[-\pi, \pi]$. [*Hint:* See Exercise 12.]

14. Prove that the polynomials
$$1,\ \ x,\ \ \tfrac{3}{2}x^2 - \tfrac{1}{2},\ \ \tfrac{5}{2}x^3 - \tfrac{3}{2}x$$
   form a basis for $\mathcal{P}_4$.
15. Express x^2 and x^3 as linear combinations of the basis vectors for $\mathcal{P}_4$ given in Exercise 14.

16. Assume that $\mathbf{x}_1$, $\mathbf{x}_2$, and $\mathbf{x}_3$ are linearly independent vectors in $\mathcal{V}$. Prove that $\mathbf{x}_1 + \mathbf{x}_2$, $\mathbf{x}_1 + \mathbf{x}_3$, and $\mathbf{x}_2 + \mathbf{x}_3$ are linearly independent.

17. Prove that the functions $1, e^x, e^{2x}$ are linearly independent in $\mathcal{C}[0, 1]$. [*Hint:* Differentiate the expression $\alpha + \beta e^x + \gamma e^{2x} = 0$.]

18. Prove that the functions $1, e^x, xe^x$ are linearly independent in $\mathcal{C}[0, 1]$.

19. Show that the polynomials

$$1, \quad x - a, \quad (x - a)^2, \quad \ldots, \quad (x - a)^{n-1},$$

where a is an arbitrary real number, form a basis for $\mathcal{P}_n$. [*Hint:* Consider the Taylor series expansion of a polynomial about the point $x = a$.]

*20. Show that the polynomials

$$1, \quad x, \quad x(x - 1), \quad x(x - 1)(x - 2), \quad \ldots, \quad x(x - 1)(x - 2) \cdots (x - n + 1)$$

form a basis for $\mathcal{P}_{n+1}$. [*Hint:* Use mathematical induction.]

21. Express the polynomials x^3 and $x^3 + 3x - 1$ as linear combinations of the basis vectors in $\mathcal{P}_4$ described in Exercise 20.

*22. Let $\mathcal{X}_n$ be any set of n polynomials in $\mathcal{P}_n$, one of each degree $0, 1, \ldots, n - 1$. Prove that $\mathcal{X}_n$ is a basis for $\mathcal{P}_n$. [*Hint:* Use mathematical induction.]

23. Let $\mathcal{X}_4$ be the basis for $\mathcal{P}_4$ consisting of the cubic polynomial $x^3 + 2x + 5$ and its first three derivatives (see Exercise 22). Write each of the following polynomials $\mathcal{X}$ is a as a linear combination of polynomials in $\mathcal{X}_4$.
    (a) $x^3 + 2x + 5$
    (b) $x^2 + 1$
    (c) $2x^3 - x^2 + 10x + 2$

24. Let $\mathcal{X}$ be a finite linearly independent subset of a vector space $\mathcal{V}$, and suppose that every finite subset of $\mathcal{V}$ which properly contains $\mathcal{X}$ is linearly dependent. Prove that $\mathcal{X}$ is a basis for $\mathcal{V}$.

25. Let $\mathcal{X}$ be a finite subset of a vector space $\mathcal{V}$ which spans $\mathcal{V}$, and suppose that no *proper* subset of $\mathcal{X}$ spans $\mathcal{V}$. Prove that $\mathcal{X}$ is a basis for $\mathcal{V}$.

26. Let $\mathbf{e}_1, \ldots, \mathbf{e}_n$ be a basis for $\mathcal{V}$. Prove that $\mathbf{e}_1, \alpha\mathbf{e}_1 + \mathbf{e}_2, \ldots, \alpha\mathbf{e}_1 + \mathbf{e}_n$ is also a basis for $\mathcal{V}$ for every real number α.

*27. Prove that every basis for $\mathcal{R}^3$ contains exactly three vectors. [*Hint:* Let $\mathbf{e}_1, \ldots, \mathbf{e}_n$ be any basis in $\mathcal{R}^3$, and express each $\mathbf{e}_i$ as a linear combination of the standard basis vectors $\mathbf{i}, \mathbf{j}, \mathbf{k}$ as

$$\begin{aligned}
\mathbf{e}_1 &= \alpha_{11}\mathbf{i} + \alpha_{21}\mathbf{j} + \alpha_{31}\mathbf{k}, \\
\mathbf{e}_2 &= \alpha_{12}\mathbf{i} + \alpha_{22}\mathbf{j} + \alpha_{32}\mathbf{k}, \\
&\vdots \\
\mathbf{e}_n &= \alpha_{1n}\mathbf{i} + \alpha_{2n}\mathbf{j} + \alpha_{3n}\mathbf{k}.
\end{aligned}$$

Use the fact that none of the $\mathbf{e}_i$ are zero to successively eliminate $\mathbf{i}$, $\mathbf{j}$, and $\mathbf{k}$ from these equations, and conclude that $n \leq 3$. Now reverse the argument to show that $3 \leq n$.]

1-6 COORDINATE SYSTEMS

Definition 1-7. Let $\mathbf{e}_1, \ldots, \mathbf{e}_n$ be a basis for $\mathcal{V}$ and let

$$\mathbf{x} = \alpha_1\mathbf{e}_1 + \cdots + \alpha_n\mathbf{e}_n$$

be the unique expression for $\mathbf{x}$ in terms of this basis (Theorem 1-3). Then the scalars $\alpha_1, \ldots, \alpha_n$ are called the *coordinates* or *components* of $\mathbf{x}$ with respect to $\mathbf{e}_1, \ldots, \mathbf{e}_n$, and the basis vectors themselves are said to form a *coordinate system* for $\mathcal{V}$. Finally, the subspaces of $\mathcal{V}$ spanned by each of the $\mathbf{e}_i$ are called the *coordinate axes* of the given coordinate system.

Thus a basis is a coordinate system, and the unique expression for a vector as a linear combination of basis vectors is nothing other than the "decomposition" of the vector into its components along the various coordinate axes. In these terms the direct statement in Theorem 1-3 assumes the following eminently reasonable form: *The coordinates of a vector are uniquely determined by the coordinate system.*

At the same time, we caution the student not to expect too much from a coordinate system, and especially not to fall into the error of expecting coordinate axes to be mutually perpendicular. Strictly speaking, of course, the concept of perpendicularity in a vector space has no meaning yet, but it will be defined in Chapter 7. Nevertheless, it is common knowledge that certain coordinate axes, such as the standard ones in $\mathcal{R}^n$, are mutually perpendicular. We merely wish to emphasize the sometime nature of this phenomenon, and call attention to the existence of "oblique" coordinate systems. One such is the coordinate system $\mathbf{e}'_1, \ldots, \mathbf{e}'_n$ for $\mathcal{R}^n$ introduced in Example 2 of the preceding section.

In this connection, it is also worth mentioning explicitly that the coordinates of a vector change with a change of coordinate system. Failure to appreciate the implications of this innocent and obvious statement often causes confusion, or worse, for the unwary. For example, the vector $\mathbf{x} = (4, 2)$ has coordinates 4, 2 with respect to the standard coordinate system $\mathbf{e}_1 = (1, 0), \mathbf{e}_2 = (0, 1)$ in $\mathcal{R}^2$, since

$$(4, 2) = 4\mathbf{e}_1 + 2\mathbf{e}_2.$$

However, if we use the coordinate system $\mathbf{e}'_1 = (1, 0)$, $\mathbf{e}'_2 = (1, 1)$, then the

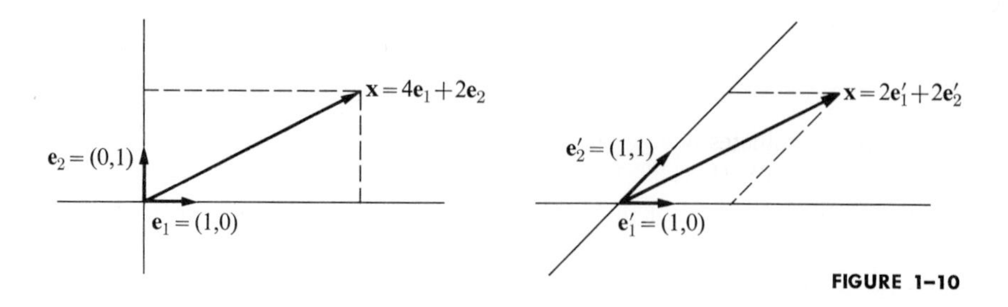

FIGURE 1-10

coordinates of $\mathbf{x}$ become 2, 2, since

$$(4, 2) = 2\mathbf{e}_1' + 2\mathbf{e}_2'.$$

Indeed, the vector in $\mathcal{R}^2$ having coordinates 4, 2 with respect to $\mathbf{e}_1'$, $\mathbf{e}_2'$ is the ordered pair (6, 2). (See Figs. 1–10 and 1–11.)

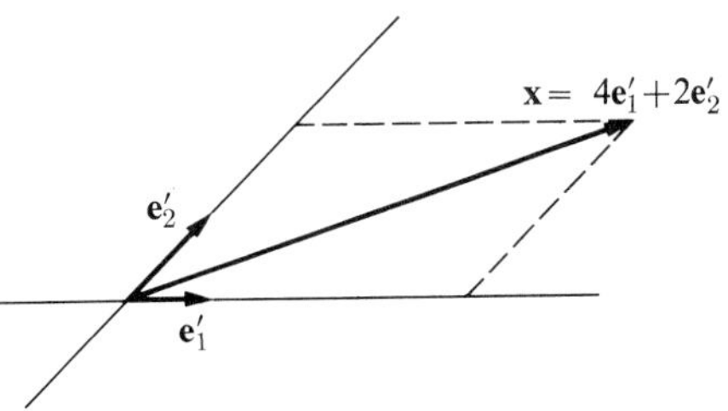

FIGURE 1–11

Finally, we call attention to the fact that the operations of vector addition and scalar multiplication are converted into ordinary addition and multiplication when they are carried out with respect to a basis. For then these operations are performed componentwise, irrespective of the nature of the vectors involved (n-tuples of real numbers, polynomials, etc.). We prove this assertion as follows:

Theorem 1–4. *Let $\mathbf{e}_1, \ldots, \mathbf{e}_n$ be any basis for a vector space $\mathcal{V}$. Then the sum of two vectors in $\mathcal{V}$ is found by adding their corresponding components, and the product of a vector and a scalar α is found by multiplying each component of the vector by α.*

Proof. If $\mathbf{x} = \alpha_1\mathbf{e}_1 + \cdots + \alpha_n\mathbf{e}_n$ and $\mathbf{y} = \beta_1\mathbf{e}_1 + \cdots + \beta_n\mathbf{e}_n$, then it follows directly from the axioms defining a vector space that

$$\mathbf{x} + \mathbf{y} = (\alpha_1 + \beta_1)\mathbf{e}_1 + \cdots + (\alpha_n + \beta_n)\mathbf{e}_n,$$

and that

$$\alpha\mathbf{x} = (\alpha\alpha_1)\mathbf{e}_1 + \cdots + (\alpha\alpha_n)\mathbf{e}_n,$$

as asserted. ▮

Among its various implications, this theorem foreshadows the use of bases in finite dimensional vector spaces whenever extensive numerical calculations are in the offing. On the other hand, as long as we are concerned with the general theory of vector spaces, bases are a distinct hindrance. This stems from the fact that whenever a basis is used in the proof of a theorem which purports to be a general statement about finite dimensional vector spaces, one must then prove that the result in question is independent of the particular basis chosen to prove it. And this is usually as difficult as it is to construct a coordinate free proof of the original statement.

EXERCISES

1. Find the coordinates of each of the following vectors in $\mathcal{R}^3$ with respect to the basis (1, 0, 0), (1, 1, 0), (1, 1, 1).
   (a) (0, 1, 0) (c) (0, 0, 1) (e) (4, −2, 2)
   (b) (−2, 1, 1) (d) $(-1, \frac{1}{2}, 1)$ (f) (1, 3, 2)
2. Repeat Exercise 1 using the basis (1, 1, 0), (1, 0, 1), (0, 1, 1).

3. Prove that the vectors (2, 1, 0), (2, 1, 1), (2, 2, 1) form a basis for $\mathcal{R}^3$. Find the vectors in $\mathcal{R}^3$ which have the following coordinates with respect to this basis.
   (a) 1, 0, 0 (c) 4, −5, 0 (e) 3, 1, −1
   (b) −1, 2, 1 (d) $\frac{1}{2}$, 2, 1 (f) 2, 2, 1
4. Find the coordinates of the standard basis vectors in $\mathcal{R}^3$ with respect to the basis given in Exercise 3.
5. Find a basis in $\mathcal{R}^4$ with respect to which the vector (−3, 1, 2, −1) has coordinates 1, 1, 1, 1.
6. Let $\mathbf{e}_1, \mathbf{e}_2, \mathbf{e}_3$ and $\mathbf{e}'_1, \mathbf{e}'_2, \mathbf{e}'_3$ be bases for a vector space $\mathcal{V}$, and suppose that

$$\mathbf{e}_1 = \alpha_1\mathbf{e}'_1 + \alpha_2\mathbf{e}'_2 + \alpha_3\mathbf{e}'_3,$$
$$\mathbf{e}_2 = \beta_1\mathbf{e}'_1 + \beta_2\mathbf{e}'_2 + \beta_3\mathbf{e}'_3,$$
$$\mathbf{e}_3 = \gamma_1\mathbf{e}'_1 + \gamma_2\mathbf{e}'_2 + \gamma_3\mathbf{e}'_3.$$

   Find the coordinates of the vector $\mathbf{x} = x_1\mathbf{e}_1 + x_2\mathbf{e}_2 + x_3\mathbf{e}_3$ with respect to the basis $\mathbf{e}'_1, \mathbf{e}'_2, \mathbf{e}'_3$.
7. Does there exist a basis for $\mathcal{R}^2$ with respect to which an arbitrary vector (x_1, x_2) has coordinates $2x_1$ and $3x_2$?
8. Find a basis for $\mathcal{R}^2$ with respect to which an arbitrary vector (x_1, x_2) has coordinates x_1 and $x_1 + 2x_2$.
9. Let $\mathbf{e}_1, \ldots, \mathbf{e}_n$ and $\mathbf{e}'_1, \ldots, \mathbf{e}'_n$ be bases for a finite dimensional vector space $\mathcal{V}$, and let

$$\mathbf{e}'_1 = \alpha_{11}\mathbf{e}_1 + \alpha_{21}\mathbf{e}_2 + \cdots + \alpha_{n1}\mathbf{e}_n,$$
$$\mathbf{e}'_2 = \alpha_{12}\mathbf{e}_1 + \alpha_{22}\mathbf{e}_2 + \cdots + \alpha_{n2}\mathbf{e}_n,$$
$$\vdots$$
$$\mathbf{e}'_n = \alpha_{1n}\mathbf{e}_1 + \alpha_{2n}\mathbf{e}_2 + \cdots + \alpha_{nn}\mathbf{e}_n.$$

   Find the coordinates of $\mathbf{x}$ with respect to $\mathbf{e}_1, \ldots, \mathbf{e}_n$ given that

$$\mathbf{x} = x'_1\mathbf{e}'_1 + x'_2\mathbf{e}'_2 + \cdots + x'_n\mathbf{e}'_n.$$

1–7 DIMENSION

Theorem 1–5. *If $\mathcal{V}$ has a basis containing n vectors, then any $n + 1$ or more vectors in $\mathcal{V}$ are linearly dependent.**

The technique used to prove this theorem has already been introduced in Exercise 27 of Section 1–5 to treat a particular case. The reader may find it helpful to keep that exercise in mind while reading the following proof.

---

* The statement of Theorem 1–5 also applies to the trivial space consisting of just the zero vector, provided we agree that the empty set of vectors is a basis for this space. Such an agreement is consistent with our definition of a basis for a vector space, and with the convention that the dimension of the trivial space is zero.

Proof. Let $\mathbf{e}_1, \ldots, \mathbf{e}_n$ be a basis for $\mathcal{V}$, and suppose, contrary to the assertion of the theorem, that $\mathcal{V}$ contains a linearly independent set $\mathbf{e}'_1, \ldots, \mathbf{e}'_m$ in which $m > n$. Express each of the $\mathbf{e}'_j$ as a linear combination of the $\mathbf{e}_i$, thereby obtaining the system of equations

$$\begin{aligned} \mathbf{e}'_1 &= \alpha_{11}\mathbf{e}_1 + \alpha_{21}\mathbf{e}_2 + \cdots + \alpha_{n1}\mathbf{e}_n, \\ \mathbf{e}'_2 &= \alpha_{12}\mathbf{e}_1 + \alpha_{22}\mathbf{e}_2 + \cdots + \alpha_{n2}\mathbf{e}_n, \\ &\vdots \\ \mathbf{e}'_m &= \alpha_{1m}\mathbf{e}_1 + \alpha_{2m}\mathbf{e}_2 + \cdots + \alpha_{nm}\mathbf{e}_n, \end{aligned} \tag{1–30}$$

in which the α_{ij} are scalars. Since none of the $\mathbf{e}'_j$ is the zero vector, at least one of the α_{ij} is different from zero in each of these equations. (Recall that the zero vector is linearly dependent on every vector in $\mathcal{V}$.) Thus, by relabeling the $\mathbf{e}_i$ if necessary, we may assume that $\alpha_{11} \neq 0$. This done, solve the first equation for $\mathbf{e}_1$, and substitute the value obtained in the remaining $m - 1$ equations. This eliminates $\mathbf{e}_1$ from (1–30), and yields a system of equations of the form

$$\begin{aligned} \mathbf{e}'_2 &= \beta_{22}\mathbf{e}_2 + \beta_{32}\mathbf{e}_3 + \cdots + \beta_{n2}\mathbf{e}_n + \beta_{12}\mathbf{e}'_1, \\ \mathbf{e}'_3 &= \beta_{23}\mathbf{e}_2 + \beta_{33}\mathbf{e}_3 + \cdots + \beta_{n3}\mathbf{e}_n + \beta_{13}\mathbf{e}'_1, \\ &\vdots \\ \mathbf{e}'_m &= \beta_{2m}\mathbf{e}_2 + \beta_{3m}\mathbf{e}_3 + \cdots + \beta_{nm}\mathbf{e}_n + \beta_{1m}\mathbf{e}'_1. \end{aligned} \tag{1–31}$$

Focusing our attention on the first of these equations we note that the linear independence of $\mathbf{e}'_1$ and $\mathbf{e}'_2$ implies that at least one of the coefficients $\beta_{22}, \beta_{32}, \ldots, \beta_{n2}$ is different from zero. Assume that the $\mathbf{e}_i$ are labeled so that $\beta_{22} \neq 0$. Then a repetition of the above argument, now applied to $\mathbf{e}_2$, reduces (1–31) to the system

$$\begin{aligned} \mathbf{e}'_3 &= \gamma_{33}\mathbf{e}_3 + \cdots + \gamma_{n3}\mathbf{e}_n + \gamma_{13}\mathbf{e}'_1 + \gamma_{23}\mathbf{e}'_2, \\ &\vdots \\ \mathbf{e}'_m &= \gamma_{3m}\mathbf{e}_3 + \cdots + \gamma_{nm}\mathbf{e}_n + \gamma_{1m}\mathbf{e}'_1 + \gamma_{2m}\mathbf{e}'_2. \end{aligned}$$

Let us now speculate on the effect of our assumption that m is greater than n. A moment's thought will reveal that by continuing the above process of elimination we will eventually find ourselves confronted with a system of $m - n$ equations expressing each of the vectors $\mathbf{e}'_{n+1}, \ldots, \mathbf{e}'_m$ as a linear combination of $\mathbf{e}'_1, \ldots, \mathbf{e}'_n$. But this cannot be. Hence $m \leq n$ after all. ▮

Corollary 1. *If $\mathcal{V}$ has a basis containing n vectors, then every basis for $\mathcal{V}$ contains n vectors.*

Proof. If $\mathbf{e}_1, \ldots, \mathbf{e}_n$ and $\mathbf{e}'_1, \ldots, \mathbf{e}'_m$ are bases for $\mathcal{V}$, then the above theorem implies that $m \leq n$, and $n \leq m$. Hence $m = n$. ▮

This result furnishes the necessary justification for the definition of the dimension of a vector space given in Section 1–5.

Theorem 1–5 also allows us to prove the reassuring fact that every subspace of a finite dimensional vector space is finite dimensional, and that its dimension does not exceed the dimension of the whole space. This is the content of

Theorem 1–6. *If $\mathcal{W}$ is a subspace of an n-dimensional vector space $\mathcal{V}$, then* $\dim \mathcal{W} \leq n$.

Proof. The theorem is obviously true if $n = 0$, or if $\mathcal{W}$ is the trivial subspace of $\mathcal{V}$.* Thus we can assume $n > 0$, and $\mathcal{W}$ nontrivial.

By virtue of this last assumption, $\mathcal{W}$ contains linearly independent sets of vectors, since any nonzero vector in $\mathcal{W}$ is, by itself, such a set. Moreover, every linearly independent set in $\mathcal{W}$ is also linearly independent as a set in $\mathcal{V}$. Thus, by Theorem 1–5, the number of vectors in such a set cannot exceed n. Finally, if $\mathbf{e}_1, \ldots, \mathbf{e}_m$ is a linearly independent set in $\mathcal{W}$ containing a *maximum* number of vectors, then $\mathcal{S}(\mathbf{e}_1, \ldots, \mathbf{e}_m) = \mathcal{W}$. Hence $\dim \mathcal{W} = m \leq n$, as advertised. ▮

This theorem may be read as asserting that every nontrivial subspace $\mathcal{W}$ of an n-dimensional space $\mathcal{V}$ has a basis $\mathbf{e}_1, \ldots, \mathbf{e}_m$ with $m \leq n$. If $m = n$, then $\mathbf{e}_1, \ldots, \mathbf{e}_m$ is also a basis for $\mathcal{V}$, and $\mathcal{W} = \mathcal{V}$. On the other hand, if $m < n$, then $\mathcal{W}$ is a *proper* subspace of $\mathcal{V}$ (i.e., $\mathcal{W} \neq \mathcal{V}$), and there exist vectors in $\mathcal{V}$ which do not belong to $\mathcal{W}$. Choose any such vector, and label it $\mathbf{e}_{m+1}$. Then it is all but obvious that $\mathbf{e}_1, \ldots, \mathbf{e}_{m+1}$ are linearly independent in $\mathcal{V}$.

To prove the truth of this observation, we apply the test for linear independence (p. 20) as follows. Suppose that

$$\alpha_1\mathbf{e}_1 + \cdots + \alpha_m\mathbf{e}_m + \alpha_{m+1}\mathbf{e}_{m+1} = \mathbf{0}. \tag{1–32}$$

Then $\alpha_{m+1} = 0$, for otherwise

$$\mathbf{e}_{m+1} = -\frac{\alpha_1}{\alpha_{m+1}}\mathbf{e}_1 - \cdots - \frac{\alpha_m}{\alpha_{m+1}}\mathbf{e}_m,$$

and $\mathbf{e}_{m+1}$ is in $\mathcal{W}$. Thus

$$\alpha_1\mathbf{e}_1 + \cdots + \alpha_m\mathbf{e}_m = \mathbf{0},$$

and it follows from the linear independence of $\mathbf{e}_1, \ldots, \mathbf{e}_m$ that $\alpha_1 = \cdots = \alpha_m = 0$. Hence all of the coefficients in (1–32) are zero, and $\mathbf{e}_1, \ldots, \mathbf{e}_{m+1}$ are linearly independent.

We now repeat the above argument, this time starting with the subspace $\mathcal{S}(\mathbf{e}_1, \ldots, \mathbf{e}_{m+1})$. If $\mathcal{S}(\mathbf{e}_1, \ldots, \mathbf{e}_{m+1})$ is a proper subspace of $\mathcal{V}$ we can enlarge $\mathbf{e}_1, \ldots, \mathbf{e}_{m+1}$ to a linearly independent set in $\mathcal{V}$ containing $m + 2$ vectors. But

---

* Recall that the dimension of the trivial space is zero.

Theorem 1–5 implies that this process must come to a halt after $n - m$ steps, at which point we will have a *basis* for $\mathcal{V}$. With this we have proved the following important and useful result.

Theorem 1–7. *Let $\mathcal{V}$ be an n-dimensional vector space, and let $\mathbf{e}_1, \ldots, \mathbf{e}_m$ be a basis for an m-dimensional subspace of $\mathcal{V}$. Then there exist $n - m$ vectors $\mathbf{e}_{m+1}, \ldots, \mathbf{e}_n$ in $\mathcal{V}$ such that $\mathbf{e}_1, \ldots, \mathbf{e}_m, \mathbf{e}_{m+1}, \ldots, \mathbf{e}_n$ is a basis for $\mathcal{V}$.*

EXERCISES

1. What is the dimension of the subspace of $\mathcal{R}^3$ spanned by
   (a) the vectors $(2, 1, -1)$, $(3, 2, 1)$, $(1, 0, -3)$?
   (b) the vectors $(1, -1, 2)$, $(0, 2, 1)$, $(-1, 0, 1)$?
2. What is the dimension of the subspace of $\mathcal{R}^4$ spanned by
   (a) the vectors $(1, 0, 2, -1)$, $(3, -1, -2, 0)$, $(1, -1, -6, 2)$, $(0, 1, 8, -3)$?
   (b) the vectors $(-\frac{1}{2}, \frac{1}{2}, 3, -1)$, $(\frac{1}{2}, 0, 1, -\frac{1}{2})$, $(1, 1, 10, -4)$?
3. Let $\mathcal{W}$ be the set of all polynomials in $\mathcal{P}_n$ whose second derivative is zero; i.e., $\mathbf{p}(x)$ belongs to $\mathcal{W}$ if and only if $(d^2/dx^2)\mathbf{p}(x) = 0$.
   (a) Prove that $\mathcal{W}$ is a subspace of $\mathcal{P}_n$, and find a basis for $\mathcal{W}$.
   (b) Extend the basis for $\mathcal{W}$ found in (a) to a basis for $\mathcal{P}_n$.
4. Let $\mathcal{W}$ be the set of all polynomials $\mathbf{p}(x)$ in $\mathcal{P}_n$ such that $\mathbf{p}(1) = \mathbf{p}'(1) = 0$.
   (a) Prove that $\mathcal{W}$ is a subspace of $\mathcal{P}_n$, and find a basis for $\mathcal{W}$.
   (b) Extend the basis for $\mathcal{W}$ found in (a) to a basis for $\mathcal{P}_n$.
5. (a) Find the dimension of the subspace of $\mathcal{C}[-\pi, \pi]$ spanned by the vectors 1, $\sin x$, $\sin^2 x$, $\cos^2 x$.
   (b) Repeat part (a) for the vectors $\sin x \cos x$, $\sin 2x$, $\cos 2x$, $\sin^2 x$, $\cos^2 x$.
6. Prove that the vector space $\mathcal{C}[a, b]$ is infinite dimensional.
7. What is the dimension of the subspace of all solutions (in $\mathcal{R}^3$) of the single linear equation
$$a_1x_1 + a_2x_2 + a_3x_3 = 0?$$
8. What is the dimension of the subspace of all solutions (in $\mathcal{R}^n$) of the single linear equation
$$a_1x_1 + a_2x_2 + \cdots + a_nx_n = 0?$$
9. Given the vectors $\mathbf{x}_1 = (2, 0, 1, 1)$ and $\mathbf{x}_2 = (1, 1, 0, 3)$ in $\mathcal{R}^4$, find vectors $\mathbf{x}_3$ and $\mathbf{x}_4$ such that $\mathbf{x}_1, \mathbf{x}_2, \mathbf{x}_3, \mathbf{x}_4$ form a basis for $\mathcal{R}^4$.
10. Let $\mathcal{V}$ be a vector space of dimension n. Prove that $\mathcal{V}$ contains a sequence of subspaces
$$\mathcal{V}_0, \mathcal{V}_1, \ldots, \mathcal{V}_n$$
    having the following two properties:
    (a) $\dim \mathcal{V}_i = i$;
    (b) $\mathcal{V}_i$ is a subspace of $\mathcal{V}_j$ whenever $i \leq j$.

11. Let $\mathcal{V}_1$ and $\mathcal{V}_2$ be finite dimensional subspaces of a vector space $\mathcal{V}$, and suppose that $\mathcal{V}_1$ and $\mathcal{V}_2$ have only the zero vector in common. Let $\mathbf{e}_1, \ldots, \mathbf{e}_m$ be a basis for $\mathcal{V}_1$, and $\mathbf{e}'_1, \ldots, \mathbf{e}'_n$ a basis for $\mathcal{V}_2$. Prove that $\mathbf{e}_1, \ldots, \mathbf{e}_m, \mathbf{e}'_1, \ldots, \mathbf{e}'_n$ is a basis for the subspace $\mathcal{W} = \mathcal{V}_1 + \mathcal{V}_2$ of $\mathcal{V}$ (cf. Exercise 22, Section 1–4). Deduce that $\dim \mathcal{W} = \dim \mathcal{V}_1 + \dim \mathcal{V}_2$.

*12. Let $\mathcal{V}_1$ and $\mathcal{V}_2$ be finite dimensional subspaces of a vector space $\mathcal{V}$. Prove that $\mathcal{V}_1 + \mathcal{V}_2$ is finite dimensional, and that

$$\dim (\mathcal{V}_1 + \mathcal{V}_2) = \dim \mathcal{V}_1 + \dim \mathcal{V}_2 - \dim (\mathcal{V}_1 \cap \mathcal{V}_2).$$

13. Let $\mathcal{V}_1$ be an m-dimensional subspace of an n-dimensional vector space $\mathcal{V}$. Prove that there exists an $(n - m)$-dimensional subspace $\mathcal{V}_2$ of $\mathcal{V}$ such that
(a) $\mathcal{V}_1 + \mathcal{V}_2 = \mathcal{V}$, and
(b) $\mathcal{V}_1 \cap \mathcal{V}_2$ contains only the zero vector. [*Hint:* Choose a basis for $\mathcal{V}_1$ and extend this to a basis for $\mathcal{V}$.]

1–8 GEOMETRIC VECTORS

Informally, geometric vectors are arrows in the plane or 3-space. As such, they are familiar to anyone who has studied elementary physics, where they appear as forces, velocities, accelerations, etc., i.e., quantities having a magnitude and direction. In this section we propose to examine some of the vague ideas associated with the use of such arrows, and make these ideas precise by constructing the space of two-dimensional geometric vectors. Besides furnishing us with still another example of a real vector space, this discussion will provide the link between our definition of the term vector and the vectors introduced in elementary calculus and physics.*

The geometric notion of an arrow in the plane finds its mathematical analogue in the concept of a *directed line segment.* Specifically, the line segment between two distinct points A and B in $\mathcal{R}^2$ is said to be *directed* if the points are given a definite order, say A, B. In this case we speak of the directed line segment *from A to B*, which we denote by $\overrightarrow{AB}$. A is then called the *initial point* of the segment, and B the *terminal point.* We also agree to regard a single point as a directed line segment, in which case the relevant symbol is $\overrightarrow{AA}$.

Intuitively the directed line segment $\overrightarrow{AB}$ has a magnitude and direction. At first sight one might be tempted to define these concepts as length and angular measure with respect to a coordinate system in the plane. However, any such definition would have the grave defect of making the magnitude and direction of $\overrightarrow{AB}$ dependent upon the coordinate system used, in conflict with the intuitive demand that they be intrinsically associated with $\overrightarrow{AB}$ itself. Unfortunately there is

* With obvious minor changes the following discussion can be adapted to 3-space or, for that matter, to n-space.

no way out of this dilemma so long as we continue to focus our attention upon a single segment. But when we turn to the set of *all* directed line segments in $\mathcal{R}^2$, we observe that it is as easy to determine when two segments have *the same* magnitude and direction as it is difficult to say what these terms mean. Indeed, using the notion of parallel translation, we can say that $\overrightarrow{AB}$ and $\overrightarrow{CD}$ have the same magnitude and direction if and only if they can be brought into coincidence by such a translation. Furthermore, the entire theory of geometric vectors can be based upon this simple observation.

We begin by giving the above discussion formal status in

Definition 1–8. Let $\overrightarrow{AB}$ and $\overrightarrow{CD}$ be directed line segments in the plane, and suppose $\overrightarrow{CD}$ is translated parallel to itself until its initial point coincides with A. Then, if the terminal points of the two segments also coincide, we say that $\overrightarrow{AB}$ and $\overrightarrow{CD}$ *have the same magnitude and direction*, and we write $\overrightarrow{AB} \sim \overrightarrow{CD}$ (read "$\overrightarrow{AB}$ is equivalent to $\overrightarrow{CD}$").

For future use we note the following simple consequences of this definition:

$$\overrightarrow{AB} \sim \overrightarrow{AB}, \tag{1–33}$$

$$\overrightarrow{AB} \sim \overrightarrow{CD} \quad \text{implies} \quad \overrightarrow{CD} \sim \overrightarrow{AB}, \tag{1–34}$$

and

$$\overrightarrow{AB} \sim \overrightarrow{CD} \quad \text{and} \quad \overrightarrow{CD} \sim \overrightarrow{EF} \quad \text{imply} \quad \overrightarrow{AB} \sim \overrightarrow{EF}. \tag{1–35}$$

And now we give the basic definition.

Definition 1–9. The collection of *all* directed line segments in the plane which have the same magnitude and direction as a given segment $\overrightarrow{AB}$ is, by definition, the *two-dimensional geometric vector* $\mathbf{v}(\overrightarrow{AB})$ determined by $\overrightarrow{AB}$. Any directed line segment in this collection will be called a *representative* of $\mathbf{v}(\overrightarrow{AB})$, and the set $\mathcal{G}^2$ consisting of all such vectors is called the *space of two-dimensional geometric vectors*.

At first sight this definition may seem somewhat bizarre, but it does in fact yield precisely the sort of quantity we want in a geometric vector. For, whatever other ideas one may have concerning geometric vectors, it is clear that every such vector must be completely determined by its magnitude and direction. In other words, it consists of *nothing but* a magnitude and direction. And when we consider equivalent directed line segments $\overrightarrow{AB}$ and $\overrightarrow{CD}$ as entities having *only* a magnitude and direction, forgetting about their initial and terminal points, it is also clear that $\overrightarrow{AB}$ and $\overrightarrow{CD}$ are then effectively identical. Thus equivalent directed line segments are *not* distinct geometric vectors; they are distinct representatives of *the same* geometric vector. This, of course, is the content of Definition 1–9. Figuratively speaking, $\mathcal{G}^2$ is the totality of directed line segments in the plane viewed so myopically that segments having the same magnitude and direction are indistinguishable.

But can we accept Definition 1–9 as it stands? Hardly; for we still lack the necessary assurance that a geometric vector is unambiguously determined by any one of its representatives. Phrased somewhat differently, the validity of Definition 1–9 depends upon the fact that no directed line segment can be a representative of more than one geometric vector. Intuitively this is clear, but the student should nevertheless appreciate the need for a proof based upon the definitions. This is accomplished by the following theorem, which actually establishes a somewhat stronger result.

Theorem 1–8. *Every directed line segment in the plane is a representative of one and only one two-dimensional geometric vector.*

Proof. If $\overrightarrow{AB}$ is a directed line segment in $\mathcal{R}^2$, then by Definition 1–8, $\mathbf{v}(\overrightarrow{AB})$ is a geometric vector having $\overrightarrow{AB}$ as a representative. Hence every directed line segment is a representative of *at least one* geometric vector, and it remains to prove that $\overrightarrow{AB}$ cannot represent more than one.

Thus suppose $\overrightarrow{AB}$ also represents the vector $\mathbf{v}(\overrightarrow{CD})$. Then, if $\overrightarrow{EF}$ is any representative of $\mathbf{v}(\overrightarrow{CD})$, $\overrightarrow{EF} \sim \overrightarrow{CD}$. But $\overrightarrow{AB} \sim \overrightarrow{CD}$, and hence by (1–34) we have $\overrightarrow{CD} \sim \overrightarrow{AB}$, and by (1–35) $\overrightarrow{EF} \sim \overrightarrow{AB}$. This shows that every representative of $\mathbf{v}(\overrightarrow{CD})$ is also a representative of $\mathbf{v}(\overrightarrow{AB})$. Now reverse the argument to conclude that every representative of $\mathbf{v}(\overrightarrow{AB})$ is also a representative of $\mathbf{v}(\overrightarrow{CD})$. Hence $\mathbf{v}(\overrightarrow{AB}) = \mathbf{v}(\overrightarrow{CD})$, as required. ▮

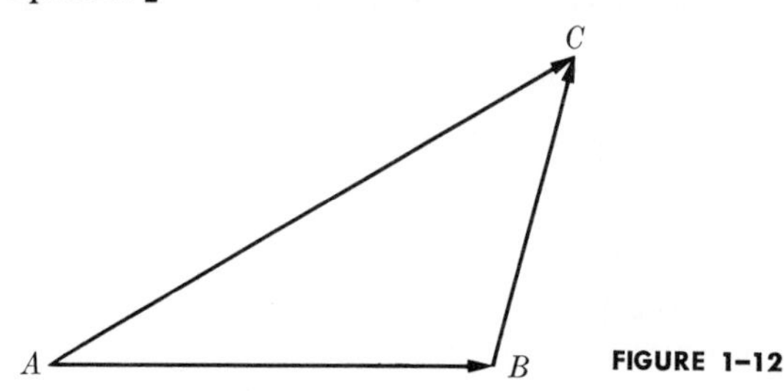

FIGURE 1–12

To define vector addition in $\mathcal{G}^2$ we make use of the simple and geometrically obvious fact that if $\mathbf{v}$ is an arbitrary geometric vector, and P is any point in the plane, then there exists precisely one representative of $\mathbf{v}$ with initial point P. (To produce this representative, select any segment belonging to $\mathbf{v}$ and translate its initial point to P.) This having been said, let $\mathbf{v}$ and $\mathbf{w}$ be vectors in $\mathcal{G}^2$, and choose a representative $\overrightarrow{AB}$ of $\mathbf{v}$. Then, if $\overrightarrow{BC}$ is that representative of $\mathbf{w}$ with initial point B, we define $\mathbf{v} + \mathbf{w}$ to be the vector $\mathbf{v}(\overrightarrow{AC})$. Geometrically, $\overrightarrow{AC}$ is the third side of the triangle formed from $\overrightarrow{AB}$ and $\overrightarrow{BC}$, as shown in Fig. 1–12.

Here again we have a definition which requires justification, for it is predicated upon the fact that $\mathbf{v} + \mathbf{w}$ remains the same regardless of which representatives of $\mathbf{v}$

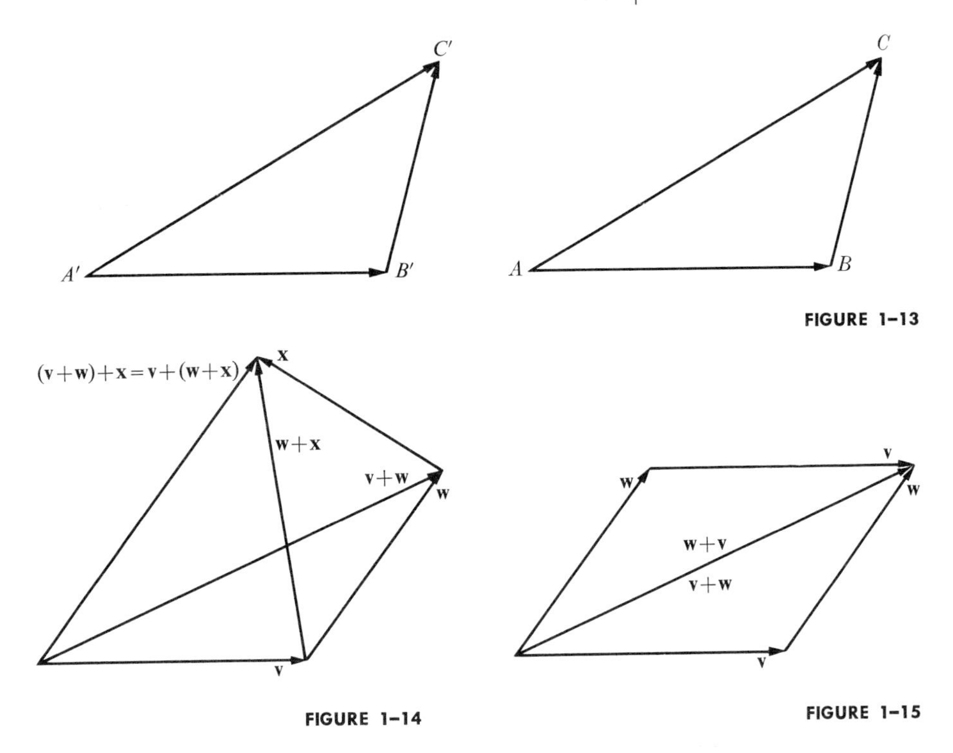

FIGURE 1-13

FIGURE 1-14

FIGURE 1-15

and **w** are used to compute it. But this, we assert, is obvious. For if $\overrightarrow{A'B'} \sim \overrightarrow{AB}$, and $\overrightarrow{B'C'} \sim \overrightarrow{BC}$, then triangles $A'B'C'$ and ABC will be congruent, and a parallel translation will bring $\overrightarrow{A'C'}$ into coincidence with $\overrightarrow{AC}$ (see Fig. 1-13).

With this difficulty out of the way, we invoke elementary geometry to prove that $\mathbf{v} + (\mathbf{w} + \mathbf{x}) = (\mathbf{v} + \mathbf{w}) + \mathbf{x}$, and that $\mathbf{v} + \mathbf{w} = \mathbf{w} + \mathbf{v}$ (appropriate diagrams appear above in Figs. 1-14 and 1-15). Moreover, if we set $\mathbf{0} = \mathbf{v}(\overrightarrow{AA})$, then $\mathbf{v} + \mathbf{0} = \mathbf{v}$ for every $\mathbf{v}$ in $\mathcal{G}^2$; while if $\mathbf{v} = \mathbf{v}(\overrightarrow{AB})$, it is clear that the vector $-\mathbf{v} = \mathbf{v}(\overrightarrow{BA})$ has the property that $\mathbf{v} + (-\mathbf{v}) = \mathbf{0}$. Thus the additive axioms for a vector space are satisfied in $\mathcal{G}^2$.

Next let $\mathbf{v}(\overrightarrow{AB})$ be an arbitrary nonzero vector in $\mathcal{G}^2$, and let α be a real number. Choose a unit of distance on the ray from A through B, and let $|\overrightarrow{AB}|$ denote the length of the segment $\overrightarrow{AB}$. Then there exists a *unique* point C on this ray such that $|\overrightarrow{AC}|/|\overrightarrow{AB}| = |\alpha|$, where $|\alpha|$ denotes the absolute value of α. If $\alpha \geq 0$, let $\alpha\mathbf{v} = \mathbf{v}(\overrightarrow{AC})$; while if $\alpha < 0$, let $\alpha\mathbf{v} = -\mathbf{v}(\overrightarrow{AC}) = \mathbf{v}(\overrightarrow{CA})$. Finally, when $\mathbf{v} = \mathbf{0}$, set $\alpha\mathbf{v} = \mathbf{0}$ for all α. (We can describe $\alpha\mathbf{v}$ as the collection of line segments in the plane whose magnitude is $|\alpha|$-times the magnitude of $\overrightarrow{AB}$, and whose direction is the same as or opposite to that of $\overrightarrow{AB}$ according as α is positive or negative.) With this we have defined a scalar multiplication in $\mathcal{G}^2$, and geometric arguments can again be used to prove that the required axioms are satisfied. Thus $\mathcal{G}^2$ is a real vector space.

It is both interesting and instructive to compare this space with the vector space $\mathcal{R}^2$ defined in Section 1–1. To effect this comparison consider the totality of directed line segments in the plane which have their initial point at the coordinate origin in $\mathcal{R}^2$. On the one hand, this collection of vectors is none other than the set of vectors which comprise $\mathcal{R}^2$. However (and this is the crux of our argument), it may also be viewed as a *complete* set of representatives of the vectors in $\mathcal{G}^2$. (Recall that every vector in $\mathcal{G}^2$ has precisely one such representative, and conversely, every such directed line segment represents a unique vector in $\mathcal{G}^2$.) In other words, the vectors in $\mathcal{R}^2$ are simply a particular set of representatives of the vectors in $\mathcal{G}^2$. From this it follows that if we agree to replace each vector in $\mathcal{G}^2$ by its unique representative emanating from the coordinate origin in $\mathcal{R}^2$, we find that $\mathcal{R}^2$ *and* $\mathcal{G}^2$ *then consist of precisely the same vectors.* Moreover, addition and scalar multiplication in $\mathcal{G}^2$ then become identical with the corresponding operations in $\mathcal{R}^2$. Thus we conclude that $\mathcal{R}^2$ and the space of two-dimensional geometric vectors are essentially identical. $\mathcal{G}^2$ is simply $\mathcal{R}^2$ stripped of its coordinate system, and conversely, $\mathcal{R}^2$ is $\mathcal{G}^2$ seen by means of a coordinate system.

EXERCISES

In Exercises 1 through 5 find two other directed line segments which have the same magnitude and direction as $\overrightarrow{AB}$ for the given values of A and B.

1. $A = (1, 1)$, $B = (3, 1)$
2. $A = (-2, 1)$, $B = (0, 0)$
3. $A = (1, -1)$, $B = (3, 1)$
4. $A = (-4, -\frac{1}{2})$, $B = (-\frac{11}{2}, \frac{3}{2})$
5. $A = (0, 2)$, $B = (-3, -1)$

In Exercises 6 through 10 find $\mathbf{v}(\overrightarrow{AB}) + \mathbf{v}(\overrightarrow{CD})$ for the given values of A, B, C, D.

6. $A = (0, 0)$, $B = (1, 2)$, $C = (-1, 2)$, $D = (3, 0)$
7. $A = (0, -1)$, $B = (-2, 1)$, $C = (3, 0)$, $D = (5, -2)$
8. $A = (-1, 1)$, $B = (2, 1)$, $C = (1, 3)$, $D = (-1, 2)$
9. $A = (4, \frac{1}{2})$, $B = (\frac{3}{2}, -\frac{3}{2})$, $C = (-2, \frac{1}{3})$, $D = (0, -\frac{2}{3})$
10. $A = (-2, -5)$, $B = (-1, -2)$, $C = (1, 4)$, $D = (6, 7)$

In Exercises 11 through 15 find the vector in $\mathcal{R}^2$ which represents the geometric vector $\mathbf{v}(\overrightarrow{AB})$ for the given values of A and B.

11. $A = (-1, 1)$, $B = (2, 1)$
12. $A = (-2, 1)$, $B = (0, 0)$
13. $A = (0, 2)$, $B = (-3, -2)$
14. $A = (\frac{2}{3}, \frac{1}{2})$, $B = (-\frac{1}{3}, 2)$
15. $A = (6, -3)$, $B = (-7, -9)$

In Exercises 16 through 20 find the value of $\alpha\mathbf{v}(\overrightarrow{AB})$ for the given values of A, B, and α.

16. $A = (-5, 7)$, $B = (3, 10)$, $\alpha = 4$
17. $A = (9, 2)$, $B = (5, -2)$, $\alpha = -\frac{1}{2}$
18. $A = (\frac{5}{3}, 6)$, $B = (4, \frac{8}{3})$, $\alpha = 3$
19. $A = (0, 5)$, $B = (11, 0)$, $\alpha = -1$
20. $A = (-3, -1)$, $B = (-4, -5)$, $\alpha = \frac{1}{3}$

21. Let $A = (x_1, y_1)$ and $B = (x_2, y_2)$ be any two points in $\mathcal{R}^2$, and suppose $\overrightarrow{AB} \sim \overrightarrow{CD}$, where $C = (x_3, y_3)$. Find the coordinates of the point D.
22. Let $A = (x_1, y_1)$ and $B = (x_2, y_2)$ be any two points in $\mathcal{R}^2$, and let α be a real number. Find a representative of the geometric vector $\alpha\mathbf{v}(\overrightarrow{AB})$.
23. Given $A = (x_1, y_1)$, $B = (x_2, y_2)$, $C = (x_3, y_3)$, $D = (x_4, y_4)$, four points in $\mathcal{R}^2$. Compute $\mathbf{v}(\overrightarrow{AB}) + \mathbf{v}(\overrightarrow{CD})$.
24. Prove that $\mathbf{v}(\overrightarrow{AB}) + \mathbf{v}(\overrightarrow{CD}) = \mathbf{v}(\overrightarrow{CD}) + \mathbf{v}(\overrightarrow{AB})$ for any pair of geometric vectors. [*Hint:* Use the result of Exercise 23.]
25. Prove that the addition of geometric vectors is associative. [*Hint:* Use the result of Exercise 23.]
26. Using the results of Exercises 22 and 23 prove that $\alpha(\mathbf{v} + \mathbf{w}) = \alpha\mathbf{v} + \alpha\mathbf{w}$ and $(\alpha + \beta)\mathbf{v} = \alpha\mathbf{v} + \beta\mathbf{v}$ for any pair of geometric vectors $\mathbf{v}$, $\mathbf{w}$, and any pair of real numbers α, β.
27. Use the result of Exercise 22 to prove that $(\alpha\beta)\mathbf{v} = \alpha(\beta\mathbf{v})$ for any geometric vector $\mathbf{v}$, and any pair of real numbers α, β.

*1-9 EQUIVALENCE RELATIONS

We have seen that a geometric vector is a collection of directed line segments mutually related by magnitude and direction. This is but one example of the method whereby a new mathematical entity is defined as a collection of related objects of some familiar type. We shall have occasion to use this technique again, and it may therefore be of some interest to present it in its general setting. As usual we begin with a definition.

Definition 1-10. An *equivalence relation* $\mathcal{R}$ on a set $\mathcal{S}$ is a set of ordered pairs (x, y) of elements of $\mathcal{S}$, subject to the following conditions:

(i) The pair (x, x) belongs to $\mathcal{R}$ for every x in $\mathcal{S}$;

(ii) If (x, y) belongs to $\mathcal{R}$, then so does (y, x);

(iii) If (x, y) and (y, z) belong to $\mathcal{R}$, then (x, z) belongs to $\mathcal{R}$.

Whenever an ordered pair (x, y) belongs to an equivalence relation on $\mathcal{S}$, one says that *x is equivalent to y*, and writes $x \sim y$. Custom then dictates that the symbol $\sim$ (usually called "tilda" or simply "wiggle") rather than $\mathcal{R}$ itself be referred to as the equivalence relation on $\mathcal{S}$. In these terms the defining conditions of an equivalence relation on $\mathcal{S}$ become

(i) $x \sim x$ for all x in $\mathcal{S}$,

(ii) $x \sim y$ implies $y \sim x$,

(iii) $x \sim y$ and $y \sim z$ imply $x \sim z$.

One also says that an equivalence relation is *reflexive*, *symmetric*, and *transitive;* these names being given respectively to properties (i), (ii), and (iii) above.

Equivalence relations crop up in every branch of mathematics, and usually in a very fundamental way, as the following examples illustrate.

EXAMPLE 1. The relation of equality applied to the elements of any set $\mathcal{S}$ is obviously an equivalence relation on $\mathcal{S}$. In fact, the notion of an equivalence relation can be viewed as a generalized form of equality.

EXAMPLE 2. Let $\mathcal{S}$ be the set of all triangles in the plane, and let $\triangle_1 \sim \triangle_2$ mean that $\triangle_1$ and $\triangle_2$ are congruent. Then $\sim$ is an equivalence relation on $\mathcal{S}$.

EXAMPLE 3. Let $\mathcal{S}$ be the set of directed line segments in the plane, and let $\overrightarrow{AB} \sim \overrightarrow{CD}$ have the meaning assigned in Definition 1–8. Then (1–33) through (1–35) imply that $\sim$ is an equivalence relation on $\mathcal{S}$.

EXAMPLE 4. Let $\mathcal{S}$ be the set of real valued functions which are continuous at all but a finite number of points in an interval $[a, b]$. If f and g are two such functions, let $f \sim g$ mean that $f(x) = g(x)$ for all but a finite number of values of x in $[a, b]$. Then $\sim$ is an equivalence relation on $\mathcal{S}$ (see Exercise 1).

EXAMPLE 5. Let $\mathcal{S}$ be the set of all "symbols" of the form a/b, where a and b are integers, and $b \neq 0$. Set $(a/b) \sim (c/d)$ if and only if $ad = bc$. Then $\sim$ is an equivalence relation on $\mathcal{S}$ (see Exercise 2). (Just as arrows in the plane *represent* geometric vectors, the symbols a/b in $\mathcal{S}$ *represent* rational numbers, and the equivalence relation introduced here allows us to *define* a rational number as a collection of mutually equivalent symbols of the form a/b. This example is discussed in greater detail in Exercise 6.)

If $\sim$ is an equivalence relation on a set $\mathcal{S}$, and x is any element in $\mathcal{S}$, then the set of all elements y in $\mathcal{S}$ such that $y \sim x$ is called the *equivalence class determined by* x. This equivalence class is denoted by $[x]$, and any element belonging to it is said to be a *representative* of $[x]$. In particular, x itself is a representative of the equivalence class $[x]$, since $x \sim x$.

We now come to the fundamental theorem concerning equivalence relations, a special case of which was proved in the last section.

Theorem 1–9. *If $\sim$ is an equivalence relation on a set $\mathcal{S}$, then every element of $\mathcal{S}$ belongs to one and only one equivalence class of elements of $\mathcal{S}$.*

For a proof, see Theorem 1–8.

It is this result which justifies using equivalence relations to define new mathematical objects. The objects in question are the equivalence classes, and the above theorem asserts that every such object is uniquely determined by any one of its representatives. This type of definition is sometimes called "definition by abstraction," since it passes from the particular to the general by regarding mutually equivalent individuals as identical.

EXERCISES

1. Prove that the relation defined in Example 4 above is an equivalence relation.
2. Prove that the relation defined in Example 5 above is an equivalence relation. What is the equivalence class determined by $\frac{1}{2}$? By $\frac{1}{4}$? By $\frac{2}{4}$?
3. Let $\mathcal{S}$ be the set of all integers, and let m be a fixed positive integer. If x and y belong to $\mathcal{S}$, set $x \sim y$ if and only if $x - y$ is divisible by m. Prove that $\sim$ is an equivalence relation on $\mathcal{S}$, and find *all* the equivalence classes for this relation when $m = 2$, 3, and 4.
4. A *partition* of a set $\mathcal{S}$ is defined to be a collection $\mathcal{P}$ of subsets of $\mathcal{S}$, each of which contains at least one element of $\mathcal{S}$, *and* such that every element of $\mathcal{S}$ belongs to one and only one of the subsets making up $\mathcal{P}$. (Informally, a partition chops $\mathcal{S}$ into pieces, and these pieces are the subsets belonging to $\mathcal{P}$.) In these terms, the fundamental theorem on equivalence relations can be restated as follows: *If $\sim$ is an equivalence relation on a set $\mathcal{S}$, then the equivalence classes determined by $\sim$ yield a partition of $\mathcal{S}$.* Prove the following converse of this theorem: *If $\mathcal{P}$ is a partition of a set $\mathcal{S}$, then there exists an equivalence relation on $\mathcal{S}$ whose equivalence classes coincide with the subsets of the given partition $\mathcal{P}$.* (Note that these two results imply that the concepts of an equivalence relation on a set and a partition of a set are identical.)

*5. Let f be a function with domain $\mathcal{D}$ and range $\mathcal{R}$ (i.e., f associates with each x in $\mathcal{D}$ a unique element y in $\mathcal{R}$ denoted $f(x)$ and called the image of x under f, *and* each y in $\mathcal{R}$ is the image of at least one x in $\mathcal{D}$).

(a) If x_1 and x_2 belong to $\mathcal{D}$, set $x_1 \sim x_2$ if and only if $f(x_1) = f(x_2)$. Prove that $\sim$ is an equivalence relation on $\mathcal{D}$.

(b) Let $\mathcal{P}$ be the partition of $\mathcal{D}$ defined by this equivalence relation (see Exercise 4). Prove that there exists a one-to-one function g with domain $\mathcal{P}$ and range $\mathcal{R}$ such that $g([x]) = f(x)$ for each x in $\mathcal{D}$. (Recall that a function f is one-to-one if $f(x_1) = f(x_2)$ implies $x_1 = x_2$.)

*6. *The rational numbers.* Let $\mathcal{S}$ be the set of all symbols a/b, $b \neq 0$, introduced in Example 5 above, and let $\sim$ be the equivalence relation that was defined on $\mathcal{S}$; i.e., $a/b \sim c/d$ if and only if $ad = bc$. Let Q be the set of all equivalence classes of the elements of $\mathcal{S}$ under this equivalence relation. Then the equivalence classes belonging to Q are, by definition, *rational numbers*, and the set Q itself is called the set of all rational numbers. Thus the rational number $[a/b]$ is the equivalence class containing the symbol a/b.

(a) To define addition of rational numbers set

$$\left[\frac{a}{b}\right] + \left[\frac{c}{d}\right] = \left[\frac{ad + bc}{bd}\right].$$

Prove that if $a'/b' \sim a/b$ and $c'/d' \sim c/d$, then

$$\frac{a'd' + b'c'}{b'd'} \sim \frac{ad + bc}{bd},$$

and thus conclude that the above equation actually defines an addition of equivalence classes.

(b) To define multiplication of rational numbers set

$$\left[\frac{a}{b}\right]\left[\frac{c}{d}\right] = \left[\frac{ac}{bd}\right].$$

Prove that if $a'/b' \sim a/b$ and $c'/d' \sim c/d$, then

$$\frac{a'c'}{b'd'} \sim \frac{ac}{bd},$$

and thus conclude that the above equation actually defines a multiplication of equivalence classes.

2 linear transformations and matrices

2–1 LINEAR TRANSFORMATIONS

Up to this point our study of real vector spaces can best be described as a modest generalization of some of the ideas implicit in analytic geometry. Although such terms as linear dependence and independence, subspaces, bases, and the like, may have been unfamiliar they actually add little to the knowledge of vector spaces taught in elementary geometry. All this changes, however, as soon as these ideas are used to study functions defined on vector spaces. Here new and important things do happen, and as the following discussion unfolds we shall find the concepts introduced earlier taking on added meaning and significance.

The simplest yet most important functions which arise in the study of vector spaces are known as linear transformations, and are defined as follows:

Definition 2–1. A *linear transformation*, or *linear operator*, from a vector space $\mathcal{V}_1$ to a vector space $\mathcal{V}_2$ is a function A which associates with each vector $\mathbf{x}$ in $\mathcal{V}_1$ a unique vector $A(\mathbf{x})$ in $\mathcal{V}_2$ in such a way that

$$A(\mathbf{x}_1 + \mathbf{x}_2) = A(\mathbf{x}_1) + A(\mathbf{x}_2) \tag{2–1}$$

and

$$A(\alpha\mathbf{x}) = \alpha A(\mathbf{x}) \tag{2–2}$$

for all vectors $\mathbf{x}_1$, $\mathbf{x}_2$, $\mathbf{x}$ in $\mathcal{V}_1$, and all scalars α.

In other words, a linear transformation is a function, or mapping, from one vector space to another which sends sums into sums and scalar products into scalar products (see Fig. 2–1). These requirements are sometimes referred to by saying that a linear transformation is "compatible" with the algebraic operations of addition and scalar multiplication defined on vector spaces, and it is just this compatibility which accounts for the importance of such functions in linear algebra.

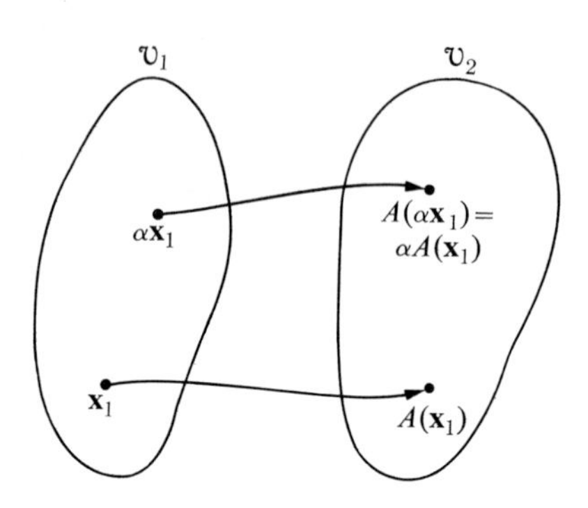

FIGURE 2–1

One consequence of Definition 2–1 is that a linear transformation *always* maps the zero vector of $\mathcal{V}_1$ onto the zero vector of $\mathcal{V}_2$; that is,

$$A(\mathbf{0}) = \mathbf{0}. \tag{2–3}$$

Another is that

$$A(\alpha_1\mathbf{x}_1 + \cdots + \alpha_n\mathbf{x}_n) = \alpha_1 A(\mathbf{x}_1) + \cdots + \alpha_n A(\mathbf{x}_n) \tag{2–4}$$

for any finite collection of vectors $\mathbf{x}_1, \ldots, \mathbf{x}_n$ in $\mathcal{V}_1$ and scalars $\alpha_1, \ldots, \alpha_n$. The first of these assertions can be established by setting $\alpha = 0$ in (2–2), the second by repeated use of (2–1) and (2–2) in the obvious fashion. In particular, when $n = 2$, (2–4) becomes

$$A(\alpha_1\mathbf{x}_1 + \alpha_2\mathbf{x}_2) = \alpha_1 A(\mathbf{x}_1) + \alpha_2 A(\mathbf{x}_2). \tag{2–5}$$

We call attention to this equation in order to remark that, by itself, it can be (and often is) taken as the definition of a linear transformation, since (2–1) and (2–2) are satisfied if and only if (2–5) is. (See Exercise 17.) From time to time we shall use this fact when proving that a function is a linear transformation.

If A is a linear transformation from $\mathcal{V}_1$ to $\mathcal{V}_2$ we write $A: \mathcal{V}_1 \to \mathcal{V}_2$ (read "A maps $\mathcal{V}_1$ into $\mathcal{V}_2$"), and refer to $\mathcal{V}_1$ as the *domain* of A. In this case the set of all vectors $\mathbf{y}$ in $\mathcal{V}_2$ such that $\mathbf{y} = A(\mathbf{x})$ for some $\mathbf{x}$ in $\mathcal{V}_1$ is called the *image* or *range* of A, and is denoted by $\mathcal{I}(A)$. Lest it be overlooked, we point out that the image of A need not be all of $\mathcal{V}_2$, a possibility which is made explicit by saying that A maps $\mathcal{V}_1$ *into* $\mathcal{V}_2$ (Fig. 2–2). Of course, it may happen that $\mathcal{I}(A) = \mathcal{V}_2$, in which case the term *onto* is used. And finally, it should be observed that there is nothing in the above definition to prevent $\mathcal{V}_1$ and $\mathcal{V}_2$ from being

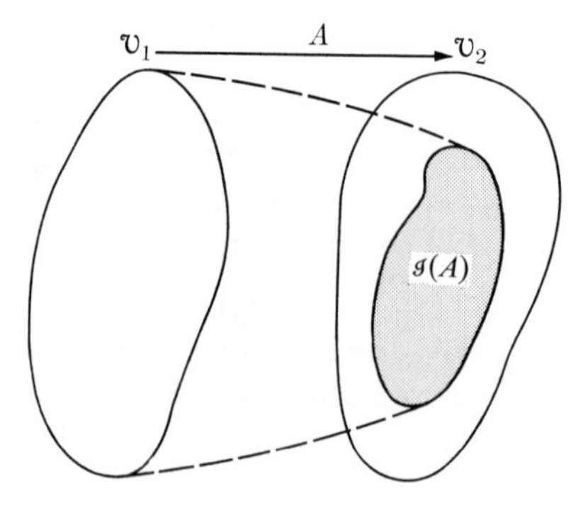

FIGURE 2–2

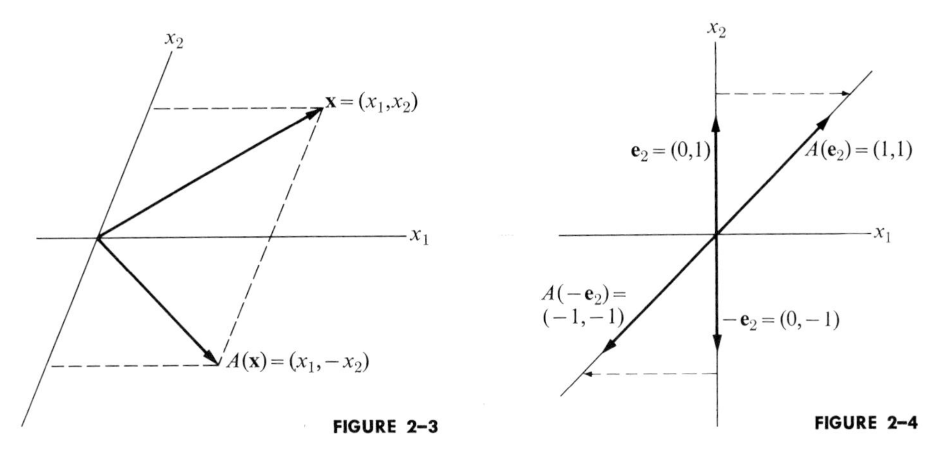

FIGURE 2–3

FIGURE 2–4

one and the same vector space. Indeed, this is one of the most fruitful settings in which to pursue the study of linear transformations.

We conclude this section by giving a number of examples, several of which will figure prominently in our later work. For the most part we simply state the definition of the function in question, and omit the routine verification of linearity in the expectation that the reader will supply the missing argument for himself.

EXAMPLE 1. Let $\mathbf{x} = (x_1, x_2)$ be an arbitrary vector in $\mathcal{R}^2$, and set

$$A(\mathbf{x}) = (x_1, -x_2).$$

Geometrically A can be described as the linear transformation mapping $\mathcal{R}^2$ onto itself by reflection across the x_1-axis. (See Fig. 2–3 where, for generality, the effect of A has been depicted relative to an oblique coordinate system.)

EXAMPLE 2. Let A be the mapping of $\mathcal{R}^2$ onto itself obtained by shearing the plane horizontally so that the x_2-axis is shifted through a 45° angle as shown in Fig. 2–4. Analytically, A is defined by the equation

$$A(x_1, x_2) = (x_1 + x_2, x_2),$$

and is clearly linear.

EXAMPLE 3. Let $\mathcal{L}$ be any line through the origin in $\mathcal{R}^3$, and let A be a fixed rotation about $\mathcal{L}$. Then, arguing geometrically, it is easy to show that A is a linear transformation mapping $\mathcal{R}^3$ onto itself.

EXAMPLE 4. The mapping which sends each vector in $\mathcal{V}_1$ onto the zero vector in $\mathcal{V}_2$ is clearly a linear transformation from $\mathcal{V}_1$ to $\mathcal{V}_2$ for all $\mathcal{V}_1$ and $\mathcal{V}_2$. It is called the *zero transformation*, and is denoted by the symbol O, irrespective of the vector spaces involved.

EXAMPLE 5. A second linear transformation for which we reserve a special symbol is the *identity transformation* I mapping a vector space $\mathcal{V}$ onto itself.

The defining equation for I is

$$I(\mathbf{x}) = \mathbf{x}$$

for all x in $\mathcal{V}$; its linearity is obvious.

EXAMPLE 6. Consider the space $\mathcal{C}[a, b]$ of all real valued continuous functions on the interval $[a, b]$, and for each $\mathbf{f}$ in $\mathcal{C}[a, b]$ set

$$A(\mathbf{f}) = \int_a^x \mathbf{f}(t)\, dt, \qquad a \leq x \leq b.$$

Then since $A(\mathbf{f})$ is continuous on $[a, b]$, A can be viewed as a mapping of $\mathcal{C}[a, b]$ into itself. As such it is linear since

$$\begin{aligned} A(\alpha_1\mathbf{f}_1 + \alpha_2\mathbf{f}_2) &= \int_a^x [\alpha_1\mathbf{f}_1(t) + \alpha_2\mathbf{f}_2(t)]\, dt \\ &= \int_a^x \alpha_1\mathbf{f}_1(t)\, dt + \int_a^x \alpha_2\mathbf{f}_2(t)\, dt \\ &= \alpha_1 \int_a^x \mathbf{f}_1(t)\, dt + \alpha_2 \int_a^x \mathbf{f}_2(t)\, dt \\ &= \alpha_1 A(\mathbf{f}_1) + \alpha_2 A(\mathbf{f}_2). \end{aligned}$$

EXAMPLE 7. For the same reasons as those just given, the mapping $A\colon \mathcal{C}[a,b] \to \mathcal{R}^1$ defined by

$$A(\mathbf{f}) = \int_a^b \mathbf{f}(x)\, dx$$

is also linear.

EXAMPLE 8. Let $\mathcal{C}^1[a, b]$ denote the space of all continuously differentiable functions on $[a, b]$ (see Example 3 in Section 1–4), and let D denote the operation of differentiation on this space; that is, $D(\mathbf{f}) = \mathbf{f}'$. Then the familiar identities

$$D(\mathbf{f}_1 + \mathbf{f}_2) = D(\mathbf{f}_1) + D(\mathbf{f}_2), \qquad D(\alpha\mathbf{f}) = \alpha D(\mathbf{f})$$

imply that D is a linear transformation from $\mathcal{C}^1[a, b]$ to $\mathcal{C}[a, b]$. More generally, the operation of taking nth derivatives is a linear transformation mapping the space of n-times continuously differentiable functions on an interval $[a, b]$ into the space $\mathcal{C}[a, b]$.

EXERCISES

Prove that each of the following equations defines a linear transformation from $\mathcal{R}^2$ into (or onto) itself, and describe the effect of the transformation in geometric terms.

1. $A(x_1, x_2) = -(x_1, x_2)$
2. $A(x_1, x_2) = (2x_1, x_2)$
3. $A(x_1, x_2) = 2(x_1, x_2/3)$
4. $A(x_1, x_2) = 3(x_1, x_2)$
5. $A(x_1, x_2) = \sqrt{2}(x_1 - x_2, x_1 + x_2)$
6. $A(x_1, x_2) = (x_2, x_1)$

7. $A(x_1, x_2) = -(x_2, x_1)$
8. $A(x_1, x_2) = (x_1 + x_2, x_1 + x_2)$
9. $A(x_1, x_2) = (0, 0)$
10. $A(x_1, x_2) = (x_1 + x_2, 0)$

Determine which of the following equations defines a linear transformation on the space of polynomials $\mathcal{P}$.

11. $A(\mathbf{p}) = \mathbf{p}(x)^2$
12. $A(\mathbf{p}) = x\ (\mathbf{p}(x))$
13. $A(\mathbf{p}) = \mathbf{p}(x + 1) - \mathbf{p}(x)$
14. $A(\mathbf{p}) = \mathbf{p}(x + 1) - \mathbf{p}(0)$
15. $A(\mathbf{p}) = \mathbf{p}''(x) - 2\mathbf{p}'(x)$
16. $A(\mathbf{p}) = \mathbf{p}(x^2)$
17. Prove that $A: \mathcal{V}_1 \to \mathcal{V}_2$ is linear if and only if

$$A(\alpha_1\mathbf{x}_1 + \alpha_2\mathbf{x}_2) = \alpha_1 A(\mathbf{x}_1) + \alpha_2 A(\mathbf{x}_2)$$

for all $\mathbf{x}_1, \mathbf{x}_2$ in $\mathcal{V}_1$, and all scalars α_1, α_2.

18. Prove Eq. (2–4). [*Hint:* Use mathematical induction.]
19. Let $\mathbf{e}_1, \ldots, \mathbf{e}_n$ be a basis for a finite dimensional vector space $\mathcal{V}$, and for each index i, $1 \leq i \leq n$, let η_i be an arbitrary real number. Prove that the function $A: \mathcal{V} \to \mathcal{R}^1$ defined by

$$A(\alpha_1\mathbf{e}_1 + \cdots + \alpha_n\mathbf{e}_n) = \alpha_1\eta_1 + \cdots + \alpha_n\eta_n$$

for each vector $\alpha_1\mathbf{e}_1 + \cdots + \alpha_n\mathbf{e}_n$ in $\mathcal{V}$ is a linear transformation.

20. Let $\mathcal{V}_1$ be a finite dimensional vector space with basis $\mathbf{e}_1, \ldots, \mathbf{e}_n$, let $\mathcal{V}_2$ be an arbitrary vector space, let $\mathbf{y}_1, \ldots, \mathbf{y}_n$ be vectors in $\mathcal{V}_2$, and for each $\mathbf{x} = \alpha_1\mathbf{e}_1 + \cdots + \alpha_n\mathbf{e}_n$ in $\mathcal{V}_1$ set

$$A(\mathbf{x}) = \alpha_1\mathbf{y}_1 + \cdots + \alpha_n\mathbf{y}_n.$$

Prove that A is a linear transformation from $\mathcal{V}_1$ to $\mathcal{V}_2$.

*21. Find all linear transformations mapping $\mathcal{R}^1$ into (or onto) itself. [*Hint:* 1 is a basis for $\mathcal{R}^1$.]

2–2 ADDITION AND SCALAR MULTIPLICATION OF TRANSFORMATIONS

We begin the systematic study of linear transformations by describing several ways in which new transformations can be formed from old ones. Of these the simplest is the addition of two transformations, each of which maps a given vector space $\mathcal{V}_1$ into *the same* space $\mathcal{V}_2$. The definition reads as follows:

Definition 2–2. Let A and B be linear transformations from $\mathcal{V}_1$ to $\mathcal{V}_2$. Then their *sum*, $A + B$, is the transformation from $\mathcal{V}_1$ to $\mathcal{V}_2$ defined by

$$(A + B)(\mathbf{x}) = A(\mathbf{x}) + B(\mathbf{x}) \tag{2–6}$$

for all $\mathbf{x}$ in $\mathcal{V}_1$.

This, of course, is just the familiar addition of functions here applied to linear transformations, and it is an easy matter to show that $A + B$ is again a linear transformation from $\mathcal{V}_1$ to $\mathcal{V}_2$. Indeed, if $\mathbf{x}_1$ and $\mathbf{x}_2$ belong to $\mathcal{V}_1$, and α_1 and α_2

are scalars, then

$$\begin{aligned}(A + B)(\alpha_1\mathbf{x}_1 + \alpha_2\mathbf{x}_2) &= A(\alpha_1\mathbf{x}_1 + \alpha_2\mathbf{x}_2) + B(\alpha_1\mathbf{x}_1 + \alpha_2\mathbf{x}_2)\\ &= \alpha_1 A(\mathbf{x}_1) + \alpha_2 A(\mathbf{x}_2) + \alpha_1 B(\mathbf{x}_1) + \alpha_2 B(\mathbf{x}_2)\\ &= \alpha_1[A(\mathbf{x}_1) + B(\mathbf{x}_1)] + \alpha_2[A(\mathbf{x}_2) + B(\mathbf{x}_2)]\\ &= \alpha_1(A + B)(\mathbf{x}_1) + \alpha_2(A + B)(\mathbf{x}_2).\end{aligned}$$

Thus $A + B$ satisfies Eq. (2–5), and is therefore linear, as asserted.

EXAMPLE 1. Let D and D^2 denote, respectively, the operations of taking first and second derivatives in $\mathcal{C}^2[a, b]$. Then the sum $D^2 + D$ is the linear transformation from $\mathcal{C}^2[a, b]$ to $\mathcal{C}[a, b]$ which sends each function $\mathbf{y}$ in $\mathcal{C}^2[a, b]$ onto the continuous function $\mathbf{y}'' + \mathbf{y}'$; that is,

$$(D^2 + D)(\mathbf{y}) = D^2\mathbf{y} + D\mathbf{y}.$$

EXAMPLE 2. Let $\mathbf{K}(t)$ be a fixed function in $\mathcal{C}[a, b]$, and let A be the linear transformation mapping $\mathcal{C}[a, b]$ into itself given by

$$A(\mathbf{f}) = \int_a^x \mathbf{K}(t)\mathbf{f}(t)\,dt, \qquad a \leq x \leq b.$$

Then the sum $A + I$, I the identity transformation on $\mathcal{C}[a, b]$ (see Example 5, Section 2–1), is the linear transformation mapping $\mathcal{C}[a, b]$ into itself whose defining equation is

$$(A + I)(\mathbf{f}) = \int_a^x \mathbf{K}(t)\mathbf{f}(t)\,dt + \mathbf{f}.$$

The addition of linear transformations defined above has a number of familiar and suggestive properties. In the first place, it is clear that

$$A + (B + C) = (A + B) + C \tag{2–7}$$

and

$$A + B = B + A \tag{2–8}$$

whenever A, B, and C are linear transformations from $\mathcal{V}_1$ to $\mathcal{V}_2$ (see Exercise 2). Secondly, the zero mapping from $\mathcal{V}_1$ to $\mathcal{V}_2$ defined in Example 4 of the preceding section acts as a "zero" for this addition since

$$A + O = O + A = A \tag{2–9}$$

for all $A: \mathcal{V}_1 \to \mathcal{V}_2$. And finally, if A is any linear transformation from $\mathcal{V}_1$ to $\mathcal{V}_2$, and if we define $-A$ by the equation

$$(-A)(\mathbf{x}) = -A(\mathbf{x}) \tag{2–10}$$

for all $\mathbf{x}$ in $\mathcal{V}_1$, we obtain a linear transformation from $\mathcal{V}_1$ to $\mathcal{V}_2$ with the property that

$$A + (-A) = -A + A = O. \tag{2–11}$$

In short, the addition of linear transformations from $\mathcal{V}_1$ to $\mathcal{V}_2$ satisfies all of the axioms postulated for addition in a vector space.

To complete what should by now be an obvious sequence of ideas we introduce a scalar multiplication on the set of linear transformations from $\mathcal{V}_1$ to $\mathcal{V}_2$. The relevant definition is as follows:

Definition 2–3. The *product* of a real number α and a linear transformation $A: \mathcal{V}_1 \to \mathcal{V}_2$ is the mapping αA from $\mathcal{V}_1$ to $\mathcal{V}_2$ given by

$$(\alpha A)(\mathbf{x}) = \alpha A(\mathbf{x}) \tag{2–12}$$

for all $\mathbf{x}$ in $\mathcal{V}_1$. In other words, αA is the function whose value at $\mathbf{x}$ is computed by forming the scalar product of α and the vector $A(\mathbf{x})$.

We omit the proof that αA is linear, as well as the easy sequence of arguments required to show that the remaining axioms in the definition of a real vector space are now satisfied. Granting the truth of these facts, we have

Theorem 2–1. *The set of linear transformations from $\mathcal{V}_1$ to $\mathcal{V}_2$ is itself a real vector space under the definitions of addition and scalar multiplication given above.*

EXERCISES

1. Cite the relevant axiom or definition needed to justify each step in the proof of the linearity of $A + B$.
2. Prove that addition of linear transformations is associative and commutative.
3. Prove that the mapping $-A$ defined in (2–10) is linear, and that

$$A + (-A) = -A + A = O.$$

4. (a) Prove that αA as defined in the text is linear.

   (b) Prove Theorem 2–1.
5. (a) Let $\mathfrak{X}$ be a nonempty subset of a vector space $\mathcal{V}_1$, and let $\mathcal{A}(\mathfrak{X})$ denote the set of all linear transformations A from $\mathcal{V}_1$ to $\mathcal{V}_2$ with the property that $A(\mathbf{x}) = \mathbf{0}$ for all $\mathbf{x}$ in $\mathfrak{X}$. Prove that $\mathcal{A}(\mathfrak{X})$ is a subspace of the space of all linear transformations from $\mathcal{V}_1$ to $\mathcal{V}_2$.

   (b) What is $\mathcal{A}(\mathfrak{X})$ when $\mathfrak{X}$ consists of just the zero vector? When $\mathfrak{X} = \mathcal{V}_1$?

   (c) Prove that $\mathcal{A}(\mathfrak{X}) = \mathcal{A}(\mathcal{S}(\mathfrak{X}))$.

*6. Let $\mathcal{V}_1$ and $\mathcal{V}_2$ be finite dimensional vector spaces with bases $\mathbf{e}_1, \ldots, \mathbf{e}_n$ and $\mathbf{e}'_1, \ldots, \mathbf{e}'_m$, respectively. For each pair of integers i, j, $1 \le i \le n$, $1 \le j \le m$, define $A_{ij}: \mathcal{V}_1 \to \mathcal{V}_2$ by first defining A_{ij} on the basis vectors of $\mathcal{V}_1$ according to the formula

$$A_{ij}(\mathbf{e}_k) = \begin{cases} \mathbf{e}'_j & \text{if } k = i, \\ \mathbf{0} & \text{if } k \neq i, \end{cases}$$

and then use (2–4) to obtain the value of A_{ij} for each $\mathbf{x}$ in $\mathcal{V}_1$. (See Exercise 20, Section 2–1.)

(a) Prove that the A_{ij} are linear transformations from $\mathcal{V}_1$ to $\mathcal{V}_2$, and that they are linearly independent in the vector space of all such transformations.

(b) Prove that the A_{ij} span the space of linear transformations from $\mathcal{V}_1$ to $\mathcal{V}_2$, and hence deduce that this space is finite dimensional with dimension mn. [*Hint:* Two linear transformations from $\mathcal{V}_1$ to $\mathcal{V}_2$ are identical if and only if they coincide on a basis for $\mathcal{V}_1$.]

2–3 PRODUCTS OF LINEAR TRANSFORMATIONS

The theorem established at the end of the last section would seem to imply that the study of linear transformations can be subsumed within the general theory of vector spaces. Such would indeed be the case were addition and scalar multiplication the only algebraic operations that could be performed on linear transformations. However, under suitable hypotheses, it is also possible to define a multiplication of transformations. And, as we shall see, this single fact makes their study much richer in content and quite different in spirit from that of vector spaces alone.

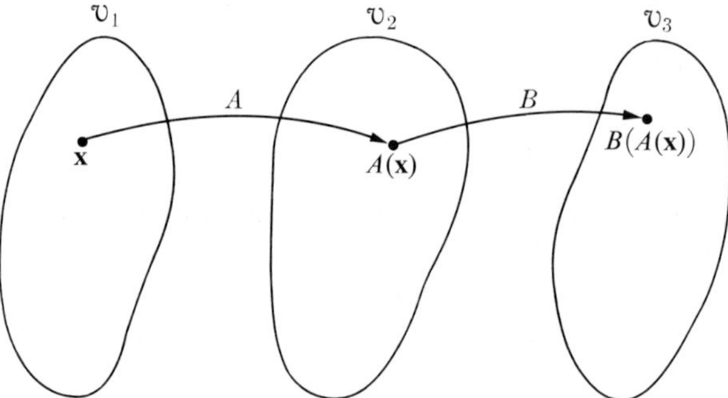

FIGURE 2–5

To introduce this multiplication, let $\mathcal{V}_1$, $\mathcal{V}_2$, $\mathcal{V}_3$ be vector spaces, and consider a pair of linear transformations

$$A: \mathcal{V}_1 \to \mathcal{V}_2, \qquad B: \mathcal{V}_2 \to \mathcal{V}_3.$$

Then, for each $\mathbf{x}$ in $\mathcal{V}_1$, $A(\mathbf{x})$ is a vector in $\mathcal{V}_2$, and it therefore makes sense to speak of applying B to $A(\mathbf{x})$ to obtain the vector $B(A(\mathbf{x}))$ in $\mathcal{V}_3$ (see Fig. 2–5). Thus A and B can be combined, or multiplied, to produce a function from $\mathcal{V}_1$ to $\mathcal{V}_3$ which will be denoted by BA, and called the *product* of A and B in that order, viz., *first A, then B*. This is the content of

Definition 2–4. If $A: \mathcal{V}_1 \to \mathcal{V}_2$ and $B: \mathcal{V}_2 \to \mathcal{V}_3$ are linear transformations, then their *product*, BA, is the mapping from $\mathcal{V}_1$ to $\mathcal{V}_3$ defined by the equation

$$BA(\mathbf{x}) = B(A(\mathbf{x})) \tag{2–13}$$

for all $\mathbf{x}$ in $\mathcal{V}_1$.

The essential fact about such products is that they are always linear. Indeed, if $\mathbf{x}_1$ and $\mathbf{x}_2$ belong to $\mathcal{V}_1$, and α_1 and α_2 are arbitrary real numbers, then

$$\begin{aligned} BA(\alpha_1\mathbf{x}_1 + \alpha_2\mathbf{x}_2) &= B[A(\alpha_1\mathbf{x}_1 + \alpha_2\mathbf{x}_2)] \\ &= B[\alpha_1 A(\mathbf{x}_1) + \alpha_2 A(\mathbf{x}_2)] \\ &= \alpha_1 B(A(\mathbf{x}_1)) + \alpha_2 B(A(\mathbf{x}_2)) \\ &= \alpha_1 BA(\mathbf{x}_1) + \alpha_2 BA(\mathbf{x}_2). \end{aligned}$$

Hence BA satisfies (2–5), and its linearity has been established.

Before going any further, a comment on notation is in order. At first sight it might seem more reasonable to denote the product of A and B by AB rather than BA as above. The explanation for not adopting this notation is quite simple. Were it used, (2–13) would have to be changed to read $AB(\mathbf{x}) = B(A(\mathbf{x}))$ and the writing of equations would then be an open invitation to error.

Having established the convention that the symbol BA stands for the product of A and B, in that order, we observe that this product is defined *only* when the image of A is contained in the domain of B. Thus one of the products AB or BA may exist and the other not, a phenomenon which will reappear later when we introduce the subject of matrices. But even when both A and B map a given vector space into itself, in which case AB and BA are linear transformations on the same space, it is by no means true that they must be equal. A simple example of this disturbing fact can be given in $\mathcal{R}^2$ by letting A be a counterclockwise rotation of 90° about the origin, and B a reflection across any line through the origin, say the x-axis. Then, with $\mathbf{e}_1$ and $\mathbf{e}_2$ the standard basis vectors, $AB(\mathbf{e}_1) = \mathbf{e}_2$ while $BA(\mathbf{e}_1) = -\mathbf{e}_2$, and $AB \neq BA$. (Picture?) In short, *the multiplication of linear transformations is noncommutative.*

The foregoing example illustrates one of the ways in which this multiplication differs from "ordinary" multiplication. Why then call it multiplication at all? The answer is provided by the following identities which show that *most* of the properties usually associated with the term multiplication are still valid when phrased in terms of linear transformations. Specifically, assuming that all of the indicated products are defined, we have

$$A(BC) = (AB)C, \tag{2–14}$$

$$(A_1 + A_2)B = A_1B + A_2B, \qquad A(B_1 + B_2) = AB_1 + AB_2, \tag{2–15}$$

$$(\alpha A)B = A(\alpha B) = \alpha(AB), \quad \alpha \text{ a scalar}, \tag{2–16}$$

$$\left.\begin{aligned} AI &= A, \\ IA &= A, \end{aligned}\right. \quad I \text{ the identity map.} \tag{2–17}$$

The first of these identities asserts that the multiplication of linear transformations is *associative*, the next two that it is *distributive* over addition, and the fourth that it commutes with the operation of scalar multiplication. Finally, (2–17) implies that the identity transformation plays the same role in operator multiplication that the number one plays in arithmetic. The reader should note, however, that two different identity maps are usually involved here, and, strictly speaking, if $A: \mathcal{V}_1 \to \mathcal{V}_2$, then (2–17) ought to be written

$$AI_{\mathcal{V}_1} = A,$$
$$I_{\mathcal{V}_2}A = A,$$

where $I_{\mathcal{V}_1}$ denotes the identity map on $\mathcal{V}_1$, $I_{\mathcal{V}_2}$ the identity map on $\mathcal{V}_2$. But this notation is rarely used since the meaning of the unidentified symbol I is always clear from the context.

The proof of each of the above identities is an easy exercise in the definitions of the operations involved. Thus to establish (2–14) suppose that $C: \mathcal{V}_1 \to \mathcal{V}_2$, $B: \mathcal{V}_2 \to \mathcal{V}_3$, and $A: \mathcal{V}_3 \to \mathcal{V}_4$. Then each of the products $A(BC)$ and $(AB)C$ is a linear transformation from $\mathcal{V}_1$ to $\mathcal{V}_4$, and to prove their equality we simply apply Definition 2–4, twice for each product. This gives

$$[A(BC)](\mathbf{x}) = A(BC(\mathbf{x})) = A\big(B(C(\mathbf{x}))\big),$$

and

$$[(AB)C](\mathbf{x}) = AB(C(\mathbf{x})) = A\big(B(C(\mathbf{x}))\big),$$

and (2–14) now follows from the equality of the right-hand sides of these expressions. The remaining proofs have been left to the exercises.

EXAMPLE 1. *Powers of a linear transformation.* If A is a linear transformation on a fixed vector space $\mathcal{V}$ (i.e., $A: \mathcal{V} \to \mathcal{V}$) we can form the product of A with itself any finite number of times, thereby obtaining a sequence of linear transformations on $\mathcal{V}$ known as the *powers* of A. The associativity of operator multiplication implies that each of these powers is independent of the grouping of its factors and hence can be denoted without ambiguity by A^n, n a positive integer. Thus

$$A^1 = A,\ A^2 = AA,\ A^3 = AA^2, \ldots .$$

In addition, it is customary to let A^0 denote the identity transformation on $\mathcal{V}$, i.e., $A^0 = I$, so that all of the familiar rules for manipulating (nonnegative) exponents become valid. In particular, we have

$$A^mA^n = A^nA^m = A^{m+n},$$
$$(A^m)^n = A^{mn}$$

for all nonnegative integers m and n.

EXAMPLE 2. Let D be the differentiation operator on the space of polynomials $\mathcal{P}_n$. Then D is a linear transformation mapping $\mathcal{P}_n$ into itself, and its powers are simply the derivatives of orders two, three, etc. Since differentiation lowers the degree of every nonzero polynomial by one, the nth power of D maps *every* polynomial in $\mathcal{P}_n$ onto zero, and D^n is the zero transformation on $\mathcal{P}_n$. However, if $n > 1$, then D^{n-1}, and hence D itself, is certainly different from zero, and we have therefore shown that *a power of a nonzero linear transformation may be zero.*

In general, a nonzero linear transformation $A: \mathcal{V} \to \mathcal{V}$ with the property that $A^n = 0$ for some $n > 1$ is said to be *nilpotent* on $\mathcal{V}$, and the smallest integer n such that $A^n = 0$ is called the *degree of nilpotence* of A. We call attention to the fact that the property of being nilpotent actually depends upon the vector space under consideration as well as the linear transformation involved. For instance, D is nilpotent of degree n on $\mathcal{P}_n$, but is not nilpotent on $\mathcal{P}$. (Why?)

EXAMPLE 3. *Polynomials in A.* If A is a linear transformation on a vector space $\mathcal{V}$, we can use the powers of A together with the operations of addition and scalar multiplication to form polynomials in A. Thus if

$$p(x) = a_0 + a_1 x + \cdots + a_n x^n$$

is a polynomial in x with real coefficients, we define $p(A)$ to be the linear transformation on $\mathcal{V}$ obtained by substituting A for x in $p(x)$. In other words,

$$p(A) = a_0 I + a_1 A + \cdots + a_n A^n,$$

or

$$p(A) = a_0 + a_1 A + \cdots + a_n A^n,$$

the factor I being understood in the first term of this expression just as $x^0 = 1$ is understood in $p(x)$. Hence if $\mathbf{x}$ is any vector in $\mathcal{V}$,

$$p(A)(\mathbf{x}) = a_0 \mathbf{x} + a_1 A(\mathbf{x}) + \cdots + a_n A^n(\mathbf{x}).$$

Multiplicatively, these polynomials obey all of the familiar rules of polynomial algebra with the single exception that products can sometimes vanish without any of their factors vanishing, as was shown in the example above. In particular, *the multiplication of polynomials in a linear transformation is commutative* since the identity $p(x)q(x) = q(x)p(x)$ for "ordinary" polynomials p and q implies that $p(A)q(A) = q(A)p(A)$. This, in turn, implies that such polynomials can be factored, for, as the reader will remember, factorization of polynomials depends only on the commutativity of multiplication and its distributivity over addition.

EXAMPLE 4. Let $\mathcal{C}^\infty[a, b]$ denote the space of all infinitely differentiable functions defined on the interval $[a, b]$, and again let D be differentiation. Then D maps $\mathcal{C}^\infty[a, b]$ into itself, and we can therefore form polynomials in D, which in this setting are expressions of the type

$$a_n D^n + a_{n-1} D^{n-1} + \cdots + a_1 D + a_0,$$

$a_0, \ldots, a_n$ real numbers. (Such expressions are known as *constant coefficient linear differential operators*, and it should be observed that they can also be interpreted as linear transformations from $\mathcal{C}^n[a, b]$ to $\mathcal{C}[a, b]$.) The polynomial $D^2 + D - 2$ is a typical example, and if y is any function in $\mathcal{C}^\infty[a, b]$ (or $\mathcal{C}^2[a, b]$), then

$$(D^2 + D - 2)y = \frac{d^2y}{dx^2} + \frac{dy}{dx} - 2y.$$

By virtue of the remarks made in the preceding example, we know that $D^2 + D - 2$ may be rewritten in either of the equivalent forms $(D + 2)(D - 1)$ or $(D - 1)(D + 2)$, and in this case it is easy to verify directly that these factorizations are correct. Indeed,

$$\begin{aligned}(D + 2)(D - 1)y &= (D + 2)[(D - 1)y] \\ &= (D + 2)\left(\frac{dy}{dx} - y\right) \\ &= \frac{d}{dx}\left(\frac{dy}{dx} - y\right) + 2\left(\frac{dy}{dx} - y\right) \\ &= \frac{d^2y}{dx^2} + \frac{dy}{dx} - 2y \\ &= (D^2 + D - 2)y,\end{aligned}$$

while a similar calculation yields the equality $(D - 1)(D + 2) = D^2 + D - 2$.

EXAMPLE 5. Let A and B denote the linear transformations mapping $\mathcal{C}^\infty[a, b]$ into itself defined by

$$A = xD + 1, \qquad B = D - x;$$

that is,

$$A(y) = x\frac{dy}{dx} + y, \qquad B(y) = \frac{dy}{dx} - xy$$

for each y in $\mathcal{C}^\infty[a, b]$ (see Exercise 4). Then

$$\begin{aligned}AB(y) &= (xD + 1)[(D - x)y] \\ &= (xD + 1)\left(\frac{dy}{dx} - xy\right) \\ &= x\frac{d}{dx}\left(\frac{dy}{dx} - xy\right) + \frac{dy}{dx} - xy \\ &= x\frac{d^2y}{dx^2} + (1 - x^2)\frac{dy}{dx} - 2xy \\ &= [xD^2 + (1 - x^2)D - 2x]y,\end{aligned}$$

and hence

$$(xD + 1)(D - x) = xD^2 + (1 - x^2)D - 2x. \tag{2–18}$$

On the other hand,

$$\begin{aligned} BA(y) &= (D - x)\left(x\frac{dy}{dx} + y\right) \\ &= \frac{d}{dx}\left(x\frac{dy}{dx} + y\right) - x\left(x\frac{dy}{dx} + y\right) \\ &= x\frac{d^2y}{dx^2} + (2 - x^2)\frac{dy}{dx} - xy \\ &= [xD^2 + (2 - x^2)D - x]y, \end{aligned}$$

and hence

$$(D - x)(xD + 1) = xD^2 + (2 - x^2)D - x. \tag{2–19}$$

Comparing these results, we see that

$$(xD + 1)(D - x) \neq (D - x)(xD + 1),$$

and we have another illustration of the noncommutativity of operator multiplication. The reader should note that in this case neither the product AB nor BA can be evaluated by using the rules of elementary algebra. This is another of the unpleasant consequences of a noncommutative multiplication, and, as we shall see, has a decisive effect upon the study of linear differential equations.

In view of the examples just given it is clear that the time has come for us to discuss the general problem of functional notation. All of the functions which we shall encounter in this text will be elements in one of a number of vector spaces, and hence should be denoted by such symbols as **f** or **f**(x) were we to be inflexible in our use of bold face type. This, however, would ultimately involve us in such unsightly (and confusing) expressions as $x^{\mathbf{n}}$, **sin** x, $d^2\mathbf{y}/dx^2$, $\int_a^b \mathbf{f}(x)\mathbf{g}(x)\,dx$. Such pedantry is pointless, and so we shall use the symbols f and $f(x)$ when in our opinion the printed page or its reader would suffer from the use of bold face type. As a general rule, when we wish to call attention to the fact that a function is a vector, we shall emphasize it; otherwise, not.

Finally, to settle notational matters once and for all, we comment on our intended use (and mild misuse) of the symbols f and $f(x)$. Strictly speaking, f should be used to denote a function, and $f(x)$ its value at the point x. But here again strict adherence to the letter of the law violates the spirit of clarity of exposition, for then we would be forced, for example, to write ^{n}, sin , and e where everyone expects x^n, sin x, and e^x. We shall not disappoint the reader's expectations on this score either.

EXERCISES

1. Cite the relevant axiom or definition needed to justify each step in the proof of the linearity of BA as given in the text.
2. Prove the distributivity formulas (2–15).

3. Let D and L denote, respectively, the operations of differentiation and integration on $\mathcal{C}^\infty[a, b]$; that is,

$$D(y) = \frac{dy}{dx}, \qquad L(y) = \int_a^x y(t)\,dt$$

for all y in $\mathcal{C}^\infty[a, b]$.

(a) Compute the value of LD and DL.

(b) Compute the value of L^nD^n and D^nL^n, n a non-negative integer.

4. (a) Prove that the mappings A and B defined in Example 5 above are linear transformations on $\mathcal{C}^\infty[a, b]$.

(b) Prove that every expression of the form

$$a_n(x)D^n + a_{n-1}(x)D^{n-1} + \cdots + a_1(x)D + a_0(x)$$

defines a linear transformation from $\mathcal{C}^n[a, b]$ to $\mathcal{C}[a, b]$ whenever $a_0(x), \ldots, a_n(x)$ are continuous on $[a, b]$.

5. Write each of the following products in the form $a_2(x)D^2 + a_1(x)D + a_0(x)$.

(a) $(xD + 1)^2$

(b) $(2xD + 1)(D - 1)$

(c) $(D - 1)(2xD + 1)$

(d) $(x^2D + 2x)(D - 2x)$

(e) $(D - 2x)(x^2D + 2x)$

6. Let $A, B\colon \mathcal{V} \to \mathcal{V}$ be linear, and suppose that $AB = BA$. Find a formula for $(A + B)^n$, n a non-negative integer.

7. In each of the following find the result obtained by applying the given polynomial in D to the indicated functions.

(a) $D^2 - 1$; $2e^x$, e^{-x}, $e^x + e^{-x}$

(b) $D^2 + 1$; $\sin x + \cos x$, $2 \sin 2x$, e^x

(c) $(D + 1)(D - 2)$; $\sin x + e^{-x} + e^{2x}$, e^x, x^2

(d) $(D + 2)^2$; e^{-2x}, xe^{-2x}, x^2e^{-2x}

(e) $(xD - 1)(xD + 2)$; x^2, $(x^3 + 1)/x$, e^x.

8. (a) Prove that $AB = BA$ for any pair of linear transformations A, B mapping a one-dimensional vector space $\mathcal{V}$ into itself. [*Hint:* Choose a basis for $\mathcal{V}$.]

(b) Let $\mathcal{V}$ be finite dimensional, with $\dim \mathcal{V} > 1$. Prove that there always exist linear transformations A and B mapping $\mathcal{V}$ into itself such that $AB \neq BA$.

9. Let A, B, C be linear transformations each mapping a given (unspecified) vector space into another vector space, and suppose that each of the products AB and BC is defined. Prove that $(AB)C$ and $A(BC)$ are also defined.

*10. (a) Let A and B be nilpotent linear transformations on a vector space $\mathcal{V}$, and suppose that $AB = BA$. Prove that AB is nilpotent on $\mathcal{V}$.

(b) Give an example to show that the conclusion in (a) may be false if $AB \neq BA$.

11. Let B be a fixed linear transformation on a vector space $\mathcal{V}$. Show that the set of all linear transformations A on $\mathcal{V}$ such that $AB = O$ is a subspace of the vector space of all linear transformations on $\mathcal{V}$. What is the subspace if $B = O$? If $B = I$?

2-4 THE NULL SPACE AND IMAGE; INVERSES

Let A be a linear transformation from $\mathcal{V}_1$ to $\mathcal{V}_2$, and let $\mathfrak{N}(A)$ denote the set of all $\mathbf{x}$ in $\mathcal{V}_1$ such that $A(\mathbf{x}) = \mathbf{0}$. Then, as we have already observed, $\mathfrak{N}(A)$ always contains the zero vector of $\mathcal{V}_1$. Actually we can say much more than this, for if $A(\mathbf{x}_1) = A(\mathbf{x}_2) = \mathbf{0}$, then

$$A(\alpha_1\mathbf{x}_1 + \alpha_2\mathbf{x}_2) = \alpha_1 A(\mathbf{x}_1) + \alpha_2 A(\mathbf{x}_2) = \mathbf{0}$$

for all scalars α_1, α_2, and it follows that $\mathfrak{N}(A)$ *is a subspace of* $\mathcal{V}_1$. This subspace is called the *null space* or *kernel* of A, and is of fundamental importance in studying the behavior of A on $\mathcal{V}_1$.

Of equal importance with the null space of A is its image, $\mathfrak{I}(A)$, which, we recall, is the set of all $\mathbf{y}$ in $\mathcal{V}_2$ such that $\mathbf{y} = A(\mathbf{x})$ for some $\mathbf{x}$ in $\mathcal{V}_1$. It too is a subspace—this time in $\mathcal{V}_2$—since if $\mathbf{y}_1$ and $\mathbf{y}_2$ belong to $\mathfrak{I}(A)$ with $\mathbf{y}_1 = A(\mathbf{x}_1)$, $\mathbf{y}_2 = A(\mathbf{x}_2)$, then

$$A(\alpha_1\mathbf{x}_1 + \alpha_2\mathbf{x}_2) = \alpha_1 A(\mathbf{x}_1) + \alpha_2 A(\mathbf{x}_2) = \alpha_1\mathbf{y}_1 + \alpha_2\mathbf{y}_2,$$

and $\alpha_1\mathbf{y}_1 + \alpha_2\mathbf{y}_2$ is also in the image of A, as required.

EXAMPLE 1. Let $I: \mathcal{V} \to \mathcal{V}$ be the identity transformation. Then $\mathfrak{N}(I) = \mathcal{O}$, the trivial subspace of $\mathcal{V}$, while $\mathfrak{I}(I) = \mathcal{V}$.

EXAMPLE 2. If $O: \mathcal{V}_1 \to \mathcal{V}_2$ is the zero transformation, then, by its very definition, $\mathfrak{N}(O) = \mathcal{V}_1$, $\mathfrak{I}(O) = \mathcal{O}$.

EXAMPLE 3. Let D be the differentiation operator on the space of polynomials $\mathcal{P}_n$. Then the null space of D consists of all polynomials of degree zero together with the zero polynomial, while its image consists of the zero polynomial and all polynomials of degree $< n - 1$.

EXAMPLE 4. Let $\mathcal{C}^2(-\infty, \infty)$ denote the space of all twice continuously differentiable functions on $(-\infty, \infty)$, and let $A: \mathcal{C}^2(-\infty, \infty) \to \mathcal{C}(-\infty, \infty)$ be the linear transformation $D^2 - I$. Then

$$A(y) = \frac{d^2y}{dx^2} - y,$$

and the null space of A is the set of all functions y in $\mathcal{C}^2(-\infty, \infty)$ for which

$$\frac{d^2y}{dx^2} - y = 0.$$

Thus, $\mathfrak{N}(A)$ is the set of solutions of a certain differential equation, and the problem of finding all solutions of this equation is identical with that of finding the null space of $D^2 - I$.

EXAMPLE 5. Let $\mathcal{R}^\infty$ be the space of all infinite sequences $\{x_1, x_2, x_3, \ldots\}$ of real numbers, with addition and scalar multiplication defined termwise (see

Exercise 8, Section 1–2), and let A and B be the linear transformations on $\mathcal{R}^\infty$ defined by

$$A\{x_1, x_2, x_3, \ldots\} = \{x_2, x_3, x_4, \ldots\},$$
$$B\{x_1, x_2, x_3, \ldots\} = \{0, x_1, x_2, \ldots\}.$$

Then $\mathfrak{N}(A)$ is the subspace of $\mathcal{R}^\infty$ consisting of all sequences of the form $\{x_1, 0, 0, \ldots\}$, with x_1 arbitrary, while $\mathfrak{N}(B) = \mathcal{O}$. On the other hand, $\mathcal{I}(A) = \mathcal{R}^\infty$, while, by definition, $\mathcal{I}(B)$ consists of all sequences whose first entry is zero.

Now that we have introduced the null space and image of a linear transformation we propose to take a closer look at those transformations $A: \mathcal{V}_1 \to \mathcal{V}_2$ for which either

$$\text{(i)} \quad \mathfrak{N}(A) = \mathcal{O}, \qquad \text{or} \qquad \text{(ii)} \quad \mathcal{I}(A) = \mathcal{V}_2,$$

or both. The second of these equations asserts that A maps $\mathcal{V}_1$ *onto* $\mathcal{V}_2$, and implies that for each $\mathbf{y}$ in $\mathcal{V}_2$ there exists at least one $\mathbf{x}$ in $\mathcal{V}_1$ such that $\mathbf{y} = A(\mathbf{x})$. The first, which says that the null space of A contains only the zero vector, turns out to be equivalent to the assertion that A is *one-to-one* in the sense of the following definition.

Definition 2–5. A linear transformation $A: \mathcal{V}_1 \to \mathcal{V}_2$ is said to be *one-to-one* if and only if $A(\mathbf{x}_1) = A(\mathbf{x}_2)$ implies that $\mathbf{x}_1 = \mathbf{x}_2$.

In other words, A is one-to-one if and only if A maps *distinct* vectors in $\mathcal{V}_1$ onto *distinct* vectors in $\mathcal{V}_2$; whence the name. (See Fig. 2–6.) This said, we now prove

Theorem 2–2. *A linear transformation $A: \mathcal{V}_1 \to \mathcal{V}_2$ is one-to-one if and only if $\mathfrak{N}(A) = \mathcal{O}$.*

Proof. Let A be one-to-one, and suppose that $A(\mathbf{x}) = \mathbf{0}$. Then $A(\mathbf{x}) = A(\mathbf{0})$, and Definition 2–5 implies that $\mathbf{x} = \mathbf{0}$. Thus $\mathfrak{N}(A) = \mathcal{O}$. Conversely, if $\mathfrak{N}(A) = \mathcal{O}$ and $A(\mathbf{x}_1) = A(\mathbf{x}_2)$, then $A(\mathbf{x}_1) - A(\mathbf{x}_2) = \mathbf{0}$, or $A(\mathbf{x}_1 - \mathbf{x}_2) = \mathbf{0}$. Thus $\mathbf{x}_1 - \mathbf{x}_2 = \mathbf{0}$, and $\mathbf{x}_1 = \mathbf{x}_2$, as asserted. ▌

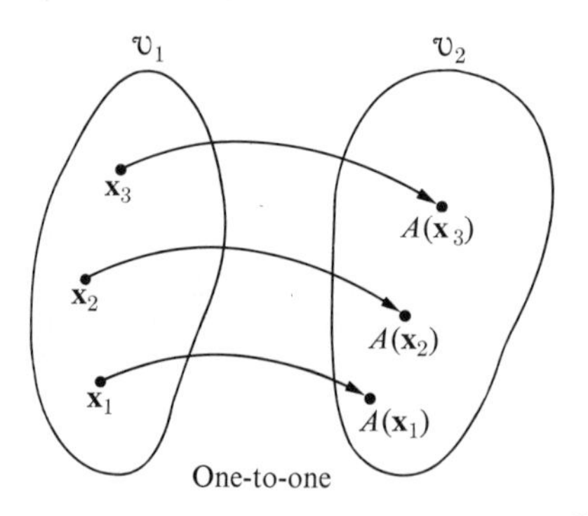

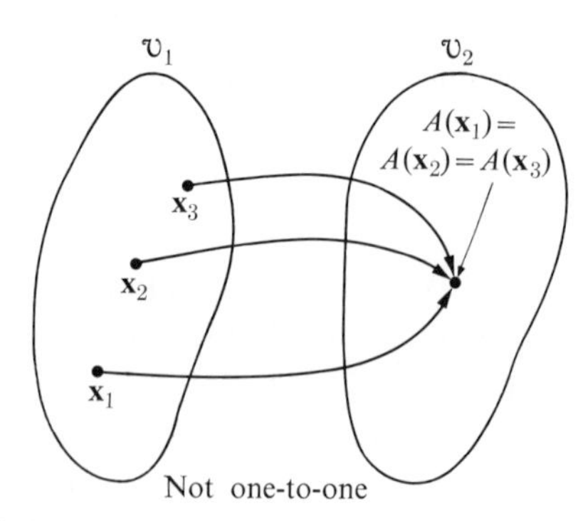

FIGURE 2–6

Among the various transformations appearing in the examples above, only $B: \mathcal{R}^\infty \to \mathcal{R}^\infty$ and I were one-to-one, since only they had trivial null spaces. Additional examples of such transformations are provided by rotations of $\mathcal{R}^2$ about the origin, reflections across any line through the origin, etc. The reader should have no difficulty in augmenting this list indefinitely.

Linear transformations which are both one-to-one and onto are called *isomorphisms*, and are said to be *invertible*. They are of particular importance since, just as with ordinary one-to-one onto functions, they have inverses, and all of the standard facts concerning inverse functions can then be established. Indeed, if $A: \mathcal{V}_1 \to \mathcal{V}_2$ is one-to-one and onto, *each* vector $\mathbf{y}$ in $\mathcal{V}_2$ is paired with a *unique* vector $\mathbf{x}$ in $\mathcal{V}_1$, and A can therefore be used to define a function from $\mathcal{V}_2$ to $\mathcal{V}_1$. This function is called the *inverse* of A, and is denoted by A^{-1} (read "A inverse"). It can be described explicitly as *the* function from $\mathcal{V}_2$ to $\mathcal{V}_1$ such that

$$A^{-1}(\mathbf{y}) = \mathbf{x}, \qquad \text{where} \quad A(\mathbf{x}) = \mathbf{y} \tag{2–20}$$

for each $\mathbf{y}$ in $\mathcal{V}_2$. Loosely speaking, A^{-1} is obtained from A by reading the definition of A from right to left, as suggested in Fig. 2–7, and is clearly a one-to-one map of $\mathcal{V}_2$ onto $\mathcal{V}_1$. Moreover, it is linear, since if $\mathbf{y}_1$ and $\mathbf{y}_2$ belong to $\mathcal{V}_2$ with $\mathbf{y}_1 = A(\mathbf{x}_1)$, $\mathbf{y}_2 = A(\mathbf{x}_2)$, then

$$A(\alpha_1\mathbf{x}_1 + \alpha_2\mathbf{x}_2) = \alpha_1\mathbf{y}_1 + \alpha_2\mathbf{y}_2,$$

and (2–20) implies that

$$A^{-1}(\alpha_1\mathbf{y}_1 + \alpha_2\mathbf{y}_2) = \alpha_1\mathbf{x}_1 + \alpha_2\mathbf{x}_2 = \alpha_1A^{-1}(\mathbf{y}_1) + \alpha_2A^{-1}(\mathbf{y}_2).$$

Having observed that A^{-1} is one-to-one, onto, and linear, it follows that it too is invertible, and if we simply parrot the construction given above, this time starting with A^{-1}, we find that $(A^{-1})^{-1} = A$. Finally, if we form the products $A^{-1}A$ and AA^{-1}, each of which is certainly defined, an easy argument reveals that they both reduce to the identity; that is,

$$A^{-1}A(\mathbf{x}) = \mathbf{x} \qquad \text{and} \qquad AA^{-1}(\mathbf{y}) = \mathbf{y}$$

for all $\mathbf{x}$ in $\mathcal{V}_1$ and all $\mathbf{y}$ in $\mathcal{V}_2$.* And with this we have proved the following theorem.

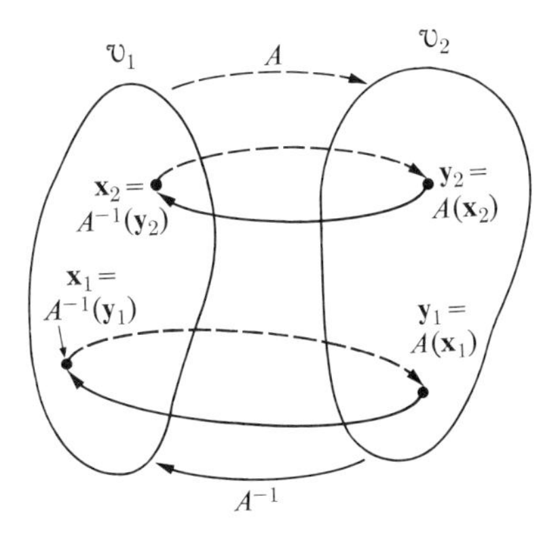

FIGURE 2–7

* These equations are the vector space analogs of such pairs of statements as

$$\sin^{-1}(\sin x) = x, \qquad \sin(\sin^{-1} x) = x, \qquad -\pi/2 \le x \le \pi/2,$$

or

$$e^{\ln x} = x, \qquad \ln(e^x) = x,$$

which are familiar from calculus.

Theorem 2–3. *Every one-to-one linear transformation A mapping $\mathcal{V}_1$ onto $\mathcal{V}_2$ has a unique inverse from $\mathcal{V}_2$ to $\mathcal{V}_1$ defined by*

$$A^{-1}(\mathbf{y}) = \mathbf{x},$$

where $A(\mathbf{x}) = \mathbf{y}$ for all $\mathbf{y}$ in $\mathcal{V}_2$. A^{-1} is also one-to-one, onto, and linear, with $(A^{-1})^{-1} = A$, and

$$A^{-1}A = I_{\mathcal{V}_1}, \qquad AA^{-1} = I_{\mathcal{V}_2}, \tag{2–21}$$

where $I_{\mathcal{V}_1}$ and $I_{\mathcal{V}_2}$ denote, respectively, the identity maps on $\mathcal{V}_1$ and $\mathcal{V}_2$.

These last equations actually serve to characterize invertible linear transformations—a fact which when stated precisely reads as follows:

Theorem 2–4. *Let $A: \mathcal{V}_1 \to \mathcal{V}_2$ and $B: \mathcal{V}_2 \to \mathcal{V}_1$ be linear, and suppose that BA and AB are, respectively, the identity maps on $\mathcal{V}_1$ and $\mathcal{V}_2$. Then A is one-to-one and onto, and $B = A^{-1}$.*

Proof. Let $\mathbf{x}$ in $\mathcal{V}_1$ be such that $A(\mathbf{x}) = \mathbf{0}$. Then on the one hand,

$$B\big(A(\mathbf{x})\big) = B(\mathbf{0}) = \mathbf{0},$$

and on the other,

$$B\big(A(\mathbf{x})\big) = BA(\mathbf{x}) = I(\mathbf{x}) = \mathbf{x}.$$

Thus $\mathbf{x} = \mathbf{0}$, and $\mathfrak{N}(A) = \mathcal{O}$.

Now let $\mathbf{y}$ be an arbitrary vector in $\mathcal{V}_2$. Then

$$\mathbf{y} = I(\mathbf{y}) = AB(\mathbf{y}) = A\big(B(\mathbf{y})\big),$$

and it follows that $\mathbf{y}$ is the image under A of the vector $B(\mathbf{y})$ in $\mathcal{V}_1$. Thus $\mathfrak{I}(A) = \mathcal{V}_2$, and we are done. ▌

Example 6. If A is any rotation of $\mathbb{R}^2$ about the origin through an angle θ, then A is invertible with A^{-1} the rotation through $-\theta$, since $A^{-1}A = AA^{-1} = I$.

Example 7. Let $A: \mathbb{R}^3 \to \mathbb{R}^3$ be defined by

$$A(x_1, x_2, x_3) = (x_1 + x_2, x_2, x_3).$$

Then A is invertible, with A^{-1} given by

$$A^{-1}(x_1, x_2, x_3) = (x_1 - x_2, x_2, x_3),$$

since

$$A^{-1}A(x_1, x_2, x_3) = (x_1, x_2, x_3) = AA^{-1}(x_1, x_2, x_3).$$

Theorem 2–4 suggests a natural and valuable generalization of the notion of the inverse of a linear transformation $A: \mathcal{V}_1 \to \mathcal{V}_2$; to wit, a linear transformation

$B: \mathcal{V}_2 \to \mathcal{V}_1$ such that

$$AB = I, \qquad BA \neq I.$$

The fact that such transformations do exist can be seen by looking at Example 5 above where

$$AB\{x_1, x_2, x_3, \ldots\} = \{x_1, x_2, x_3, \ldots\},$$

and

$$BA\{x_1, x_2, x_3, \ldots\} = \{0, x_2, x_3, \ldots\}.$$

Transformations of this sort are encountered fairly often in certain types of problems and are therefore distinguished by name according to the following definition.

Definition 2–6. A linear transformation $B: \mathcal{V}_2 \to \mathcal{V}_1$ is said to be a *right inverse* for $A: \mathcal{V}_1 \to \mathcal{V}_2$ if the product AB is the identity map on $\mathcal{V}_2$. Similarly, B is said to be a *left inverse* for A if BA is the identity map on $\mathcal{V}_1$.

Remark. If B is a right (left) inverse for A, then A is a left (right) inverse for B.

The example given a moment ago shows that a linear transformation may have a right or left inverse without having an inverse. It is easy to show, however, that if A has *both* a right inverse B and a left inverse C, then A is invertible, and $B = C = A^{-1}$. For then

$$AB = I \quad \text{and} \quad CA = I,$$

and it follows that

$$C(AB) = CI = C, \qquad (CA)B = IB = B,$$

and hence that $B = C$. Thus $AB = BA = I$, and the assertion that A is invertible with $B = C = A^{-1}$ now follows from Theorem 2–4.

EXAMPLE 8. Let $\mathcal{C}^\infty[a, b]$ be the space of infinitely differentiable functions on $[a, b]$, and let D and L be differentiation and integration, respectively; that is,

$$Dy = \frac{dy}{dx}, \qquad Ly = \int_a^x y(t)\, dt.$$

Then

$$LD(y) = \int_a^x y'(t)\, dt = y(x) - y(a),$$

while

$$DL(y) = \frac{d}{dx} \int_a^x y(t)\, dt = y(x),$$

and it follows that $DL = I$, $LD \neq I$. In other words, the operation of integration on function spaces is only a right inverse, and not an inverse, for differentiation. It is this fact more than any other which motivated us to introduce the notions of right and left inverses in the first place.

EXERCISES

1. Find the null space and, where applicable, the inverse of each of the following linear transformations on $\mathcal{R}^2$.

   (a) $A(x_1, x_2) = 2(x_1, -x_2)$

   (b) $A(x_1, x_2) = (x_2, 0)$

   (c) $A(x_1, x_2) = (x_1 + x_2, x_1 + x_2)$

   (d) $A(x_1, x_2) = (x_1 + x_2, x_1 - x_2)$

2. Repeat Exercise 1 for the following transformations on $\mathcal{R}^3$.

   (a) $A(x_1, x_2, x_3) = (x_1 + x_2, x_2 + x_3, x_1)$

   (b) $A(x_1, x_2, x_3) = (2x_1, -x_3, x_1 + x_3)$

   (c) $A(x_1, x_2, x_3) = (x_2 + 2x_3, x_1 - x_2, 0)$

   (d) $A(x_1, x_2, x_3) = (x_1 - x_2, x_1 + x_2 + x_3, x_2 + x_3)$

3. Repeat Exercise 1 for the following transformations on $\mathcal{P}$.

   (a) $A(p) = \dfrac{d^2p}{dx^2} - 2\dfrac{dp}{dx}$

   (b) $A(p) = xp(x)$

   (c) $A(p) = p(x) - p(0)$

   (d) $A(p) = p(x)q(x)$, $q(x)$ a fixed polynomial in $\mathcal{P}$

4. In giving the definition of A^{-1} we insisted that A be one-to-one. Why?

5. Let $A: \mathcal{R}^2 \to \mathcal{R}^2$ be defined by

$$A(x_1, x_2) = (\alpha_1 x_1 + \alpha_2 x_2, \beta_1 x_1 + \beta_2 x_2),$$

   where $\alpha_1, \alpha_2, \beta_1, \beta_2$ are real numbers. Prove that A is linear, and find a necessary and sufficient condition in terms of $\alpha_1, \alpha_2, \beta_1, \beta_2$ for A to be invertible.

6. Let $A: \mathcal{V}_1 \to \mathcal{V}_2$ be linear, let $\mathcal{W}$ be a subspace of $\mathcal{V}_1$, and let $A(\mathcal{W})$ denote the image of $\mathcal{W}$ in $\mathcal{V}_2$; i.e., $A(\mathcal{W})$ is the set of all vectors $\mathbf{y}$ in $\mathcal{V}_2$ such that $\mathbf{y} = A(\mathbf{w})$ for some $\mathbf{w}$ in $\mathcal{W}$. Prove that $A(\mathcal{W})$ is a subspace of $\mathcal{V}_2$.

7. Let $A: \mathcal{V}_1 \to \mathcal{V}_2$ be linear, and let $\mathcal{W}$ be a subspace of $\mathcal{V}_2$. Prove that the set of all $\mathbf{x}$ in $\mathcal{V}_1$ such that $A(\mathbf{x})$ belongs to $\mathcal{W}$ is a subspace of $\mathcal{V}_1$.

8. Let $A: \mathcal{V}_1 \to \mathcal{V}_2$ be a one-to-one linear transformation, and let $\mathbf{e}_1, \ldots, \mathbf{e}_n$ be linearly independent vectors in $\mathcal{V}_1$. Prove that $A(\mathbf{e}_1), \ldots, A(\mathbf{e}_n)$ are linearly independent in $\mathcal{V}_2$.

9. Let $A: \mathcal{V}_1 \to \mathcal{V}_2$ be linear, and suppose that

$$\dim \mathcal{V}_1 = \dim \mathcal{V}_2 = n < \infty.$$

   Prove that A is one-to-one if and only if it is onto. [*Hint:* See Exercise 8.]

10. Let A and B be invertible linear transformations mapping $\mathcal{V}$ onto itself. Prove that AB and BA are also invertible, and that

$$(AB)^{-1} = B^{-1}A^{-1}, \qquad (BA)^{-1} = A^{-1}B^{-1}.$$

11. Let $A: \mathcal{V} \to \mathcal{V}$ be linear, and suppose that $A^2 + I = A$. Prove that A is invertible.

*12. (a) Let $\mathcal{V}_2$ be finite dimensional, and let $A: \mathcal{V}_1 \to \mathcal{V}_2$ be one-to-one. Prove that A has a left inverse but not a right inverse whenever $\mathcal{I}(A) \neq \mathcal{V}_2$. [*Hint:* Choose an appropriate basis in $\mathcal{V}_2$.]

(b) Now let $\mathcal{V}_1$ be finite dimensional, and suppose that A is onto. Prove that A has a right inverse but no left inverse whenever $\mathcal{N}(A) \neq \mathcal{O}$.

13. A linear transformation $P: \mathcal{V} \to \mathcal{V}$ is said to be *idempotent* if and only if $P^2 = P$.

(a) Prove that P is idempotent if and only if $I - P$ is.

(b) Prove that $\mathcal{N}(P) = \mathcal{I}(I - P)$, and that $\mathcal{I}(P) = \mathcal{N}(I - P)$ whenever P is idempotent. [*Hint:* The image of P consists of all $\mathbf{x}$ in $\mathcal{V}$ such that $P(\mathbf{x}) = \mathbf{x}$.]

(c) Use the results of (b) to show that every $\mathbf{x}$ in $\mathcal{V}$ can be written *uniquely* in the form

$$\mathbf{x} = \mathbf{x}_1 + \mathbf{x}_2$$

with $\mathbf{x}_1$ in $\mathcal{N}(P)$, $\mathbf{x}_2$ in $\mathcal{I}(P)$ whenever P is idempotent.

14. Let $\mathcal{V}$ be finite dimensional, and let $\mathcal{W}$ be a subspace of $\mathcal{V}$.

(a) Prove that there exists a linear transformation $P: \mathcal{V} \to \mathcal{W}$ such that $\mathcal{I}(P) = \mathcal{W}$, and $P(\mathbf{x}) = \mathbf{x}$ for all $\mathbf{x}$ in $\mathcal{W}$. A linear transformation of this type is said to be a *projection* of $\mathcal{V}$ onto $\mathcal{W}$. [*Hint:* Choose a basis for $\mathcal{W}$, and extend to a basis for $\mathcal{V}$.]

(b) Prove that a linear transformation P on $\mathcal{V}$ is a projection onto a subspace $\mathcal{W}$ if and only if P is idempotent and $\mathcal{W} = \mathcal{I}(P)$. (See Exercise 13.)

15. Two vector spaces $\mathcal{V}_1$ and $\mathcal{V}_2$ are said to be *isomorphic* if and only if there exists an isomorphism $A: \mathcal{V}_1 \to \mathcal{V}_2$.

(a) Prove that two finite dimensional spaces $\mathcal{V}_1$ and $\mathcal{V}_2$ are isomorphic if and only if $\dim \mathcal{V}_1 = \dim \mathcal{V}_2$.

(b) Let $\mathcal{R}^+$ denote the space of Exercise 10, Section 1–2. Prove that $\mathcal{R}^+$ and $\mathcal{R}^1$ are isomorphic by exhibiting an isomorphism $A: \mathcal{R}^+ \to \mathcal{R}^1$.

2–5 LINEAR TRANSFORMATIONS AND BASES

Until now we have carefully refrained from using bases and coordinates in our study of linear transformations in order to avoid subverting our results by tying them to a particular choice of coordinate system or suggesting that they are valid only for finite dimensional spaces. But as everyone knows full well, coordinate systems are invaluable in computations, and we therefore propose to devote the remainder of this chapter to exploring some of the connections between linear transformations and bases. In so doing we will be led to the notion of a matrix and to the subject of matrix algebra, which can best be described as the arithmetic of coordinatized linear transformations. Thus we now impose the restriction that every one of the vector spaces encountered in the several sections which follow is *finite dimensional* unless explicitly stated otherwise.

With this restriction in force, let A be a linear transformation from $\mathcal{V}_1$ to $\mathcal{V}_2$, and let $\mathbf{e}_1, \ldots, \mathbf{e}_n$ be a basis for $\mathcal{V}_1$. Then if

$$\mathbf{x} = x_1\mathbf{e}_1 + \cdots + x_n\mathbf{e}_n$$

is any vector in $\mathcal{V}_1$,

$$A(\mathbf{x}) = x_1 A(\mathbf{e}_1) + \cdots + x_n A(\mathbf{e}_n), \tag{2–22}$$

and it follows that the value of $A(\mathbf{x})$ is completely determined by the vectors $A(\mathbf{e}_1), \ldots, A(\mathbf{e}_n)$ in $\mathcal{V}_2$; i.e., A is uniquely determined by its values on a basis for $\mathcal{V}_1$. Moreover, if $\mathbf{y}_1, \ldots, \mathbf{y}_n$ are *arbitrary* vectors in $\mathcal{V}_2$, the mapping $A: \mathcal{V}_1 \to \mathcal{V}_2$ defined by setting

$$A(\mathbf{e}_1) = \mathbf{y}_1, \ \ldots \ , A(\mathbf{e}_n) = \mathbf{y}_n,$$

and then using (2–22) to compute the value of $A(\mathbf{x})$ for every $\mathbf{x}$ in $\mathcal{V}_1$ is clearly linear. Thus Eq. (2–22) also tells us how to construct *all* linear transformations from $\mathcal{V}_1$ to $\mathcal{V}_2$, and we have proved

Theorem 2–5. *Every linear transformation from $\mathcal{V}_1$ to $\mathcal{V}_2$ is uniquely determined by its values on a basis for $\mathcal{V}_1$. These values can be chosen arbitrarily in $\mathcal{V}_2$, different choices yielding different transformations, and every linear transformation from $\mathcal{V}_1$ to $\mathcal{V}_2$ can be obtained in this way.*

EXAMPLE 1. Let $\mathbf{e}_1$ and $\mathbf{e}_2$ be the standard basis vectors in $\mathcal{R}^2$, and let A be a linear transformation from $\mathcal{R}^2$ to $\mathcal{R}^1$. Then A is completely determined by the pair of real numbers $A(\mathbf{e}_1)$, $A(\mathbf{e}_2)$, and can therefore be represented by the ordered pair $\big(A(\mathbf{e}_1), A(\mathbf{e}_2)\big)$. Since distinct ordered pairs define distinct linear transformations, it follows that there are exactly as many linear transformations from $\mathcal{R}^2$ to $\mathcal{R}^1$ as there are vectors in $\mathcal{R}^2$.

EXAMPLE 2. Let $A: \mathcal{R}^2 \to \mathcal{R}^2$ be the linear transformation defined by

$$\begin{aligned} A(\mathbf{e}_1) &= (\alpha_1, \alpha_2), \\ A(\mathbf{e}_2) &= (\beta_1, \beta_2), \end{aligned}$$

where $\mathbf{e}_1$ and $\mathbf{e}_2$ are the standard basis vectors, and let

$$\mathbf{x} = x_1\mathbf{e}_1 + x_2\mathbf{e}_2$$

be any vector in $\mathcal{R}^2$. Then

$$\begin{aligned} A(\mathbf{x}) &= x_1 A(\mathbf{e}_1) + x_2 A(\mathbf{e}_2) \\ &= x_1(\alpha_1, \alpha_2) + x_2(\beta_1, \beta_2) \\ &= (\alpha_1 x_1 + \beta_1 x_2, \ \alpha_2 x_1 + \beta_2 x_2). \end{aligned}$$

In this case A can be represented by the array of scalars

$$\begin{bmatrix} \alpha_1 & \beta_1 \\ \alpha_2 & \beta_2 \end{bmatrix},$$

and every such array can be viewed as the definition of a linear transformation

A of $\mathcal{R}^2$ into itself where

$$A(x_1, x_2) = (\alpha_1 x_1 + \beta_1 x_2, \alpha_2 x_1 + \beta_2 x_2).$$

EXAMPLE 3. Let $A: \mathcal{R}^2 \to \mathcal{R}^1$ be the linear transformation described by the ordered pair of real numbers $(2, -1)$ with respect to the standard basis in $\mathcal{R}^2$; that is,

$$A(\mathbf{e}_1) = 2, \qquad A(\mathbf{e}_2) = -1,$$

and let $\mathbf{e}_1' = \mathbf{e}_1 + \mathbf{e}_2$, $\mathbf{e}_2' = -\mathbf{e}_1$. Then $\mathbf{e}_1'$, $\mathbf{e}_2'$ is also a basis for $\mathcal{R}^2$, and we have

$$A(\mathbf{e}_1') = A(\mathbf{e}_1) + A(\mathbf{e}_2) = 1,$$
$$A(\mathbf{e}_2') = -A(\mathbf{e}_1) = -2.$$

Thus the ordered pair which describes A with respect to the basis $\mathbf{e}_1'$, $\mathbf{e}_2'$ is $(1, -2)$, and we see that *the description of a linear transformation by means of its values on a basis changes with a change of basis.*

EXERCISES

1. Let $A: \mathcal{R}^2 \to \mathcal{R}^1$ be represented by the ordered pair of real numbers (α_1, α_2) with respect to the standard basis in $\mathcal{R}^2$, and let $B: \mathcal{R}^2 \to \mathcal{R}^1$ be represented by (β_1, β_2). Show that αA is represented by $(\alpha\alpha_1, \alpha\alpha_2)$, and that $A + B$ is represented by $(\alpha_1 + \beta_1, \alpha_2 + \beta_2)$.
2. Let $A: \mathcal{R}^2 \to \mathcal{R}^1$ be represented by the ordered pair of real numbers (α_1, α_2) with respect to the standard basis in $\mathcal{R}^2$. Find a necessary and sufficient condition that there exist a basis $\mathbf{e}_1'$, $\mathbf{e}_2'$ in $\mathcal{R}^2$ with respect to which A is represented by $(1, 1)$. Assume that this condition is satisfied, and find $\mathbf{e}_1'$, $\mathbf{e}_2'$.
3. Let $A: \mathcal{R}^2 \to \mathcal{R}^2$ be represented by the array

$$\begin{bmatrix} \alpha_1 & \beta_1 \\ \alpha_2 & \beta_2 \end{bmatrix},$$

and let $B: \mathcal{R}^2 \to \mathcal{R}^2$ be represented by

$$\begin{bmatrix} \alpha_1' & \beta_1' \\ \alpha_2' & \beta_2' \end{bmatrix}.$$

Find the representation of αA, $A + B$, and AB.
4. Let $\mathcal{L}(\mathcal{R}^2, \mathcal{R}^1)$ denote the vector space of all linear transformations from $\mathcal{R}^2$ to $\mathcal{R}^1$, and let

$$T: \mathcal{L}(\mathcal{R}^2, \mathcal{R}^1) \to \mathcal{R}^2$$

be defined by

$$T(A) = (\alpha_1, \alpha_2),$$

where (α_1, α_2) is the ordered pair of real numbers which describes A with respect to the standard basis in $\mathcal{R}^2$.

(a) Prove that T is a one-to-one linear transformation mapping $\mathcal{L}(\mathcal{R}^2, \mathcal{R}^1)$ onto $\mathcal{R}^2$, and hence deduce that $\dim \mathcal{L}(\mathcal{R}^2, \mathcal{R}^1) = 2$. [*Hint:* See Exercise 1.]

(b) Use the result in (a) to find a basis A_1, A_2 in $\mathcal{L}(\mathcal{R}^2, \mathcal{R}^1)$ which corresponds to the standard basis in $\mathcal{R}^2$. What is the effect of A_1 on the vector (1, 1) in $\mathcal{R}^2$? Of A_2?

*5. Generalize the technique used in the preceding exercise and show that there exists a one-to-one linear transformation mapping $\mathcal{L}(\mathcal{R}^2, \mathcal{R}^2)$ onto $\mathcal{R}^4$. What is the dimension of $\mathcal{L}(\mathcal{R}^2, \mathcal{R}^2)$?

2–6 MATRICES

We have seen that *every* linear transformation $A: \mathcal{V}_1 \to \mathcal{V}_2$ can be obtained from the formula

$$A(\mathbf{x}) = x_1 A(\mathbf{e}_1) + \cdots + x_n A(\mathbf{e}_n) \tag{2–23}$$

by suitably choosing the $A(\mathbf{e}_j)$ in $\mathcal{V}_2$, and that (2–23) defines a linear transformation from $\mathcal{V}_1$ to $\mathcal{V}_2$ for *every* choice of these vectors. (Recall that $\mathbf{e}_1, \ldots, \mathbf{e}_n$ is a basis for $\mathcal{V}_1$, and that $x_1, \ldots, x_n$ are the coordinates of $\mathbf{x}$ with respect to this basis.) We now use this observation to define the notion of a matrix for a linear transformation, as follows.

Let $\mathbf{f}_1, \ldots, \mathbf{f}_m$ be a basis for $\mathcal{V}_2$, and let $A: \mathcal{V}_1 \to \mathcal{V}_2$ be given. Then, for each integer j, $1 \le j \le n$, there exist scalars α_{ij} such that

$$A(\mathbf{e}_j) = \sum_{i=1}^{m} \alpha_{ij}\mathbf{f}_i, \tag{2–24}$$

i.e., such that

$$\begin{aligned} A(\mathbf{e}_1) &= \alpha_{11}\mathbf{f}_1 + \alpha_{21}\mathbf{f}_2 + \cdots + \alpha_{m1}\mathbf{f}_m, \\ A(\mathbf{e}_2) &= \alpha_{12}\mathbf{f}_1 + \alpha_{22}\mathbf{f}_2 + \cdots + \alpha_{m2}\mathbf{f}_m, \\ &\vdots \\ A(\mathbf{e}_n) &= \alpha_{1n}\mathbf{f}_1 + \alpha_{2n}\mathbf{f}_2 + \cdots + \alpha_{mn}\mathbf{f}_m. \end{aligned} \tag{2–25}$$

For computational purposes it turns out to be convenient to display these scalars in the rectangular array

$$\begin{bmatrix} \alpha_{11} & \alpha_{12} & \cdots & \alpha_{1n} \\ \alpha_{21} & \alpha_{22} & \cdots & \alpha_{2n} \\ \vdots & & & \vdots \\ \alpha_{m1} & \alpha_{m2} & \cdots & \alpha_{mn} \end{bmatrix}, \tag{2–26}$$

whose *columns* are the coefficients of the various equations in (2–25). The reader should note that the first subscript on an entry in (2–26) indicates the *row* in which that entry appears, and the second indicates the *column*. With this convention in force the entire array can be abbreviated (α_{ij}), it being understood that i and j range independently over the integers $1, \ldots, m$ and $1, \ldots, n$, respectively. When displayed as above, this set of scalars is called *the matrix of A with respect to the bases* $\mathcal{B}_1 = \{\mathbf{e}_1, \ldots, \mathbf{e}_n\}$ *and* $\mathcal{B}_2 = \{\mathbf{f}_1, \ldots, \mathbf{f}_m\}$ and is denoted by $[A: \mathcal{B}_1, \mathcal{B}_2]$, or simply by $[A]$. (In the special case where A maps $\mathcal{V}$ into itself and $\mathcal{B}_1 = \mathcal{B}_2 = \mathcal{B}$, the notation $[A: \mathcal{B}]$ is also used.) When $m = n$, we say that (2–26) is a *square* matrix; otherwise, *rectangular*. In general, a matrix consisting of m rows and n columns will be referred to as an $m \times n$-matrix (read "m by n").

The argument just given shows that *every* linear transformation from $\mathcal{V}_1$ to $\mathcal{V}_2$ determines a *unique* $m \times n$-matrix with respect to $\mathcal{B}_1$ and $\mathcal{B}_2$. But since the α_{ij} in (2–25) uniquely determine the $A(\mathbf{e}_j)$ and hence, by (2–23), $A(\mathbf{x})$ for all $\mathbf{x}$ in $\mathcal{V}_1$, it follows that *every* $m \times n$-matrix determines a *unique* linear transformation from $\mathcal{V}_1$ to $\mathcal{V}_2$ in terms of $\mathcal{B}_1$ and $\mathcal{B}_2$, and we therefore have

Theorem 2–6. *Let $\mathcal{V}_1$ and $\mathcal{V}_2$ be finite dimensional vector spaces with* $\dim \mathcal{V}_1 = n$, $\dim \mathcal{V}_2 = m$, *and let $\mathcal{B}_1$ and $\mathcal{B}_2$ be bases for $\mathcal{V}_1$ and $\mathcal{V}_2$, respectively. Then every linear transformation from $\mathcal{V}_1$ to $\mathcal{V}_2$ determines a unique $m \times n$-matrix with respect to $\mathcal{B}_1$ and $\mathcal{B}_2$, and, conversely, every such matrix determines a unique linear transformation from $\mathcal{V}_1$ to $\mathcal{V}_2$ defined by* (2–23) *and* (2–24).

It is important to realize that this theorem does *not* assert that every linear transformation has a unique matrix. Indeed, any such assertion would be patently false, for, as we have already seen, the matrix of a linear transformation can change with a change of basis (Example 3, Section 2–5). Thus the several references to bases which appear in Theorem 2–6 cannot under any circumstances be deleted.

EXAMPLE 1. Let $\mathbf{e}_1$ and $\mathbf{e}_2$ be the standard basis vectors in $\mathcal{R}^2$, and let $A: \mathcal{R}^2 \to \mathcal{R}^2$ denote the reflection across the $\mathbf{e}_1$-axis. Then

$$A(\mathbf{e}_1) = 1 \cdot \mathbf{e}_1 + 0 \cdot \mathbf{e}_2, \qquad A(\mathbf{e}_2) = 0 \cdot \mathbf{e}_1 - 1 \cdot \mathbf{e}_2,$$

and the matrix of A with respect to the basis $\mathbf{e}_1, \mathbf{e}_2$ is

$$\begin{bmatrix} 1 & 0 \\ 0 & -1 \end{bmatrix}.$$

EXAMPLE 2. Let $A: \mathcal{R}^2 \to \mathcal{R}^2$ be the (counterclockwise) rotation about the origin through an angle θ, and again let $\mathbf{e}_1$ and $\mathbf{e}_2$ be the standard basis vectors. Then

$$A(\mathbf{e}_1) = (\cos\theta)\mathbf{e}_1 + (\sin\theta)\mathbf{e}_2,$$
$$A(\mathbf{e}_2) = -(\sin\theta)\mathbf{e}_1 + (\cos\theta)\mathbf{e}_2$$

(see Fig. 2–8), and the matrix of A with respect to this basis is

$$\begin{bmatrix} \cos\theta & -\sin\theta \\ \sin\theta & \cos\theta \end{bmatrix}.$$

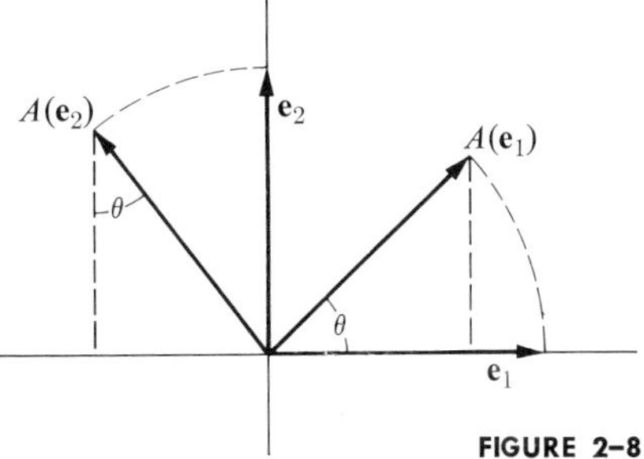

FIGURE 2–8

141564

EXAMPLE 3. If I is the identity map on $\mathcal{V}$, then, regardless of the basis used, the matrix of I is the $n \times n$-matrix

$$\begin{bmatrix} 1 & 0 & \cdots & 0 \\ 0 & 1 & \cdots & 0 \\ \vdots & & & \vdots \\ 0 & 0 & \cdots & 1 \end{bmatrix}$$

with ones along its *principal diagonal* and zeros elsewhere. For obvious reasons this matrix is called the *$n \times n$-identity matrix.*

Similarly the matrix of the zero transformation from $\mathcal{V}_1$ to $\mathcal{V}_2$ is always the the *$m \times n$-zero matrix*

$$\begin{bmatrix} 0 & 0 & \cdots & 0 \\ 0 & 0 & \cdots & 0 \\ \vdots & & & \vdots \\ 0 & 0 & \cdots & 0 \end{bmatrix}.$$

EXAMPLE 4. Let

$$D: \mathcal{P}_n \to \mathcal{P}_n$$

be differentiation, and let

$$\mathcal{B} = \{1, x, \ldots, x^{n-1}\}$$

be the "standard" basis. Then

$$\begin{aligned} D(1) &= 0 \cdot 1 + 0 \cdot x + \cdots + 0 \cdot x^{n-2} + 0 \cdot x^{n-1}, \\ D(x) &= 1 \cdot 1 + 0 \cdot x + \cdots + 0 \cdot x^{n-2} + 0 \cdot x^{n-1}, \\ D(x^2) &= 0 \cdot 1 + 2 \cdot x + \cdots + 0 \cdot x^{n-2} + 0 \cdot x^{n-1}, \\ &\vdots \\ D(x^{n-1}) &= 0 \cdot 1 + 0 \cdot x + \cdots + (n-1)x^{n-2} + 0 \cdot x^{n-1}, \end{aligned}$$

and

$$[D: \mathcal{B}] = \begin{bmatrix} 0 & 1 & 0 & \cdots & 0 \\ 0 & 0 & 2 & \cdots & 0 \\ \vdots & & & & \vdots \\ 0 & 0 & 0 & \cdots & n-1 \\ 0 & 0 & 0 & \cdots & 0 \end{bmatrix}.$$

EXAMPLE 5. Let

$$A: \mathcal{P}_3 \to \mathcal{P}_5$$

be the linear transformation defined by

$$A(p(x)) = (2x^2 - 3)p(x)$$

for all $p(x)$ in $\mathcal{P}_3$, and let $\mathcal{B}_1$ and $\mathcal{B}_2$ be the standard bases in $\mathcal{P}_3$ and $\mathcal{P}_5$, respectively. Then

$$\begin{aligned} A(1) &= 2x^2 - 3 = -3\cdot 1 + 0\cdot x + 2\cdot x^2 + 0\cdot x^3 + 0\cdot x^4, \\ A(x) &= 2x^3 - 3x = 0\cdot 1 - 3\cdot x + 0\cdot x^2 + 2\cdot x^3 + 0\cdot x^4, \\ A(x^2) &= 2x^4 - 3x^2 = 0\cdot 1 + 0\cdot x - 3\cdot x^2 + 0\cdot x^3 + 2\cdot x^4, \end{aligned}$$

and

$$[A\colon \mathcal{B}_1, \mathcal{B}_2] = \begin{bmatrix} -3 & 0 & 0 \\ 0 & -3 & 0 \\ 2 & 0 & -3 \\ 0 & 2 & 0 \\ 0 & 0 & 2 \end{bmatrix}.$$

EXERCISES

In Exercises 1 through 6 find the matrix of the given linear transformation with respect to each of the given pairs of bases.

1. $A\colon \mathcal{R}^3 \to \mathcal{R}^3$; $A(x_1, x_2, x_3) = (x_1 + x_2, x_1 + x_3, 0)$
   (a) $\mathcal{B}_1$ and $\mathcal{B}_2$ the standard basis
   (b) $\mathcal{B}_1$ the standard basis, $\mathcal{B}_2 = \begin{cases}(1, 1, 0)\\(1, 0, 1)\\(0, 0, 1)\end{cases}$
   (c) $\mathcal{B}_1 = \begin{cases}(1, 1, 0)\\(1, 0, 1)\\(0, 1, 1)\end{cases}$, $\mathcal{B}_2 = \begin{cases}(1, 0, 0)\\(1, 1, 0)\\(1, 1, 1)\end{cases}$
2. $A\colon \mathcal{R}^3 \to \mathcal{R}^2$; $A(x_1, x_2, x_3) = (x_1 - x_2, 2x_2 - 3x_3)$
   (a) $\mathcal{B}_1$ and $\mathcal{B}_2$ the standard bases
   (b) $\mathcal{B}_1 = \begin{cases}(1, 1, 0)\\(1, 0, 1)\\(0, 1, 1)\end{cases}$, $\mathcal{B}_2 =$ the standard basis
   (c) $\mathcal{B}_1 = \begin{cases}(1, 1, \frac{2}{3})\\(2, -1, -1),\\(3, 2, \frac{1}{2})\end{cases}$ $\mathcal{B}_2 = \begin{cases}(3, 1)\\(1, \frac{5}{2})\end{cases}$
3. $A\colon \mathcal{P}_3 \to \mathcal{R}^1$; $A(p(x)) = \int_0^1 p(x)\,dx$
   (a) $\mathcal{B}_1 = \{1, x, x^2\}$, $\mathcal{B}_2 = \{1\}$
   (b) $\mathcal{B}_1 = \{1, x - 1, x(x - 1)\}$, $\mathcal{B}_2 = \{1\}$
   (c) $\mathcal{B}_1 = (1, x - 1, x(x - 1)\}$, $\mathcal{B}_2 = \{2\}$

4. $A: \mathcal{P}_3 \to \mathcal{R}^1$; $A(p(x)) = \int_{-1}^{1} p(x)\, dx$
   (a) $\mathcal{B}_1 = \{1, x, x^2\}$, $\mathcal{B}_2 = \{1\}$
   (b) $\mathcal{B}_1 = \{1, x, x^2 - \frac{1}{3}\}$, $\mathcal{B}_2 = \{1\}$
   (c) $\mathcal{B}_1 = \{1, x + 1, x^2 - 1\}$, $\mathcal{B}_2 = \{-2\}$

5. $A: \mathcal{P}_3 \to \mathcal{P}_4$; $A(p(x)) = (x^2 - 1)p'(x)$
   (a) $\mathcal{B}_1 = \{1, x, x^2\}$, $\mathcal{B}_2 = \{1, x, x^2, x^3\}$
   (b) $\mathcal{B}_1 = \{1, x, x^2\}$, $\mathcal{B}_2 = \{1, x - 1, (x - 1)^2, (x - 1)^3\}$
   (c) $\mathcal{B}_1 = \{1, x - 1, (x - 1)^2\}$, $\mathcal{B}_2 = \{1, x - 2, (x - 2)^2, (x - 1)(x - 2)^2\}$

6. $A: \mathcal{P}_3 \to \mathcal{P}_4$; $A(p(x)) = \int_0^x p(t)\, dt$
   (a) $\mathcal{B}_1 = \{1, x, x^2\}$, $\mathcal{B}_2 = \{1, x, x^2, x^3\}$
   (b) $\mathcal{B}_1 = \{1, x, x^2\}$, $\mathcal{B}_2 = \{1, x, x^2/2, x^3/3\}$
   (c) $\mathcal{B}_1 = \{2x - 1, 2x + 1, x(x + 1)\}$, $\mathcal{B}_2 = \{x - 1, x + 1, x^2, x^3/3\}$

7. Find the value of $A(\mathbf{x})$ for any $\mathbf{x}$ in $\mathcal{V}_1$, given that $A: \mathcal{V}_1 \to \mathcal{V}_2$ is linear, and $[A: \mathcal{B}_1, \mathcal{B}_2] = (\alpha_{ij})$.

8. A subspace $\mathcal{W}$ of a vector space $\mathcal{V}$ is said to be *invariant* under a linear transformation $A: \mathcal{V} \to \mathcal{V}$ if and only if $A(\mathbf{w})$ belongs to $\mathcal{W}$ for every $\mathbf{w}$ in $\mathcal{W}$.
   (a) Every linear transformation $A: \mathcal{V} \to \mathcal{V}$ has at least two invariant subspaces. What are they?
   (b) Give an example of a linear transformation on a finite dimensional vector space $\mathcal{V}$ which has *exactly* two invariant subspaces. (Do *not* assume $\dim \mathcal{V} = 1$.)
   (c) Let $\dim \mathcal{V} = n$, $\dim \mathcal{W} = m$, and suppose that $\mathcal{W}$ is invariant under $A: \mathcal{V} \to \mathcal{V}$. Prove that there exists a basis $\mathcal{B}$ for $\mathcal{V}$ such that

$$[A: \mathcal{B}] = \begin{bmatrix} \alpha_{11} & \cdots & \alpha_{1m} & \alpha_{1,m+1} & \cdots & \alpha_{1n} \\ \vdots & & & & & \vdots \\ \alpha_{m1} & \cdots & \alpha_{mm} & \alpha_{m,m+1} & \cdots & \alpha_{mn} \\ 0 & \cdots & 0 & \alpha_{m+1,m+1} & \cdots & \alpha_{m+1,n} \\ \vdots & & & & & \vdots \\ 0 & \cdots & 0 & \alpha_{n,m+1} & \cdots & \alpha_{nn} \end{bmatrix}.$$

   [*Hint:* See Theorem 1–7.]
   (d) Let $\mathcal{W}_1$ and $\mathcal{W}_2$ be invariant subspaces of $\mathcal{V}$ under A, and suppose that
      (i) $\dim \mathcal{W}_1 = m$, $\dim \mathcal{W}_2 = n - m$, where $n = \dim \mathcal{V}$,
      (ii) $\mathcal{W}_1$ and $\mathcal{W}_2$ have only the zero vector in common.
   Prove that there exists a basis $\mathcal{B}$ for $\mathcal{V}$ such that

$$[A: \mathcal{B}] = \begin{bmatrix} \alpha_{11} & \cdots & \alpha_{1m} & 0 & \cdots & 0 \\ \vdots & & & & & \vdots \\ \alpha_{m1} & \cdots & \alpha_{mm} & 0 & \cdots & 0 \\ 0 & \cdots & 0 & \alpha_{m+1,m+1} & \cdots & \alpha_{m+1,n} \\ \vdots & & & & & \vdots \\ 0 & \cdots & 0 & \alpha_{n,m+1} & \cdots & \alpha_{nn} \end{bmatrix}.$$

2–7 ADDITION AND SCALAR MULTIPLICATION OF MATRICES

Let $\mathcal{L}(\mathcal{V}_1, \mathcal{V}_2)$ denote the set of all linear transformations from $\mathcal{V}_1$ to $\mathcal{V}_2$, and let $\mathfrak{M}_{mn}$ denote the set of all $m \times n$-matrices. Then if $\dim \mathcal{V}_1 = n$ and $\dim \mathcal{V}_2 = m$, Theorem 2–6 asserts that the function which associates each A in $\mathcal{L}(\mathcal{V}_1, \mathcal{V}_2)$ with its matrix $[A]$ with respect to a fixed pair of bases $\mathcal{B}_1$ and $\mathcal{B}_2$ is a *one-to-one* mapping of $\mathcal{L}(\mathcal{V}_1, \mathcal{V}_2)$ *onto* $\mathfrak{M}_{mn}$. This simple fact allows us to translate algebraic statements concerning linear transformations into statements concerning matrices, and leads to the subject of matrix algebra. In particular, it allows us to convert $\mathfrak{M}_{mn}$ into a real vector space by using the matrix analogs of the addition and scalar multiplication in $\mathcal{L}(\mathcal{V}_1, \mathcal{V}_2)$ to define an addition and scalar multiplication for matrices. The argument goes as follows.

Let (α_{ij}) and (β_{ij}) be arbitrary $m \times n$-matrices, and let σ be a real number. Then by choosing bases in $\mathcal{V}_1$ and $\mathcal{V}_2$ we find *unique* linear transformations A and B in $\mathcal{L}(\mathcal{V}_1, \mathcal{V}_2)$ such that

$$[A] = (\alpha_{ij}), \qquad [B] = (\beta_{ij}).$$

Thus, using the notation of the preceding section, we find

$$A(\mathbf{e}_j) = \sum_{i=1}^{m} \alpha_{ij}\mathbf{f}_i, \qquad B(\mathbf{e}_j) = \sum_{i=1}^{m} \beta_{ij}\mathbf{f}_i,$$

and it follows that

$$\begin{aligned}(A + B)(\mathbf{e}_j) &= A(\mathbf{e}_j) + B(\mathbf{e}_j)\\ &= \sum_{i=1}^{m} \alpha_{ij}\mathbf{f}_i + \sum_{i=1}^{m} \beta_{ij}\mathbf{f}_i\\ &= \sum_{i=1}^{m} (\alpha_{ij} + \beta_{ij})\mathbf{f}_i,\end{aligned}$$

and

$$\begin{aligned}(\sigma A)(\mathbf{e}_j) &= \sigma\left(\sum_{i=1}^{m} \alpha_{ij}\mathbf{f}_i\right)\\ &= \sum_{i=1}^{m} (\sigma\alpha_{ij})\mathbf{f}_i.\end{aligned}$$

Hence

$$\begin{aligned}[A + B] &= (\alpha_{ij} + \beta_{ij}),\\ [\sigma A] &= (\sigma\alpha_{ij}),\end{aligned}$$

and if we now require, as reason dictates we must, that

$$[A] + [B] = [A + B], \qquad \text{and} \qquad \sigma[A] = [\sigma A],$$

we are *forced* to give the following definition.

Definition 2–7. The *sum* $(\alpha_{ij}) + (\beta_{ij})$ of two $m \times n$-matrices is by definition the $m \times n$-matrix $(\alpha_{ij} + \beta_{ij})$; the *product* $\sigma(\alpha_{ij})$ of a real number σ and an $m \times n$-matrix is by definition the $m \times n$-matrix $(\sigma\alpha_{ij})$. In other words,

$$(\alpha_{ij}) + (\beta_{ij}) = (\alpha_{ij} + \beta_{ij}) \tag{2–27}$$

and

$$\sigma(\alpha_{ij}) = (\sigma\alpha_{ij}) \tag{2–28}$$

for all $m \times n$-matrices (α_{ij}), (β_{ij}), and all real numbers σ.

When expressed as rectangular arrays these equations become

$$\begin{bmatrix} \alpha_{11} & \alpha_{12} & \cdots & \alpha_{1n} \\ \alpha_{21} & \alpha_{22} & \cdots & \alpha_{2n} \\ \vdots & & & \vdots \\ \alpha_{m1} & \alpha_{m2} & \cdots & \alpha_{mn} \end{bmatrix} + \begin{bmatrix} \beta_{11} & \beta_{12} & \cdots & \beta_{1n} \\ \beta_{21} & \beta_{22} & \cdots & \beta_{2n} \\ \vdots & & & \vdots \\ \beta_{m1} & \beta_{m2} & \cdots & \beta_{mn} \end{bmatrix}$$
$$= \begin{bmatrix} \alpha_{11} + \beta_{11} & \alpha_{12} + \beta_{12} & \cdots & \alpha_{1n} + \beta_{1n} \\ \alpha_{21} + \beta_{21} & \alpha_{22} + \beta_{22} & \cdots & \alpha_{2n} + \beta_{2n} \\ \vdots & & & \vdots \\ \alpha_{m1} + \beta_{m1} & \alpha_{m2} + \beta_{m2} & \cdots & \alpha_{mn} + \beta_{mn} \end{bmatrix},$$

and

$$\sigma \begin{bmatrix} \alpha_{11} & \alpha_{12} & \cdots & \alpha_{1n} \\ \alpha_{21} & \alpha_{22} & \cdots & \alpha_{2n} \\ \vdots & & & \vdots \\ \alpha_{m1} & \alpha_{m2} & \cdots & \alpha_{mn} \end{bmatrix} = \begin{bmatrix} \sigma\alpha_{11} & \sigma\alpha_{12} & \cdots & \sigma\alpha_{1n} \\ \sigma\alpha_{21} & \sigma\alpha_{22} & \cdots & \sigma\alpha_{2n} \\ \vdots & & & \vdots \\ \sigma\alpha_{m1} & \sigma\alpha_{m2} & \cdots & \sigma\alpha_{mn} \end{bmatrix},$$

and assert that matrix addition and scalar multiplication are performed entry by entry, or termwise. Moreover, we now have

Theorem 2–7. *The set $\mathfrak{M}_{mn}$ of all $m \times n$-matrices is a real vector space under the above definition of addition and scalar multiplication.*

The student should appreciate that there is no need for a formal proof at this point since the asserted result follows automatically from Theorem 2–6, the fact that $\mathcal{L}(\mathcal{V}_1, \mathcal{V}_2)$ is a real vector space, and the way in which addition and scalar multiplication were defined in $\mathfrak{M}_{mn}$. Indeed, we can now assert that $\mathcal{L}(\mathcal{V}_1, \mathcal{V}_2)$ and $\mathfrak{M}_{mn}$ are *algebraically identical* (or *isomorphic*), and that the function which sends each linear transformation $A: \mathcal{V}_1 \to \mathcal{V}_2$ onto its matrix $[A: \mathcal{B}_1, \mathcal{B}_2]$ with respect to a fixed pair of bases $\mathcal{B}_1$ and $\mathcal{B}_2$ is an *isomorphism* of $\mathcal{L}(\mathcal{V}_1, \mathcal{V}_2)$ onto $\mathfrak{M}_{mn}$.

As an illustration of the way in which this fact can be used to establish results which are not otherwise obvious, we now propose to show that $\mathcal{L}(\mathcal{V}_1, \mathcal{V}_2)$ is finite dimensional and to compute its dimension. For this purpose we introduce the special matrices (e_{ij}), $1 \leq i \leq m$, $1 \leq j \leq n$, each of which has the entry 1 at the intersection of the ith row and jth column and zeros elsewhere:

$$i\begin{bmatrix} 0 & \overset{j}{\vdots} & 0 \\ \cdots & 1 & \cdots \\ 0 & \vdots & 0 \end{bmatrix} = (e_{ij}). \tag{2–29}$$

Then, for each (α_{ij}) in $\mathfrak{M}_{mn}$ we have

$$\begin{aligned}(\alpha_{ij}) = {} & \alpha_{11}(e_{11}) + \cdots + \alpha_{1n}(e_{1n}) \\ & + \alpha_{21}(e_{21}) + \cdots + \alpha_{2n}(e_{2n}) \\ & \vdots \\ & + \alpha_{m1}(e_{m1}) + \cdots + \alpha_{mn}(e_{mn}),\end{aligned}$$

or

$$(\alpha_{ij}) = \sum_{i,j=1}^{m,n} \alpha_{ij}(e_{ij}). \tag{2–30}$$

Thus the (e_{ij}) span $\mathfrak{M}_{mn}$, and since it is clear that (2–30) is the only possible way of writing (α_{ij}) as a linear combination of the (e_{ij}), it follows that these matrices are a *basis* for $\mathfrak{M}_{mn}$. (This particular basis is called the *standard basis* for $\mathfrak{M}_{mn}$.) Hence $\mathcal{L}(\mathcal{V}_1, \mathcal{V}_2)$ is also finite dimensional with dimension mn, and we have proved

Theorem 2–8. *If $\mathcal{V}_1$ and $\mathcal{V}_2$ are finite dimensional vector spaces, then so is $\mathcal{L}(\mathcal{V}_1, \mathcal{V}_2)$, and*

$$\dim \mathcal{L}(\mathcal{V}_1, \mathcal{V}_2) = (\dim \mathcal{V}_1)(\dim \mathcal{V}_2). \tag{2–31}$$

EXAMPLE 1. The set of all $1 \times n$-matrices $\mathfrak{M}_{1n}$ is an n-dimensional vector space with

$$\begin{aligned}(e_{11}) &= (1, 0, \ldots, 0) \\ (e_{12}) &= (0, 1, \ldots, 0) \\ &\vdots \\ (e_{1n}) &= (0, 0, \ldots, 1)\end{aligned}$$

as a basis. In this case the (e_{ij}) can be identified with the standard basis vectors in $\mathcal{R}^n$, and when this identification is made $\mathfrak{M}_{1n}$ becomes identical with $\mathcal{R}^n$. Thus $\mathcal{L}(\mathcal{V}, \mathcal{R}^1)$ is isomorphic with $\mathcal{R}^n$, whenever $\dim \mathcal{V} = n$, and we conclude that there are exactly as many linear transformations from $\mathcal{V}$ to $\mathcal{R}^1$ as there are vectors in $\mathcal{R}^n$ (cf. Example 1, Section 2–5).

EXAMPLE 2. Let

$$\mathcal{V}_1 = \mathcal{V}_2 = \mathcal{R}^2.$$

Then $\dim \mathcal{L}(\mathcal{R}^2, \mathcal{R}^2) = 4$, and the (e_{ij}) are four in number:

$$(e_{11}) = \begin{bmatrix} 1 & 0 \\ 0 & 0 \end{bmatrix}, \quad (e_{12}) = \begin{bmatrix} 0 & 1 \\ 0 & 0 \end{bmatrix},$$

$$(e_{21}) = \begin{bmatrix} 0 & 0 \\ 1 & 0 \end{bmatrix}, \quad (e_{22}) = \begin{bmatrix} 0 & 0 \\ 0 & 1 \end{bmatrix}.$$

Moreover, if E_{ij} denotes the linear transformation corresponding to (e_{ij}) with respect to a fixed basis in $\mathcal{R}^2$, then every linear transformation mapping $\mathcal{R}^2$ into itself can be written uniquely in the form

$$\alpha_{11}E_{11} + \alpha_{12}E_{12} + \alpha_{21}E_{21} + \alpha_{22}E_{22}$$

for suitable scalars $\alpha_{11}, \alpha_{12}, \alpha_{21}, \alpha_{22}$.

EXAMPLE 3. If A is a nonzero linear transformation mapping a finite dimensional vector space $\mathcal{V}$ into itself, then

$$I, A, A^2, \ldots, A^k, \ldots$$

also map $\mathcal{V}$ into itself, and thus belong to $\mathcal{L}(\mathcal{V}, \mathcal{V})$. But by Theorem 2–8 this set is linearly *dependent* in $\mathcal{L}(\mathcal{V}, \mathcal{V})$. Hence there exists a *smallest* positive integer k such that A^k is linearly dependent on $I, A, \ldots, A^{k-1}$, and it is now easy to show that these transformations are a *basis* for the subspace of $\mathcal{L}(\mathcal{V}, \mathcal{V})$ spanned by the powers of A (see Exercise 14). In particular, we can write A^k in the form

$$A^k = a_{k-1}A^{k-1} + \cdots + a_1A + a_0I,$$

or

$$A^k - a_{k-1}A^{k-1} - \cdots - a_1A - a_0 = O, \tag{2–32}$$

where O is the zero transformation on $\mathcal{V}$, and it follows that A is a root of the polynomial

$$m_A(x) = x^k - a_{k-1}x^{k-1} - \cdots - a_1x - a_0. \tag{2–33}$$

Since k was chosen as small as possible in this argument there is no polynomial of lower degree having A as a root. For this reason $m_A(x)$ is called the *minimum polynomial* of A. It can be characterized as *the* polynomial of *least degree with leading coefficient* 1 which has A as a root, and is clearly of degree $\leq n^2$ when $\dim \mathcal{V} = n$. Actually, it can be shown that the degree of $m_A(x)$ does not exceed the dimension of $\mathcal{V}$ for any nonzero transformation $A: \mathcal{V} \to \mathcal{V}$. The proof, however, is not easy.

EXERCISES

In Exercises 1 through 5 compute the value of $\alpha[A] + \beta[B]$ for the given scalars α, β and matrices $[A]$, $[B]$.

1. $\alpha = 2$, $\beta = -1$,

$$[A] = \begin{bmatrix} 1 & 0 & -2 \\ 3 & 1 & -5 \\ 2 & -2 & 3 \end{bmatrix}, \quad [B] = \begin{bmatrix} 4 & -1 & 2 \\ 3 & -5 & 1 \\ 0 & 4 & -2 \end{bmatrix}$$

2. $\alpha = -\frac{1}{2}$, $\beta = 4$,

$$[A] = \begin{bmatrix} 2 & -3 \\ 1 & 0 \\ 4 & -2 \end{bmatrix}, \quad [B] = \begin{bmatrix} -1 & 2 \\ 3 & 1 \\ 4 & -5 \end{bmatrix}$$

3. $\alpha = \frac{2}{3}$, $\beta = -2$,

$$[A] = \begin{bmatrix} 6 & 2 & -3 \\ 1 & 0 & 4 \end{bmatrix}, \quad [B] = \begin{bmatrix} -8 & -\frac{2}{3} & 1 \\ -\frac{1}{3} & 0 & -\frac{4}{3} \end{bmatrix}$$

4. $\alpha = \frac{1}{3}$, $\beta = \frac{2}{3}$,

$$[A] = \begin{bmatrix} 6 & 2 & -3 \\ 0 & 4 & 1 \\ 3 & -1 & 3 \end{bmatrix}, \quad [B] = \begin{bmatrix} -\frac{3}{2} & -1 & \frac{3}{2} \\ 0 & -\frac{1}{2} & -\frac{1}{2} \\ -\frac{3}{2} & \frac{1}{2} & 0 \end{bmatrix}$$

5. $\alpha = \frac{1}{2}$, $\beta = -\frac{1}{3}$,

$$[A] = \begin{bmatrix} 0 & \frac{5}{2} & -2 & 6 \\ \frac{2}{3} & -1 & -\frac{1}{3} & 0 \\ -3 & 1 & \frac{1}{2} & \frac{4}{3} \end{bmatrix}, \quad [B] = \begin{bmatrix} \frac{2}{3} & 3 & -4 & 1 \\ 0 & 6 & -2 & \frac{1}{2} \\ \frac{4}{3} & \frac{1}{3} & 0 & 2 \end{bmatrix}$$

6. Let A denote the counterclockwise rotation of $\mathcal{R}^2$ about the origin through the angle $\pi/4$, and let B denote the reflection across the origin. Find the matrix of $A + B$ with respect to the standard basis $\mathbf{e}_1 = (1, 0)$, $\mathbf{e}_2 = (0, 1)$, and with respect to the basis $\mathbf{e}_1' = (1, 1)$, $\mathbf{e}_2' = (-1, 1)$.

7. Let $A, B\colon \mathcal{P}_3 \to \mathcal{P}_4$ be defined by

$$A(p(x)) = xp(x) - p(1), \qquad B(p(x)) = (x - 1)p(x).$$

Find the matrix of $2A - B$ with respect to the standard bases in $\mathcal{P}_3$ and $\mathcal{P}_4$, and with respect to the bases

$$\mathcal{B}_1 = \{1, x - 1, (x - 1)^2\}, \quad \mathcal{B}_2 = \{1, x - 1, (x - 1)^2, (x - 1)^3\}.$$

8. Compute the dimension of the subspace of $\mathfrak{M}_{22}$ spanned by the matrices

$$\begin{bmatrix} 2 & 1 \\ 0 & -1 \end{bmatrix}, \quad \begin{bmatrix} 1 & -\frac{9}{2} \\ 0 & \frac{3}{4} \end{bmatrix}, \quad \begin{bmatrix} 3 & -6 \\ 0 & -\frac{3}{2} \end{bmatrix}.$$

9. Let $\mathbf{e}_1, \ldots, \mathbf{e}_n$ and $\mathbf{f}_1, \ldots, \mathbf{f}_m$ be bases for $\mathcal{V}_1$ and $\mathcal{V}_2$, respectively, and for each pair of integers i,j with $1 \leq i \leq m$, $1 \leq j \leq n$, let $E_{ij}: \mathcal{V}_1 \to \mathcal{V}_2$ be defined by

$$E_{ij}(\mathbf{e}_k) = \begin{cases} \mathbf{0} & \text{if } j \neq k, \\ \mathbf{f}_i & \text{if } j = k. \end{cases}$$

Prove that the E_{ij} are a basis for $\mathcal{L}(\mathcal{V}_1, \mathcal{V}_2)$.

10. Let $A: \mathfrak{M}_{22} \to \mathfrak{M}_{22}$ be the mapping defined by

$$A \begin{bmatrix} \alpha_{11} & \alpha_{12} \\ \alpha_{21} & \alpha_{22} \end{bmatrix} = \begin{bmatrix} \alpha_{11} & \alpha_{12} \\ 0 & \alpha_{22} \end{bmatrix}.$$

Prove that A is linear, and find the matrix of A with respect to the standard basis in $\mathfrak{M}_{22}$.

11. Repeat Exercise 10 for the mapping $A: \mathfrak{M}_{22} \to \mathfrak{M}_{23}$ defined by

$$A \begin{bmatrix} \alpha_{11} & \alpha_{12} \\ \alpha_{21} & \alpha_{22} \end{bmatrix} = \begin{bmatrix} \alpha_{11} & \alpha_{12} & 0 \\ \alpha_{21} & \alpha_{22} & 0 \end{bmatrix}.$$

12. What is the dimension of the vector space $\mathcal{L}(\mathfrak{M}_{m1}, \mathfrak{M}_{1n})$? Of $\mathcal{L}(\mathfrak{M}_{mn}, \mathfrak{M}_{pq})$?

13. (a) Prove that the functions $\sin x$, $\cos x$, $\sin x \cos x$, $\sin^2 x$, $\cos^2 x$ are linearly independent in $\mathcal{C}(-\infty, \infty)$.

(b) Let $\mathcal{V}$ denote the subspace of $\mathcal{C}(-\infty, \infty)$ spanned by the functions in (a). Prove that D^n, the nth power of the differentiation operator, maps $\mathcal{V}$ into itself for all n, and find the matrix of $D^2 - 2D + 1$ with respect to the given basis for $\mathcal{V}$.

14. Let $A: \mathcal{V} \to \mathcal{V}$ be linear, with $A \neq O$, and let k be the smallest positive integer such that A^k is linearly dependent on $I, A, \ldots, A^{k-1}$. Prove that $I, A, \ldots, A^{k-1}$ is a basis for the subspace of $\mathcal{L}(\mathcal{V}, \mathcal{V})$ spanned by the powers of A.

15. Find the minimum polynomial of the linear transformation $D: \mathcal{P}_3 \to \mathcal{P}_3$.

2–8 MATRIX MULTIPLICATION

Continuing in the spirit of the last section we now define a multiplication for matrices by rewriting the definition of the product of two linear transformations in matrix form. To this end let

$$B: \mathcal{V}_1 \to \mathcal{V}_2, \qquad A: \mathcal{V}_2 \to \mathcal{V}_3$$

be given, let

$$\mathcal{B}_1 = \{\mathbf{e}_1, \ldots, \mathbf{e}_r\}, \qquad \mathcal{B}_2 = \{\mathbf{f}_1, \ldots, \mathbf{f}_n\}, \qquad \mathcal{B}_3 = \{\mathbf{g}_1, \ldots, \mathbf{g}_m\}$$

be bases for $\mathcal{V}_1$, $\mathcal{V}_2$, $\mathcal{V}_3$, respectively, and let

$$B(\mathbf{e}_j) = \sum_{k=1}^{n} \beta_{kj}\mathbf{f}_k,$$

$$A(\mathbf{f}_k) = \sum_{i=1}^{m} \alpha_{ik}\mathbf{g}_i.$$

Then we have

$$\begin{aligned} AB(\mathbf{e}_j) &= A\left(\sum_{k=1}^{n} \beta_{kj}\mathbf{f}_k\right) = \sum_{k=1}^{n} \beta_{kj}A(\mathbf{f}_k) \\ &= \sum_{k=1}^{n} \beta_{kj}\left(\sum_{i=1}^{m} \alpha_{ik}\mathbf{g}_i\right) \\ &= \sum_{i=1}^{m}\left(\sum_{k=1}^{n} \alpha_{ik}\beta_{kj}\right)\mathbf{g}_i, \end{aligned}$$

and the matrix of AB with respect to $\mathcal{B}_1$, $\mathcal{B}_3$ is the $m \times r$-matrix whose ijth entry is $\sum_{k=1}^{n} \alpha_{ik}\beta_{kj}$. But since

$$[A\colon \mathcal{B}_2, \mathcal{B}_3] = (\alpha_{ik}), \qquad 1 \leq i \leq m, \quad 1 \leq k \leq n,$$

and

$$[B\colon \mathcal{B}_1, \mathcal{B}_2] = (\beta_{kj}), \qquad 1 \leq k \leq n, \quad 1 \leq j \leq r,$$

the requirement that

$$[AB\colon \mathcal{B}_1, \mathcal{B}_3] = [A\colon \mathcal{B}_2, \mathcal{B}_3][B\colon \mathcal{B}_1, \mathcal{B}_2]$$

leads to the following definition of matrix multiplication.

Definition 2–8. The *product* $(\alpha_{ik})(\beta_{kj})$ of an $m \times n$-matrix (α_{ik}) and an $n \times r$-matrix (β_{kj}) is by definition the $m \times r$-matrix

$$(\alpha_{ik})(\beta_{kj}) = \left(\sum_{k=1}^{n} \alpha_{ik}\beta_{kj}\right). \tag{2–34}$$

It is important to notice that the product of two matrices is defined *only when the number of columns in the first matrix is equal to the number of rows in the second;* a restriction which is the matrix analog of the fact that the product of two linear transformations is defined only when the image of the first is contained in the domain of the second. When written in greater detail, Eq. (2–34) becomes

$$\begin{bmatrix} \alpha_{11} & \alpha_{12} & \cdots & \alpha_{1n} \\ \alpha_{21} & \alpha_{22} & \cdots & \alpha_{2n} \\ \vdots & & & \vdots \\ \alpha_{m1} & \alpha_{m2} & \cdots & \alpha_{mn} \end{bmatrix} \begin{bmatrix} \beta_{11} & \beta_{12} & \cdots & \beta_{1r} \\ \beta_{21} & \beta_{22} & \cdots & \beta_{2r} \\ \vdots & \vdots & & \vdots \\ \beta_{n1} & \beta_{n2} & \cdots & \beta_{nr} \end{bmatrix}$$

$$= \begin{bmatrix} \alpha_{11}\beta_{11} + \cdots + \alpha_{1n}\beta_{n1} & \cdots & \alpha_{11}\beta_{1r} + \cdots + \alpha_{1n}\beta_{nr} \\ \alpha_{21}\beta_{11} + \cdots + \alpha_{2n}\beta_{n1} & \cdots & \alpha_{21}\beta_{1r} + \cdots + \alpha_{2n}\beta_{nr} \\ \vdots & & \vdots \\ \alpha_{m1}\beta_{11} + \cdots + \alpha_{mn}\beta_{n1} & \cdots & \alpha_{m1}\beta_{1r} + \cdots + \alpha_{mn}\beta_{nr} \end{bmatrix},$$

and is easily remembered in terms of the kinesthetic relationship between the rows of the first matrix and the columns of the second.

EXAMPLE 1. If

$$(\alpha_{ik}) = \begin{bmatrix} 2 & -1 & 0 \\ 1 & 2 & -3 \end{bmatrix}, \qquad (\beta_{kj}) = \begin{bmatrix} 1 & 3 \\ -2 & 1 \\ 0 & 4 \end{bmatrix},$$

then $(\alpha_{ik})(\beta_{kj})$ is defined, and we have

$$\begin{aligned}(\alpha_{ik})(\beta_{kj}) &= \begin{bmatrix} 2\cdot 1 + (-1)(-2) + 0\cdot 0 & 2\cdot 3 + (-1)\cdot 1 + 0\cdot 4 \\ 1\cdot 1 + 2\cdot(-2) + (-3)\cdot 0 & 1\cdot 3 + 2\cdot 1 + (-3)\cdot 4 \end{bmatrix} \\ &= \begin{bmatrix} 4 & 5 \\ -3 & -7 \end{bmatrix}.\end{aligned}$$

On the other hand, $(\beta_{kj})(\alpha_{ik})$ is not defined since the number of columns in (β_{kj}) is not equal to the number of rows in (α_{ik}).

EXAMPLE 2. Let

$$(\alpha_{ij}) = \begin{bmatrix} 3 & 1 \\ -1 & 2 \end{bmatrix}, \qquad (\beta_{ij}) = \begin{bmatrix} -1 & 2 \\ 1 & 0 \end{bmatrix}.$$

Then

$$(\alpha_{ij})(\beta_{ij}) = \begin{bmatrix} 3 & 1 \\ -1 & 2 \end{bmatrix}\begin{bmatrix} -1 & 2 \\ 1 & 0 \end{bmatrix} = \begin{bmatrix} -2 & 6 \\ 3 & -2 \end{bmatrix},$$

$$(\beta_{ij})(\alpha_{ij}) = \begin{bmatrix} -1 & 2 \\ 1 & 0 \end{bmatrix}\begin{bmatrix} 3 & 1 \\ -1 & 2 \end{bmatrix} = \begin{bmatrix} -5 & 3 \\ 3 & 1 \end{bmatrix},$$

and we see that *the multiplication of square matrices is noncommutative.*

EXAMPLE 3. Let $[A]$ be the matrix of a linear transformation $A: \mathcal{V} \to \mathcal{V}$ with respect to a basis $\mathcal{B}$ in $\mathcal{V}$, and set $[A]^0 = [I]$ where $[I]$ is the appropriate identity matrix. Then the various powers $[A]^k$, k a nonnegative integer, are all defined, and are simply the powers of A with respect to $\mathcal{B}$. (Why?) This in turn implies that the matrix of a polynomial

$$p(A) = a_k A^k + a_{k-1}A^{k-1} + \cdots + a_1 A + a_0 I$$

in A is

$$p([A]) = a_k[A]^k + a_{k-1}[A]^{k-1} + \cdots + a_1[A] + a_0[I].$$

Such expressions are called polynomials in $[A]$, and are defined whenever $[A]$ is a square matrix. In particular, if

$$m_A(x) = x^k - a_{k-1}x^{k-1} - \cdots - a_1 x - a_0$$

is the minimum polynomial of A defined in Example 3 of the preceding section, then

$$m_A([A]) = [A]^k - a_{k-1}[A]^{k-1} - \cdots - a_1[A] - a_0[I] = [O],$$

and it follows that $[A]$ is a root of $m_A(x)$. For obvious reasons this polynomial is also called the *minimum polynomial* of the matrix $[A]$.

Since Definition 2–8 is simply a coordinatized version of the multiplication of linear transformations, all of the results proved in Section 2–3 remain true when phrased in terms of matrices. This allows us to assert (without proof) that *matrix multiplication is associative, and is distributive over addition;* that is,

$$[(\alpha_{ik})(\beta_{kj})](\gamma_{jl}) = (\alpha_{ik})[(\beta_{kj})(\gamma_{jl})] \tag{2–35}$$

and

$$\begin{aligned}(\alpha_{ik})[(\beta_{kj}) + (\gamma_{kj})] &= (\alpha_{ik})(\beta_{kj}) + (\alpha_{ik})(\gamma_{kj}),\\ [(\alpha_{ik}) + (\beta_{ik})](\gamma_{kj}) &= (\alpha_{ik})(\gamma_{kj}) + (\beta_{ik})(\gamma_{kj})\end{aligned} \tag{2–36}$$

whenever the sums and products appearing in these equations are defined. Similarly, the identity matrices introduced in Example 3 of Section 2–6 play the same role in matrix multiplication that the identity transformations play in operator multiplication. And finally, we can use these identity matrices to define right, left, and two-sided inverses for matrices by rewriting Eq. (2–21) and Definition 2–6 in matrix terms. The details have been left to the reader.

EXERCISES

1. Evaluate each of the following products.

(a) $\begin{bmatrix} 1 & -3 & 2 \\ 4 & 1 & -1 \\ 2 & -5 & 3 \end{bmatrix} \begin{bmatrix} 6 & 1 \\ -2 & 3 \\ 3 & -4 \end{bmatrix}$

(b) $\begin{bmatrix} 2 & 0 & -1 & 1 \\ 4 & 5 & 3 & -2 \\ -1 & 0 & 2 & 4 \end{bmatrix} \begin{bmatrix} 1 & 3 & -2 \\ 0 & 1 & 0 \\ 5 & -1 & 3 \end{bmatrix}$

(c) $\begin{bmatrix} \frac{1}{3} & \frac{1}{6} & 0 \\ \frac{1}{6} & \frac{1}{2} & -\frac{1}{6} \\ 0 & \frac{1}{6} & \frac{1}{3} \end{bmatrix} \begin{bmatrix} \frac{7}{2} & -1 & -\frac{1}{2} \\ -1 & 2 & 1 \\ \frac{1}{2} & -1 & \frac{5}{2} \end{bmatrix}$

(d) $[2 \quad \frac{1}{2} \quad -1 \quad 3] \begin{bmatrix} 4 \\ 0 \\ 2 \\ -1 \end{bmatrix}$

(e) $\begin{bmatrix} 4 \\ 0 \\ 2 \\ -1 \end{bmatrix} [2 \quad \frac{1}{2} \quad -1 \quad 3]$

2. Find all 2×2-matrices which commute with

$$\begin{bmatrix} 1 & -1 \\ 0 & 2 \end{bmatrix}.$$

3. Find necessary and sufficient conditions for the matrices

$$\begin{bmatrix} \alpha_{11} & \alpha_{12} \\ \alpha_{21} & \alpha_{22} \end{bmatrix} \quad \text{and} \quad \begin{bmatrix} \alpha_{11} & \alpha_{21} \\ \alpha_{12} & \alpha_{22} \end{bmatrix}$$

to commute.

4. An $n \times n$-matrix $[A]$ is said to be *invertible* with inverse $[A]^{-1}$ if and only if

$$[A][A]^{-1} = [A]^{-1}[A] = [I],$$

where $[I]$ is the $n \times n$-identity matrix.

(a) Show that if $[A]$ has a right inverse $[B]$ and a left inverse $[C]$ then

$$[B] = [C] = [A]^{-1}.$$

(b) Show that the matrix

$$\begin{bmatrix} 1 & 0 & 1 \\ 0 & 1 & 1 \end{bmatrix}$$

has no left inverse in $\mathfrak{M}_{32}$, but has infinitely many right inverses in $\mathfrak{M}_{32}$.

5. Find the inverse of each of the following matrices.

(a) $\begin{bmatrix} 1 & -1 \\ 4 & 1 \end{bmatrix}$ (b) $\begin{bmatrix} 1 & 0 & -1 \\ 2 & 1 & 0 \\ 1 & 0 & 1 \end{bmatrix}$ (c) $\begin{bmatrix} 0 & 1 & -3 \\ 2 & 1 & 0 \\ 1 & -2 & 1 \end{bmatrix}$

6. Show that the matrix

$$\begin{bmatrix} 2 & 1 & -1 \\ \alpha & 0 & 1 \\ \beta & 0 & \alpha \end{bmatrix}$$

is invertible if and only if $\beta \neq \alpha^2$, and find its inverse when this condition is satisfied.

7. Find all values of α and β for which each of the following matrices is invertible, and compute their inverses.

(a) $\begin{bmatrix} 0 & \alpha \\ \beta & 2 \end{bmatrix}$ (b) $\begin{bmatrix} \alpha & \beta \\ \alpha^2 & \beta^2 \end{bmatrix}$ (c) $\begin{bmatrix} 1 & 0 & \alpha \\ 0 & -1 & 0 \\ \alpha & \beta & 0 \end{bmatrix}$

8. Compute $[A]^n$ when

(a) $[A] = \begin{bmatrix} 1 & 1 \\ 1 & 1 \end{bmatrix}$ (b) $[A] = \begin{bmatrix} 1 & 1 \\ 0 & 0 \end{bmatrix}$

(c) $[A] = \begin{bmatrix} 1 & -1 \\ -1 & 1 \end{bmatrix}$ (d) $[A] = \begin{bmatrix} 0 & 1 \\ 1 & 0 \end{bmatrix}$ (e) $[A] = \begin{bmatrix} \alpha & 1 \\ 0 & \alpha \end{bmatrix}$

9. Find all 2×2-matrices

$$\begin{bmatrix} \alpha & \beta \\ \gamma & \delta \end{bmatrix}$$

such that

$$\begin{bmatrix} \alpha & \beta \\ \gamma & \delta \end{bmatrix}^2 = \begin{bmatrix} 1 & 0 \\ 0 & 1 \end{bmatrix}.$$

10. An $n \times n$-matrix (α_{ij}) is said to be *nilpotent* if and only if there exists an integer $k > 0$ such that $(\alpha_{ij})^k$ is the $n \times n$-zero matrix. The smallest positive integer for which this is true is called the *degree of nilpotence* of (α_{ij}). Show that each of the following matrices is nilpotent, and find their degree of nilpotence.

(a) $\begin{bmatrix} 0 & 1 & -1 & 2 \\ 0 & 0 & 3 & -2 \\ 0 & 0 & 0 & 4 \\ 0 & 0 & 0 & 0 \end{bmatrix}$ (b) $\begin{bmatrix} 1 & 1 & 3 \\ 5 & 2 & 6 \\ -2 & -1 & -3 \end{bmatrix}$ (c) $\begin{bmatrix} 1 & -3 & -4 \\ -1 & 3 & 4 \\ 1 & -3 & -4 \end{bmatrix}$

11. Find all 2×2 nilpotent matrices with degree of nilpotence two. (See Exercise 10.)

12. Determine which of the following matrices are roots of the polynomial $p(x) = x^3 - x^2 + x - 1$.

(a) $\begin{bmatrix} 0 & 0 \\ 0 & 1 \end{bmatrix}$ (b) $\begin{bmatrix} 0 & 1 & 0 \\ 0 & 0 & 1 \\ 1 & 0 & 0 \end{bmatrix}$

(c) $\begin{bmatrix} 1 & 0 & 0 \\ 0 & 1 & 0 \\ 1 & 0 & -1 \end{bmatrix}$ (d) $\begin{bmatrix} 1 & 0 & 0 \\ 0 & 1 & 0 \\ 1 & 0 & -1 \end{bmatrix}$

13. Let $A: \mathfrak{M}_{22} \to \mathfrak{M}_{22}$ be defined by

$$A\begin{bmatrix} \alpha_{11} & \alpha_{12} \\ \alpha_{21} & \alpha_{22} \end{bmatrix} = \begin{bmatrix} 2 & 1 \\ 0 & -1 \end{bmatrix} \begin{bmatrix} \alpha_{11} & \alpha_{12} \\ \alpha_{21} & \alpha_{22} \end{bmatrix}.$$

Prove that A is linear, and find its matrix with respect to the standard basis in $\mathfrak{M}_{22}$.

14. Repeat Exercise 13 for the mapping $A: \mathfrak{M}_{22} \to \mathfrak{M}_{23}$ defined by

$$A\begin{bmatrix} \alpha_{11} & \alpha_{12} \\ \alpha_{21} & \alpha_{22} \end{bmatrix} = \begin{bmatrix} \alpha_{11} & \alpha_{12} \\ \alpha_{21} & \alpha_{22} \end{bmatrix} \begin{bmatrix} 1 & 0 & -1 \\ 2 & 1 & 1 \end{bmatrix}.$$

15. Solve each of the following matrix equations for $[X]$, given that

$$[A] = \begin{bmatrix} 2 & 1 \\ 0 & 3 \end{bmatrix}, \quad [B] = \begin{bmatrix} -1 & 1 \\ 2 & 1 \end{bmatrix}, \quad [C] = \begin{bmatrix} 0 & 1 \\ -1 & 1 \end{bmatrix}, \quad [O] = \begin{bmatrix} 0 & 0 \\ 0 & 0 \end{bmatrix}.$$

(a) $2[X] - [A][C] = [B]$. (b) $[C][X] - [A][B] = [O]$.
(c) $[X]^2 = [A]([B] - [C])$.

16. Prove the associativity of matrix multiplication directly from Definition 2–8.

17. Find 2×2-matrices $[A]$, $[B]$ such that

$$[A][B] = [O], \qquad [B][A] \neq [O],$$

where $[O]$ is the 2×2-zero matrix.

18. Does $[A][B] = [A][C]$ for $n \times n$-matrices $[A]$, $[B]$, $[C]$ necessarily imply that $[B] = [C]$? Why?

19. An $n \times n$-matrix (α_{ij}) is said to be a *diagonal matrix* if and only if $\alpha_{ij} = 0$ whenever $i \neq j$.

(a) Show that the product of diagonal matrices is diagonal.

(b) Under what conditions is an $n \times n$-diagonal matrix invertible? What is its inverse?

(c) Let $T: \mathcal{R}^1 \to \mathfrak{M}_{nn}$ be defined by

$$T(\alpha) = \begin{bmatrix} \alpha & 0 & \cdots & 0 \\ 0 & \alpha & \cdots & 0 \\ \vdots & & & \vdots \\ 0 & 0 & \cdots & \alpha \end{bmatrix}$$

for all α in $\mathcal{R}^1$. Prove that T is one-to-one, linear, and that $T(\alpha)T(\beta) = T(\alpha\beta) = T(\beta)T(\alpha)$ for all α, β.

(d) Find the matrix of T with respect to the standard bases in $\mathcal{R}^1$ and $\mathfrak{M}_{nn}$.

20. Prove that the only $n \times n$-matrices which commute with *every* matrix in $\mathfrak{M}_{nn}$ are the *scalar matrices*

$$\begin{bmatrix} \alpha & 0 & \cdots & 0 \\ 0 & \alpha & \cdots & 0 \\ \vdots & & & \vdots \\ 0 & 0 & \cdots & \alpha \end{bmatrix}.$$

2–9 OPERATOR EQUATIONS

Much of the study of linear transformations is given over to devising methods for solving equations of the form

$$A\mathbf{x} = \mathbf{y}, \tag{2–37}$$

where $\mathbf{y}$ is known, $\mathbf{x}$ unknown, and A a linear transformation from $\mathcal{V}_1$ to $\mathcal{V}_2$. Such equations are known under the generic name of *operator equations*, and will appear throughout this book in a variety of forms. In general, of course, the technique for solving a particular operator equation depends upon the operator involved, and also upon the underlying vector spaces. Nevertheless there are a number of facts concerning such equations which can be proved without using anything other than the linearity of A, and we propose to get them on record here before going on to more specialized topics.

A vector $\mathbf{x}_0$ in $\mathcal{V}_1$ is said to be a *solution* of (2–37) if $A(\mathbf{x}_0) = \mathbf{y}$. The totality of such vectors is called the *solution set* of the equation. In the special case of a *homogeneous equation*

$$A\mathbf{x} = \mathbf{0} \tag{2–38}$$

whose right-hand side is zero, we know that this set is a *subspace* of $\mathcal{V}_1$. It is called the *solution space* of the equation. One of the most important properties of operator equations is that the problem of solving a *nonhomogeneous* equation $A\mathbf{x} = \mathbf{y}, \mathbf{y} \neq \mathbf{0}$, can all but be reduced to that of solving its *associated homogeneous equation* $A\mathbf{x} = \mathbf{0}$. In fact, if $\mathbf{x}_p$ is a fixed solution of $A\mathbf{x} = \mathbf{y}$, and if $\mathbf{x}_h$ is any solution whatever of $A\mathbf{x} = \mathbf{0}$, then $\mathbf{x}_p + \mathbf{x}_h$ is also a solution of $A\mathbf{x} = \mathbf{y}$, since

$$\begin{aligned} A(\mathbf{x}_p + \mathbf{x}_h) &= A(\mathbf{x}_p) + A(\mathbf{x}_h) \\ &= \mathbf{y} + \mathbf{0} \\ &= \mathbf{y}. \end{aligned}$$

Moreover, *every* solution $\mathbf{x}_0$ of $A\mathbf{x} = \mathbf{y}$ can be written in this form for a suitable $\mathbf{x}_h$, since from

$$\begin{aligned} A(\mathbf{x}_0 - \mathbf{x}_p) &= A(\mathbf{x}_0) - A(\mathbf{x}_p) \\ &= \mathbf{y} - \mathbf{y} \\ &= \mathbf{0} \end{aligned}$$

it follows that $\mathbf{x}_0 - \mathbf{x}_p = \mathbf{x}_h$ is a solution of $A\mathbf{x} = \mathbf{0}$, and hence that $\mathbf{x}_0 = \mathbf{x}_p + \mathbf{x}_h$, as asserted.

The solution $\mathbf{x}_p$ appearing in this argument is frequently called a *particular solution* of $A\mathbf{x} = \mathbf{y}$, and in these terms we can state the above result as follows:

Theorem 2–9. *If $\mathbf{x}_p$ is a particular solution of $A\mathbf{x} = \mathbf{y}$, then the solution set of this equation consists of all vectors of the form $\mathbf{x}_p + \mathbf{x}_h$, where $\mathbf{x}_h$ is an arbitrary solution of the associated homogeneous equation $A\mathbf{x} = \mathbf{0}$.*

Geometrically this theorem asserts that the solution set of a nonhomogeneous operator equation can be obtained from the solution space of its associated homogeneous equation by *translating* that subspace by a particular solution $\mathbf{x}_p$ as shown in Fig. 2–9. Algebraically it gives us a prescription for solving $A\mathbf{x} = \mathbf{y}$; viz., find *all* solutions of $A\mathbf{x} = \mathbf{0}$, *one* solution

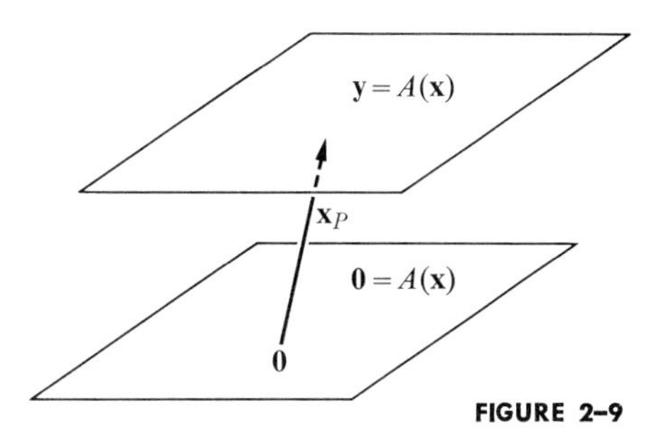

FIGURE 2–9

of $A\mathbf{x} = \mathbf{y}$, and add. The reader will do well to keep both of these points of view in mind as we continue.

EXAMPLE 1. Let $A: \mathcal{C}^2(-\infty, \infty) \to \mathcal{C}(-\infty, \infty)$ be the linear transformation $D^2 - I$ introduced in Example 4, Section 2–4. Then the operator equation

$$Ay = 1,$$

y in $\mathcal{C}^2(-\infty, \infty)$, assumes the form

$$\frac{d^2y}{dx^2} - y = 1, \tag{2–39}$$

and its solution set consists of all functions in $\mathcal{C}^2(-\infty, \infty)$ which satisfy this equation on the entire real line. In this case it is obvious that $y = -1$ is one such function. Thus, to complete the solution of (2–39) it suffices to find *all* solutions of the homogeneous equation

$$\frac{d^2y}{dx^2} - y = 0.$$

In Chapter 3 we will prove that the solution space of this equation has the functions e^x and e^{-x} as a basis, and hence as a corollary, that the solution set of (2–39) is the totality of functions in $\mathcal{C}^2(-\infty, \infty)$ of the form

$$y = -1 + c_1e^x + c_2e^{-x},$$

where c_1 and c_2 are arbitrary constants.

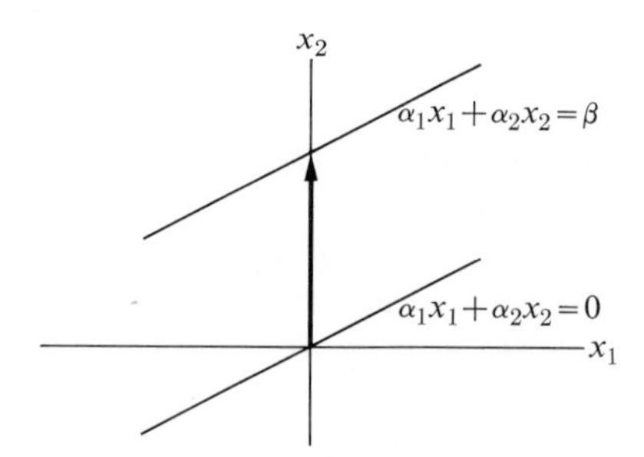

FIGURE 2–10

EXAMPLE 2. Let A be a linear transformation from $\mathcal{R}^2$ to $\mathcal{R}^1$, let $\mathbf{e}_1$ and $\mathbf{e}_2$ be the standard basis vectors in $\mathcal{R}^2$, and let $A(\mathbf{e}_1) = \alpha_1$, $A(\mathbf{e}_2) = \alpha_2$, α_1 and α_2 real numbers. Then if $\mathbf{x} = x_1\mathbf{e}_1 + x_2\mathbf{e}_2$ is any vector in $\mathcal{R}^2$,

$$\begin{aligned} A(\mathbf{x}) &= x_1A(\mathbf{e}_1) + x_2A(\mathbf{e}_2) \\ &= \alpha_1x_1 + \alpha_2x_2, \end{aligned}$$

and the operator equation $A\mathbf{x} = \mathbf{0}$ is an abbreviated version of

$$\alpha_1 x_1 + \alpha_2 x_2 = 0. \tag{2-40}$$

Since (2-40) is the equation of the line through the origin in $\mathcal{R}^2$ with slope $-\alpha_1/\alpha_2$, the solution space of the equation $A\mathbf{x} = \mathbf{0}$ is just the set of points in $\mathcal{R}^2$ which comprise that line. In this case the solution set of the nonhomogeneous equation $A\mathbf{x} = \beta$, β a real number, can be interpreted as a translation of the line described by (2-40), as shown in Fig. 2-10.

It goes without saying that in particular instances the solution set of $A\mathbf{x} = \mathbf{y}$ may be empty, in which case the equation has no solutions. In fact, one of the major problems in the study of operator equations (or arbitrary equations for that matter) is to determine conditions under which the equation will have solutions. This is the so-called *existence problem* for operator equations, and theorems which establish such conditions are called *existence theorems.*

Of equal, or even greater importance is the problem of ascertaining when $A\mathbf{x} = \mathbf{y}$ admits *at most one* solution for any given $\mathbf{y}$ in $\mathcal{V}_2$. This problem is known as the *uniqueness problem* for operator equations, and can always be answered by examining the homogeneous equation $A\mathbf{x} = \mathbf{0}$ and using the following theorem.

Theorem 2-10. *An operator equation $A\mathbf{x} = \mathbf{y}$ will have a unique solution (provided it has any solutions at all) if and only if its associated homogeneous equation $A\mathbf{x} = \mathbf{0}$ has no nonzero solutions, i.e., if and only if $\mathfrak{N}(A) = \mathcal{O}$.*

The student should have no difficulty in convincing himself that this result is an immediate consequence of Theorem 2-9 and the description of the solution set of $A\mathbf{x} = \mathbf{y}$ given there.

In the case where A admits an inverse of one of the various types discussed in Section 2-4, the equation $A\mathbf{x} = \mathbf{y}$ can be immediately solved. If, for instance, A is invertible, then from $A\mathbf{x} = \mathbf{y}$ we deduce that

$$A^{-1}(A\mathbf{x}) = A^{-1}\mathbf{y},$$

or

$$\mathbf{x} = A^{-1}\mathbf{y},$$

and the solution (which in this case must be unique) has been described in terms of A^{-1}. Similarly, if B is either a right or left inverse for A we find that the solution set of $A\mathbf{x} = \mathbf{y}$ is the set of all $\mathbf{x}$ in $\mathcal{V}_1$ such that $\mathbf{x} = B\mathbf{y}$. This technique for solving an operator equation is known as *inverting the operator*, and is used whenever an explicit formula for an inverse can be deduced from the definition of A.

EXAMPLE 3. *Systems of Linear Equations.* As our final example we apply the above ideas to the study of systems of linear equations. Our motive for presenting this somewhat extended example is twofold. First, the theory of linear equations is important in its own right, and the results we are about to obtain will be needed from time to time in our later work. Second, this material provides perhaps the easiest application of linear transformations to the solution of a nontrivial problem, and should therefore help the student become familiar with such transformations.

We begin by introducing some standard terminology. A system of m linear equations in the n *unknowns* $x_1, x_2, \ldots, x_n$ is a set of equations of the form

$$\begin{aligned} \alpha_{11}x_1 + \alpha_{12}x_2 + \cdots + \alpha_{1n}x_n &= \beta_1, \\ \alpha_{21}x_1 + \alpha_{22}x_2 + \cdots + \alpha_{2n}x_n &= \beta_2, \\ &\vdots \\ \alpha_{m1}x_1 + \alpha_{m2}x_2 + \cdots + \alpha_{mn}x_n &= \beta_m, \end{aligned} \tag{2-41}$$

in which the α_{ij} and β_i are real numbers. The α_{ij} are called the *coefficients* of the system, and have been so indexed that the first subscript on any coefficient indicates the equation in which it appears, and the second the unknown with which it is associated. Such a system is said to be *homogeneous* if all of the β_i are zero; *nonhomogeneous* otherwise. Finally, a *solution* of (2-41) is an n-tuple of real numbers $(c_1, c_2, \ldots, c_n)$ with the property that when c_1 is substituted for x_1, c_2 for x_2, etc., each of the equations in the system becomes an identity. A system without solutions is said to be *incompatible.*

Let

$$[A] = \begin{bmatrix} \alpha_{11} & \alpha_{12} & \cdots & \alpha_{1n} \\ \alpha_{21} & \alpha_{22} & \cdots & \alpha_{2n} \\ \vdots & & & \vdots \\ \alpha_{m1} & \alpha_{m2} & \cdots & \alpha_{mn} \end{bmatrix}$$

be the $m \times n$-matrix formed from the coefficients of (2-41), and let $A: \mathcal{R}^n \to \mathcal{R}^m$ be the linear transformation defined by this matrix relative to the standard bases. Then if $\mathbf{x}$ denotes the vector $(x_1, \ldots, x_n)$ in $\mathcal{R}^n$, and $\mathbf{y}$ the vector $(\beta_1, \ldots, \beta_m)$ in $\mathcal{R}^m$, (2-41) can be rewritten in operator form as

$$A\mathbf{x} = \mathbf{y}. \tag{2-42}$$

Conversely, if A is an arbitrary linear transformation from $\mathcal{R}^n$ to $\mathcal{R}^m$, and if $[A]$ is the matrix of A with respect to the standard bases in these spaces, Eq. (2-42) is equivalent to a system of m linear equations in n unknowns for each fixed $\mathbf{y}$ in $\mathcal{R}^m$. We now propose to use these observations to establish several important facts concerning solutions of systems of linear equations.

In the first place, we know that (2-42) will have solutions if and only if the vector $\mathbf{y} = (\beta_1, \ldots, \beta_m)$ belongs to the image of the linear transformation A. Furthermore, if

$$\mathbf{y}_1 = (\alpha_{11}, \alpha_{21}, \ldots, \alpha_{m1}), \quad \ldots, \quad \mathbf{y}_n = (\alpha_{1n}, \alpha_{2n}, \ldots, \alpha_{mn})$$

denote the vectors in $\mathcal{R}^m$ formed from the *columns* of (2–41) (or, equivalently, from the columns of the matrix $[A]$), then

$$A\mathbf{x} = x_1\mathbf{y}_1 + x_2\mathbf{y}_2 + \cdots + x_n\mathbf{y}_n$$

for each $\mathbf{x} = (x_1, \ldots, x_n)$ in $\mathcal{R}^n$. Thus $A\mathbf{x}$ is a linear combination of $\mathbf{y}_1, \ldots, \mathbf{y}_n$, and it follows that these vectors span the image of A. This, combined with the observation made a moment ago, yields our first existence theorem.

Theorem 2–11. *The system of linear equations* (2–41) *has a solution if and only if the vector* $\mathbf{y} = (\beta_1, \ldots, \beta_m)$ *in* $\mathcal{R}^m$ *is linearly dependent on the vectors* $\mathbf{y}_1, \ldots, \mathbf{y}_n$ *formed from the columns of the system.*

To answer the uniqueness problem for (2–41) we pass to the associated homogeneous system $A\mathbf{x} = \mathbf{0}$, and apply Theorem 2–10. We again use the fact that if $\mathbf{c} = (c_1, \ldots, c_n)$ is any vector in $\mathcal{R}^n$, then

$$A\mathbf{c} = c_1\mathbf{y}_1 + \cdots + c_n\mathbf{y}_n.$$

Hence $\mathbf{c}$ will be a solution of the homogeneous equation $A\mathbf{x} = \mathbf{0}$ if and only if

$$c_1\mathbf{y}_1 + \cdots + c_n\mathbf{y}_n = \mathbf{0}.$$

Thus $A\mathbf{x} = \mathbf{0}$ has a *nontrivial* solution (i.e., a solution in which at least one of the $c_i \neq 0$) if and only if the vectors $\mathbf{y}_1, \ldots, \mathbf{y}_n$ are linearly *dependent* in $\mathcal{R}^m$, and we have

Theorem 2–12. *The system of linear equations* (2–41) *has a unique solution (provided it has any solutions at all) if and only if the vectors formed from the columns of the system are linearly independent in* $\mathcal{R}^m$.

Finally, since $\mathbf{y}_1, \ldots, \mathbf{y}_n$ are always linearly dependent if $n > m$, this result also yields

Corollary 2–1. *A homogeneous system of linear equations has nontrivial solutions whenever the number of unknowns exceeds the number of equations.*

3 the general theory of linear differential equations

3–1 LINEAR DIFFERENTIAL OPERATORS

We have already had occasion to remark that the operator D which maps a differentiable function onto its derivative is a linear transformation (see Section 2–1). The same is true of polynomials in D, and of even more complicated expressions such as $xD^2 + D + x$. Linear transformations of this sort which involve D and its powers are called *linear differential operators.* The study of such operators leads naturally to the theory of linear differential equations, the subject matter of this chapter and the chapters which follow.

To give a precise meaning to the term "linear differential operator," let I be an arbitrary interval of the real line, and for each non-negative integer n, let $\mathcal{C}^n(I)$ denote the vector space of all real-valued functions which have a continuous nth derivative everywhere in I. [Recall that the vectors in $\mathcal{C}^n(I)$ are real-valued functions whose first n derivatives exist and are continuous throughout I, and that vector addition and scalar multiplication in this space are defined by the equations

$$(f + g)(x) = f(x) + g(x), \qquad (\alpha f)(x) = \alpha f(x),$$

for all x in I. By agreement, $\mathcal{C}^0(I) = \mathcal{C}(I)$.]

In these terms we now state

Definition 3–1. A linear transformation $L: \mathcal{C}^n(I) \to \mathcal{C}(I)$ is said to be a *linear differential operator of order n on the interval I* if it can be expressed in the form

$$L = a_n(x)D^n + a_{n-1}(x)D^{n-1} + \cdots + a_1(x)D + a_0(x), \tag{3–1}$$

where the *coefficients* $a_0(x), \ldots, a_n(x)$ are continuous everywhere in I, and $a_n(x)$ is not identically zero on I. In addition, the transformation which maps every function in $\mathcal{C}^n(I)$ onto the zero function is also considered to be a linear differential operator. It, however, is not assigned an order.

Thus the image of a function f in $\mathcal{C}^n(I)$ under the linear differential operator described above is the function in $\mathcal{C}(I)$ defined by the identity

$$Lf(x) = a_n(x)\frac{d^n}{dx^n}f(x) + \cdots + a_1(x)\frac{d}{dx}f(x) + a_0(x)f(x), \tag{3–2}$$

or, more simply, by

$$Ly = a_n(x)y^{(n)} + \cdots + a_1(x)y' + a_0(x)y, \tag{3–3}$$

where $y', \ldots, y^{(n)}$ are the first n derivatives of the function $y = f(x)$. Strictly speaking, the left-hand side of (3–2) is the value of Lf at the point x, and a scrupulous regard for accuracy would require that it be written $\big(L(f)\big)(x)$. For obvious reasons the extra parentheses are almost always omitted. Moreover, we shall occasionally refer to $Lf(x)$ as "the linear transformation L applied to the function $f(x)$," thereby following the familiar custom of confusing a function with its value at a point. This, of course, is just a linguistic convenience, and once understood as such causes no difficulty.

EXAMPLE 1. The nth derivative operator, D^n, is the simplest example of a linear differential operator of order n on an arbitrary interval I. When $n = 0$, D^0 is just the identity transformation, and, in general, D^n can be viewed as the nth power of the linear transformation D (see Section 2–3).

EXAMPLE 2. Any polynomial in D of degree n, with real coefficients, is a linear differential operator of order n on every interval of the real line.

EXAMPLE 3. A linear differential operator of order zero on I has the form

$$L = a_0(x), \tag{3–4}$$

where $a_0(x)$ is continuous and not identically zero on I. Thus if f is any function in $\mathcal{C}(I)$,

$$Lf(x) = a_0(x)f(x),$$

which, of course, is just the product of the functions a_0 and f. Occasionally one finds (3–4) written in the form $a_0(x)D^0$ to emphasize the fact that $a_0(x)$ is being viewed as an operator and not as a function in $\mathcal{C}(I)$.*

EXAMPLE 4. The linear differential operator

$$xD^2 + 3\sqrt{x}\,D - 1$$

---

* Note that the expression $a_0(x)f(x)$ actually admits three different interpretations. It can be viewed as the product of the functions $a_0(x)$ and $f(x)$, *or* as the value of the operator $a_0(x)$ applied to the function $f(x)$, *or* as the product of the operators $a_0(x)$ and $f(x)$. The particular interpretation chosen, however, is usually a matter of indifference.

is of order 2 in $[0, \infty)$ or any of its subintervals.* By way of contrast

$$(x + |x|)D^2 - \sqrt{x + 1}\, D + \ln(x + 1)$$

is of order 2 on $(-1, 1)$, but of order 1 on the subinterval $(-1, 0]$ since $x + |x|$ vanishes identically there. Thus the order of a linear differential operator may depend upon the interval in which it is being considered, as well as on the algebraic form of the operator itself.

A linear differential operator is, by definition, a linear transformation, and hence, under suitable hypotheses, it makes sense to talk about the product of two such operators. Such products are again linear differential operators although it is impossible to say very much about their order or domain of definition (see Exercise 7). We remind the reader that the usual precautions arising from the noncommutativity of operator multiplication must also be observed in this setting. For instance, a product such as $(xD + 2)(2xD + 1)$ *cannot* be computed by multiplying the expressions $xD + 2$ and $2xD + 1$ according to the usual rules of algebra. Indeed, if it could, we would have $(xD + 2)(2xD + 1) = 2x^2D^2 + 5xD + 2$, when, in fact, the correct answer is $2x^2D^2 + 7xD + 2$, as can be seen from the following computation:

$$\begin{aligned}(xD + 2)(2xD + 1)y &= (xD + 2)(2xy' + y) \\ &= xD(2xy' + y) + 2(2xy' + y) \\ &= x(2xy'' + 3y') + 4xy' + 2y \\ &= 2x^2y'' + 7xy' + 2y.\end{aligned}$$

However, in the special case of operators with *constant coefficients* products *can* be computed as though the operators were ordinary polynomials in D (see Section 2–3, and Exercise 12 below). As we shall see, this fact will ultimately enable us to solve *all* linear differential equations with constant coefficients.

EXERCISES

1. Evaluate each of the following expressions.
   (a) $(D^2 + D)e^{2x}$ (b) $(3D^2 + 2D + 2)\sin x$
   (c) $(xD - x)(2\ln x)$ (d) $(D + 1)(D - x)(2e^x + \cos x)$
2. Repeat Exercise 1 for each of the following expressions.
   (a) $(aD^2 + bD + c)e^{kx}$, a, b, c, k constants
   (b) $(x^2D^2 - 2xD + 4)x^k$, k a constant

---

* The intervals $[a, b)$ and $(a, b]$ are defined, respectively, by the inequalities $a \leq x < b$, $a < x \leq b$. The first is said to be open on the right and closed on the left, the second closed on the right and open on the left.

(c) $(4x^2D^2 + 4xD + 4x^2 + 1)\dfrac{1}{\sqrt{x}}\sin x$

3. Find constants a, b, c such that $a + b + c = 1$, and

$$[(1 - x^2)D^2 - 2xD + 6](ax^2 + bx + c) = 0.$$

4. Write each of the following linear differential operators in the standard form

$$a_n(x)D^n + \cdots + a_1(x)D + a_0(x).$$

(a) $(D^2 + 1)(D - 1)$ (b) $xD(D - x)$
(c) $(xD^2 + D)^2$ (d) $D^2(xD - 1)D$
(e) $D(De^x + 1) + e^x$

5. Show that $D(xD) \neq (xD)D$.
6. (a) Prove that a linear differential operator of order n is a linear transformation from $\mathcal{C}^n(I)$ to $\mathcal{C}(I)$.
(b) Is this linear transformation one-to-one when $n > 0$? Why?
7. (a) Compute the product of the linear differential operators $a_1(x)D + 1$ and $b_1(x)D + 1$ when

$$a_1(x) = \begin{cases} 0, & x \leq 0, \\ x^2/2, & x \geq 0, \end{cases} \qquad b_1(x) = \begin{cases} x^2/2, & x \leq 0, \\ 0, & x \geq 0, \end{cases}$$

and thus deduce that the order of the product of two such operators need not be the sum of the orders of the factors.
(b) Give an example to show that the product of two linear differential operators on an interval I need not be defined on the same interval.
8. Prove that $D^m(a(x)D^n)$ is a linear differential operator of order $m + n$ by expressing this product in standard form as a "polynomial" in D. [Assume the existence and continuity of all the necessary derivatives of $a(x)$.]
9. Find the sum $L_1 + L_2$ of each of the following pairs of linear differential operators.
(a) $L_1 = 2xD + 3$, $L_2 = xD - 1$
(b) $L_1 = e^xD^2 + D$, $L_2 = e^{-x}D^2 - D$
(c) $L_1 = xD + 1$, $L_2 = Dx$
10. Prove that the sum of two linear differential operators defined on an interval I is the linear differential operator on I obtained by adding the corresponding coefficients in the standard "polynomial" representation (3–1) of the given operators.
11. Let

$$L_1 = \sum_{k=0}^{m} a_k(x)D^k \quad \text{and} \quad L_2 = \sum_{k=0}^{n} b_k(x)D^k$$

be linear differential operators on an interval I. Prove that $L_1 = L_2$ if and only if $m = n$, and $a_k(x) \equiv b_k(x)$ for all k.
12. (a) Prove that

$$(aD^m)(bD^n) = (bD^n)(aD^m) = abD^{m+n}$$

whenever a and b are constants.

(b) Use (a) and the general distributivity formula for linear transformations that was established in Section 2–3 to prove that the multiplication of constant coefficient linear differential operators is commutative. Deduce from this that the product of two such operators can be obtained by treating them as ordinary polynomials in D and using the usual rules of elementary algebra.

13. Factor each of the following linear differential operators into a product of irreducible factors of lower order.

(a) $D^2 - 3D + 2$
(b) $2D^2 + 5D + 2$
(c) $4D^2 + 4D + 1$
(d) $D^3 - 3D^2 + 4$
(e) $4D^4 + 4D^3 - 7D^2 + D - 2$
(f) $D^4 - 1$
(g) $D^4 + 1$
(h) $D^5 - 1$

14. Prove that

(a) $D^2[f(x)g(x)] = f''(x)g(x) + 2f'(x)g'(x) + f(x)g''(x)$,
(b) $D^3[f(x)g(x)] = f'''(x)g(x) + 3f''(x)g'(x) + 3f'(x)g''(x) + f(x)g'''(x)$.

Can you make a conjecture as to the form of $D^n[f(x)g(x)]$? ("No" is not an acceptable answer.)

*15. Use mathematical induction to prove *Leibnitz's rule:*

$$D^n[f(x)g(x)] = \sum_{k=0}^{n} \binom{n}{k} \left(D^{n-k} f(x)\right)\left(D^k g(x)\right),$$

where

$$\binom{n}{k} = \frac{n!}{k!(n-k)!} = \frac{n(n-1)\cdots(n-k+1)}{k!}.$$

16. Use the result of the preceding exercise to express each of the following linear differential operators in the form $a_n(x)D^n + \cdots + a_1(x)D + a_0(x)$.

(a) $D^3(xD)$
(b) $D^m(xD)$
(c) $D^5(xD^2 + e^x)$

17. Prove that for any pair of non-negative integers k and m,

$$D^m x^k = \begin{cases} m! \binom{k}{m} x^{k-m}, & m \le k, \\ 0, & m > k. \end{cases}$$

*18. (a) Prove that

$$(x^m D^m)x^k = k(k-1)\cdots(k-m+1)x^k$$

for any real number k.

(b) Prove that

$$(a_2 x^2 D^2 + a_1 x D + a_0)x^k = [a_2 k(k-1) + a_1 k + a_0]x^k$$

for any real number k (a_0, a_1, a_2 constants).

(c) Prove that $(xD)(x^3D^3) = (x^3D^3)(xD)$.

*19. A linear differential operator is sometimes said to be *equidimensional* or an *Euler* operator if it can be written in the form $a_n x^n D^n + \cdots + a_1 x D + a_0$, where $a_0, \ldots, a_n$ are constants.

(a) Compute the value of Lx^k, k an arbitrary real number, when L is equidimensional.

(b) Prove that $(x^m D^m)(x^n D^n) = (x^n D^n)(x^m D^m)$ for any pair of non-negative integers m, n, and hence deduce that the multiplication of equidimensional operators is commutative. [*Hint:* Use the results of Exercises 15 and 17.]

3–2 LINEAR DIFFERENTIAL EQUATIONS

An nth-*order linear differential equation* on an interval I is, by definition, an operator equation of the form

$$Ly = h(x), \tag{3–5}$$

in which h is continuous on I, and L is an nth-order linear differential operator defined on I. Such an equation is said to be *homogeneous* if h is identically zero on I, *nonhomogeneous* otherwise, and *normal* whenever the leading coefficient $a_n(x)$ of the operator L does not vanish anywhere on I. Finally, a function $y(x)$ is said to be a *solution* of (3–5) if and only if $y(x)$ belongs to $\mathcal{C}^n(I)$ and satisfies the equation identically on I.

Thus an nth order linear differential equation is simply an equation of the form

$$a_n(x)\frac{d^n y}{dx^n} + \cdots + a_1(x)\frac{dy}{dx} + a_0(x)y = h(x), \tag{3–6}$$

whose coefficients $a_0(x), \ldots, a_n(x)$ and right-hand side $h(x)$ are continuous on an interval I in which $a_n(x)$ is not identically zero. Typical examples are provided by the equations

$$\frac{d^2 y}{dx^2} + y = 0,$$

which is homogeneous, normal, and of order 2 on $(-\infty, \infty)$ or any of its subintervals, and

$$x^3 \frac{d^3 y}{dx^3} + x\frac{dy}{dx} = 3,$$

which is nonhomogeneous, normal, and of order 3 on $(0, \infty)$ and $(-\infty, 0)$, but is non-normal on any interval containing the origin.

The primary objective in the study of linear differential equations is to find *all* solutions of any given equation on an interval I. As might be expected, this is a difficult problem, and a complete answer is known only for certain special types of equations. However, there exists a considerable body of knowledge concerning the general behavior of solutions of linear differential equations, and in this respect the theory of such equations stands in refreshing contrast to that of nonlinear equations. This, of course, is due to the fact that the techniques of linear algebra can be used in this context, and the present chapter constitutes our first substantial application of these ideas to the study of a problem in analysis.

As an illustration of the way in which linear algebra intervenes in the study of differential equations, let

$$Ly = 0 \tag{3-7}$$

be a *normal, homogeneous* linear differential equation of order n on an interval I of the x-axis. In this case the solution set of the equation is none other than the *null space* of the linear transformation L, and hence is a *subspace* of $\mathcal{C}^n(I)$. Out of deference to the problem at hand this subspace is called the *solution space* of the equation, and the task of solving (3–7) has been reduced to that of finding a *basis* for its solution space, provided, of course, the solution space is finite dimensional. It is; and later in this chapter we shall in fact prove that *the solution space of any normal* nth *order homogeneous linear differential equation is an n-dimensional subspace of* $\mathcal{C}^n(I)$. Thus if L is normal, and if $y_1(x), \ldots, y_n(x)$ are *n linearly independent* solutions of (3–7), then *every* solution of that equation must be of the form

$$y(x) = c_1y_1(x) + \cdots + c_ny_n(x) \tag{3-8}$$

for suitable real numbers c_i. Conversely, every function of this type is certainly a solution of (3–7) whenever $y_1(x), \ldots, y_n(x)$ are, and for this reason (3–8), with the c_i arbitrary, is called the *general solution* of (3–7). Finally, any function obtained from the general solution by assigning definite values to the c_i is called a *particular solution.* We leave the reader to reflect upon the merits and shortcomings of this somewhat unfortunate choice of terminology.

By a familiar line of reasoning, these results are also pertinent to the study of nonhomogeneous equations. Indeed, in Section 2–9 we saw that if $y_p(x)$ is any solution of the nonhomogeneous equation

$$Ly = h(x) \tag{3-9}$$

and if $y_h(x)$ is the general solution of the *associated homogeneous equation* $Ly = 0$, then the expression $y_p(x) + y_h(x)$ is the general solution of (3–9). In other words, *the solution set of a nonhomogeneous linear differential equation can be found by adding all solutions of the associated homogeneous equation to any particular solution of the given equation.* Needless to say, this argument effectively reduces the problem of solving a nonhomogeneous equation to that of finding the general solution of its associated homogeneous equation. And in the next chapter we shall complete this reduction by giving a method whereby a particular solution $y_p(x)$ of (3–9) can always be found once the general solution of the associated homogeneous equation is known.

EXAMPLE 1. The functions $\sin x$ and $\cos x$ are easily seen to be solutions of the second-order equation

$$y'' + y = 0 \tag{3-10}$$

on the interval $(-\infty, \infty)$. Moreover, these functions are linearly independent in $\mathcal{C}^2(-\infty, \infty)$, since

$$c_1 \sin x + c_2 \cos x = 0$$

for *all* x implies, by setting $x = \pi/2$ and $x = 0$, that $c_1 = 0$ and $c_2 = 0$. Thus the general solution of (3–10) is

$$y = c_1 \sin x + c_2 \cos x, \tag{3–11}$$

where c_1 and c_2 are arbitrary constants. The reader should note that without a theorem such as the one cited above there would be no guarantee whatever that (3–11) includes *every* solution of the given equation.

EXAMPLE 2. The function $y_p(x) = x$ is obviously a solution of the nonhomogeneous equation

$$y'' + y = x \tag{3–12}$$

on $(-\infty, \infty)$. Hence, since $c_1 \sin x + c_2 \cos x$ is the general solution of the associated homogeneous equation $y'' + y = 0$, the general solution of (3–12) is

$$y = x + c_1 \sin x + c_2 \cos x.$$

Before leaving this section it may be instructive to compare the solution set of a *nonlinear* differential equation with that of a linear equation. To this end we consider

$$y' - 3y^{2/3} = 0, \tag{3–13}$$

which, as is easily seen, has the family of cubic curves

$$y = (x + c)^3 \tag{3–14}$$

as its "general" solution on the interval $(-\infty, \infty)$. (See Fig. 3–1.) In particular, the functions x^3 and $(x + 1)^3$ are solutions of (3–13). But their sum is not, and hence the solution set of this equation is *not* a subspace of $\mathcal{C}^1(-\infty, \infty)$, even though the equation appears to be homogeneous. Moreover, all of the various solutions obtained from (3–14) by assigning different values to c are linearly independent in $\mathcal{C}^1(-\infty, \infty)$, and we conclude that a *first-order* nonlinear differential equation can actually have *infinitely many* linearly independent solutions. Finally, (3–13) also admits an infinite number of solutions which cannot be obtained from $(x + c)^3$ by specializing the constant c. All of these somewhat peculiar solutions have the property that they are zero along an interval of the x-axis and are of the following three forms:

$$y = \begin{cases} (x - a)^3, & x \le a, \\ 0, & x > a, \end{cases} \qquad y = \begin{cases} 0, & x < b, \\ (x - b)^3, & x \ge b, \end{cases}$$

$$y = \begin{cases} (x - a)^3, & x \le a, \\ 0, & a < x < b, \\ (x - b)^3, & x \ge b. \end{cases}$$

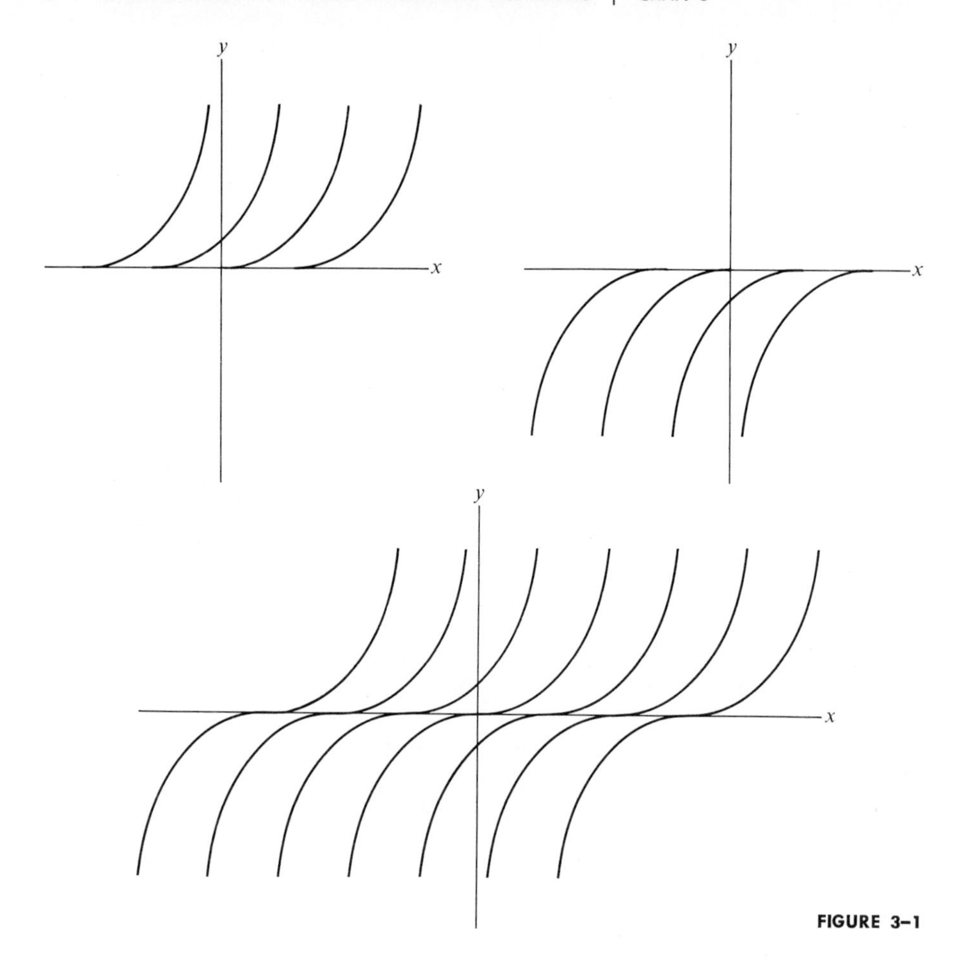

FIGURE 3–1

Thus, to use the term "general solution" in reference to (3–14) is in this case a genuine misnomer. In short, every single one of the properties enjoyed by the solution set of a *linear* differential equation fails to hold here—a fact which, if it does nothing else, should convince the student that linear differential equations are rather more pleasant to encounter than nonlinear equations.

EXERCISES

1. Determine the order of each of the following linear differential equations on the indicated intervals.

   (a) $xy'' - (2x + 1)y = 3$, on $(-\infty, \infty)$ (b) $(D + 1)^3 y = 0$, on $(0, 1)$

   (c) $(x + |x|)y''' + (\sin x)y' = 2e^x$, on $(-1, 1)$; on $(0, \infty)$

   (d) $\sqrt{x}\, y'' - 2y' + (\sin x)y = \ln x$, on $(1, \infty)$

   (e) $(x + 1 + |x + 1|)y''' + (x + |x|)y' + 2y = 0$, on $(-\infty, \infty)$; on $(0, \infty)$; on $(-1, 0)$

2. In each of the following show that the given function is a solution of the associated linear differential equation, and find the interval (or intervals) in which this is the case.
   (a) $xy'' + y' = 0$; $\ln(1/x)$
   (b) $4x^2y'' + 4xy' + (4x^2 - 1)y = 0$; $\sqrt{2/(\pi x)}\sin x$
   (c) $(1 - x^2)y'' - 2xy' + 6y = 0$; $3x^2 - 1$
   (d) $x^2y'' - xy' + y = 1$; $1 + 2x\ln x$
   (e) $(1 - x^2)y'' - 2xy' + 2y = 2$; $x\tanh^{-1} x$
3. (a) Show that $e^{ax}\cos bx$ and $e^{ax}\sin bx$ are linearly independent solutions of the equation
$$(D^2 - 2aD + a^2 + b^2)y = 0, \quad b \neq 0,$$
on $(-\infty, \infty)$.
   (b) What is the general solution of this equation?
   (c) Find the particular solution of the equation in (a) which satisfies the "initial" conditions $y(0) = b$, $y'(0) = -a$.
4. (a) Show that e^{ax} and xe^{ax} are linearly independent solutions of the equation
$$(D - a)^2y = 0.$$
   (b) Find the particular solution of this equation which satisfies the "initial" conditions $y(0) = 1$, $y'(0) = 2$.
5. (a) Verify that $\sin^3 x$ and $\sin x - \frac{1}{3}\sin 3x$ are solutions of
$$y'' + (\tan x - 2\cot x)y' = 0$$
on any interval where $\tan x$ and $\cot x$ are both defined. Are these solutions linearly independent?
   (b) Find the general solution of this equation.
6. Show that $\frac{1}{9}x^3$ and $\frac{1}{9}(x^{3/2} + 1)^2$ are solutions of the nonlinear differential equation $(dy/dx)^2 - xy = 0$ on $(0, \infty)$. Is the sum of these functions a solution?
7. In each of the following show that the given functions span the solution space of the associated differential equation. Find a basis for the solution space in each case, and use it to obtain the general solution of the equation in question.
   (a) $y'' - y = 0$; $\sinh x, 2e^{-x}, -\cosh x$, on $(-\infty, \infty)$
   (b) $x^2y'' - 5xy' + 9y = 0$; $2x^3\ln x, x^3, x^3(2\ln x - 1)$, on $(0, \infty)$
   (c) $y'' + 4y = 0$; $\sin 2x, -2\cos 2x, -\cos(2x - 3)$, on $(-\infty, \infty)$
   (d) $(1 - x^2)y'' - 2xy' + 2y = 0$; $3x, \dfrac{x}{2}\ln\dfrac{1 + x}{1 - x} - 1, \dfrac{x}{2}$, on $(-1, 1)$

3–3 FIRST-ORDER EQUATIONS

Let
$$a_1(x)\frac{dy}{dx} + a_0(x)y = h(x) \tag{3–15}$$
be a *normal* first-order linear differential equation defined on an interval I of the x-axis. Then, as we know, the general solution of this equation can be expressed

in the form

$$y = y_p(x) + y_h(x), \tag{3-16}$$

where $y_p(x)$ is any "particular" solution, and $y_h(x)$ is the general solution of the homogeneous equation

$$a_1(x)\frac{dy}{dx} + a_0(x)y = 0. \tag{3-17}$$

Since $a_1(x) \neq 0$ everywhere in I, (3–17) may be rewritten

$$\frac{1}{y}\frac{dy}{dx} = -\frac{a_0(x)}{a_1(x)}, \qquad y \neq 0,$$

and integrated to yield

$$\ln|y| = -\int \frac{a_0(x)}{a_1(x)}\,dx,$$

or

$$|y| = e^{-\int [a_0(x)/a_1(x)]dx}.$$

Hence, by the theorem cited in the preceding section, the general solution of (3–17) is

$$y = ce^{-\int [a_0(x)/a_1(x)]dx},$$

where c is an arbitrary constant.

To obtain a particular solution of (3–15), we rewrite the equation as

$$\frac{dy}{dx} + \frac{a_0(x)}{a_1(x)}y = \frac{h(x)}{a_1(x)}, \tag{3-18}$$

and multiply by $e^{\int [a_0(x)/a_1(x)]dx}$ to obtain

$$\left(\frac{dy}{dx} + \frac{a_0(x)}{a_1(x)}y\right)e^{\int [a_0(x)/a_1(x)]dx} = \frac{h(x)}{a_1(x)}e^{\int [a_0(x)/a_1(x)]dx}. \tag{3-19}$$

But

$$\frac{d}{dx}\left(ye^{\int [a_0(x)/a_1(x)]dx}\right) = \left(\frac{dy}{dx} + \frac{a_0(x)}{a_1(x)}y\right)e^{\int [a_0(x)/a_1(x)]dx},$$

and so (3–19) may be replaced by the equivalent equation

$$\frac{d}{dx}\left(ye^{\int [a_0(x)/a_1(x)]dx}\right) = \frac{h(x)}{a_1(x)}e^{\int [a_0(x)/a_1(x)]dx}.$$

Thus

$$y = e^{-\int [a_0(x)/a_1(x)]dx}\int \frac{h(x)}{a_1(x)}e^{\int [a_0(x)/a_1(x)]dx}\,dx,$$

and it follows from (3–16) that the general solution of (3–15) is

$$y = \left[c + \int \frac{h(x)}{a_1(x)} e^{\int [a_0(x)/a_1(x)]dx}\, dx\right] e^{-\int [a_0(x)/a_1(x)]dx}, \tag{3–20}$$

c an arbitrary constant.

Considering the simplicity of the technique underlying this result, it is not recommended that the student commit (3–20) to memory. Instead he should remember the general method, which can be described as follows: *To find the general solution of a normal first-order linear differential equation, rewrite the equation in the form*

$$\frac{dy}{dx} + \frac{a_0(x)}{a_1(x)} y = \frac{h(x)}{a_1(x)},$$

multiply by $e^{\int [a_0(x)/a_1(x)]dx}$, *and integrate.*

EXAMPLE 1. Find the general solution of

$$\frac{dy}{dx} + 2xy = x.$$

In this case we multiply the equation by $e^{\int 2x\, dx} = e^{x^2}$ to obtain

$$\left(\frac{dy}{dx} + 2xy\right) e^{x^2} = xe^{x^2}.$$

Thus

$$ye^{x^2} = \int xe^{x^2}\, dx + c,$$

and it follows that

$$y = \left(\frac{e^{x^2}}{2} + c\right) e^{-x^2} = \frac{1}{2} + ce^{-x^2}.$$

EXAMPLE 2. Solve the equation

$$x\frac{dy}{dx} + y = x. \tag{3–21}$$

Since the leading coefficient of this equation vanishes when $x = 0$, the above method applies only on the intervals $(0, \infty)$ and $(-\infty, 0)$. There, however, (3–21) may be rewritten

$$\frac{dy}{dx} + \frac{1}{x} y = 1, \tag{3–22}$$

and solved by introducing the "integrating factor"

$$e^{\int dx/x} = e^{\ln|x|} = |x|.$$

Multiplying (3–22) by $|x|$ and integrating, we obtain

$$y|x| = \int |x|\,dx + c$$

$$= \begin{cases} \dfrac{x^2}{2} + c, & x > 0, \\ -\dfrac{x^2}{2} + c, & x < 0. \end{cases}$$

Thus

$$y = \begin{cases} \dfrac{c}{x} + \dfrac{x}{2}, & x > 0, \\ -\dfrac{c}{x} + \dfrac{x}{2}, & x < 0, \end{cases}$$

and since c is arbitrary we have

$$y = \frac{c}{x} + \frac{x}{2}.$$

We call the reader's attention to the fact that $x/2$ is the only solution of the given equation defined on the entire real line. Nevertheless it is common practice to call $c/x + x/2$ the "general solution" of (3–21) without specifying the interval in question—a practice which is admittedly convenient, but potentially misleading.

EXAMPLE 3. We now use the above technique to solve the nonlinear equation

$$a_1(x)\frac{dy}{dx} + a_0(x)y = h(x)y^n, \tag{3–23}$$

where n is an arbitrary real number. This equation is known as *Bernoulli's equation*, and here, as always, we assume that $a_0(x)$, $a_1(x)$, and $h(x)$ are continuous on an interval I, and that $a_1(x) \neq 0$ on I.

Dismissing the cases $n = 0$ and $n = 1$ which have been treated above, we rewrite (3–23) as

$$a_1(x)y^{-n}\frac{dy}{dx} + a_0(x)y^{1-n} = h(x), \tag{3–24}$$

and make the change of variable $u = y^{1-n}$. Then

$$\frac{du}{dx} = (1 - n)y^{-n}\frac{dy}{dx},$$

and (3–24) becomes

$$\frac{a_1(x)}{1 - n}\frac{du}{dx} + a_0(x)u = h(x),$$

which is a normal first-order equation on the interval I. We now solve this equation for u, and then express the general solution of (3–23) as $y = u^{1/(1-n)}$. Finally, if $n > 0$, we add the solution $y = 0$ "suppressed" in passing from (3–23) to (3–24), and we are done.

Thus, to solve

$$\frac{dy}{dx} + y = (xy)^2, \tag{3–25}$$

we first rewrite the equation as

$$y^{-2}\frac{dy}{dx} + y^{-1} = x^2,$$

and then make the change of variable $u = y^{-1}$. This gives

$$-\frac{du}{dx} + u = x^2,$$

from which we obtain

$$ue^{-x} = -\int x^2 e^{-x}\,dx + c.$$

Hence

$$u = 2 + 2x + x^2 + ce^x,$$

and the solutions of (3–25) are

$$y = (2 + 2x + x^2 + ce^x)^{-1}, \quad c \text{ arbitrary},$$

and $y = 0$.

EXERCISES

Find the general solution of each of the following equations.

1. $xy' + 2y = 0$
2. $(1 - x^2)y' - y = 0$
3. $(\sin x)y' + (\cos x)y = 0$
4. $3y' + ky = 0$, k a constant
5. $2y' + 3y = e^{-x}$
6. $3xy' - y = \ln x + 1$
7. $L\dfrac{di}{dt} + Ri = E$, L, R, E constants, $L, R \neq 0$
8. $(3x^2 + 1)y' - 2xy = 6x$
9. $(x^2 + 1)y' - (1 - x)^2 y = xe^{-x}$
10. $(x^2 + 1)y' + xy = (1 - 2x)\sqrt{x^2 + 1}$
11. $x \sin x \dfrac{dy}{dx} + (\sin x + x \cos x)y = xe^x$
12. $x\dfrac{dy}{dx} + \dfrac{y}{\sqrt{2x + 1}} = 1 + \sqrt{2x + 1}$

13. $x\dfrac{dy}{dx} + \dfrac{y}{\sqrt{1-x^2}} = (1 + \sqrt{1-x^2})e^x$

14. $\sin x \cos x \dfrac{dy}{dx} + y = \tan^2 x$

15. $(1 + \sin x)\dfrac{dy}{dx} + (2\cos x)y = \tan x$

16. $2(1 - x^2)y' - (1 - x^2)y = xy^3e^{-x}$

17. $y' = \dfrac{y^2 \sin x - y\cos^2 x}{\sin x \cos x}$

18. $yy' + xy^2 - x = 0$

19. $(x^2 + 1)\sqrt{y}\,y' = xe^{3x/2} + (1 - x)^2 y\sqrt{y}$

20. $(x^2 + x + 1)yy' + (2x + 1)y^2 = 2x - 1$

21. $xy' + \dfrac{y}{\ln x} = \dfrac{x(x + \ln x)}{y^2 \ln x}$

22. $\dfrac{\sin 2x}{6}\, y' + y = (1 + \cos x)y^{2/3}$

*23. $(x - 1)y' - 2y = \sqrt{(x^2 - 1)y}$

24. $y' = \dfrac{(x + 1)\ln x - x(3x + 4)y^3}{(x^3 + 2x^2 - 1)y^2}$

25. $(xy^2)' = (xy)^3(x^2 + 1)$

26. Find the particular solution of the equation $xy' - (\sin x)y = 0$ on the interval $(0, \infty)$ which passes through the point $(1, -1)$. [*Hint:* Show that the general solution of this equation on $(0, \infty)$ may be written in the form $y = ce^{\int_1^x (\sin t)/t\, dt}$, $x > 0$.]

27. (a) Find the solution curve of the equation

$$x\frac{dy}{dx} + y = e^{-x^2/2}$$

which passes through the point $(2, -3)$. [*Hint:* Find the general solution and show that it can be written in the form

$$y = \frac{c}{x} + \frac{1}{x}\int_2^x e^{-x^2/2}\,dx.]$$

(b) What is the ordinate of the point on the solution curve found in (a) corresponding to the point $x = 1$? (Consult a table of values for $(1/\sqrt{2\pi})\int_{-\infty}^x e^{-t^2/2}\,dt$.) Find the slope of the solution curve at this point.

Riccati's equation. Any first-order differential equation of the form

$$\frac{dy}{dx} + a_2(x)y^2 + a_1(x)y + a_0(x) = 0, \tag{3-26}$$

in which $a_0(x)$, $a_1(x)$, $a_2(x)$ are continuous on an interval I and $a_2(x) \not\equiv 0$ on I, is called a *Riccati equation.* A number of elementary facts concerning the solutions of such equations are given in the exercises which follow.

28. Let $y_1(x)$ be a particular solution of (3-26). Make the change of variable $y = y_1 + 1/z$ to reduce (3-26) to a first-order *linear* equation in z, and hence deduce that the general solution of a Riccati equation can be found as soon as a particular solution is known.

Use the technique suggested in the preceding exercise to find the general solution of each of the following Riccati equations.

29. $y' - xy^2 + (2x - 1)y = x - 1$; particular solution $y = 1$

30. $y' + xy^2 - 2x^2y + x^3 = x + 1$; particular solution $y = x - 1$

31. $2y' - (y/x)^2 - 1 = 0$; particular solution $y = x$

32. $y' + y^2 - (1 + 2e^x)y + e^{2x} = 0$; particular solution $y = e^x$

33. $y' - (\sin^2 x)y^2 + \dfrac{1}{\sin x \cos x} y + \cos^2 x = 0$; particular solution $y = \dfrac{\cos x}{\sin x}$

34. (a) Let $y_1(x)$ and $y_2(x)$ be two particular solutions of Eq. (3–26). Show that the general solution of the equation is

$$\frac{y - y_1}{y - y_2} = ce^{\int a_2(x)(y_2 - y_1)\,dx},$$

c an arbitrary constant. [*Hint:* Consider the expression

$$\frac{y' - y_1'}{y - y_1} - \frac{y' - y_2'}{y - y_2}.]$$

(b) Let $y_1(x)$, $y_2(x)$, and $y_3(x)$ be distinct particular solutions of Eq. (3–26). Use the result established in (a) to prove that the general solution of the equation is

$$\frac{(y - y_1)(y_3 - y_2)}{(y - y_2)(y_3 - y_1)} = c,$$

c an arbitrary constant.

35. (a) Show that a constant coefficient Riccati equation

$$\frac{dy}{dx} + ay^2 + by + c = 0$$

has a solution of the form $y = m$, m a constant, if and only if m is a root of the quadratic equation

$$am^2 + bm + c = 0.$$

(b) Use this result, together with Exercise 28 or Exercise 34(a), as appropriate, to find the general solution of each of the following Riccati equations.

(i) $y' + y^2 + 3y + 2 = 0$ (ii) $y' + 4y^2 - 9 = 0$

(iii) $y' + y^2 - 2y + 1 = 0$ (iv) $6y' + 6y^2 + y - 1 = 0$

36. (a) Prove that the change of variable $v = y'/y$ reduces the second-order homogeneous linear differential equation

$$y'' + a_1(x)y' + a_0(x)y = 0 \tag{3–27}$$

to the Riccati equation

$$v' + v^2 + a_1(x)v + a_0(x) = 0, \tag{3–28}$$

and hence deduce that the problem of solving (3–27) is equivalent to that of solving the simultaneous pair of first-order equations

$$\frac{dy}{dx} = vy, \qquad \frac{dv}{dx} = -v^2 - a_1(x)v - a_0(x). \tag{3–29}$$

[Equation (3–28) is called the Riccati equation *associated with* (3–27).]

(b) What conditions ought one impose on (3–29) to correspond to the conditions $y(0) = y_0$, $y'(0) = y_1$ on (3–27)?

(c) Prove that every Riccati equation (3–26) in which $a_2(x) \not\equiv 0$ can be converted to a second-order homogeneous linear differential equation by making the change of variable $y = v'/(a_2 v)$.

37. Find the Riccati equation associated with $y'' - y = 0$. Solve this equation, and hence find the general solution of $y'' - y = 0$.

38. Prove that whenever m_1 and m_2 are distinct real roots of the quadratic equation

$$am^2 + bm + c = 0, \quad a, b, c \text{ constants},$$

then $e^{m_1 x}$ and $e^{m_2 x}$ are linearly independent solutions in $\mathcal{C}(-\infty, \infty)$ of the second-order homogeneous linear differential equation

$$ay'' + by' + c = 0.$$

(See Exercises 35(a) and 36(a).)

39. Use the result of the preceding exercise to find the general solution of each of the following second-order linear differential equations.

(a) $y'' - 5y' + 6y = 0$ (b) $2y'' + y' - 3 = 0$

(c) $(D + 1)(D - 2)y = 0$ (d) $(12D^2 - D - 20)y = 0$

(e) $(2D^2 - 3)y = 0$

40. Prove that e^{mx} and xe^{mx} are linearly independent solutions in $\mathcal{C}(-\infty, \infty)$ of the second-order homogeneous linear differential equation

$$y'' - 2my' + m^2 y = 0,$$

m a constant.

41. Use the result of the preceding exercise to find the general solution of each of the following second-order linear differential equations.

(a) $y'' + 2y' + 1 = 0$ (b) $4y'' - 12y' + 9y = 0$

(c) $(D - \frac{3}{2})^2 y = 0$ (d) $(36D^2 - 12D + 1)y = 0$

(e) $(2D^2 - 2\sqrt{2}\, D + 1)y = 0$

3–4 EXISTENCE AND UNIQUENESS OF SOLUTIONS; INITIAL-VALUE PROBLEMS

On the strength of the results of the preceding section we can assert that every first-order linear differential equation which is normal on an interval I has solutions. In fact, it has infinitely many, one for each value of c in the expression

$$y = \left[c + \int \frac{h(x)}{a_1(x)}\, e^{\int [a_0(x)/a_1(x)]dx}\, dx\right] e^{-\int [a_0(x)/a_1(x)]dx}, \tag{3–30}$$

and the general solution of such an equation therefore is a one-parameter family of plane curves which traverse the strip of the xy-plane determined by I, as shown in Fig. 3-2. Even more important, it is easy to see that there is a solution curve passing through any preassigned point (x_0, y_0) in this strip, since (3-30) can be solved for c when $x = x_0$, $y = y_0$.

The problem of finding a function $y = y(x)$ which is a solution of a normal first-order linear differential equation and which also satisfies the condition $y(x_0) = y_0$ is called an *initial-value problem* for the given equation. This terminology is designed to serve as a reminder of the physical interpretation which views such a solution as the path, or trajectory, of a moving particle which started at the point (x_0, y_0), and whose subsequent motion was governed by the equation in question. In these terms our earlier results can be summarized by saying that every initial-value problem involving a normal first-order linear differential equation has *at least one* solution.

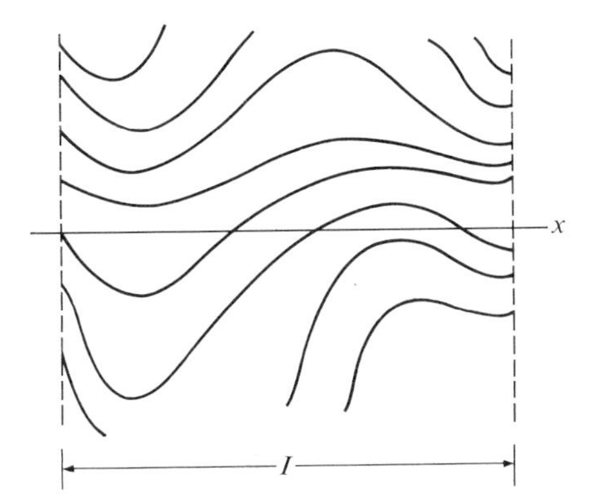

FIGURE 3-2

At this point it is only natural to ask whether or not such a problem can admit more than one solution. This is the so-called *uniqueness problem* for first-order linear differential equations, and is anything but an idle question. Indeed, in applications of differential equations to the natural sciences it is often essential to be able to guarantee that the problem being investigated has a unique solution, since any attempt to predict the future behavior of a physical system governed by an initial value problem relies upon this knowledge. In the case at hand, it is not difficult to show that the desired uniqueness obtains (see Exercise 14 below), and hence the above assertion can be amended to read as follows:

Theorem 3-1. *Every initial-value problem involving a normal first-order linear differential equation has precisely one solution.*

The general theory of linear differential equations can properly be said to begin with the theorem which generalizes this result to nth-order equations. In the special case treated above, the theorem was proved by the simple expedient of exhibiting all of the solutions at issue. Unfortunately, it is impossible to give an argument of this type for equations of higher order, and though the asserted theorem is true, its proof is not conspicuously easy. Thus, rather than become involved in a long and somewhat arid discussion at this time, we content ourselves with a formal statement of the result.*

Theorem 3–2. (The existence and uniqueness theorem for linear differential equations.) *Let*

$$a_n(x)\frac{d^n y}{dx^n} + \cdots + a_0(x)y = h(x) \tag{3-31}$$

be a normal n*th-order linear differential equation defined on an interval* I, *and let* x_0 *be any point in* I. *Then if* $y_0, \ldots, y_{n-1}$ *are arbitrary real numbers there exists* **one and only one** *solution* $y(x)$ *of* (3–31) *with the property that*

$$y(x_0) = y_0,\ y'(x_0) = y_1, \ldots, y^{(n-1)}(x_0) = y_{n-1}. \tag{3-32}$$

As in the case of first-order equations, the problem of finding a solution of (3–31) which satisfies the n additional conditions given in (3–32) is called an *initial-value problem* with *initial conditions*

$$y(x_0) = y_0,\ y'(x_0) = y_1, \ldots, y^{(n-1)}(x_0) = y_{n-1}.$$

It is also worth noting that Theorem 3–2 can be phrased in the language of linear operators, in which case it assumes the following suggestive form:

If $L\colon \mathcal{C}^n(I) \to \mathcal{C}(I)$ *is a normal* n*th-order linear differential operator, there exists a unique inverse operator* $G\colon \mathcal{C}(I) \to \mathcal{C}^n(I)$ *such that*

(i) $L[G(h)] = h$, *for all* h *in* $\mathcal{C}(I)$,

and

(ii) $G(h)(x_0) = y_0,\ G(h)'(x_0) = y_1, \ldots, G(h)^{(n-1)}(x_0) = y_{n-1}$.

When stated in these terms it is clear that the task of solving an initial-value problem for a normal linear differential equation comes down to finding an explicit form for the inverse operator G, since once G is known the problem

$$Ly = h;$$

$$y(x_0) = y_0, \ldots, y^{(n-1)}(x_0) = y_{n-1},$$

can be solved by computing the value of $G(h)$. This point of view will be exploited in later chapters where much of our work will be directed toward finding G for specific classes of linear differential operators. As we shall see, G will turn out to be an integral operator of the type considered in Example 2 of Section 2–2.

---

* For a proof the reader should consult E. Coddington, *An Introduction to Ordinary Differential Equations*, Prentice-Hall, Englewood Cliffs, N.J., 1961.

EXERCISES

Find the solution of each of the following initial-value problems and specify the domain of the solution.

1. $xy' + 2y = 0, \quad y(1) = -1$
2. $(\sin x)y' + (\cos x)y = 0, \quad y\left(\frac{3\pi}{4}\right) = 2$
3. $2y' + 3y = e^{-x}, \quad y(-3) = -3$
4. $(x^2 + 1)y' - (1 - x^2)y = e^{-x}, \quad y(-2) = 0$
5. $x \sin x \dfrac{dy}{dx} + (\sin x + \cos x)y = \dfrac{e^x}{x}, \quad y(-\frac{1}{2}) = 0$
6. $x^2y' + \dfrac{xy}{\sqrt{1 - x^2}} = 1 + \sqrt{1 - x^2}, \quad y(\frac{1}{2}) = 0$
7. $(1 + \sin x)\dfrac{dy}{dx} + (\cot x)y = \cos x, \quad y\left(\frac{\pi}{2}\right) = 1$

Use the given general solution to solve each of the following initial-value problems.

8. $y'' - k^2y = 0, \quad y(0) = y'(0) = 1; \quad y = c_1 \sinh kx + c_2 \cosh kx, \ k \neq 0$
9. $(1 - x^2)y'' - 2xy' = 0, \quad y(-2) = 0, \quad y'(-2) = 1; \quad y = c_1 + c_2 \ln \left|\dfrac{x - 1}{x + 1}\right|$
10. $xy'' + y' + xy = 0, \ y(1) = y'(1) = 1; \ y = c_1J_0(x) + c_2Y_0(x)$, where J_0 and Y_0 are linearly independent solutions of the equation on $(0, \infty)$.
11. $4x^2y'' + 4xy' + (4x^2 - 1)y = 0, \quad y\left(\frac{\pi}{6}\right) = -1, y'\left(\frac{\pi}{6}\right) = 0;$

    $y = \sqrt{\dfrac{2}{\pi x}}\,(c_1 \sin x + c_2 \cos x)$
12. $y'' + (\tan x - 2 \cot x)y' = 0, \quad y\left(-\frac{\pi}{4}\right) = y'\left(-\frac{\pi}{4}\right) = 1;$

    $y = c_1 \sin^3 x + c_2$
13. $xy'' + y' = 0, \quad y(-2) = y'(-2) = 1; \quad y = c_1 + c_2 \ln |x|$
14. (a) Let y_1 and y_2 be solutions of a normal first-order linear differential equation on an interval I. Prove that $y_1 - y_2$ is either identically zero or is different from zero everywhere on I.

    (b) Use the result in (a) to deduce that every initial-value problem involving a normal first-order linear differential equation has at most one solution.
15. Give an example to show that the conclusion of Theorem 3–2 fails when the hypothesis of normality is not satisfied.
16. Let y_1 and y_2 be *distinct* solutions of a normal first-order linear differential equation on an interval I. Prove that the general solution of the equation on I is

$$\frac{y - y_1}{y_1 - y_2} = c,$$

where c is an arbitrary constant. [*Hint:* See Exercise 14(a).]

17. Prove that a nontrivial solution of a homogeneous first-order linear differential equation cannot intersect the x-axis. [*Hint:* Use Theorem 3–1.]
18. Use the results of this section to prove that $y = c_1 \sin x + c_2 \cos x$ is the *general solution* of $y'' + y = 0$ on $(-\infty, \infty)$. [*Hint:* If $u(x)$ is any solution on $(-\infty, \infty)$, show that c_1 and c_2 can be chosen so that $y(0) = u(0)$, $y'(0) = u'(0)$, and then apply Theorem 3–2.]
19. Show that two distinct solutions of a normal first-order linear differential equation cannot have a point of intersection.
20. Prove that every nontrivial solution $u(x)$ of a normal second-order linear differential equation

$$a_2(x)y'' + a_1(x)y' + a_0(x)y = 0$$

has only simple zeros. [A point x_0 is said to be a *zero* of a function $u(x)$ if and only if $u(x_0) = 0$. A zero of $u(x)$ is *simple* if and only if $u'(x_0) \neq 0$.]
21. Prove that distinct solutions of a normal second-order linear differential equation have no point of mutual tangency.
22. Given the linear differential operator D, find the inverse operator G which satisfies $G(h)(x_0) = y_0$.
23. Given the linear differential operator $D - k$ (k a constant), find the inverse operator G which satisfies $G(h)(x_0) = y_0$.

3–5 DIMENSION OF THE SOLUTION SPACE

In this section we shall use the existence and uniqueness theorem stated above to give a simple yet elegant proof of the fact that the dimension of the solution space of every *normal* homogeneous linear differential equation is equal to the order of the equation. The reader should note, however, that this result fails in the case of an equation whose leading coefficient vanishes somewhere in the interval under consideration (witness the equation $xy' + y = 0$ on an interval containing the origin).

This said, we now prove

Theorem 3–3. *The solution space of any normal* nth*-order homogeneous linear differential equation*

$$a_n(x)\frac{d^n y}{dx^n} + \cdots + a_0(x)y = 0 \tag{3–33}$$

defined on an interval I is an n-dimensional subspace of $\mathcal{C}(I)$.

Proof. Let x_0 be a fixed point in I. Then by Theorem 3–2 we know that this equation admits solutions $y_1(x), \ldots, y_n(x)$ which, respectively, satisfy the initial conditions

$$\begin{aligned} y_1(x_0) &= 1, y_1'(x_0) = 0, \ldots, y_1^{(n-1)}(x_0) = 0, \\ y_2(x_0) &= 0, y_2'(x_0) = 1, \ldots, y_2^{(n-1)}(x_0) = 0, \\ &\vdots \\ y_n(x_0) &= 0, y_n'(x_0) = 0, \ldots, y_n^{(n-1)}(x_0) = 1. \end{aligned} \tag{3–34}$$

In other words, $y_1(x), \ldots, y_n(x)$ have the property that the vectors

$$\left(y_i(x_0), y_i'(x_0), \ldots, y_i^{(n-1)}(x_0)\right), \quad i = 1, \ldots, n,$$

are the standard basis vectors in $\mathcal{R}^n$.* (See Fig. 3–3.) We assert that these solutions are a basis for the solution space of (3–33).

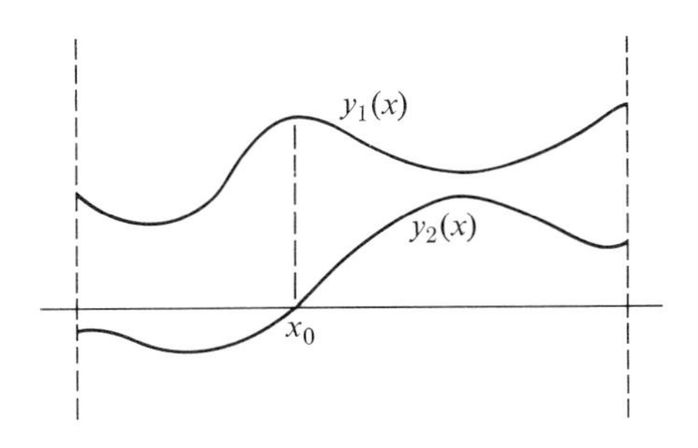

FIGURE 3–3

Indeed, suppose that $c_1, \ldots, c_n$ are real numbers such that

$$c_1y_1(x) + \cdots + c_ny_n(x) \equiv 0$$

on I. Then this identity, together with its first $n - 1$ derivatives, yields the system

$$\begin{aligned} c_1y_1(x) + c_2y_2(x) + \cdots + c_ny_n(x) &\equiv 0, \\ c_1y_1'(x) + c_2y_2'(x) + \cdots + c_ny_n'(x) &\equiv 0, \\ &\vdots \\ c_1y_1^{(n-1)}(x) + c_2y_2^{(n-1)}(x) + \cdots + c_ny_n^{(n-1)}(x) &\equiv 0. \end{aligned} \tag{3–35}$$

Setting $x = x_0$, we obtain

$$\begin{aligned} c_1y_1(x_0) + c_2y_2(x_0) + \cdots + c_ny_n(x_0) &= 0, \\ c_1y_1'(x_0) + c_2y_2'(x_0) + \cdots + c_ny_n'(x_0) &= 0, \\ &\vdots \\ c_1y_1^{(n-1)}(x_0) + c_2y_2^{(n-1)}(x_0) + \cdots + c_ny_n^{(n-1)}(x_0) &= 0, \end{aligned} \tag{3–36}$$

and (3–34) now implies that $c_1 = c_2 = \cdots = c_n = 0$. Thus $y_1(x), \ldots, y_n(x)$ are *linearly independent* in $\mathcal{C}(I)$.

It remains to prove that *every* solution of (3–33) can be written as a linear combination of $y_1(x), \ldots, y_n(x)$. To this end let $y(x)$ be an arbitrary solution of the equation and suppose that

$$y(x_0) = a_0, \; y'(x_0) = a_1, \; \ldots, \; y^{(n-1)}(x_0) = a_{n-1}. \tag{3–37}$$

---

* This choice of solutions has been illustrated in Fig. 3–3 for a second-order equation, in which case

$$\begin{aligned} (y_1(x_0), y_1'(x_0)) &= (1, 0), \\ (y_2(x_0), y_2'(x_0)) &= (0, 1). \end{aligned}$$

Then by the uniqueness statement in Theorem 3–2 we know that $y(x)$ is *the* solution of (3–33) which satisfies these particular initial conditions. But, using (3–34) again, we see that the function

$$a_0 y_1(x) + a_1 y_2(x) + \cdots + a_{n-1} y_n(x)$$

also satisfies this initial-value problem. Hence

$$y(x) = a_0 y_1(x) + a_1 y_2(x) + \cdots + a_{n-1} y_n(x),$$

and it follows that $y_1(x), \ldots, y_n(x)$ *span* the solution space of (3–33). With this, the proof is complete. ▮

We call the reader's attention to the fact that the particular numerical values used to fix the solutions $y_1(x), \ldots, y_n(x)$ did not really play an essential role in the argument given above. Indeed, the success of the proof depended only on the linear independence of the vectors

$$\left(y_i(x_0), y_i'(x_0), \ldots, y_i^{(n-1)}(x_0)\right), \quad i = 1, \ldots, n,$$

in $\mathcal{R}^n$, and the choice made in (3–34) merely served to simplify our computations. For as long as these vectors are linearly independent, the system of homogeneous linear equations (3–36) will have a *unique* solution, and the c_i will be zero, as required.

EXAMPLE 1. The second-order equation

$$\frac{d^2y}{dx^2} - y = 0 \tag{3–38}$$

is normal on the entire x-axis, and thus its solution space is a 2-dimensional subspace of $\mathcal{C}(-\infty, \infty)$. Moreover, it is easy to show that the functions

$$y_1(x) = \tfrac{1}{2}(e^x + e^{-x}) = \cosh x,$$
$$y_2(x) = \tfrac{1}{2}(e^x - e^{-x}) = \sinh x$$

are solutions of (3–38) on $(-\infty, \infty)$, and since

$$y_1(0) = 1, \qquad y_1'(0) = 0,$$
$$y_2(0) = 0, \qquad y_2'(0) = 1,$$

the argument given above implies that $\cosh x$ and $\sinh x$ are a basis for the solution space of this equation. Thus the general solution of (3–38) is

$$y = c_1 \cosh x + c_2 \sinh x,$$

where c_1 and c_2 are arbitrary constants.

EXAMPLE 2. The functions

$$y_1(x) = e^x, \qquad y_2(x) = e^{-x}$$

provide a second pair of solutions of Eq. (3–38). In this case

$$\begin{aligned} y_1(0) &= 1, & y_1'(0) &= 1, \\ y_2(0) &= 1, & y_2'(0) &= -1, \end{aligned}$$

and since the vectors $(1, 1)$ and $(1, -1)$ are linearly independent in $\mathcal{R}^2$, e^x and e^{-x} also form a basis for the solution space of the equation. It follows that the general solution of (3–38) may also be written

$$y = c_1 e^x + c_2 e^{-x},$$

which, of course, is a variant of the solution obtained above.

EXAMPLE 3. The functions

$$y_1(x) = \sin 2x, \qquad y_2(x) = \cos 2x$$

are solutions of the normal second-order equation

$$\frac{d^2y}{dx^2} + 4y = 0 \tag{3–39}$$

on $(-\infty, \infty)$. Furthermore, $y_1(0) = 0$, $y_1'(0) = 2$, while $y_2(0) = 1$, $y_2'(0) = 0$, and since $(0, 2)$ and $(1, 0)$ are linearly independent vectors in $\mathcal{R}^2$, we conclude that $\sin 2x$ and $\cos 2x$ are linearly independent in $\mathcal{C}(-\infty, \infty)$. Hence they are a basis for the solution space of (3–39), and the general solution of that equation is

$$y = c_1 \sin 2x + c_2 \cos 2x.$$

At this point it is impossible to escape the conclusion that in proving Theorem 3–3 we also established a method for testing functions for linear independence. This fact is well worth bringing out into the open, since it will be used in the following sections to obtain a number of important results concerning (normal) linear differential equations. Specifically, we have

Corollary 3–1. *Let $y_1(x), \ldots, y_n(x)$ be functions in $\mathcal{C}(I)$, each of which possesses derivatives up to and including those of order $n - 1$ everywhere in I, and suppose that at some point x_0 in I the vectors*

$$\left(y_i(x_0), y_i'(x_0), \ldots, y_i^{(n-1)}(x_0)\right), \quad i = 1, \ldots, n, \tag{3–40}$$

are linearly independent in $\mathcal{R}^n$. Then $y_1(x), \ldots, y_n(x)$ are linearly independent in $\mathcal{C}(I)$.

EXAMPLE 4. The functions

$$e^x, \; xe^x, \; x^2e^x$$

are linearly independent in $\mathcal{C}(-\infty, \infty)$ since the above test applied at $x = 0$ yields the linearly independent vectors $(1, 1, 1)$, $(0, 1, 2)$, $(0, 0, 2)$ in $\mathcal{R}^3$.

EXERCISES

For each of the following homogeneous linear differential equations, (a) show that the given functions span the solution space of the equation on an appropriate interval, (b) choose a basis for the solution space from among these functions and use it to express the general solution of the equation, and (c) find a basis for the solution space of the equation which satisfies initial conditions of the form (3–34) at the given point x_0.

1. $y''' - y'' - 2y' = 0;\quad e^{-x}, \sinh x - \frac{1}{2}e^x, 2e^{2x}, 1;\quad x_0 = 0$
2. $y''' - y' = 0;\quad \cosh x, e^{-x} + \sinh x, e^x + \sinh x, \cosh x - 1;\quad x_0 = 0$
3. $(D^4 - 4D^3 + 7D^2 - 6D + 2)y = 0;$
   $xe^x,\ (1 + x)e^x,\ e^x(1 - \sin x),\ e^x(x + \sin x),\ e^x \cos x;\quad x_0 = 0$
4. $x^3y''' + 2x^2y'' - xy' + y = 0;$
   $x + \dfrac{1}{x},\ x + x \ln x,\ \dfrac{1}{x} + x \ln \dfrac{1}{x},\ x(1 - \ln x);\quad x_0 = e$
5. $(x^2D^2 - 2)y = 0;\quad 2x^2 - \dfrac{1}{x},\ \dfrac{x^3 + 1}{x},\ 3x^2;\quad x_0 = 1$
6. $y'' - \dfrac{2x}{1 - x^2}\, y' = 0;\quad 1 - \tanh^{-1} x,\ 1 + \ln \dfrac{1 - x}{1 + x},\ 2;\quad x_0 = 0$
7. Use the equation $xy' + y = 0$ to show that Theorem 3–3 fails if the hypothesis of normality is not satisfied.
8. Show that the following functions are linearly independent in $\mathcal{C}(I)$ for the given interval I.
   (a) $e^{ax}, e^{bx}\ (a \neq b);\quad (-\infty, \infty)$
   (b) $1, x, x^2;\quad (-\infty, \infty)$
   (c) $1, \ln \dfrac{x - 1}{x + 1};\quad (1, \infty)$
   (d) $x, x \ln x;\quad (0, \infty)$
   (e) $e^{ax} \sin bx, e^{ax} \cos bx\ (b \neq 0);\quad (-\infty, \infty)$
9. Prove that the functions $1, x, x^2, \ldots, x^n$ are linearly independent in $\mathcal{C}(I)$ for *any* interval I.

*10. Suppose that $y_1(x), \ldots, y_n(x)$ are solutions of a normal homogeneous linear differential equation of order $m \geq n$ on an interval I. Prove that if $y_1(x), \ldots, y_n(x)$ are linearly independent in $\mathcal{C}(I)$, they are linearly independent in $\mathcal{C}(J)$ for any subinterval J of I. Give an example to show that this result fails if $y_1(x), \ldots, y_n(x)$ do not satisfy such an equation.

3–6 THE WRONSKIAN

In the preceding section we proved that $y_1, \ldots, y_n$ are linearly independent functions in $\mathcal{C}^{n-1}(I)$ whenever there exists a point x_0 in I such that the vectors

$$\big(y_i(x_0), y_i'(x_0), \ldots, y_i^{(n-1)}(x_0)\big), \quad i = 1, \ldots, n, \tag{3–41}$$

are linearly independent in $\mathcal{R}^n$. For our present purposes this result can be stated more conveniently in terms of the determinant of a certain matrix, as follows:

Let $y_1, \ldots, y_n$ be arbitrary functions in $\mathcal{C}^{n-1}(I)$, and for each x in I consider the matrix

$$\begin{bmatrix} y_1(x) & y_2(x) & \cdots & y_n(x) \\ y_1'(x) & y_2'(x) & \cdots & y_n'(x) \\ \vdots & & & \vdots \\ y_1^{(n-1)}(x) & y_2^{(n-1)}(x) & \cdots & y_n^{(n-1)}(x) \end{bmatrix}. \tag{3–42}$$

Then (3–42) defines a function on the interval I whose value at x is the indicated matrix, and by forming the determinant of this matrix we obtain a *real valued* function on I known as the *Wronskian* of $y_1, \ldots, y_n$. This function will be denoted by $W[y_1, \ldots, y_n]$ to indicate its dependence on $y_1, \ldots, y_n$, and its value at x by $W[y_1(x), \ldots, y_n(x)]$. In short, the Wronskian of $y_1, \ldots, y_n$ is the (real valued) function whose defining equation is

$$W[y_1(x), \ldots, y_n(x)] = \begin{vmatrix} y_1(x) & y_2(x) & \cdots & y_n(x) \\ y_1'(x) & y'(x) & \cdots & y_n'(x) \\ \vdots & & & \vdots \\ y_1^{(n-1)}(x) & y_2^{(n-1)}(x) & \cdots & y_n^{(n-1)}(x) \end{vmatrix}. \tag{3–43}$$

For example,

$$W[x, \sin x] = \begin{vmatrix} x & \sin x \\ 1 & \cos x \end{vmatrix} = x\cos x - \sin x,$$

and

$$W[x, 2x] = \begin{vmatrix} x & 2x \\ 1 & 2 \end{vmatrix} = 0.$$

We now recall that the determinant of an $n \times n$ matrix is nonzero if and only if the columns of the matrix are linearly independent vectors in $\mathcal{R}^n$ (see Theorem III–8). Thus the Wronskian of $y_1, \ldots, y_n$ is different from zero at x_0 if and only if the columns of (3–42) are linearly independent when $x = x_0$. But for each x_0 in I the columns of (3–42) are none other than the vectors in (3–41), and we therefore have the following theorem.

Theorem 3–4. *The functions $y_1, \ldots, y_n$ are linearly independent in $\mathcal{C}^{n-1}(I)$, and hence also in $\mathcal{C}(I)$, whenever their Wronskian is* **not** *identically zero on I.*

EXAMPLE 1. Since

$$W[e^x, e^{-x}] = \begin{vmatrix} e^x & e^{-x} \\ e^x & -e^{-x} \end{vmatrix} = -2,$$

the functions e^x and e^{-x} are linearly independent in $\mathcal{C}(I)$ for any interval I.

EXAMPLE 2. The functions x, $x^{1/2}$, $x^{3/2}$ are linearly independent in $\mathcal{C}(I)$ for any subinterval I of the positive x-axis since

$$W[x, x^{1/2}, x^{3/2}] = \begin{vmatrix} x & x^{1/2} & x^{3/2} \\ 1 & \frac{1}{2}x^{-1/2} & \frac{3}{2}x^{1/2} \\ 0 & -\frac{1}{4}x^{-3/2} & \frac{3}{4}x^{-1/2} \end{vmatrix} = -\tfrac{1}{4}.$$

More generally, x^α, x^β, x^γ are linearly independent in $\mathcal{C}(I)$, I as above, if and only if α, β, γ are distinct real numbers (see Exercises 13 and 14 below).

EXAMPLE 3. The functions x^3 and $|x|^3$ are linearly independent in $\mathcal{C}(-\infty, \infty)$, for if $c_1x^3 + c_2|x|^3 \equiv 0$, then

$$\begin{aligned} c_1(1)^3 + c_2|1|^3 &= 0, \\ c_1(-1)^3 + c_2|-1|^3 &= 0, \end{aligned}$$

and $c_1 = c_2 = 0$. On the other hand, the Wronskian of x^3 and $|x|^3$ is identically zero on $(-\infty, \infty)$ since

$$W[x^3, |x|^3] = \begin{vmatrix} x^3 & x^3 \\ 3x^2 & 3x^2 \end{vmatrix} = 0,$$

if $x \geq 0$, and

$$W[x^3, |x|^3] = \begin{vmatrix} x^3 & -x^3 \\ 3x^2 & -3x^2 \end{vmatrix} = 0,$$

if $x < 0$. Thus the converse of Theorem 3–4 is *false*, and one cannot deduce the linear *dependence* of a set of functions in $\mathcal{C}(I)$ from the fact that their Wronskian vanishes identically on I. (See Fig. 3–4).

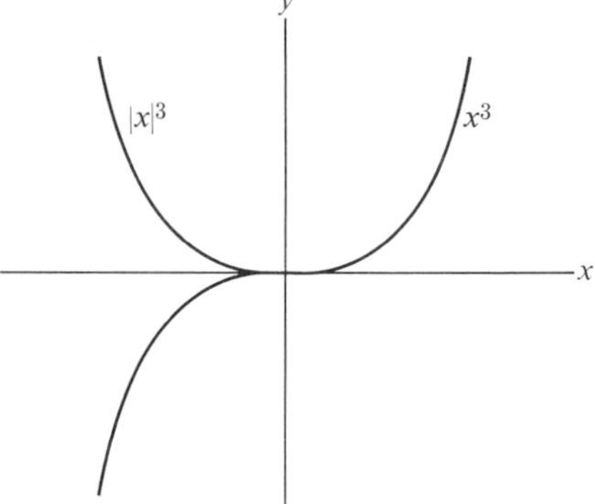

FIGURE 3–4

This example notwithstanding, it *is* true that the Wronskian of a linearly dependent set of functions in $\mathcal{C}(I)$ vanishes identically on I, provided, of course, that the Wronskian exists in the first place. Hence, rather than abandon the search for a converse to Theorem 3–4, we weaken our requirements, and ask whether it is possible to impose additional conditions on a set of functions which, together with the vanishing of their Wronskian, will imply linear dependence.

This can in fact be done, simply by requiring that the functions be solutions of a homogeneous linear differential equation. We prove this assertion as

Theorem 3–5. *Let $y_1, \ldots, y_n$ be solutions of a normal* nth*-order homogeneous linear differential equation*

$$a_n(x) \frac{d^n y}{dx^n} + \cdots + a_0(x)y = 0 \tag{3–44}$$

on an interval I, and suppose that $W[y_1, \ldots, y_n]$ is identically zero on I. Then $y_1, \ldots, y_n$ are linearly dependent in $\mathcal{C}(I)$.

Proof. Let x_0 be any point in I, and consider the system of equations

$$\begin{aligned} c_1 y_1(x_0) + \cdots + c_n y_n(x_0) &= 0, \\ c_1 y_1'(x_0) + \cdots + c_n y_n'(x_0) &= 0, \\ &\vdots \\ c_1 y_1^{(n-1)}(x_0) + \cdots + c_n y_n^{(n-1)}(x_0) &= 0, \end{aligned} \tag{3–45}$$

in the unknowns $c_1, \ldots, c_n$. Since the Wronskian of $y_1, \ldots, y_n$ vanishes identically on I the determinant of (3–45) is zero, and the system has a *nontrivial* solution $(\bar{c}_1, \ldots, \bar{c}_n)$. (See Theorem III–9.) Thus the function

$$y(x) = \sum_{i=1}^{n} \bar{c}_i y_i(x)$$

is a solution of the initial-value problem consisting of (3–44) and initial conditions

$$y(x_0) = 0,\ y'(x_0) = 0, \ldots, y^{(n-1)}(x_0) = 0.$$

But the zero function is also a solution of this problem, and hence Theorem 3–2 implies that

$$\bar{c}_1 y_1(x) + \cdots + \bar{c}_n y_n(x) = 0$$

for all x in I. The linear dependence of $y_1, \ldots, y_n$ now follows from the fact that the $\bar{c}_i$ are not all zero. ▌

Once again we have established a result which is stronger than the one advertised. For the above proof only made use of the fact that the Wronskian of $y_1, \ldots, y_n$ vanished *at a single point* in I, and hence the conclusion remains true under this more restrictive hypothesis. Combined with Theorem 3–4 this observation yields

Theorem 3–6. *A set of solutions of a normal* nth*-order homogeneous linear differential equation is linearly independent in $\mathcal{C}(I)$, and hence is a basis for the solution space of the equation, if and only if its Wronskian* **never** *vanishes on I.*

EXAMPLE 4. By direct substitution the student can verify that $\sin^3 x$ and $1/\sin^2 x$ are solutions of

$$\frac{d^2y}{dx^2} + \tan x \frac{dy}{dx} - 6(\cot^2 x)y = 0 \tag{3–46}$$

on any interval I in which $\tan x$ and $\cot x$ are both defined. Moreover,

$$W\left[\sin^3 x, \frac{1}{\sin^2 x}\right] = \begin{vmatrix} \sin^3 x & \dfrac{1}{\sin^2 x} \\ 3\sin^2 x \cos x & -\dfrac{2\cos x}{\sin^3 x} \end{vmatrix} = 5\cos x.$$

Since $\cos x$ is never zero on I, the above theorem implies that $\sin^3 x$ and $1/\sin^2 x$ are linearly independent in $\mathcal{C}(I)$, and the general solution of (3–46) therefore is

$$y = c_1 \sin^3 x + \frac{c_2}{\sin^2 x}\cdot$$

(Note that this result can also be obtained directly from Theorem 3–4.)

EXAMPLE 5. The functions

$$y_1(x) = \sin^3 x, \qquad y_2(x) = \sin x - \tfrac{1}{3}\sin 3x$$

are solutions of

$$\frac{d^2y}{dx^2} + (\tan x - 2\cot x)\frac{dy}{dx} = 0 \tag{3–47}$$

on any interval I in which $\tan x$ and $\cot x$ are defined. But

$$\begin{aligned} W[y_1(x), y_2(x)] &= \begin{vmatrix} \sin^3 x & \sin x - \frac{1}{3}\sin 3x \\ 3\sin^2 x\cos x & \cos x - \cos 3x \end{vmatrix} \\ &= \sin^3 x(\cos x - \cos 3x) - 3\sin^2 x\cos x(\sin x - \tfrac{1}{3}\sin 3x) \\ &= \sin^2 x(\sin 3x\cos x - \sin x\cos 3x) - \sin^2 x(2\sin x\cos x) \\ &= \sin^2 x\sin 2x - \sin^2 x\sin 2x = 0. \end{aligned}$$

Hence y_1 and y_2 are linearly dependent in $\mathcal{C}(I)$, and do *not* form a basis for the solution space of (3–47).* In this case, however, it is clear that any constant c is a solution of (3–47), and since c and $\sin^3 x$ are obviously linearly independent in $\mathcal{C}(I)$, the general solution of the equation is

$$y = c_1 + c_2\sin^3 x.$$

Of course, this expression may also be written

$$y = c_1 + c_2(\sin x - \tfrac{1}{3}\sin 3x).$$

---

* In fact, $\sin^3 x = \frac{3}{4}\sin x - \frac{1}{4}\sin 3x$.

EXERCISES

By computing Wronskians, show that each of the following sets of functions is linearly independent in $\mathcal{C}(I)$ for the indicated interval I.

1. $1,\ e^{-x},\ 2e^{2x}$ on any interval I
2. $e^x,\ \sin 2x$ on any interval I
3. $1,\ x,\ x^2, \ldots, x^n$ on any interval I
4. $\ln x,\ x \ln x$ on $(0, \infty)$
5. $x^{1/2},\ x^{1/3}$ on $(0, \infty)$
6. $e^{ax} \sin bx,\ e^{ax} \cos bx \quad (b \neq 0)$ on any interval I
7. $e^x,\ e^x \sin x$ on any interval I
8. $e^{-x},\ xe^{-x},\ x^2e^{-x}$ on any interval I
9. $1,\ \sin^2 x,\ 1 - \cos x$ on any interval I
10. $\ln \dfrac{x-1}{x+1},\ 1$ on $(-\infty, -1)$
11. $\sqrt{1 - x^2},\ x$ on $(-1, 1)$
12. $\sin \dfrac{x}{2},\ \cos^2 x$ on any interval I
13. Show that x^α, x^β, x^γ are linearly independent in $\mathcal{C}(0, \infty)$ if and only if α, β, γ are distinct real numbers. [*Hint:* If α, β, γ are distinct and $c_1x^\alpha + c_2x^\beta + c_3x^\gamma \equiv 0$ on $(0, \infty)$, show that $c_1 = c_2 = c_3 = 0$ by considering what happens as x tends to ∞.]
14. Show that x^α, x^β, x^γ are linearly independent in $\mathcal{C}(0, \infty)$ if and only if they are linearly independent in $\mathcal{C}(I)$ for *every* subinterval I of $(0, \infty)$. [*Hint:* First establish both of the following assertions, and then use Theorem 3–6:

    (a) x^α, x^β, x^γ satisfy the linear differential equation

    $$x^3y''' + a_2x^2y'' + a_1xy' + a_0y = 0,$$

    where $a_2 = 3 - \alpha - \beta - \gamma$, $a_1 = 1 - \alpha - \beta - \gamma + \alpha\beta + \alpha\gamma + \beta\gamma$, $a_0 = -\alpha\beta\gamma$.

    (b) $$W(x^\alpha, x^\beta, x^\gamma) = x^{\alpha+\beta+\gamma-3} \begin{vmatrix} 1 & 1 & 1 \\ \alpha & \beta & \gamma \\ \alpha(\alpha-1) & \beta(\beta-1) & \gamma(\gamma-1) \end{vmatrix},$$

    and hence $W(x^\alpha, x^\beta, x^\gamma)$ either vanishes nowhere in $(0, \infty)$ or vanishes identically.]
15. Generalize the results of Exercises 13 and 14(b) to show that $x^{\alpha_1}, \ldots, x^{\alpha_n}$ are linearly independent in $\mathcal{C}(I)$ for any subinterval I of $(0, \infty)$ if and only if $\alpha_1, \ldots, \alpha_n$ are distinct real numbers.
16. Let f belong to $\mathcal{C}^1[a, b]$, and suppose that f is not the zero function. By computing their Wronskian show that $f(x)$ and $xf(x)$ are linearly independent in $\mathcal{C}^1[a, b]$. (Also see Exercise 17.)
17. Show that $f(x)$ and $xf(x)$ are linearly independent in $\mathcal{C}[a, b]$ if f is continuous and not identically zero on $[a, b]$.

18. Suppose that f is an odd function in $\mathcal{C}^1[-a, a]$ (that is, $f(-x) = -f(x)$) and that $f(0) = f'(0) = 0$. Show that

$$W[f(x), |f(x)|] = 0$$

for all x in $[-a, a]$, but that f and $|f|$ are linearly independent in $\mathcal{C}^1[-a, a]$ unless f is identically zero. Compare this result with Example 3 in the text.

19. Let f, g be any two functions in $\mathcal{C}^1(I)$, and suppose that g never vanishes in I. Prove that if $W[f(x), g(x)] \equiv 0$ on I, then f and g are linearly dependent in $\mathcal{C}^1(I)$. [*Hint:* Calculate $d/dx(f(x)/g(x))$.]

*20. Let f, g be any two functions in $\mathcal{C}^1(I)$ which have only finitely many zeros in I and have no common zeros. Prove that if $W[f(x), g(x)] \equiv 0$ on I, then f, g are linearly dependent in $\mathcal{C}^1(I)$. [*Hint:* Apply the result of Exercise 19 to the finite number of subintervals of I on which f or g never vanishes.]

*21. (a) Show that

$$W[e^{a_1 x}, \ldots, e^{a_n x}] = e^{(a_1 + \cdots + a_n)x} \begin{vmatrix} 1 & 1 & \cdots & 1 \\ a_1 & a_2 & \cdots & a_n \\ a_1^2 & a_2^2 & \cdots & a_n^2 \\ \vdots & & & \vdots \\ a_1^{n-1} & a_2^{n-1} & \cdots & a_n^{n-1} \end{vmatrix}.$$

(b) The determinant appearing in (a) is known as a *Vandermonde* determinant. Show that it is zero if and only if $a_i = a_j$ for some pair of indices i, j with $i \neq j$. [*Hint:* Expand the determinant by cofactors of the 1st column to obtain a polynomial in a_1. Is a_2 a root of this polynomial?]

*22. Prove that if $u_1, \ldots, u_n$ is a basis for the solution space of the linear differential equation

$$y^{(n)} + a_{n-1}(x)y^{(n-1)} + \cdots + a_0(x)y = 0,$$

then the $a_i(x)$, $0 \leq i \leq n - 1$, are *uniquely determined* by $u_1, \ldots, u_n$. [*Hint:* For each index i, let $u_{i,x_0}^{(j)}$ be *the* solution of the given equation which satisfies the initial conditions $u_{i,x_0}^{(j)}(x_0) = 0$, $0 \leq j \leq n - 1$, $j \neq i$, and $u_{i,x_0}^{(i)}(x_0) = -1$. Then $a_i(x_0) = u_{i,x_0}^{(n)}(x_0)$.]

23. Let u_1 and u_2 be linearly independent solutions of the normal second-order linear differential equation

$$y'' + a_1(x)y' + a_0(x)y = 0.$$

Express the coefficients $a_0(x)$, $a_1(x)$ in terms of u_1 and u_2. [*Hint:* Let y be an arbitrary solution of the equation, and consider the Wronskian of y, u_1, u_2.]

24. Generalize the result of the preceding exercise to an nth-order equation

$$y^{(n)} + a_{n-1}(x)y^{(n-1)} + \cdots + a_0(x)y = 0.$$

25. Use the results of Exercise 23 to find a homogeneous second-order linear differential equation whose solution space has the following functions as a basis.

(a) x, xe^x (b) x, x^2
(c) $\sin x$, $\cos x$ (d) x, $\sin x$
(e) x, $\ln x$

3–7 ABEL'S FORMULA

According to Theorem 3–6 the Wronskian of a set of solutions of a normal homogeneous linear differential equation either vanishes identically or not at all. This fact can also be deduced from the following theorem, which gives an explicit formula for the Wronskian in this case.

Theorem 3–7. *Let $y_1, \ldots, y_n$ be solutions on an interval I of the nth-order equation*

$$a_n(x)\frac{d^ny}{dx^n} + a_{n-1}(x)\frac{d^{n-1}y}{dx^{n-1}} + \cdots + a_0(x)y = 0, \tag{3–48}$$

and suppose that $a_n(x) \neq 0$ everywhere in I. Then

$$W[y_1(x), \ldots, y_n(x)] = ce^{-\int[a_{n-1}(x)/a_n(x)]dx} \tag{3–49}$$

for an appropriate constant c. (This result is known as *Abel's formula* for the Wronskian.)

Proof. In order to avoid using general properties of determinants, we shall prove (3–49) only in the case $n = 2$. The general proof is identical, except that it uses the formula for the derivative of an nth-order determinant (see Exercise 9).

Thus suppose that y_1 and y_2 are solutions of

$$a_2(x)\frac{d^2y}{dx^2} + a_1(x)\frac{dy}{dx} + a_0(x)y = 0$$

on an interval I in which $a_2(x)$ does not vanish. Then

$$\begin{aligned}
\frac{d}{dx}W[y_1(x), y_2(x)] &= \frac{d}{dx}\begin{vmatrix} y_1(x) & y_2(x) \\ y_1'(x) & y_2'(x) \end{vmatrix} \\
&= \frac{d}{dx}[y_1(x)y_2'(x) - y_2(x)y_1'(x)] \\
&= y_1(x)y_2''(x) - y_1''(x)y_2(x) \\
&= y_1(x)\left[-\frac{a_1(x)}{a_2(x)}y_2'(x) - \frac{a_0(x)}{a_2(x)}y_2(x)\right] \\
&\quad - y_2(x)\left[-\frac{a_1(x)}{a_2(x)}y_1'(x) - \frac{a_0(x)}{a_2(x)}y_1(x)\right] \\
&= -\frac{a_1(x)}{a_2(x)}[y_1(x)y_2'(x) - y_2(x)y_1'(x)] \\
&= -\frac{a_1(x)}{a_2(x)}W[y_1(x), y_2(x)].
\end{aligned}$$

Thus $W[y_1(x), y_2(x)]$ is differentiable on I, and satisfies the first-order linear differential equation

$$\frac{dy}{dx} + \frac{a_1(x)}{a_2(x)} y = 0.*$$

But the general solution of this equation is

$$y = ce^{-\int [a_1(x)/a_2(x)]dx},$$

and hence, for an appropriate value of c, we have

$$W[y_1(x), y_2(x)] = ce^{-\int [a_1(x)/a_2(x)]dx},$$

as asserted. ▮

If x_0 is any fixed point in I, and if $W(x_0)$ denotes the value of the Wronskian of $y_1, \ldots, y_n$ at x_0, then Abel's formula may be written in the form

$$W[y_1(x), \ldots, y_n(x)] = W(x_0)e^{-\int_{x_0}^{x} [a_{n-1}(x)/a_n(x)]dx}. \tag{3-50}$$

This formula shows that the Wronskian of any basis for the solution space of a normal homogeneous linear differential equation is determined up to a multiplicative constant by the equation itself, and does not depend upon the particular basis used to compute it. This simple observation will be important later on.

EXAMPLE 1. Since

$$x\frac{d^2y}{dx^2} + \frac{dy}{dx} + xy = 0$$

is normal on $(0, \infty)$, the Wronskian of any two solutions y_1, y_2 of this equation must be of the form

$$W[y_1(x), y_2(x)] = ce^{-\int (dx/x)} = \frac{c}{x}.$$

If, in addition, y_1 and y_2 satisfy the initial conditions

$$y_1(x_0) = a_0, \; y_1'(x_0) = a_1,$$
$$y_2(x_0) = b_0, \; y_2'(x_0) = b_1,$$

at some point $x_0 > 0$, then

$$c = x_0 \begin{vmatrix} a_0 & b_0 \\ a_1 & b_1 \end{vmatrix} = x_0(a_0b_1 - a_1b_0),$$

and

$$W[y_1(x), y_2(x)] = \frac{x_0(a_0b_1 - a_1b_0)}{x}.$$

---

* This result actually holds for all $n \geq 2$, provided that $a_1(x)$ and $a_2(x)$ are replaced by $a_{n-1}(x)$ and $a_n(x)$, respectively.

As our first substantial application of Theorem 3–7 we shall use Abel's formula to find the general solution of a second-order homogeneous linear differential equation given one nontrivial solution of the equation. Thus let $y_1 \not\equiv 0$ be a solution of

$$a_2(x)y'' + a_1(x)y' + a_0(x)y = 0 \tag{3–51}$$

on an interval I in which $a_2(x) \neq 0$. Then every solution y_2 of (3–51) must satisfy the equation

$$W[y_1(x), y_2(x)] = ce^{-\int [a_1(x)/a_2(x)]dx}, \tag{3–52}$$

and y_2 can thus be found by solving the nonhomogeneous *first-order* equation

$$y_1(x)\frac{dy_2}{dx} - y_1'(x)y_2 = ce^{-\int [a_1(x)/a_2(x)]dx}.$$

By (3–20) the general solution of this equation is

$$y_2 = cy_1(x) \int \frac{e^{-\int [a_1(x)/a_2(x)]dx}}{y_1(x)^2}\, dx + ky_1(x),$$

where k is an arbitrary constant, and since this formula is valid on any subinterval of I in which $y_1 \neq 0$, it can be used to determine a second solution of (3–51) on such a subinterval. In particular, the function

$$y_2(x) = y_1(x) \int \frac{e^{-\int [a_1(x)/a_2(x)]dx}}{y_1(x)^2}\, dx$$

will be such a solution, and is clearly linearly independent of y_1, as desired. Thus we have proved the following theorem.

Theorem 3–8. *If y_1 is a nontrivial solution of Eq.* (3–51) *on an interval in which $a_2(x)$ does not vanish, then*

$$y_2(x) = y_1(x) \int \frac{e^{-\int [a_1(x)/a_2(x)]dx}}{y_1(x)^2}\, dx \tag{3–53}$$

is a solution of the equation on any subinterval of I in which $y_1 \neq 0$. Moreover, y_2 is linearly independent of y_1, and the general solution of (3–51) *is*

$$y = c_1y_1 + c_2y_2,$$

where c_1 and c_2 are arbitrary constants.

Example 2. By direct substitution we find that x^2 is a solution on $(0, \infty)$ of the second-order equation

$$x^2y'' + x^3y' - 2(1 + x^2)y = 0. \tag{3–54}$$

Hence a second linearly independent solution in $\mathcal{C}(0, \infty)$ can be found by solving the first-order equation

$$x^2y' - 2xy = e^{-x^2/2}.$$

Using Formula (3–53), we obtain

$$\begin{aligned} y_2 &= x^2 \int \frac{e^{-\int x\,dx}}{x^4}\,dx \\ &= x^2 \int x^{-4}e^{-x^2/2}\,dx, \end{aligned}$$

and the general solution of (3–54) on $(0, \infty)$ is

$$y = x^2\left(c_1 + c_2 \int x^{-4}e^{-x^2/2}\,dx\right).$$

Here the solution must be left in integral form since $\int x^{-4}e^{-x^2/2}\,dx$ cannot be expressed in terms of elementary functions.

EXAMPLE 3. The function $y_1 = 1$ is a solution of

$$y'' + (\tan x - 2\cot x)y' = 0 \tag{3–55}$$

on any interval in which $\tan x$ and $\cot x$ are both defined. Applying the above result we obtain a second solution

$$\begin{aligned} y_2(x) &= \int e^{-\int(\tan x - 2\cot x)dx}\,dx \\ &= \int e^{\ln(\cos x \sin^2 x)dx}\,dx \\ &= \int \sin^2 x \cos x\,dx \\ &= \tfrac{1}{3}\sin^3 x. \end{aligned}$$

Thus the general solution of (3–55) is

$$y = c_1 + c_2 \sin^3 x,$$

which agrees with the result obtained at the end of the last section.

EXERCISES

1. For each of the following differential equations find the Wronskian of the solutions y_1, y_2 which satisfy the given initial conditions.
   (a) $x^2y'' + xy' + (x^2 + 1)y = 0$; $y_1(1) = 0$, $y_1'(1) = 1$, $y_2(1) = y_2'(1) = 1$
   (b) $(1 - x^2)y'' - 2xy' + n(n + 1)y = 0$ (n a positive integer);
   $y_1(0) = y_1'(0) = 2$, $y_2(0) = 1$, $y_2'(0) = -1$

(c) $x^2y'' - 3xy' + y = 0$; $y_1(-1) = y_1'(-1) = 2$, $y_2(-1) = 0$, $y_2'(-1) = -1$

(d) $y'' + 2xy = 0$; $y_1(0) = y_1'(0) = 1$, $y_2(0) = 1$, $y_2'(0) = 0$

(e) $y'' - (\sin x)y' + 3(\tan x)y = 0$;
$y_1(0) = 1$, $y_1'(0) = 0$, $y_2(0) = 0$, $y_2'(0) = 1$

(f) $\sqrt{1 + x^3}\, y'' - x^2y' + y = 0$; $y_1(1) = 1$, $y_1'(1) = 0$, $y_2(1) = -1$, $y_2'(1) = 1$

In Exercises 2 through 8, one solution of the differential equation is given. Find a second linearly independent solution using the method of this section.

2. $y'' - 4y' + 4y = 0$; e^{2x}
3. $y'' - 2ay' + a^2y = 0$ (a constant); e^{ax}
4. $3xy'' - y' = 0$; 1
5. $y'' + (\tan x)y' - 6(\cot^2 x)y = 0$; $\sin^3 x$
6. $(1 - x^2)y'' - 2xy' = 0$; 1
7. $(1 - x^2)y'' - 2xy' + 2y = 0$; x
8. $2xy'' - (e^x)y' = 0$; 1
9. (a) Prove that if f_{ij} is in $\mathcal{C}^1(I)$, $1 \le i, j \le 2$, then so is the function F defined by

$$F(x) = \begin{vmatrix} f_{11}(x) & f_{12}(x) \\ f_{21}(x) & f_{22}(x) \end{vmatrix},$$

and that

$$\frac{d}{dx}\begin{vmatrix} f_{11}(x) & f_{12}(x) \\ f_{21}(x) & f_{22}(x) \end{vmatrix} = \begin{vmatrix} f_{11}'(x) & f_{12}(x) \\ f_{21}'(x) & f_{22}(x) \end{vmatrix} + \begin{vmatrix} f_{11}(x) & f_{12}'(x) \\ f_{21}(x) & f_{22}'(x) \end{vmatrix}.$$

(b) Generalize the result in (a) to the case of nth-order determinants; and show, in particular, that

$$\frac{d}{dx}\begin{vmatrix} f_{11}(x) & f_{12}(x) & \cdots & f_{1n}(x) \\ f_{21}(x) & f_{22}(x) & \cdots & f_{2n}(x) \\ \vdots & & & \vdots \\ f_{n1}(x) & f_{n2}(x) & \cdots & f_{nn}(x) \end{vmatrix}$$

can be expressed as the sum of n determinants, the ith of which is obtained from $|f_{ij}(x)|$ by differentiating the functions in the ith column.

*3-8 THE EQUATION $y'' + y = 0$

By now it should be obvious that the theorems we have in hand are powerful tools for studying linear differential equations. What is not so obvious, however, is that they can also be used to obtain detailed information about the solutions of an individual equation, and before going any further we propose to illustrate this aspect of our results. Since this method is applied almost exclusively in rather complicated situations we have chosen to introduce it by means of an example which, though somewhat artificial, has the merit of absolute clarity. In the process we will undoubtedly give the impression of someone who is resolutely shutting his eyes to the obvious (which, in fact, is what we shall be doing), but the reader should nevertheless appreciate that the technique in question and the spirit underlying its application are of considerable importance.

The problem we set for ourselves is to study the solutions of the second-order equation

$$y'' + y = 0 \tag{3–56}$$

without using any information other than that provided by the equation itself and the general theorems proved above.

In the first place, this equation is normal on $(-\infty, \infty)$, and we can therefore apply Theorem 3–3 to assert that its solution space is spanned by two linearly independent functions, $C(x)$ and $S(x)$, which are defined for all x and satisfy the initial conditions

$$\begin{aligned} C(0) &= 1, \qquad & C'(0) &= 0, \\ S(0) &= 0, \qquad & S'(0) &= 1. \end{aligned} \tag{3–57}$$

Moreover, from the identities

$$C''(x) + C(x) = 0, \qquad S''(x) + S(x) = 0, \tag{3–58}$$

valid for all x, we conclude that both $C(x)$ and $S(x)$ are infinitely differentiable on $(-\infty, \infty)$, and that all of their derivatives are solutions of (3–56). For example, the identity $C''(x) + C(x) = 0$ implies that $C''(x)$ is differentiable, since $C(x)$ is, and that $C'''(x) + C'(x) = 0$. But this is just (3–56) again with $C'(x)$ in place of y; $S'(x)$ is treated similarly, and the argument can be repeated to establish the assertion for still higher derivatives. In particular,

$$\begin{aligned} C'''(x) &= -C'(x), \qquad & S'''(x) &= -S'(x), \\ C^{(\text{iv})}(x) &= C(x), \qquad & S^{(\text{iv})}(x) &= S(x), \end{aligned}$$

and the derivatives of $C(x)$ and $S(x)$ repeat in cycles of four, as expected. Finally, (3–57) and (3–58) imply that

$$\begin{aligned} C'(0) &= 0, \qquad & C''(0) &= -1, \\ S'(0) &= 1, \qquad & S''(0) &= 0, \end{aligned}$$

and it follows that $C'(x)$ is *the* solution of (3–56) which satisfies the initial conditions $C'(0) = 0$, $C''(0) = -1$, while $S'(x)$ is *the* solution which satisfies $S'(0) = 1$, $S''(0) = 0$. Thus

$$C'(x) = -S(x), \qquad S'(x) = C(x),$$

and we have evaluated *all* of the derivatives of $C(x)$ and $S(x)$.

Using these results we can now prove the familiar identity

$$S(x)^2 + C(x)^2 = 1. \tag{3–59}$$

Indeed, since

$$\frac{d}{dx}[S(x)^2 + C(x)^2] = 2S(x)S'(x) + 2C(x)C'(x) = 0,$$

it follows that

$$S(x)^2 + C(x)^2 = k,$$

k a constant. Setting $x = 0$ gives $k = 1$, as asserted. Among other things, (3–59) implies that $|C(x)| \leq 1$, $|S(x)| \leq 1$ for all x.

Next, we establish the important addition formulas

$$\begin{aligned} C(a + x) &= C(a)C(x) - S(a)S(x), \\ S(a + x) &= S(a)C(x) + C(a)S(x), \end{aligned} \tag{3–60}$$

a an arbitrary real number. To do so, we note that

$$\frac{d^2}{dx^2} C(a + x) = -C(a + x) \quad \text{and} \quad \frac{d^2}{dx^2} S(a + x) = -S(a + x).$$

Thus $C(a + x)$ and $S(a + x)$ are solutions of (3–56), and as such must be linear combinations of $C(x)$ and $S(x)$; that is,

$$\begin{aligned} C(a + x) &= \alpha_1 C(x) + \alpha_2 S(x), \\ S(a + x) &= \beta_1 C(x) + \beta_2 S(x) \end{aligned} \tag{3–61}$$

for suitable constants $\alpha_1, \alpha_2, \beta_1, \beta_2$. To obtain the values of these constants we set $x = 0$ in (3–61) and the identities obtained from it by differentiation. This gives $\alpha_1 = C(a)$, $\beta_1 = S(a)$, $\alpha_2 = -S(a)$, $\beta_2 = C(a)$, and (3–60) has been proved.

In much the same fashion it can be shown that

$$C(-x) = C(x), \qquad S(-x) = -S(x), \tag{3–62}$$

and we conclude that the graph of $C(x)$ is symmetric about the y-axis, while that of $S(x)$ is symmetric about the origin (see Exercise 1).

At this point we could derive those long and all too familiar lists of trigonometric identities involving $C(x)$ and $S(x)$. However, it is much more instructive to prove that these functions are periodic with period 2π. Here we begin by *defining* $\pi/2$ to be the smallest positive real number such that $C(x) = 0$. (The proof that such a number exists has been left to the student in Exercise 2.) Then $C(x)$ is positive on the interval $(0, \pi/2)$, and since $S'(x) = C(x)$, we conclude that $S(x)$ is *increasing* on that interval. But $S(0) = 0$, and hence $S(x)$ is also positive on $(0, \pi/2)$.

Thus by (3–59), $S(\pi/2) = 1$, and the addition formulas now give

$$C(\pi) = -1, \qquad C(3\pi/2) = 0, \qquad C(2\pi) = 1,$$
$$S(\pi) = 0, \qquad S(3\pi/2) = -1, \qquad S(2\pi) = 0.$$

Hence

$$C(x + 2\pi) = C(x)C(2\pi) - S(x)S(2\pi) = C(x),$$
$$S(x + 2\pi) = S(x)C(2\pi) + C(x)S(2\pi) = S(x),$$

and the periodicity of $C(x)$ and $S(x)$ has been established.

To complete the discussion it remains to show that 2π is the *smallest* period for each of these functions. For $C(x)$ the argument goes as follows: From (3–61) we obtain

$$C(x + \pi/2) = -S(x),$$
$$C(x + \pi) = -S(x),$$
$$C(x + 3\pi/2) = S(x),$$

and it follows that $C(x)$ is negative on the interval $(\pi/2, 3\pi/2)$, positive on $(3\pi/2, 2\pi)$. For similar reasons $S(x)$ is positive on $(0, \pi)$, negative on $(\pi, 2\pi)$. But since $C'(x) = -S(x)$, $C(x)$ is decreasing on $(0, \pi)$ and increasing on $(\pi, 2\pi)$. This, together with $C(3\pi/2) = 0$ and $C(2\pi) = 1$, implies that 2π is the smallest positive real number such that $C(x) = 1$, and we are done.

EXERCISES

1. Establish the identities (3–62) by showing that $C(-x)$, $S(-x)$ are solutions of (3–56) and expressing them in terms of the basis $C(x)$, $S(x)$ for the solution space.

*2. Show that there is a least positive real number α such that $C(\alpha) = 0$. [*Hint:* Assume the contrary; then argue that
   (a) $C(x) > 0$ for $0 < x < \infty$,
   (b) $S(x) > 0$ for $0 < x < \infty$,
   (c) $C(x)$ is strictly decreasing and concave downwards on $(0, \infty)$,
   and derive a contradiction from (a) and (c).* Having established that $C(x)$ has positive zeros, consider the greatest lower bound α of the set of positive zeros of $C(x)$.]

3. Let $E(x)$ be the unique solution on $(-\infty, \infty)$ of the initial value problem $y' - y = 0$, $y(0) = 1$. Establish the following properties of $E(x)$:
   (a) $E(x)$ has derivations of all orders, and $E^{(n)}(x) = E(x)$;
   (b) $E(x) > 0$ on $(0, \infty)$; [*Hint:* Otherwise let x_0 be the smallest positive real number for which $E(x_0) = 0$, and derive a contradiction by applying the mean value theorem. (Why would such a smallest number exist?)]

---

* To establish these facts rigorously, the intermediate value theorem and mean value theorem from calculus must be applied. The student, however, may give an intuitive argument based on a consideration of the graphs of $C(x)$ and $S(x)$.

(c) $E(x)$ is strictly increasing and concave upwards on $(0, \infty)$;

(d) $E(a + x) = E(a)E(x)$ for all real numbers a, x;

(e) $E(-x) = \dfrac{1}{E(x)}$ for every real number x; [*Hint:* Apply (d).]

(f) $0 < E(x) < 1$ on $(-\infty, 0)$;

(g) $\lim_{x \to \infty} E(x) = \infty$; [*Hint:* Use (c).]

(h) $\lim_{x \to -\infty} E(x) = 0$;

(i) Set $E(1) = e$, and show that $E(n) = e^n$.

4. (a) Prove that every solution of a homogeneous linear differential equation with constant coefficients has derivatives of all orders at every point on the x-axis.

(b) Generalize this result to homogeneous equations with variable coefficients and to nonhomogeneous equations.

4 equations with constant coefficients

4–1 INTRODUCTION

Linear differential equations with constant coefficients, that is, equations of the form

$$a_n y^{(n)} + a_{n-1} y^{(n-1)} + \cdots + a_0 y = h(x) \tag{4–1}$$

in which $a_0, \ldots, a_n \neq 0$ are (real) constants, are in many respects the simplest of all differential equations. For one thing, they can be discussed *entirely* within the context of linear algebra, and form the only substantial class of equations of order greater than one which can be explicitly solved. This, plus the fact that such equations arise in a surprisingly wide variety of physical problems, accounts for the special place they occupy in the theory of linear differential equations.

We shall begin the discussion of this chapter by considering the homogeneous version of Eq. (4–1), which can be written in normal form as

$$(D^n + a_{n-1}D^{n-1} + \cdots + a_0)y = 0, \tag{4–2}$$

or as

$$Ly = 0, \tag{4–3}$$

where L is the constant coefficient linear differential operator $D^n + a_{n-1}D^{n-1} + \cdots + a_0$. Algebraically such operators behave *exactly* as if they were ordinary polynomials in D, and can therefore be factored according to the rules of elementary algebra. In particular, it follows that *every linear differential operator with constant coefficients can be expressed as a product of constant coefficient operators of degrees one and two* (see Exercise 2 below). As we shall see, this essentially reduces the task of solving (4–2) to the second-order case where complete results can be obtained with relative ease.

This done, we will take up the problem of finding a particular solution of $Ly = h$ given the general solution of the associated homogeneous equation

$Ly = 0$ (see Section 3–2). Here the restriction on the coefficients of L will be dropped and much more far reaching results obtained. The language of operator theory and the ideas of linear algebra will dominate this portion of our discussion and furnish just that measure of insight needed to make it intelligible. Finally, we shall conclude the chapter with some special results involving constant coefficient equations and a number of applications to problems in elementary physics.

EXERCISES

1. (a) Prove that the product of two complex numbers $a + bi$ and $c + di$ is real if and only if either

   (i) $b = d = 0$, or

   (ii) $a = c$ and $b = -d$.

   [*Hint:* Recall that a complex number $a + bi$ is real if and only if $b = 0$, and that a product of the form $(a + bi)(c + di)$ is computed by using the distributive law and the rule $i^2 = -1$.]

   (b) Let $P(x)$ be a polynomial with *real* coefficients, and suppose that $P(x)$ has $a + bi$, $b > 0$, as a root; that is, $P(a + bi) = 0$. Prove that $a - bi$ is also a root of $P(x)$.
2. Let $P(x)$ be a polynomial of degree n, $n > 0$, with real coefficients. Use the fact that $P(x)$ has *exactly* n roots in the complex number system to prove that $P(x)$ can be factored into a product of linear and quadratic factors with real coefficients. [*Hint:* See Exercise 1(b) above.]
3. Find the second-degree polynomial which has $a + bi$ and $a - bi$, $b > 0$, as roots.
4. Prove that every polynomial of odd degree with real coefficients has at least one real root. [*Hint:* See Exercises 1(b) and 2 above.]
5. Write each of the following linear differential operators as a product of operators of degrees one and two.

   (a) $D^3 + 4D^2 + 5D + 2$ (b) $D^3 - D^2 + D - 1$

   (c) $D^4 + 2D^3 - 10D - 25$ (d) $D^4 - 5D^2 + 4$

   (e) $D^4 + 2D^2 + 10$
6. Factor the operator $D^4 + 1$ into a product of operators with real coefficients.

4–2 HOMOGENEOUS EQUATIONS OF ORDER TWO

We have already emphasized that the technique for solving constant coefficient linear differential equations depends upon the commutativity of the operator multiplication involved. To make this dependence explicit, and at the same time phrase it in the form best suited to our immediate needs, we begin by establishing

Lemma 4–1. *If $L_1, \ldots, L_n$ are constant coefficient linear differential operators, then the null space of each of them is contained in the null space of their product.*

Proof. To prove this assertion we must show that $(L_1 \dots L_n)y = 0$ whenever $L_i y = 0$. But this is a triviality since

$$\begin{aligned}(L_1 \dots L_n)y &= (L_1 \dots L_{i-1}L_{i+1} \dots L_n L_i)y \\ &= (L_1 \dots L_{i-1}L_{i+1} \dots L_n)(L_i y) \\ &= (L_1 \dots L_{i-1}L_{i+1} \dots L_n)0 \\ &= 0. \blacksquare\end{aligned}$$

EXAMPLE. The second-order equation

$$(D^2 - 4)y = 0 \tag{4–4}$$

may be rewritten

$$(D + 2)(D - 2)y = 0.$$

Hence e^{2x} and e^{-2x} are solutions of (4–4) since they are, respectively, solutions of the first-order equations $(D - 2)y = 0$ and $(D + 2)y = 0$. Furthermore, these functions are linearly independent in $\mathcal{C}(-\infty, \infty)$ (compute their Wronskian!), and it therefore follows from Theorem 3–3 that the general solution of (4–4) is

$$y = c_1 e^{2x} + c_2 e^{-2x},$$

c_1 and c_2 arbitrary constants.

This simple example suggests that we attempt to solve the general second-order equation

$$(D^2 + a_1 D + a_0)y = 0 \tag{4–5}$$

by decomposing the operator $D^2 + a_1 D + a_0$ into linear factors. To this end we first find the roots α_1, α_2 of the quadratic equation

$$m^2 + a_1 m + a_0 = 0 \tag{4–6}$$

known as the *auxiliary* or *characteristic equation* of (4–5), and then rewrite (4–5) as

$$(D - \alpha_1)(D - \alpha_2)y = 0. \tag{4–7}$$

This done, the argument falls into cases depending on the nature of α_1 and α_2, as follows:

Case 1. *α_1 and α_2 real and unequal.* Here the reasoning used in the above example carries over without change; the functions $e^{\alpha_1 x}$ and $e^{\alpha_2 x}$ are linearly independent solutions of (4–7), and

$$y = c_1 e^{\alpha_1 x} + c_2 e^{\alpha_2 x}$$

is the general solution.

Case 2. $\alpha_1 = \alpha_2 = \alpha$. In this case (4–7) becomes

$$(D - \alpha)^2 y = 0, \tag{4–8}$$

and our earlier argument yields but one solution of the equation, namely $e^{\alpha x}$. Using it, however, we can apply the method introduced in Section 3-7 to find a second linearly independent solution by solving the first-order equation

$$W[e^{\alpha x}, y(x)] = e^{2\alpha x}.$$

An easy computation reveals that up to multiplicative constants, $y(x) = xe^{\alpha x}$, and hence that the general solution of (4-8) is

$$y = (c_1 + c_2 x)e^{\alpha x}.$$

Case 3. *α_1 and α_2 complex.* Here $\alpha_1 = a + bi$, $\alpha_2 = a - bi$, a and b real, $b > 0$, and the above method apparently breaks down. Nevertheless, if we pretend that $e^{\alpha_1 x}$ and $e^{\alpha_2 x}$ continue to make sense when α_1 and α_2 are complex,* the discussion under Case 1 would imply that the general solution of (4-7) is

$$\begin{aligned} y &= c_1 e^{\alpha_1 x} + c_2 e^{\alpha_2 x} \\ &= c_1 e^{(a+bi)x} + c_2 e^{(a-bi)x} \\ &= e^{ax}(c_1 e^{ibx} + c_2 e^{-ibx}). \end{aligned}$$

At this point we invoke Euler's famous formula

$$e^{ix} = \cos x + i \sin x$$

(see Exercise 34) to rewrite this expression as

$$\begin{aligned} y &= e^{ax}[c_1(\cos bx + i \sin bx) + c_2(\cos bx - i \sin bx)] \\ &= e^{ax}[(c_1 + c_2)\cos bx + i(c_1 - c_2)\sin bx] \\ &= c_3 e^{ax}\cos bx + c_4 e^{ax}\sin bx. \end{aligned}$$

Thus, on purely formal grounds we are led to $e^{ax}\cos bx$ and $e^{ax}\sin bx$ as a basis for the solution space of (4-7) when $\alpha_1 = a + bi$ and $\alpha_2 = a - bi$. Of course, we must now verify that these functions actually are solutions of the given equation, and that they are linearly independent in $\mathcal{C}(-\infty, \infty)$. But this is routine and has been left as an exercise for the reader.

Since these three cases include all possible combinations of α_1 and α_2, we have completed the task of solving the general second-order homogeneous linear differential equation with constant coefficients. For convenience of reference we conclude by summarizing our results.

To solve a second-order homogeneous linear differential equation of the form

$$(D^2 + a_1 D + a_0)y = 0,$$

---

* Properly interpreted they do, as the reader may verify by consulting any text on the theory of functions of a complex variable.

first find the roots α_1 *and* α_2 *of the auxiliary equation*

$$m^2 + a_1 m + a_0 = 0.$$

Then the general solution of the given equation can be expressed in terms of α_1 *and* α_2 *as follows:*

α_1, α_2	*General solution*
Real, $\alpha_1 \neq \alpha_2$	$c_1 e^{\alpha_1 x} + c_2 e^{\alpha_2 x}$
Real, $\alpha_1 = \alpha_2 = \alpha$	$(c_1 + c_2 x)e^{\alpha x}$
Complex, $\alpha_1 = a + bi$, $\alpha_2 = a - bi$	$e^{ax}(c_1 \cos bx + c_2 \sin bx)$

EXERCISES

Find the general solution of each of the following differential equations.

1. $y'' + y' - 2y = 0$
2. $3y'' - 5y' + 2y = 0$
3. $8y'' + 14y' - 15y = 0$
4. $y'' - 2y' = 0$
5. $y'' + 4y = 0$
6. $3y'' + 2y = 0$
7. $y'' + 4y' + 8y = 0$
8. $4y'' - 4y' + 3y = 0$
9. $y'' - 2y' + 2y = 0$
10. $9y'' - 12y' + 4y = 0$
11. $y'' + 2y' + 4y = 0$
12. $2y'' - 2\sqrt{2}\, y' + y = 0$
13. $2y'' - 5\sqrt{3}\, y' + 6y = 0$
14. $9y'' + 6y' + y = 0$
15. $64y'' - 48y' + 17y = 0$

In Exercises 16 through 25 find the solutions of the given initial-value problems.

16. $2y'' - y' - 3y = 0;\ y(0) = 2,\ y'(0) = -\frac{7}{2}$
17. $y'' - 8y' + 16y = 0;\ y(0) = \frac{1}{2},\ y'(0) = -\frac{1}{3}$
18. $4y'' - 12y' + 9y = 0;\ y(0) = 1,\ y'(0) = \frac{7}{2}$
19. $y'' + 2y = 0;\ y(0) = 2,\ y'(0) = 2\sqrt{2}$
20. $4y'' - 4y' + 5y = 0;\ y(0) = \frac{1}{2},\ y'(0) = 1$
21. $y'' + 4y' + 13y = 0;\ y(0) = 0,\ y'(0) = -2$
22. $9y'' - 3y' - 2y = 0;\ y(0) = 3,\ y'(0) = 1$
23. $y'' - 2\sqrt{5}\, y' + 5y = 0;\ y(0) = 0,\ y'(0) = 3$
24. $16y'' + 8y' + 5y = 0;\ y(0) = 4,\ y'(0) = -1$
25. $y'' - \sqrt{2}\, y' + y = 0;\ y(0) = \sqrt{2},\ y'(0) = 0$
26. Prove that $e^{\alpha_1 x}$ and $e^{\alpha_2 x}$ are linearly independent in $\mathcal{C}(-\infty, \infty)$ whenever α_1 and α_2 are distinct real numbers.
27. Verify that $xe^{\alpha x}$ is a solution of the second-order equation $(D - \alpha)^2 y = 0$. Prove that this solution and $e^{\alpha x}$ are linearly independent in $\mathcal{C}(-\infty, \infty)$.

28. Verify that $e^{ax}\cos bx$ and $e^{ax}\sin bx$ are linearly independent solutions of the equation

$$(D - \alpha_1)(D - \alpha_2)y = 0$$

when $\alpha_1 = a + bi$ and $\alpha_2 = a - bi$, $b \neq 0$.

29. Find a constant coefficient linear differential equation whose general solution is

(a) $(c_1 + c_2x)e^{-3x}$ (b) $c_1e^x \sin 2x + c_2e^x \cos 2x$

(c) $(c_1 + c_2x)e^{-2x} + 1$ (d) $c_1e^{-x} + c_2e^{-3x} + x + 4$

(e) $c_1 \sin 3x + c_2 \cos 3x + x/3$

30. For each of the following functions find a second-order linear differential equation with constant coefficients which has the given function as a particular solution.

(a) $x(1 + e^x)$ (b) $4 \sin x \cos x$

(c) $(1 + 2e^x)e^{2x} + 6x + 5$ (d) $\cos x(1 - 4\sin^2 x)$

(e) $e^{3x} + e^{2x} + xe^{3x}$

31. (a) Show that the general solution of the second-order equation

$$[D^2 - 2aD + (a^2 + b^2)]y = 0$$

can be written in the form

$$y = c_1e^{ax}\cos(bx + c_2),$$

c_1 and c_2 arbitrary constants. This form is frequently called the *phase-amplitude* form of the solution. Why?

(b) Write the general solution of $(D^2 + 4)y = 0$ in phase-amplitude form.

32. If $L = (D - \alpha)^2$, α real, show that

$$Le^{kx} = (k - \alpha)^2e^{kx}.$$

Differentiate both sides of this identity with respect to k to prove that

$$Lxe^{kx} = (k - \alpha)[2e^{kx} + k(k - \alpha)e^{kx}],$$

and then show that $xe^{\alpha x}$ is a solution of $Ly = 0$.

33. (a) Find the solution of

$$(D^2 - 2D + 26)y = 0$$

whose graph passes through the point $(0, 1)$ with slope 2.

(b) Solve the problem given in (a) again, this time writing the general solution in the form

$$y = c_1e^{(a+bi)x} + c_2e^{(a-bi)x}.$$

Evaluate c_1 and c_2 formally, and then use Euler's formula to show that the resulting solution can be transformed into the solution found in (a).

*34. The function e^z, z a complex number, is defined by the infinite series

$$e^z = 1 + z + \frac{z^2}{2!} + \cdots + \frac{z^n}{n!} + \cdots,$$

and it can be shown that this series converges absolutely for all values of z.* Set $z = ix$ in this series, and use the fact that $i^2 = -1$ to prove Euler's formula. [*Hint:* Since the series is absolutely convergent for all z, its terms may be rearranged at will.]

4–3 HOMOGENEOUS EQUATIONS OF ARBITRARY ORDER

The technique for solving homogeneous constant coefficient linear differential equations is now all but complete. For instance, to solve

$$(D^4 - 2D^3 + 2D^2 - 2D + 1)y = 0 \tag{4–9}$$

we first decompose the operator into linear and quadratic factors, as suggested in Section 4–1, to obtain the equivalent equation

$$(D - 1)^2(D^2 + 1)y = 0,$$

and then invoke Lemma 4–1 to assert that the solution space of each of the second-order equations $(D - 1)^2y = 0$ and $(D^2 + 1)y = 0$ is contained in the solution space of (4–9). Thus e^x, xe^x, $\sin x$, and $\cos x$ are solutions of (4–9), and since these functions are linearly independent in $\mathcal{C}(-\infty, \infty)$, the general solution of the given equation is

$$y = (c_1 + c_2x)e^x + c_3 \sin x + c_4 \cos x.$$

This, in brief, is how all homogeneous constant coefficient equations are solved, and save for the difficulty occasioned by equations such as

$$(D - 1)^4y = 0$$

and

$$(D^2 + 1)^2y = 0,$$

where the above argument fails to yield the required number of linearly independent solutions, we are done. But, recalling our experience with the equation $(D - \alpha)^2y = 0$, it is not difficult to guess that the missing solutions for the above equations are, respectively, x^2e^x, x^3e^x, and $x \sin x$, $x \cos x$. Both of these conjectures are correct, and we shall now prove the relevant generalization of this fact for arbitrary equations of the form

$$(D^n + a_{n-1}D^{n-1} + \cdots + a_0)y = 0, \tag{4–10}$$

where $a_0, \ldots, a_{n-1}$ are constants.

---

* By definition the absolute value, or *modulus*, of a complex number $z = a + bi$ is the real number $\sqrt{a^2 + b^2}$. A series $\sum_{n=1}^{\infty} z_n$ of complex numbers is said to converge absolutely if the series $\sum_{n=1}^{\infty} |z_n|$ of real numbers converges in the usual sense.

We begin by decomposing the operator into linear and quadratic factors, the linear factors being determined by the real roots of the *auxiliary* or *characteristic equation*

$$m^n + a_{n-1}m^{n-1} + \cdots + a_0 = 0, \tag{4–11}$$

the quadratic factors by its complex roots. Then, by Lemma 4–1, we can find solutions of (4–10) by finding the null space of each factor of the form $(D - \alpha)^m$ corresponding to the real root α, and of each factor of the form $[D^2 - 2aD + (a^2 + b^2)]^m$ corresponding to the pair of complex roots $a \pm bi$, $b > 0$. This we accomplish in the following theorem, which, it should be noted, also includes the case $(D - \alpha)^2$ discussed earlier.

Theorem 4–1. *If $y(x)$ belongs to the null space of a constant coefficient linear differential operator L, then $x^{m-1}y(x)$ belongs to the null space of L^m.*

Proof. To establish this result we must compute the value of the linear differential operator L^m applied to the product $x^{m-1}y$, and we therefore begin by giving a formula for evaluating all such expressions. Specifically, if L is any linear differential operator whatever, and if L', L'', ... denote the formal derivatives of L *with respect to D*,* then

$$L(uv) = (Lu)v + (L'u)Dv + \frac{1}{2!}(L''u)D^2v + \frac{1}{3!}(L'''u)D^3v + \cdots \tag{4–12}$$

whenever u and v are functions for which $L(uv)$ is defined.†

Accepting the validity of (4–12), let L and $y(x)$ be as in the statement of the theorem; i.e., $Ly = 0$. We must prove that

$$L^m(yx^{m-1}) = 0.$$

To this end we set $L^m = M$, and apply the above formula to obtain

$$M(yx^{m-1}) = (My)x^{m-1} + (M'y)Dx^{m-1} + \frac{1}{2!}(M''y)D^2x^{m-1} + \cdots.$$

To show that this expression is zero, we first note that $D^rx^{m-1} = 0$ whenever $r \geq m$. Moreover, when $r < m$, the rth formal derivative of M with respect to D, $M^{(r)}$, consists of a sum of terms each of which contains the factor L^{m-r} (see Exercise 20 below). Hence, since $Ly = 0$ and since L is a constant coefficient operator, Lemma 4–1 applies, and we find that $M^{(r)}y = 0$. Thus all of the terms in the above expression are zero, and the theorem is proved. ▮

---

* Thus, if $L = 3D^2 + 4D - 1$, then $L' = 6D + 4$, $L'' = 6$, $L''' = 0$, etc.

† This formula is really no more than a generalized version of Leibnitz's formula $D^n(uv) = \sum_{k=0}^{n} \binom{n}{k}(D^ku)(D^{n-k}v)$, introduced in Exercise 15, Section 3–1. Its proof has been left as an exercise (see Exercise 18).

As a consequence of this theorem we can now assert that the null space of the operator $(D - \alpha)^m$ contains the functions

$$e^{\alpha x}, xe^{\alpha x}, \ldots, x^{m-1}e^{\alpha x},$$

and that the null space of $[D^2 - 2aD + (a^2 + b^2)]^m$ contains

$$\begin{aligned} &e^{ax}\sin bx, \quad xe^{ax}\sin bx, \ldots, \quad x^{m-1}e^{ax}\sin bx,\\ &e^{ax}\cos bx, \quad xe^{ax}\cos bx, \ldots, \quad x^{m-1}e^{ax}\cos bx. \end{aligned}$$

And it is out of just such functions that the general solution of every homogeneous constant coefficient linear differential equation is constructed. The construction depends, of course, upon the fact that the various functions obtained in this way are linearly independent in $\mathcal{C}(-\infty, \infty)$. They are, but unfortunately there is no really brief proof of this assertion. One particularly elegant proof will be given in Section 7–7 as an illustration of the ideas introduced there, and in the meantime we will content ourselves with indicating an alternate approach in Example 5 below.

EXAMPLE 1. Find the general solution of

$$(D^3 + 1)y = 0. \tag{4–13}$$

Here the factorization of the operator is $(D + 1)(D^2 - D + 1)$, and it follows that the roots of the auxiliary equation

$$m^3 + 1 = 0$$

are -1, $\frac{1}{2} + (\sqrt{3}/2)i$, and $\frac{1}{2} - (\sqrt{3}/2)i$. Thus the general solution of (4–13) is

$$y = c_1e^{-x} + e^{x/2}\left(c_2 \sin \frac{\sqrt{3}}{2}x + c_3 \cos \frac{\sqrt{3}}{2}x\right).$$

EXAMPLE 2. Solve

$$y^{(7)} - 2y^{(5)} + y^{(3)} = 0.$$

In operator notation this equation becomes

$$(D^7 - 2D^5 + D^3)y = 0,$$

and since

$$D^7 - 2D^5 + D^3 = D^3[(D - 1)(D + 1)]^2,$$

the roots of the auxiliary equation are 0 (a root of multiplicity 3), 1 (a root of multiplicity 2), -1 (a root of multiplicity 2),* $y = c_1 + c_2x + c_3x^2 + (c_4 + c_5x)e^x + (c_6 + c_7x)e^{-x}$.

* The *multiplicity* of a root α of the equation $m^n + a_{n-1}m^{n-1} + \cdots + a_0 = 0$ is the largest integer k such that $(m - \alpha)^k$ is a factor of the operator.

EXAMPLE 3. Find a constant coefficient linear differential equation which has e^{2x} and xe^{-3x} among its solutions.

In this case we must find a constant coefficient linear differential operator L with the property that $Le^{2x} = 0$ and $Lxe^{-3x} = 0$. (In more picturesque terminology it is sometimes said that L *annihilates* these functions.) Clearly any operator which contains the factors $D - 2$ and $(D + 3)^2$ will answer the problem, and hence

$$(D - 2)(D + 3)^2 y = 0$$

will be an equation of the required type.

EXAMPLE 4. Find a constant coefficient linear differential operator L which, when applied to the equation

$$(D^2 + 1)(D - 1)y = e^x + 2 - 7x \sin x,$$

produces the homogeneous equation

$$L(D^2 + 1)(D - 1)y = 0.$$

Since L must annihilate the functions e^x, 2, and $x \sin x$, it must contain the factors $D - 1$, D, and $(D^2 + 1)^2$. By Lemma 4–1 we can therefore set

$$L = D(D - 1)(D^2 + 1)^2.$$

EXAMPLE 5. As our final example we shall prove that the various solutions of the equation

$$(D - 2)(D + 5)^3(D^2 - 4D + 13)y = 0$$

obtained using Theorem 4–1 are linearly independent in $\mathcal{C}(-\infty, \infty)$. In this case the solutions are

$$\begin{aligned} e^{2x} &\quad \text{corresponding to the factor } D - 2, \\ e^{-5x}, xe^{-5x}, x^2e^{-5x} &\quad \text{corresponding to the factor } (D + 5)^3, \\ e^{2x} \cos 3x, e^{2x} \sin 3x &\quad \text{corresponding to the factor } D^2 - 4D + 13, \end{aligned}$$

and it is obvious that they are somewhat too numerous to permit their Wronskian to be computed easily. Instead we reason as follows:

Let $c_1, \ldots, c_6$ be constants such that

$$c_1e^{2x} + (c_2 + c_3x + c_4x^2)e^{-5x} + e^{2x}(c_5 \cos 3x + c_6 \sin 3x) = 0 \qquad (4\text{–}14)$$

for all x. Apply the operator $(D + 5)^3(D^2 - 4D + 13)$ to this expression to annihilate every term but the first, thereby obtaining

$$c_1(D + 5)^3(D^2 - 4D + 13)e^{2x} \equiv 0.$$

But since $(D + 5)^3(D^2 - 4D + 13)$ does *not* annihilate e^{2x}, it follows that

$c_1 = 0$ and that (4–14) reduces to

$$(c_2 + c_3x + c_4x^2)e^{-5x} + e^{2x}(c_5 \cos 3x + c_6 \sin 3x) \equiv 0. \qquad (4\text{–}15)$$

Next apply the operator $(D + 5)^2(D^2 - 4D + 13)$ to annihilate every term in (4–15) except the one arising from $c_4x^2e^{-5x}$. This gives

$$c_4(D + 5)^2(D^2 - 4D + 13)x^2e^{-5x} \equiv 0,$$

and since $(D + 5)^2(D^2 - 4D + 13)x^2e^{-5x}$ is not identically zero, we conclude that $c_4 = 0$. Hence (4–15) becomes

$$(c_2 + c_3x)e^{-5x} + e^{2x}(c_5 \cos 3x + c_6 \sin 3x) \equiv 0.$$

Continue the argument, using the operators $(D + 5)(D^2 - 4D + 13)$ and $D^2 - 4D + 13$ in turn to deduce that $c_3 = 0$ and $c_2 = 0$. Finally, from the identity

$$e^{2x}(c_5 \cos 3x + c_6 \sin 3x) \equiv 0$$

we conclude directly that $c_5 = c_6 = 0$, and the linear independence of this particular set of solutions has been proved.

In point of fact, this argument can be refined to give a general proof of the linear independence of the functions appearing in the solutions of constant coefficient homogeneous linear differential equations. We refrain from doing so, however, since the problem will be considered in a later chapter where an entirely different proof will be given.

EXERCISES

Find the general solution of each of the following differential equations.

1. $y''' + 3y'' - y' - 3y = 0$
2. $y''' + 5y'' - 8y' - 12y = 0$
3. $4y''' + 12y'' + 9y' = 0$
4. $y''' + 6y'' + 13y' = 0$
5. $2y''' + y'' - 8y' - 4y = 0$
6. $y''' + 3y'' + y' + 3y = 0$
7. $y^{(iv)} - y'' = 0$
8. $y^{(iv)} - 8y'' + 16y = 0$
9. $y^{(iv)} + 18y'' + 81y = 0$
10. $4y^{(iv)} - 8y''' - y'' + 2y' = 0$
11. $y^{(iv)} + y''' + y'' = 0$
12. $y^{(iv)} = 0$
13. $y^{(iv)} - 4y''' + 6y'' - 4y' + y = 0$
14. $y^{(v)} + 2y''' + y' = 0$
15. $y^{(v)} + 6y^{(iv)} + 15y''' + 26y'' + 36y' + 24y = 0$
16. Find the general solution of $(D^4 + 1)y = 0$.
17. Find linear differential operators which annihilate each of the following functions.
    (a) $x^2e^{(x+1)}$
    (b) $3e^{2x} \cos 2x$
    (c) $x(2x + 1) \sin x$
    (d) $3 + 4x - 2e^{-2x}$
    (e) $x^2 \sin x \cos x$
    (f) $x^2e^x \sin^2 x$

(g) $x \sin(x + 1)$ (h) $(x^2 - 1)(\cos x + 1)e^{3x}$

(i) $(xe^x + 1)^3$ (j) $(x + e^x)^n$

18. Use Leibnitz's formula and mathematical induction to establish (4–12) for any linear differential operator L.

19. Verify Formula (4–12) when

(a) $L = D$

(b) $L = D^2 + 2D + 1$

(c) $L = D^n$

20. (a) Let $P = Q^nR$, where Q and R are polynomials, and n is a positive integer. Show that P' can be written in the form $Q^{n-1}S$, where S is again a polynomial. More generally, prove that if Q^k divides a polynomial P, then Q^{k-r} divides $D^{(r)}P$.

(b) Use the result in (a) to prove the assertion made during the proof of Theorem 4–1 to the effect that $M^{(r)}y = 0$ whenever $r < m$.

*21. (a) Show that the Wronskian of the functions $e^{k_1x}, e^{k_2x}, \ldots, e^{k_nx}$ is

$$e^{(k_1+\cdots+k_n)x}\begin{vmatrix} 1 & 1 & \cdots & 1 \\ k_1 & k_2 & \cdots & k_n \\ k_1^2 & k_2^2 & \cdots & k_n^2 \\ \vdots & & & \vdots \\ k_1^{n-1} & k_2^{n-1} & \cdots & k_n^{n-1} \end{vmatrix}.$$

(b) The determinant appearing in (a) is known as a *Vandermonde* determinant. Prove that every such determinant is nonzero whenever $k_1, \ldots, k_n$ are distinct. [*Hint:* Let V denote the determinant in question, and consider k_1 as a variable. Use an inductive argument to show that for each n, V can be viewed as a polynomial of degree $n - 1$ in k_1 which has $k_2, \ldots, k_n$ as roots.]

(c) By direct computation prove that every 3×3 Vandermonde determinant is different from zero if k_1, k_2, k_3 are distinct.

(d) What conclusion can be drawn about solutions of homogeneous constant coefficient linear differential equations from the results of (a) and (b)?

22. Prove that the solutions $y_1(x), \ldots, y_n(x)$ of a constant coefficient linear differential equation are linearly independent in $\mathcal{C}(-\infty, \infty)$ if and only if they are linearly independent in $\mathcal{C}(I)$ for *every* interval I. [*Hint:* Assume linear dependence on I and use the uniqueness theorem.]

23. Prove that the functions $e^{2x}, xe^{2x}, e^{2x}\sin x, e^{2x}\cos x$ are linearly independent in $\mathcal{C}(I)$ for any I. [*Hint:* Show that this assertion is equivalent to asserting the linear independence of $1, x, \sin x, \cos x$ in $\mathcal{C}(-\infty, \infty)$ (see Exercise 22), and then study the behavior of $c_1 + c_2x + c_3 \sin x + c_4 \cos x$ as $x \to \infty$.]

24. Establish the linear independence of the functions given in Exercise 23 by applying the annihilator of $\sin x$ and $\cos x$ to the identity $c_1 + c_2x + c_3 \sin x + c_4 \cos x \equiv 0$.

25. Prove that the functions $e^{\alpha x}, xe^{\alpha x}, \ldots, x^{m-1}e^{\alpha x}$, α real, are linearly independent in $\mathcal{C}(I)$ for any I. [*Hint:* A polynomial of positive degree has only a finite number of zeros.]

26. (a) Let $P(x)$ be a polynomial with real coefficients, and let $L = P(D)$ be the associated constant coefficient linear differential operator. Prove that

$$Le^{\alpha x} = P(\alpha)e^{\alpha x}.$$

(b) Use the result in (a) to show that for any constant coefficient linear differential operator L, $Le^{\alpha x} \equiv 0$ if and only if L has $D - \alpha$ as a factor.

27. Prove that

$$(D - \alpha)^k x^m e^{\alpha x} = \begin{cases} 0 & \text{if } k > m, \\ m(m-1)\cdots(m-k+1)x^{m-k}e^{\alpha x} & \text{if } k \leq m. \end{cases}$$

In particular, deduce that

$$(D - \alpha)^m x^m e^{\alpha x} = m!e^{\alpha x}.$$

28. (a) Let $L = (D - \alpha)^m(D - \beta)^n$, α and β real, $\alpha \neq \beta$. Prove that the functions

$$e^{\alpha x}, \ldots, x^{m-1}e^{\alpha x}, e^{\beta x}, \ldots, x^{n-1}e^{\beta x}$$

in the null space of L are linearly independent in $\mathcal{C}(I)$ for any I. [*Hint:* Apply the operators $(D - \beta)^n(D - \alpha)^{m-1}$, $(D - \beta)^n(D - \alpha)^{m-2}, \ldots$, in succession to the identity

$$c_1e^{\alpha x} + \cdots + c_mx^{m-1}e^{\alpha x} + d_1e^{\beta x} + \cdots + d_nx^{n-1}e^{\beta x} \equiv 0,$$

and use the results of Exercises 26 and 27.]

(b) Generalize the result in (a) to operators of the form

$$L = (D - \alpha_1)^{m_1} \cdots (D - \alpha_k)^{m_k}$$

where $\alpha_1, \ldots, \alpha_k$ are distinct real numbers.

Remark. The results of the last three exercises, when generalized to include complex functions, furnish a proof of the linear independence of the solutions of the equation $Ly = 0$ obtained in this section.

4–4 NONHOMOGENEOUS EQUATIONS: VARIATION OF PARAMETERS AND GREEN'S FUNCTIONS

In Section 3–2 we observed that the general solution of a nonhomogeneous linear differential equation

$$a_n(x)\frac{d^ny}{dx^n} + \cdots + a_0(x)y = h(x) \tag{4–16}$$

may be written in the form

$$y = y_p + y_h, \tag{4–17}$$

where y_p is any *particular solution* of (4–16), and y_h is the *general solution* of the associated homogeneous equation

$$a_n(x)\frac{d^ny}{dx^n} + \cdots + a_0(x)y = 0. \tag{4–18}$$

Using the language of linear operators, the problem of finding a particular solution of (4–16)—which we assume defined and normal on an interval I—consists of finding *exactly one* function in $\mathcal{C}^n(I)$ which satisfies the equation

$$Ly = h, \tag{4–19}$$

where L is the linear differential operator $a_n(x)D^n + \cdots + a_0(x)$. And this, as we know, is equivalent to the problem of constructing a *right inverse* for L; meaning, of course, a linear transformation $G\colon \mathcal{C}(I) \to \mathcal{C}^n(I)$ such that $L(G(h)) = h$ for all h in $\mathcal{C}(I)$ (cf., Section 2–5 and Section 3–4).

The existence of such inverses is guaranteed by the fact that Eq. (4–19) has solutions for every h in $\mathcal{C}(I)$, and the only open question is how to go about selecting a particular inverse for L from the infinitely many that exist. In other words, how do we impose conditions on Eq. (4–19) to ensure that it has a *unique* solution for each h in $\mathcal{C}(I)$? When asked in these terms an answer is obvious: We simply require that the solution satisfy a "complete" set of initial conditions at some point x_0 in the interval I.* Since the particular solution obtained is quite immaterial we choose the simplest of all possible initial conditions, namely

$$y(x_0) = 0,\ y'(x_0) = 0, \ldots, y^{(n-1)}(x_0) = 0. \tag{4–20}$$

And with this we have in fact defined a right inverse G for the operator L. Specifically, G can be described as *the* (linear) mapping from $\mathcal{C}(I)$ to $\mathcal{C}^n(I)$ which sends each function h in $\mathcal{C}(I)$ onto *the* solution of (4–19) which satisfies the initial conditions given above. In this section we shall obtain an explicit formula for G in terms of a basis for the solution space of the homogeneous equation $Ly = 0$ when L is an operator of order two. In the next section these results will be generalized to operators of arbitrary order, and once this has been done the study of linear differential equations will have been reduced to the homogeneous case. (The reader should note that this portion of our discussion is *not* restricted to constant coefficient operators.)

Thus we begin by considering a normal second-order linear differential equation

$$\frac{d^2y}{dx^2} + a_1(x)\frac{dy}{dx} + a_0(x)y = h(x), \tag{4–21}$$

defined on an interval I of the x-axis, and the general solution

$$y_h = c_1y_1(x) + c_2y_2(x) \tag{4–22}$$

of its associated homogeneous equation. We seek a particular solution y_p of (4–21) such that

$$y_p(x_0) = 0, \qquad y_p'(x_0) = 0, \tag{4–23}$$

where x_0 is a fixed but otherwise arbitrary point in I.

---

* This requirement can be viewed as restricting the domain of L in such a way that L becomes one-to-one and has an inverse.

The construction of y_p is begun by making the unjustified but not unreasonable assumption that any particular solution of (4–21) ought to be related to the expression for y_h, and we therefore attempt to alter the latter in such a way that it becomes a solution of the given equation. One way of doing this is to allow the *parameters* c_1 and c_2 in (4–22) to vary with x in the hope of finding a solution of (4–21) of the form

$$y_p = c_1(x)y_1(x) + c_2(x)y_2(x).^* \tag{4–24}$$

If (4–24) is substituted in (4–21), and the notation simplified by suppressing mention of the variable x, we obtain

$$\begin{aligned} c_1(y_1'' + a_1y_1' + a_0y_1) + c_2(y_2'' + a_1y_2' + a_0y_2) + (c_1'y_1 + c_2'y_2)' \\ + a_1(c_1'y_1 + c_2'y_2) + (c_1'y_1' + c_2'y_2') = h. \end{aligned} \tag{4–25}$$

Moreover, since y_1 and y_2 are solutions of the homogeneous equation $y'' + a_1y' + a_0y = 0$, the first two terms in (4–25) vanish, and we have

$$(c_1'y_1 + c_2'y_2)' + a_1(c_1'y_1 + c_2'y_2) + (c_1'y_1' + c_2'y_2') = h.$$

This identity, which must hold if (4–24) is to be a solution of (4–21), will obviously be satisfied if c_1 and c_2 can be chosen so that

$$\begin{aligned} c_1'(x)y_1(x) + c_2'(x)y_2(x) &= 0, \\ c_1'(x)y_1'(x) + c_2'(x)y_2'(x) &= h(x), \end{aligned} \tag{4–26}$$

for all x in I. Thus it remains to show that these equations serve to determine $c_1(x)$ and $c_2(x)$, and that this can be done in such a way that the function

$$y_p = c_1(x)y_1(x) + c_2(x)y_2(x)$$

satisfies the initial conditions given in (4–23).

But, for each x in I, (4–26) may be viewed as a pair of linear equations in the unknowns $c_1'(x)$ and $c_2'(x)$. As such, the determinant of its coefficients is

$$\begin{vmatrix} y_1(x) & y_2(x) \\ y_1'(x) & y_2'(x) \end{vmatrix},$$

which we recognize as the Wronskian of the linearly independent solutions $y_1(x)$ and $y_2(x)$ of the homogeneous equation associated with (4–21). We now recall (Theorem 3–6) that this determinant is a continuous function of x which *never vanishes* on I. Hence (4–26) has a unique solution for $c_1'(x)$ and $c_2'(x)$, and, once this solution is known, $c_1(x)$ and $c_2(x)$ can be found by integration. Moreover, by

---

* The term "parameter" is frequently used as a synonym for "constant," particularly when, as in this case, the constant is allowed to assume arbitrary values.

suitably choosing the limits of integration, the required initial conditions can also be satisfied, and the argument is complete. For obvious reasons, this method of constructing a particular solution for a nonhomogeneous linear differential equation out of the general solution of its associated homogeneous equation is known as the method of *variation of parameters.*

Starting with (4–26), an easy calculation gives

$$c_1'(x) = -\frac{h(x)y_2(x)}{W[y_1(x), y_2(x)]}, \qquad c_2'(x) = \frac{h(x)y_1(x)}{W[y_1(x), y_2(x)]}.$$

Thus

$$c_1(x) = -\int_{x_0}^{x} \frac{h(t)y_2(t)}{W[y_1(t), y_2(t)]}\, dt, \quad c_2(x) = \int_{x_0}^{x} \frac{h(t)y_1(t)}{W[y_1(t), y_2(t)]}\, dt, \tag{4–27}$$

and if these values are substituted in (4–24), and terms combined, we find that y_p can be written in integral form as

$$y_p(x) = \int_{x_0}^{x} \frac{y_2(x)y_1(t) - y_1(x)y_2(t)}{W[y_1(t), y_2(t)]}\, h(t)\, dt. \tag{4–28}$$

One reason for calling attention to this expression is that it can be read as the definition of a right inverse for the linear differential operator $L = D^2 + a_1(x)D + a_0(x)$, and, in fact, is the particular right inverse discussed earlier in this section. For if h is any function in $\mathcal{C}(I)$, and if we set

$$G(h) = \int_{x_0}^{x} K(x, t)h(t)\, dt, \tag{4–29}$$

where

$$K(x, t) = \frac{y_2(x)y_1(t) - y_1(x)y_2(t)}{W[y_1(t), y_2(t)]}, \tag{4–30}$$

then G maps $\mathcal{C}(I)$ to $\mathcal{C}^2(I)$, acts as a right inverse for L, and has the further property that $G(h)$ satisfies the initial conditions $G(h)(x_0) = G(h)'(x_0) = 0$ (see Exercise 29). It should be pointed out that the function $K(x, t)$ defined by (4–30) is independent of the particular choice of x_0 in the interval I, and is completely determined by the operator L.* As such it is referred to as the *Green's function for L for initial value problems on the internal I*, or, more simply, as the *Green's function for L.*

EXAMPLE 1. Find the general solution of the second-order equation

$$y'' + y = \tan x. \tag{4–31}$$

---

* In the next section it will be shown that $K(x, t)$ is also independent of the particular basis chosen for the solution space of $Ly = 0$.

In this case the associated homogeneous equation has

$$y_h = c_1 \sin x + c_2 \cos x$$

as its general solution. Thus we seek a particular solution of (4–31) in the form

$$y_p = c_1(x) \sin x + c_2(x) \cos x,$$

where $c_1(x)$ and $c_2(x)$ are determined from the pair of equations

$$\begin{aligned} c_1'(x) \sin x + c_2'(x) \cos x &= 0, \\ c_1'(x) \cos x - c_2'(x) \sin x &= \tan x \end{aligned}$$

(see Eq. 4–26). This gives

$$\begin{aligned} c_1'(x) &= \sin x, \\ c_2'(x) &= -\sin x \tan x, \end{aligned}$$

and it follows that

$$\begin{aligned} c_1(x) &= \int \sin x \, dx = -\cos x, \\ c_2(x) &= -\int \frac{\sin^2 x}{\cos x} \, dx = -\int (\sec x - \cos x) \, dx \\ &= -\ln |\sec x + \tan x| + \sin x.^* \end{aligned}$$

Thus

$$\begin{aligned} y_p &= -\cos x \sin x + [\sin x - \ln |\sec x + \tan x|] \cos x \\ &= -\cos x \ln |\sec x + \tan x|, \end{aligned}$$

and the general solution of (4–31) is

$$y = -\cos x \ln |\sec x + \tan x| + c_1 \sin x + c_2 \cos x.$$

An alternate method of solving (4–31) relies upon determining the Green's function $K(x, t)$ for the operator $L = D^2 + 1$. According to (4–30)

$$K(x, t) = \frac{\sin x \cos t - \cos x \sin t}{W[\cos t, \sin t]} = \sin (x - t),$$

and if we set $x_0 = 0$ in (4–29), the expression

$$G(h) = \int_0^x \sin (x - t) h(t) \, dt$$

defines a right inverse for L. (The reader should note that at this point we are in a position to solve *all* linear differential equations involving $D^2 + 1$, a fact which

---

* It is standard practice to omit the constants of integration at this point since without them we still obtain a particular solution of the given equation.

vividly illustrates the economy of using Green's functions.) It now follows that a particular solution of (4–31) can be obtained by applying G to the function $\tan x$; i.e., by computing

$$y_p(x) = \int_0^x \sin(x - t)\tan t\, dt. \tag{4–32}$$

This gives

$$\begin{aligned} y_p(x) &= \int_0^x (\sin x \cos t - \cos x \sin t)\tan t\, dt \\ &= \sin x \int_0^x \sin t\, dt - \cos x \int_0^x \frac{\sin^2 t}{\cos t}\, dt \\ &= \sin x - \cos x \ln|\sec x + \tan x|. \end{aligned}$$

EXAMPLE 2. In Section 4–8 we will show that the general solution of

$$xy'' + y' = 0$$

on $(0, \infty)$ and $(-\infty, 0)$ is $c_1 + c_2 \ln|x|$. Hence the nonhomogeneous equation

$$xy'' + y' = x + 1 \tag{4–33}$$

has a particular solution of the form $y_p = c_1(x) + c_2(x)\ln|x|$, where $c_1(x)$ and $c_2(x)$ are determined from the equations

$$\begin{aligned} c_1'(x) + c_2'(x)\ln|x| &= 0, \\ c_2'(x) \cdot \frac{1}{x} &= \frac{x+1}{x}. \end{aligned}$$

[Here $h(x) = (x + 1)/x$ since we must divide by x to put (4–33) in normal form.] Thus

$$\begin{aligned} c_1'(x) &= -(x + 1)\ln|x|, \\ c_2'(x) &= x + 1, \end{aligned}$$

and

$$\begin{aligned} c_1(x) &= \frac{x^2}{4} + x - \left(\frac{x^2}{2} + x\right)\ln|x|, \\ c_2(x) &= \frac{x^2}{2} + x. \end{aligned}$$

This gives

$$y_p = \frac{x^2}{4} + x - \left(\frac{x^2}{2} + x\right)\ln|x| + \left(\frac{x^2}{2} + x\right)\ln|x| = \frac{x^2}{4} + x,$$

and the general solution of (4–33) is

$$y = \frac{x^2}{4} + x + c_1 + c_2 \ln|x|.$$

EXERCISES

Find the general solution of each of the following differential equations.

1. $(D^2 + 1)y = \dfrac{1}{\cos x}$
2. $(D^2 - D - 2)y = e^{-x} \sin x$
3. $(D^2 + 4D + 4)y = xe^{2x}$
4. $(D^2 + 3D - 4)y = x^2 e^x$
5. $(4D^2 + 4D + 1)y = xe^{-x/2} \sin x$
6. $(D^2 + 4)y = \dfrac{e^{2x}}{2}$
7. $(D^2 + 10D - 12)y = \dfrac{(e^{2x} + 1)^2}{e^{2x}}$
8. $(D + 3)^2 y = (x + 1)e^x$
9. $(D^2 - 2D + 2)y = e^{2x} \sin x$
10. $(4D^2 - 8D + 5)y = e^x \tan^2 \dfrac{x}{2}$

For each of the following differential equations verify that the given expression is the general solution of the associated homogeneous equation and then find a particular solution of the equation.

11. $x^2y'' - 2xy' + 2y = x^3 \ln x, \quad x > 0; \qquad y_h = c_1x + c_2x^2$
12. $x^2y'' - xy' + y = x(x + 1); \qquad y_h = (c_1 + c_2 \ln |x|)x$
13. $(\sin 4x)y'' - 4(\cos^2 2x)y' = \tan x; \qquad y_h = c_1 + c_2 \cos 2x$
14. $xy'' - (1 + 2x^2)y' = x^5e^{x^2}; \qquad y_h = c_1 + c_2e^{x^2}$
15. $(1 - x^2)y'' - 2xy' = 2x, \quad -1 < x < 1; \qquad y_h = c_1 + c_2 \ln \dfrac{1 + x}{1 - x}$

In each of the following exercises find a Green's function for the given linear differential operator.

16. $D^2 + 3$
17. $D^2 - D - 2$
18. $D^2 + 4D + 4$
19. $4D^2 - 8D + 5$
20. $D^2 + 3D - 4$
21. $x^2D^2 - 2xD + 2$
22. $xD^2 - (1 + 2x)D$
23. $(1 - x^2)D^2 - 2xD$
24. Solve the initial value problem

$$(D^2 + 2aD + b^2)y = \sin \omega t, \qquad y(0) = y'(0) = 0,$$

where a, b, ω are (real) constants, $a < b$. Consider separately the cases $\omega \neq \sqrt{b^2 - a^2}$ and $\omega = \sqrt{b^2 - a^2}$, and sketch the solution curve in each case.

25. Find a Green's function for the operator in Exercise 24, and use this function to obtain the desired solution as $G(\sin \omega t)$.
26. The second-order equation

$$(D - 1)(xD + 3)y = e^x$$

may be solved by setting $(xD + 3)y = u$, and then successively solving the first-order equations

$$(D - 1)u = e^x \qquad \text{and} \qquad (xD + 3)y = u.$$

Use this technique to show that the general solution of this equation is

$$y = \frac{c_1}{x^3} + \frac{c_2e^x}{x^3}(x^2 - 2x + 2) + e^x.$$

27. Use the technique introduced in the preceding exercise to show once more that $xe^{\alpha x}$ is a solution of $(D - \alpha)^2 y = 0$.

28. Let $K(x, t)$ denote the Green's function (4–30) for initial value problems involving the operator $L = D^2 + a_1(x)D + a_0(x)$, and assume that L is defined on an interval I of the x-axis.

 (a) What is the domain of $K(x, t)$ in the xt-plane?

 (b) Prove that $K(x, x) = 0$ and $K_x(x, x) = 1$ for all x in I. [*Note:* K_x denotes the partial derivative of $K(x, t)$ with respect to x.]

 (c) Show that for each fixed t in I the function $\varphi(x) = K(x, t)$ is a solution on I of the initial value problem $Ly = 0$; $\varphi(t) = 0$, $\varphi'(t) = 1$.

 (d) Use the results of (b) and (c) to deduce that $K(x, t)$ is independent of the particular basis $y_1(x)$ and $y_2(x)$ chosen for the solution space of the homogeneous equation $Ly = 0$.

29. With $K(x, t)$ as in the preceding exercise, show that the function y_p defined by

$$y_p(x) = \int_{x_0}^{x} K(x, t)h(t)\, dt$$

satisfies the initial conditions $y_p(x_0) = y_p'(x_0) = 0$ for all h in $\mathcal{C}(I)$. [*Hint:* Use Leibnitz's formula for differentiating integrals,† namely,

$$\frac{d}{dx}\int_{a(x)}^{b(x)} f(x, t)\, dt = \int_{a(x)}^{b(x)} f_x(x, t)\, dt + f\big(x, b(x)\big)b'(x) - f\big(x, a(x)\big)a'(x).]$$

*30. Find the Green's function for initial value problems on $(0, \infty)$ for the operator

$$L = D^2 + \frac{1}{x} D + \left(1 - \frac{p^2}{x^2}\right),$$

p a non-negative real number.

*4–5 VARIATION OF PARAMETERS; GREEN'S FUNCTIONS (*continued*)

The method of variation of parameters can easily be extended to equations of arbitrary order. In this case we begin with a normal equation

$$y^{(n)} + a_{n-1}(x)y^{(n-1)} + \cdots + a_0(x)y = h(x) \tag{4–34}$$

defined on an interval I, and again assume that the general solution

$$y_h = c_1 y_1(x) + \cdots + c_n y_n(x) \tag{4–35}$$

of the associated homogeneous equation is known. Then, following the argument given in the second-order case, we seek a particular solution of the form

$$y_p = c_1(x)y_1(x) + \cdots + c_n(x)y_n(x), \tag{4–36}$$

† See Theorem I–36.

where, in addition to the requirement that y_p satisfy (4–34), we impose the following $n - 1$ conditions on the unknown functions $c_1(x), \ldots, c_n(x)$:

$$\begin{aligned} c_1'y_1 + \cdots + c_n'y_n &= 0, \\ c_1'y_1' + \cdots + c_n'y_n' &= 0, \\ &\vdots \\ c_1'y_1^{(n-2)} + \cdots + c_n'y_n^{(n-2)} &= 0 \end{aligned} \tag{4–37}$$

for all x in I. If the expression for y_p is now substituted in (4–34) and the above conditions are used, we obtain the additional equation

$$c_1'y_1^{(n-1)} + \cdots + c_n'y_n^{(n-1)} = h(x), \tag{4–38}$$

and, for each x in I, (4–37) and (4–38) may be viewed as a system of n linear equations in the unknowns $c_1', \ldots, c_n'$ whose determinant is $W[y_1(x), \ldots, y_n(x)]$. Our earlier reasoning still applies, and we can obtain a particular solution for (4–34) by solving this system for $c_1', \ldots, c_n'$, integrating, and then substituting the resulting functions in (4–36).

In fact, if $V_k(x)$ denotes the determinant obtained from $W[y_1(x), \ldots, y_n(x)]$ by replacing its kth column with

$$\begin{bmatrix} 0 \\ \vdots \\ 0 \\ 1 \end{bmatrix},$$

then a straightforward computation gives

$$c_k'(x) = \frac{V_k(x)h(x)}{W[y_1(x), \ldots, y_n(x)]}. \tag{4–39}$$

(See Exercise 17.) Hence, just as in the second-order case, the particular solution may be written in integral form as

$$y_p(x) = \int_{x_0}^{x} \frac{y_1(x)V_1(t) + \cdots + y_n(x)V_n(t)}{W[y_1(t), \ldots, y_n(t)]} h(t)\, dt, \tag{4–40}$$

where x_0 is any point in I, or as

$$y_p(x) = \int_{x_0}^{x} K(x, t)h(t)\, dt, \tag{4–41}$$

where

$$K(x, t) = \frac{y_1(x)V_1(t) + \cdots + y_n(x)V_n(t)}{W[y_1(t), \ldots, y_n(t)]}, \tag{4–42}$$

or, for the reader who prefers determinant notation,

$$K(x,t) = \frac{\begin{vmatrix} y_1(t) & \cdots & y_n(t) \\ y_1'(t) & \cdots & y_n'(t) \\ \vdots & & \vdots \\ y_1^{(n-2)}(t) & \cdots & y_n^{(n-2)}(t) \\ y_1(x) & \cdots & y_n(x) \end{vmatrix}}{\begin{vmatrix} y_1(t) & \cdots & y_n(t) \\ y_1'(t) & \cdots & y_n'(t) \\ \vdots & & \vdots \\ y_1^{(n-2)}(t) & \cdots & y_n^{(n-2)}(t) \\ y_1^{(n-1)}(t) & \cdots & y_n^{(n-1)}(t) \end{vmatrix}}. \tag{4–43}$$

The function $K(x, t)$ defined here is called the *Green's function for the operator* $L = D^n + a_{n-1}(x)D^{n-1} + \cdots + a_0(x)$ (for initial value problems in the interval I), and the expression

$$G(h) = \int_{x_0}^{x} K(x,t)h(t)\,dt \tag{4–44}$$

defines a right inverse $G\colon \mathcal{C}(I) \to \mathcal{C}^n(I)$ for the operator L. In fact, G is *the* inverse for L such that $G(h)$ satisfies the initial conditions

$$G(h)(x_0) = G(h)'(x_0) = \cdots = G(h)^{(n-1)}(x_0) = 0 \tag{4–45}$$

for all h in $\mathcal{C}(I)$.

EXAMPLE. Find a particular solution y_p for the equation

$$3y''' + 5y'' - 2y' = r(x), \tag{4–46}$$

$r(x)$ continuous on $(-\infty, \infty)$.

Here the general solution of the associated homogeneous equation is $c_1 + c_2e^{-2x} + c_3e^{x/3}$. Hence

$$y_p = c_1(x) + c_2(x)e^{-2x} + c_3(x)e^{x/3},$$

where $c_1(x)$, $c_2(x)$, $c_3(x)$ satisfy the identities

$$\begin{aligned} c_1'(x) + c_2'(x)e^{-2x} + c_3'(x)e^{x/3} &= 0, \\ -2c_2'(x)e^{-2x} + \tfrac{1}{3}c_3'(x)e^{x/3} &= 0, \\ 4c_2'(x)e^{-2x} + \tfrac{1}{9}c_3'(x)e^{x/3} &= \frac{r(x)}{3}. \end{aligned}$$

Thus

$$c_1'(x) = -\frac{r(x)}{2},$$
$$c_2'(x) = \tfrac{1}{14}e^{2x}r(x),$$
$$c_3'(x) = \tfrac{3}{7}e^{-x/3}r(x),$$

and it follows that

$$\begin{aligned} y_p &= -\tfrac{1}{2}\int r(x)\,dx + \frac{e^{-2x}}{14}\int e^{2x}r(x)\,dx + \frac{3e^{x/3}}{7}\int e^{-x/3}r(x)\,dx \\ &= \int [-\tfrac{1}{2} + \tfrac{1}{14}e^{-2(x-t)} + \tfrac{3}{7}e^{(x-t)/3}]r(t)\,dt. \end{aligned}$$

Alternately, we could have computed the Green's function $K(x, t)$ for the (normal) operator $D^3 + \frac{5}{3}D^2 - \frac{2}{3}D$, and used (4–41) to express y_p as an integral involving $K(x, t)$. Starting with the basis $1, e^{-2x}, e^{x/3}$ for the solution space of the associated homogeneous equation, we then obtain

$$\begin{aligned} K(x, t) &= \frac{\begin{vmatrix} 1 & e^{-2t} & e^{t/3} \\ 0 & -2e^{-2t} & \frac{1}{3}e^{t/3} \\ 1 & e^{-2x} & e^{x/3} \end{vmatrix}}{\begin{vmatrix} 1 & e^{-2t} & e^{t/3} \\ 0 & -2e^{-2t} & \frac{1}{3}e^{t/3} \\ 0 & 4e^{-2t} & \frac{1}{9}e^{t/3} \end{vmatrix}} \\ &= \frac{\frac{7}{3}e^{-5t/3} - \frac{1}{3}e^{-2x}e^{t/3} - 2e^{x/3}e^{-2t}}{-\frac{2}{9}e^{-5t/3} - \frac{4}{3}e^{-5t/3}} \\ &= -\tfrac{3}{2} + \tfrac{3}{14}e^{-2(x-t)} + \tfrac{9}{7}e^{(x-t)/3}. \end{aligned}$$

Thus

$$\begin{aligned} y_p &= G\left(\frac{r(x)}{3}\right) \\ &= \int_{x_0}^{x} [-\tfrac{1}{2} + \tfrac{1}{14}e^{-2(x-t)} + \tfrac{3}{7}e^{(x-t)/3}]r(t)\,dt, \end{aligned}$$

which agrees with the result obtained above.

The remainder of this section will be devoted to taking a closer look at the Green's functions for initial-value problems, and to establishing some of their more important properties. Throughout this discussion we shall assume that

$$L = D^n + a_{n-1}(x)D^{n-1} + \cdots + a_0(x)$$

is a fixed linear differential operator on $\mathcal{C}^n(I)$, and that $K(x, t)$ is the function obtained above from the general solution of the equation $Ly = 0$. It then follows from the very way in which $K(x, t)$ was constructed (see Exercises 18–21 below) that

(1) $K(x, t)$ is defined throughout the region R of the xt-plane consisting of all points (x, t) with x and t in I (see Fig. 4–1);

(2) $K(x, t)$ and $\partial K/\partial x$, $\partial^2 K/\partial x^2, \ldots,$ $\partial^n K/\partial x^n$ are continuous everywhere in R;

(3) For *every* x_0 in I, and *every* h in $\mathcal{C}(I)$, the function

$$y(x) = \int_{x_0}^{x} K(x, t)h(t)\,dt$$

is a solution of the initial-value problem

$$Ly = h;$$
$$y(x_0) = y'(x_0) = \cdots = y^{(n-1)}(x_0) = 0$$

on I.

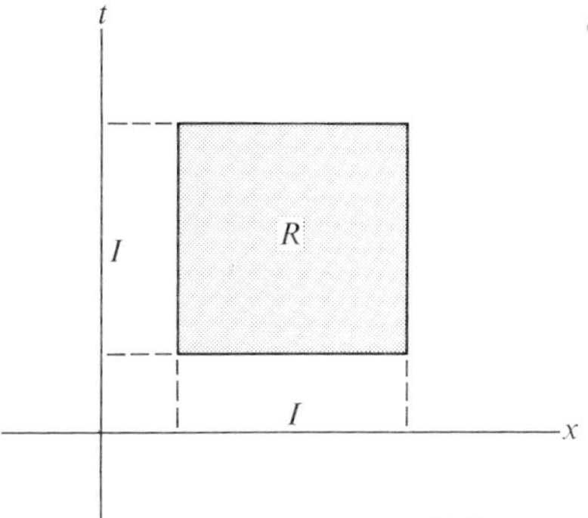

FIGURE 4–1

These properties are actually sufficient to characterize the Green's function for initial-value problems involving L in the sense that $K(x, t)$ is the *unique* function defined in R which satisfies (1), (2), and (3). This assertion will be proved below as Theorem 4–3, and is mentioned here only to motivate the following definition.

Definition 4–1. A function $H(x, t)$ is said to be a *Green's function for initial value problems involving the linear differential operator L* if and only if $H(x, t)$ enjoys the three properties listed above for the function $K(x, t)$.

This said, we proceed at once to give an alternate, and for our purposes much more useful description of a Green's function for L. For convenience, we shall denote the various derivatives $\partial H/\partial x$, $\partial^2 H/\partial x^2, \ldots$ appearing in the following argument by $H_1, H_2, \ldots$. And with this notation in effect, we have

Theorem 4–2. *Let $H(x, t)$ be defined throughout the region R described above, and suppose that H and its partial derivatives $H_1, H_2, \ldots, H_n$ are continuous everywhere in R. Then $H(x, t)$ is a Green's function for the linear differential operator $L = D^n + a_{n-1}(x)D^{n-1} + \cdots + a_0(x)$ if and only if the following identities are satisfied throughout R:*

$$\begin{aligned} H(x, x) &\equiv 0, \\ H_1(x, x) &\equiv 0, \\ &\vdots \\ H_{n-2}(x, x) &\equiv 0, \\ H_{n-1}(x, x) &\equiv 1, \end{aligned} \tag{4–47}$$

and

$$H_n(x, t) + a_{n-1}(x)H_{n-1}(x, t) + \cdots + a_0(x)H(x, t) \equiv 0. \tag{4–48}$$

Proof. First assume that $H(x, t)$ is a Green's function for L. Then, by definition, the function

$$y(x) = \int_{x_0}^{x} H(x, t)h(t)\, dt \tag{4–49}$$

is a solution of the initial value problem

$$Ly = h; \quad y(x_0) = y'(x_0) = \cdots = y^{(n-1)}(x_0) = 0$$

for every x_0 in I, and every h in $\mathcal{C}(I)$; see (3) above. We now differentiate (4–49) using Leibnitz's formula* to obtain

$$y'(x) = \int_{x_0}^{x} H_1(x, t)h(t)\, dt + H(x, x)h(x), \tag{4–50}$$

which reduces to

$$H(x_0, x_0)h(x_0) = 0$$

when $x = x_0$ (recall that $y'(x_0) = 0$). But, by assumption, this expression is valid for all h in $\mathcal{C}(I)$, and hence, in particular, for $h \equiv 1$. Thus $H(x_0, x_0) = 0$, and since x_0 can be chosen arbitrarily in I, we have

$$H(x, x) \equiv 0,$$

and

$$y'(x) = \int_{x_0}^{x} H_1(x, t)h(t)\, dt. \tag{4–51}$$

We now repeat the argument, starting with (4–51), to get, first

$$y''(x) = \int_{x_0}^{x} H_2(x, t)h(t)\, dt + H_1(x, x)h(x),$$

then

$$H_1(x, x) \equiv 0,$$

and finally

$$y''(x) = \int_{x_0}^{x} H_2(x, t)h(t)\, dt.$$

Continuing in this fashion we eventually arrive at the situation

$$H_{n-2}(x, x) \equiv 0,$$

and

$$y^{(n-1)}(x) = \int_{x_0}^{x} H_{n-1}(x, t)h(t)\, dt.$$

---

* Leibnitz's formula is

$$\frac{d}{dx}\int_{a(x)}^{b(x)} f(x, t)\, dt = \int_{a(x)}^{b(x)} \frac{\partial f}{\partial x}(x, t)\, dx + f\big(x, b(x)\big)b'(x) - f\big(x, a(x)\big)a'(x).$$

Differentiating once more, we obtain

$$y^{(n)}(x) = \int_{x_0}^{x} H_n(x, t)h(t)\,dt + H_{n-1}(x, x)h(x),$$

whence

$$y^{(n)}(x_0) = H(x_0, x_0)h(x_0). \tag{4-52}$$

But since $y(x)$ is a solution of $Ly = h$,

$$y^{(n)}(x) + a_{n-1}(x)y^{(n-1)}(x) + \cdots + a_0(x)y(x) \equiv h(x),$$

and the initial conditions in effect imply that

$$y^{(n)}(x_0) = h(x_0).$$

This, together with (4–52) and the fact that h and x_0 are still arbitrary, implies that

$$H_{n-1}(x, x) \equiv 1,$$

and

$$y^{(n)}(x) = \int_{x_0}^{x} H_n(x, t)h(t)\,dt + h(x).$$

Thus, in particular, we have established the several identities listed in (4–47).

To prove (4–48) we substitute the formulas obtained above for $y, y', \ldots, y^{(n-1)}$ in $Ly = h$. After the various terms are collected we have

$$\int_{x_0}^{x} [H_n(x, t) + a_{n-1}(x)H_{n-1}(x, t) + \cdots + a_0(x)H(x, t)]h(t)\,dt \equiv 0, \tag{4-53}$$

and the fact that this expression holds for all x_0 in I and all h in $\mathcal{C}(I)$ allows us to conclude that the bracketed portion of the integrand is identically zero (see Exercise 23). And with this, the first part of the proof is complete.

As for the remainder, the argument needed to show that (4–47) and (4–48) imply that $H(x, t)$ is a Green's function for L is an even more elementary computation than the one just given, and has therefore been relegated to the exercises (see Exercise 24). ▌

Among other things, this theorem asserts that for *any* fixed t_0 in the interval I, the function

$$k(x) = H(x, t_0)$$

is a solution of the initial-value problem

$$Ly = 0; \quad y(t_0) = y'(t_0) = \cdots = y^{(n-2)}(t_0) = 0, \quad y^{(n-1)}(t_0) = 1.$$

But, as we know, the solution of this problem is unique. Thus the values of $H(x, t)$ are *uniquely determined* by the operator L on the line segment consisting of those points (x, t) in R with $t = t_0$ (see Fig. 4–2). However, t_0 can be chosen arbitrarily in I, and we therefore have our main result.

Theorem 4–3. *The Green's function for initial value problems on I involving a linear differential operator L is uniquely determined by L, and hence must coincide with the function $K(x, t)$ defined by* (4–42) *or* (4–43). *In particular, $K(x, t)$ is independent of the basis for the solution space of $Ly = 0$ used in computing it.*

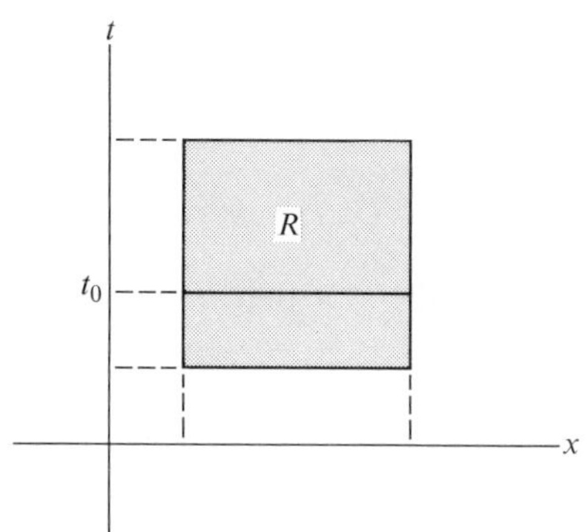

FIGURE 4–2

Everything that has been said up to this point in our discussion applies to arbitrary linear differential operators. As might be expected, much more precise information can be given in the case of operators with constant coefficients, and we conclude this section with a theorem which describes the Green's functions obtained in this special case.

Theorem 4–4. *The Green's function for a constant coefficient linear differential operator L can be written in the form $k(x - t)$, where $k(x)$ is the solution on $(-\infty, \infty)$ of the initial value problem*

$$Ly = 0; \qquad y(0) = y'(0) = \cdots = y^{(n-2)}(0) = 0, \quad y^{(n-1)}(0) = 1.$$

Proof. The function $H(x, t) = k(x - t)$ clearly satisfies the identities (4–47) and (4–48) of Theorem 4–2, and hence, by Theorem 4–3, is *the* Green's function for L.* ▮

EXERCISES

Use the method of variation of parameters (without employing Formulas (4–41) through (4–43)) to find the general solution of each of the following differential equations.

1. $y''' - y'' - y' + y = 4xe^x$
2. $y''' - y' = \sin x$
3. $y''' - 2y'' = 4(x + 1)$
4. $y''' - 3y'' - y' + 3y = 1 + e^x$

---

* Note that the assumption of constant coefficients is needed to verify (4–48); see Exercise 25.

5. $y''' - 7y' + 6y = 2 \sin x$

6. $y''' - 3y' - 2y = 9e^{-x}$

7. $y''' - y' = \sin x$

8. $y''' + y'' + y' + y = 2(\sin x + \cos x)$

9. $y^{(\text{iv})} - y'' = 2xe^x$

10. $y^{(\text{iv})} - y = x^2 + 1$

In Exercises 11 through 16 compute the Green's function $K(x, t)$ for the given operator (a) by using Formula (4–42) or (4–43) and (b) by applying Theorem 4–4.

11. $D^2(D - 1)$

12. $D(D^2 - 4)$

13. $D^3 - 6D^2 + 11D - 6$

14. $D^3 + \frac{5}{2}D^2 - \frac{3}{2}$

15. $D^2(D^2 - 1)$

16. $D^4 - 1$

17. Verify Formula (4–39). [*Hint:* Use Cramer's rule.]

Exercises 18 through 21 concern properties of the function $K(x, t)$ defined by Formula (4–42) or (4–43). Establish each of them.

18. When $n = 2$, Formula (4–43) reduces to (4–30) of the preceding section.

19. $K(x, t)$ is defined and has continuous derivatives through order n in the region R described in the text.

20. The partial derivatives with respect to x of $K(x, t)$ satisfy the identities

$$K(x, x) \equiv K_1(x, x) \equiv \cdots \equiv K_{n-2}(x, x) \equiv 0, \quad K_{n-1}(x, x) \equiv 1,$$

in the interval I. Here

$$K_i(x, t) = \frac{\partial^i}{\partial x^i} K(x, t).$$

21. For each x_0 in I and each h in $\mathcal{C}(I)$ the function

$$y(x) = \int_{x_0}^{x} K(x, t)h(t)\, dt$$

satisfies the initial conditions $y(x_0) = \cdots = y^{(n-1)}(x_0) = 0$ and the equation $Ly = h(x)$. [*Hint:* Use Leibnitz's formula and the result of Exercise 20.]

22. If $f(x)$ is continuous on the interval $[a, b]$ and if

$$\int_a^b f(x)g(x)dx = 0$$

for *every* g in $\mathcal{C}[a, b]$, then $f \equiv 0$ on $[a, b]$. [*Hint:* Assume $f(x_0) \not\equiv 0$ and use the continuity of f to obtain an interval $(x_0 - \delta, x_0 + \delta)$ in which $|f(x)| \geq |f(x_0)|/2$. Then find a function g in $\mathcal{C}(I)$ for which the above integral is different from zero.]

23. Use the result of Exercise 22 to prove the assertion made in the text concerning the bracketed term in (4–53).

24. Let $H(x, t)$ satisfy the hypotheses of Theorem 4–2, and the identities given in (4–47) and (4–48). Prove that for every x_0 in I and every h in $\mathcal{C}(I)$ the function

$$y(x) = \int_{x_0}^{x} H(x, t)h(t)\, dt$$

satisfies the initial value problem

$$Ly = 0; \qquad y(x_0) = y'(x_0) = \cdots = y^{(n-1)}(x_0) = 0.$$

[*Hint:* Use Leibnitz's formula.]

25. Show that the function $H(x, t) = k(x - t)$ appearing in the proof of Theorem 4–4 satisfies (4–47) and (4–48).

The only values of $K(x, t)$ which enter into the integration in (4–44) are those for which the point (x, t) lies in the subregion R_{x_0} of R shaded in Fig. 4–3. This suggests the possibility of generalizing the notion of a Green's function as follows:

Definition 4–2. A function $\overline{K}(x, t)$ is called a *Green's function for the operator* $L = D^n + a_{n-1}(x)D^{n-1} + \cdots + a_0(x)$ *for initial value problems at the point* x_0 if it is defined and continuous in R_{x_0}, and if for *every* h in $\mathcal{C}(I)$ the function

$$y(x) = \int_{x_0}^{x} \overline{K}(x, t)h(t)\,dt$$

is the solution of the initial value problem

$$Ly = h;$$

$$y(x_0) = \cdots = y^{(n-1)}(x_0) = 0.$$

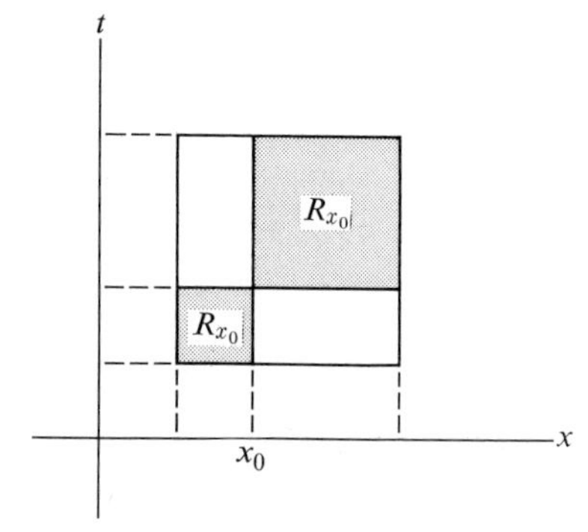

FIGURE 4–3

In the exercises which follow we explore some of the properties of these functions, and, in particular, show that under certain additional assumptions they coincide with $K(x, t)$.

26. Let $\overline{K}(x, t)$ be as described above, and assume that $L = D^2 + a_1(x)D + a_0(x)$. Suppose, in addition, that

    (i) $\overline{K}(x, t)$, $\overline{K}_1(x, t)$, $\overline{K}_2(x, t)$ are continuous in R_{x_0},

    (ii) $\overline{K}(x, x) \equiv 0$ on I,

    (iii) $\overline{K}_1(x, x) \equiv 1$ on I.

    Prove that

    $$\overline{K}_2(x, t) + a_1(x)\overline{K}_1(x, t) + a_0(x)\overline{K}(x, t) \equiv 0.$$

    [*Hint:* Follow the proof of Theorem 4–2.]

27. Let $\overline{K}(x, t)$ be as described in Exercise 26. Prove that $\overline{K}(x, t) \equiv K(x, t)$ in the region R_{x_0}.

28. Generalize the results of Exercises 26 and 27 to the nth-order case.

4–6 REDUCTION OF ORDER

One of the remarkable properties of linear differential equations is that we can simplify (and sometimes solve) the equation $Ly = h$ even when we do not have a complete basis for the solution space of $Ly = 0$. Again the technique is variation of parameters, but this time it leads to a *reduction in the order* of the equation. The following example will serve to introduce the method.

EXAMPLE 1. Consider the second-order equation

$$x^2 \frac{d^2y}{dx^2} + x^3 \frac{dy}{dx} - 2(1 + x^2)y = x \tag{4–54}$$

on the interval $(0, \infty)$. Here none of our earlier techniques is sufficient to obtain the general solution of the associated homogeneous equation

$$x^2 \frac{d^2y}{dx^2} + x^3 \frac{dy}{dx} - 2(1 + x^2)y = 0. \tag{4–55}$$

However, the solution $y = x^2$ of (4–55) is easily discovered by inspection,* and we now proceed to seek solutions of (4–54) in the form

$$y = x^2c(x).$$

Then

$$\begin{aligned} y' &= x^2c'(x) + 2xc(x), \\ y'' &= x^2c''(x) + 4xc'(x) + 2c(x), \end{aligned}$$

and (4–54) yields

$$x^2(x^2c'' + 4xc' + 2c) + x^3(x^2c' + 2xc) - 2(1 + x^2)x^2c = x,$$

which, upon simplification, becomes

$$x^4c'' + (4x^3 + x^5)c' = x,$$

or

$$c'' + \frac{4 + x^2}{x} c' = \frac{1}{x^3} \cdot \tag{4–56}$$

But (4–56) may be viewed as a *first-order* equation in c', and as such can be solved by the technique introduced in Section 3–3. Indeed, using the integrating factor

$$e^{\int (4+x^2)/x \, dx} = x^4e^{x^2/2},$$

we obtain

$$\begin{aligned} c' &= \left[k_1 + \int xe^{x^2/2} \, dx\right]x^{-4}e^{-x^2/2} \\ &= [k_1 + e^{x^2/2}]x^{-4}e^{-x^2/2} \\ &= \frac{1}{x^4} + k_1x^{-4}e^{-x^2/2}, \end{aligned}$$

where k_1 is an arbitrary constant. Hence

$$c(x) = -\frac{1}{3x^3} + k_1\int x^{-4}e^{-x^2/2} \, dx + k_2,$$

where k_2 is also arbitrary, and it follows that

$$y = x^2c(x) = \frac{1}{3x} + k_1x^2\int x^{-4}e^{-x^2/2} \, dx + k_2x^2$$

---

* The phrase "discovered by inspection" is just a dodge to hide the fact that the process was one of trial and error.

is a solution of (4–54). In fact, since x^2 and $x^2 \int x^{-4} e^{-x^2/2}\, dx$ are linearly independent in $\mathcal{C}(0, \infty)$, this expression is actually the general solution of (4–54) on the interval $(0, \infty)$.

The preceding example is representative of the technique whereby the order of a linear differential equation can be reduced by one as soon as a single (nontrivial) solution of its associated homogeneous equation is known. To establish this assertion in general, let $Ly = h$ be an equation of order n, and suppose that $u(x)$ is a nontrivial solution of $Ly = 0$. Then, evaluating the left-hand side of the expression $L[u(x)c(x)] = h(x)$ by means of Formula (4–12), we obtain

$$(Lu)c + (L'u)c' + \frac{1}{2!}(L''u)c'' + \cdots + \frac{1}{n!}(L^{(n)}u)c^{(n)} = h. \tag{4–57}$$

But since $Lu = 0$, the first term in this equation vanishes, and we may therefore view (4–57) as a linear equation of order $n - 1$ in c'. This is the asserted reduction in order. In particular, this technique can always be used as it was above to find the general solution of a *second-order* equation whenever one nontrivial solution of its associated homogeneous equation is known.

EXAMPLE 2. The second-order equation

$$(D - \alpha)^2 y = 0 \tag{4–58}$$

has $e^{\alpha x}$ as a solution. Using the above technique to find the general solution we set

$$y = c(x)e^{\alpha x},$$

substitute in the equation, and obtain

$$c''e^{\alpha x} + 2\alpha c'e^{\alpha x} + \alpha^2 ce^{\alpha x} - 2\alpha(c'e^{\alpha x} + \alpha ce^{\alpha x}) + \alpha^2 ce^{\alpha x} = 0.$$

This simplifies to $c'' = 0$, which, by two integrations gives

$$c = k_1 x + k_2.$$

Thus

$$y = (k_1 x + k_2)e^{\alpha x},$$

as expected.

EXERCISES

In Exercises 1 through 6, find the general solution of each differential equation using the given solution of the associated homogeneous equation.

1. $y'' + xy' = 3x, \quad 1$
2. $xy'' - (x + 2)y' + 2y = x^3 + x, \quad e^x$
3. $xy'' - y' = 0, \quad 1$
4. $xy'' + (2 + x)y' + y = e^{-x}, \quad 1/x$

5. $4x^2y'' - 8xy' + 9y = 0, \quad x^{3/2}$
6. $xy'' + (x - 1)y' - y = 0, \quad e^{-x}$
7. Starting with the solution $y = e^{-\int [a_0(x)/a_1(x)]dx}$ of the normal first-order equation

$$a_1(x)\frac{dy}{dx} + a_0(x)y = 0,$$

use the method of reduction of order to find the general solution of

$$a_1(x)\frac{dy}{dx} + a_0(x)y = h(x).$$

8. The second-order equation

$$(1 - x^2)y'' - 2xy' + 2y = 0$$

has $y = x$ as a particular solution. Use the method of variation of parameters to reduce the order of this equation, and then find a second linearly independent solution in each of the intervals $(-\infty, -1)$, $(-1, 1)$, $(1, \infty)$ in which the equation is normal.

9. It can be shown that the equation

$$xy'' + y' + xy = 0$$

has a solution, $J_0(x)$, on the interval $(0, \infty)$ which can be expanded in a power series as

$$J_0(x) = 1 - \frac{x^2}{2^2} + \frac{x^4}{2^4(2!)^2} - \frac{x^6}{2^6(3!)^2} + \cdots.$$

Find a second solution, linearly independent of $J_0(x)$ in $\mathcal{C}(0, \infty)$, of the form

$$J_0(x) = \int \frac{dx}{x[J_0(x)]^2},$$

and then prove that this solution can be written in the form

$$J_0(x) \ln x + (\text{power series in } x).$$

What is the behavior of this solution as $x \to 0$?

10. Let y_1 and y_2 be linearly independent solutions of a normal third-order homogeneous linear differential equation $Ly = 0$ on an interval I, and let $My = 0$ be the second-order equation obtained by using the solution y_1 to reduce the order of $Ly = 0$ by one. Prove that $(y_2/y_1)'$ is a solution of $My = 0$ on any subinterval of I in which y_1 has no zeros.

4–7 THE METHOD OF UNDETERMINED COEFFICIENTS

The method of variation of parameters enables us to find a particular solution of a nonhomogeneous linear differential equation whenever the general solution of its associated homogeneous equation is known. However, it is not always the

---

* This equation is known as *Bessel's equation of order zero* and will be studied in detail in Chapter 15.

most efficient way of producing such a solution, and for certain equations the work involved can be considerably lightened. For instance, it would be pointless to use variation of parameters to find a particular solution of $(D^2 - D + 5)y = 3$ since the solution $y_p = \frac{3}{5}$ is immediately evident. And even for an equation such as

$$(D^2 + 3)y = e^x$$

it is obvious that a solution of the form

$$y_p = Ae^x$$

must exist for a suitable value of A. Moreover, by substituting Ae^x in the equation we obtain

$$Ae^x + 3Ae^x = e^x,$$

from which it follows that $A = \frac{1}{4}$, and $y_p = e^x/4$.

This method, wherein a particular solution is known up to certain undetermined constants, and the values of these constants are found by using the differential equation, is known as the *method of undetermined coefficients*. It is clear that this method depends for its success upon the ability to recognize the form of a particular solution, and for this reason lacks the generality possessed by the method of variation of parameters. Nevertheless, it can be used often enough to merit some attention.

One type of equation for which this method always works is a *constant coefficient* equation

$$Ly = h \tag{4–59}$$

in which h itself is a solution of a linear differential equation with constant coefficients. For then we can find a linear differential operator L_1 which annihilates h (i.e., such that $L_1h = 0$), and it follows that every solution of (4–59) is also a solution of the homogeneous equation

$$L_1Ly = 0. \tag{4–60}$$

Thus we can obtain a particular solution for (4–59) by appropriately determining the constants in the general solution of (4–60). A few examples will suffice to illustrate this technique.

EXAMPLE 1. Since D^3 annihilates the right-hand side of the equation

$$(D^2 + 1)y = 3x^2 + 4, \tag{4–61}$$

a particular solution of (4–61) can be found among the solutions of the homogeneous equation

$$D^3(D^2 + 1)y = 0. \tag{4–62}$$

In other words, (4–61) has a particular solution of the form

$$y_p = c_1 + c_2x + c_3x^2 + c_4 \sin x + c_5 \cos x \tag{4–63}$$

for suitable values of $c_1, \ldots, c_5$. In fact, we can say even more than this if we observe that $c_4 \sin x + c_5 \cos x$ is the general solution of the homogeneous equation $(D^2 + 1)y = 0$. For then it is clear that the last two terms in (4–63) will be annihilated when substituted in (4–61), and so, rather than mindlessly dragging them through our computations only to see them disappear in the process, we can begin by setting

$$y_p = c_1 + c_2x + c_3x^2.$$

Substituting this expression in (4–61) we obtain

$$2c_3 + c_1 + c_2x + c_3x^2 = 3x^2 + 4,$$

from which it follows that

$$c_1 + 2c_3 = 4, \qquad c_2 = 0, \qquad c_3 = 3.$$

Thus $c_1 = -2$, $c_2 = 0$, $c_3 = 3$, and

$$y_p = 3x^2 - 2.$$

Example 2. To find a particular solution of

$$(D^2 - 4D + 4)y = 2e^{2x} + \cos x, \tag{4–64}$$

we apply the operator $(D - 2)(D^2 + 1)$ to the equation and obtain

$$(D - 2)^3(D^2 + 1)y = 0.$$

The general solution of this last equation is

$$y = c_1e^{2x} + c_2xe^{2x} + c_3x^2e^{2x} + c_4 \sin x + c_5 \cos x,$$

and since the first two terms are annihilated by the operator $D^2 - 4D + 4$, we look for a particular solution of (4–64) of the form

$$y_p = c_3x^2e^{2x} + c_4 \sin x + c_5 \cos x.$$

In this case

$$y_p' = 2c_3xe^{2x} + 2c_3x^2e^{2x} + c_4 \cos x - c_5 \sin x,$$
$$y_p'' = 2c_3e^{2x} + 8c_3xe^{2x} + 4c_3x^2e^{2x} - c_4 \sin x - c_5 \cos x,$$

and substitution in (4–64) yields

$$2c_2e^{2x} + (3c_4 + 4c_5) \sin x + (3c^5 - 4c_4) \cos x = 2e^{2x} + \cos x.$$

Hence

$$2c_3 = 2, \qquad 3c_4 + 4c_5 = 0, \qquad 3c_5 - 4c_4 = 1,$$

and

$$c_3 = 1, \qquad c_4 = -\tfrac{4}{25}, \qquad c_5 = \tfrac{3}{25}.$$

Thus the desired particular solution is

$$y_p = x^2e^{2x} - \tfrac{4}{25}\sin x + \tfrac{3}{25}\cos x.$$

EXERCISES

Use the method of undetermined coefficients to find a particular solution for each of the following differential equations.

1. $D(D+1)y = 2x + 3e^x$
2. $D(D+1)y = 2 + e^{-x}$
3. $D(D-1)y = \sin x$
4. $(D^2+1)y = 3\cos x$
5. $(D^2+4D+2)y = xe^{-2x}$
6. $(D^2+D-6)y = 6(x+1)$
7. $(6D^2+2D-1)y = 7x(x+1)e^x$
8. $(D^2-5D+6)y = -2+36x^2+e^x$
9. $(D^2-4D+5)y = (x+1)^3$
10. $(D^2-4D+8)y = e^{2x}(1+\sin 2x)$
11. $(D^2+6D+10)y = x^4+2x^2+2$
12. $(D^2-D+\frac{1}{4})y = xe^{x/2}$
13. $D(D^2-2D+10)y = 3xe^x$
14. $(D^3+3D^2+3D+1)y = x^4+4x^3+10x^2+20x+1$
15. $(D^3-D^2-D+1)y = 2(x+2e^{-x})$
16. $(D^3-3D-2)y = e^x(1+xe^x)$
17. $(D^3+D-1)y = \sin x + \cos x$
18. $D^2(D^2+1)y = 1+2xe^x$
19. $D(D^2-1)(D-2)y = x^2+2x+3-2e^x$
20. $(D^4+5D^2+4)y = 2\cos x$

Give the form of a particular solution for each of the following equations. The coefficients need not be evaluated.

21. $(D^2-4D+4)y = x(2e^{2x}+x\sin x)$
22. $(D^2+2D+2)y = x^2-3xe^{-2x}\cos 5x$
23. $(D^2+1)^3(D-1)y = 3e^{-x}+5x^2\cos x$
24. $D^2(D^4-4D^3+6D^2-4D+1)y = (x^2+1)(1-e^x)$
25. $(D^8-2D^4+1)y = (2x-1)\cosh x + x^3\sin x$
26. $(D^3-1)(D^2+D-2)y = e^{x/2}\sin\sqrt{3}\,x - x\cos\sqrt{3}\,x$
27. $(D^3-1)^3y = (2x+1)^2e^x + \dfrac{x\sin x}{2}$
28. $(D^2-2D+1)(D^2-4)^2y = x\sinh x + \cosh 2x$
29. $[(4D^2-4D+5)(D^2+2D+1)]^2y = x(1+e^{x/2}\sin x - x\cos x)$
30. $D(D^2-4)^5y = (x+1)^2[(x+1)+\sinh 2x]$

4-8 THE EULER EQUATION

A linear differential equation of the form

$$x^n \frac{d^n y}{dx^n} + a_{n-1} x^{n-1} \frac{d^{n-1} y}{dx^{n-1}} + \cdots + a_1 x \frac{dy}{dx} + a_0 y = 0, \tag{4-65}$$

with $a_0, \ldots, a_{n-1}$ constants, is called a (homogeneous) *Euler equation* of order n. The reader should note that this equation is defined on the entire x-axis, but is normal only on intervals which do not contain the point $x = 0$. It is one of the relatively few equations with variable coefficients that can be solved in closed form in terms of elementary functions, and is important because its solutions are, to some extent, typical of those of a large class of linear differential equations whose leading coefficient vanishes at the origin.

As we shall see, Eq. (4-65) can be converted into a (linear) equation with constant coefficients by making the change of variable $u = \ln x$, and hence can be solved by the methods introduced in this chapter.* Although this reduction can be effected in a routine fashion, it is illuminating to consider it from the point of view of linear operators by introducing the transformation $T: \mathcal{C}^n(-\infty, \infty) \to \mathcal{C}^n(0, \infty)$ defined by

$$(Tg)(x) = g(\ln x) \tag{4-66}$$

for all g in $\mathcal{C}^n(-\infty, \infty)$. Thus T maps the function x onto $\ln x$, $\sin x$ onto $\sin(\ln x)$, etc., and is obviously linear. More important, it is invertible, with $(T^{-1}f)(x) = f(e^x)$ for all f in $\mathcal{C}^n(0, \infty)$, and hence is a *one-to-one* linear transformation mapping $\mathcal{C}^n(-\infty, \infty)$ *onto* $\mathcal{C}^n(0, \infty)$. From this it follows that the problem of solving the equation $Ly = 0$ on the interval $(0, \infty)$, with $L = x^n D^n + a_{n-1} x^{n-1} D^{n-1} + \cdots + a_0$, is equivalent to the problem of finding all functions g in $\mathcal{C}^n(-\infty, \infty)$ such that $LTg = 0$. In other words, we must find the null space of the transformation LT.

To this end we begin by computing the various products DT, $D^2T, \ldots$, as follows:

$$DTg = Dg(\ln x) = \frac{1}{x} g'(\ln x) = \frac{1}{x} TDg,$$

$$\begin{aligned}
D^2Tg &= D(DTg) = D\left(\frac{1}{x} TDg\right) \\
&= -\frac{1}{x^2} TDg + \frac{1}{x} D(TDg) \\
&= -\frac{1}{x^2} TDg + \frac{1}{x}\left(\frac{1}{x} TD^2 g\right) \\
&= \frac{1}{x^2} TD(D-1)g,
\end{aligned}$$

* In the following discussion we shall restrict our attention to the interval $(0, \infty)$. On $(-\infty, 0)$ the change of variable $u = \ln(-x)$ must be used.

and, in general,

$$D^kTg = \frac{1}{x^k} TD(D-1)\cdots(D-k+1)g, \qquad k = 1, 2, \ldots. \tag{4-67}$$

Thus

$$x^kD^kT = TD(D-1)\cdots(D-k+1),$$

and hence when

$$L = x^nD^n + a_{n-1}x^{n-1}D^{n-1} + \cdots + a_1xD + a_0, \tag{4-68}$$

we have

$$LT = T\hat{L}, \tag{4-69}$$

where $\hat{L}$ is the *constant coefficient* linear differential operator

$$\begin{aligned}\hat{L} = {} & D(D-1)\cdots(D-n+1) + a_{n-1}D(D-1)\cdots(D-n+2) \\ & + \cdots + a_1D + a_0.\end{aligned} \tag{4-70}$$

Equation (4–69), together with the fact that T is one-to-one, implies that the null space of the transformation LT coincides with the null space of $\hat{L}$. This establishes our contention that (4–65) can be reduced to an equation with constant coefficients and also allows us to describe the solution of any (homogeneous) Euler equation on the interval $(0, \infty)$ as follows:

> *The general solution on* $(0, \infty)$ *of the equation* $Ly = 0$, *L as in* (4–68), *is*
>
> $$y = c_1y_1(\ln x) + \cdots + c_ny_n(\ln x), \tag{4-71}$$
>
> *where* $y_1(u), \ldots, y_n(u)$ *are a basis for the solution space of the constant coefficient equation* $\hat{L}y = 0$, $\hat{L}$ *deduced from L by* (4–70), *and* $c_1, \ldots, c_n$ *are arbitrary constants.*

Finally, to remove the restriction on the interval, we note that for each nonnegative integer k,

$$x^kD^kg(-x) = x^k(-1)^kg^{(k)}(-x) = (-x)^kD^kg(-x).$$

Hence if $y(x)$ is a solution of the Euler equation $Ly = 0$ on $(0, \infty)$, $y(-x)$ is a solution on $(-\infty, 0)$, and $y(|x|)$ on $(0, \infty)$ *and* $(-\infty, 0)$ (see Exercise 12).

EXAMPLE 1. *The Euler equation of order two.* Let

$$x^2\frac{d^2y}{dx^2} + a_1x\frac{dy}{dx} + a_0y = 0 \tag{4-72}$$

be a (homogeneous) Euler equation of order two. Then the constant coefficient linear differential operator $\hat{L}$ appearing above is $D(D-1) + a_1D + a_0$, and the general solution $y(x)$ of (4–72) is determined by the roots α_1, α_2 of the quadratic equation

$$m(m - 1) + a_1 m + a_0 = 0, \tag{4–73}$$

known as the *indicial equation* associated with (4–72). Thus

(i) if α_1 and α_2 are real, $\alpha_1 \neq \alpha_2$, then

$$y(x) = c_1|x|^{\alpha_1} + c_2|x|^{\alpha_2};$$

(ii) if $\alpha_1 = \alpha_2 = \alpha$, then

$$y(x) = |x|^{\alpha}(c_1 + c_2 \ln |x|);$$

(iii) if $\alpha_1 = a + bi$, $\alpha_2 = a - bi$, $b > 0$, then

$$y(x) = |x|^{a}[c_1 \sin (b \ln |x|) + c_2 \cos (b \ln |x|)].$$

We shall have occasion to refer to these results in a later chapter.

EXAMPLE 2. Find the solution of

$$x^2y'' + 2xy' - 6y = 0 \tag{4–74}$$

which passes through the point (1, 1) with slope zero.

The indicial equation associated with (4–74) is

$$m(m - 1) + 2m - 6 = 0,$$

and has 2 and -3 as roots. Thus the general solution of (4–74) on $(0, \infty)$ is

$$y = c_1x^2 + c_2x^{-3},$$

and since the given initial conditions $y(1) = 1$, $y'(1) = 0$ imply that $c_1 = \frac{3}{5}$, $c_2 = \frac{2}{5}$, the required solution is

$$y = \tfrac{3}{5}x^2 + \tfrac{2}{5}x^{-3}.$$

EXAMPLE 3. In Example 2 of Section 4–4 we said that the general solution of the equation

$$xy'' + y' = 0 \tag{4–75}$$

is

$$y = c_1 + c_2 \ln |x|. \tag{4–76}$$

To prove this assertion we multiply (4–75) by x, and obtain the second order Euler equation $x^2y'' + xy' = 0$. Here the indicial equation is $m^2 = 0$, and (4–74) now follows from the results of Example 1.

The general solution of a nonhomogeneous Euler equation $Ly = h$, defined on an interval I not including the origin, can be obtained by using variation of parameters or the Green's function for L (see Example 2, Section 4–4). But here too the problem can be reduced to one involving a constant coefficient equation, and is usually easier to solve in this form. To do so we again use the linear transformation T, this time to rewrite the equation

$$Ly = h \tag{4–77}$$

as

$$LT(T^{-1}y) = T(T^{-1}h).$$

But, by (4–69), $LT = T\hat{L}$. Thus

$$T\hat{L}(T^{-1}y) = T(T^{-1}h),$$

and since T is one-to-one, this last equation may be rewritten

$$\hat{L}(T^{-1}y) = T^{-1}h, \tag{4–78}$$

in which form it is a constant coefficient equation. Moreover, we can now assert that *if* $y(u)$ *is the general solution of* (4–78) *on* I, *then* $y(\ln|x|)$ *is the general solution of* (4–77) *on* I.

EXAMPLE 4. Find the general solution of

$$x^2y'' + xy' = x + x^2 \tag{4–79}$$

on $(0, \infty)$.

In this case $\hat{L} = D^2$, $T^{-1}(x + x^2) = e^u + e^{2u}$, and the transformed version of (4–79) is

$$D^2y = e^u + e^{2u}. \tag{4–80}$$

The general solution of $D^2y = 0$ is $c_1 + c_2u$, and a particular solution of (4–80) may be found by using the method of undetermined coefficients on the expression $Ae^u + Be^{2u}$. A simple computation gives $A = 1$, $B = \frac{1}{4}$. Thus the general solution of (4–80) is

$$y = e^u + \tfrac{1}{4}e^{2u} + c_1 + c_2u,$$

and it follows that the general solution of (4–79) on $(0, \infty)$ is

$$y = x + \frac{x^2}{4} + c_1 + c_2 \ln x.$$

EXERCISES

Find the general solution of each of the following Euler equations.

1. $x^2y'' + 2xy' - 2y = 0$
2. $4x^2y'' - 8xy' + 9y = 0$
3. $x^2y'' + xy' + 9y = 0$
4. $x^2y'' - 3xy' + 7y = 0$
5. $x^2y'' + xy' - p^2y = 0$, p a constant
6. $2x^2y'' + xy' - y = 0$

7. $x^3y''' - 2x^2y'' - 17xy' - 7y = 0$
8. $x^3y''' - 3x^2y'' + 6xy' - 6y = 0$
9. $x^3y''' + 4x^2y'' - 2y = 0$
10. $x^3y''' + 4x^2y'' - 8xy' + 8y = 0$
11. $x^4y^{(iv)} + 6x^3y''' + 7x^2y'' + xy' - y = 0$
12. Let S be the linear transformation from $\mathcal{C}^n(0, \infty)$ to $\mathcal{C}^n(-\infty, 0)$ defined by

$$(Sf)(x) = f(-x).$$

(a) Show that S is one-to-one, onto, and that $SL = LS$ whenever

$$L = x^nD^n + a_{n-1}x^{n-1}D^{n-1} + \cdots + a_1xD + a_0.$$

(b) Use the results of (a) to prove that $L(Sy) = 0$ if and only if $Ly = 0$, L as above, and hence deduce that $y(-x)$ is a solution of $Ly = 0$ on $(-\infty, 0)$ if and only if $y(x)$ is a solution on $(0, \infty)$.

13. Compute the Green's function for the linear differential operator $x^2D^2 + a_1xD + a_0$ on $\mathcal{C}(0, \infty)$.
14. Prove that (4–67) is valid for all positive integers k. [*Hint:* Use mathematical induction.]
15. (a) Let T be the linear transformation defined in (4–66), and let $\hat{L} = D + a_0$. Prove that

$$T\hat{L}T^{-1} = xD + a_0.$$

(b) Let $\hat{L} = D^2 + a_1D + a_0$. Prove that

$$T\hat{L}T^{-1} = x^2D^2 + (a_1 + 1)xD + a_0.$$

16. Prove that every linear differential operator of the form

$$L = x^nD^n + a_{n-1}x^{n-1}D^{n-1} + \cdots + a_1xD + a_0$$

can be written as a product of operators of this form of orders one and two. [*Hint:* Let $\hat{L}$ be the constant coefficient operator associated with L, and write

$$\hat{L} = \hat{L}_1\hat{L}_2\cdots\hat{L}_k,$$

where $\hat{L}_i$, $i = 1, \ldots, k$, is a constant coefficient operator of order one or two. Show that

$$L = (T\hat{L}_1T^{-1})(T\hat{L}_2T^{-1})\cdots(T\hat{L}_kT^{-1}),$$

and use the results of Exercise 15.]

17. Use the results of Exercises 15 and 16 to factor each of the following operators.
(a) $x^2D^2 + 2xD - 2$
(b) $x^2D^2 + D - 9$
(c) $x^3D^3 - 2x^2D^2 - 17xD - 7$
(d) $x^3D^3 - 3x^2D^2 + 6xD - 6$
(e) $x^4D^4 + 6x^3D^3 + 7x^2D^2 + xD - 1$

Find a particular solution of each of the following Euler equations on $(-\infty, \infty)$.

18. $x^2y'' + xy' - 9 = x^3 + 1$
19. $x^2y'' + xy' + 9 = \sin(\ln x^3)$
20. $x^2y'' + 4xy' + 2y = 2\ln x$
21. $x^3y''' + 4x^2y'' + xy' + y = x$
22. $x^2y''' + xy'' + 4y' = 1 + \cos(2\ln x)$

4–9 ELEMENTARY APPLICATIONS

We conclude this chapter with several examples illustrating the way in which linear differential equations arise in the study of natural phenomena. Although the problems we discuss are rather simple—at least from a physical point of view—they are not entirely without interest, and should be construed as reasonable, albeit elementary, applications of differential equations to biology and physics. Some of a more substantial nature will be considered in later chapters.

I. *The growth of populations.* Problems of this type consist of determining the future size of a population under the assumption that its rate of growth is known, and arise in such diverse situations as the radioactive decay of matter and the increase of bacteria in a culture. In such problems it is frequently assumed that the rate of increase (or decrease) in population at time t is proportional to the number of individuals present at that time.* Then if $y(t)$ denotes the number of individuals present at time t,

$$\frac{dy}{dt} = ky \tag{4–81}$$

for an appropriate constant k, and it follows that y obeys the well known *law of exponential growth*

$$y = ce^{kt}, \tag{4–82}$$

c a positive constant.

Hypotheses which result in something more realistic than prolonged exponential growth are, of course, also used. One of the simpliest of these is the assumption that the supply of necessities for life is constant and sufficient to support a total population P. This implies that the factor of proportionality in (4–81) depends upon y and P, and approaches zero as y approaches P. Thus y must now satisfy a nonlinear equation of the form

$$\frac{dy}{dt} = f(y, P)y, \tag{4–83}$$

where f has to be determined experimentally. If, for instance, $f(y, P) = P - y$, (4–83) becomes

$$\frac{dy}{dt} - Py = -y^2,$$

and can be solved as a Bernoulli equation. Indeed, making the substitution $u = y^{-1}$, we find that

$$\frac{du}{dt} + Pu = 1, \qquad \text{and} \qquad u = \frac{1}{P}(1 + ce^{-Pt}).$$

---

* In the case of living organisms it has been found that this assumption is fairly accurate when the population is small in comparison with the availability of such necessities as food, living space, etc. For radioactive decay, short of an atomic explosion, the assumption is in complete accord with experimental fact.

Thus

$$y = \frac{P}{1 + ce^{-Pt}},$$

and if we assume that $y(0) = P_0$, we have

$$y = \frac{PP_0}{P_0 + (P - P_0)e^{-Pt}}.$$

In Fig. 4–4 we have sketched a number of these curves for various values of P_0.

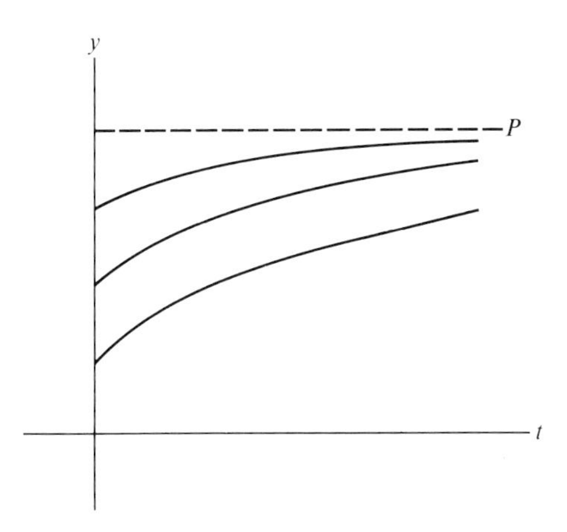

FIGURE 4–4

II. *The simple pendulum.* An overwhelming majority of the problems encountered in elementary physics are solved by invoking *Newton's second law of motion* which, in crude form, asserts that the (vector) sum of the forces acting on a moving object is proportional to the product of the mass m of the object and its acceleration $\mathbf{a}$; that is,

$$\mathbf{F} = km\mathbf{a}. \tag{4–84}$$

The reader should note that in 3-space this equation can be rewritten as a system of three scalar equations

$$F_x = kma_x,$$
$$F_y = kma_y,$$
$$F_z = kma_z,$$

where F_x, a_x, etc., denote, respectively, the x-, y-, z-components of $\mathbf{F}$ and $\mathbf{a}$. For convenience the physical units are usually chosen so that $k = 1$ in these equations, and in the future we shall always assume this choice has been made.

As it stands, Eq. (4–84) assumes that m is constant. A slightly more sophisticated formula which avoids this assumption is

$$\mathbf{F} = \frac{d}{dt}(m\mathbf{v}), \tag{4–85}$$

where **v** denotes velocity. The quantity $m\mathbf{v}$ appearing in (4–85) is known as the *momentum* of the object, and in this form Newton's second law says that the (vector) sum of the forces acting on a moving particle is equal to the time derivative of its momentum. This, for example, would be the equation used to study the flight of a rocket whose mass decreases as its fuel is consumed.

As an illustration of how Newton's second law is used to solve an elementary problem we consider a pendulum bob of mass m supported at the end of a string as shown in Fig. 4–5. The forces acting on the bob are the tension T in the string and the vertical force mg due to gravity. Let φ denote the angular displacement of the string from the vertical. Then the component of the gravitational force parallel to the string balances the tension, while the component perpendicular to the string provides the tangential restoring force which causes the pendulum to oscillate. Since the magnitude of this tangential force is $mg \sin \varphi$, and since the momentum of the bob is $mv = mL(d\varphi/dt)$, Newton's second law gives

$$\frac{d}{dt}\left(mL\frac{d\varphi}{dt}\right) = -mg \sin \varphi, \qquad -\frac{\pi}{2} \le \varphi \le \frac{\pi}{2}. \tag{4–86}$$

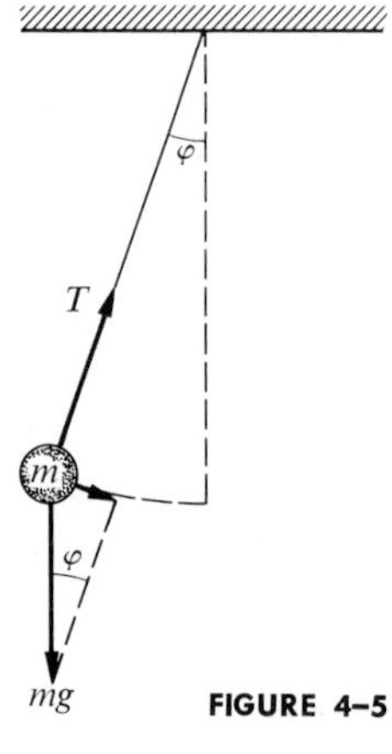

FIGURE 4–5

If we now assume that φ is small so that $\sin \varphi \approx \varphi$, the above equation may be replaced by the linear equation

$$\frac{d^2\varphi}{dt^2} + \frac{g}{L}\varphi = 0, \tag{4–87}$$

whose general solution is

$$\varphi(t) = c_1 \sin \sqrt{g/L}\, t + c_2 \cos \sqrt{g/L}\, t.$$

Finally, if we assume that the pendulum was initially released from rest at an angle φ_0 from the vertical, then $\varphi(0) = \varphi_0$, $\varphi'(0) = 0$, and the corresponding particular solution of (4–87) is $\varphi(t) = \varphi_0 \cos \sqrt{g/L}t$. The student should recognize this as the equation for *simple harmonic motion* whose period of oscillation is $2\pi\sqrt{L/g}$.

EXERCISES

1. The *half-life* of a radioactive substance is defined as the length of time required for half of the atoms in any sample of the substance to decay.

   (a) The half-life of radioactive carbon 14 is 5600 years. Find the amount of carbon 14 remaining in a sample of amount x_0 at the end of t years. (Assume that the rate of decay is proportional to the amount present.)

   (b) If 90% of the carbon 14 in a given sample of carbon has decayed, find the age of the sample.

2. If 20% of a certain radioactive substance disintegrates in 50 years, find the half-life of the substance.
3. A radioactive substance with a half-life of 50 years lies exposed to the weather, and erodes at a constant rate of k pounds per year.

   (a) Find the formula for the amount of material remaining after t years in a sample which originally contained x_0 pounds.

   (b) How long will it take for the substance to disappear entirely?
4. Bacteria in a certain colony are born and die at rates proportional to the number present, so that the equation governing the growth of the colony is

$$\frac{dy}{dt} = (k_1 - k_2)y.$$

   Determine k_1 and k_2 if it is known that the colony doubles in size every 24 hours, and would have its size halved in 8 hours were there no births.
5. If the population of a certain colony of bacteria doubles in 40 hours, how long will it take for the population to increase 10-fold?
6. The population of a country increases 3% per year; its present population is 190 million.

   (a) How many years will elapse before the population reaches 250 million?

   (b) What will the total population be in 5 years? In 50 years?
7. Solve Exercise 6 under the additional assumption that the country admits 200,000 immigrants each year.
8. What is the annual rate of interest being paid on an account where interest is continuously credited at the rate of 5%?
9. Assume that evaporation causes a spherical raindrop to decrease in volume at a rate proportional to its surface area, and find the length of time it will take a drop of radius r_0 to evaporate entirely.
10. It can be shown that a body inside the earth is attracted toward the center by a force which is directly proportional to the distance from the center. Find the equation of motion of a ball dropped into a hole bored through the center of the earth. When will the ball reach the opposite end of the hole?
11. Liquid is oscillating without friction in a U-tube as shown in Figure 4–6. If the liquid was initially at rest with one side h_0 inches higher than the other, determine the subsequent motion of the liquid by finding h as a function of time. Show that the period of oscillation is $\pi\sqrt{2L/g}$, where L is the total length of liquid in the tube, and g the acceleration of gravity.

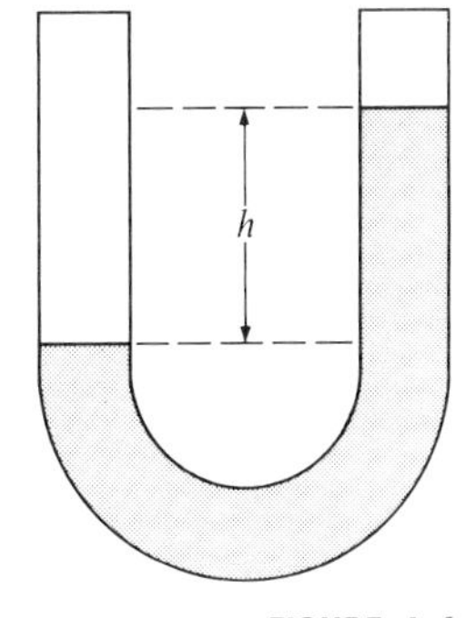

FIGURE 4–6

12. The angular momentum of a rotating body is given by the formula $T = I\alpha$, where T is the torque applied, I the moment of intertia of the body about its axis of rotation, and $\alpha = d^2\varphi/dt^2$ is the angular acceleration. Suppose that a bob on a twisted wire resists the twisting force with a torque $k\varphi$. Find I as a function of time if $k = 1$ and the period of rotation is half a second.

4–10 SIMPLE ELECTRICAL CIRCUITS

The flow of current in an electrical network consisting of a finite number of closed loops, or *circuits*, is governed by the following rules known as *Kirchhoff's laws:*

(a) The algebraic sum of the currents flowing into any point in the network is zero;

(b) The algebraic sum of the voltage drops across the various electrical components in any *oriented* closed loop in the network is zero.*

In this discussion we shall restrict our attention to networks consisting of a single circuit made up of a voltage source E, a resistance R, a capacitance C, and an inductance L, the last in the form, say, of a coil of copper wire. The formulas relating the flow of current i to the voltage drop across each of these components are

$$\begin{aligned} E_R &= iR && \text{for a resistance,} \\ E_L &= L\frac{di}{dt} && \text{for an inductance,} \\ i &= C\frac{dE}{dt} && \text{for a capacitance.} \end{aligned} \tag{4–88}$$

We begin by considering the R-L circuit shown in Fig. 4–7, where the symbol ⊣|▮⊢ denotes a constant source of voltage E such as might be supplied by a battery, and the arrows indicate the direction of the flow of current. By Kirchhoff's second law we have

$$E_L + E_R - E = 0,$$

the negative sign being due to the fact that the voltage *rises* across the battery. Hence, by (4–88),

$$L\frac{di}{dt} + Ri = E,$$

and if we now assume that the circuit was energized at time $t = 0$, the flow of current is obtained as the solution of the initial value problem

$$L\frac{di}{dt} + Ri = E, \qquad i(0) = 0. \tag{4–89}$$

An easy computation reveals that

$$i = \frac{E}{R}(1 - e^{-(R/L)t}), \tag{4–90}$$

and we see that the current flow in this circuit is a sum of two terms, a time

---

* A closed curve is said to be *oriented* if a positive direction has been assigned for traversing it.

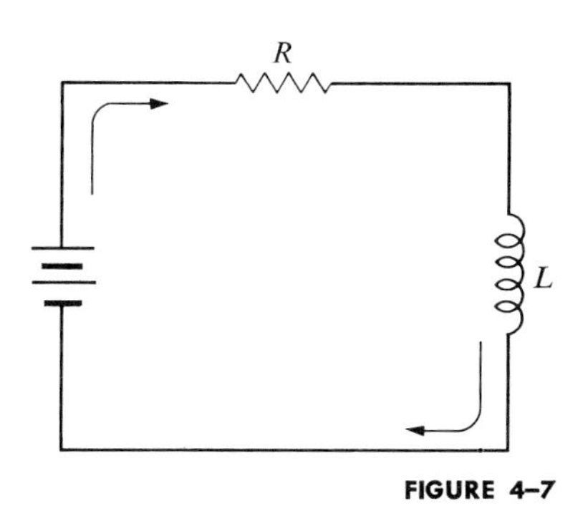

FIGURE 4-7

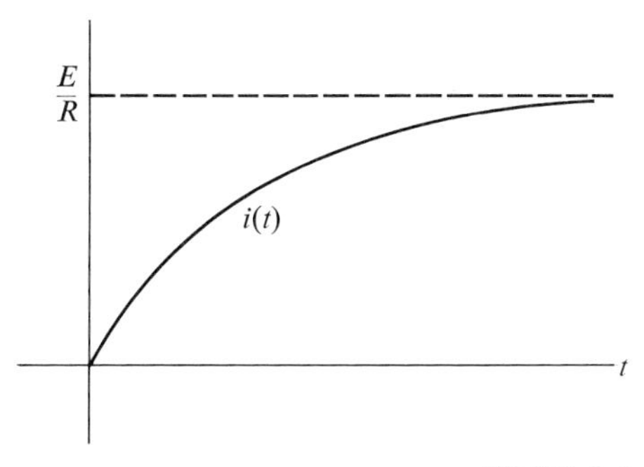

FIGURE 4-8

independent *steady-state* term E/R, and a *transient term* $-(E/R)e^{-(R/L)t}$, whose effect diminishes with time. (See Fig. 4-8.) Since the inductance L appears only in the latter term it follows that a simple R-L circuit operating under a constant impressed voltage will eventually behave very much as if the circuit were noninductive. The length of time required for the transient term to become negligible is sometimes called the *delay time* of the circuit, and furnishes a measure of its sensitivity in responding to the voltage source E.

If we replace the battery in the above circuit by an alternating current source $E \sin \omega t$, (4-89) becomes

$$L\frac{di}{dt} + Ri = E \sin \omega t, \qquad i(0) = 0. \tag{4-91}$$

This time the solution assumes the more complicated form

$$i = \frac{\omega EL}{R^2 + \omega^2 L^2} e^{-(R/L)t} + \frac{E}{\sqrt{R^2 + \omega^2 L^2}} (R \sin \omega t - \omega L \cos \omega t) \tag{4-92}$$

(see Exercise 2), which may be rewritten

$$i = \frac{\omega EL}{Z^2} e^{-(R/L)t} + \frac{E}{Z} \sin (\omega t - \alpha) \tag{4-93}$$

by setting $R = Z \cos \alpha$ and $\omega L = Z \sin \alpha$, where $Z = \sqrt{R^2 + \omega^2 L^2}$. Again the flow is the sum of two terms, i_t and i_s, the first of which is transient and dies out as t increases. The second,

$$i_s = \frac{E}{Z} \sin (\omega t - \alpha),$$

is the steady-state current, and is sinusoidal in nature as one might expect. It differs from the impressed voltage $E \sin \omega t$ by the *phase angle* α and the multiplicative factor $1/Z$. (The quantity $Z = \sqrt{R^2 + \omega^2 L^2}$ is called the *steady-state impedance* of the circuit.) Thus the graph of the steady-state current can be

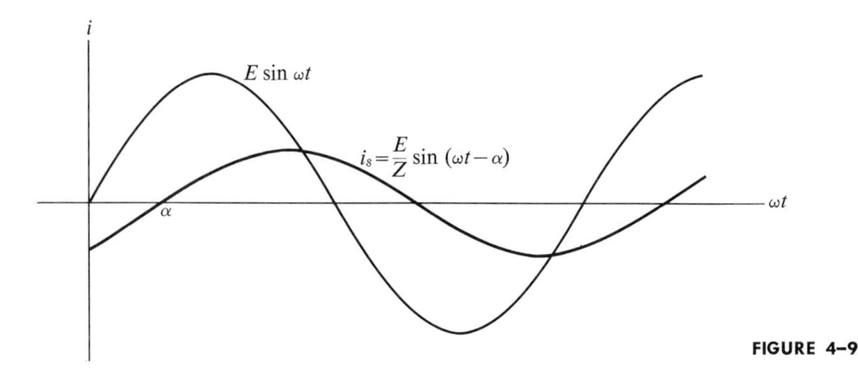

FIGURE 4–9

obtained by multiplying the amplitude of the graph of the impressed voltage by $1/Z$, and translating the result α units to the right (see Fig. 4–9). Since $0 \leq \alpha \leq \pi/2$, the current in such a circuit is said to *lag* the voltage by the phase angle α. Finally, since

$$\sin \alpha = \frac{\omega L}{Z} \quad \text{and} \quad \cos \alpha = \frac{R}{Z},$$

we see that $\alpha = 0$ in a purely resistive circuit ($L = 0$), and that $\alpha = \pi/2$ in a purely inductive circuit ($R = 0$). Hence when $L = 0$ the current and voltage are *in phase*, while when $R = 0$ they are 90° out of phase. In either case the reader will note that the steady-state impedance plays exactly the same role that the resistance plays in an R-L circuit under a constant voltage, a fact which explains the use of the term "impedance" here.

We now consider a simple R-L-C circuit under a sinusoidal impressed voltage $E \sin \omega t$ (see Fig. 4–10). In this case it is somewhat simpler to describe the state of the system in terms of the charge q on the capacitor as a function of time, rather than in terms of the current i. Since

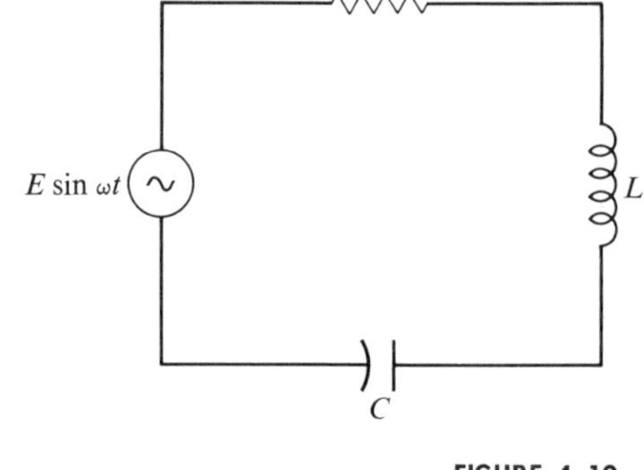

FIGURE 4–10

$$i = \frac{dq}{dt}, \tag{4–94}$$

the voltage drop across a capacitor is $(1/C)q$, and Kirchhoff's second law leads to the equation

$$L \frac{d^2q}{dt^2} + R \frac{dq}{dt} + \frac{1}{C} q = E \sin \omega t \tag{4–95}$$

governing the accumulation of charge $q(t)$ on the capacitor in the circuit. Thus the behavior of this circuit can be determined as soon as the initial values of q and $i = dq/dt$ are known.

The solution q_h of the homogeneous equation associated with (4–95) can be expressed in terms of the roots

$$-\frac{R}{2L} \pm \sqrt{\left(\frac{R}{2L}\right)^2 - \frac{1}{LC}}$$

of the auxiliary equation

$$Lm^2 + Rm + \frac{1}{C} = 0.$$

Thus q_h assumes different forms depending on whether

$$\left(\frac{R}{2L}\right)^2 - \frac{1}{LC}$$

is positive, negative, or zero. However, it is easily seen that no matter what initial conditions are imposed on the circuit, $q_h \to 0$ as $t \to \infty$. In other words, q_h represents a *transient charge* on the capacitor, and as t increases, the time variation $q(t)$ of the charge on the capacitor approaches a *steady-state* value $q_s(t)$, which can be determined by finding a particular solution of (4–95). Furthermore, it is interesting to note that this steady-state charge is independent of the initial conditions $q(0)$ and $i(0)$ imposed on the circuit, and depends only on R, L, C, and $E \sin \omega t$. [Why?] This observation is in agreement with our intuition which suggests that the long-range behavior of such a circuit ought to depend only upon its components, and not upon their state at time $t = 0$.

We now find the steady-state behavior of this system under the assumption that $R \neq 0$. In this case $E \sin \omega t$ cannot be a solution of (4–95) for any value of E, and hence the equation has a particular solution of the form

$$q_s = c_1 \sin \omega t + c_2 \cos \omega t.$$

When this expression is substituted in (4–95) and the resulting identity solved for c_1 and c_2, we obtain

$$c_1 = -\frac{E[\omega L - (1/\omega C)]}{\omega\{R^2 + [\omega L - (1/\omega C)]^2\}},$$

$$c_2 = -\frac{ER}{\omega\{R^2 + [\omega L - (1/\omega C)]^2\}}.$$

If we introduce the abbreviations

$$\gamma = \omega L - \frac{1}{\omega C}, \qquad Z^2 = R^2 + \left(\omega L - \frac{1}{\omega C}\right)^2,$$

the particular solution q_s may be written

$$q_s = -\frac{E\gamma}{\omega Z^2} \sin \omega t - \frac{ER}{\omega Z^2} \cos \omega t,$$

or even more simply

$$q_s = -\frac{E}{\omega Z}\cos(\omega t - \alpha), \tag{4-96}$$

where $\sin\alpha = \gamma/Z$ and $\cos\alpha = R/Z$. Finally, by differentiating this expression we find that the *steady-state* current for an *R-L-C* circuit is

$$i_s = \frac{E}{Z}\sin(\omega t - \alpha). \tag{4-97}$$

As in the case of an *R-L* circuit with impressed electromotive force $E\sin\omega t$, the constants α and Z are called the *phase angle* and *steady-state impedance* of the circuit. In fact, it is customary to view an *R-L* circuit as the limiting case of an *R-L-C* circuit obtained by setting $C = \infty$.

Since the steady-state impedance

$$Z = \sqrt{R^2 + [\omega L - (1/\omega C)]^2}$$

in an *R-L-C* circuit depends upon ω, the maximum amplitude E/Z of the steady-state current i_s also depends on ω. It is clear that for fixed values of L and C the quantity E/Z is a maximum when $\omega L - 1/\omega C$ vanishes, i.e., when $\omega = 1/\sqrt{LC}$, and for this value of ω, $E/Z = E/R$. Thus, if we plot E/Z, the amplitude of i_s, as a function of ω, the graph will attain its maximum value when $\omega = 1/\sqrt{LC}$. Furthermore, this maximum increases with decreasing R, as shown in Fig. 4–11. Physically, these observations tell us that the circuit may react quite differently to input voltages with different frequencies ω. The more the frequency differs from $1/\sqrt{LC}$, the smaller the amplitude of the steady-state current becomes, and hence the voltage drop across the various components of the circuit is small. By minimizing its resistance, such a circuit can be made highly selective, in the sense that it discriminates very sharply against inputs whose frequency differs from the circuit's *natural* or *resonating frequency* $1/\sqrt{LC}$. Thus if a mixture of sinusoidal input voltages of the same magnitude is applied to an *R-L-C* circuit of high selectivity, so that the impressed electromotive force is a sum of the form

$$\sum_{i=1}^{n} E\sin\omega_i t,$$

the steady-state current will depend almost exclusively on those terms whose frequencies ω_i are very close to the natural frequency $1/\sqrt{LC}$. The use of such circuits as tuning circuits or filters in electronic equipment is obvious.

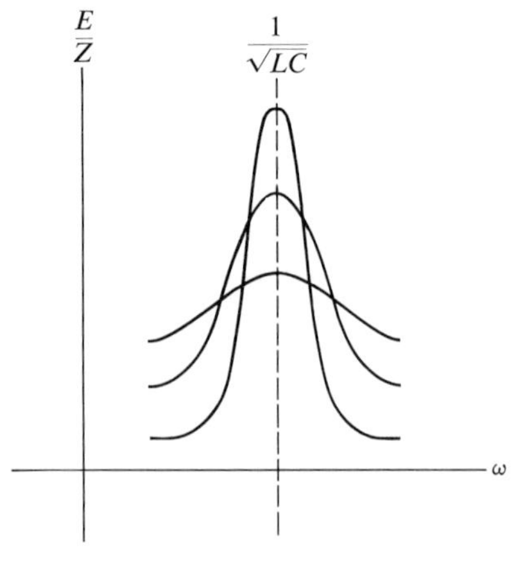

FIGURE 4–11

EXERCISES*

1. Find the current flow in a simple *R-L* circuit under a damped sinusoidal voltage $E_0e^{-at}\sin bt$, with E_0, a, b constants, $a > 0$. [Assume that at time $t = 0$, $i(0) = i_0$.]
2. Verify Eq. (4–92) of the text.
3. A simple circuit consisting of a condenser and a resistor is connected as shown in Fig. 4–12. Suppose that C initially carries a charge of 0.03 coulombs and that the switch is closed at time $t = 0$.

   (a) Find $q(t)$ as a function of time.

   (b) Sketch the graph of $q(t)$.

   (c) Find the voltage drop across the resistor after 10 seconds have elapsed if $C = 300$ microfarads (300×10^{-6} farad) and $R =$ 10,000 ohms.

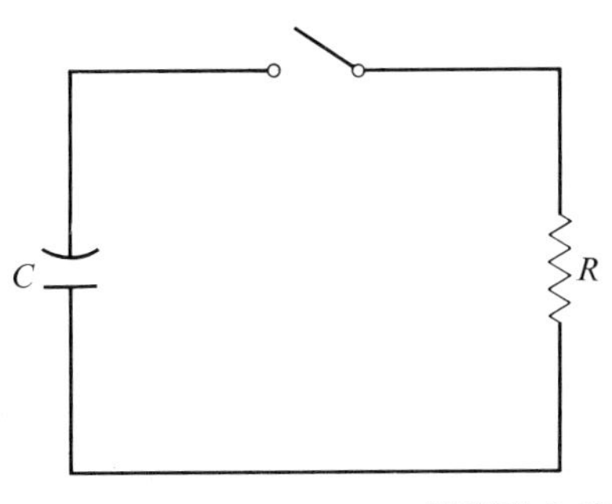

FIGURE 4–12

4. Replace the resistor in the circuit of Exercise 3 by an inductance L, and show that the charge q on the condenser satisfies the equation

$$L\frac{d^2q}{dt^2} + \frac{1}{C}q = 0.$$

   Solve this equation under the initial conditions $i(0) = 0$, $q(0) = 0.03$ coulombs. Sketch the graph of the solution.

5. Add an inductance to the circuit in Exercise 3 and solve the resulting differential equation, considering separately the three cases

$$R > 2\sqrt{L/C}, \qquad R = 2\sqrt{L/C}, \qquad R < 2\sqrt{L/C}.$$

   Sketch the graphs of the three types of solutions obtained. (In this situation the resistance R corresponds to a mechanical damping force, and these three cases yield, respectively, overdamping, critical damping, and underdamping.)

6. Find the current flow in a simple *R-L* circuit under a constant voltage given that $R = 40$ ohms, $L = 8$ henries, and that $E = 0$ volts and $i = 10$ amperes when $t = 0$. At what time will $i = 5$ amperes?
7. If the resistance is removed from the *R-L-C* circuit discussed in the text, q satisfies the differential equation

$$L\frac{d^2q}{dt^2} + \frac{1}{C}q = E\sin\omega t.$$

   Solve this equation in the two cases

$$\text{(i)}\quad \omega \neq \frac{1}{\sqrt{LC}}, \qquad \text{(ii)}\quad \omega = \frac{1}{\sqrt{LC}},$$

---

* The units of measurement used in the following exercises are resistance in *ohms*, inductance in *henries*, capacitance in *farads*, current in *amperes*, charge in *coulombs*.

and discuss the behavior of the solution in each case. Which of these cases exhibits the phenomenon of *resonance*? [Note that in the resonant case the voltage drop across the capacitor oscillates with the same frequency as the input voltage $(E \sin t)/\sqrt{LC}$. Its amplitude, however, is unbounded. This is the limiting case of an R-L-C circuit when $R = 0$, but, of course, can only be approximated in practice.]

8. Suppose that $q(0) = q'(0) = 0$, and that $R = 5$ ohms, $C = 300$ microfarads (300×10^{-6} farad), and that $L = 0.1$ henries in the R-L-C circuit discussed in the text. If the impressed electromotive force is the standard 110-volt, 60-cycle alternating current in common use (i.e., $E \sin \omega t = 110\sqrt{2} \sin (120\pi t)$, since it is the *effective* and not the *peak* voltage which is 110 volts), find q as a function of time. What is the voltage drop across R after 10 seconds? Across L?

5

the laplace transform

5–1 INTRODUCTION

In this chapter we shall for the first time make full use of the idea of a linear operator and its inverse in solving initial-value problems involving linear differential equations. By contrast to the rather pedestrian methods developed in the last chapter, our present investigations will yield an extremely efficient technique for handling such problems. In addition, they will give a much deeper insight into the role which operator theory plays in applied mathematics, and will serve as an excellent introduction to a general method which will be used later to analyze more difficult problems.

The particular linear transformation which we now intend to study is an integral operator $\mathcal{L}$ known as the *Laplace transform*. Before giving the definition, however, we introduce the notion of a *piecewise continuous function*, which will be needed when we describe the domain of $\mathcal{L}$.

Informally, a real valued function is said to be *piecewise continuous* on a closed interval if its graph consists of a finite number of continuous pieces. More precisely, f is piecewise continuous on $[a, b]$ if it is continuous at all but a finite number of points of this interval and if at each point x_0 of discontinuity both the right and left limits of f exist; that is, $f(x_0 + h)$ and $f(x_0 - h)$ both tend to a finite limit as h tends to zero through positive values.* Thus such functions as t^2, $\sin t$, and the "square wave" function shown in Fig. 5–1 are piecewise continuous on any finite interval of the t-axis. On the other hand, neither $\tan t$ nor $\sin 1/t$ is piecewise continuous on $[0, \pi/2]$; the first because of its behavior near $\pi/2$, the second because it oscillates in such a manner that it does not approach a limit as $t \to 0$ (see Exercise 10).

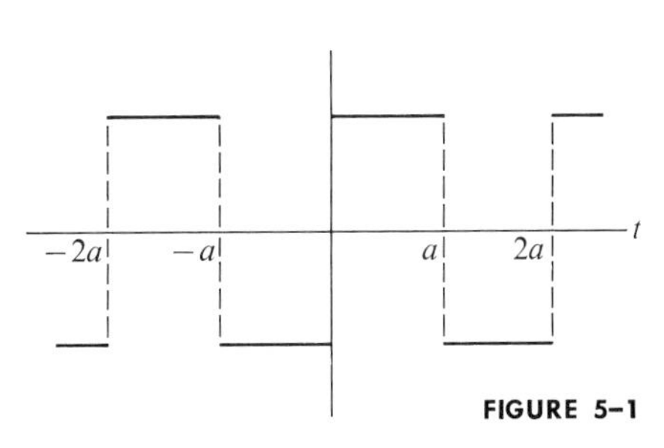

FIGURE 5–1

* Note that only one of these limits is relevant when x_0 is an endpoint of the interval.

For our purposes the essential fact concerning piecewise continuous functions on a finite interval is that they are integrable. Indeed, if f is piecewise continuous on $[a, b]$, with discontinuities at $x_0, x_1, \ldots, x_n$, and possibly at a and b as well, then $\int_a^b f(t)\,dt$ is defined and evaluated as

$$\lim_{h\to 0^+}\left[\int_{a+h}^{x_0-h} f(t)\,dt + \int_{x_0+h}^{x_1-h} f(t)\,dt + \cdots + \int_{x_n+h}^{b-h} f(t)\,dt\right],$$

where the notation $h \to 0^+$ means that h approaches zero through positive values only. (See Fig. 5–2.) It is known that this limit always exists, and the student presumably has had practice heretofore in evaluating such integrals.

At this point a number of trivial remarks are in order. The first concerns our use of the letter t in place of x for the independent variable. This is nothing more than a convention which is all but universal when discussing the Laplace transform, and stems from the fact that in most initial-value problems the independent variable is time. Moreover, since negative values of time are usually excluded, it is also standard practice to restrict attention to the non-negative t-axis $[0, \infty)$. In this context we shall be interested in functions which are piecewise continuous on *every* finite interval $[0, t_0]$, $t_0 > 0$, and for the sake of brevity shall say that such functions are piecewise continuous on $[0, \infty)$. Finally, the set of all such functions is obviously a real vector space under the definitions of vector addition and scalar multiplication given in Chapter 1 (see Exercise 5).* We shall have more to say about this space in the sections which follow.

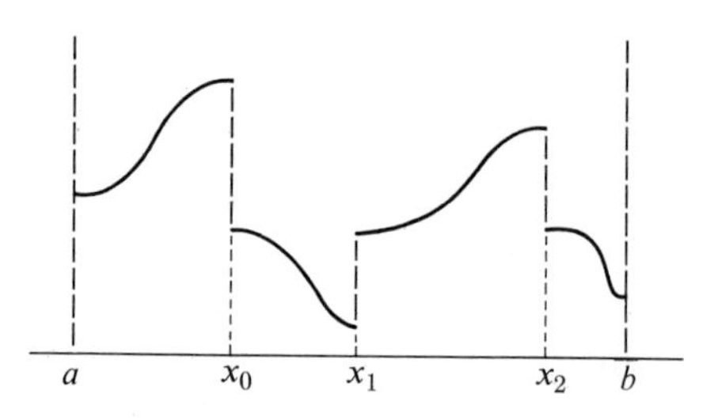

FIGURE 5–2

EXERCISES

1. Which of the following functions are piecewise continuous on $[0, \infty)$? Give reasons for your answers.

   (a) e^{t^2} (b) $\ln(t^2 + 1)$ (c) $\dfrac{t+1}{t-1}$ (d) $\dfrac{t-2}{t^2 - t - 2}$ (e) $e^{1/t}$

2. Repeat Exercise 1 for the following functions.

   (a) $\dfrac{\sin t}{t^n}$, n a positive integer

   (b) t^t (c) $f(t) = [t]$, the greatest integer less than t

   (d) $f(t) = \begin{cases} 0 \text{ if } t = \dfrac{1}{n}, \; n = 1, 2, \ldots \\ 1 \text{ otherwise} \end{cases}$ (e) $f(t) = \begin{cases} 0, \; t \text{ an integer} \\ 1 \text{ otherwise} \end{cases}$

---

* That is, $(f + g)(t) = f(t) + g(t)$ and $(\alpha f)(t) = \alpha f(t)$.

3. Prove that for any real number α, $e^{\alpha t}f(t)$ is piecewise continuous on $[0, \infty)$ whenever f is.
4. Prove that the product of two piecewise continuous functions on $[0, \infty)$ is piecewise continuous.
5. Prove that the set of all piecewise continuous functions on an interval I is a real vector space under the "usual" definitions of addition and scalar multiplication.

Evaluate $\int_0^2 f(t)\,dt$ for each of the following functions.

6. $f(t) = \begin{cases} t, & 0 < t < 1 \\ t - 1, & 1 < t < 2 \end{cases}$

7. $f(t) = \cos|\pi t|$

8. $f(t) = \begin{cases} 1 - t, & 0 < t < 1 \\ t - 1, & 1 < t < 2 \end{cases}$

9. $f(t) = \begin{cases} \frac{1}{2} - t, & 0 < t < \frac{1}{2} \\ 4t^2 - 8t + 3, & \frac{1}{2} < t < \frac{3}{2} \\ \frac{3}{2} - t, & \frac{3}{2} < t < 2 \end{cases}$

10. Show that $\lim_{t \to 0^+} (\sin 1/t)$ does not exist. [*Hint:* Sketch the graph of $\sin 1/t$ in the interval $(0, \infty)$.]
11. Does $\lim_{t \to 0^+} (t \sin 1/t)$ exist? Is $t \sin 1/t$ piecewise continuous on $[0, \infty)$?

5-2 DEFINITION OF THE LAPLACE TRANSFORM

Let $f(t)$ be a real valued function on the interval $(0, \infty)$ and consider

$$\int_0^\infty e^{-st} f(t)\,dt, \tag{5-1}$$

where s is a real variable.* Whenever f is sufficiently well behaved this integral will converge for certain values of s, in which case it defines a function of s called the *Laplace transform* of f, and denoted $\mathcal{L}[f]$, or $\mathcal{L}[f](s)$. Thus if $f(t) = \cos at$, where a is a constant, then

$$\begin{aligned}
\mathcal{L}[\cos at](s) &= \int_0^\infty e^{-st} \cos at\,dt \\
&= \lim_{t_0 \to \infty} \int_0^{t_0} e^{-st} \cos at\,dt \\
&= \lim_{t_0 \to \infty} \left\{ \frac{e^{-st}}{s^2 + a^2} (a \sin at - s \cos at) \Big|_0^{t_0} \right\} \\
&= \lim_{t_0 \to \infty} \left[\frac{e^{-st_0}}{s^2 + a^2} (a \sin at_0 - s \cos at_0) + \frac{s}{s^2 + a^2} \right].
\end{aligned}$$

---

* We recall that an integral of this sort is evaluated according to the rule

$$\int_0^\infty e^{-st} f(t)\,dt = \lim_{t_0 \to \infty} \int_0^{t_0} e^{-st} f(t)\,dt,$$

and is said to converge for a particular value of s if and only if this limit exists.

Since this limit exists if and only if $s > 0$, in which case it has the value $s/(s^2 + a^2)$, it follows that the Laplace transform of $\cos at$ is the function $s/(s^2 + a^2)$ *restricted to the interval* $(0, \infty)$. In other words,

$$\mathcal{L}[\cos at] = \frac{s}{s^2 + a^2}, \qquad s > 0. \tag{5–2}$$

In most of our applications it will be possible to compute $\mathcal{L}[f]$ by direct evaluation of (5–1) as was done above. This, however, does not obviate the need for determining a "reasonable" set of conditions which will insure the existence of the Laplace transform of a given function f, particularly since we wish to view $\mathcal{L}$ as a linear transformation defined on a suitable vector space. If we examine (5–1) from this point of view it is clear that f must be so chosen that

$$\int_0^{t_0} e^{-st} f(t)\, dt \tag{5–3}$$

exists for all $t_0 > 0$. This can be accomplished by demanding that f be *piecewise continuous* on every interval of the form $[0, t_0]$, $t_0 > 0$, for then the integrand in (5–3) will be piecewise continuous, and the integral will exist. (See Exercise 4, Section 5–1.) Piecewise continuity by itself, however, is not enough to guarantee the existence of $\mathcal{L}[f]$ since (5–3) must also *converge* as $t_0 \to \infty$ for at least one value of s. One way of assuring this convergence is to require that $f(t)$ be "dominated" by some exponential function, thus in effect demanding that $e^{-st}f(t)$ approaches zero rapidly as t increases. To make this notion precise we lay down the following definition.

Definition 5–1. A function f is said to be of *exponential order* on $[0, \infty)$ if there exist constants C and α, $C > 0$, such that

$$|f(t)| \le Ce^{\alpha t} \tag{5–4}$$

for all $t > 0$.*

In a moment we shall prove that the Laplace transform of any piecewise continuous function of exponential order does in fact exist, but first some examples.

The constant function $f(t) = 1$ is of exponential order, as can be seen by setting $\alpha = 0$, $C = 1$ in (5–4). So too—and this is important—are the functions

$$t^n, \; e^{at}, \; \sin bt, \; \cos bt, \; t^n e^{at} \sin bt, \; t^n e^{at} \cos bt,$$

familiar from the study of constant coefficient linear differential equations. For instance, the proof that $t^n e^{at} \cos bt$ is of exponential order goes as follows: If $a > 0$,

$$\left|\frac{t^n e^{at} \cos bt}{e^{2at}}\right| \le \frac{t^n e^{at}}{e^{2at}} = \frac{t^n}{e^{at}},$$

* This inequality need only be satisfied, of course, at those points of the non-negative t-axis where f is defined.

and L'Hôpital's rule shows that this expression tends to zero as $t \to \infty$. In particular, it is eventually less than 1, and hence $|t^n e^{at} \cos bt| \leq e^{2at}$ for sufficiently large values of t. Thus there exists a constant $C > 0$ such that $|t^n e^{at} \cos bt| \leq Ce^{2at}$ for *all* $t > 0$ (see Exercise 14 below). If $a \leq 0$ the proof is even easier, for then

$$|t^n e^{at} \cos bt| \leq t^n,$$

and the inequality $t^n < e^t$ for large values of t implies the existence of a constant $C > 0$ such that $|t^n e^{at} \cos bt| \leq Ce^t$ for all $t > 0$.

On the other hand, the function e^{t^2} is *not* of exponential order, since

$$\lim_{t\to\infty} \frac{e^{t^2}}{e^{\alpha t}} = \lim_{t\to\infty} e^{t(t-\alpha)} = \infty$$

for all α.

This having been said, we now prove the theorem which justifies introducing functions of exponential order in the first place.

Theorem 5–1. *If f is a piecewise continuous function of exponential order, there exists a real number α such that*

$$\int_0^\infty e^{-st} f(t)\, dt$$

converges for all values of $s > \alpha$.

Proof. This assertion is an immediate consequence of a well-known comparison theorem from analysis; viz., if f and g are integrable on every interval of the form $[a, b]$, where a is fixed and $b > a$ is arbitrary, and if $|f(t)| \leq g(t)$ for all $t \geq a$, then $\int_a^\infty f(t)\, dt$ exists whenever $\int_a^\infty g(t)\, dt$ exists.*

Granting the truth of this result, choose C and α so that $|f(t)| \leq Ce^{\alpha t}$ for all $t > 0$ (recall that f is of exponential order). Then

$$\begin{aligned}\int_0^\infty e^{-st}(Ce^{\alpha t})\, dt &= C\int_0^\infty e^{-(s-\alpha)t}\, dt \\ &= \lim_{t_0\to\infty} \frac{C}{s-\alpha}[1 - e^{-(s-\alpha)t_0}] \\ &= \frac{C}{s-\alpha} \quad \text{if} \quad s > \alpha,\end{aligned}$$

and the comparison theorem implies that $\int_0^\infty e^{-st} f(t)\, dt$ exists for all $s > \alpha$. ▮

On the strength of this result we can assert that the domain of definition of the Laplace transform of a piecewise continuous function of exponential order always

* The student who has not already met this theorem may be willing to accept it when we point out that it is the analog for integrable functions of the comparison test for the convergence of infinite series. For a proof see Appendix I.

includes a semi-infinite interval of the form (α, ∞). In point of fact, if s_0 denotes the greatest lower bound of the set of real numbers α such that $\mathcal{L}[f](s)$ exists for all $s > \alpha$, it can be shown that $\mathcal{L}[f](s)$ does not converge for *any* $s < s_0$.* Thus, with the possible exception of the point s_0 itself, the domain of definition of $\mathcal{L}[f]$ is the open interval (s_0, ∞), and for this reason s_0 is known as the *abscissa of convergence* of the function f. (Note that for certain functions such as e^{-t^2} or 0, s_0 may be $-\infty$.)

The Laplace transform of any function of exponential order exists; this much we know. But what about the converse? Is it true that a function whose Laplace transform exists is necessarily of exponential order? The answer is no, and, as a matter of fact, the function $1/\sqrt{t}$ has a Laplace transform even though it is not of exponential order. All of this is by way of saying that the set of functions possessing Laplace transforms is larger than the set $\mathcal{E}$ of functions of exponential order. How much larger we shall not say, for it is no easy matter to specify the domain of $\mathcal{L}$ precisely. Fortunately the set of functions of exponential order contains all of the functions which arise in applications, and as such is large enough for most purposes.

EXERCISES

Compute the Laplace transform and abscissa of convergence of each of the following functions.

1. t

2. e^{at}

3. $f(t) = \begin{cases} 1, & 0 < t \leq 1 \\ 0, & t > 1 \end{cases}$

4. 1

5. $\sin at$

6. $f(t) = \begin{cases} 0, & 0 < t \leq 1 \\ t, & t > 1 \end{cases}$

Prove that each of the following functions is of exponential order.

7. t^n, n a positive integer

8. e^{at}

9. $\sin bt$

10. $\cos bt$

11. $\ln(1 + t)$

12. $\sqrt{t}$

*13. Show that the Laplace transform of a function f may exist even though f "grows too fast" to be of exponential order.

14. Let f be piecewise continuous on $[0, \infty)$, and suppose that there exist constants C and α such that $|f(t)| \leq Ce^{\alpha t}$ whenever $t > t_0 > 0$. Prove that f is of exponential order.

15. Prove that the product of two functions of exponential order is of exponential order.

---

* Recall that b is a *lower bound* of a nonempty set S of real numbers if and only if $b \leq s$ for all s in S, and that B is a *greatest lower bound* of S if and only if B is a lower bound of S and $B \geq b$ for every lower bound b of S. One of the most important properties of the real number system is that every nonempty set S of real numbers has a unique greatest lower bound B (provided B is allowed to assume the value $-\infty$ wherever S has no finite lower bound).

16. Let f be piecewise continuous on $[0, \infty)$.

 (a) Prove that f is of exponential order whenever there exists a constant α such that

$$\lim_{t \to \infty} \frac{f(t)}{e^{\alpha t}} = 0.$$

 (b) Prove that f is *not* of exponential order if

$$\lim_{t \to \infty} \frac{f(t)}{e^{\alpha t}} = \infty$$

 for *all* real numbers α.

17. Use the results of the preceding exercise to prove that $e^{\alpha t^{\alpha}}$ is of exponential order if $\alpha \leq 1$, and not if $\alpha > 1$.

18. Is the function t^t of exponential order on $[0, \infty)$? [*Hint:* Use Exercise 16 and the identity $t^t = e^{t \ln t}$.]

19. Another version of the integral comparison theorem stated in the proof of Theorem 5–1 is the following. If f and g are integrable on $[a, 1]$, $0 < a < 1$, and if $|f(t)| \leq g(t)$ whenever $0 < t \leq 1$, then $\int_0^1 f(t)\, dt$ exists whenever $\int_0^1 g(t)\, dt$ exists. Use this result to prove that $1/\sqrt{t}$ has a Laplace transform.

$$[\textit{Hint:} \quad \int_0^{\infty} \frac{e^{-st}}{\sqrt{t}}\, dt = \int_0^1 \frac{e^{-st}}{\sqrt{t}}\, dt + \int_1^{\infty} \frac{e^{-st}}{\sqrt{t}}\, dt.]$$

20. Let f be a function of exponential order, and let α_0 be the least real number such that for some constant C,

$$|f(t)| \leq Ce^{\alpha t}$$

 for all $\alpha > \alpha_0$.

 (a) Show that $\alpha_0 \geq s_0$, the abscissa of convergence of f.

 (b) Show that there exist functions for which $\alpha_0 > s_0$. [*Hint:* Consider the function

$$f(t) = \begin{cases} e^t & \text{if } t \text{ is an integer,} \\ 0 & \text{otherwise.} \end{cases}]$$

21. Let f be piecewise continuous and bounded on $[0, \infty)$; i.e., there exists a constant M such that $|f(t)| \leq M$ for all $t > 0$. Prove that f is a function of exponential order with abscissa of convergence ≤ 0.

5–3 THE LAPLACE TRANSFORM AS A LINEAR TRANSFORMATION

Let $\mathcal{E}$ denote the set of all piecewise continuous functions of exponential order, viewed as a real vector space under the usual definitions of addition and scalar multiplication, and let $\mathfrak{F}$ denote the set of all real valued functions defined on intervals of the form (s_0, ∞) or $[s_0, \infty)$, $s_0 \geq -\infty$. Then $\mathfrak{F}$ too can be made into a real vector space provided we modify the addition used heretofore in function spaces to accommodate the fact that the members of $\mathfrak{F}$ are not all defined on the same interval. Specifically, if f and g are any two functions in $\mathfrak{F}$, $f + g$ is defined to be the function whose domain is the *intersection* of the domains of f and g, and whose value at any point s in that intersection is $f(s) + g(s)$. Then, with scalar multiplication as usual, $\mathfrak{F}$ is a real vector space (see Exercise 1).

By virtue of the observations made following the proof of Theorem 5–1, we can assert that the Laplace transform $\mathcal{L}$ maps the vector space $\mathcal{E}$ into the vector space $\mathfrak{F}$, and it is only natural to ask if this mapping is linear. It might seem that the answer to this question is obvious, but, unfortunately, the obvious answer is wrong. The difficulty arises from the fact that $\mathcal{L}[f + g]$ need not be the same function as $\mathcal{L}[f] + \mathcal{L}[g]$, as can be seen by considering the case where $f(t) = \cos at$ and $g(t) = -\cos at$. For then $\mathcal{L}[f] + \mathcal{L}[g]$ is the function which is zero on the interval $(0, \infty)$, but is not defined for $s \leq 0$, while $\mathcal{L}[f + g] = \mathcal{L}[0]$ is the zero function on the entire s-axis $(-\infty, \infty)$. From this it is clear that we are only entitled to say that $\mathcal{L}[f + g]$ and $\mathcal{L}[f] + \mathcal{L}[g]$ are identical for those values of s where *both* of these functions are defined, a statement which is not at all the same as asserting their equality.*

But once this difficulty has been recognized, it is clear that it can be circumvented simply by agreeing to regard two functions in $\mathfrak{F}$ as *identical* whenever they coincide on an interval of the form (a, ∞). (Thus, for example, the two functions encountered above are "identified," meaning that they are considered as one and the same.) Enforcing this identification, it is now an easy matter to prove that $\mathcal{L}[f + g] = \mathcal{L}[f] + \mathcal{L}[g]$ for any two functions f and g in $\mathcal{E}$, and since, in any event, $\mathcal{L}[\alpha f] = \alpha\mathcal{L}[f]$ whenever α is a real number, we have succeeded in interpreting $\mathcal{L}$ as a linear transformation from $\mathcal{E}$ to $\mathfrak{F}$.†

This done, we now ask if $\mathcal{L}$ is *one-to-one*; i.e., does $\mathcal{L}[f] = \mathcal{L}[g]$ imply that $f = g$? The reader should recognize that this is just another way of asking if an operator equation of the form

$$\mathcal{L}[y] = \varphi(s)$$

can be solved uniquely for y when φ is given, and by now he should realize that this is not an idle question. As in the discussion of the linearity of $\mathcal{L}$, there is a trivial difficulty which prevents us from giving an affirmative answer. For if f and g are functions in $\mathcal{E}$ which differ *only* at their points of discontinuity, then $\mathcal{L}[f] = \mathcal{L}[g]$ even though $f \neq g$. But two such functions are "very nearly" identical, and should this be the worst that can happen we would certainly be justified in asserting that for all practical purposes $\mathcal{L}$ *is* one-to-one. The following theorem guarantees that it is, and for this reason is one of the most important results in the theory of the Laplace transform.

---

* Recall that two functions are *equal* if and only if they have the same domain and take the same value at each point in this domain.

† Strictly speaking, we should at this point replace $\mathfrak{F}$ by the vector space $\mathfrak{F}^*$ whose elements are equivalence classes of functions in $\mathfrak{F}$ as determined by the above process of identification (see Exercise 3). Then the mapping $\mathcal{L}^* : \mathcal{E} \to \mathfrak{F}^*$ defined by $\mathcal{L}^*[f] = \mathcal{L}[f]^*$, where $\mathcal{L}[f]^*$ denotes the equivalence class in $\mathfrak{F}^*$ containing the function $\mathcal{L}[f]$, is the linear transformation in question. It is common practice however to ignore this distinction and simply speak of $\mathcal{L}$ itself as being linear.

Theorem 5–2. (Lerch's theorem.)* *Let f and g be piecewise continuous functions of exponential order, and suppose there exists a real number s_0 such that $\mathcal{L}[f](s) = \mathcal{L}[g](s)$ for all $s > s_0$. Then, with the possible exception of points of discontinuity, $f(t) = g(t)$ for all $t > 0$.*

Thus whenever an equation

$$\mathcal{L}[y] = \varphi(s) \tag{5–5}$$

can be solved for y, the solution is "essentially" unique. In fact, if we agree to identify any two functions in $\mathcal{E}$ which coincide except at their points of discontinuity, we can then speak of *the* solution of such an equation.† This solution is called the *inverse Laplace transform* of the function φ, and is denoted by $\mathcal{L}^{-1}[\varphi]$. It is characterized by the property

$$\mathcal{L}^{-1}[\varphi] = y \quad \text{if and only if} \quad \mathcal{L}[y] = \varphi. \tag{5–6}$$

At this point in our discussion only one general question remains unanswered; viz., does $\mathcal{L}$ map $\mathcal{E}$ *onto* $\mathfrak{F}$? In terms of operator equations this is equivalent to asking if (5–5) has a solution for *every* function φ in $\mathfrak{F}$. And this time the answer is an honest no, since we have

Theorem 5–3. *If f is a function of exponential order, then* $\lim_{s\to\infty} \mathcal{L}[f] = 0$.

Proof. Indeed, in proving Theorem 5–1 we saw that there exist constants C and α such that

$$|\mathcal{L}[f]| \leq \frac{C}{s - \alpha}$$

for all $s > \alpha$, and the desired result follows by taking the limit as $s \to \infty$. ▮

On the strength of this theorem we can assert that such functions as 1, s, $\sin s$, and $s/(s + 1)$ do not have inverse transforms in $\mathcal{E}$, since none of them approaches zero as $s \to \infty$.

EXERCISES

1. (a) Prove that the set $\mathcal{E}$ of piecewise continuous functions of exponential order is a real vector space under the usual definitions of addition and scalar multiplication.

   (b) Using the definition of addition given in the text, prove that the set $\mathfrak{F}$ is a real vector space.

---

* See Appendix II for a proof.

† The knowledgeable reader will recognize that we are again defining an equivalence relation.

*2. Let f and g belong to $\mathfrak{F}$, and define $f \sim g$ if and only if $f(s) = g(s)$ on some interval of the form $s_0 < s$.

(a) Prove that $\sim$ is an equivalence relation on $\mathfrak{F}$.

(b) Exhibit an equivalence class of functions in $\mathfrak{F}$ which does not contain the Laplace transform of any function in $\mathcal{E}$.

(c) Can an equivalence class in $\mathfrak{F}$ contain two different functions which are both Laplace transforms of functions in $\mathcal{E}$? Why?

*3. Let $\mathfrak{F}^*$ denote the set of all equivalence classes of functions in $\mathfrak{F}$ under the equivalence relation defined in Exercise 2.

(a) Give an appropriate definition of addition and scalar multiplication for elements of $\mathfrak{F}^*$ so that $\mathfrak{F}^*$ becomes a vector space.

(b) Define $\mathcal{L}^*: \mathcal{E} \to \mathfrak{F}^*$ by $\mathcal{L}^*[f] = \mathcal{L}[f]^*$, where $\mathcal{L}[f]^*$ is the equivalence class in $\mathfrak{F}^*$ containing the function $\mathcal{L}[f]$. Prove that $\mathcal{L}^*$ is a linear transformation.

4. Prove that $\lim_{s\to\infty} s\mathcal{L}[f]$ is bounded for any function f of exponential order, and then use this result together with Theorem 5–3 to deduce that $\mathcal{L}^{-1}[s^\alpha]$ does not exist for any $\alpha > -1$. [*Hint:* See the proof of Theorem 5–3.]

5–4 ELEMENTARY FORMULAS

In Section 5–2 we used the definition of the Laplace transform to prove that

$$\mathcal{L}[\cos at] = \frac{s}{s^2 + a^2}, \qquad s > 0. \tag{5–7}$$

In like fashion one can produce an almost endless list of elementary formulas, among which are

$$\mathcal{L}[1] = \frac{1}{s}, \qquad s > 0, \tag{5–8}$$

$$\mathcal{L}[e^{at}] = \frac{1}{s - a}, \qquad s > a, \tag{5–9}$$

$$\mathcal{L}[\sin at] = \frac{a}{s^2 + a^2}, \qquad s > 0, \tag{5–10}$$

$$\mathcal{L}[t^n] = \frac{n!}{s^{n+1}}, \qquad s > 0,\ n \text{ a non-negative integer.}^* \tag{5–11}$$

For example,

$$\begin{aligned}
\mathcal{L}[e^{at}] &= \int_0^\infty e^{-st}e^{at}\,dt \\
&= \lim_{t_0\to\infty} \int_0^{t_0} e^{-(s-a)t}\,dt \\
&= \lim_{t_0\to\infty} \left[\frac{1}{s - a}(1 - e^{-(s-a)t_0})\right] \\
&= \frac{1}{s - a}, \qquad s > a.
\end{aligned}$$

* Recall that $0! = 1$.

This proves (5–9), and (5–8) can be obtained from it by setting $a = 0$. Formulas (5–10) and (5–11) will be established presently, and a more comprehensive list is found in the table of transforms on p. 228.

Although these simple formulas are not without significance, it is clear that any applications of the Laplace transform must rest upon more substantial results. One of the most important of these is a formula which expresses the transform of the derivative of f in terms of $\mathcal{L}[f]$ and the behavior of f at 0. This result in turn depends upon the elementary fact that *any function which is continuous on* $(0, \infty)$, *and has a piecewise continuous derivative which is of exponential order, is itself of exponential order* (see Exercise 15). In particular, this allows us to deduce the existence of $\mathcal{L}[f]$ from the continuity of f and the existence of $\mathcal{L}[f']$, and this is precisely what we need to prove

Theorem 5–4. *Let f be continuous on* $(0, \infty)$, *and suppose that f' is piecewise continuous and of exponential order on* $[0, \infty)$. *Then*

$$\mathcal{L}[f'] = s\mathcal{L}[f] - f(0^+), \tag{5–12}$$

where $f(0^+) = \lim_{t\to 0^+} f(t)$. More generally, if $f, f', \ldots, f^{(n-1)}$ are continuous for all $t > 0$, and if $f^{(n)}$ is piecewise continuous and of exponential order on $[0, \infty)$, *then*

$$\mathcal{L}[f''] = s^2\mathcal{L}[f] - sf(0^+) - f'(0^+), \tag{5–13}$$

$$\vdots$$

$$\mathcal{L}[f^{(n)}] = s^n\mathcal{L}[f] - s^{n-1}f(0^+) - s^{n-2}f'(0^+) - \cdots - f^{(n-1)}(0^+). \tag{5–14}$$

Proof. To establish (5–12) we use integration by parts to evaluate $\mathcal{L}[f']$ as follows:

$$\begin{aligned}\mathcal{L}[f'] &= \int_0^\infty e^{-st}f'(t)\,dt \\ &= e^{-st}f(t)\big|_0^\infty + s\int_0^\infty e^{-st}f(t)\,dt \\ &= s\mathcal{L}[f] + e^{-st}f(t)\big|_0^\infty,\end{aligned}$$

and the proof will be complete if we can show that $e^{-st}f(t)\big|_0^\infty = -f(0^+)$.

To this end we note that since f is of exponential order, $e^{-st}f(t) \to 0$ as $t \to \infty$ whenever s is sufficiently large. Thus $e^{-st}f(t)\big|_0^\infty$ vanishes at its upper limit, and, taking account of the fact that f may have a jump discontinuity at the origin, we have

$$\begin{aligned}e^{-st}f(t)\big|_0^\infty &= \lim_{t\to\infty} e^{-st}f(t) - \lim_{t\to 0^+} e^{-st}f(t) \\ &= -\lim_{t\to 0^+} e^{-st}f(t) \\ &= -f(0^+),\end{aligned}$$

as required.

Formulas (5–13) and (5–14) can now be established by repeated use of (5–12), and we are done (see Exercise 28).* ▮

EXAMPLE 1. Let $f(t) = -(1/a)\cos at$. Then $f'(t) = \sin at$, and so, using (5–12) and (5–7),

$$\begin{aligned} \mathcal{L}[\sin at] &= -\frac{s}{a}\mathcal{L}[\cos at] + \frac{1}{a} \\ &= -\frac{s}{a}\left(\frac{s}{s^2 + a^2}\right) + \frac{1}{a} \\ &= \frac{a}{s^2 + a^2}, \qquad s > 0. \end{aligned}$$

This proves (5–10).

EXAMPLE 2. Since $D^n t^n = n!$, (5–8) implies that

$$\mathcal{L}[D^n t^n] = \mathcal{L}[n!] = n!\mathcal{L}[1] = \frac{n!}{s}, \qquad s > 0.$$

On the other hand, (5–14) yields

$$\begin{aligned} \mathcal{L}[D^n t^n] &= s^n \mathcal{L}[t^n] - s^{n-1} \cdot 0 - \cdots - 0 \\ &= s^n \mathcal{L}[t^n]. \end{aligned}$$

Hence $s^n \mathcal{L}[t^n] = n!/s$ for every non-negative integer n, and

$$\mathcal{L}[t^n] = \frac{n!}{s^{n+1}}, \qquad s > 0.$$

This proves (5–11).

With the mechanics of using the differentiation formulas now out of the way, we are in a position to illustrate the use of Laplace transforms in the solution of an initial-value problem. Our example must perforce be a simple one since our list of transforms is still rather limited. Nevertheless it will illustrate all of the essential steps of the technique.

EXAMPLE 3. Use Laplace transforms to solve the initial-value problem

$$\frac{d^2y}{dx^2} - y = 1; \qquad y(0) = 0, \quad y'(0) = 1.$$

We begin by applying the operator $\mathcal{L}$ to both sides of the given equation to obtain

$$\mathcal{L}[y''] - \mathcal{L}[y] = \mathcal{L}[1].$$

---

* A generalization of (5–12) to the case where f has jump discontinuities is given in the exercises below.

(Note that this step depends upon the linearity of $\mathcal{L}$.) Using (5–13) and (5–8), this equation may be rewritten

$$s^2\mathcal{L}[y] - 1 - \mathcal{L}[y] = \frac{1}{s}.$$

Thus

$$\mathcal{L}[y] = \frac{1}{s(s-1)}, \tag{5–15}$$

and to complete the solution we must find a function whose Laplace transform is given by this equation. To do so we use the method of partial fractions to rewrite (5–15) as

$$\mathcal{L}[y] = \frac{1}{s-1} - \frac{1}{s},$$

from which it follows at once that

$$y = e^t - 1.$$

The above example illustrates the way Laplace transforms are used to solve initial-value problems. In general, if we are given an nth-order linear differential equation $Ly = h(t)$ *with constant coefficients* and initial conditions $y(0) = y_0$, $y'(0) = y_1, \ldots, y^{(n-1)}(0) = y_{n-1}$, then (5–14) can be used to convert this initial-value problem into an operator equation of the form

$$\mathcal{L}[y] = \varphi(s),$$

whenever $h(t)$ is of exponential order. It then follows that

$$y = \mathcal{L}^{-1}[\varphi],$$

and the problem has been solved, provided $\mathcal{L}^{-1}[\varphi]$ can be explicitly found. The reader should appreciate that this argument depends upon two facts: first, that the unique solution of *every* such problem is of exponential order, so that the existence of $\mathcal{L}[y]$ is assured from the outset, and second, that the equation $y = \mathcal{L}^{-1}[\varphi]$ has a unique *continuous* solution. But both of these facts have already been established; the first in the paragraph following Definition 5–1, the second by Lerch's theorem.

We conclude this section by establishing the integral analogs of Formulas (5–12) and (5–14) which are frequently useful in computing transforms and their inverses.

Theorem 5–5. *If f is a function of exponential order on* $[0, \infty)$, *and a is a non-negative real number, then*

$$\mathcal{L}\left[\int_a^t f(x)\,dx\right] = \frac{1}{s}\mathcal{L}[f] - \frac{1}{s}\int_0^a f(x)\,dx. \tag{5–16}$$

More generally,

$$\mathcal{L}\Big[\underbrace{\int_a^t \cdots \int_a^t}_{n \text{ times}} f(x)\,dx \cdots dx\Big] = \frac{1}{s^n}\mathcal{L}[f] - \frac{1}{s^n}\int_0^a f(x)\,dx$$

$$- \frac{1}{s^{n-1}}\int_0^a \int_a^t f(x)\,dx\,dx - \cdots - \frac{1}{s}\underbrace{\int_0^a \int_a^t \cdots \int_a^t}_{(n-1) \text{ times}} f(x)\,dx \cdots dx. \tag{5–17}$$

Proof. The proof is based upon the observation that if f is of exponential order then so is $\int_a^t f(x)\,dx$ (see Exercise 19). For using integration by parts with

$$u(t) = \int_a^t f(x)\,dx \qquad \text{and} \qquad dv = e^{-st}\,dt,$$

we have

$$\mathcal{L}\Big[\int_a^t f(x)\,dx\Big] = \int_0^\infty e^{-st}\int_a^t f(x)\,dx\,dt$$

$$= -\frac{1}{s}e^{-st}\int_a^t f(x)\,dx\Big|_0^\infty + \frac{1}{s}\int_0^\infty e^{-st}f(t)\,dt.$$

But since $\int_a^t f(x)\,dx$ is of exponential order, the first term in this expression tends to zero as $t \to \infty$ provided s is sufficiently large, and hence

$$\mathcal{L}\Big[\int_a^t f(x)\,dx\Big] = \frac{1}{s}\mathcal{L}[f] + \frac{1}{s}\int_a^0 f(x)\,dx.$$

Except for obvious notational changes this is (5–16). Equation (5–17) is established by iterating this result. ▌

In practice the integration formulas usually arise with $a = 0$, in which case they assume the much simpler forms

$$\mathcal{L}\Big[\int_0^t f(x)\,dx\Big] = \frac{1}{s}\mathcal{L}[f], \tag{5–18}$$

$$\mathcal{L}\Big[\int_0^t \cdots \int_0^t f(x)\,dx \cdots dx\Big] = \frac{1}{s^n}\mathcal{L}[f]. \tag{5–19}$$

EXAMPLE 4. Since $\int_0^t \cos ax\,dx = (1/a)\sin at$, (5–18) gives

$$\frac{1}{a}\mathcal{L}[\sin at] = \frac{1}{s}\mathcal{L}[\cos at].$$

Thus

$$\begin{aligned}\mathcal{L}[\sin at] &= \frac{a}{s}\,\mathcal{L}[\cos at]\\ &= \frac{a}{s}\left(\frac{s}{s^2 + a^2}\right)\\ &= \frac{a}{s^2 + a^2}, \quad s > 0.\end{aligned}$$

EXAMPLE 5. Use the integration formulas to compute $\mathcal{L}[te^t]$.

Since

$$\int_0^t xe^x\,dx = te^t - e^t + 1,$$

(5–18) gives

$$\mathcal{L}[te^t - e^t + 1] = \frac{1}{s}\,\mathcal{L}[te^t].$$

Using the linearity of $\mathcal{L}$ we then have

$$\mathcal{L}[te^t] - \mathcal{L}[e^t] + \mathcal{L}[1] = \frac{1}{s}\,\mathcal{L}[te^t].$$

Hence

$$\mathcal{L}[te^t] - \frac{1}{s-1} + \frac{1}{s} = \frac{1}{s}\,\mathcal{L}[te^t],$$

and it follows that

$$\mathcal{L}[te^t] = \frac{1}{(s-1)^2}, \quad s > 1.$$

(The reader should note that this result can also be obtained by using the differentiation formula.)

EXERCISES

Find the Laplace transform of each of the following functions. (In those problems where it appears, a denotes a constant.)

1. $\sin(t + a)$
2. $(t + a)^2$
3. $(t + a)^n$, n a positive integer
4. $\sinh at$
5. $\cosh at$
6. t^2e^{at}
7. $\sin^2 at$
8. $t^2 \cos t$
9. $t \sin 2t$
10. $(2t - 3)e^{(t+2)/3}$
11. Show that

$$\mathcal{L}[\cos^3 t] = \frac{s(s^2 + 7)}{(s^2 + 9)(s^2 + 1)}.$$

12. Compute $\mathcal{L}[e^t \sin t]$. [*Hint:* Use the differentiation formula.]
13. Compute the Laplace transform of the function $u_a(t)$ defined by the formula

$$u_a(t) = \begin{cases} 0, & t \le a,\\ 1, & t > a.\end{cases}$$

*14. Compute the Laplace transform of the function whose graph is given in Fig. 5–3. [*Hint:* Recall the formula for the sum of a geometric series.]

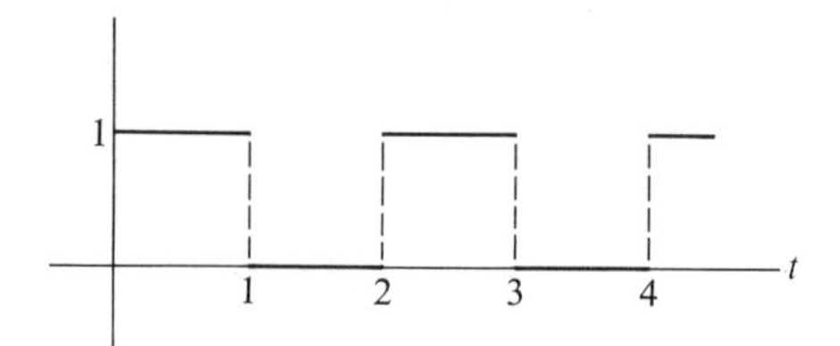

FIGURE 5–3

15. Suppose that the function f is continuous on $(0, \infty)$, and that its derivative f' is of exponential order. Integrate the inequality

$$-Ce^{\alpha x} \leq f'(x) \leq Ce^{\alpha x}$$

to deduce that f is of exponential order.

16. In the proof of Theorem 5–4 we used integration by parts to evaluate the integral

$$\int_0^\infty e^{-st} f'(t)\, dt,$$

where f' was of exponential order. Prove that this was legitimate. (The problem consists of investigating the behavior of this integral at zero and at the points of discontinuity of f'.)

17. During the proof of Theorem 5–4 it was asserted that $e^{-st}f(t)|_0^\infty$ vanishes at its upper limit provided s is sufficiently large. How large is "sufficiently large"?

18. Suppose that the function f in Theorem 5–4 has a jump discontinuity at $t_0 > 0$. Prove that

$$\mathcal{L}[f'] = s\mathcal{L}[f] - f(0^+) - e^{-st_0}[f(t_0^+) - f(t_0^-)],$$

where

$$f(t_0^+) = \lim_{h\to 0} f(t_0 + h), \qquad f(t_0^-) = \lim_{h \to 0} f(t_0 - h), \quad h > 0.$$

19. Prove that $\int_a^t f(x)\, dx$, $a \geq 0$, is of exponential order whenever f is of exponential order.

20. Use (5–16) to prove that

$$\mathcal{L}\left[\int_a^t \int_a^t f(x)\, dx\, dx\right] = \frac{1}{s^2}\mathcal{L}[f] - \frac{1}{s^2}\int_0^a f(x)\, dx - \frac{1}{s}\int_0^a \int_a^t f(x)\, dx\, dx.$$

Solve each of the following initial-value problems using Laplace transforms.

21. $y'' - 3y' + 2y = 0;\ y(0) = 3,\ y'(0) = 4$
22. $y'' + y = t;\ y(0) = -1,\ y'(0) = 3$
23. $y''' + y'' + 4y' + 4y = -2;\ y(0) = 0,\ y'(0) = 1,\ y''(0) = -1$
24. $y''' - y'' + 9y' - 9y = 0;\ y(0) = 0,\ y'(0) = 3,\ y''(0) = 0$
25. $y'' - y' - 6y = 3t^2 + t - 1;\ y(0) = -1,\ y'(0) = 6$
26. $y'' + 6y' + 8y = 4(4t + 3);\ y(0) = 2,\ y'(0) = -6$
27. $4y'' + y = -2;\ y(0) = 0,\ y'(0) = \frac{1}{2}$
28. Derive Formula (5–14).

*29. If $Ly = h(t)$ is a linear differential equation with constant coefficients and if $h(t)$ is of exponential order, show that *every* solution of this equation is of exponential order. [*Hint:* Each solution is of the form $y_h(t) + y_p(t)$, where $y_h(t)$ satisfies the homogeneous equation $Ly = 0$ and $y_p(t) = \int_0^t G(t, \xi)h(\xi)\, d\xi$. Recall how the Green's function $G(t, \xi)$ is constructed from solutions of $Ly = 0$ (see Section 4–5).]

5–5 FURTHER PROPERTIES OF THE LAPLACE TRANSFORM

As we have seen, the solution of an initial-value problem by Laplace transform methods comes down to finding the inverse transform of a function $\varphi(s)$. In practice such inverses are obtained by using the method of partial fractions to convert $\varphi(s)$ to a form in which its inverse can be recognized—usually with the aid of certain *special formulas* and a table such as the one given at the end of this chapter. In this section we shall derive a number of the above mentioned formulas, and illustrate their use in computations.

We begin with a very simple result which permits us to compute the Laplace transform of $e^{at}f(t)$ whenever the transform of f is already known.

Theorem 5–6. *If* $\mathcal{L}[f] = \varphi(s)$, *then*

$$\mathcal{L}[e^{at}f(t)] = \varphi(s - a). \tag{5–20}$$

Proof.

$$\begin{aligned}\mathcal{L}[e^{at}f(t)] &= \int_0^\infty e^{-st}e^{at}f(t)\, dt\\ &= \int_0^\infty e^{-(s-a)t}f(t)\, dt\\ &= \mathcal{L}[f(t)](s - a)\\ &= \varphi(s - a). \blacksquare\end{aligned}$$

This result is sometimes known as the *first shifting theorem* (the second will be given below), and may be written in terms of inverse transforms as

$$\mathcal{L}^{-1}[\varphi(s - a)] = e^{at}\mathcal{L}^{-1}[\varphi(s)], \tag{5–21}$$

or as

$$\mathcal{L}^{-1}[\varphi(s)] = e^{at}\mathcal{L}^{-1}[\varphi(s + a)]. \tag{5–22}$$

EXAMPLE 1. Since $\mathcal{L}[\cos 3t] = s/(s^2 + 9)$, (5–20) yields

$$\mathcal{L}[e^{-2t}\cos 3t] = \frac{s + 2}{(s + 2)^2 + 9}.$$

EXAMPLE 2. Compute

$$\mathcal{L}^{-1}\left[\frac{2s + 3}{s^2 - 4s + 20}\right].$$

Keeping (5–20) or (5–21) in mind, we write

$$\frac{2s+3}{s^2-4s+20} = \frac{2s+3}{(s-2)^2+16}$$
$$= \frac{2(s-2)+7}{(s-2)^2+16}$$
$$= 2\left[\frac{s-2}{(s-2)^2+16}\right] + \frac{7}{4}\left[\frac{4}{(s-2)^2+16}\right].$$

But

$$\mathcal{L}^{-1}\left[\frac{s-2}{(s-2)^2+16}\right] = e^{2t}\cos 4t,$$

and

$$\mathcal{L}^{-1}\left[\frac{4}{(s-2)^2+16}\right] = e^{2t}\sin 4t.$$

Hence

$$\mathcal{L}^{-1}\left[\frac{2s+3}{(s-2)^2+16}\right] = 2e^{2t}\cos 4t + \tfrac{7}{4}e^{2t}\sin 4t.$$

In order to state our next result we must introduce the so called *unit step function* $u_a(t)$, which is defined by the formula

$$u_a(t) = \begin{cases} 0, & t \le a, \\ 1, & t > a. \end{cases} \tag{5–23}$$

(See Fig. 5–4.) (For our purposes we shall always assume $a \ge 0$.) This function enables us to write the formula for a function such as the sinusoidal curve

$$f(t) = \begin{cases} 0, & t \le a, \\ \sin(t-a), & t > a, \end{cases} \tag{5–24}$$

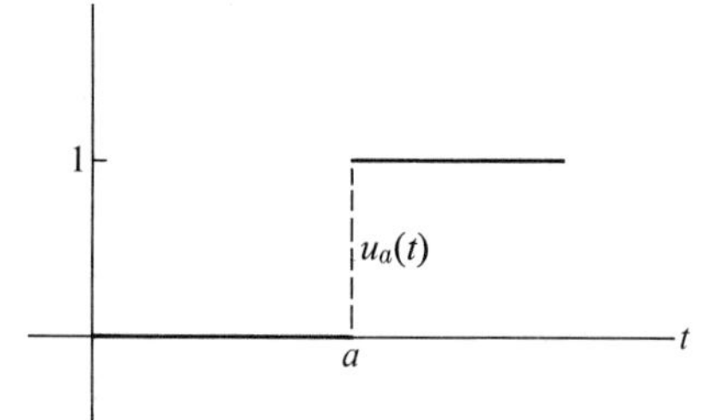

FIGURE 5–4

shown in Fig. 5–5, in very simple form. Indeed, since $u_a(t)$ is zero for $t \le a$, we have

$$f(t) = u_a(t)\sin(t-a), \tag{5–25}$$

an expression which is much better adapted to computation than (5–24).* More generally, the expression

$$f(t) = u_a(t)g(t-a)$$
$$= \begin{cases} 0, & t \le a, \\ g(t-a), & t > a, \end{cases}$$

* Some authors restrict the term "unit step function" to mean what we have called $u_0(t)$. In that case (5–25) becomes $f(t) = u_0(t-a)\sin(t-a)$.

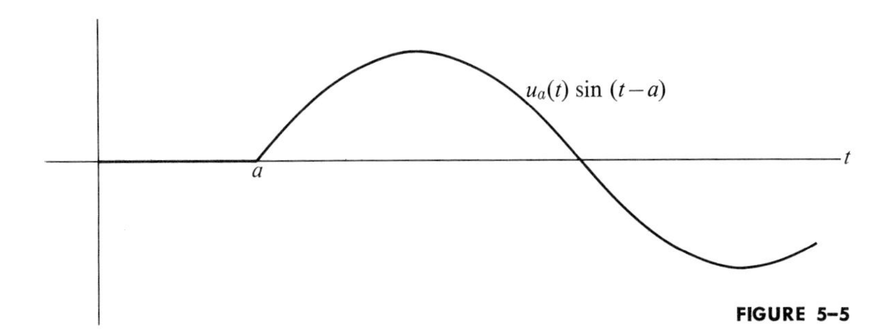

FIGURE 5–5

describes the function obtained by translating or "shifting" $g(t)$ a units to the right and then annihilating that portion to the left of a (Fig. 5–6). Such functions arise in practice as time delayed inputs to physical systems (i.e., inputs occurring at time $t = a > 0$) and are of considerable practical importance. The next theorem, known as the *second shifting theorem*, gives a formula for the Laplace transform of such a function.

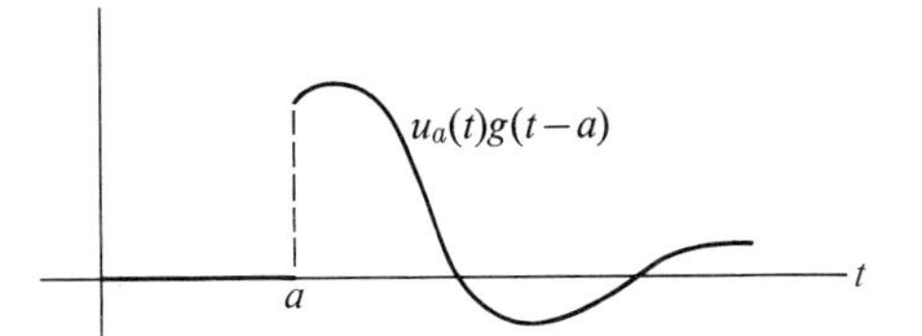

FIGURE 5–6

Theorem 5–7. *Let*

$$f(t) = u_a(t)g(t - a), \quad a \geq 0$$

be a piecewise continuous function of exponential order. Then

$$\mathcal{L}[f] = e^{-as}\mathcal{L}[g]. \tag{5–26}$$

(For physical reasons the factor e^{-as} in this formula is known as a *delaying factor.*)

Proof.

$$\begin{aligned}\mathcal{L}[f] &= \int_0^\infty e^{-st}f(t)\,dt \\ &= \int_a^\infty e^{-st}g(t - a)\,dt.\end{aligned}$$

Thus if we make the substitution $x = t - a$, we obtain

$$\begin{aligned}\mathcal{L}[f] &= \int_0^\infty e^{-s(x+a)}g(x)\,dx \\ &= e^{-as}\int_0^\infty e^{-sx}g(x)\,dx \\ &= e^{-as}\mathcal{L}[g],\end{aligned}$$

and the theorem is proved. ▮

To apply (5–26) in the computation of inverse transforms we rewrite it as

$$\mathcal{L}^{-1}[e^{-as}\mathcal{L}[g(t)]] = u_a(t)g(t - a),$$

or as

$$\mathcal{L}^{-1}[e^{-as}\varphi(s)] = u_a(t)g(t - a), \tag{5–27}$$

where $\varphi(s) = \mathcal{L}[g(t)]$.

EXAMPLE 3. If $f(t) = u_a(t) \sin t$, then

$$\begin{aligned}\mathcal{L}[f] &= \mathcal{L}[u_a(t) \sin (t + a - a)] \\ &= e^{-as}\mathcal{L}[\sin (t + a)] \\ &= e^{-as}\mathcal{L}[\sin t \cos a + \cos t \sin a] \\ &= e^{-as}\{\cos a\mathcal{L}[\sin t] + \sin a\mathcal{L}[\cos t]\} \\ &= \frac{e^{-as}(\cos a + s \sin a)}{s^2 + 1}.\end{aligned}$$

EXAMPLE 4. Let f be the function whose graph is shown in Fig. 5–7. To compute $\mathcal{L}[f]$ it is convenient to think of f as the sum of the functions

$$f_1(t) = 1, \quad t \geq 0,$$

$$f_2(t) = \begin{cases} 0, & t \leq 1, \\ 1 - t, & t > 1, \end{cases}$$

$$f_3(t) = \begin{cases} 0, & t \leq 2, \\ -(2 - t), & t > 2. \end{cases}$$

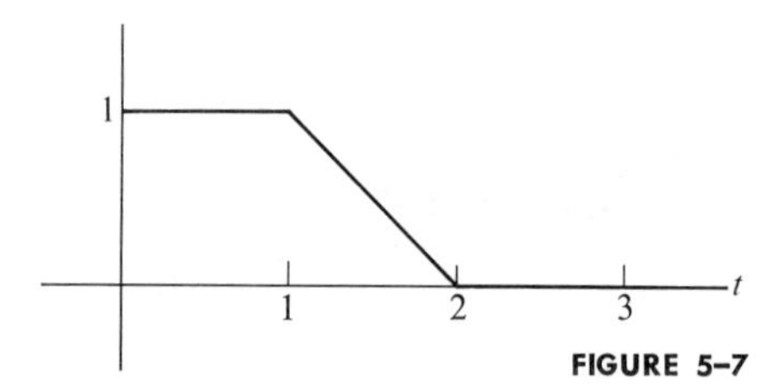

FIGURE 5–7

Then

$$\begin{aligned}\mathcal{L}[f] &= \mathcal{L}[f_1] + \mathcal{L}[f_2] + \mathcal{L}[f_3] \\ &= \frac{1}{s} + e^{-s}\mathcal{L}[-t] + e^{-2s}\mathcal{L}[t] \\ &= \frac{s + e^{-2s} - e^{-s}}{s^2}.\end{aligned}$$

EXAMPLE 5. Find $\mathcal{L}^{-1}[e^{-3s}/(s^2 + 6s + 10)]$.

Using (5–27) we have

$$\mathcal{L}^{-1}\left[\frac{e^{-3s}}{s^2 + 6s + 10}\right] = u_3(t)g(t - 3),$$

where $g(t) = \mathcal{L}^{-1}[1/(s^2 + 6s + 10)]$. But, by (5–21),

$$\mathcal{L}^{-1}\left[\frac{1}{s^2 + 6s + 10}\right] = \mathcal{L}^{-1}\left[\frac{1}{(s + 3)^2 + 1}\right] = e^{-3t}\mathcal{L}^{-1}\left[\frac{1}{s^2 + 1}\right] = e^{-3t} \sin t.$$

Hence

$$\mathcal{L}^{-1}\left[\frac{e^{-3s}}{s^2 + 6s + 10}\right] = u_3(t)e^{-3(t-3)} \sin (t - 3).$$

Our next result is somewhat akin to Theorem 5–6 in that it permits us to compute the Laplace transform of $t^n f(t)$ when $\mathcal{L}[f]$ is known. Specifically, we have

Theorem 5–8. *If* $\mathcal{L}[f] = \varphi(s)$, *then*

$$\mathcal{L}[t^n f(t)] = (-1)^n \frac{d^n}{ds^n} \varphi(s). \tag{5–28}$$

Proof. This result is established by differentiating both sides of the equation

$$\varphi(s) = \int_0^\infty e^{-st} f(t)\, dt$$

n times with respect to s. Thus

$$\begin{aligned}\frac{d}{ds}\varphi(s) &= \frac{d}{ds}\int_0^\infty e^{-st} f(t)\, dt \\ &= \int_0^\infty \frac{\partial}{\partial s}[e^{-st} f(t)]\, dt \\ &= -\int_0^\infty e^{-st} t f(t)\, dt \\ &= -\mathcal{L}[t f(t)],\end{aligned}$$

and so forth. (For a justification of this differentiation under the integral sign see Appendix I.) ▮

This time the companion formula phrased in terms of $\mathcal{L}^{-1}$ is

$$\mathcal{L}^{-1}\left[\frac{d^n}{ds^n}\varphi(s)\right] = (-1)^n t^n \mathcal{L}^{-1}[\varphi(s)]. \tag{5–29}$$

EXAMPLE 6.

$$\begin{aligned}\mathcal{L}[t \sin t] &= -\frac{d}{ds}\mathcal{L}[\sin t] \\ &= -\frac{d}{ds}\left(\frac{1}{s^2 + 1}\right) \\ &= \frac{2s}{(s^2 + 1)^2}\,.\end{aligned}$$

EXAMPLE 7.

$$\begin{aligned}\mathcal{L}[t^n] &= (-1)^n \frac{d^n}{ds^n} \mathcal{L}[1] \\ &= (-1)^n \frac{d^n}{ds^n}\left(\frac{1}{s}\right) \\ &= (-1)^n(-1)^n \frac{n!}{s^{n+1}} \\ &= \frac{n!}{s^{n+1}},\end{aligned}$$

which again proves (5–11).

EXAMPLE 8. Suppose we wish to compute $\mathcal{L}^{-1}[1/(s^2+1)^2]$. By comparing $1/(s^2+1)^2$ with $(d/ds)\big(1/(s^2+1)\big)$ we see that

$$\frac{1}{(s^2+1)^2} = -\frac{1}{2s}\frac{d}{ds}\left(\frac{1}{s^2+1}\right),$$

and applying Formula (5–18) we have

$$\mathcal{L}^{-1}\left[\frac{1}{(s^2+1)^2}\right] = -\tfrac{1}{2}\int_0^t \mathcal{L}^{-1}\left[\frac{d}{ds}\left(\frac{1}{s^2+1}\right)\right] dt.$$

But (5–29), with $n = 1$, yields

$$\mathcal{L}^{-1}\left[\frac{d}{ds}\left(\frac{1}{s^2+1}\right)\right] = -t\mathcal{L}^{-1}\left[\frac{1}{s^2+1}\right] = -t \sin t.$$

Hence

$$\begin{aligned}\mathcal{L}^{-1}\left[\frac{1}{(s^2+1)^2}\right] &= -\tfrac{1}{2}\int_0^t (-t\sin t)\, dt \\ &= \tfrac{1}{2}\int_0^t t \sin t \, dt \\ &= -\tfrac{1}{2}t\cos t + \tfrac{1}{2}\sin t.\end{aligned}$$

Our final formula is designed to reduce the computation of the Laplace transform of a periodic function* to the evaluation of an integral over a finite interval, and reads as follows:

Theorem 5–9. *If f is of exponential order and is periodic with period p, then*

$$\mathcal{L}[f] = \frac{\int_0^p e^{-st} f(t)\, dt}{1 - e^{-ps}}. \tag{5–30}$$

---

* A piecewise continuous function f is said to be *periodic* with period $p > 0$ if $f(t+p) = f(t)$ for all values of t.

Proof. By definition,

$$\begin{aligned}\mathcal{L}[f] &= \int_0^\infty e^{-st} f(t)\,dt \\ &= \int_0^p e^{-st} f(t)\,dt + \int_p^{2p} e^{-st} f(t)\,dt + \cdots \\ &\qquad + \int_{np}^{(n+1)p} e^{-st} f(t)\,dt + \cdots.\end{aligned}$$

We now set $x + np = t$ in the $(n + 1)$st integral of the above series to obtain

$$\begin{aligned}\int_{np}^{(n+1)p} e^{-st} f(t)\,dt &= \int_0^p e^{-s(x+np)} f(x + np)\,dx \\ &= e^{-nps} \int_0^p e^{-sx} f(x)\,dx,\end{aligned}$$

the last step following from the periodicity of f. Hence

$$\begin{aligned}\mathcal{L}[f] &= \int_0^p e^{-sx} f(x)\,dx + e^{-ps} \int_0^p e^{-sx} f(x)\,dx + \cdots \\ &\qquad + e^{-nps} \int_0^p e^{-sx} f(x)\,dx + \cdots \\ &= [1 + e^{-ps} + e^{-2ps} + \cdots] \int_0^p e^{-sx} f(x)\,dx.\end{aligned}$$

But the sum of the geometric series

$$1 + e^{-ps} + \cdots + e^{-nps} + \cdots$$

is $1/(1 - e^{-ps})$, and it follows that

$$\mathcal{L}[f] = \frac{\int_0^p e^{-st} f(t)\,dt}{1 - e^{-ps}},$$

as asserted. ▮

EXAMPLE 9. Find the Laplace transform of the function whose graph is shown in Fig. 5–8. In this case f is periodic with period 2, whence

$$\begin{aligned}\mathcal{L}[f] &= \frac{\int_0^2 e^{-st} f(t)\,dt}{1 - e^{-2s}} \\ &= \frac{\int_0^1 e^{-st}\,dt}{1 - e^{-2s}} \\ &= \frac{1 - e^{-s}}{s(1 - e^{-2s})} \\ &= \frac{1}{s(1 + e^{-s})}.\end{aligned}$$

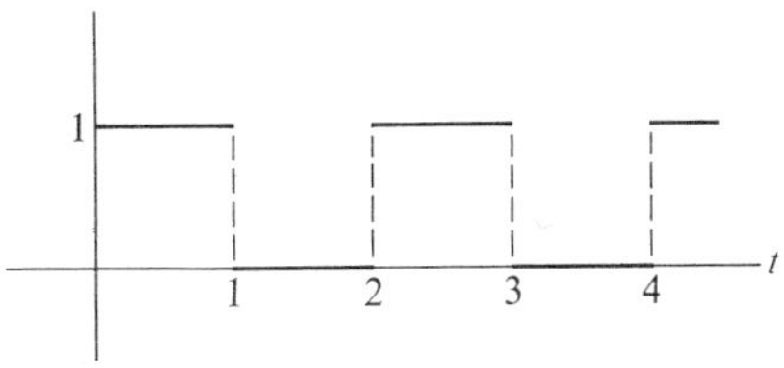

FIGURE 5–8

EXERCISES

Find the Laplace transform of each of the following functions.

1. $e^{2t}\sin 3t$
2. $3e^{-t}\cos 2t$
3. $t^3\sin 3t$
4. $t^2e^t\cos t$
5. $e^{-3t}\cos(2t+4)$
6. $te^t(d/dt)(\sin 2t)$
7. $te^{2t}f'(t)$
8. $(D^3+1)f(t)$, where $D = d/dt$
9. $f(t) = \begin{cases} 0, & t < \frac{1}{2} \\ 1+t, & t \ge \frac{1}{2} \end{cases}$
10. $f(t) = \begin{cases} t, & t < 2 \\ 2, & t > 2 \end{cases}$
11. $f(t) = \begin{cases} \sin t, & t < 2\pi \\ 0, & t > 2\pi \end{cases}$
12. $f(t) = \begin{cases} 0, & t < \dfrac{\pi}{2} \\ \cos t, & \dfrac{\pi}{2} < t < \dfrac{3\pi}{2} \\ 0, & t > \dfrac{3\pi}{2} \end{cases}$
13. $f(t) = \begin{cases} t, & t < 2 \\ 8-3t, & 2 \le t \le 3 \\ t-4, & 3 < t \le 4 \\ 0, & t > 4 \end{cases}$
14. $\sin t\cos t$
15. $2e^{2t}\sin t\cos t$
16. $\sin^2 t$ [*Hint:* What is the period of $\sin^2 t$?]
17. $|\sin t|$
18. $f(t)$, as shown in Fig.5–9
19. $\displaystyle\int_0^t te^{2t}\sin t\,dt$
20. $\displaystyle e^{-3t}\int_0^t t\cos 4t\,dt$
21. $\displaystyle t^2\int_1^t t\sin t\,dt$
22. $\displaystyle\frac{d^2}{dt^2}\int_0^t e^{-t}\cos 3t\,dt$
23. $\displaystyle te^t\int_a^t t\frac{d}{dt}(e^{2t}\sin t)\,dt$
24. Find the Laplace transform of the *staircase function*

$$f(t) = n+1,$$
$$n < t < n+1, \quad n = 0, 1, 2, \ldots,$$

shown in Fig. 5–10.

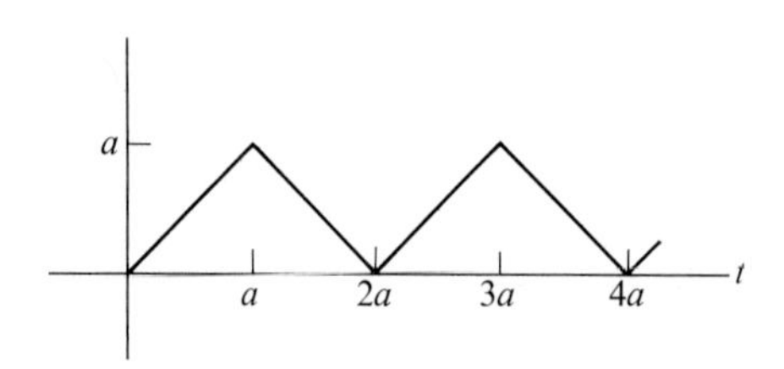

FIGURE 5–9

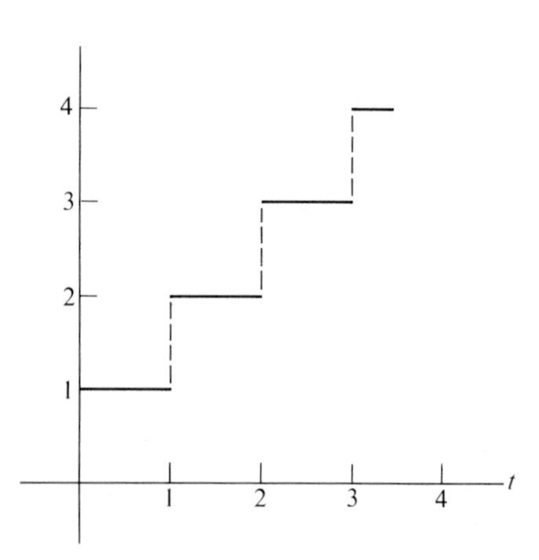

FIGURE 5–10

Find the inverse Laplace transform of each of the following functions.

25. $\dfrac{1}{s(s+1)}$ 26. $\dfrac{3}{(s-1)^2}$ 27. $\dfrac{1}{s(s+2)^2}$

28. $\dfrac{5}{s^2(s-5)^2}$ 29. $\dfrac{1}{(s-a)^n}$, $n \geq 1$

30. $\dfrac{1}{(s-a)(s-b)}$, a, b constants 31. $\dfrac{1}{s^2+4s+29}$

32. $\dfrac{2s}{2s^2+1}$ 33. $\dfrac{2s}{(s^2+1)^2}$ [*Hint*: See Example 8.]

*34. $\dfrac{1}{(s^2+4)^3}$ 35. $\dfrac{3s^2}{(s^2+1)^2}$ [*Hint*: First expand in partial fractions.]

*36. $\dfrac{2s^3}{(s^2+1)^3}$ 37. $\dfrac{3e^{-2s}}{3s^2+1}$ 38. $\dfrac{e^{-s}}{s^4+1}$

39. $\dfrac{1}{s^4+1}$ [*Hint*: $s^4 + 1 \equiv (s^4 + 2s^2 + 1) - 2s^2$.]

40. $\dfrac{3s}{(s+1)^4}$ [*Hint*: See Exercise 35.]

41. $\dfrac{1+e^{-s}}{s}$ 42. $\dfrac{e^{-s}}{(s-1)(s-2)}$

43. $\dfrac{1}{s^3+a^3}$ 44. $\dfrac{s^2}{s^3+a^3}$

45. $\ln \dfrac{s+3}{s+2}$ [*Hint*: $\dfrac{d}{ds} \ln \dfrac{s+3}{s+2} = -\dfrac{1}{(s+2)(s+3)}$. Apply Formula (5–29).]

46. $\ln \dfrac{s^2+1}{s(s+3)}$

*47. Describe a procedure for finding the inverse Laplace transform of any rational function $P(s)/Q(s)$, P, Q polynomials with degree $P <$ degree Q. In particular, show that it is sufficient to consider the special cases

$$\frac{1}{(s^2+a^2)^n}, \qquad \frac{s}{(s^2+a^2)^n}, \qquad \frac{1}{(s+a)^n},$$

where a is a constant and n a positive integer.

*48. Show that for any integer $n \geq 1$, $a \neq 0$,

$$\mathcal{L}^{-1}\left[\frac{1}{(s^2+a^2)^{n+1}}\right] = \frac{1}{2n}\int_0^t t\mathcal{L}^{-1}\left[\frac{1}{(s^2+a^2)^n}\right] dt.$$

*49. (a) Use the result of Exercise 48 to show that

$$\mathcal{L}^{-1}\left[\frac{1}{(s^2+a^2)^{n+1}}\right] = \frac{1}{2^n a n!}\underbrace{\int_0^t t \int_0^t t \cdots \int_0^t}_{n \text{ times}} t \sin at\, dt\, dt \cdots dt.$$

(b) Derive a similar formula for $\mathcal{L}^{-1}[s/(s^2+a^2)^{n+1}]$.

5-6 THE LAPLACE TRANSFORM AND DIFFERENTIAL EQUATIONS

The use of the Laplace transform in solving initial-value problems was introduced and justified in Section 5-4. In this section we shall illustrate how the formulas just derived allow us to solve more elaborate problems.

EXAMPLE 1. Find the solution of

$$\begin{aligned} y'' + 4y' + 13y &= 2t + 3e^{-2t}\cos 3t, \\ y(0) = 0, \quad & y'(0) = -1. \end{aligned} \tag{5-31}$$

Taking the Laplace transform of both sides of this equation, and applying the given initial conditions, we obtain

$$s^2\mathcal{L}[y] + 1 + 4s\mathcal{L}[y] + 13\mathcal{L}[y] = \frac{2}{s^2} + \frac{3(s+2)}{(s+2)^2 + 9}.$$

Hence

$$\mathcal{L}[y] = -\frac{1}{s^2 + 4s + 13} + \frac{2}{s^2(s^2 + 4s + 13)} + \frac{3(s+2)}{(s^2 + 4s + 13)^2},$$

and we must now find the inverse transform of the various terms on the right-hand side of this equation. The first can be disposed of without difficulty since

$$\frac{1}{s^2 + 4s + 13} = \frac{1}{(s+2)^2 + 9} = \frac{1}{3}\left(\frac{3}{(s+2)^2 + 9}\right).$$

Hence

$$\mathcal{L}^{-1}\left[-\frac{1}{s^2 + 4s + 13}\right] = -\tfrac{1}{3}e^{-2t}\sin 3t.$$

To handle the second, we use the method of partial fractions, as follows:

$$\frac{2}{s^2(s^2 + 4s + 13)} = \frac{A}{s} + \frac{B}{s^2} + \frac{Cs + D}{s^2 + 4s + 13},$$

whence

$$As(s^2 + 4s + 13) + B(s^2 + 4s + 13) + (Cs + D)s^2 = 2.$$

In order that this equation hold identically in s, we must have

$$\begin{aligned} A + C &= 0, \\ 4A + B + D &= 0, \\ 13A + 4B &= 0, \\ 13B &= 2, \end{aligned}$$

and it follows that $A = -\frac{8}{169}$, $B = \frac{2}{13}$, $C = \frac{8}{169}$, $D = \frac{6}{169}$. Thus

$$\begin{aligned}\frac{2}{s^2(s^2+4s+13)} &= -\frac{8}{169}\left(\frac{1}{s}\right) + \frac{2}{13}\left(\frac{1}{s^2}\right) + \frac{1}{169}\left(\frac{8s+6}{s^2+4s+13}\right)\\ &= -\frac{8}{169}\left(\frac{1}{s}\right) + \frac{2}{13}\left(\frac{1}{s^2}\right) + \frac{8}{169}\left(\frac{s+2}{(s+2)^2+9}\right)\\ &\qquad - \frac{10}{3(169)}\left(\frac{3}{(s+2)^2+9}\right),\end{aligned}$$

and

$$\mathcal{L}^{-1}\left[\frac{2}{s^2(s^2+4s+13)}\right] = -\tfrac{8}{169} + \tfrac{2}{13}t + \tfrac{8}{169}e^{-2t}\cos 3t - \tfrac{10}{507}e^{-2t}\sin 3t.$$

Finally, since

$$\begin{aligned}\frac{3(s+2)}{(s^2+4s+13)^2} &= -\frac{3}{2}\frac{d}{ds}\left(\frac{1}{s^2+4s+13}\right)\\ &= -\frac{1}{2}\frac{d}{ds}\left(\frac{3}{(s+2)^2+9}\right),\end{aligned}$$

we can apply Formulas 5–21 and 5–29 to obtain

$$\mathcal{L}^{-1}\left[\frac{3(s+2)}{(s^2+4s+13)^2}\right] = \tfrac{1}{2}te^{-2t}\sin 3t.$$

Combining these results we see that the solution of (5–31) is

$$y = -\tfrac{179}{507}e^{-2t}\sin 3t + \tfrac{8}{169}e^{-2t}\cos 3t + \tfrac{1}{2}te^{-2t}\sin 3t + \tfrac{2}{13}t - \tfrac{8}{169}.$$

(The student who feels overwhelmed by the details of these computations is urged to compare them with those involved in the method of variation of parameters or undetermined coefficients. He will find that the argument given above is simpler and much more straightforward than either of the other two.)

EXAMPLE 2. In this example we show how the method of Laplace transforms can be used to find the *general solution* of a differential equation. As an illustration we again solve the equation

$$(D - \alpha)^2 y = 0, \tag{5–32}$$

where α is an arbitrary real number.

In the absence of specific initial conditions we choose them arbitrarily as $y(0) = c_1$, $y'(0) = c_2$. Taking Laplace transforms of both sides of (5–32) and applying our initial conditions, we have

$$s^2\mathcal{L}[y] - c_1 s - c_2 - 2\alpha(s\mathcal{L}[y] - c_1) + \alpha^2\mathcal{L}[y] = 0.$$

Hence

$$\begin{aligned}\mathcal{L}[y] &= \frac{c_1 s + c_2 + 2\alpha c_1}{s^2 - 2\alpha s + \alpha^2} \\ &= \frac{c_1 s - c_1\alpha}{(s-\alpha)^2} + \frac{c_2 + 3c_1\alpha}{(s-\alpha)^2} \\ &= \frac{c_1}{s-\alpha} + (c_2 + 3c_1\alpha)\frac{1}{(s-\alpha)^2}.\end{aligned}$$

But

$$\mathcal{L}^{-1}\left[\frac{1}{s-\alpha}\right] = e^{\alpha t}, \quad \text{and} \quad \mathcal{L}^{-1}\left[\frac{1}{(s-\alpha)^2}\right] = te^{\alpha t}.$$

Hence

$$\begin{aligned} y &= c_1 e^{\alpha t} + (c_2 + 3c_1\alpha)te^{\alpha t} \\ &= c_3 e^{\alpha t} + c_4 t e^{\alpha t}, \end{aligned}$$

where c_3, c_4 are still arbitrary, is the general solution of (5–32).

EXERCISES

Solve each of the following initial-value problems using Laplace transforms.

1. $(D^2 + 2D + 1)y = e^t$; $y(0) = y'(0) = 0$.
2. $\dfrac{dy}{dt} + 3y = t \sin at$; $y(0) = -1$.
3. $\dfrac{d^2y}{dt^2} + 2\dfrac{dy}{dt} + 3y = 3t$; $y(0) = 0$, $y'(0) = 1$.
4. $(D^2 - 4D + 4)y = 2e^{2t} + \cos t$; $y(0) = \frac{3}{25}$, $y'(0) = -\frac{4}{25}$.

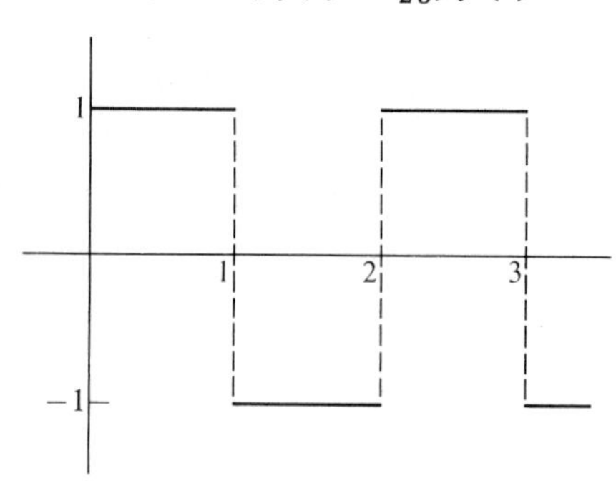

FIGURE 5–11

5. $\dfrac{dy}{dt} + ky = h(t)$; $y(0) = 0$, with k a constant and the graph of $h(t)$ given by Fig. 5–11.
6. $\dfrac{d^2y}{dt^2} - 2\dfrac{dy}{dt} + y = te^t \sin t$; $y(0) = y'(0) = 0$.

7. $\dfrac{d^3y}{dt^3} - \dfrac{d^2y}{dt^2} + 4\dfrac{dy}{dt} - 4y = -3e^t + 4e^{2t}$; $y(0) = 0$, $y'(0) = 5$, $y''(0) = 3$.

8. $\dfrac{d^4y}{dt^4} + 3\dfrac{d^3y}{dt^3} + \dfrac{d^2y}{dt^2} - 3\dfrac{dy}{dt} - 2y = t$; $y(0) = y'(0) = y''(0) = y'''(0) = 0$.

9. $\dfrac{d^2y}{dt^2} + 4\dfrac{dy}{dt} + 4y = \begin{cases} 0, & t \leq 2; \\ e^{-(t-2)}, & t > 2; \end{cases}$ $y(0) = 1$, $y'(0) = -1$.

10. $\dfrac{d^2y}{dt^2} + y = t^2 + 1$; $y(\pi) = \pi^2$, $y'(\pi) = 2\pi$. [*Hint*: First make the substitution $x = t - \pi$.]

11. $\dfrac{d^2y}{dt^2} - y = -10\sin 2t$; $y(\pi) = -1$, $y'(\pi) = 0$. [*Hint*: See Exercise 10.]

12. $\dfrac{d^4y}{dt^4} + y = \begin{cases} 0, & t \leq 1; \\ t - 1, & t > 1; \end{cases}$ $y(1) = y'(1) = 1$, $y''(1) = y'''(1) = 0$.

[*Hint*: See Exercise 10.]

13. Use Laplace transforms to solve the equation

$$\frac{dy}{dt} + 2y + \int_0^t y(t)\,dt = \begin{cases} t, & t < 1, \\ 2 - t, & 1 \leq t \leq 2, \\ 0, & t > 2, \end{cases}$$

subject to the initial condition $y(0) = 1$.

14. (a) Suppose that $y(t)$ is of exponential order and is a solution of the Euler equation

$$t^2\frac{d^2y}{dt^2} + at\frac{dy}{dt} + by = 0,$$

a, b constants. Show that $\mathcal{L}[y(t)]$ also satisfies an equation of the Euler type.

(b) Prove that the result in (a) is also valid for any solution (of exponential order) of an Euler equation of order n.

15. Show that if $f(t)$ and $f'(t)$ are of exponential order, and if $f(t)$ is continuous for all $t > 0$, then

$$\lim_{s\to\infty} s\mathcal{L}[f] = f(0^+).$$

*16. Bessel's function of the first kind of order zero, denoted J_0, is by definition that solution of the differential equation

$$t\frac{d^2y}{dt^2} + \frac{dy}{dt} + ty = 0$$

which is defined for $t = 0$ and satisfies $J_0(0) = 1$. Prove that

$$\mathcal{L}[J_0(t)] = \frac{1}{\sqrt{1 + s^2}}.$$

[*Hint:* Show that $\mathcal{L}[J_0]$ is a solution of the differential equation

$$(1 + s^2)\varphi'(s) + s\varphi(s) = 0,$$

and use Exercise 15. Assume J_0 is of exponential order.]

*17. By expanding $s\mathcal{L}[J_0] = s(s^2 + 1)^{-1/2}$ (see Exercise 16) in a binomial series, express $J_0(t)$ as a power series.

5–7 THE CONVOLUTION THEOREM

In this section we shall establish the most important single property of the Laplace transform, the so-called *convolution formula.* Far from being just a computational device, as were the results of Section 5–5, the convolution formula plays an important role in certain theoretical investigations in advanced analysis. And in the following pages we shall find that it is also ideally suited to the task of constructing inverses for linear differential operators with constant coefficients.

We prove the formula in question as

Theorem 5–10. *Let f and g be piecewise continuous functions of exponential order, and suppose that*

$$\mathcal{L}[f] = \varphi(s), \qquad \mathcal{L}[g] = \psi(s).$$

Then

$$\mathcal{L}\left[\int_0^t f(t - \xi)g(\xi)\,d\xi\right] = \varphi(s)\psi(s). \tag{5–33}$$

When written in terms of inverse transforms, (5–33) becomes

$$\mathcal{L}^{-1}[\varphi(s)\psi(s)] = \int_0^t f(t - \xi)g(\xi)\,d\xi, \tag{5–34}$$

and in this form asserts that if we know the inverse transforms, f and g, of the functions φ and ψ, we can express the inverse transform of the product $\varphi(s)\psi(s)$ as an integral involving f and g. The integral in question is called the *convolution* of f and g and is denoted $f * g$; that is,

$$(f * g)(t) = \int_0^t f(t - \xi)g(\xi)\,d\xi. \tag{5–35}$$

Using this notation, (5–34) can be written

$$f * g = \mathcal{L}^{-1}[\varphi(s)\psi(s)],$$

or, even more suggestively,

$$\mathcal{L}^{-1}[\varphi(s)\psi(s)] = \mathcal{L}^{-1}[\varphi(s)] * \mathcal{L}^{-1}[\psi(s)]. \tag{5–36}$$

Proof of the theorem. Using the definition of $\mathcal{L}$ we have

$$\mathcal{L}\left[\int_0^t f(t-\xi)g(\xi)\,d\xi\right] = \int_0^\infty e^{-st}\int_0^t f(t-\xi)g(\xi)\,d\xi\,dt$$
$$= \int_0^\infty \int_0^t e^{-st} f(t-\xi)g(\xi)\,d\xi\,dt,$$

where the integration is being performed over the region of the $t\xi$-plane described by the inequalities

$$0 \le \xi \le t, \qquad 0 \le t < \infty$$

(see Fig. 5–12). But this region is also described by

$$\xi \le t < \infty, \qquad 0 \le \xi < \infty,$$

and hence the above iterated integral may be written

$$\int_0^\infty \int_\xi^\infty e^{-st} f(t-\xi)g(\xi)\,dt\,d\xi,$$

or

$$\int_0^\infty g(\xi)\left(\int_\xi^\infty e^{-st} f(t-\xi)\,dt\right)d\xi.\dagger$$

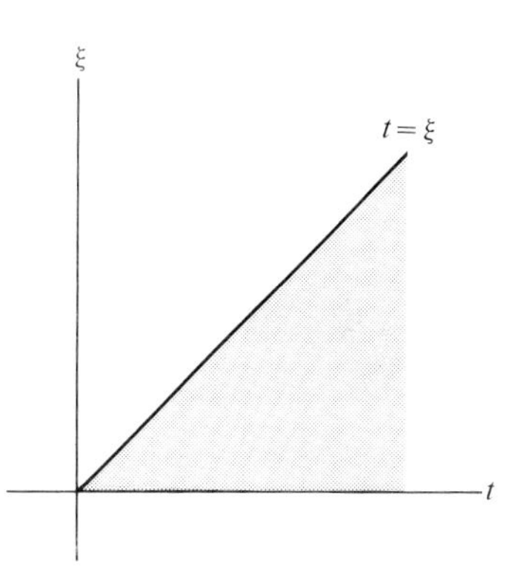

FIGURE 5–12

We now make the change of variable $u = t - \xi$ in $\int_\xi^\infty e^{-st} f(t-\xi)\,dt$ and obtain

$$\int_\xi^\infty e^{-st} f(t-\xi)\,dt = \int_0^\infty e^{-s(u+\xi)} f(u)\,du.$$

Hence

$$\mathcal{L}\left[\int_0^t f(t-\xi)g(\xi)\,d\xi\right] = \int_0^\infty g(\xi)\left(\int_0^\infty e^{-s(u+\xi)} f(u)\,du\right)d\xi$$
$$= \int_0^\infty e^{-s\xi} g(\xi)\left(\int_0^\infty e^{-su} f(u)\,du\right)d\xi$$
$$= \int_0^\infty e^{-su} f(u)\,du \int_0^\infty e^{-s\xi} g(\xi)\,d\xi$$
$$= \mathcal{L}[f]\mathcal{L}[g],$$

and the proof is complete. ▌

The operation of convolution can be viewed as a multiplication on the vector space $\mathcal{E}$, with $f * g$ as the product of f and g, and it is of some interest to determine its properties. For instance, Formula (5–34) and the equality $\mathcal{L}[f]\mathcal{L}[g] = \mathcal{L}[g]\mathcal{L}[f]$

† It can be shown that an interchange of the order of integration is permissible at this point. See Appendix I.

imply at once that

$$\int_0^t f(t - \xi)g(\xi)\,d\xi = \int_0^t g(t - \xi)f(\xi)\,d\xi.$$

Hence $f * g = g * f$, and convolution is a commutative operation. It is also associative and distributive; that is

$$f * (g * h) = (f * g) * h \qquad \text{and} \qquad f * (g + h) = f * g + f * h$$

(see Exercises 14 and 15 below), and thus defines a very reasonable multiplication on $\mathcal{E}$. Much of the advanced work connected with the convolution integral is devoted to studying the behavior of $\mathcal{E}$ under this multiplication.

EXAMPLE 1. Find

$$\mathcal{L}^{-1}\left[\frac{1}{s(s^2 + 1)}\right].$$

It is clear that this problem can be solved by separating $1/(s(s^2 + 1))$ into partial fractions as $(1/s) - (s/(s^2 + 1))$ and applying our earlier formulas. But it can also be handled just as well by using (5–35) and (5–36) as follows:

$$\begin{aligned}\mathcal{L}^{-1}\left[\frac{1}{s(s^2 + 1)}\right] &= \mathcal{L}^{-1}\left[\frac{1}{s}\right] * \mathcal{L}^{-1}\left[\frac{1}{s^2 + 1}\right]\\ &= 1 * \sin t\\ &= \int_0^t \sin \xi\,d\xi\\ &= 1 - \cos t.\end{aligned}$$

EXAMPLE 2. Find the solution of the initial-value problem

$$\begin{aligned}(D^2 + D - 6)y &= h(t),\\ y(0) = y'(0) &= 0,\end{aligned} \tag{5–37}$$

given that $h(t)$ is a function of exponential order.

We apply the operator $\mathcal{L}$ to the given equation and obtain

$$(s^2 + s - 6)\mathcal{L}[y] = \mathcal{L}[h],$$

or

$$\mathcal{L}[y] = \frac{\mathcal{L}[h]}{s^2 + s - 6},$$

from which it follows that

$$y = \mathcal{L}^{-1}\left[\frac{1}{(s - 2)(s + 3)}\right] * h(t).$$

But since

$$\mathcal{L}^{-1}\left[\frac{1}{(s-2)(s+3)}\right] = \mathcal{L}^{-1}\left[\frac{1}{5(s-2)} - \frac{1}{5(s+3)}\right]$$
$$= \tfrac{1}{5}e^{2t} - \tfrac{1}{5}e^{-3t},$$

we have

$$y = \int_0^t [\tfrac{1}{5}e^{2(t-\xi)} - \tfrac{1}{5}e^{-3(t-\xi)}]h(\xi)\,d\xi, \tag{5–38}$$

and the problem has been solved pending explicit knowledge of h.

If, for example, $h(t) = 1$, then

$$y = \int_0^t [\tfrac{1}{5}e^{2(t-\xi)} - \tfrac{1}{5}e^{-3(t-\xi)}]\,d\xi$$
$$= \tfrac{1}{5}e^{2t}\int_0^t e^{-2\xi}\,d\xi - \tfrac{1}{5}e^{-3t}\int_0^t e^{3\xi}\,d\xi$$
$$= \tfrac{1}{10}e^{2t} + \tfrac{1}{15}e^{-3t} - \tfrac{1}{6}.$$

The reader will recall from Section 4–5 that the function $K(t, \xi) = \frac{1}{5}e^{2(t-\xi)} - \frac{1}{5}e^{-3(t-\xi)}$ appearing in (5–38) is called the Green's function for the linear differential operator $L = D^2 + D - 6$ (for initial-value problems at $t = 0$), and as such completely determines an inverse for L. In the next section we shall have more to say about the simple formula

$$G(t, \xi) = \mathcal{L}^{-1}\left[\frac{1}{s^2 + s - 6}\right](t - \xi),$$

which allows us to compute this Green's function by Laplace transform methods.

EXERCISES

Use the convolution formula to find the inverse Laplace transform of each of the following functions.

1. $\dfrac{\mathcal{L}[f]}{s^2 + 1}$
2. $\dfrac{e^{-3s}\mathcal{L}[f]}{s^3}$
3. $\dfrac{1}{s^2(s + 1)}$
4. $\dfrac{s}{(s^2 + 1)^2}$
5. $\dfrac{3s^2}{(s^2 + 1)^2}$
6. $\dfrac{1}{(s - a)(s - b)}$, $a \neq b$

Evaluate each of the following.

7. $e^{at} * e^{bt}$
8. $t * \cos at$
9. $\sin at * \cos bt$
10. $t * e^{at}$
11. $f(t - 1) * e^{-t}g(t + 1)$
12. $f(-t) * (\sin t)g(t^2)$

13. Prove directly that $f * g = g * f$, that is,

$$\int_0^t f(t - \xi)g(\xi)\, d\xi = \int_0^t g(t - \xi)f(\xi)\, d\xi.$$

[*Hint:* Make the substitution $u = t - \xi$.]

14. Prove that $f * (g + h) = f * g + f * h$.
15. Prove that $f * (g * h) = (f * g) * h$.
16. Find $1 * 1$ and $1 * 1 * 1$.
17. Derive a formula for $1 * 1 * 1 * \cdots * 1$ (n factors).
18. Prove that

$$\int_0^t \sin a(t - u_1) \int_0^{u_1} \sin a(u_1 - u_2) \int_0^{u_2} \sin a(u_2 - u_3) \cdots$$
$$\int_0^{u_{n-1}} \sin a(u_{n-1} - u_n) \sin au_n\, du_n \cdots du_1$$
$$= \frac{a^n}{2^n n!} \underbrace{\int_0^t t \int_0^t t \int_0^t t \cdots \int_0^t}_{n \text{ times}} t \sin at\, dt \cdots dt.$$

[*Hint:* Compare $\mathcal{L}^{-1}[1/(s^2 + a^2)^{n+1}]$ as computed by the formula of Exercise 49, Section 5–5, and as computed by repeated use of the convolution integral.]

19. Suppose that $f(t)$ is of exponential order and that $\lim_{t \to 0^+} f(t)/t$ exists. Assuming that the order of integration may be reversed in the computation, prove that

$$\mathcal{L}\left[\frac{f(t)}{t}\right] = \int_s^\infty \mathcal{L}[f]\, ds.$$

Use the result of Exercise 19 to compute the Laplace transform of each of the following functions.

20. $t \left(= \frac{t^2}{t}, \text{ with } \mathcal{L}[t^2] = \frac{2}{s^3} \right)$

21. $\frac{\sin at}{t}$

22. $\frac{e^t - 1}{t}$

23. $\frac{1 - \cos 3t}{t^2}$

24. (a) It can be proved that whenever $\int_0^\infty [f(t)/t]\, dt$ exists, the formula in Exercise 19 remains valid with s set equal to zero. Show that we then obtain

$$\int_0^\infty \frac{f(t)}{t}\, dt = \int_0^\infty \mathcal{L}[f]\, ds.$$

(b) Use this formula to prove that

$$\int_0^\infty \frac{\sin t}{t}\, dt = \frac{\pi}{2}.$$

The *Gamma function* $\Gamma(x)$ and *Beta function* $\mathrm{B}(x, y)$ are introduced in Exercises 25–33.

25. The gamma function is defined by the equation

$$\Gamma(x) = \int_0^\infty e^{-t} t^{x-1}\, dt. \tag{5-39}$$

(It can be proved that this integral converges for all $x > 0$.) Use integration by parts to show that

$$\Gamma(x + 1) = x\Gamma(x) \quad \text{for all } x > 0.$$

26. Use the result of Exercise 25 together with the value of $\Gamma(1)$ to prove that

$$\Gamma(n + 1) = n!, \quad n = 0, 1, 2, \ldots.$$

(Because of this property the gamma function is also called the *generalized factorial function.*)

27. By differentiation of (5-39), show that $\Gamma''(x)$ is non-negative on $(0, \infty)$. Where, approximately, does the minimum of $\Gamma(x)$ lie? Draw an accurate graph of $\Gamma(x)$.

28. Let n be an arbitrary real number greater than -1. Prove that

$$\mathcal{L}[t^n] = \frac{\Gamma(n + 1)}{s^{n+1}}.$$

[*Hint:* Set $t = u/s$ in the integral defining $\mathcal{L}[t^n]$.]

29. Prove that $\Gamma(\frac{1}{2}) = \sqrt{\pi}$. [*Hint*: Show that

$$\Gamma(\tfrac{1}{2}) = 2\int_0^\infty e^{-u^2}\, du,$$

and hence that

$$\left[\Gamma\left(\frac{1}{2}\right)\right]^2 = 4\int_0^\infty \int_0^\infty e^{-(u^2+v^2)}\, du\, dv.$$

Now evaluate this integral by changing to polar coordinates.]

30. Find the value of $\Gamma(\frac{5}{2})$. (See Exercise 29.)

31. Evaluate each of the following integrals.

(a) $\displaystyle\int_0^\infty \frac{e^{-x}}{\sqrt{x}}\, dx$ (b) $\displaystyle\int_0^\infty e^{-\sqrt{x}}\, dx$

32. Prove that

$$\frac{\Gamma(x)\Gamma(y)}{\Gamma(x + y)} = \int_0^1 u^{x-1}(1 - u)^{y-1}\, du$$

whenever $x, y > 0$. This integral is known as the *Beta function* of x and y and is denoted $B(x, y)$; that is,

$$B(x, y) = \int_0^1 u^{x-1}(1 - u)^{y-1}\, du.$$

[*Hint:* Use the result of Exercise 28 and the convolution formula to evaluate $\mathcal{L}^{-1}[1/s^{x+y}]$ in two ways.]

33. Use Exercise 32 to prove that

$$\int_0^1 \frac{dx}{\sqrt{1 - x^n}} = \frac{\sqrt{\pi}}{n} \frac{\Gamma(1/n)}{\Gamma[(n + 2)/(2n)]}.$$

*5–8 GREEN'S FUNCTIONS FOR CONSTANT COEFFICIENT LINEAR DIFFERENTIAL OPERATORS

We begin this section by recalling a number of facts concerning linear differential operators which were established in Chapters 3 and 4.

Let $L = D^n + a_{n-1}(t)D^{n-1} + \cdots + a_0(t)$ be a (normal) linear differential operator whose coefficients are continuous on an interval I, let h be continuous on I, and consider the equation

$$Ly = h(t). \tag{5–40}$$

Then L can be viewed as a linear transformation from $\mathcal{C}^n(I)$ to $\mathcal{C}(I)$, and a solution of (5–40) can be described as a linear transformation G from $\mathcal{C}(I)$ to $\mathcal{C}^n(I)$ which acts as a *right inverse* for L, i.e., is such that $L(G(h)) = h$ for all h in $\mathcal{C}(I)$. In the absence of any further conditions on the unknown y, G is not uniquely determined by L since L is not one-to-one. Thus to construct G we must impose a sufficient number of additional restrictions on y so as to make the solution of (5–40) unique. For example, we might require, as in Section 4–4, that y be *the* solution of (5–40) which satisfies the complete set of initial conditions

$$y(t_0) = y'(t_0) = \cdots = y^{(n-1)}(t_0) = 0 \tag{5–41}$$

at some point t_0 in the interval I. Then the corresponding inverse G can be expressed as an integral operator

$$G(h) = \int_{t_0}^{t} K(t, \xi)h(\xi)\, d\xi, \tag{5–42}$$

where the function $K(t, \xi)$ may be constructed by the method of variation of parameters from a basis for the solution space of $Ly = 0$ [see Formulas (4–30) and (4–42)]. This function is called the *Green's function for L for initial value problems on I*, and is completely determined by L since we saw that it is independent of the point t_0 at which the initial conditions were imposed and of the particular basis chosen for the solution space of $Ly = 0$.

In this section we shall rederive several of the above results in the case of a *constant coefficient operator L* by means of the convolution formula. One of the principal advantages of this method is that it allows us to bypass the explicit use of a basis for the solution space of $Ly = 0$, and eliminates the tedious computations involved in the method of variation of parameters. As such it furnishes an excellent example of the efficiency that accrues from using the "coordinate free" ideas of linear algebra.

Thus, for the remainder of this discussion we shall consider the constant coefficient linear differential operator

$$L = D^n + a_{n-1}D^{n-1} + \cdots + a_0, \tag{5–43}$$

and we begin by solving the initial-value problem

$$\begin{gathered} Ly = h(t), \\ y(0) = y'(0) = \cdots = y^{(n-1)}(0) = 0, \end{gathered} \tag{5–44}$$

where $h(t)$ is a function of exponential order. Applying the Laplace transform $\mathcal{L}$ to (5–44), we obtain the equation

$$p(s)\mathcal{L}[y] = \bar{h}(s),$$

where $p(s) = s^n + a_{n-1}s^{n-1} + \cdots + a_0$, and $\bar{h}(s) = \mathcal{L}[h]$. Thus

$$\mathcal{L}[y] = \frac{\bar{h}(s)}{p(s)},$$

and hence if

$$\mathcal{L}^{-1}\left[\frac{1}{p(s)}\right] = g(t),$$

the convolution formula yields

$$y(t) = \int_0^t g(t - \xi)h(\xi)\, d\xi \tag{5–45}$$

as the desired solution of (5–44).

We can now view (5–45) as determining, in the usual way, a mapping from $\mathcal{C}[0, \infty)$ to $\mathcal{C}^n[0, \infty)$, and we have therefore obtained a right inverse G for the operator (5–43) on this interval.* The defining equation for G is

$$G(h) = \int_0^t g(t - \xi)h(\xi)\, d\xi, \tag{5–46}$$

and the function $g(t - \xi)$ appearing in this integral is then, by definition, a Green's function for L for initial-value problems at $t = 0$ (see Definition 4–2).

Fortunately we can say much more about $g(t - \xi)$. For the function

$$g(t) = \mathcal{L}^{-1}\left[\frac{1}{p(s)}\right]$$

---

* In Exercise 14 it is shown that (5–46) can be derived, using Laplace transforms, without the assumption that $h(t)$ is of exponential order.

is the *unique* solution on $[0, \infty)$ of the initial-value problem

$$\begin{gathered} Ly = 0, \\ y(0) = y'(0) = \cdots = y^{(n-2)}(0) = 0, \\ y^{(n-1)}(0) = 1, \end{gathered} \tag{5–47}$$

as is easily verified by using Laplace transform methods to solve (5–47). (See Exercise 12.) Moreover, the usual techniques of Sections 5–4 and 5–5 for computing $\mathcal{L}^{-1}[1/p(s)]$ lead to a function $K(t)$, which extends $g(t)$ to the unique solution of (5–47) on all of $(-\infty, \infty)$. Thus by Theorem 4–3, $K(t - \xi)$ is *the* Green's function for L (for initial-value problems) on the entire interval $(-\infty, \infty)$. When Laplace transforms are used in a purely computational fashion it is common practice to ignore the distinction between $K(t)$ and $g(t)$, as we do, for example, in stating the next theorem which summarizes the above results.

Theorem 5–11. *Let L be the constant coefficient operator* (5–43), *and let*

$$p(s) = s^n + a_{n-1}s^{n-1} + \cdots + a_0$$

be the auxiliary polynomial of L. Then if

$$g(t) = \mathcal{L}^{-1}\left[\frac{1}{p(s)}\right],$$

the function $g(t - \xi)$ *is the Green's function for L on the interval* $(-\infty, \infty)$.

EXAMPLE 1. To find a Green's function for the operator

$$L = D^2 - 2aD + a^2 + b^2, \quad b \neq 0,$$

we first set

$$\begin{aligned} g(t) &= \mathcal{L}^{-1}\left[\frac{1}{s^2 - 2as + a^2 + b^2}\right] \\ &= \mathcal{L}^{-1}\left[\frac{1}{(s - a)^2 + b^2}\right] \\ &= \frac{1}{b} e^{at} \sin bt. \end{aligned}$$

Then, by the above theorem, the desired Green's function is

$$g(t - \xi) = \frac{1}{b} e^{a(t-\xi)} \sin b(t - \xi).$$

EXAMPLE 2. Find the particular solution of the differential equation

$$(4D^4 - 4D^3 + 5D^2 - 4D + 1)y = \ln t$$

which satisfies the initial conditions $y(1) = y'(1) = y''(1) = y'''(1) = 0$.

We first use Theorem 5–11 to compute the Green's function for

$$L = D^4 - D^3 + \tfrac{5}{4}D^2 - D + \tfrac{1}{4}.$$

Since

$$\begin{aligned} g(t) &= \mathcal{L}^{-1}\left[\frac{1}{s^4 - s^3 + \frac{5}{4}s^2 - s + \frac{1}{4}}\right] \\ &= \mathcal{L}^{-1}\left[\frac{4}{(2s-1)^2(s^2+1)}\right] \\ &= \mathcal{L}^{-1}\left[\frac{\frac{16}{5}}{(2s-1)^2} - \frac{\frac{32}{25}}{2s-1} + \frac{\frac{16}{25}s}{s^2+1} - \frac{\frac{12}{25}}{s^2+1}\right] \\ &= \tfrac{4}{5}te^{t/2} - \tfrac{16}{25}e^{t/2} + \tfrac{16}{25}\cos t - \tfrac{12}{25}\sin t, \end{aligned}$$

the desired Green's function is

$$g(t-\xi) = \tfrac{4}{5}(t-\xi)e^{(t-\xi)/2} - \tfrac{16}{25}e^{(t-\xi)/2} + \tfrac{16}{25}\cos(t-\xi) - \tfrac{12}{25}\sin(t-\xi).$$

The solution of the given initial-value problem can now be written in the form

$$y(t) = \int_1^t g(t-\xi)\frac{\ln \xi}{4}\,d\xi,$$

since $h(t) = (\ln t)/4$ is the right-hand side of the *normalized* differential equation $(D^4 - D^3 + \frac{5}{4}D^2 - D + \frac{1}{4})y = \frac{1}{4}\ln t$.

As a final example let us use Green's functions and Laplace transform methods to solve an initial-value problem with *nonhomogeneous* initial conditions, say,

$$\begin{gathered} Ly = h(t); \\ y(t_0) = c_0, \quad \ldots, \quad y^{(n-1)}(t_0) = c_{n-1}. \end{gathered} \tag{5–48}$$

Following the methods of Chapter 4 we can write the solution of (5–48) in the form

$$y = y_h + G(h),$$

with $G(h)$ as above, and y_h the solution of the *homogeneous* equation $Ly = 0$ which satisfies the initial conditions of (5–48).

If, for example, (5–48) is

$$\begin{gathered} (D^2 - 2aD + a^2 + b^2)y = h(t), \\ y(\pi) = 2, \qquad y'(\pi) = -3, \end{gathered} \tag{5–49}$$

then, using the Green's function for $D^2 - 2aD + a^2 + b^2$ obtained in Example 1, we have

$$G(h) = \frac{1}{b}\int_\pi^t e^{a(t-\xi)}\sin b(t-\xi)h(\xi)\,d\xi.$$

Thus it remains to find the solution $y_h(t)$ of the initial-value problem

$$(D^2 - 2aD + a^2 + b^2)y = 0,$$
$$y(\pi) = 2, \qquad y'(\pi) = -3.$$

This, of course, may be done with the methods of Chapter 4. But we can also use Laplace transforms if we note that $y_h(t) = y(t - \pi)$, where $y(t)$ satisfies

$$\begin{aligned}(D^2 - 2aD + a^2 + b^2)y &= 0,\\ y(0) = 2, \qquad y'(0) &= -3.\end{aligned} \tag{5–50}$$

Applying $\mathcal{L}$ to (5–50) we obtain

$$(s^2 - 2as + a^2 + b^2)\mathcal{L}[y] = 2s - 3 - 4a,$$

or

$$\begin{aligned}\mathcal{L}[y] &= \frac{2s - 3 - 4a}{s^2 - 2as + a^2 + b^2}\\ &= \frac{2(s - a)}{(s - a)^2 + b^2} - \frac{3 + 2a}{(s - a)^2 + b^2}\cdot\end{aligned}$$

Thus

$$y(t) = 2e^{at}\cos bt - \frac{3 + 2a}{b}e^{at}\sin bt,$$

and

$$y_h(t) = y(t - \pi) = e^{a(t-\pi)}\left[2\cos b(t - \pi) - \frac{3 + 2a}{b}\sin b(t - \pi)\right]\cdot$$

The desired solution of the original problem is now completely determined as

$$y = e^{a(t-\pi)}\left[2\cos b(t - \pi) - \frac{3 + 2a}{b}\sin b(t - \pi)\right]$$
$$+ \frac{1}{b}\int_{\pi}^{t} e^{a(t-\xi)}\sin b(t - \xi)h(\xi)\,d\xi.$$

EXERCISES

Determine Green's functions for each of the following linear differential operators in three ways: (a) by applying Formula (4–42) or (4–43); (b) by applying Theorem 4–4; and (c) by applying Theorem 5–11.

1. $D^2 - 4D + 4$
2. $D^2 + D$
3. $D^2 + 6D + 13$
4. $D^2 + \frac{1}{2}D - \frac{1}{2}$
5. $D^2 - \frac{1}{3}D + \frac{1}{36}$
6. $4D^3 - D$
7. $D^3 + 1$
8. $(D^2 + 1)^2$
9. $D^4 + 1$
10. $(D^2 - 4D + 20)^2$

11. Solve the initial-value problem

$$(D^2 - 1)y = \begin{cases} 0, & 0 \le t \le 1, \\ t - 1, & t > 1, \end{cases}$$
$$y(0) = y'(0) = 0,$$

in three different ways: (a) by using Laplace transforms directly; (b) by first determining a Green's function for $D^2 - 1$; and (c) by solving the initial-value problem

$$(D^2 - 1)y = t - 1; \quad y(1) = a, \quad y'(1) = b,$$

with an appropriate choice of constants a, b.

12. Prove that the function $g(t) = \mathcal{L}^{-1}[1/p(s)]$ defined in the text satisfies the initial-value problem

$$Ly = 0;$$
$$y(0) = \cdots = y^{(n-2)}(0) = 0, \quad y^{(n-1)}(0) = 1,$$

on the interval $[0, \infty)$.

13. Show that the methods of Sections 5–4 and 5–5, when used to compute $\mathcal{L}^{-1}[1/p(s)]$, lead to the unique solution on the entire interval $(-\infty, \infty)$ of the initial-value problem of Exercise 12. (This justifies the statement that $g(t - \xi)$ is *the* Green's function for L on $(-\infty, \infty)$, where $g(t) = \mathcal{L}^{-1}[1/p(s)]$.)

14. Use Laplace transforms to derive (5–45) for any function $h(t)$ which is piecewise continuous on $[0, \infty)$. [*Hint:* First consider the solution of (5–44) with $h(t)$ replaced by

$$H(t) = \begin{cases} h(t), & 0 \le t \le a, \\ 0, & t > a, \end{cases}$$

where a is a constant > 0.]

15. If L is a constant coefficient operator, show that $y_p(t)$ is a solution of the initial-value problem

$$Ly = h(t);$$
$$y(t_0) = c_0, \quad y'(t_0) = c_1, \quad \ldots, \quad y^{(n-1)}(t_0) = c_{n-1},$$

if and only if $y_p(t) = Y(t - t_0)$, where $Y(t)$ satisfies

$$Ly = h(t + t_0);$$
$$y(0) = c_0, \quad y'(0) = c_1, \quad \ldots, \quad y^{(n-1)}(0) = c_{n-1}.$$

Solve each of the following initial-value problems using the method given at the end of this section.

16. $(D^2 + 1)y = e^{t-1}; \quad y(1) = y'(1) = 0$
17. $(2D^2 + D - 1)y = \sin t; \quad y(\pi) = y'(\pi) = 0$
18. $(4D^2 + 16D + 17)y = t^2 - 1; \quad y(a) = y'(a) = 0$
19. $(D^2 - 3D - 4)y = e^{-t}; \quad y(2) = 3, \ y'(2) = 0$
20. $(2D^2 - 3D + 1)y = t; \quad y(1) = 0, \ y'(1) = -1$

21. $(4D^2 - 4D + 37)y = e^{t/2}\cos 3t;\quad y(a) = 7,\ y'(a) = -2,\ a > 0$
22. $(D^3 + 1)y = te^t;\quad y(1) = y'(1) = 0,\ y''(1) = 1$
23. $D^2(D^2 + 1)^2y = h(t);\quad y(a) = y'(a) = y''(a) = y'''(a) = y^{(\text{iv})}(a) = 0,\ y^{(\text{v})}(a) = 1$

5-9 THE VIBRATING SPRING; IMPULSE FUNCTIONS

Consider an elastic spring which is fixed at one end and is free to vibrate in a vertical direction as shown in Fig. 5-13. Suppose a weight of mass m is attached to the spring and the entire system comes into equilibrium with the weight at the point $y = 0$ located y_0 units below the natural length of the spring. Then, by Hooke's law, the weight experiences an upward force of magnitude ky_0.* Since the system is in equilibrium this force is exactly counteracted by the force of gravity acting on the weight, and so we have

$$ky_0 = mg. \tag{5-51}$$

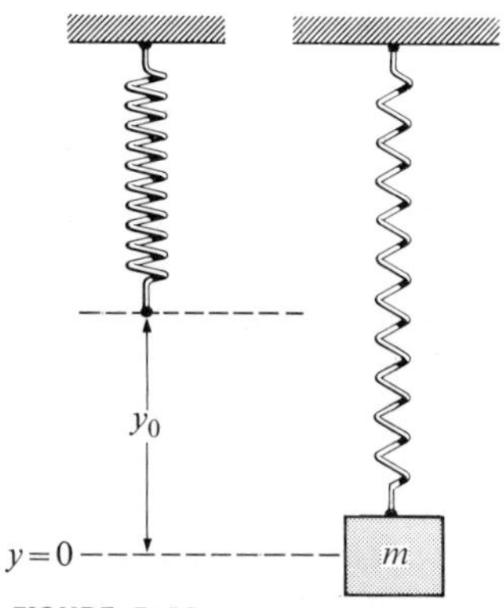

FIGURE 5-13

Suppose that the spring-weight system is in equilibrium, and that the weight is now subjected to an additional vertical force $h(t)$ which may vary with time. Then at time t, with the weight a distance $y(t)$ from the equilibrium position, and the positive y-direction measured downward, the forces on the weight are mg due to gravity, $-k(y - y_0)$ due to the restoring force in the spring, and $h(t)$. Hence, by Newton's second law, we have

$$m\frac{d^2y}{dt^2} = mg - k(y - y_0) + h(t),$$

or, using (5-51),

$$m\frac{d^2y}{dt^2} + ky = h(t). \tag{5-52}$$

Furthermore, since the system was initially at rest in equilibrium, $y(t)$ must satisfy

$$y(0) = 0, \qquad y'(0) = 0, \tag{5-53}$$

and the motion of the weight is thus obtained as the solution of an initial-value problem.

To solve this problem we use Laplace transforms and obtain

$$ms^2\mathcal{L}[y] + k\mathcal{L}[y] = \mathcal{L}[h(t)].$$

---

* The positive constant k is known as the *spring constant*. It depends upon the material from which the spring is made, the dimensions of the spring, etc.

Hence

$$\mathcal{L}[y] = \frac{\mathcal{L}[h]}{ms^2 + k}, \tag{5-54}$$

and

$$\begin{aligned} y(t) &= \mathcal{L}^{-1}\left[\frac{1}{ms^2 + k}\right] * h(t) \\ &= \mathcal{L}^{-1}\left[\frac{\sqrt{k/m}}{\sqrt{mk}\,(s^2 + (k/m))}\right] * h(t) \\ &= \left(\frac{1}{\sqrt{mk}} \sin \sqrt{\frac{k}{m}}\, t\right) * h(t).\dagger \end{aligned}$$

Thus the equation of motion for this system under the action of an arbitrary external force h can be expressed in integral form as

$$y(t) = \frac{1}{\sqrt{mk}} \int_0^t \sin\,[\sqrt{k/m}\,(t - \xi)]h(\xi)\,d\xi. \tag{5-55}$$

In applications the impressed force $h(t)$ is often of the form

$$h(t) = A \sin \omega t, \tag{5-56}$$

where A and ω are positive constants, in which case the equation of motion becomes

$$y(t) = \frac{A}{\sqrt{mk}} \int_0^t \sin\,[\sqrt{k/m}\,(t - \xi)] \sin \omega\xi\,d\xi.$$

Although this integral can be evaluated by elementary techniques it is instructive to begin again with (5-54) and solve the problem directly. Thus

$$\begin{aligned} \mathcal{L}[y] &= \frac{1}{ms^2 + k} \mathcal{L}[A \sin \omega t] \\ &= \frac{1}{ms^2 + k} \cdot \frac{A\omega}{s^2 + \omega^2}. \end{aligned}$$

We now consider two cases according as ω is or is not equal to $\sqrt{k/m}$.

Case 1. $\omega \neq \sqrt{k/m}$. Then $A\omega/[(ms^2 + k)(s^2 + \omega^2)]$ can be rewritten by the method of partial fractions as

$$\frac{A\omega}{k - m\omega^2}\left(\frac{1}{s^2 + \omega^2} - \frac{m}{ms^2 + k}\right),$$

† Those of our readers who are familiar with the material of the preceding section will recognize that we have now computed the Green's function $g(t - \xi)$ for initial-value problems involving the *normalized* operator $L = D^2 + (k/m)$, and that

$$g(t - \xi) = \sqrt{m/k} \sin \sqrt{k/m}\,(t - \xi).$$

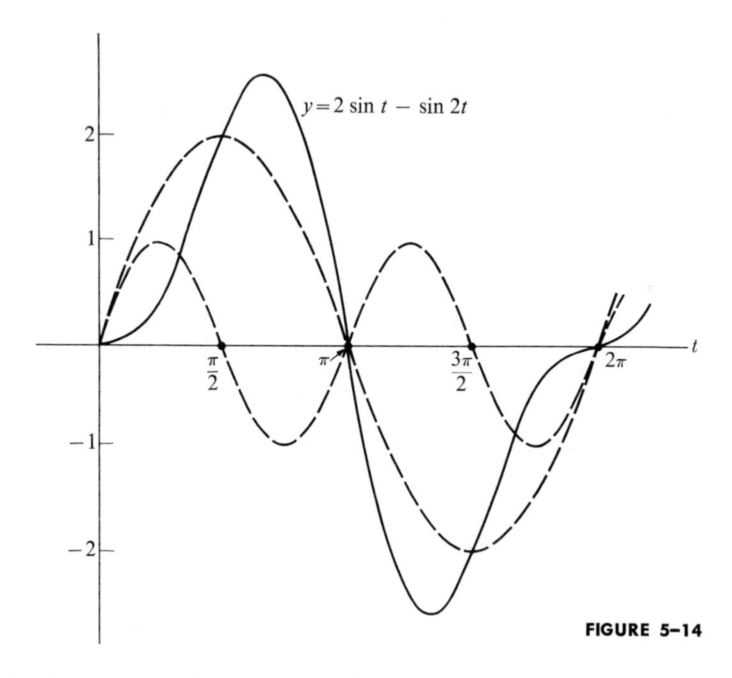

FIGURE 5–14

and, taking inverse transforms, we obtain

$$y(t) = \frac{A\omega}{k - m\omega^2}\left[\frac{1}{\omega}\sin \omega t - \sqrt{m/k}\sin \sqrt{k/m}\, t\right]. \tag{5–57}$$

This function may be interpreted as the superposition of oscillations of two different frequencies, $\sqrt{k/m}\,2\pi$ and $\omega\, 2\pi$. The first of these is the so-called *natural frequency* of the system, while the second is the frequency of the impressed force (5–56). In Fig. 5–14 we have sketched the graph of $y(t)$ when $m = k = 1, A = 3$, and $\omega = 2$.

Case 2. $\omega = \sqrt{k/m}$. Then

$$\begin{aligned}\mathcal{L}[y] &= \frac{A\sqrt{km}}{(ms^2 + k)^2} \\ &= \frac{A\sqrt{km}}{s}\cdot\frac{s}{(ms^2 + k)^2}.\end{aligned}$$

But since

$$\begin{aligned}\mathcal{L}^{-1}\left[\frac{s}{(ms^2 + k)^2}\right] &= \mathcal{L}^{-1}\left[-\frac{1}{2m}\frac{d}{ds}\left(\frac{1}{ms^2 + k}\right)\right] \\ &= \frac{1}{2m}\, t\mathcal{L}^{-1}\left[\frac{1}{ms^2 + k}\right] \\ &= \frac{t}{2m\sqrt{km}}\sin\sqrt{k/m}\, t,\end{aligned}$$

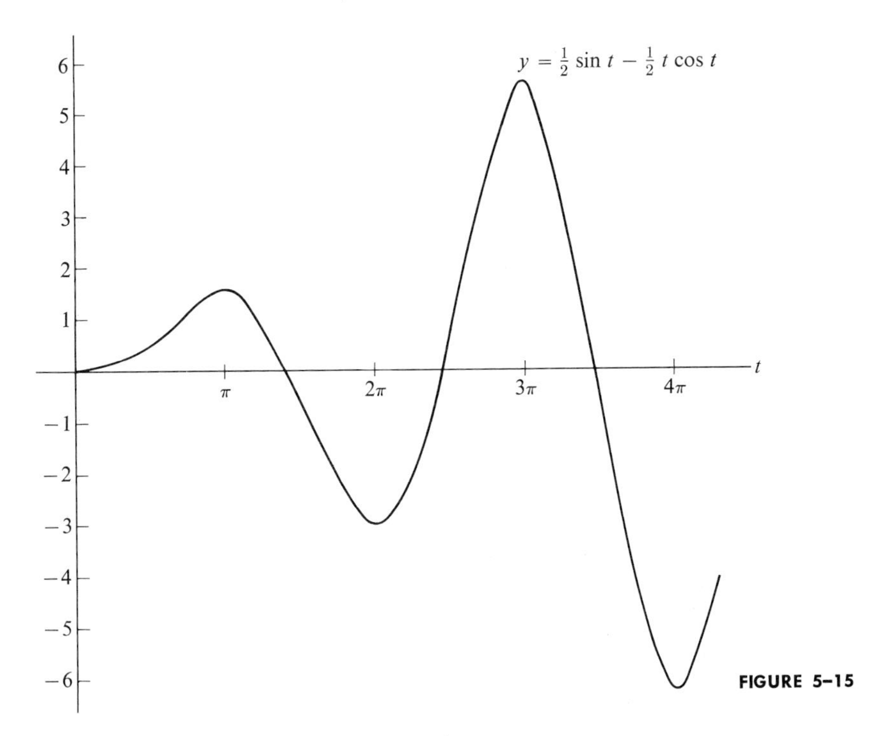

FIGURE 5–15

the convolution formula can be used to obtain

$$y = \frac{A}{2m}\int_0^t \xi \sin\sqrt{k/m}\,\xi\,d\xi$$

$$= \frac{A}{2m}\left[\frac{m}{k}\sin\sqrt{k/m}\,t - \sqrt{m/k}\,t\cos\sqrt{k/m}\,t\right],$$

and we have

$$y(t) = \frac{A}{2k}\sin\sqrt{k/m}\,t - \frac{A}{2\sqrt{km}}\,t\cos\sqrt{k/m}\,t. \tag{5–58}$$

Therefore when the impressed frequency is equal to the natural frequency, the amplitude of the oscillations increases with time and the spring is eventually stretched beyond its elastic limit (see Fig. 5–15, sketched for $A = k = m = 1$). This phenomenon is known as *resonance*, and is important in various physical problems.†

A rather different situation arises if we attempt to find the response of this system when the weight is struck a sharp blow in the vertical direction at time $t = a$,

† Also see the discussion of resonance in connection with electrical networks in Section 4–10.

$a \geq 0$. To obtain the equation of motion in this case we introduce the function h defined by

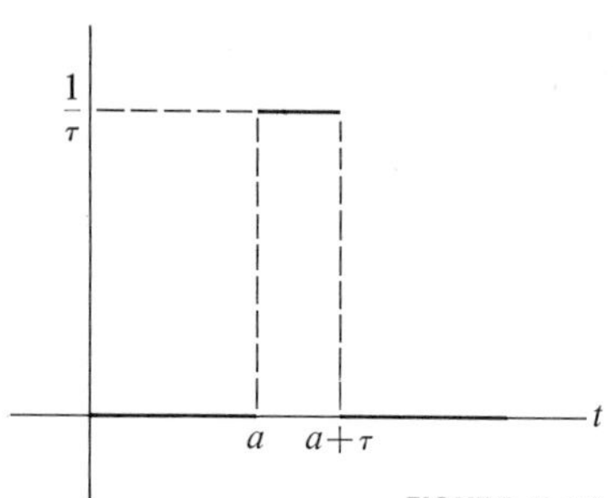

FIGURE 5–16

$$h(t) = \begin{cases} 0, & 0 \leq t \leq a, \\ \dfrac{1}{\tau}, & a < t < a + \tau, \\ 0, & a + \tau \leq t, \end{cases} \tag{5–59}$$

where τ is an arbitrary positive constant (see Fig. 5–16). Physically, h represents a force of magnitude $1/\tau$ acting on the system for a time τ, and hence h imparts a total *impulse* of 1 to the system.* We now agree that the mathematical description of the physical situation described by the words "sharp blow" is obtained by using a force which acts throughout an arbitrarily short interval of time but imparts a predetermined impulse, or change of momentum, to the system. Our problem then becomes that of determining the behavior of the solution $y(t)$ in (5–55) when h is as above, and $\tau \to 0$.

Substitution of (5–59) in (5–55) gives

$$y(t) = \begin{cases} 0, \quad 0 \leq t \leq a, \\ \dfrac{1}{\tau} \displaystyle\int_a^t \left[\frac{1}{\sqrt{mk}} \sin \sqrt{k/m}\,(t - \xi) \right] d\xi, \quad a < t < a + \tau, \\ \dfrac{1}{\tau} \displaystyle\int_a^{a+\tau} \left[\frac{1}{\sqrt{mk}} \sin \sqrt{k/m}\,(t - \xi) \right] d\xi, \quad a + \tau \leq t. \end{cases} \tag{5–60}$$

Hence, passing to the limit as $\tau \to 0$, we obtain the solution

$$y_0(t) = \begin{cases} 0, \quad t \leq a, \\ \dfrac{1}{\sqrt{mk}} \sin \sqrt{k/m}\,(t - a), \quad t > a, \end{cases} \tag{5–61}$$

(see Exercise 2). But $y_0(t)$ is also the solution of the initial-value problem

$$m \frac{d^2y}{dt^2} + ky = 0;$$

$$y(a) = 0, \qquad y'(a) = \frac{1}{m},$$

---

* A constant force of magnitude F acting on an object of mass m for t seconds is said to impart an *impulse* $I = F \cdot t$ to the object. Since $F = (d/dt)(mv)$, where v is the magnitude of the velocity of the object (Newton's second law), it follows that when F is constant the *total* change in the momentum mv of the object is equal to the impulse. Throughout this discussion we shall assume for the sake of convenience that $I = 1$.

and as such can be interpreted as the response of a weighted spring which is given unit momentum at $t = a$ [i.e., $my'(a) = 1$], and left undisturbed thereafter. It follows (see the preceding footnote) that in this situation $y_0(t)$ may be interpreted as arising from an instantaneous unit impulse imparted to the system at $t = a$, and the problem has been solved.

In certain circumstances it is convenient to think of such a unit impulse at $t = 0$ as arising from a *fictitious* function $\delta(t)$, called the *Dirac Delta Function*, whose defining properties are

$$\delta(t) = 0 \quad \text{for all } t \neq 0, \qquad \int_{-\infty}^{\infty} \delta(t)\, dt = 1. \tag{5–62}$$

In these terms a unit impulse at $t = a$ would be provided by the "function" $\delta(t - a)$. The initial-value problem leading to (5–61) then becomes

$$m\frac{d^2y}{dt^2} + ky = \delta(t - a);$$
$$y(0) = y'(0) = 0,$$

and, as we have seen, has $y_0(t)$ as its solution.

The foregoing discussion can easily be generalized to initial-value problems of the form

$$\begin{gathered} Ly = \delta(t - a); \\ y(0) = y'(0) = \cdots = y^{(n-1)}(0) = 0, \end{gathered} \tag{5–63}$$

where $L = D^n + a_{n-1}D^{n-1} + \cdots + a_0$. To solve this problem we again use Laplace transforms, first replacing $\delta(t - a)$ by the function h defined in (5–59), and then passing to the limit as $\tau \to 0$. Thus

$$\mathcal{L}[y] = \frac{1}{p(s)}\,\mathcal{L}[h]$$

with $p(s) = s^n + a_{n-1}s^{n-1} + \cdots + a_0$, and hence

$$\begin{aligned} y &= \mathcal{L}^{-1}\left[\frac{1}{p(s)}\right] * h(t) \\ &= \int_0^t g(t - \xi)h(\xi)\, d\xi, \end{aligned}$$

where $g(t) = \mathcal{L}^{-1}[1/p(s)]$. Using the given value of $h(t)$ we thus obtain

$$y(t) = \begin{cases} 0, & t \leq a, \\ \dfrac{1}{\tau}\displaystyle\int_a^t g(t - \xi)\, d\xi, & a < t < a + \tau, \\ \dfrac{1}{\tau}\displaystyle\int_a^{a+\tau} g(t - \xi)\, d\xi, & a + \tau \leq t, \end{cases} \tag{5–64}$$

and it follows that the solution of (5–63) is

$$y_0(t) = \begin{cases} 0, & t \leq a, \\ g(t-a), & t > a. \end{cases} \tag{5–65}$$

This result is valuable because it can be used to *define* the Laplace transform of $\delta(t-a)$. Indeed, by Formula (5–26) we have

$$\begin{aligned} \mathcal{L}[y_0] &= e^{-as}\mathcal{L}[g] \\ &= e^{-as}\frac{1}{p(s)}. \end{aligned}$$

On the other hand, if we apply $\mathcal{L}$ to (5–63), under the unjustified assumption that $\mathcal{L}[\delta(t-a)]$ exists, we find that

$$\mathcal{L}[y_0] = \frac{1}{p(s)}\mathcal{L}[\delta(t-a)].$$

Thus if $\mathcal{L}[\delta(t-a)]$ exists at all it must have the value e^{-as}, and hence we are forced to the *definition*

$$\mathcal{L}[\delta(t-a)] = e^{-as}. \tag{5–66}$$

(For a different approach to this formula see Exercise 1 below.) In particular, for $a = 0$ we have the curious formula

$$\mathcal{L}[\delta(t)] = 1. \tag{5–67}$$

This apparent contradiction of Theorem 5–3, which asserts that $\mathcal{L}[f] \to 0$ as $s \to \infty$, is merely a reflection of the fact that $\delta(t)$ does not belong to the space of piecewise continuous functions of exponential order. But this is hardly surprising since, as was pointed out above, $\delta(t)$ is not even a function in the usual sense of the term.

EXAMPLE. Consider a spring in equilibrium with a body of mass m attached. Suppose that at time $t = 0$ the system is subjected to a sinusoidal force $h(t) = A \sin \omega t$, and that at time $t = 10$ it is struck a sharp blow from below which instantaneously imparts 2 units of momentum to the mass. Problem: Find the motion of the system from $t = 0$ onward.

In this case we have to solve the initial-value problem

$$m\frac{d^2y}{dt^2} + ky = A \sin \omega t - 2\,\delta(t-10),$$
$$y(0) = y'(0) = 0.$$

Then

$$(ms^2 + k)\mathcal{L}[y] = \frac{A\omega}{s^2 + \omega^2} - 2e^{-10s},$$

and

$$\mathcal{L}[y] = \frac{A\omega}{(ms^2 + k)(s^2 + \omega^2)} - \frac{2}{ms^2 + k} e^{-10s}.$$

Referring to our earlier calculations, the inverse transform $d(t)$ of $A\omega/[(ms^2 + k)(s^2 + \omega^2)]$ is given by either (5–57) or (5–58) depending on whether $\omega \neq \sqrt{k/m}$ or not. As for the second term, its inverse transform is

$$\begin{aligned} e(t) &= \begin{cases} 0, & t \leq 10, \\ -\dfrac{2}{\sqrt{km}} \sin \sqrt{k/m}\,(t - 10), & t > 10, \end{cases} \\ &= -\frac{2}{\sqrt{km}} u_{10}(t) \sin \sqrt{k/m}\,(t - 10). \end{aligned}$$

Hence $y(t) = d(t) + e(t)$, and the description of the motion is complete.

EXERCISES

1. Show that the definition

$$\mathcal{L}[\delta(t - a)] = \lim_{\tau \to 0} \mathcal{L}[h(t)],$$

where $h(t)$ is defined by (5–59), leads to the same result as Formula (5–66).

2. Let $y(t)$ be defined as in (5–64). Prove that

$$\lim_{\tau \to 0} y(t) = \begin{cases} 0, & t \leq a, \\ g(t - a), & t > a. \end{cases}$$

Use this result to verify that (5–61) is correct.

3. Solve the initial value problem

$$y''' + y = e^t + \delta(t - 1);$$
$$y(0) = 1, \quad y'(0) = 1, \quad y''(0) = 2.$$

4. Find the equation of motion of a weight having mass m which is initially in equilibrium at the end of a spring, and which, at time $t = 0$, is struck a sharp blow from above which instantaneously imparts 1 unit of momentum to the system.

5. A weight of mass 1 is attached to a spring whose spring constant is 4, and at time $t = 0$ the weight is struck a blow from above which instantaneously imparts 1 unit of momentum to the system. At time $t = \pi/2$ a sinusoidal force of magnitude $-\sin(t - (\pi/2))$ begins to act vertically on the system. Find the equation of motion of the mass.

6. A mass m, hanging in equilibrium at the end of a spring, is struck from below and instantaneously given two units of momentum. At time $t = a$ the mass is subjected to the external force $\sin(t - a)$. Assuming that the spring constant is different from m, find the equation of motion of the mass m.

7. A unit mass is attached to a rigid spring whose spring constant is 3, and is then mounted in an elevator as shown in Fig. 5–17. At time $t = 0$ the elevator begins to descend with a constant velocity of 2 ft/sec, and at that moment the mass is struck a blow from above which instantaneously gives it one unit of momentum. Find the equation of motion of the mass as a function of time.

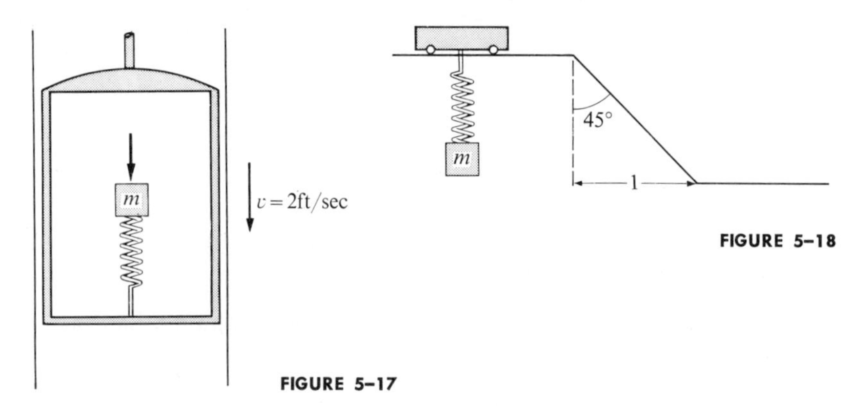

FIGURE 5–17

FIGURE 5–18

8. A mass m is suspended on a spring beneath a car which moves with constant velocity v along the track shown in Fig. 5–18. Suppose that at time $t = 0$ the mass is struck from below and instantaneously given one additional unit of momentum. Find the equation for the motion of m in the vertical direction as a function of time.
9. Find the equation of motion of a mass m bobbing at the end of a spring with spring constant k if the mass was originally displaced a units from equilibrium and given initial velocity b. (This is an example of what is known as *simple harmonic motion.*)
10. It is observed that the spring of Exercise 9 oscillates with a period of 2 seconds when m is 1 gram.

    (a) Determine the spring constant k.

    (b) Now suppose the mass on this spring is 9 grams and that the oscillations start at $a = 4$ with initial velocity $b = \pi$. Find the period and amplitude of the oscillations.
11. Suppose that a spring is suspended in a resisting medium which opposes any motion through it with a force proportional to the velocity of the moving object.

    (a) Show that the equation governing the motion of a mass m bobbing at the end of such a spring is

$$my'' + \lambda y' + ky = 0,$$

    where λ is a positive constant and k is the spring constant.

    (b) Let $y(0) = a$, $y'(0) = b$ for the mass in (a). Show that

$$y = \mathcal{L}^{-1}\left[\frac{mas + mb + \lambda a}{ms^2 + \lambda s + k}\right].$$

12. Assume that the resisting force in Exercise 11 is sufficiently large to ensure that $\lambda^2 > 4km$.

(a) Show that the quadratic equation $ms^2 + \lambda s + k = 0$ then has distinct negative roots, $-\sigma_1$, $-\sigma_2$ with $\sigma_1 < \sigma_2$, and that

$$y = e^{-\sigma_1 t}[\alpha + \beta e^{-(\sigma_2 - \sigma_1)t}],$$

where

$$\alpha = \frac{a}{2} + \frac{2bm + a\lambda}{2\sqrt{\lambda^2 - 4km}}, \qquad \beta = \frac{a}{2} - \frac{2bm + a\lambda}{2\sqrt{\lambda^2 - 4km}},$$

$$\sigma_1 = \frac{\lambda - \sqrt{\lambda^2 - 4km}}{2m}, \qquad \sigma_2 = \frac{\lambda + \sqrt{\lambda^2 - 4km}}{2m}.$$

(b) Show that except in the trivial case where $y \equiv 0$, the mass in this problem is in its equilibrium position $y = 0$ for at most one value of t. (The effect of the resistance in this problem is often described by the word "damping," and the resulting motion is known as *damped harmonic motion.*)

13. Suppose that the spring in Exercises 11 and 12 has spring constant 9, and that a unit mass is attached to the spring, displaced one unit in the positive y-direction, and given an initial velocity $b = -10$. Suppose in addition that the medium offers a resistance corresponding to $\lambda = 10$.

    (a) Find the equation of motion of the mass.

    (b) Show that the mass is in its equilibrium position for exactly one instant. When is this?

    (c) What is the maximum height which the mass attains? (Recall that the positive y-direction is downward.)

    (d) Sketch the graph of the solution curve for this problem.

14. Show that a necessary and sufficient condition that the mass of Exercises 11 and 12 be in its equilibrium position at some one instant after the motion begins is that $0 < -\alpha/\beta < 1$, and hence deduce that if $a \neq 0$ and $b = 0$ the mass approaches its equilibrium position without ever passing through it.

15. Show that the quadratic equation $ms^2 + \lambda s + k = 0$ for the vibrating spring problem described in Exercise 11 has a double root $s = -\sigma = -\lambda/(2m) < 0$ when $\lambda^2 = 4km$. Find the equation of motion of the mass in this case. (This is an example of what is known as *critical damping.*)

16. Let $y = y(t)$ be the solution obtained in Exercise 15.

    (a) Show that $y \to 0$ as $t \to \infty$, and that the mass passes through its equilibrium position at most once during the motion.

    (b) Show that if $b/a < -\sigma$, then $y = 0$ when $t = -a/(a\sigma + b) > 0$, but that otherwise the mass approaches its equilibrium position without ever passing through it. (Assume $y \not\equiv 0$.)

17. Find the equation of motion for the system in Exercise 11 in the case where $\lambda^2 < 4km$. What is the behavior of y as $t \to \infty$?

A Short Table of Laplace Transforms†

	Function	Transform
	$f(t)$	$\mathcal{L}[f] = \int_0^\infty e^{-st} f(t)\, dt$
	$\alpha f(t) + \beta g(t)$	$\alpha\mathcal{L}[f] + \beta\mathcal{L}[f]$
	$f'(t)$	$s\mathcal{L}[f] - f(0^+)$
	$f''(t)$	$s^2\mathcal{L}[f] - sf(0^+) - f'(0^+)$
	$f^{(n)}(t)$	$s^n\mathcal{L}[f] - s^{n-1}f(0^+) - s^{n-2}f'(0^+) - \cdots - f^{(n-1)}(0^+)$
*	$\int_0^t f(t)\, dt$	$\frac{1}{s}\mathcal{L}[f]$
	$\int_a^t f(t)\, dt$	$\frac{1}{s}\mathcal{L}[f] - \frac{1}{s}\int_0^a f(t)\, dt$
*	$\underbrace{\int_0^t \cdots \int_0^t}_{n \text{ times}} f(t)\, dt \ldots dt$	$\frac{1}{s^n}\mathcal{L}[f]$
	$\underbrace{\int_a^t \cdots \int_a^t}_{n \text{ times}} f(t)\, dt \ldots dt$	$\frac{1}{s^n}\mathcal{L}[f] - \frac{1}{s^n}\int_0^a f(t)\, dt - \frac{1}{s^{n-1}}\int_0^a \int_a^t f(t)\, dt\, dt - \cdots - \frac{1}{s}\int_0^a \underbrace{\int_a^t \cdots \int_a^t}_{n-1 \text{ times}} f(t)\, dt \ldots dt$
*	$e^{at} f(t)$	$\bar{f}(s - a)$, where $\bar{f}(s) = \mathcal{L}[f]$
*	$t^n f(t)$	$(-1)^n \frac{d^n}{ds^n}\mathcal{L}[f]$
	$u_a(t)g(t) = \begin{cases} 0, & t \le a \\ g(t), & t > a \end{cases}$	$e^{-as}\mathcal{L}[g(t + a)]$

† The table consists of three parts. The first contains formulas of a general nature, and the second the transforms of a small number of selected functions. When used in conjunction, these two parts of the table will yield the Laplace transforms of most functions which occur in practice. The third part of the table is primarily intended for computing inverse transforms, and hence is designed to be read from right to left. The methods of partial fractions, completing the square, etc., are, of course, indispensible in such computations. Finally, those formulas in the first part of the table which are particularly well suited to evaluating inverse transforms have been marked with an asterisk.

A SHORT TABLE OF LAPLACE TRANSFORMS (*Continued*)

Function	Transform
* $u_a(t)g(t-a) = \begin{cases} 0, & t \leq a \\ g(t-a), & t > a \end{cases}$	$e^{-as}\mathcal{L}[g]$
* $\int_0^t f(t-\xi)g(\xi)\,d\xi$	$\mathcal{L}[f]\mathcal{L}[g]$
$f(t)$ periodic with period p ($p > 0$)	$\dfrac{\int_0^p e^{-st} f(t)\,dt}{1 - e^{-ps}}$
$\dfrac{f(t)}{t}$ if $\lim\limits_{t \to 0+} \dfrac{f(t)}{t}$ exists	$\int_s^\infty \mathcal{L}[f]\,ds$
1	$\dfrac{1}{s}$
e^{at}	$\dfrac{1}{s-a}$
t^n	$\dfrac{n!}{s^{n+1}}$
$\sin at$	$\dfrac{a}{s^2+a^2}$
$\cos at$	$\dfrac{s}{s^2+a^2}$
$\sinh at$	$\dfrac{a}{s^2-a^2}$
$\cosh at$	$\dfrac{s}{s^2-a^2}$
$\delta(t)$	1
$\delta(t-a)$	e^{-as}
$\dfrac{t^{n-1}e^{at}}{(n-1)!}$	$\dfrac{1}{(s-a)^n}$ $(n \geq 1)$
$\dfrac{1}{2a^3}(\sin at - at\cos at)$	$\dfrac{1}{(s^2+a^2)^2}$
$\dfrac{t}{2a}\sin at$	$\dfrac{s}{(s^2+a^2)^2}$
$\int_0^t \dfrac{t}{2n}\mathcal{L}^{-1}\left[\dfrac{1}{(s^2+a^2)^n}\right]dt$	$\dfrac{1}{(s^2+a^2)^{n+1}}$
$\dfrac{t}{2n}\mathcal{L}^{-1}\left[\dfrac{1}{(s^2+a^2)^n}\right]$	$\dfrac{s}{(s^2+a^2)^{n+1}}$

A SHORT TABLE OF LAPLACE TRANSFORMS (*Continued*)

Function	Transform
$\dfrac{1}{2^n a n!} \underbrace{\int_0^t t \int_0^t t \cdots \int_0^t}_{n \text{ times}} t \sin at\, dt \cdots dt$	$\dfrac{1}{(s^2 + a^2)^{n+1}}$
$\dfrac{t}{2^n a n!} \underbrace{\int_0^t t \int_0^t t \cdots \int_0^t}_{n-1 \text{ times}} t \sin at\, dt \cdots dt$	$\dfrac{s}{(s^2 + a^2)^{n+1}}$

6

further topics in the theory of linear differential equations*

6–1 THE SEPARATION AND COMPARISON THEOREMS

In this chapter we resume the general discussion of linear differential equations begun in Chapter 3, and establish a number of extremely valuable results concerning equations with variable coefficients. The first two are almost immediate consequences of Abel's formula (Theorem 3–7), and describe the behavior of the zeros of solutions of second-order homogeneous linear differential equations.

> **Theorem 6–1.** (Sturm separation theorem.) *If y_1 and y_2 are linearly independent solutions of the second-order homogeneous equation*
>
> $$a_2(x)y'' + a_1(x)y' + a_0(x)y = 0 \tag{6–1}$$
>
> *on an interval I in which a_2 does not vanish, then the zeros of y_1 and y_2 alternate on I.*†

Proof. Let a and b ($a < b$) be two points of I such that $y_2(a) = y_2(b) = 0$, and suppose that y_2 is nonzero everywhere between a and b. We must show that there exists *exactly one* point c between a and b such that $y_1(c) = 0$. (See Fig. 6–1.)

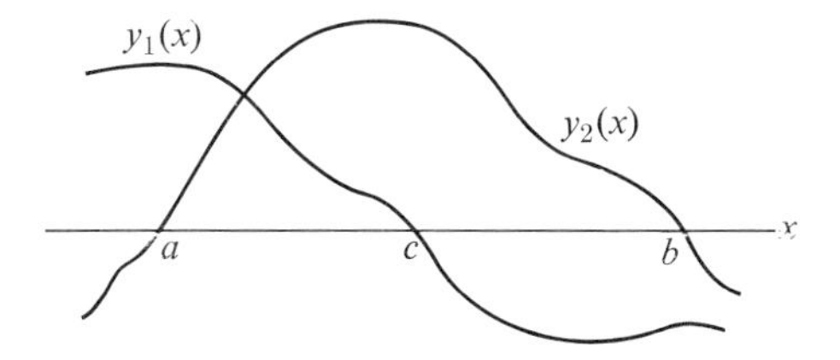

FIGURE 6–1

---

* None of the material in this chapter will be used in any essential way until Chapter 15 where equations with regular singular points are studied.

† A *zero* of a function y is a point x_0 at which $y(x_0) = 0$. It may happen, of course, that y_1 and y_2 have no zeros on I, in which case the conclusion of the theorem is vacuously satisfied.

Since y_1 and y_2 are linearly independent solutions of (6–1), their Wronskian never vanishes on I (Theorem 3–6). Hence the constant in Abel's formula is nonzero, and

$$W[y_1(x), y_2(x)] = y_1(x)y_2'(x) - y_2(x)y_1'(x)$$

has the same algebraic sign everywhere in I. Moreover, the values of the Wronskian at $x = a$ and $x = b$ are, respectively,

$$y_1(a)y_2'(a) \qquad \text{and} \qquad y_1(b)y_2'(b),$$

and hence $y_1(a)$, $y_1(b)$, $y_2'(a)$, and $y_2'(b)$ are all different from zero. But since a and b are successive zeros of y_2, the derivative of y_2 must have opposite signs at a and b (i.e., the graph of y_2 must be rising at a and falling at b, or *vice versa*). Hence $y_1(a)$ and $y_1(b)$ have opposite signs, and it follows that $y_1(c) = 0$ for *at least one* point c between a and b.*

Finally, by reversing the roles of y_1 and y_2 in the above argument, we conclude that y_2 has at least one zero between every pair of successive zeros of y_1 on I, and we are done. ▮

EXAMPLE 1. On the strength of this theorem we can deduce the well-known fact that the zeros of $\sin x$ and $\cos x$ alternate on $(-\infty, \infty)$, since these functions are linearly independent solutions of the equation

$$y'' + y = 0. \tag{6–2}$$

A somewhat less obvious consequence is that *any* two functions of the form

$$a_1 \sin x + a_2 \cos x, \qquad b_1 \sin x + b_2 \cos x$$

have alternating zeros whenever $a_1b_2 \neq a_2b_1$, for all such pairs of functions are also linearly independent solutions of (6–2). (See Exercise 1.)

EXAMPLE 2. The functions $\sinh x$ and $\cosh x$ are linearly independent solutions of $y'' - y = 0$, and hence their zeros alternate on $(-\infty, \infty)$. This, of course, is obvious since $\sinh x$ has but a single zero, namely, $x = 0$, while $\cosh x$ has no zeros at all, and is cited merely to emphasize that the separation theorem says nothing at all about the *number* of zeros of a solution of (6–1).

Questions of the latter sort can sometimes be answered by using the following theorem, also due to Sturm.

Theorem 6–2. (Sturm comparison theorem.) *Let y_1 and y_2 be, respectively, nontrivial solutions of the differential equations*

$$y'' + p_1(x)y = 0 \qquad \text{and} \qquad y'' + p_2(x)y = 0$$

---

* The validity of this assertion depends upon the fact that a continuous function assumes *all* values between its maximum and minimum on a *closed* interval $[a, b]$.

on an interval I, and suppose that $p_1(x) \geq p_2(x)$ everywhere in I. Then between any two zeros of y_2 there is **at least one** *zero of y_1, unless $p_1(x) \equiv p_2(x)$ and y_1 and y_2 are linearly dependent in $\mathcal{C}(I)$.*

Proof. Let a and b be adjacent zeros of y_2, with $a < b$, and suppose that y_1 does *not* vanish in the interval (a, b).* Since the zeros of a function y are the same as those of $-y$, we may assume that y_1 and y_2 are both positive throughout (a, b). Then, arguing as in the preceding proof, we have

$$\begin{aligned} W(a) &= y_1(a)y_2'(a) \geq 0, \\ W(b) &= y_1(b)y_2'(b) \leq 0, \end{aligned} \tag{6-3}$$

where $W(a)$ and $W(b)$ denote the values of the Wronskian of y_1 and y_2 at a and b, respectively. But

$$\begin{aligned} \frac{d}{dx} W[y_1(x), y_2(x)] &= \frac{d}{dx}[y_1(x)y_2'(x) - y_2(x)y_1'(x)] \\ &= y_1(x)y_2''(x) - y_2(x)y_1''(x) \\ &= -y_1(x)p_2(x)y_2(x) + y_2(x)p_1(x)y_1(x) \\ &= y_1(x)y_2(x)[p_1(x) - p_2(x)] \geq 0, \end{aligned}$$

and it follows that the Wronskian of y_1 and y_2 is a *nondecreasing* function on I. This, however, contradicts (6–3) unless $p_1(x) \equiv p_2(x)$, in which case $W[y_1(x), y_2(x)] \equiv 0$ for all x in I. Thus y_1 must have a zero between a and b whenever $p_1(x) \not\equiv p_2(x)$. Moreover, the separation theorem implies that this result continues to hold when $p_1(x) \equiv p_2(x)$ unless y_1 and y_2 are linearly dependent in $\mathcal{C}(I)$, and the proof is complete. ▌

EXAMPLE 3. Every nontrivial solution of the equation

$$y'' + p(x)y = 0 \tag{6-4}$$

has at most one zero on any interval in which $p(x) \leq 0$. For if we apply the comparison theorem to this equation and

$$y'' = 0, \tag{6-5}$$

we conclude that on such an interval *every* solution of (6–5) must vanish at least once between successive zeros of a solution of (6–4). The assertion now follows from the fact that $y'' = 0$ has solutions (namely $y = c$) which do not vanish on *any* interval.

---

* At this point we are making use of the fact that a nontrivial solution of a second-order linear differential equation cannot have infinitely many zeros in any finite *closed* interval of the x-axis. (See Exercise 6.)

If we agree to say that a nontrivial solution of a linear differential equation *oscillates* on an interval I if and only if it has at least two zeros in I, we can restate the conclusion of this example by saying that no solution of (6–4) oscillates on any interval in which $p(x) \leq 0$. In particular, this result holds for the nontrivial solutions of

$$y'' - ky = 0$$

on $(-\infty, \infty)$ whenever k is a positive constant (cf. Example 2).

EXERCISES

1. Show that the functions $a_1 \sin x + a_2 \cos x$ and $b_1 \sin x + b_2 \cos x$ are linearly independent solutions of $y'' + y = 0$ whenever $a_1 b_2 \neq a_2 b_1$, and thus prove that the zeros of these functions alternate on $(-\infty, \infty)$.
2. Prove that $\sin k_1 x$ has at least one zero between any two zeros of $\sin k_2 x$ whenever $k_1 > k_2 > 0$.
3. (a) Show that every nontrivial solution of the equation $y'' + (\sinh x)y = 0$ has at most one zero in $(-\infty, 0)$, and infinitely many zeros in $(0, \infty)$.

   (b) Prove that the distance between successive zeros of any nontrivial solution of this equation tends to zero as $x \to \infty$.
4. Every solution of the equation $y'' + xy = 0$ has infinitely many zeros in $(0, \infty)$. True or false? Why?
5. (a) Let f be continuous on $(0, \infty)$, and suppose that $f(x) \geq \epsilon > 0$ for all $x > 0$. Prove that every solution of $y'' + f(x)y = 0$ has infinitely many zeros in $(0, \infty)$.

   (b) Does the conclusion in (a) remain true if $f(x) \geq 0$ in $(0, \infty)$? Why?

*6. Prove that no nontrivial solution $y(x)$ of a normal second-order linear differential equation can have infinitely many zeros in a closed interval $I = [a, b]$. [*Hint:* Assume the contrary, and then show by partitioning I into arbitrarily small subintervals that there would exist a point x_0 in I with the property that *every* open interval centered at I contains infinitely many zeros of $y(x)$. Use this fact to prove that

$$y(x_0) = y'(x_0) = 0,$$

and then invoke the uniqueness theorem.]

6–2 THE ZEROS OF SOLUTIONS OF BESSEL'S EQUATION

The second-order linear differential equation

$$x^2 y'' + xy' + (x^2 - p^2)y = 0, \tag{6–6}$$

p a non-negative real number, is known as *Bessel's equation of order p*. It is easily one of the most important differential equations in mathematical physics, and will be studied in some detail in Chapter 15. For the present we restrict ourselves to investigating the behavior of the zeros of its solutions on the interval $(0, \infty)$.

Before the comparison theorem can be applied to (6–6) the term involving y' must be eliminated. This can be done by making the change of variable $y = u/\sqrt{x}$ which transforms (6–6) into

$$u'' + \left(1 + \frac{1 - 4p^2}{4x^2}\right) u = 0 \tag{6–7}$$

without disturbing the zeros of its solutions. Thus it suffices to study the solutions of (6–7). There are three cases to be considered.

Case 1. $0 \le p < \frac{1}{2}$. Then $1 + (1 - 4p^2)/4x^2 > 1$, and the comparison theorem implies that every solution of Bessel's equation vanishes at least once between any two zeros of a nontrivial solution of $u'' + u = 0$. But, as we know, the general solution of this equation is $c_1 \sin x + c_2 \cos x$. Moreover, if α is an arbitrary real number, c_1 and c_2 can be chosen in such a way that the zeros of $c_1 \sin x + c_2 \cos x$ are $\alpha, \alpha \pm \pi, \alpha \pm 2\pi, \ldots$ (See Exercise 2.) Thus we conclude that *every solution of Bessel's equation of order p, $p < \frac{1}{2}$, has* **at least one** *zero in every subinterval of the positive x-axis of length* π; i.e., the distance between successive zeros of these solutions does not exceed π. Finally, it is not difficult to show that this distance is always less than π, and approaches π as $x \to \infty$ (Exercise 3).

Case 2. $p = \frac{1}{2}$. Here Eq. (6–7) reduces to $u'' + u = 0$, and the general solution of (6–6) can be written explicitly as

$$y = \frac{1}{\sqrt{x}}(c_1 \sin x + c_2 \cos x).$$

The remark made a moment ago concerning the choice of c_1 and c_2 again applies, and we can assert that the *zeros of every nontrivial solution of Bessel's equation of order $\frac{1}{2}$ are equally spaced along the positive x-axis, successive zeros separated by an interval of length π.*

Case 3. $p > \frac{1}{2}$. In this case $1 + (1 - 4p^2)/4x^2 < 1$, and comparison with $u'' + u = 0$ implies that every nontrivial solution of (6–7) has *at most* one zero in any subinterval of the positive x-axis of length π. To prove the existence of zeros here we reason as follows.

For any fixed value of p, $(1 - 4p^2)/4x^2 \to 0$, as $x \to \infty$. Hence there exists an $x_0 > 0$ such that $1 + (1 - 4p^2)/4x^2 > \frac{1}{2}$ whenever $x > x_0$, and we now apply the comparison theorem on the interval (x_0, ∞) to (6–7) and the equation $u'' + \frac{1}{2}u = 0$ with general solution $c_1 \sin (\sqrt{2}/2)x + c_2 \cos (\sqrt{2}/2)x$. Since this last function has infinitely many zeros in (x_0, ∞) with successive zeros separated by an interval of length $\sqrt{2}\,\pi$, it follows that *every solution of Bessel's equation of order $p > \frac{1}{2}$ has infinitely many zeros*, and that the distance between successive zeros eventually becomes less than $\sqrt{2}\pi$. By modifying the argument in the obvious way, it can be shown that here too the distance between successive zeros approaches π as $x \to \infty$.

EXERCISES

1. Verify that the substitution $y = u/\sqrt{x}$ reduces Bessel's equation of order p to

$$u'' + \left(1 + \frac{1 - 4p^2}{4x^2}\right) u = 0.$$

2. Prove that for any real number α there exist constants c_1 and c_2 such that the zeros of $c_1 \sin x + c_2 \cos x$ are $\alpha, \alpha \pm \pi, \alpha \pm 2\pi, \ldots$.
3. (a) Prove that the distance between successive zeros of any nontrivial solution of Bessel's equation of order $p < \frac{1}{2}$ is always less than π. [*Hint:* For any fixed $p < \frac{1}{2}$, and any $x_0 > 0$, $1 < 1 + \epsilon < 1 + (1 - 4p^2)/(4x^2)$ on $(0, x_0)$.]
   (b) Prove that the distance in (a) approaches π as $x \to \infty$. [*Hint:* $(1 - 4p^2)/(4x^2) \to 0$ as $x \to \infty$.]
4. Prove that the distance between successive zeros of any nontrivial solution of Bessel's equation of order $p > \frac{1}{2}$ approaches π as $x \to \infty$.

*5. (a) Let $a_0(x)$ be continuous on $(0, \infty)$, and suppose that there exist positive numbers b, B such that $b^2 \leq a_0(x) \leq B^2$ for all $x > 0$. Prove that every nontrivial solution of

$$y'' + a_0(x)y = 0$$

has infinitely many zeros on $(0, \infty)$, and that the distance d between successive zeros can be estimated as

$$\frac{\pi}{B} \leq d \leq \frac{\pi}{b}.$$

[*Hint:* Use the comparison theorem on the given equation and each of $y'' + b^2y = 0$ and $y'' + B^2y = 0$.]

(b) Use the results in (a) to deduce the facts proved in this section concerning the zeros of the solutions of Bessel's equation.

6. Let

$$a_2(x)\frac{d^2y}{dx^2} + a_1(x)\frac{dy}{dx} + a_0(x)y = 0$$

be normal on an interval I. Show that there exists a function $v(x)$ defined on I with the property that the substitution $y = uv$ reduces this equation to

$$u'' + i(x)u = 0.$$

(The function $i(x)$ is called the *invariant* of the equation.)

7. Use the method suggested in Exercise 6 to reduce the following equations to a form in which the first derivative does not appear. (In each of these equations p is a non-negative constant.)
   (a) $x^2y'' + xy' + (x^2 - p^2)y = 0$
   (b) $(1 - x^2)y'' - 2xy' + p(p + 1)y = 0$
   (c) $(1 - x^2)y'' - xy' + p^2y = 0$ (d) $y'' - 2xy' + 2py = 0$
8. Prove that every nontrivial solution of the *Hermite equation*

$$y'' - 2xy' + 2py = 0,$$

p a non-negative constant, has at most finitely many zeros on $(-\infty, \infty)$. [*Hint:* See Example 3 and Exercise 6 of the preceding section, and Exercise 7 (d) above.]

6-3 SELF-ADJOINT FORM; THE SONIN-POLYA THEOREM

Let

$$a_2(x)\frac{d^2y}{dx^2} + a_1(x)\frac{dy}{dx} + a_0(x)y = 0 \tag{6-8}$$

be normal on an interval I, and let

$$p(x) = e^{\int [a_1(x)/a_2(x)]\,dx}. \tag{6-9}$$

Then since

$$\frac{d}{dx}\left(p(x)\frac{dy}{dx}\right) = p(x)\frac{d^2y}{dx^2} + \frac{a_1(x)}{a_2(x)}\,p(x)\frac{dy}{dx},$$

Eq. (6–8) may be rewritten

$$\frac{d}{dx}\left(p(x)\frac{dy}{dx}\right) + \frac{a_0(x)}{a_2(x)}\,p(x)y = 0,$$

or, more simply,

$$\frac{d}{dx}\left(p(x)\frac{dy}{dx}\right) + q(x)y = 0, \tag{6-10}$$

where $q(x) = [a_0(x)/a_2(x)]p(x)$.

Equation (6–10) is known as the *self-adjoint* form of (6–8), and enjoys certain advantages over the original version of the equation. Not the least of these is that it provides a "standard form" for all normal second-order linear differential equations which is easy to derive, and, as we shall see, well suited for computations. For future reference we note that the function $p(x)$ appearing in (6–10) is always *positive* throughout the interval I.

In addition to the self-adjoint form, there are a number of other special forms for normal second-order equations which are sometimes useful. One of them involving the invariant of the equation was given in the exercises at the end of the last section. Another is a normalized version of the self-adjoint form in which $p(x) \equiv 1$. It can be deduced from (6–10) by making the change of variable

$$t = \int \frac{dx}{p(x)}.$$

For then

$$\frac{dy}{dx} = \frac{dy}{dt}\frac{dt}{dx} = \frac{1}{p(x)}\frac{dy}{dt},$$

$$\frac{d}{dx}\left(p(x)\frac{dy}{dx}\right) = \frac{d}{dx}\left(\frac{dy}{dt}\right) = \frac{d}{dt}\left(\frac{dy}{dt}\right)\frac{dt}{dx} = \frac{1}{p(x)}\frac{d^2y}{dt^2},$$

and (6–10) becomes

$$\frac{d^2y}{dt^2} + Q(t)y = 0, \tag{6-11}$$

where $Q(t) = p(x)q(x)$.

This argument proves that *every* normal homogeneous second-order linear differential equation can be written in the form required by the comparison theorem, and thus considerably extends the usefulness of that result. Actually, it proves even more. For if the points x_0 and t_0 correspond under the change of variable

$$t = \int \frac{dx}{p(x)}$$

introduced in passing from (6–10) to (6–11), and if

$$\begin{aligned} Q_1(t) &= p(x)q_1(x), \\ Q_2(t) &= p(x)q_2(x), \end{aligned}$$

then the fact that $p(x)$ is positive throughout I implies that $Q_1(t_0) \geq Q_2(t_0)$ if and only if $q_1(x_0) \geq q_2(x_0)$. Thus if

$$\frac{d^2y}{dt^2} + Q_1(t)y = 0 \qquad \text{and} \qquad \frac{d^2y}{dt^2} + Q_2(t)y = 0$$

are deduced, respectively, from the self-adjoint equations

$$\frac{d}{dx}\left(p(x)\frac{dy}{dx}\right) + q_1(x)y = 0 \qquad \text{and} \qquad \frac{d}{dx}\left(p(x)\frac{dy}{dx}\right) + q_2(x)y = 0, \tag{6–12}$$

then $Q_1(t) \geq Q_2(t)$ if and only if $q_1(x) \geq q_2(x)$. Hence the comparison theorem stated in Section 6–1 is also valid for a pair of self-adjoint equations of the form (6–12) whenever $q_1(x) \geq q_2(x)$ on I.*

As a final application of the ideas we have been exploring here we prove a rather surprising result concerning the zeros of the *derivative* of any solution of a certain class of self-adjoint equations.

Theorem 6–3. (Sonin-Polya theorem.) *Let $p(x) > 0$ and $q(x) \neq 0$ be continuously differentiable on an interval I, and suppose that $p(x)q(x)$ is nonincreasing (nondecreasing) on I. Then the absolute values of the relative*

---

* The reader should note that we have proved this assertion under the assumption that the function $p(x)$ is *the same* in both equations in (6–12). It can be shown, however, that the conclusion of the comparison theorem continues to hold exactly as stated earlier for the solutions of a pair of self-adjoint equations

$$\frac{d}{dx}\left(p_1(x)\frac{dy}{dx}\right) + q_1(x)y = 0, \qquad \frac{d}{dx}\left(p_2(x)\frac{dy}{dx}\right) + q_2(x)y = 0,$$

in which $q_1(x) \geq q_2(x)$ and $0 < p_1(x) \leq p_2(x)$. For a proof of this more general theorem see G. Birkhoff and G. C. Rota, *Ordinary Differential Equations*, Ginn, Boston, 1962.

maxima and minima of every nontrivial solution of the equation

$$\frac{d}{dx}\left(p(x)\frac{dy}{dx}\right) + q(x)y = 0 \tag{6–13}$$

*are nondecreasing (nonincreasing) as x increases.**

Proof. Let y be a nontrivial solution of (6–13), and consider the function

$$F(x) = [y(x)]^2 + \frac{1}{p(x)q(x)}[p(x)y'(x)]^2.$$

Then

$$F' = 2yy' + \frac{2py'}{pq}(py')' - \frac{(pq)'}{(pq)^2}(py')^2,$$

and since, by assumption, $(py')' = -qy$,

$$F'(x) = -\left(\frac{y'}{q}\right)^2 \frac{d}{dx}(pq).$$

Now suppose that pq is nonincreasing on I. Then $d/dx(pq) \leq 0$ and F is nondecreasing on I (i.e., $F' \geq 0$ on I). Hence the same is true of any sequence of values of F computed at points $x_1 < x_2 < \cdots$. If, in particular, $x_1, x_2, \ldots$ are the points at which y has a relative maximum or minimum, then $y'(x_i) = 0$, $F(x_i) = y(x_i)^2$, and we have

$$y(x_1)^2 \leq y(x_2)^2 \leq \cdots.$$

Thus

$$|y(x_1)| \leq |y(x_2)| \leq \cdots,$$

as asserted. A similar argument applies when pq is nondecreasing, and the proof is complete. ▌

Example. In the preceding section we proved that every nontrivial solution of Bessel's equation of order p has infinitely many relative maxima and minima on $(0, \infty)$. Since the self-adjoint form of Bessel's equation is

$$\frac{d}{dx}\left(x\frac{dy}{dx}\right) + \left(\frac{x^2 - p^2}{x}\right)y = 0, \tag{6–14}$$

and since $p(x)q(x) = x^2 - p^2$ is increasing and positive on the interval (p, ∞), the Sonin-Polya theorem implies that the magnitude of the oscillations of such a

* Note that the relative maxima and minima of a solution y of (6–13) are the zeros of y'.

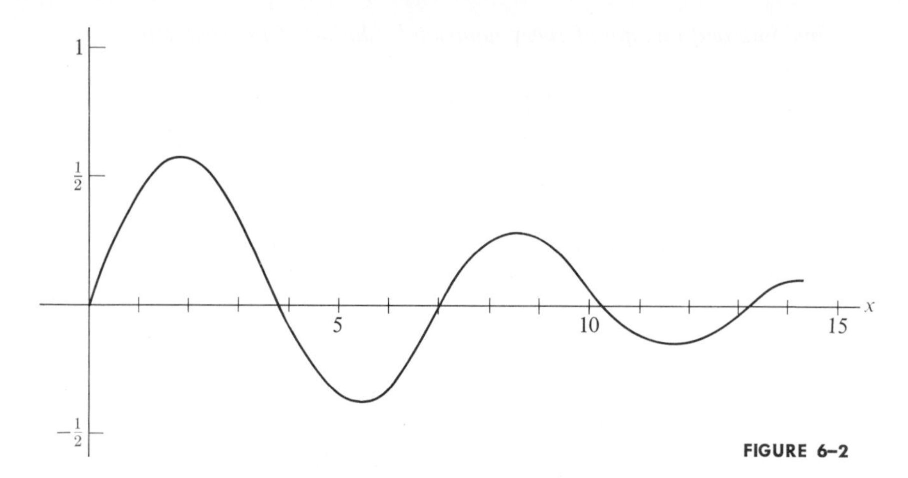

FIGURE 6–2

solution is nonincreasing on this interval. In fact, in this case it can be proved (see Exercise 6) that the oscillations actually *decrease* as shown in Fig. 6–2 where the graph of a solution of Bessel's equation of order 1 has been sketched.

EXERCISES

1. Write each of the following equations in self-adjoint form.
   (a) $(1 - x^2)y'' - 2xy' + 6y = 0$ (b) $x^2y'' - 2x^3y' - (4 - x^2)y = 0$
   (c) $(x^3 - 2)y'' - x^2y' - 3y = 0$ (d) $2x(\ln x)y'' + 3y' - (\sin x)y = 0$
   (e) $(x + 1)y'' - y' + 2xy = 0$
2. Prove that if y_1 and y_2 are solutions of the self-adjoint equation

$$\frac{d}{dx}\left(p(x)\frac{dy}{dx}\right) + q(x)y = 0,$$

   then $p[y_1y_2' - y_1'y_2]$ is a constant. (This result is known as *Abel's identity.*)
3. Write Bessel's equation in the form

$$\frac{d^2y}{dt^2} + Q(t)y = 0$$

   by using the substitution suggested in the text.
4. (a) Prove that no solution of a self-adjoint equation

$$\frac{d}{dx}\left(p(x)\frac{dy}{dx}\right) + q(x)y = 0$$

   can oscillate in any interval where $q(x) \leq 0$.
   (b) For each of the equations in Exercise 1 determine the intervals of the x-axis in which a nontrivial solution can have at most one zero.

5. Discuss the oscillatory behavior of the nontrivial solutions of each of the following equations.

   (a) $2x^2y'' + 6xy' + \left(1 - \frac{1}{x^2}\right)y = 0$

   (b) $y'' - y' + e^x(1 - x)y = 0$ (c) $x^2y'' + xy' + (x^3 - 1)y = 0$

6. Prove that the sequence of absolute values obtained in the Sonin-Polya theorem is strictly increasing (decreasing) when pq is strictly increasing (decreasing).

7. (a) Discuss the oscillatory behavior of the nontrivial solutions of *Airy's differential equation*

$$y'' + xy = 0$$

   on $(-\infty, \infty)$.

   (b) Prove that if $y(x)$ is a solution of Bessel's equation of order $\frac{1}{3}$ on $(0, \infty)$, then $x^{1/2}y(\frac{2}{3}x^{3/2})$ is a solution of Airy's equation.

8. Let $p(x)$ and $q(x) > 0$ satisfy the hypotheses of the Sonin-Polya theorem, and let $y(x)$ be a nontrivial solution of the equation

$$\frac{d}{dx}\left(p(x)\frac{dy}{dx}\right) + q(x)y = 0.$$

   Prove that the values of $|p(x)q(x)|^{1/2}|y(x)|$ at the points where $y' = 0$ form an increasing (decreasing) sequence if $p(x)q(x)$ is an increasing (decreasing) function. [*Hint:* Argue as in the proof of Theorem 6–3, starting with the function

$$F(x) = p(x)q(x)y(x)^2 + [p(x)y'(x)]^2.]$$

6–4 POWER SERIES AND ANALYTIC FUNCTIONS

In an earlier section we introduced the equation $x^2y'' + xy' + (x^2 - p^2)y = 0$ to illustrate how the comparison theorem is used to obtain information about the solutions of a differential equation. This particular equation is but one of a number of linear differential equations with variable coefficients which arise repeatedly in mathematics and mathematical physics, and whose study has had a decisive influence on the development of the theory of differential equations. One of the distinctive features of these equations is that their solutions cannot, in general, be expressed in closed form in terms of elementary functions. Thus it is quite impossible to "solve" them within the context of the naive interpretation which considers a solution of a differential equation as a neat little formula involving familiar functions. Nevertheless, it *is* possible to gain information about the solutions of such equations (witness our discussion of the oscillatory behavior of the solutions of Bessel's equation) and, in fact, enough information to be able to use these solutions effectively in the analysis of other problems. This suggests that we ought to drop the artificial restriction of seeking solutions within some preassigned collection of "known" functions, and adopt instead the larger point of view which sees the solutions of a differential equation as functions *defined* by the equation itself.

One example of the way in which this point of view can be exploited was given earlier when we derived all of the familiar properties of the functions $\sin x$ and $\cos x$ from the fact that they satisfy the equation $y'' + y = 0$ (see Section 3–8). Another is furnished by that treatment of the natural logarithm which sees this function as *the* solution of the first-order equation $xy' = 1$ subject to the initial condition $y(1) = 0$. Together these examples pointedly illustrate the fact that one of the most satisfactory ways of defining new functions in mathematics is as solutions of differential equations.

In the remaining sections of this chapter we shall use the method of power series expansions to study the solutions of a certain large class of linear differential equations, or, rather, to study the functions defined by this class of equations. Since this discussion depends upon certain results in the theory of power series expansions of analytic functions we begin by reviewing some of the basic facts concerning such series. It is assumed that the reader is already familiar with these results, and the real purpose of this resumé is to fix terminology and provide formulas for convenient reference.

In the first place, we recall that an expression of the form

$$\sum_{k=0}^{\infty} a_k(x - x_0)^k, \tag{6–15}$$

x_0 and a_k constants, is called a power series in $x - x_0$, or, alternately, a power series about the point x_0, and is said to converge at x_1 if and only if the series obtained from it by setting $x = x_1$ is a convergent series of real numbers. It is well known that every power series about x_0 either converges at the single point $x = x_0$, or else throughout an interval centered at x_0 of the form $|x - x_0| < r$ with $0 < r \leq \infty$. If, in the latter case, r is chosen as large as possible, the series diverges when $|x - x_0| > r$, and for this reason r is called the radius of convergence of the series.*

Every power series with radius of convergence r defines a function f in the interval $|x - x_0| < r$, and we write

$$f(x) = \sum_{k=0}^{\infty} a_k(x - x_0)^k. \tag{6–16}$$

It can be shown that such functions are infinitely differentiable throughout $|x - x_0| < r$, and that

$$\begin{aligned} f'(x) &= \sum_{k=1}^{\infty} ka_k(x - x_0)^{k-1}, \\ f''(x) &= \sum_{k=2}^{\infty} k(k - 1)a_k(x - x_0)^{k-2}, \\ &\vdots \end{aligned} \tag{6–17}$$

* The behavior of a power series at the endpoints of its interval of convergence cannot be predicted in advance.

where the radius of convergence of each of these derived series is identical with the radius of convergence of (6–16). In short, a power series may be differentiated term-by-term without changing its radius of convergence.

Any function which can be represented by a convergent power series of the form (6–16) in an open interval I about the point x_0 is said to be *analytic* at x_0. We have just seen that such a function must have derivatives of all orders everywhere in I, and, as might be expected, is actually analytic at each point of I. Thus it is customary to speak of functions as being analytic on an interval, and the phrase "analytic at x_0" is used primarily to direct attention to the point about which the series is expanded. Finally, the reader should be aware of the fact that all of the elementary functions in mathematics—polynomials, exponentials, trigonometric functions, etc.—are analytic. Indeed, this is one of the major results of that chapter of calculus which deals with Taylor series expansions of (analytic) functions, and will be used throughout the following discussion.

EXERCISES

1. Let $\mathcal{A}(I)$ denote the set of all functions analytic on an open interval I of the x-axis. Prove that $\mathcal{A}(I)$ is a real vector space under the "usual" definitions of addition and scalar multiplication.
2. Let $y(x)$ be a solution of

$$\frac{d^n y}{dx^n} + a_{n-1}(x)\frac{d^{n-1} y}{dx^{n-1}} + \cdots + a_0(x)y = h(x)$$

on an interval I, and suppose that $a_0(x), \ldots, a_{n-1}(x)$, and $h(x)$ are analytic at a point x_0 in I. Prove that $y(x)$ is infinitely differentiable at x_0.

3. Prove that the power series expansion of an analytic function about the point x_0 is unique, i.e., that such a function has precisely one such expansion.

6–5 ANALYTIC SOLUTIONS OF LINEAR DIFFERENTIAL EQUATIONS

Analytic functions arise in the study of differential equations for the very simple reason that the solutions of any *normal* linear differential equation whose coefficients and right-hand side are analytic on an interval I are themselves analytic on I. This result is known as the existence theorem for equations with analytic coefficients, and in contrast to most existence theorems in the theory of differential equations immediately leads to an explicit technique for computing solutions. Before introducing this technique, however, we give a formal statement of the theorem in question.

Theorem 6–4. (The existence theorem for equations with analytic coefficients.) *Let*

$$\frac{d^n y}{dx^n} + a_{n-1}(x)\frac{d^{n-1} y}{dx^{n-1}} + \cdots + a_0(x)y = h(x) \qquad (6\text{–}18)$$

be a normal nth-order linear differential equation whose coefficients $a_0(x), \ldots, a_{n-1}(x)$ *and right-hand side* $h(x)$ *are analytic in an open interval I. Let* x_0 *be an arbitrary point in I, and suppose that the power series expansions of* $a_0(x), \ldots, h(x)$ *all converge in the interval* $|x - x_0| < r$, $r > 0$. *Then every solution of* (6–18) *which is defined at the point* x_0 *is analytic at that point, and its power series expansion also converges in the interval* $|x - x_0| < r$.

It is important to note that this theorem not only asserts the analyticity of solutions of an equation of the type described, but also specifies an interval in which the power series expansions of these solutions converge. At the same time, it should be observed that we make no statement concerning the behavior of such series *outside* this interval. As we shall see, this is in the nature of things, since it is possible to give examples of series solutions which converge in a larger interval than the one described above.*

EXAMPLE 1. Find the general solution of

$$y'' + xy' + y = 0 \tag{6–19}$$

on $(-\infty, \infty)$.

Since this equation satisfies the hypotheses of Theorem 6–4 on the entire x-axis, each of its solutions has a power series expansion about the point $x_0 = 0$ which converges *for all values of* x.† To compute these solutions we use the so-called *method of undetermined coefficients*, as follows.

Let $y(x)$ be an arbitrary solution of (6–19). Then for suitable constants a_k we have

$$y(x) = \sum_{k=0}^{\infty} a_k x^k, \tag{6–20}$$

and

$$y'(x) = \sum_{k=1}^{\infty} k a_k x^{k-1}, \qquad y''(x) = \sum_{k=2}^{\infty} k(k-1) a_k x^{k-2}.$$

Substituting these series in (6–19), we obtain

$$\sum_{k=2}^{\infty} k(k-1) a_k x^{k-2} + \sum_{k=1}^{\infty} k a_k x^k + \sum_{k=0}^{\infty} a_k x^k = 0, \tag{6–21}$$

and it follows that (6–20) will be a solution of the given equation if and only if the a_k are chosen so that the sum of the coefficients of like powers of x in this expression are all zero (see Exercise 3 of the preceding section).

---

* For a proof of Theorem 6–4 the reader should consult E. A. Coddington, *An Introduction to Ordinary Differential Equations*, Prentice-Hall, Englewood Cliffs, N. J., 1961.

† The student should appreciate that the choice $x_0 = 0$ is one of convenience, not necessity.

To facilitate collecting terms in (6–21) we now replace the index of summation in the first series by $k + 2$, which can certainly be done without prejudice to the sum of the series. This gives

$$\sum_{k=2}^{\infty} k(k-1)a_k x^{k-2} = \sum_{k+2=2}^{\infty} (k+2)(k+1)a_{k+2}x^k$$
$$= \sum_{k=0}^{\infty} (k+2)(k+1)a_{k+2}x^k,$$

and (6–21) becomes

$$\sum_{k=0}^{\infty} (k+2)(k+1)a_{k+2}x^k + \sum_{k=1}^{\infty} ka_k x^k + \sum_{k=0}^{\infty} a_k x^k = 0.$$

But since the order of summation in a power series is a matter of indifference, we can rewrite this expression as

$$a_0 + 2a_2 + \sum_{k=1}^{\infty} (k+1)[(k+2)a_{k+2} + a_k]x^k = 0;$$

whence

$$a_0 + 2a_2 = 0,$$
$$(k+2)a_{k+2} + a_k = 0, \quad k \geq 1.$$

The second of these equations is known as a *recurrence relation* (or *finite difference equation*), and can be used to express the a_k from $k = 3$ onward in terms of the preceding ones. Moreover, since $a_2 = -(a_0/2)$, it follows that *all of the a_k for $k \geq 2$ are uniquely determined by the values of a_0 and a_1*. They fall into two distinct sets depending on the parity of k, as follows:

k even	k odd
a_0	a_1
$a_2 = -\dfrac{a_0}{2}$	$a_3 = -\dfrac{a_1}{3}$
$a_4 = \dfrac{a_0}{4 \cdot 2}$	$a_5 = \dfrac{a_1}{5 \cdot 3}$
$a_6 = -\dfrac{a_0}{6 \cdot 4 \cdot 2}$	$a_7 = -\dfrac{a_1}{7 \cdot 5 \cdot 3}$
$\vdots$	$\vdots$
$a_{2k} = (-1)^k \dfrac{a_0}{(2k)(2k-2)\cdots 4 \cdot 2}$	$a_{2k+1} = (-1)^k \dfrac{a_1}{(2k+1)(2k-1)\cdots 5 \cdot 3}$

Substituting these values in (6–20), we obtain

$$y(x) = a_0\left[1 - \frac{x^2}{2} + \frac{x^4}{4 \cdot 2} - \frac{x^6}{6 \cdot 4 \cdot 2} + \cdots\right] + a_1\left[x - \frac{x^3}{3} + \frac{x^5}{5 \cdot 3} - \frac{x^7}{7 \cdot 5 \cdot 3} + \cdots\right], \tag{6–22}$$

where a_0 and a_1 are arbitrary constants.

To complete the problem it remains to show that (6–22) is the *general solution* of the given equation on $(-\infty, \infty)$. To this end we introduce the series

$$y_0(x) = 1 + \sum_{k=1}^{\infty} (-1)^k \frac{x^{2k}}{(2k)(2k-2)\cdots 4\cdot 2},$$
$$y_1(x) = x + \sum_{k=1}^{\infty} (-1)^k \frac{x^{2k+1}}{(2k+1)(2k-1)\cdots 5\cdot 3}, \tag{6–23}$$

and rewrite (6–22) as

$$y(x) = a_0 y_0(x) + a_1 y_1(x). \tag{6–24}$$

An easy computation using the ratio test now shows that y_0 and y_1 both converge for all values of x. Hence so does (6–24), and we conclude that this expression is a solution of (6–19) on the entire x-axis, just as predicted by Theorem 6–4. Finally, we note that y_0 and y_1 are themselves solutions of (6–19), and that

$$y_0(0) = 1, \qquad y_0'(0) = 0,$$
$$y_1(0) = 0, \qquad y_1'(0) = 1.$$

Hence y_0 and y_1 are linearly independent in $\mathcal{C}(-\infty, \infty)$ and therefore span the solution space of (6–19) (cf. Theorem 3–3). This implies that (6–24) is the general solution of the given equation, and we are done.

EXAMPLE 2. The differential equation

$$(1 - x^2)y'' - 2xy' + \lambda(\lambda + 1)y = 0, \tag{6–25}$$

λ a non-negative constant, is known as *Legendre's equation of order* λ, and will play an important role in much of our later work. In this section we seek the power series expansion

$$y(x) = \sum_{k=0}^{\infty} a_k x^k \tag{6–26}$$

of its general solution about the point $x = 0$. Before we begin, however, we note that since the leading coefficient of (6–25) vanishes when $x = \pm 1$, Theorem 6–4 only guarantees that the series in question will converge in the interval $(-1, 1)$.

This said, we substitute (6–26) and its first two derivatives in (6–25) to obtain

$$(1 - x^2)\sum_{k=2}^{\infty} k(k-1)a_k x^{k-2} - 2x\sum_{k=1}^{\infty} k a_k x^{k-1} + \lambda(\lambda+1)\sum_{k=0}^{\infty} a_k x^k = 0,$$

or

$$\sum_{k=2}^{\infty} k(k-1)a_k x^{k-2} - \sum_{k=2}^{\infty} k(k-1)a_k x^k - \sum_{k=1}^{\infty} 2ka_k x^k + \sum_{k=0}^{\infty} \lambda(\lambda+1)a_k x^k = 0.$$

By shifting the index of summation on the first series and consolidating the last three, this expression may be rewritten

$$-2a_1 x + \lambda(\lambda+1)[a_0 + a_1 x] + \sum_{k=0}^{\infty} (k+2)(k+1)a_{k+2}x^k + \sum_{k=2}^{\infty} [-k(k-1) - 2k + \lambda(\lambda+1)]a_k x^k = 0. \tag{6–27}$$

Finally, we use the identity

$$-k(k-1) - 2k + \lambda(\lambda+1) = (\lambda + k + 1)(\lambda - k)$$

to put (6–27) in the simpler form

$$2a_2 + \lambda(\lambda+1)a_0 + [(\lambda+2)(\lambda-1)a_1 + (3 \cdot 2)a_3]x + \sum_{k=2}^{\infty} [(k+2)(k+1)a_{k+2} + (\lambda+k+1)(\lambda-k)a_k]x^k = 0.$$

To complete the solution we set the various coefficients in this last series equal to zero and solve the resulting equations. This gives

$$2a_2 + \lambda(\lambda+1)a_0 = 0, \qquad (\lambda+2)(\lambda-1)a_1 + (3 \cdot 2)a_3 = 0,$$
$$(k+2)(k+1)a_{k+2} + (\lambda+k+1)(\lambda-k)a_k = 0, \quad k \geq 2,$$

and it is now an easy matter to express all of the a_k in terms of a_0 and a_1. Indeed,

$$a_2 = -\frac{(\lambda+1)\lambda}{2} a_0,$$

$$a_4 = -\frac{(\lambda+3)(\lambda-2)}{4 \cdot 3} a_2 = \frac{(\lambda+3)(\lambda+1)\lambda(\lambda-2)}{4!} a_0,$$

etc., while

$$a_3 = -\frac{(\lambda+2)(\lambda-1)}{3!} a_1,$$

$$a_5 = -\frac{(\lambda+4)(\lambda-3)}{5 \cdot 4} a_3 = \frac{(\lambda+4)(\lambda+2)(\lambda-1)(\lambda-3)}{5!} a_1,$$

etc. In general,

$$a_{2k} = (-1)^k \times \frac{(\lambda + 2k - 1)(\lambda + 2k - 3) \cdots (\lambda + 1)\lambda(\lambda - 2) \cdots (\lambda - 2k + 2)}{(2k)!} a_0,$$

$$a_{2k+1} = (-1)^k \frac{(\lambda + 2k)(\lambda + 2k - 2) \cdots (\lambda + 2)(\lambda - 1) \cdots (\lambda - 2k + 1)}{(2k + 1)!} a_1$$

for all $k > 0$, and it follows that (6–26) may be written

$$y(x) = a_0 y_0(x) + a_1 y_1(x), \tag{6–28}$$

where

$$y_0(x) = 1 - \frac{(\lambda + 1)\lambda}{2!} x^2 + \frac{(\lambda + 3)(\lambda + 1)\lambda(\lambda - 2)}{4!} x^4 - \cdots,$$

$$y_1(x) = x - \frac{(\lambda + 2)(\lambda - 1)}{3!} x^3 + \frac{(\lambda + 4)(\lambda + 2)(\lambda - 1)(\lambda - 3)}{5!} x^5 - \cdots, \tag{6–29}$$

and a_0 and a_1 are arbitrary constants. Moreover, since

$$y_0(0) = 1, \qquad y_0'(0) = 0,$$
$$y_1(0) = 0, \qquad y_1'(0) = 1,$$

we see that (6–28) is the general solution of Legendre's equation of order λ on $(-1, 1)$, as required.

When λ is a (non-negative) integer the solutions of Legendre's equation are of particular interest. For instance, when $\lambda = 2n$, $n = 0, 1, 2, \ldots$, the series for y_0 given above has only a finite number of nonzero terms, and hence is a polynomial. In fact, it is a polynomial of degree $2n$ involving only *even* powers of x. Similarly, when $\lambda = 2n + 1$, $n = 0, 1, 2, \ldots, y_1$ is a polynomial of degree $2n + 1$ involving only *odd* powers of x. But since y_0 and y_1 are themselves solutions of (6–25) we conclude that *Legendre's equation has polynomial solutions for each non-negative integral value of the parameter* λ. These polynomials will be studied in considerable detail in Chapter 11, and we therefore say no more about them here other than to remark that they provide examples of power series solutions of a linear differential equation whose radius of convergence exceeds that predicted by Theorem 6–4.*

* In fact, Legendre's equation has polynomial solutions *only* when λ is a non-negative integer, and these polynomial solutions are the only ones which are continuous on the *closed* interval $[-1, 1]$. For a proof see F. Tricomi, *Differential Equations*, Hafner, New York, 1961, p. 192.

EXERCISES

Express the general solution of each of the following equations as a power series about the point $x = 0$.

1. $y'' + y = 0$
2. $y'' - y = 0$
3. $y'' - 3xy = 0$
4. $y'' - 3xy' - y = 0$
5. $(x^2 + 1)y'' - 6y = 0$
6. $(x^2 + 1)y'' - 8xy' + 15y = 0$
7. $(2x^2 + 1)y'' + 2xy' - 18y = 0$
8. $2y'' + 2x^2y' - xy = 0$
9. $y'' + x^2y' + 2xy = 0$
10. $y'' - x^2y = 6x$
11. $3y'' - xy' + y = x^2 + 2x + 1$
12. $y''' + 3x^2y'' - 2y = 0$
13. $y''' - 3xy' - y = 0$
14. $y''' + x^2y' - xy = 0$

Use the method of undetermined coefficients to express the general solution of each of the following equations as a power series about the point $x = 0$, and specify an interval in which the solution is valid.

15. $(x^2 - 1)y'' + xy' - 4y = 0$
16. $(x^2 - 2)y'' + xy' - y = x^2$
17. $y'' + \dfrac{x}{x^2 - 4}y' - \dfrac{25y}{x^2 - 4} = \dfrac{1 + 2x}{x^2 - 4}$
18. $(x^3 - 8)y'' + x^2y' + xy = 16$
19. $(x^3 + 2)y'' + 6x^2y' - 6xy = 0$
20. $(x^4 - 4)y''' - 36x^2y' - 48xy = 0$
21. Use the method of undetermined coefficients to solve the initial-value problem

$$y'' + xy' - 2y = e^x;$$
$$y(0) = y'(0) = 0.$$

[*Hint:* Expand e^x as a power series about $x = 0$.]

22. Find the general solution of Airy's equation $y'' + xy = 0$.
23. Find a necessary and sufficient condition that a differential equation of the form

$$(x^2 + \alpha)y'' + \beta xy' + \gamma y = 0$$

has a polynomial solution of degree n.

24. Let y_0 and y_1 be the series solutions of Legendre's equation given in (6–29).

(a) Find the radius of convergence of these series. [*Hint:* Use the ratio test.]

(b) Prove that up to constant multiples Legendre's equation of order n (n a non-negative integer) has only one polynomial solution.

(c) Prove that the function

$$\frac{x}{2}\ln\left(\frac{1 + x}{1 - x}\right) - 1$$

is a solution of Legendre's equation of order one on the interval $(-1, 1)$, and use this fact to write the general solution of the equation in closed form.

25. The second-order linear differential equation

$$y'' - 2xy' + 2\lambda y = 0,$$

λ a non-negative constant, is known as *Hermite's equation of order* λ.

(a) Prove that y is a solution of this equation if and only if $u = e^{-x^2/2}y$ is a solution of

$$u'' + (2\lambda + 1 - x^2)u = 0.$$

(b) Use the method of undetermined coefficients to find a basis for the solution space of Hermite's equation.

(c) Show that Hermite's equation of order n has polynomial solutions of degree n for each integer $n \geq 0$, and that up to constant multiples there is precisely one such solution for each n.

*26. Use the method of undetermined coefficients to show that Bessel's equation of order zero has a solution J_0 which is analytic on the entire x-axis and satisfies the condition $J_0(0) = 1$.

6–6 FURTHER EXAMPLES

In certain respects the exercises and examples in the preceding section were somewhat too special to be completely representative of the power series method for solving linear differential equations. For one thing, all of the coefficients in each of these equations were polynomials—which hardly qualify as typical analytic functions. And for another, each of the recurrence relations we obtained led immediately to a general formula for the coefficients of the series being sought. In many cases neither of these simplifications occur, and in order to put this discussion in a more reasonable perspective we now present a number of less elementary examples.

EXAMPLE 1. Solve the initial-value problem

$$\begin{aligned} 3y'' - y' + (x + 1)y &= 1; \\ y(0) = y'(0) &= 0. \end{aligned} \tag{6–30}$$

By Theorem 6–4 we know that the desired solution can be expressed in the form

$$y(x) = \sum_{k=0}^{\infty} a_k x^k \tag{6–31}$$

for suitable constants a_k, and that the resulting series will converge on $(-\infty, \infty)$. Thus we substitute (6–31) and its first two derivatives in the given equation to obtain

$$\sum_{k=2}^{\infty} 3k(k - 1)a_k x^{k-2} - \sum_{k=1}^{\infty} ka_k x^{k-1} + \sum_{k=0}^{\infty} a_k x^{k+1} + \sum_{k=0}^{\infty} a_k x^k = 1,$$

or, after shifting indices of summation,

$$\sum_{k=0}^{\infty} 3(k+2)(k+1)a_{k+2}x^k - \sum_{k=0}^{\infty} (k+1)a_{k+1}x^k + \sum_{k=1}^{\infty} a_{k-1}x^k + \sum_{k=0}^{\infty} a_k x^k = 1.$$

Collecting terms we have

$$6a_2 - a_1 + a_0 + \sum_{k=1}^{\infty} [3(k+2)(k+1)a_{k+2} - (k+1)a_{k+1} + a_k + a_{k-1}]x^k = 1,$$

and it now follows that

$$6a_2 - a_1 + a_0 = 1,$$
$$3(k+2)(k+1)a_{k+2} - (k+1)a_{k+1} + a_k + a_{k-1} = 0, \quad k \geq 1.$$

In addition, by setting $x = 0$ in (6–31) and its first derivative, and using the given initial conditions, we find that $a_0 = a_1 = 0$. Hence

$$a_0 = 0, \qquad a_1 = 0, \qquad a_2 = \tfrac{1}{6},$$

and, in general,

$$a_{k+2} = \frac{a_{k+1}}{3(k+2)} - \frac{a_k + a_{k-1}}{3(k+1)(k+2)}, \quad k \geq 1.$$

Here, for the first time, we are confronted with a recurrence relation which cannot be solved for a_k as a function of k alone. As suggested above, this is not at all uncommon, and when it occurs we have no choice but to compute a few terms of the series involved and then use Theorem 6–4 to determine a (minimal) interval of convergence. In point of fact, this is usually sufficient for most purposes, since it is always possible to develop the series to the point where it can be used in numerical work.

In the present case we have

$$a_3 = \tfrac{1}{54}, \qquad a_4 = -\tfrac{1}{324}, \qquad a_5 = -\tfrac{4}{1215},$$

and

$$y(x) = \tfrac{1}{6}x^2 + \tfrac{1}{54}x^3 - \tfrac{1}{324}x^4 - \tfrac{4}{1215}x^5 + \cdots,$$

an expression which already furnishes an excellent approximation to the required solution in the interval $(-1, 1)$.

EXAMPLE 2. Solve the initial-value problem

$$\begin{gathered} xy'' + y' + xy = 0; \\ y(1) = 0, \qquad y'(1) = -1. \end{gathered} \tag{6–32}$$

Here we begin by making the change of variable $u = x - 1$ to shift the computations to the origin. Then

$$\frac{dy}{du} = \frac{dy}{dx}, \qquad \frac{d^2y}{du^2} = \frac{d^2y}{dx^2}, \tag{6-33}$$

and (6-32) becomes

$$(u + 1)\frac{d^2y}{du^2} + \frac{dy}{du} + (u + 1)y = 0;$$
$$y(0) = 0, \qquad y'(0) = -1.$$

Substituting

$$y(u) = \sum_{k=0}^{\infty} a_k u^k$$

in this equation we obtain

$$(u + 1)\sum_{k=2}^{\infty} k(k - 1)a_k u^{k-2} + \sum_{k=1}^{\infty} k a_k u^{k-1} + (u + 1)\sum_{k=0}^{\infty} a_k u^k = 0,$$

or, after the usual simplifications,

$$a_0 + a_1 + 2a_2 + \sum_{k=1}^{\infty} [(k + 2)(k + 1)a_{k+2} + (k + 1)^2 a_{k+1} + a_k + a_{k-1}]u^k = 0.$$

Thus

$$a_0 + a_1 + 2a_2 = 0,$$
$$(k + 2)(k + 1)a_{k+2} + (k + 1)^2 a_{k+1} + a_k + a_{k-1} = 0, \quad k \geq 1,$$

and since the initial conditions imply that $a_0 = 0$, $a_1 = -1$, we have

$$a_2 = \tfrac{1}{2}, \qquad a_3 = -\tfrac{1}{6}, \qquad a_4 = \tfrac{1}{6}.$$

Hence

$$y(u) = -u + \tfrac{1}{2}u^2 - \tfrac{1}{6}u^3 + \tfrac{1}{6}u^4 + \cdots,$$

and, setting $u = x - 1$,

$$y(x) = 1 - x + \tfrac{1}{2}(x - 1)^2 - \tfrac{1}{6}(x - 1)^3 + \tfrac{1}{6}(x - 1)^4 + \cdots.$$

From Theorem 6-4 we conclude that this series converges at least in the interval $0 < x < 2$, since this is the largest interval centered at $x_0 = 1$ in which the equation is normal.

Example 3. For our final example we solve the initial value problem

$$y'' - e^x y = 0,$$
$$y(0) = y'(0) = 1, \tag{6-34}$$

involving an equation with nonpolynomial coefficients.

As usual we start by substituting

$$y(x) = \sum_{k=0}^{\infty} a_k x^k$$

in the given equation. This yields

$$\sum_{k=2}^{\infty} k(k-1)a_k x^{k-2} - e^x \sum_{k=0}^{\infty} a_k x^k = 0,$$

which we rewrite as

$$\sum_{k=0}^{\infty} (k+2)(k+1)a_{k+2}x^k - \left(\sum_{k=0}^{\infty} \frac{x^k}{k!}\right)\left(\sum_{k=0}^{\infty} a_k x^k\right) = 0.^* \qquad (6\text{–}35)$$

In order to put this expression in the form required by the method of undetermined coefficients (i.e., a power series in x), we now use the theorem which asserts that power series may be multiplied according to the usual rules of algebra within their common intervals of convergence. Thus

$$\begin{aligned}
(a_0 + a_1 x + a_2 x^2 + \cdots)&\left(1 + x + \frac{x^2}{2!} + \cdots\right) \\
&= a_0 + (a_0 + a_1)x + \left(\frac{a_0}{2!} + a_1 + a_2\right)x^2 \\
&\quad + \left(\frac{a_0}{3!} + \frac{a_1}{2!} + a_2 + a_3\right)x^3 + \cdots \\
&= \sum_{k=0}^{\infty}\left(\sum_{j=0}^{k} \frac{a_j}{(k-j)!}\right)x^k.
\end{aligned}$$

Substituting this expression in (6–35) we obtain

$$\sum_{k=0}^{\infty}\left[(k+2)(k+1)a_{k+2} - \sum_{j=0}^{k} \frac{a_j}{(k-j)!}\right]x^k = 0,$$

and it follows that

$$a_{k+2} = \frac{1}{(k+2)(k+1)} \sum_{j=0}^{k} \frac{a_j}{(k-j)!}, \quad k \geq 0.$$

In particular,

$$\begin{aligned}
a_2 &= \frac{a_0}{2}, \\
a_3 &= \frac{a_0 + a_1}{6}, \\
a_4 &= \frac{1}{12}\left(\frac{a_0}{2} + a_1 + a_2\right) = \frac{a_0 + a_1}{12},
\end{aligned}$$

* Recall that $e^x = 1 + x + x^2/2! + x^3/3! + \cdots$.

etc., and in principle all of the a_k can now be computed in terms of a_0 and a_1. Finally, since the initial conditions in effect imply that $a_0 = a_1 = 1$, we have

$$\begin{aligned} a_2 &= \tfrac{1}{2}, \\ a_3 &= \tfrac{1}{3}, \\ a_4 &= \tfrac{1}{6}, \\ &\vdots \end{aligned}$$

and

$$y(x) = 1 + x + \frac{x^2}{2} + \frac{x^3}{3} + \frac{x^4}{6} + \cdots.$$

The validity of the above computations obviously depends upon the fact that convergent power series can be multiplied as though they were polynomials. Since we shall have occasion to refer to this result again we now state it precisely and formally as

Theorem 6–5. *Let*

$$\sum_{k=0}^{\infty} a_k x^k \quad \textit{and} \quad \sum_{k=0}^{\infty} b_k x^k \tag{6–36}$$

be convergent in the interval $|x| < r$, $r > 0$. *Then the series*

$$\sum_{k=0}^{\infty} c_k x^k, \tag{6–37}$$

with

$$c_k = \sum_{j=0}^{k} a_j b_{k-j}, \tag{6–38}$$

known as the **Cauchy product** *of the series in* (6–36), *also converges for* $|x| < r$, *and*

$$\left(\sum_{k=0}^{\infty} a_k x^k\right)\left(\sum_{k=0}^{\infty} b_k x^k\right) = \sum_{k=0}^{\infty} c_k x^k \tag{6–39}$$

for all x in this interval.

When phrased in terms of analytic functions this theorem asserts that the product of two functions f and g which are analytic on an interval I is itself analytic on I, and that its power series expansion about any point x_0 in I is the Cauchy product of the power series expansions of f and g about x_0.*

---

* For a proof the reader should consult Buck, *Advanced Calculus*, 2nd Ed., McGraw-Hill, New York, 1965.

EXERCISES

Find the first four nonzero terms in the series expansion of the solution of each of the following initial value problems, and determine a (minimal) interval of convergence for the series.

1. $y'' + (\sin x)y = 0;$
   $y(0) = 1, y'(0) = 0.$
2. $2y'' - y' + (x + 1)y = 0;$
   $y(0) = 0, y'(0) = 1.$
3. $(x + 1)y'' + y' + xy = 0;$
   $y(0) = y'(0) = -1.$
4. $2y'' - xy = \cos x;$
   $y(0) = y'(0) = 0.$
5. $(1 - 4x^2)y'' + x^2y' - 2y = 0;$
   $y(0) = y'(0) = 1.$
6. $(\cos x)y'' + 2xe^x y = 0;$
   $y(0) = 0, y'(0) = 1.$
7. $(x - 3)y'' + x^2y' + y = 0;$
   $y(0) = 0, y'(0) = 6.$
8. $[1 + \ln(1 + x)]y'' - xy' + y = \sin x;$
   $y(0) = 0, y'(0) = 1.$
9. $xy''' - y = 0;$
   $y(3) = y'(3) = 0, y''(3) = 9.$
10. $xy'' + y' + xy = 0;$
    $y(1) = 0, y'(1) = -1.$
11. $3y''' - xy' + x^2y = e^x;$
    $y(0) = y'(0) = 0, y''(0) = \frac{1}{4}.$
12. $3xy'' - y' = 0;$
    $y(-2) = 1, y'(-2) = -1.$
13. Find the first four nonzero terms in the power series expansion of

$$\int_0^x e^{-t^2}\, dt.$$

What is the interval of convergence of the power series expansion of this integral?

14. Let

$$y(x) = \sum_{k=0}^{\infty} a_k x^k$$

be a solution of the equation

$$y'' + p(x)y' + q(x)y = 0$$

in the interval $|x| < r$, $r > 0$, and suppose that

$$p(x) = \sum_{k=0}^{\infty} p_k x^k, \qquad q(x) = \sum_{k=0}^{\infty} q_k x^k$$

in this interval. Prove that

$$a_{k+2} = -\frac{1}{(k + 1)(k + 2)} \sum_{j=0}^{k} [(j + 1)p_{k-j}a_{j+1} + q_{k-j}a_j],$$

for $k = 0, 1, 2, \ldots$.

7 euclidean spaces

7-1 INNER PRODUCTS

Much of the content of elementary geometry depends upon the ability to measure distance between points. In this chapter we shall show how distance, together with such related concepts as length and angular measure, can be generalized to arbitrary real vector spaces. These so-called "metric" concepts are the foundation of Euclidean geometry, and from them flow a wealth of results in both geometry and analysis.

In order to introduce a metric into a real vector space we must first choose a unit of distance for our measurements. This can be done most easily by defining what is known as an *inner product* on a vector space. The logic behind taking the notion of inner product as primitive will become compelling once we have used it to define length, angular measure, and distance. For each of these concepts then appears as a natural consequence of the notion of inner product, and the student is led to appreciate them as an elaboration of a single idea.

Definition 7-1. An *inner product* is said to be defined on a real vector space $\mathcal{V}$ if with each pair of vectors $\mathbf{x}$, $\mathbf{y}$ in $\mathcal{V}$ there is associated a real number $\mathbf{x} \cdot \mathbf{y}$ in such a way that

$$\mathbf{x} \cdot \mathbf{y} = \mathbf{y} \cdot \mathbf{x}, \tag{7-1}$$

$$(\alpha \mathbf{x}) \cdot \mathbf{y} = \alpha(\mathbf{x} \cdot \mathbf{y}) \quad \text{for every real number } \alpha, \tag{7-2}$$

$$(\mathbf{x}_1 + \mathbf{x}_2) \cdot \mathbf{y} = \mathbf{x}_1 \cdot \mathbf{y} + \mathbf{x}_2 \cdot \mathbf{y}, \tag{7-3}$$

$$\mathbf{x} \cdot \mathbf{x} \geq 0, \quad \text{and} \quad \mathbf{x} \cdot \mathbf{x} = 0 \quad \text{if and only if } \mathbf{x} = \mathbf{0}. \tag{7-4}$$

A vector space with an inner product is known as a *Euclidean* or *inner product space*, and the real number $\mathbf{x} \cdot \mathbf{y}$ (read "$\mathbf{x}$ dot $\mathbf{y}$") is called the inner product of $\mathbf{x}$ and $\mathbf{y}$.*

---

* Some authors call $\mathbf{x} \cdot \mathbf{y}$ the "scalar product" of $\mathbf{x}$ and $\mathbf{y}$. We shall avoid this terminology, however, because of the possibility of confusing it with scalar multiplication as introduced in Chapter 1.

If we apply Eq. (7–1) to (7–2) and (7–3), we see that an inner product also satisfies

$$\mathbf{x} \cdot (\mathbf{y}_1 + \mathbf{y}_2) = \mathbf{x} \cdot \mathbf{y}_1 + \mathbf{x} \cdot \mathbf{y}_2 \quad \text{and} \quad \mathbf{x} \cdot (\alpha \mathbf{y}) = \alpha(\mathbf{x} \cdot \mathbf{y}). \tag{7–5}$$

Even more generally, we have

$$\begin{aligned}(\alpha_1 \mathbf{x}_1 + \alpha_2 \mathbf{x}_2) \cdot (\beta_1 \mathbf{y}_1 + \beta_2 \mathbf{y}_2) = \alpha_1\beta_1(\mathbf{x}_1 \cdot \mathbf{y}_1) + \alpha_1\beta_2(\mathbf{x}_1 \cdot \mathbf{y}_2) \\ + \alpha_2\beta_1(\mathbf{x}_2 \cdot \mathbf{y}_1) + \alpha_2\beta_2(\mathbf{x}_2 \cdot \mathbf{y}_2),\end{aligned} \tag{7–6}$$

where $\alpha_1, \alpha_2, \beta_1, \beta_2$ are arbitrary scalars. For future reference we also note that

$$\mathbf{x} \cdot \mathbf{y} = 0 \quad \text{whenever} \quad \mathbf{x} = \mathbf{0} \quad \text{or} \quad \mathbf{y} = \mathbf{0}. \tag{7–7}$$

[The proofs of (7–6) and (7–7) have been left to the reader in Exercises 6 and 7 below.]

Equation (7–1) in the above definition asserts that an inner product is a commutative or *symmetric* operation on pairs of vectors. Equation (7–2) may be interpreted as an associativity requirement, this time with respect to scalars, while (7–3) requires that the operation be distributive. These two conditions, together with their analogs given in (7–5), are said to make the inner product *bilinear*. Finally, (7–4) is referred to by saying that an inner product is *positive definite*, the allusion here being to the fact that the inner product of a vector with itself is always greater than zero unless the vector involved is the zero vector. Thus one frequently hears an inner product called a real-valued, symmetric, bilinear, positive definite operation on pairs of vectors.

EXAMPLE 1. Let $\mathbf{x} = (x_1, \ldots, x_n)$ and $\mathbf{y} = (y_1, \ldots, y_n)$ be vectors in $\mathcal{R}^n$, and define $\mathbf{x} \cdot \mathbf{y}$ by

$$\mathbf{x} \cdot \mathbf{y} = x_1 y_1 + \cdots + x_n y_n. \tag{7–8}$$

Then $\mathcal{R}^n$ becomes a Euclidean space, and as such is called *Euclidean n-space*. In $\mathcal{R}^2$ and $\mathcal{R}^3$ this inner product is none other than the familiar "dot product" of physics, where the definition is usually phrased geometrically as the product of the length of $\mathbf{x}$, the length of $\mathbf{y}$, and the cosine of the angle between them. The equivalence of these definitions will become apparent in the next section (see Eq. 7–21).

EXAMPLE 2. This example furnishes the first intimation of things to come in analysis. The vector space in question is $\mathcal{C}[a, b]$, the space of all continuous functions on the interval $[a, b]$, with $\mathbf{f} \cdot \mathbf{g}$ defined by

$$\mathbf{f} \cdot \mathbf{g} = \int_a^b f(x)g(x)\,dx. \tag{7–9}$$

It is not difficult to show that (7–9) satisfies all of the requirements for an inner product. Indeed, it is perfectly obvious that $\mathbf{f} \cdot \mathbf{g}$ is both real-valued and symmetric, while bilinearity follows from the equations

$$\int_a^b \alpha f(x)g(x)\,dx = \alpha \int_a^b f(x)g(x)\,dx, \quad \alpha \text{ a real number},$$

and

$$\int_a^b [f_1(x) + f_2(x)]g(x)\,dx = \int_a^b f_1(x)g(x)\,dx + \int_a^b f_2(x)g(x)\,dx.$$

Finally, if we recall that the integral of a non-negative function is non-negative, and that the integral of a *continuous* non-negative function is zero if and only if the function is identically zero (see Exercise 12), we see that

$$\mathbf{f} \cdot \mathbf{f} = \int_a^b f(x)^2\,dx \geq 0,$$

and

$$\mathbf{f} \cdot \mathbf{f} = 0 \quad \text{if and only if} \quad \mathbf{f} = \mathbf{0}.$$

Hereafter whenever we refer to $\mathcal{R}^n$ or $\mathcal{C}[a, b]$ as Euclidean spaces, we shall assume that we are using the inner products defined in the above examples unless express mention is made to the contrary.

EXAMPLE 3. Let $\mathbf{r}$ be a non-negative function in $\mathcal{C}[a, b]$ which vanishes at most at finitely many points in the interval $[a, b]$. Then

$$\mathbf{f} \cdot \mathbf{g} = \int_a^b f(x)g(x)r(x)\,dx \tag{7–10}$$

also defines an inner product on $\mathcal{C}[a, b]$. The function $\mathbf{r}$ is called the *weight function* for this inner product, and we note that when $\mathbf{r}$ is identically equal to 1, (7–10) reduces to the inner product defined in Example 2. We shall meet this inner product again when we study boundary value problems for differential equations.

EXAMPLE 4. Let $\mathcal{P}$ be the vector space consisting of all polynomials in x with real coefficients. In most of our prior discussions of $\mathcal{P}$ we have considered its vectors as objects in and of themselves, and have ignored their interpretation as real-valued continuous functions defined on the entire real line or any of its subintervals. From now on, however, we shall consider the members of $\mathcal{P}$ as polynomial *functions* in order that $\mathcal{P}$ may be viewed as a subspace of $\mathcal{C}[a, b]$ for any pair of real numbers $a < b$. In this case we must define the inner product on $\mathcal{P}$ by (7–9) or (7–10) according as one or the other of these products is being used in $\mathcal{C}[a, b]$. (Similar remarks apply to each of the vector spaces $\mathcal{P}_n$.)

In the preceding example we mentioned the notion of a subspace of a Euclidean space. The relevant definition is obvious: A Euclidean space $\mathcal{W}$ is a subspace of a Euclidean space $\mathcal{V}$ if $\mathcal{W}$ is a subspace of $\mathcal{V}$ as defined in Chapter 1, *and* if the inner product defined on $\mathcal{W}$ coincides with the inner product defined on $\mathcal{V}$. It is clear that an arbitrary subspace $\mathcal{W}$ of a Euclidean space $\mathcal{V}$ is always a subspace of $\mathcal{V}$ in this sense provided we use as the inner product on $\mathcal{W}$ the one that is defined on $\mathcal{V}$. However, it is equally clear that it is possible to furnish $\mathcal{W}$ with an inner product which differs from that defined on $\mathcal{V}$, in which case $\mathcal{W}$, *as a Euclidean space*, is not a subspace of $\mathcal{V}$. The reader should be able to construct examples involving $\mathcal{P}$ and $\mathcal{C}[a, b]$. We furnish one in Exercise 10 below.

EXERCISES

1. Find $\mathbf{x} \cdot \mathbf{y}$ for each of the following pairs of vectors in $\mathcal{R}^3$.

   (a) $\mathbf{x} = (\frac{1}{2}, 2, -1)$, $\mathbf{y} = (4, -2, 3)$
   (b) $\mathbf{x} = (\frac{2}{3}, \frac{1}{2}, 1)$, $\mathbf{y} = (-\frac{1}{2}, 4, 2)$
   (c) $\mathbf{x} = (1, \frac{2}{5}, -3)$, $\mathbf{y} = (1, \frac{1}{2}, \frac{2}{5})$
   (d) $\mathbf{x} = (-5, 0, 1)$, $\mathbf{y} = (1, 0, -5)$
   (e) $\mathbf{x} = (2, -2, 1)$, $\mathbf{y} = (\frac{1}{2}, 1, 3)$

2. Find $\mathbf{f} \cdot \mathbf{g}$ for each of the following pairs of vectors in $\mathcal{C}[0, 1]$. (Recall that the inner product is defined by (7–9).)

   (a) $\mathbf{f}(x) = x$, $\mathbf{g}(x) = 1 - x^2$
   (b) $\mathbf{f}(x) = x$, $\mathbf{g}(x) = 1 - x$
   (c) $\mathbf{f}(x) = \sin \pi x/2$, $\mathbf{g}(x) = \cos \pi x/2$
   (d) $\mathbf{f}(x) = e^x$, $\mathbf{g}(x) = \sin x$
   (e) $\mathbf{f}(x) = |x - \frac{1}{2}|$, $\mathbf{g}(x) = \frac{1}{2} - |x - \frac{1}{2}|$

3. Find $\mathbf{f} \cdot \mathbf{g}$ for each of the following pairs of vectors in $\mathcal{C}[0, 1]$ when the inner product is defined with respect to the weight function $\mathbf{r}(x) = e^x$ (see (7–10)).

   (a) $\mathbf{f}(x) = 1 - 2x$, $\mathbf{g}(x) = e^{-x}$
   (b) $\mathbf{f}(x) = x^2$, $\mathbf{g}(x) = e^x$
   (c) $\mathbf{f}(x) = x$, $\mathbf{g}(x) = 1 - x$
   (d) $\mathbf{f}(x) = e^{-x/2} \sin \pi x/2$, $\mathbf{g}(x) = e^{-x/2} \sin 3\pi x/2$
   (e) $\mathbf{f}(x) = \cos \pi x/2$, $\mathbf{g}(x) = 1$

4. Prove that (7–8) defines an inner product on $\mathcal{R}^n$.

5. Let $\mathbf{x} = (x_1, x_2)$, $\mathbf{y} = (y_1, y_2)$ be arbitrary vectors in $\mathcal{R}^2$. Determine which of the following define an inner product on $\mathcal{R}^2$.

   (a) $\mathbf{x} \cdot \mathbf{y} = x_1 y_1$
   (b) $\mathbf{x} \cdot \mathbf{y} = 2(x_1 y_1 + x_2 y_2)$
   (c) $\mathbf{x} \cdot \mathbf{y} = -2(x_1 y_1 + x_2 y_2)$
   (d) $\mathbf{x} \cdot \mathbf{y} = (x_1 y_1)^2 + (x_2 y_2)^2$
   (e) $\mathbf{x} \cdot \mathbf{y} = x_1 y_1 + x_1 y_2 + x_2 y_1 + 2x_2 y_2$
   (f) $\mathbf{x} \cdot \mathbf{y} = x_1 y_2 + x_2 y_1$

6. Prove that Eq. (7–6) holds in any Euclidean space.

7. Prove (7–7). [*Hint:* Consider $(\mathbf{0} + \mathbf{0}) \cdot \mathbf{y}$.]

8. Let $\mathcal{V}$ be a real vector space, and set $\mathbf{x} \cdot \mathbf{y} = 0$ for every pair of vectors $\mathbf{x}$, $\mathbf{y}$ in $\mathcal{V}$. Is $\mathcal{V}$ a Euclidean space? Why?
9. (a) Let $a < a_1 < b_1 < b$ be real numbers, and define $\mathbf{f} \cdot \mathbf{g}$ in $\mathcal{C}[a, b]$ by

$$\mathbf{f} \cdot \mathbf{g} = \int_{a_1}^{b_1} f(x)g(x)\,dx.$$

Is $\mathcal{C}[a, b]$ then a Euclidean space? Why?

(b) Answer the same question for $\mathcal{P}$. Explain fully.

10. In the space of polynomials $\mathcal{P}$ let

$$\mathbf{p} \cdot \mathbf{q} = a_0b_0 + a_1b_1 + \cdots + a_nb_n,$$

where $\mathbf{p}(x) = a_0 + a_1x + \cdots + a_nx^n$, and $\mathbf{q}(x) = b_0 + b_1x + \cdots + b_nx^n$. (Note that by adding terms with zero coefficients we can make any two polynomials in $\mathcal{P}$ have the same *apparent* degree, as above.)

(a) Prove that this definition yields an inner product on $\mathcal{P}$.

(b) Is $\mathcal{P}$ with this inner product a subspace of the Euclidean space $\mathcal{C}[a, b]$? Why?

11. Let $\mathcal{V}$ be a Euclidean space with inner product $\mathbf{x} \cdot \mathbf{y}$.

(a) For each pair of vectors $\mathbf{x}$, $\mathbf{y}$ in $\mathcal{V}$ let $\mathbf{x} \circ \mathbf{y}$ be defined by

$$\mathbf{x} \circ \mathbf{y} = 2(\mathbf{x} \cdot \mathbf{y}).$$

Prove that this definition yields an inner product on $\mathcal{V}$.

(b) Let α be an arbitrary real number, and define $\mathbf{x} \circ \mathbf{y}$ by

$$\mathbf{x} \circ \mathbf{y} = \alpha(\mathbf{x} \cdot \mathbf{y}).$$

Determine those values of α for which this definition yields an inner product on $\mathcal{V}$.

*12. (a) Let f be a continuous function on the interval $[a, b]$, and suppose $f(x_0) > 0$ for some x_0 in this interval. Use the definition of continuity to prove that $f(x) > 0$ for all values of x in some *subinterval* of $[a, b]$ containing the point x_0.

(b) Use the result in (a) to prove that in $\mathcal{C}[a, b]$ $\mathbf{f} \cdot \mathbf{f} \geq 0$, and that $\mathbf{f} \cdot \mathbf{f} = 0$ if and only if $\mathbf{f} = \mathbf{0}$.

13. Prove that (7–10) defines an inner product on $\mathcal{C}[a, b]$.
14. Will (7–10) define an inner product on $\mathcal{C}[a, b]$ if we merely require $\mathbf{r}$ to be nonnegative on $[a, b]$? Why?
15. Let $\mathbf{r}$ be any function in $\mathcal{C}[a, b]$ which vanishes for at most finitely many values in the interval $[a, b]$. Prove that

$$\mathbf{f} \cdot \mathbf{g} = \int_a^b f(x)g(x)|r(x)|\,dx$$

defines an inner product on $\mathcal{C}[a, b]$.

*16. Let $\mathbf{x} = (x_1, x_2)$ and $\mathbf{y} = (y_1, y_2)$ be vectors in $\mathcal{R}^2$, and let

$$(a_{ij}) = \begin{pmatrix} a_{11} & a_{12} \\ a_{21} & a_{22} \end{pmatrix}$$

be a 2×2 matrix whose entries are real numbers. Set

$$\mathbf{x} \cdot \mathbf{y} = a_{11}x_1y_1 + a_{12}x_1y_2 + a_{21}x_2y_1 + a_{22}x_2y_2. \tag{7-11}$$

(a) Show that this definition satisfies Eqs. (7-2) and (7-3) of Definition 7-1 for every 2×2 matrix (a_{ij}).

(b) Show that Eq. (7-1) is satisfied if and only if $a_{12} = a_{21}$ [i.e., if and only if (a_{ij}) is a *symmetric* matrix], and hence deduce that (7-11) defines an inner product on $\mathcal{R}^2$ if and only if (a_{ij}) is a 2×2 symmetric matrix such that

$$a_{11}x_1^2 + (a_{12} + a_{21})x_1x_2 + a_{22}x_2^2$$

is non-negative for every choice of x_1 and x_2, and is zero if and only if $x_1 = x_2 = 0$.

(c) Find a matrix (a_{ij}) which reduces (7-11) to the ordinary inner product on $\mathcal{R}^2$.

(d) Determine which of the following matrices can be used to define an inner product on $\mathcal{R}^2$:

$$\begin{pmatrix} 2 & 1 \\ 1 & 1 \end{pmatrix}, \begin{pmatrix} -1 & 0 \\ 1 & 0 \end{pmatrix}, \begin{pmatrix} 1 & 1 \\ 1 & 1 \end{pmatrix}.$$

*17. Generalize the preceding exercise to $\mathcal{R}^n$.

7-2 LENGTH, ANGULAR MEASURE, DISTANCE

In this section we fulfill our promise to define length, angular measure, and distance in terms of the inner product on a Euclidean space. Each of these concepts has a well-defined meaning in Euclidean 2-space, and it is reasonable to demand that any definition we adopt reduce to the familiar one in $\mathcal{R}^2$. Thus we can obtain acceptable definitions by rewriting the relevant formulas from analytic geometry in terms of the inner product on $\mathcal{R}^2$, and then adopting the results as definitions for arbitrary Euclidean spaces.

Turning first to the notion of length, let $\mathbf{x} = (x_1, x_2)$ be any vector in $\mathcal{R}^2$. Then the length of $\mathbf{x}$, denoted by $\|\mathbf{x}\|$, is the non-negative real number

$$\|\mathbf{x}\| = \sqrt{x_1^2 + x_2^2}$$

(see Fig. 7-1). But this expression may be rewritten in terms of the inner product on $\mathcal{R}^2$ (Formula 7-8) as

$$\|\mathbf{x}\| = \sqrt{\mathbf{x} \cdot \mathbf{x}},$$

and we have our first definition.

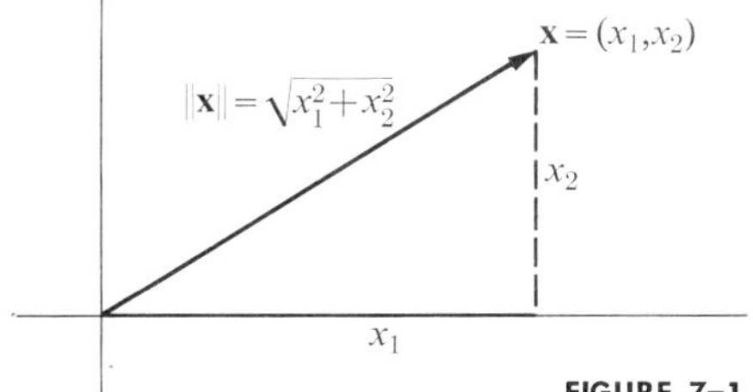

FIGURE 7-1

Definition 7-2. The *length* (or *norm*) of a vector $\mathbf{x}$ in a Euclidean space is defined to be the *non-negative* real number

$$\|\mathbf{x}\| = \sqrt{\mathbf{x} \cdot \mathbf{x}}. \tag{7-12}$$

Thus, in particular, the length of a vector $\mathbf{x} = (x_1, \ldots, x_n)$ in $\mathcal{R}^n$ is

$$\|\mathbf{x}\| = \sqrt{x_1^2 + \cdots + x_n^2}, \tag{7–13}$$

while the length of a vector $\mathbf{f}$ in $\mathcal{C}[a, b]$ is

$$\|\mathbf{f}\| = \left(\int_a^b f(x)^2\,dx\right)^{1/2}. \tag{7–14}$$

Next, we observe that if $\mathbf{x}$ and $\mathbf{y}$ are any two nonzero vectors in $\mathcal{R}^2$, the formula

$$\cos\theta = \frac{\mathbf{x}\cdot\mathbf{y}}{\|\mathbf{x}\|\,\|\mathbf{y}\|}, \quad 0 \le \theta \le \pi, \tag{7–15}$$

is an immediate consequence of the law of cosines (see Exercise 6). But the expression

$$\frac{\mathbf{x}\cdot\mathbf{y}}{\|\mathbf{x}\|\,\|\mathbf{y}\|}$$

is also meaningful in an arbitrary Euclidean space, a fact which suggests considering (7–15) as a reasonable candidate for the general definition of $\cos\theta$. Before acting on this suggestion, however, we must establish the inequality

$$-1 \le \frac{\mathbf{x}\cdot\mathbf{y}}{\|\mathbf{x}\|\,\|\mathbf{y}\|} \le 1 \tag{7–16}$$

for every pair of nonzero vectors in a Euclidean space, since, of course, any definition of $\cos\theta$ must satisfy the inequality $-1 \le \cos\theta \le 1$. This fact will emerge as a consequence of the following important result, known as the *Schwarz* or *Cauchy-Schwarz* inequality.

Theorem 7–1. (Schwarz inequality.) *If* $\mathbf{x}$ *and* $\mathbf{y}$ *are any two vectors in a Euclidean space, then*

$$(\mathbf{x}\cdot\mathbf{y})^2 \le (\mathbf{x}\cdot\mathbf{x})(\mathbf{y}\cdot\mathbf{y}). \tag{7–17}$$

Proof. We first observe that this inequality is immediate if either $\mathbf{x}$ or $\mathbf{y}$ is the zero vector, since then both sides of (7–17) are zero [see (7–7)]. Thus it suffices to consider the case in which $\mathbf{x}$ and $\mathbf{y}$ are nonzero. Here we use (7–6) to expand $(\alpha\mathbf{x} - \beta\mathbf{y})\cdot(\alpha\mathbf{x} - \beta\mathbf{y})$ where α and β are arbitrary real numbers. By (7–4) we have

$$\begin{aligned} 0 &\le (\alpha\mathbf{x} - \beta\mathbf{y})\cdot(\alpha\mathbf{x} - \beta\mathbf{y}) \\ &= \alpha^2(\mathbf{x}\cdot\mathbf{x}) - 2\alpha\beta(\mathbf{x}\cdot\mathbf{y}) + \beta^2(\mathbf{y}\cdot\mathbf{y}), \end{aligned}$$

whence

$$2\alpha\beta(\mathbf{x}\cdot\mathbf{y}) \le \alpha^2(\mathbf{x}\cdot\mathbf{x}) + \beta^2(\mathbf{y}\cdot\mathbf{y}).$$

We now set

$$\alpha = \sqrt{\mathbf{y}\cdot\mathbf{y}} \quad \text{and} \quad \beta = \sqrt{\mathbf{x}\cdot\mathbf{x}}.$$

This gives

$$2\sqrt{\mathbf{x}\cdot\mathbf{x}}\sqrt{\mathbf{y}\cdot\mathbf{y}}\,(\mathbf{x}\cdot\mathbf{y}) \leq 2(\mathbf{x}\cdot\mathbf{x})(\mathbf{y}\cdot\mathbf{y}),$$

or

$$\mathbf{x}\cdot\mathbf{y} \leq \sqrt{\mathbf{x}\cdot\mathbf{x}}\sqrt{\mathbf{y}\cdot\mathbf{y}}.$$

Squaring, we obtain (7–17). ▌

In $\mathcal{R}^n$ the Schwarz inequality assumes the form

$$\left(\sum_{i=1}^{n} x_i y_i\right)^2 \leq \left(\sum_{i=1}^{n} x_i^2\right)\left(\sum_{i=1}^{n} y_i^2\right), \tag{7–18}$$

while in $\mathcal{C}[a, b]$ it becomes

$$\left(\int_a^b f(x)g(x)\,dx\right)^2 \leq \left(\int_a^b f(x)^2\,dx\right)\left(\int_a^b g(x)^2\,dx\right)\cdot \tag{7–19}$$

The first of these inequalities is valid for any collection $x_1, \ldots, x_n, y_1, \ldots, y_n$ of real numbers, and is usually called *Cauchy's inequality.* It is worth remembering since it is often useful in deducing other arithmetic inequalities (see Exercises 9 through 11 below).

In the notation of Definition 7–2 the Schwarz inequality becomes

$$|\mathbf{x}\cdot\mathbf{y}| \leq \|\mathbf{x}\|\,\|\mathbf{y}\|, \tag{7–20}$$

and asserts that the absolute value of the inner product of two vectors does not exceed the product of the lengths of the vectors. Thus

$$\frac{|\mathbf{x}\cdot\mathbf{y}|}{\|\mathbf{x}\|\,\|\mathbf{y}\|} \leq 1,$$

whenever $\mathbf{x}$ and $\mathbf{y}$ are nonzero. But this is just another way of writing (7–16), the inequality needed to justify using (7–15) as a definition of $\cos\theta$, and we can now state

Definition 7–3. If $\mathbf{x}$ and $\mathbf{y}$ are nonzero vectors in a Euclidean space we define the *cosine of the angle between them* to be

$$\cos\theta = \frac{\mathbf{x}\cdot\mathbf{y}}{\|\mathbf{x}\|\,\|\mathbf{y}\|}\cdot \tag{7–21}$$

If, on the other hand, one of the vectors is zero, we set $\cos\theta = 0$.

It goes without saying that in defining the cosine of the angle between $\mathbf{x}$ and $\mathbf{y}$ we have, by implication, also defined the angle in question; just take the principal value of the inverse cosine.

At this point all that remains of our original program is to define the distance between any two points (i.e., vectors) in a Euclidean space. Again this is done simply by copying the definition from $\mathcal{R}^2$ where the distance between $\mathbf{x}$ and $\mathbf{y}$ is the length of the vector $\mathbf{x} - \mathbf{y}$ (Fig. 7–2). Thus

Definition 7–4. The *distance* between two vectors $\mathbf{x}$ and $\mathbf{y}$ in a Euclidean space is, by definition,

$$d(\mathbf{x}, \mathbf{y}) = \|\mathbf{x} - \mathbf{y}\|. \tag{7–22}$$

But is this a reasonable definition of the term "distance"? In order to answer this question we must first decide what properties we require of distance in general.

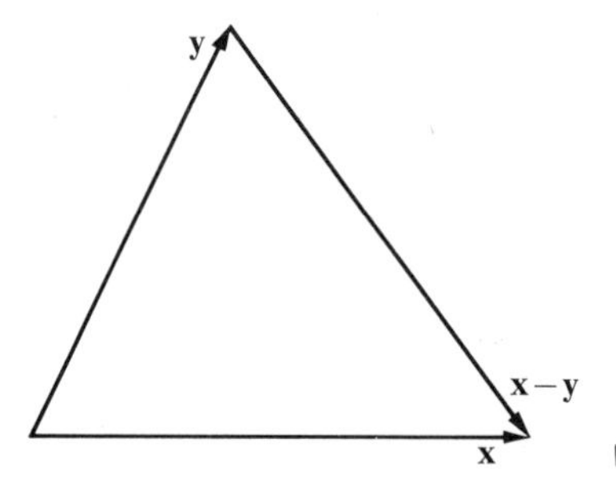

FIGURE 7–2

On this score mathematicians are in agreement, having decided as follows. The distance between two points must be a non-negative real number which is zero if and only if the points coincide. It must be independent of the order in which the points are considered, and finally the *triangle inequality*, famous from plane geometry, must be satisfied. Thus in order to justify using the term "distance" in Definition 7–4 we must show that $d(\mathbf{x}, \mathbf{y})$ is a real number satisfying

$$d(\mathbf{x}, \mathbf{y}) \geq 0, \tag{7–23}$$

$$d(\mathbf{x}, \mathbf{y}) = 0 \quad \text{if and only if } \mathbf{x} = \mathbf{y}, \tag{7–24}$$

$$d(\mathbf{x}, \mathbf{y}) = d(\mathbf{y}, \mathbf{x}), \tag{7–25}$$

$$d(\mathbf{x}, \mathbf{y}) + d(\mathbf{y}, \mathbf{z}) \geq d(\mathbf{x}, \mathbf{z}) \quad \text{for any three vectors } \mathbf{x}, \mathbf{y}, \mathbf{z}. \tag{7–26}$$

The first three of these properties follow immediately from the definition of length and the axioms governing an inner product. The last, however, is not quite so obvious. To prove it we first establish an inequality which is of some importance in its own right.

Lemma 7–1. *If* $\mathbf{x}$ *and* $\mathbf{y}$ *are arbitrary vectors in a Euclidean space, then*

$$\|\mathbf{x} + \mathbf{y}\| \leq \|\mathbf{x}\| + \|\mathbf{y}\|. \tag{7–27}$$

Proof.

$$\begin{aligned}\|\mathbf{x}+\mathbf{y}\| &= [(\mathbf{x}+\mathbf{y})\cdot(\mathbf{x}+\mathbf{y})]^{1/2}\\ &= [(\mathbf{x}\cdot\mathbf{x}) + 2(\mathbf{x}\cdot\mathbf{y}) + (\mathbf{y}\cdot\mathbf{y})]^{1/2}\\ &\le [(\mathbf{x}\cdot\mathbf{x}) + 2\sqrt{\mathbf{x}\cdot\mathbf{x}}\sqrt{\mathbf{y}\cdot\mathbf{y}} + (\mathbf{y}\cdot\mathbf{y})]^{1/2} \quad \text{(by the Schwarz inequality)}\\ &= [(\sqrt{\mathbf{x}\cdot\mathbf{x}} + \sqrt{\mathbf{y}\cdot\mathbf{y}})^2]^{1/2}\\ &= \sqrt{\mathbf{x}\cdot\mathbf{x}} + \sqrt{\mathbf{y}\cdot\mathbf{y}}\\ &= \|\mathbf{x}\| + \|\mathbf{y}\|. \blacksquare\end{aligned}$$

The triangle inequality follows at once from this result. Indeed,

$$\|\mathbf{x}-\mathbf{z}\| = \|(\mathbf{x}-\mathbf{y}) + (\mathbf{y}-\mathbf{z})\| \le \|\mathbf{x}-\mathbf{y}\| + \|\mathbf{y}-\mathbf{z}\|, \tag{7–28}$$

which is precisely what we had to show.

Finally, we note that the distance function defined above also enjoys the following agreeable properties:

$$d(\alpha\mathbf{x}, \alpha\mathbf{y}) = |\alpha|\, d(\mathbf{x}, \mathbf{y}) \quad \text{for any real number } \alpha, \tag{7–29}$$

and

$$d(\mathbf{x}+\mathbf{z}, \mathbf{y}+\mathbf{z}) = d(\mathbf{x}, \mathbf{y}). \tag{7–30}$$

The proofs and geometric interpretations are left to the reader.

EXERCISES

1. Find the length of each of the following vectors in $\mathcal{R}^4$.
   (a) $(1, -2, 2, 0)$
   (b) $(\frac{1}{2}, -\sqrt{3}, \sqrt{2}, 1)$
   (c) $(3, 4, -3, 1)$
   (d) $(\frac{2}{3}, \frac{1}{2}, -\frac{2}{3}, -\frac{1}{2})$
   (e) $(2, -1, 5, 3)$
2. Find the distance between each of the following pairs of points in $\mathcal{R}^4$.
   (a) $\mathbf{x} = (3, 0, -1, 5)$, $\mathbf{y} = (2, 2, -1, 3)$
   (b) $\mathbf{x} = (7, -4, 1, 3)$, $\mathbf{y} = (2, 1, -4, 8)$
   (c) $\mathbf{x} = (\frac{2}{3}, 1, 0, 2)$, $\mathbf{y} = (\frac{1}{6}, \frac{3}{2}, -\frac{1}{2}, \frac{3}{2})$
   (d) $\mathbf{x} = (1, -1, 0, 2)$, $\mathbf{y} = (2, 1, 1, 0)$
   (e) $\mathbf{x} = (\frac{1}{3}, \frac{1}{2}, 2, 1)$, $\mathbf{y} = (-\frac{2}{3}, \frac{1}{4}, \frac{3}{2}, 1)$
3. Compute $\|\mathbf{f}\|$ for each of the following vectors in $\mathcal{C}[0, 1]$.
   (a) $\mathbf{f}(x) = x$
   (b) $\mathbf{f}(x) = e^{x/2}$
   (c) $\mathbf{f}(x) = 1 - x^2$
   (d) $\mathbf{f}(x) = \sin \pi x/2$
   (e) $\mathbf{f}(x) = \ln(x + 1)$

4. Find the angle between each of the following pairs of vectors in $\mathcal{R}^3$.
   (a) $\mathbf{x} = (1, 1, 1)$, $\mathbf{y} = (\frac{1}{2}, -1, \frac{1}{2})$
   (b) $\mathbf{x} = (\frac{1}{2}, 1, -2)$, $\mathbf{y} = (2, 4, -8)$
   (c) $\mathbf{x} = (1, -1, 0)$, $\mathbf{y} = (2, -1, 2)$
   (d) $\mathbf{x} = (0, \sqrt{39}, \sqrt{11})$, $\mathbf{y} = (1, 5, 0)$
   (e) $\mathbf{x} = (-3, -1, 0)$, $\mathbf{y} = (1, 2, -\sqrt{5})$
5. (a) Find the cosine of the angle between each of the following pairs of vectors in $\mathcal{P}_3$ if the inner product is $\mathbf{p} \cdot \mathbf{q} = \int_{-1}^{1} p(x)q(x)\,dx$.

$$\text{(i) } 1, x \qquad \text{(ii) } x, x^2 \qquad \text{(iii) } x,\ 1 - x$$

   (b) Repeat (a), this time using the inner product $\mathbf{p} \cdot \mathbf{q} = \int_0^1 p(x)q(x)\,dx$.
6. Use the law of cosines to establish that

$$\cos\theta = \frac{\mathbf{x} \cdot \mathbf{y}}{\|\mathbf{x}\|\,\|\mathbf{y}\|}$$

   for any pair of nonzero vectors $\mathbf{x}$, $\mathbf{y}$ in $\mathcal{R}^2$ (see Fig. 7–3).

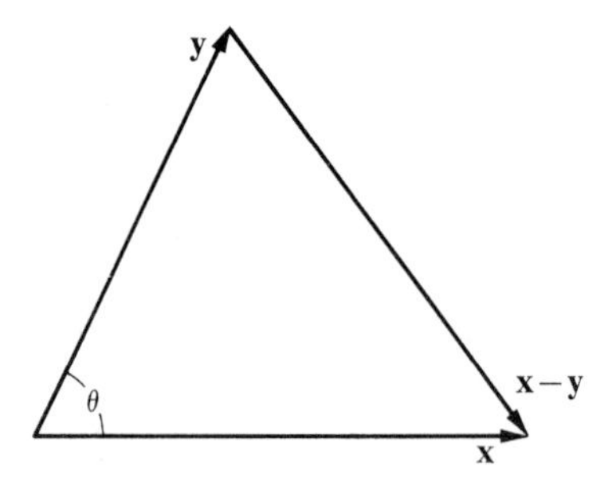

FIGURE 7–3

7. (a) Compute $\mathbf{f} \cdot \mathbf{g}$ for each of the following pairs of vectors in $\mathcal{C}[-\pi, \pi]$:
   (i) $\mathbf{f}(x) = \sin mx$, $\mathbf{g}(x) = \sin nx$,
   (ii) $\mathbf{f}(x) = \sin mx$, $\mathbf{g}(x) = \cos nx$,
   (iii) $\mathbf{f}(x) = \cos mx$, $\mathbf{g}(x) = \cos nx$,

   where m and n are arbitrary non-negative integers.
   (b) What can you say about the functions

$$1, \sin x, \cos x, \sin 2x, \cos 2x, \ldots$$

   in $\mathcal{C}[-\pi, \pi]$ on the basis of the results in (*a*)?
8. Prove that the Schwarz inequality becomes an equality if and only if $\mathbf{x}$ and $\mathbf{y}$ are linearly dependent.
9. Let $a_1, \ldots, a_n$ be positive real numbers. Prove that

$$(a_1 + \cdots + a_n)\left(\frac{1}{a_1} + \cdots + \frac{1}{a_n}\right) \geq n^2.$$

   [*Hint:* Use Cauchy's inequality.]

*10. Let a, b, c be positive real numbers such that $a + b + c = 1$. Use Cauchy's inequality to prove that

$$\left(\frac{1}{a} - 1\right)\left(\frac{1}{b} - 1\right)\left(\frac{1}{c} - 1\right) \geq 8.$$

11. Prove that the following inequality holds for any collection of real numbers $a_1, \ldots, a_n$:

$$\left(\frac{a_1 + \cdots + a_n}{n}\right)^2 \leq \frac{a_1^2 + \cdots + a_n^2}{n}.$$

*12. Let t be an arbitrary real number, and consider the inequality

$$0 \leq (t\mathbf{x} - \mathbf{y}) \cdot (t\mathbf{x} - \mathbf{y}),$$

valid for any pair of vectors $\mathbf{x}$ and $\mathbf{y}$ in a Euclidean space. Expand this inner product, and derive the Schwarz inequality by examining the discriminant of the resulting inequality in t. (This is another very popular way of deriving the Schwarz inequality.)

13. Prove that distance as defined in (7–22) satisfies (7–23) to (7–25).

*14. Set $\mathbf{x} = \mathbf{z}$ in (7–26) and use (7–24) and (7–25) to deduce (7–23), thus showing that this relation is actually implied by the other three.

15. Show that $\|\alpha\mathbf{x}\| = |\alpha|\, \|\mathbf{x}\|$ for all real numbers α.

16. Prove that

$$d(\alpha\mathbf{x}, \alpha\mathbf{y}) = |\alpha|\, d(\mathbf{x}, \mathbf{y})$$

for all real numbers α, and that

$$d(\mathbf{x} + \mathbf{z}, \mathbf{y} + \mathbf{z}) = d(\mathbf{x}, \mathbf{y}).$$

17. Use (7–27) to deduce that

$$\|\mathbf{x} \pm \mathbf{y}\| \geq \big|\, \|\mathbf{x}\| - \|\mathbf{y}\| \,\big|$$

for any pair of vectors $\mathbf{x}, \mathbf{y}$ in a Euclidean space.

*18. Prove that $\|\mathbf{x} + \mathbf{y}\| = \|\mathbf{x}\| + \|\mathbf{y}\|$ if and only if $\mathbf{y} = \alpha\mathbf{x}$ or $\mathbf{x} = \alpha\mathbf{y}$ for some real number $\alpha \geq 0$.

*19. Let

$$(a_{ij}) = \begin{pmatrix} a_{11} & a_{12} \\ a_{21} & a_{22} \end{pmatrix}$$

be a 2×2 matrix whose entries are real numbers, and suppose that (a_{ij}) is so chosen that (7–11) is an inner product on $\mathcal{R}^2$ (see Exercise 16 in the preceding section). Use the Schwarz inequality to deduce that

$$a_{12}^2 \leq a_{11}a_{22}.$$

Conversely, show that if (a_{ij}) satisfies this inequality, and $a_{12} = a_{21}$, then (7–11) defines an inner product on $\mathcal{R}^2$.

7-3 ORTHOGONALITY

Two vectors in a Euclidean space are said to be *orthogonal* or *perpendicular* if the cosine of the angle between them is zero. Referring to Definition 7-3 we see that the zero vector is othogonal to everything, and, in general, that $\mathbf{x}$ and $\mathbf{y}$ are orthogonal if and only if $\mathbf{x} \cdot \mathbf{y} = 0$. In a moment we shall generalize the notion of orthogonality somewhat, but first we prove a particularly celebrated theorem.

Theorem 7-2. (Pythagoras.) *Two vectors* $\mathbf{x}$ *and* $\mathbf{y}$ *in a Euclidean space are orthogonal if and only if*

$$\|\mathbf{x} + \mathbf{y}\|^2 = \|\mathbf{x}\|^2 + \|\mathbf{y}\|^2.$$

Proof.

$$\begin{aligned} \|\mathbf{x} + \mathbf{y}\|^2 &= (\mathbf{x} + \mathbf{y}) \cdot (\mathbf{x} + \mathbf{y}) \\ &= \mathbf{x} \cdot \mathbf{x} + 2(\mathbf{x} \cdot \mathbf{y}) + \mathbf{y} \cdot \mathbf{y} \\ &= \|\mathbf{x}\|^2 + 2(\mathbf{x} \cdot \mathbf{y}) + \|\mathbf{y}\|^2. \end{aligned}$$

Thus $\|\mathbf{x} + \mathbf{y}\|^2 = \|\mathbf{x}\|^2 + \|\mathbf{y}\|^2$ if and only if $\mathbf{x} \cdot \mathbf{y} = 0$, as asserted. ▮

We have appended two figures in illustration of this result (Figs. 7-4 and 7-5). The first is certainly one of the most familiar and well chosen diagrams in mathematics, and needs no comment. The second, on the other hand, has the negative virtue of conveying almost no information at all, and should stand as a warning to the student to exercise geometric restraint in interpreting statements concerning orthogonality.

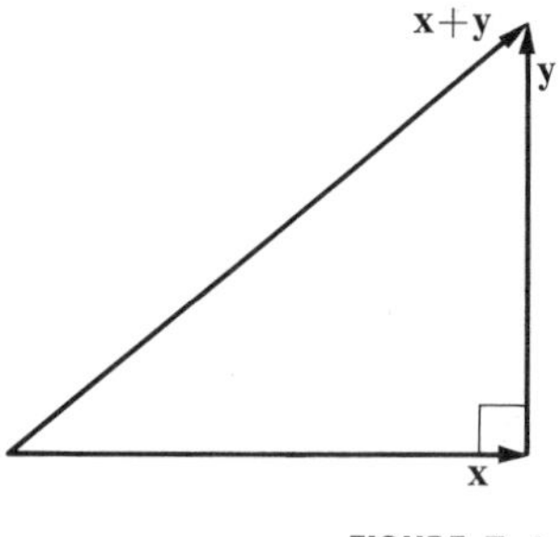

FIGURE 7-4

This said, we continue the discussion by giving the following definition.

Definition 7-5. A set of vectors $\mathbf{x}_1, \mathbf{x}_2, \ldots, \mathbf{x}_i, \ldots$ in a Euclidean space is said to be an *orthogonal set* if $\mathbf{x}_i \neq \mathbf{0}$ for all i, and

$$\mathbf{x}_i \cdot \mathbf{x}_j = 0 \tag{7-31}$$

whenever $i \neq j$. If, in addition,

$$\mathbf{x}_i \cdot \mathbf{x}_i = 1 \tag{7-32}$$

for each i, the set is said to be *orthonormal.*

Thus an orthogonal set is a set of *mutually perpendicular nonzero* vectors, while an ortho*normal* set is an orthogonal set in which each of the vectors is of unit length. For economy of notation when discussing orthonormal sets, Eqs. (7-31) and (7–32) are frequently combined by writing

$$\mathbf{x}_i \cdot \mathbf{x}_j = \delta_{ij}, \quad \text{where } \delta_{ij} = \begin{cases} 0 & \text{if } i \neq j, \\ 1 & \text{if } i = j. \end{cases} \tag{7–33}$$

The symbol δ_{ij} introduced here is called the *Kronecker delta.*

The distinction between an orthogonal and an orthonormal set is really trifling, for if we replace each vector $\mathbf{x}$ in an orthogonal set by the "normalized" vector $\mathbf{x}/\|\mathbf{x}\|$ of unit length, the resulting set is obviously orthonormal. The only reason for introducing orthonormal sets at all is for the convenience which sometimes results from working with unit vectors.

Before giving examples, we call attention to two points in the above definition. The first is that every vector in an orthogonal (or orthonormal) set is nonzero; the second is that we have placed no restriction on the number of vectors in such sets. In particular, an orthogonal set may contain an *infinite* number of vectors (see Example 3 below). Such sets will occur repeatedly in Chapters 9 and 11.

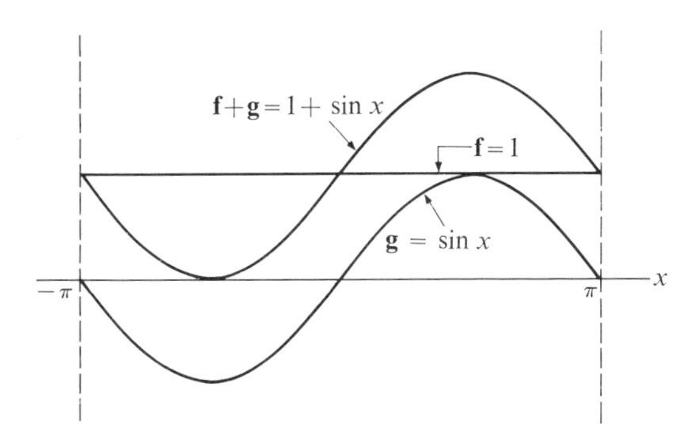

FIGURE 7–5

EXAMPLE 1. In $\mathcal{R}^3$ the vectors (1, 0, 0), (0, 2, 0), $(0, 0, -\frac{1}{2})$ form an orthogonal set, while the standard basis vectors (1, 0, 0), (0, 1, 0), (0, 0, 1) form an orthonormal set. More generally, the set consisting of the standard basis vectors in $\mathcal{R}^n$ is orthonormal.

EXAMPLE 2. We define a *trigonometric polynomial* of *degree* $2n + 1$ to be an expression of the form

$$f(x) = \frac{a_0}{2} + a_1 \cos x + a_2 \cos 2x + \cdots + a_n \cos nx$$
$$+ b_1 \sin x + b_2 \sin 2x + \cdots + b_n \sin nx, \tag{7–34}$$

where $a_0, \ldots, b_n$ are real numbers, and $a_n \neq 0$, or $b_n \neq 0$, or both. Let $\mathfrak{T}_n$ denote the set of all trigonometric polynomials of degree $\leq 2n + 1$, together with the zero polynomial. We make $\mathfrak{T}_n$ a Euclidean space by defining addition and scalar multiplication of trigonometric polynomials termwise, as with ordinary polynomials, and an inner product by

$$\mathbf{f} \cdot \mathbf{g} = \int_{-\pi}^{\pi} f(x)g(x)\,dx. \tag{7-35}$$

The set of functions

$$1, \cos x, \sin x, \ldots, \cos nx, \sin nx \tag{7-36}$$

is an orthogonal set in $\mathfrak{I}_n$ since, for non-negative integers m and n,

$$\begin{aligned} \int_{-\pi}^{\pi} \sin mx \sin nx\,dx &= 0, \quad \text{if} \quad m \neq n, \\ \int_{-\pi}^{\pi} \sin mx \cos nx\,dx &= 0, \\ \int_{-\pi}^{\pi} \cos mx \cos nx\,dx &= 0, \quad \text{if} \quad m \neq n. \end{aligned} \tag{7-37}$$

To normalize this set we observe that

$$\int_{-\pi}^{\pi} 1\,dx = 2\pi,$$

and

$$\int_{-\pi}^{\pi} \sin^2 mx\,dx = \int_{-\pi}^{\pi} \cos^2 mx\,dx = \pi, \quad \text{if} \quad m > 0. \tag{7-38}$$

Hence the functions

$$\frac{1}{\sqrt{2\pi}}, \frac{\cos x}{\sqrt{\pi}}, \ldots, \frac{\cos nx}{\sqrt{\pi}}, \frac{\sin x}{\sqrt{\pi}}, \ldots, \frac{\sin nx}{\sqrt{\pi}} \tag{7-39}$$

form an *orthonormal* set in $\mathfrak{I}_n$.

EXAMPLE 3. It follows from the preceding example that the (infinite) set

$$1, \cos x, \sin x, \ldots, \cos nx, \sin nx, \ldots$$

is orthogonal in $\mathcal{C}[-\pi, \pi]$.

The most important single property of orthogonal (and hence orthonormal) sets is that each such set in a vector space $\mathcal{V}$ is linearly independent. To prove this result in all generality we extend the definition of linear independence to include *infinite* sets by agreeing that any such set is linearly independent if and only if *every one* of its *finite* subsets is linearly independent in the earlier sense of the term. Thus, for example, the vectors $1, x, x^2, \ldots$ are linearly independent in the space of polynomials $\mathcal{P}$. (Proof?)

Theorem 7–3. *Every orthogonal set of vectors in a Euclidean space $\mathcal{V}$ is linearly independent.*

Proof. Let $\mathcal{S}$ be an orthogonal set in $\mathcal{V}$, and suppose

$$\alpha_1\mathbf{x}_1 + \cdots + \alpha_n\mathbf{x}_n = \mathbf{0}, \tag{7-40}$$

with $\mathbf{x}_1, \ldots, \mathbf{x}_n$ in $\mathcal{S}$. Then, for each index i, $1 \leq i \leq n$,

$$(\alpha_1\mathbf{x}_1 + \cdots + \alpha_n\mathbf{x}_n) \cdot \mathbf{x}_i = \mathbf{0} \cdot \mathbf{x}_i = 0.$$

But since $\mathbf{x}_i \cdot \mathbf{x}_j = 0$ whenever $i \neq j$, the above equation reduces to

$$\alpha_i(\mathbf{x}_i \cdot \mathbf{x}_i) = 0,$$

and since $\mathbf{x}_i \cdot \mathbf{x}_i \neq 0$, it follows that $\alpha_i = 0$. Thus each of the coefficients in (7–40) is zero, and the assertion follows from the test for linear independence. ▌

In particular, we can now state the following useful result.

Corollary 7–1. *An orthogonal set is a basis for an n-dimensional Euclidean space if and only if it contains n vectors.*

EXAMPLE 4. In $\mathcal{P}_3$, with inner product

$$\mathbf{p} \cdot \mathbf{q} = \int_{-1}^{1} p(x)q(x)\,dx,$$

the polynomials $1, x, x^2 - \frac{1}{3}$ are mutually orthogonal, and hence are a basis for this space.

EXAMPLE 5. We saw above that the functions $1, \cos x, \sin x, \ldots, \cos nx, \sin nx$ are mutually orthogonal in $\mathfrak{T}_n$, the space of all trigonometric polynomials of degree $\leq 2n + 1$. Hence these functions are linearly independent in $\mathfrak{T}_n$. Moreover since every vector in this space is a linear combination of these functions it follows that they form a *basis* for $\mathfrak{T}_n$, and hence that $\dim \mathfrak{T}_n = 2n + 1$.

EXAMPLE 6. The orthogonality of the set of functions

$$1, \cos x, \sin x, \ldots, \cos nx, \sin nx, \ldots$$

in $\mathcal{C}[-\pi, \pi]$ implies that this set is also linearly independent. Combined with the fact that any $n + 1$ vectors in an n-dimensional space are linearly dependent, we conclude that $\mathcal{C}[-\pi, \pi]$ is infinite dimensional.

EXERCISES

1. Verify that the Pythagorean theorem holds for the orthogonal functions $\sin x$, $\cos x$ in $\mathcal{C}[-\pi, \pi]$.

*2. Let $\mathbf{x}_1, \ldots, \mathbf{x}_n$ be mutually perpendicular vectors in a Euclidean space. Prove that

$$\|\mathbf{x}_1 + \cdots + \mathbf{x}_n\|^2 = \|\mathbf{x}_1\|^2 + \cdots + \|\mathbf{x}_n\|^2.$$

(This result is a generalized version of the Pythagorean theorem.)

3. Prove that the polynomials $1, x, x^2 - \frac{1}{3}$ are mutually orthogonal in $\mathcal{P}_3$ with the inner product defined as in Example 4 of the text.
4. Let $\mathbf{x}$ be a nonzero vector in a Euclidean space. Show that the vector $\mathbf{x}/\|\mathbf{x}\|$ is of length 1.
5. Convert the orthogonal set of Exercise 3 above into an orthonormal set.
6. Find a polynomial of unit length in $\mathcal{P}_3$ which is orthogonal to 1 and x^2. (Use the inner product of Example 4 of the text.)
7. Find a polynomial of degree 3 which is orthogonal to $1, x, x^2$ in $\mathcal{P}_4$. (Use the inner product of Example 4 of the text.)
8. Find a vector of unit length in $\mathcal{R}^3$ which is orthogonal to the vectors $\mathbf{x} = (1, -1, 0)$, $\mathbf{y} = (2, 1, -1)$.
9. Find a vector of length 2 in $\mathcal{R}^4$ which is orthogonal to the vectors $\mathbf{x} = (1, 0, 3, 1)$, $\mathbf{y} = (-1, 2, 1, 1)$, $\mathbf{z} = (2, -3, 0, -1)$, and has 0 as its second component.
10. Let $\mathbf{x}$ and $\mathbf{y}$ be linearly independent vectors in $\mathcal{R}^3$. Prove that there exist precisely two vectors of unit length which are orthogonal to $\mathbf{x}$ and $\mathbf{y}$.
11. Find a linear combination of the functions e^x and e^{-x} which is orthogonal to e^x in $\mathcal{C}[0, 1]$.
12. Let $\mathbf{x}$ and $\mathbf{y}$ be arbitrary vectors in a Euclidean space, and suppose that $\|\mathbf{x}\| = \|\mathbf{y}\|$. Prove that $\mathbf{x} + \mathbf{y}$ and $\mathbf{x} - \mathbf{y}$ are orthogonal. Interpret this result geometrically.

*13. Suppose $\mathbf{x}_1, \ldots, \mathbf{x}_n$ is a finite orthonormal set in a Euclidean space $\mathcal{V}$. Prove that for any vector $\mathbf{x}$ in $\mathcal{V}$

$$\sum_{i=1}^{n} (\mathbf{x} \cdot \mathbf{x}_i)^2 \leq \|\mathbf{x}\|^2.$$

[*Hint:* Set $\mathbf{y} = \mathbf{x} - \sum_{i=1}^{n} (\mathbf{x} \cdot \mathbf{x}_i)\mathbf{x}_i$, and compute $\|\mathbf{y}\|^2$.]
This inequality is a special case of *Bessel's inequality* which will be proved in general in Section 8–4.

*14. Let $\mathbf{x}_1, \ldots, \mathbf{x}_n$ be an orthonormal basis for a Euclidean space $\mathcal{V}$.

(a) If $\mathbf{x}$ is any vector in $\mathcal{V}$, prove that

$$\|\mathbf{x}\|^2 = \sum_{i=1}^{n} (\mathbf{x} \cdot \mathbf{x}_i)^2.$$

This result is known as *Parseval's equality*, and will be proved in more general terms in Section 8–4.

(b) If $\mathbf{x}$ and $\mathbf{y}$ are any two vectors in $\mathcal{V}$, prove that

$$\mathbf{x} \cdot \mathbf{y} = \sum_{i=1}^{n} (\mathbf{x} \cdot \mathbf{x}_i)(\mathbf{y} \cdot \mathbf{x}_i).$$

(c) Prove, conversely, that if $\mathbf{x}_1, \ldots, \mathbf{x}_n$ is an orthonormal set in a finite dimensional Euclidean space $\mathcal{V}$, and if the equality in (b) is valid for every pair of vectors $\mathbf{x}, \mathbf{y}$ in $\mathcal{V}$, then $\mathbf{x}_1, \ldots, \mathbf{x}_n$ is a basis for $\mathcal{V}$.

7-4 ORTHOGONALIZATION

We now know that every orthogonal set of vectors in a Euclidean space is linearly independent (Theorem 7-3). However, this result would be of only passing interest were it not for the fact that a Euclidean space also contains "enough" orthogonal vectors to enable us to replace a given linearly independent set by an equivalent orthogonal one. More precisely, in this section we shall prove that any (finite or infinite) linearly independent set $\mathfrak{X}$ in a Euclidean space can be converted into an orthogonal set which spans the subspace $\mathcal{S}(\mathfrak{X})$. This process of *orthogonalizing* a linearly independent set, as it is called, has a number of important and useful consequences, not the least of which are the computational simplifications which result from working with orthogonal vectors.

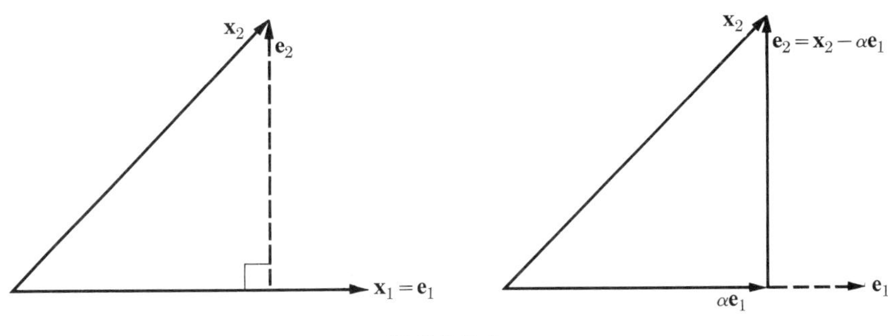

FIGURE 7-6

Rather than begin with the most general situation, we shall introduce the orthogonalization process by two examples. The first is drawn from $\mathcal{R}^2$ where we consider a pair of linearly independent vectors $\mathbf{x}_1, \mathbf{x}_2$. Then $\mathbf{x}_1$ and $\mathbf{x}_2$ form a basis for $\mathcal{R}^2$, and in this case our problem becomes that of replacing $\mathbf{x}_1$ and $\mathbf{x}_2$ with an orthogonal basis $\mathbf{e}_1, \mathbf{e}_2$ constructed out of $\mathbf{x}_1$ and $\mathbf{x}_2$ in some reasonable way. Figure 7-6 suggests the most natural solution of our problem; simply take $\mathbf{e}_1 = \mathbf{x}_1$, and then let $\mathbf{e}_2$ be the "component" of $\mathbf{x}_2$ perpendicular to $\mathbf{x}_1$. Thus we write $\mathbf{e}_2$ in the form

$$\mathbf{e}_2 = \mathbf{x}_2 - \alpha\mathbf{e}_1,$$

and then determine α so that the orthogonality condition $\mathbf{e}_2 \cdot \mathbf{e}_1 = 0$ is satisfied. This yields the equation

$$\mathbf{x}_2 \cdot \mathbf{e}_1 - \alpha(\mathbf{e}_1 \cdot \mathbf{e}_1) = 0,$$

and hence the value of α is

$$\alpha = \frac{\mathbf{x}_2 \cdot \mathbf{e}_1}{\mathbf{e}_1 \cdot \mathbf{e}_1}.$$

With this, $\mathbf{e}_2$ has been determined in terms of $\mathbf{x}_1$ ($= \mathbf{e}_1$) and $\mathbf{x}_2$, and the basis $\mathbf{x}_1, \mathbf{x}_2$ in $\mathcal{R}^2$ has been orthogonalized.

EXAMPLE 1. If $\mathbf{x}_1 = (1, 1)$ and $\mathbf{x}_2 = (0, 1)$, then

$$\alpha = \frac{(0, 1) \cdot (1, 1)}{(1, 1) \cdot (1, 1)} = \frac{1}{2},$$

and so

$$\mathbf{e}_1 = (1, 1), \qquad \mathbf{e}_2 = (-\tfrac{1}{2}, \tfrac{1}{2}).$$

Figure 7–7 shows that this is exactly the result one would expect on the basis of our earlier remarks.

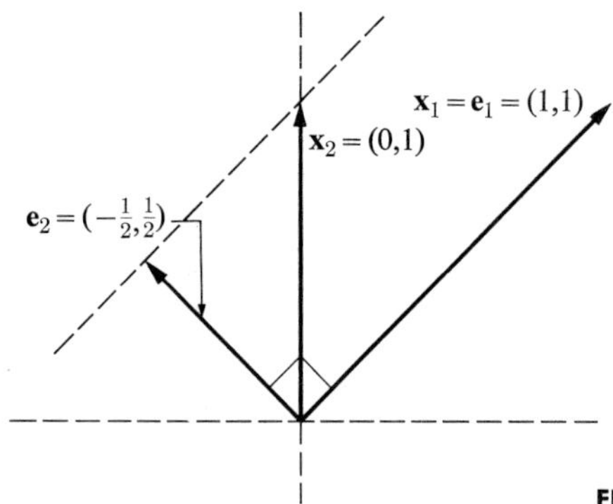

FIGURE 7–7

As our second example we orthogonalize an arbitrary basis $\mathbf{x}_1, \mathbf{x}_2, \mathbf{x}_3$ in $\mathcal{R}^3$. The procedure is essentially the same as that used above in $\mathcal{R}^2$, and is started by choosing $\mathbf{e}_1 = \mathbf{x}_1$. The second step consists of determining $\mathbf{e}_2$ according to the pair of equations

$$\mathbf{e}_2 \cdot \mathbf{e}_1 = 0, \qquad \mathbf{e}_2 = \mathbf{x}_2 - \alpha \mathbf{e}_1,$$

which gives again

$$\alpha = \frac{\mathbf{x}_2 \cdot \mathbf{e}_1}{\mathbf{e}_1 \cdot \mathbf{e}_1}.$$

It is clear that $\mathbf{e}_2$ is not the zero vector (why?), and also that $\mathbf{e}_1$ and $\mathbf{e}_2$ both belong to the subspace of $\mathcal{R}^3$ spanned by $\mathbf{x}_1$ and $\mathbf{x}_2$. Hence $\mathcal{S}(\mathbf{e}_1, \mathbf{e}_2)$ is a subspace of $\mathcal{S}(\mathbf{x}_1, \mathbf{x}_2)$. Moreover, since the orthogonal vectors $\mathbf{e}_1, \mathbf{e}_2$ are linearly independent, $\mathcal{S}(\mathbf{e}_1, \mathbf{e}_2)$ has the same dimension as $\mathcal{S}(\mathbf{x}_1, \mathbf{x}_2)$. Thus

$$\mathcal{S}(\mathbf{e}_1, \mathbf{e}_2) = \mathcal{S}(\mathbf{x}_1, \mathbf{x}_2).$$

Combined with the fact that $\mathbf{x}_1, \mathbf{x}_2, \mathbf{x}_3$ form a basis for $\mathcal{R}^3$, this equality implies that $\mathbf{x}_3$ does *not* belong to the subspace of $\mathcal{R}^3$ spanned by $\mathbf{e}_1$ and $\mathbf{e}_2$. Referring to Fig. 7–8, it is again geometrically clear that the orthogonalization process ought to be completed by letting $\mathbf{e}_3$ be the component of $\mathbf{x}_3$ perpendicular to the subspace $\mathcal{S}(\mathbf{e}_1, \mathbf{e}_2)$. Thus we set

$$\mathbf{e}_3 = \mathbf{x}_3 - \alpha_1 \mathbf{e}_1 - \alpha_2 \mathbf{e}_2,$$

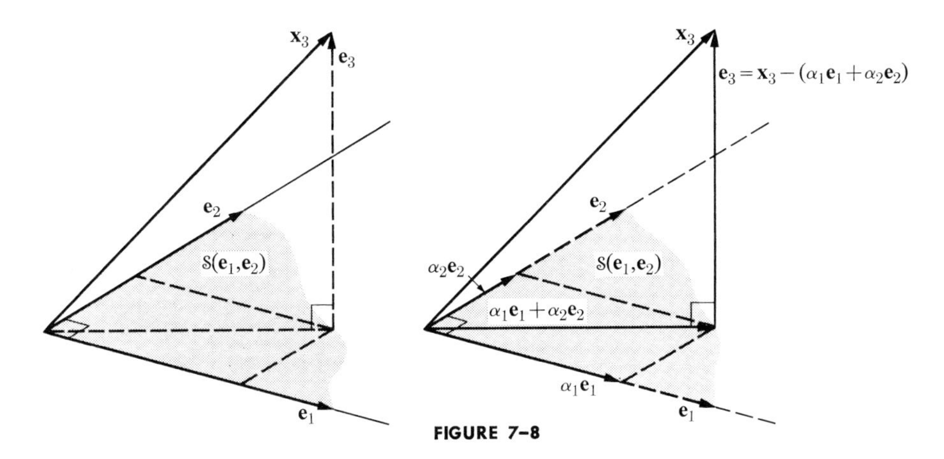

FIGURE 7–8

and find α_1 and α_2 by means of the orthogonality conditions $\mathbf{e}_1 \cdot \mathbf{e}_2 = \mathbf{e}_1 \cdot \mathbf{e}_3 = \mathbf{e}_2 \cdot \mathbf{e}_3 = 0$. They yield the pair of equations

$$0 = \mathbf{x}_3 \cdot \mathbf{e}_1 - \alpha_1(\mathbf{e}_1 \cdot \mathbf{e}_1)$$

and

$$0 = \mathbf{x}_3 \cdot \mathbf{e}_2 - \alpha_2(\mathbf{e}_2 \cdot \mathbf{e}_2),$$

whence

$$\alpha_1 = \frac{\mathbf{x}_3 \cdot \mathbf{e}_1}{\mathbf{e}_1 \cdot \mathbf{e}_1}, \qquad \alpha_2 = \frac{\mathbf{x}_3 \cdot \mathbf{e}_2}{\mathbf{e}_2 \cdot \mathbf{e}_2}.$$

This completes the orthogonalization of the basis $\mathbf{x}_1, \mathbf{x}_2, \mathbf{x}_3$.

EXAMPLE 2. Let $\mathbf{x}_1 = (1, 1, 0)$, $\mathbf{x}_2 = (0, 1, 0)$, $\mathbf{x}_3 = (1, 1, 1)$. Then $\mathbf{e}_1 = \mathbf{x}_1$, $\mathbf{e}_2 = \mathbf{x}_2 - \alpha\mathbf{e}_1$, and $\mathbf{e}_3 = \mathbf{x}_3 - \alpha_1\mathbf{e}_1 - \alpha_2\mathbf{e}_2$, where α, α_1, and α_2 are found from the equations

$$\alpha = \frac{\mathbf{x}_2 \cdot \mathbf{e}_1}{\mathbf{e}_1 \cdot \mathbf{e}_1}, \qquad \alpha_1 = \frac{\mathbf{x}_3 \cdot \mathbf{e}_1}{\mathbf{e}_1 \cdot \mathbf{e}_1}, \qquad \alpha_2 = \frac{\mathbf{x}_3 \cdot \mathbf{e}_2}{\mathbf{e}_2 \cdot \mathbf{e}_2}.$$

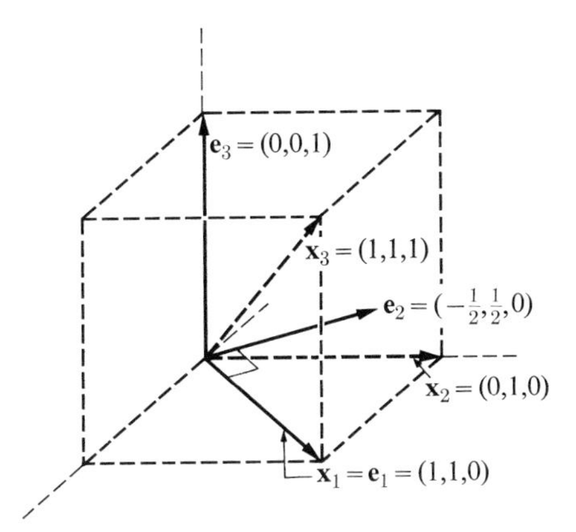

FIGURE 7–9

It follows that $\alpha = \frac{1}{2}$, $\alpha_1 = 1$, $\alpha_2 = 0$, and

$$\mathbf{e}_1 = (1, 1, 0), \qquad \mathbf{e}_2 = (-\tfrac{1}{2}, \tfrac{1}{2}, 0), \qquad \mathbf{e}_3 = (0, 0, 1)$$

(see Fig. 7–9).

We can now quickly dispose of the general situation in which we are required to orthogonalize an arbitrary set of linearly independent vectors $\mathbf{x}_1, \mathbf{x}_2, \ldots$ in a Euclidean space. First set $\mathbf{e}_1 = \mathbf{x}_1$, and then let $\mathbf{e}_2 = \mathbf{x}_2 - \alpha\mathbf{e}_1$, where α is so chosen that $\mathbf{e}_1 \cdot \mathbf{e}_2 = 0$. This determines α as $(\mathbf{x}_2 \cdot \mathbf{e}_1)/(\mathbf{e}_1 \cdot \mathbf{e}_1)$, and the linear independence of $\mathbf{x}_1$ and $\mathbf{x}_2$ implies that $\mathbf{e}_2 \neq \mathbf{0}$. Furthermore, arguing as above, we see that $\mathcal{S}(\mathbf{e}_1, \mathbf{e}_2) = \mathcal{S}(\mathbf{x}_1, \mathbf{x}_2)$.

It remains to show that this process can be continued indefinitely step by step.* To do so, suppose that we have already constructed an orthogonal set $\mathbf{e}_1, \ldots, \mathbf{e}_n$ out of $\mathbf{x}_1, \ldots, \mathbf{x}_n$ so that $\mathcal{S}(\mathbf{e}_1, \ldots, \mathbf{e}_n) = \mathcal{S}(\mathbf{x}_1, \ldots, \mathbf{x}_n)$. Then, to continue one step further, set

$$\mathbf{e}_{n+1} = \mathbf{x}_{n+1} - \alpha_1\mathbf{e}_1 - \alpha_2\mathbf{e}_2 - \cdots - \alpha_n\mathbf{e}_n,$$

and determine $\alpha_1, \ldots, \alpha_n$ so that $\mathbf{e}_{n+1}$ is orthogonal to each of $\mathbf{e}_1, \ldots, \mathbf{e}_n$. This leads to the equations

$$\begin{aligned}
\mathbf{x}_{n+1} \cdot \mathbf{e}_1 - \alpha_1(\mathbf{e}_1 \cdot \mathbf{e}_1) &= 0,\\
\mathbf{x}_{n+1} \cdot \mathbf{e}_2 - \alpha_2(\mathbf{e}_2 \cdot \mathbf{e}_2) &= 0,\\
&\vdots\\
\mathbf{x}_{n+1} \cdot \mathbf{e}_n - \alpha_n(\mathbf{e}_n \cdot \mathbf{e}_n) &= 0,
\end{aligned}$$

and so to

$$\alpha_1 = \frac{\mathbf{x}_{n+1} \cdot \mathbf{e}_1}{\mathbf{e}_1 \cdot \mathbf{e}_1}, \quad \alpha_2 = \frac{\mathbf{x}_{n+1} \cdot \mathbf{e}_2}{\mathbf{e}_2 \cdot \mathbf{e}_2}, \quad \ldots, \quad \alpha_n = \frac{\mathbf{x}_{n+1} \cdot \mathbf{e}_n}{\mathbf{e}_n \cdot \mathbf{e}_n},$$

which determines $\mathbf{e}_{n+1}$. Again the linear independence of $\mathbf{x}_1, \ldots, \mathbf{x}_{n+1}$ implies that $\mathbf{e}_{n+1} \neq 0$, and, as before, we can show that $\mathcal{S}(\mathbf{e}_1, \ldots, \mathbf{e}_{n+1}) = \mathcal{S}(\mathbf{x}_1, \ldots, \mathbf{x}_{n+1})$. (See Exercise 9 below.) Thus the orthogonalization process has been continued, as required, and we can now state the following important result.

Theorem 7–4. *Let $\mathbf{x}_1, \mathbf{x}_2, \ldots$ be a (finite or infinite) set of linearly independent vectors in a Euclidean space $\mathcal{V}$. Then there exists an orthogonal set $\mathbf{e}_1, \mathbf{e}_2, \ldots$ in $\mathcal{V}$ such that for each integer n, $\mathcal{S}(\mathbf{e}_1, \ldots, \mathbf{e}_n) = \mathcal{S}(\mathbf{x}_1, \ldots, \mathbf{x}_n)$. Moreover, the $\mathbf{e}_n$ can be chosen according to the rule*

$$\mathbf{e}_1 = \mathbf{x}_1, \tag{7–41}$$

and,

$$\mathbf{e}_{n+1} = \mathbf{x}_{n+1} - \alpha_1\mathbf{e}_1 - \cdots - \alpha_n\mathbf{e}_n, \tag{7–42}$$

---

* The knowledgeable reader will recognize that we are giving a proof by mathematical induction at this point.

where

$$\alpha_1 = \frac{\mathbf{x}_{n+1} \cdot \mathbf{e}_1}{\mathbf{e}_1 \cdot \mathbf{e}_1}, \quad \alpha_2 = \frac{\mathbf{x}_{n+1} \cdot \mathbf{e}_2}{\mathbf{e}_2 \cdot \mathbf{e}_2}, \quad \ldots, \quad \alpha_n = \frac{\mathbf{x}_{n+1} \cdot \mathbf{e}_n}{\mathbf{e}_n \cdot \mathbf{e}_n}. \tag{7-43}$$

The method of orthogonalization described in this theorem is known as the *Gram-Schmidt orthogonalization process.*

EXAMPLE 3. In this example we apply our orthogonalization process to the infinite linearly independent set of vectors

$$1, x, x^2, \ldots$$

in the space of polynomials $\mathcal{P}$ with inner product defined by

$$\mathbf{p} \cdot \mathbf{q} = \int_{-1}^{1} p(x)q(x)\,dx. \tag{7-44}$$

The orthogonalization goes as follows:

(i)

$$\boxed{\mathbf{e}_1 = 1}$$

$$\mathbf{e}_1 \cdot \mathbf{e}_1 = \int_{-1}^{1} dx = 2.$$

(ii) $\mathbf{e}_2 = x - \alpha$, where $\alpha = \frac{1}{2} \int_{-1}^{1} x\,dx = 0.$

Thus

$$\boxed{\mathbf{e}_2 = x}$$

$$\mathbf{e}_2 \cdot \mathbf{e}_2 = \int_{-1}^{1} x^2\,dx = \tfrac{2}{3}.$$

(iii) $\mathbf{e}_3 = x^2 - \alpha_1 - \alpha_2 x$, where

$$\alpha_1 = \tfrac{1}{2} \int_{-1}^{1} x^2\,dx = \tfrac{1}{3}, \qquad \alpha_2 = \tfrac{3}{2} \int_{-1}^{1} x^3\,dx = 0.$$

Thus

$$\boxed{\mathbf{e}_3 = x^2 - \tfrac{1}{3}}$$

$$\mathbf{e}_3 \cdot \mathbf{e}_3 = \int_{-1}^{1} (x^2 - \tfrac{1}{3})^2\,dx = \tfrac{8}{45}.$$

Continuing in this fashion we obtain the orthogonal sequence

$$1, x, x^2 - \tfrac{1}{3}, x^3 - \tfrac{3}{5}x, x^4 - \tfrac{6}{7}x^2 + \tfrac{3}{35}, \ldots \tag{7-45}$$

in $\mathcal{P}$.

When supplied with appropriate multiplicative constants these polynomials become the famous *Legendre polynomials* of analysis. (Also see Section 6–5.) We shall meet them again in Chapter 11, where they will be discussed in greater detail. At this point we merely ask the student to make a mental note of the fact that the Legendre polynomials form an orthogonal set in $\mathcal{P}$ when the inner product is defined by (7–44).

Theorem 7–4 has a number of useful consequences, the first of which though obvious is nonetheless useful.

Corollary 7–2. *Every finite dimensional Euclidean space has an orthonormal basis.*

This result will enable us to simplify many of our future computations. In particular, if $\mathbf{e}_1, \ldots, \mathbf{e}_n$ is an orthonormal basis in an n-dimensional Euclidean space $\mathcal{V}$, and if

$$\mathbf{x} = \alpha_1\mathbf{e}_1 + \cdots + \alpha_n\mathbf{e}_n$$

is any vector in $\mathcal{V}$, then since $\mathbf{e}_i \cdot \mathbf{e}_j = \delta_{ij}$,

$$\mathbf{x} \cdot \mathbf{e}_i = \alpha_i$$

for each integer i, $1 \leq i \leq n$. Thus *every vector* $\mathbf{x}$ *in* $\mathcal{V}$ *can be written uniquely in the form*

$$\mathbf{x} = (\mathbf{x} \cdot \mathbf{e}_1)\mathbf{e}_1 + \cdots + (\mathbf{x} \cdot \mathbf{e}_n)\mathbf{e}_n, \tag{7–46}$$

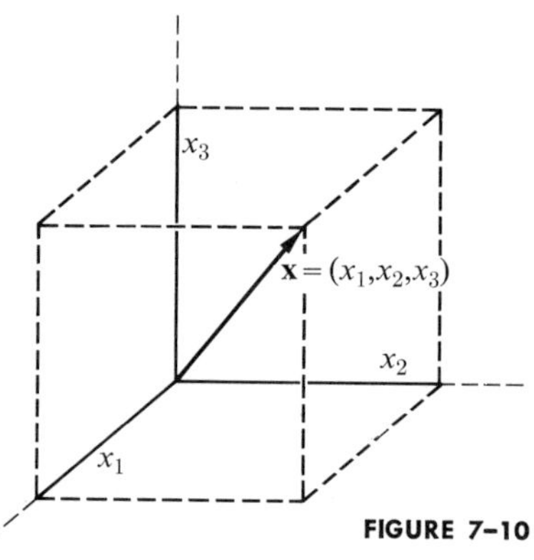

FIGURE 7–10

and it follows that the coordinates of $\mathbf{x}$ with respect to an orthonormal basis in $\mathcal{V}$ are simply the various inner products of $\mathbf{x}$ with the basis vectors. When interpreted geometrically, these coordinates are just the lengths of the projections of $\mathbf{x}$ onto the coordinate axes, as shown in Fig. 7–10.

Finally, if

$$\mathbf{x} = \alpha_1\mathbf{e}_1 + \cdots + \alpha_n\mathbf{e}_n$$

and

$$\mathbf{y} = \beta_1\mathbf{e}_1 + \cdots + \beta_n\mathbf{e}_n$$

are any two vectors in $\mathcal{V}$, their inner product is

$$\mathbf{x} \cdot \mathbf{y} = \alpha_1\beta_1 + \cdots + \alpha_n\beta_n. \tag{7–47}$$

Verbally, this result may be expressed by saying that *the inner product of two vectors in a finite dimensional Euclidean space is the sum of the products of their corresponding components when computed with respect to an orthonormal basis for the space.* The reader ought to compare this result with Example 1 and Exercise 10 of Section 7–1.

EXERCISES

1. Orthogonalize each of the following bases in $\mathcal{R}^3$.

   (a) $(1, 1, 0)$, $(-1, 1, 0)$, $(-1, 1, 1)$.
   (b) $(\frac{1}{2}, 0, 2)$, $(-1, 3, 1)$, $(2, 1, \frac{1}{2})$.
   (c) $(2, -1, 1)$, $(3, -2, \frac{1}{3})$, $(1, 0, 1)$.
   (d) $(1, 2, -1)$, $(2, 0, 3)$, $(0, 4, 1)$.
   (e) $(1, 0, 0)$, $(0, 1, 0)$, $(-1, 0, 1)$.

2. Continue the orthogonalization process in Example 3 above and show that $\mathbf{e}_4 = x^3 - \frac{3}{5}x$, and $\mathbf{e}_5 = x^4 - \frac{6}{7}x^2 + \frac{3}{35}$.

3. Orthogonalize each of the following bases in $\mathcal{R}^4$.

   (a) $(1, 0, 0, 1)$, $(-1, 0, 2, 1)$, $(0, 1, 2, 0)$, $(0, 0, -1, 1)$.
   (b) $(2, 0, \frac{1}{2}, -1)$, $(0, 0, 3, 1)$, $(1, 0, -1, 1)$, $(2, 1, 0, -3)$.
   (c) $(1, 0, 0, 0)$, $(1, 1, 0, 0)$, $(1, 1, 1, 0)$, $(1, 1, 1, 1)$.

4. (a) Orthogonalize the basis $1, x - 1, (x - 1)^2$ in $\mathcal{P}_3$ with inner product $\int_{-1}^{1} p(x)q(x)\,dx$.

   (b) Repeat (a), this time using the inner product $\int_0^1 p(x)q(x)\,dx$.

5. Orthogonalize each of the following sets of vectors in $\mathcal{C}[0, 1]$.

   (a) e^x, e^{-x}
   (b) e^x, e^{2x}
   (c) $1, 2x, e^x$

6. At one point in this section the following argument was used: If $\mathcal{W}$ is a subspace of an n-dimensional vector space $\mathcal{V}$, and $\dim \mathcal{W} = n$, then $\mathcal{W} = \mathcal{V}$. Prove this statement.

7. In $\mathcal{P}$ define $\mathbf{p} \cdot \mathbf{q}$ by

$$\mathbf{p} \cdot \mathbf{q} = a_0b_0 + a_1b_1 + \cdots + a_nb_n,$$

   where

$$\mathbf{p}(x) = a_0 + a_1x + \cdots + a_nx^n,$$
$$\mathbf{q}(x) = b_0 + b_1x + \cdots + b_nx^n.$$

   (See Exercise 10, Section 7–1.) If a is an arbitrary constant, show that the polynomials

$$1, x - a, (x - a)^2, \ldots, (x - a)^n, \ldots$$

   are linearly independent in $\mathcal{P}$, and then orthogonalize this set.

8. Orthogonalize the basis

$$\begin{aligned} \mathbf{e}_1' &= (1, 0, \ldots, 0), \\ \mathbf{e}_2' &= (1, 1, \ldots, 0), \\ &\vdots \\ \mathbf{e}_n' &= (1, 1, \ldots, 1), \end{aligned}$$

   in $\mathcal{R}^n$.

9. In passing from step n to $n + 1$ in the orthogonalization process (Theorem 7–4) we had the following situation:

   (i) $\mathcal{S}(\mathbf{e}_1, \ldots, \mathbf{e}_n) = \mathcal{S}(\mathbf{x}_1, \ldots, \mathbf{x}_n)$, with $\mathbf{e}_1, \ldots, \mathbf{e}_n$ mutually orthogonal, and

   (ii) $\mathbf{e}_{n+1} = \mathbf{x}_{n+1} - \alpha_1\mathbf{e}_1 - \cdots - \alpha_n\mathbf{e}_n$, where $\alpha_1, \ldots, \alpha_n$ are computed according to (7–43).

It was then asserted that

(a) $\mathbf{e}_{n+1} \neq \mathbf{0}$, and

(b) $\mathcal{S}(\mathbf{e}_1, \ldots, \mathbf{e}_{n+1}) = \mathcal{S}(\mathbf{x}_1, \ldots, \mathbf{x}_{n+1})$. Prove these statements.

10. Write out a proof of Eq. (7–47) of the text.

11. Let $\mathbf{e}_1, \ldots, \mathbf{e}_n$ be an orthogonal basis in a Euclidean space $\mathcal{V}$.

(a) If $\mathbf{x} = \alpha_1\mathbf{e}_1 + \cdots + \alpha_n\mathbf{e}_n$ is an arbitrary vector in $\mathcal{V}$, find the value of α_i in terms of $\mathbf{x}$ and $\mathbf{e}_i$.

(b) Compute $\mathbf{x} \cdot \mathbf{y}$ for any pair of vectors $\mathbf{x}$ and $\mathbf{y}$ in $\mathcal{V}$.

12. A vector $\mathbf{x}$ in a Euclidean space $\mathcal{V}$ is said to be *perpendicular* or *orthogonal* to a *subspace* $\mathcal{W}$ of $\mathcal{V}$ if $\mathbf{x}$ is orthogonal to every vector in $\mathcal{V}$.

(a) Show that at each step in the Gram-Schmidt orthogonalization process $\mathbf{e}_{n+1}$ is orthogonal to the subspace $\mathcal{S}(\mathbf{x}_1, \ldots, \mathbf{x}_n)$.

(b) Suppose $\mathbf{e}_1, \ldots, \mathbf{e}_n$ is a basis for $\mathcal{W}$. Prove that $\mathbf{x}$ is orthogonal to $\mathcal{W}$ if and only if $\mathbf{x}$ is orthogonal to each of the $\mathbf{e}_i$.

13. Find a vector of unit length in $\mathcal{R}^3$ which is orthogonal to the subspace spanned by the vectors $(1, 2, -1)$ and $(-1, 0, 2)$. (See Exercise 12.)

14. (a) Orthogonalize the set of vectors $1, \sin x, \sin^2 x$ in $\mathcal{C}[-\pi, \pi]$.

(b) Use the result in (a) to find a unit vector which is orthogonal to the subspace of $\mathcal{C}[-\pi, \pi]$ spanned by $1, \sin x$. (See Exercise 12.)

15. (a) Let $\mathbf{e}_1, \mathbf{e}_2$ be an orthonormal set in a Euclidean space $\mathcal{V}$, and let $\mathcal{W}$ be the subspace spanned by $\mathbf{e}_1$ and $\mathbf{e}_2$. If $\mathbf{x}$ is an arbitrary vector in $\mathcal{V}$, prove that there exists *precisely one* vector $\mathbf{y}$ in $\mathcal{W}$ such that $\mathbf{x} - \mathbf{y}$ is orthogonal to $\mathcal{W}$ (see Exercise 12).

(b) Find $\mathbf{y}$ if $\mathcal{V} = \mathcal{R}^3$, $\mathbf{e}_1 = (1, 0, 0)$, $\mathbf{e}_2 = (0, \sqrt{2}/2, \sqrt{2}/2)$, and $\mathbf{x} = (1, 1, 1)$.

16. Let $p_0(x), p_1(x), \ldots, p_n(x), \ldots$ be the sequence of polynomials obtained by orthogonalizing the sequence $1, x, x^2, \ldots$ in $\mathcal{C}[-1, 1]$.

(a) Prove that $p_n(x)$ is a polynomial of degree n, for each n.

(b) Prove that the leading coefficient of $p_n(x)$ is 1.

*(c) Prove that when n is even, $p_n(x)$ contains only terms of even degree, and when n is odd, $p_n(x)$ contains only terms of odd degree. [*Hint:* Use mathematical induction.]

17. *Legendre polynomials.* The Legendre polynomial $P_n(x)$ may be computed by the following general formula where n successively assumes the values $0, 1, 2, \ldots$:

$$P_n(x) = \frac{(2n)!}{2^n(n!)^2}\left[x^n - \frac{n(n-1)}{2(2n-1)}x^{n-2} + \frac{n(n-1)(n-2)(n-3)}{2 \cdot 4(2n-1)(2n-3)}x^{n-4} - \cdots\right] \tag{7–48}$$

(a) Show that the first five Legendre polynomials, $P_0(x), \ldots, P_4(x)$, are constant multiples of the polynomials listed in (7–45). (Recall that $0! = 1$.)

(b) Write each of the following polynomials as a linear combination of Legendre polynomials.

(i) $3x^2 - 2x + 1$

(ii) $-5x^3 + 9x^2 - 3x - 2$

(iii) $\frac{15}{2}x^3 + \frac{9}{2}x^2 - \frac{9}{2}x + 2$

7–5 PERPENDICULAR PROJECTIONS; DISTANCE TO A SUBSPACE

Let $\mathcal{W}$ be a plane through the origin in $\mathcal{R}^3$, and let $\mathbf{x}$ be an arbitrary point not on $\mathcal{W}$. Then, if $\mathbf{y}$ denotes the perpendicular projection of $\mathbf{x}$ onto $\mathcal{W}$, the distance from $\mathbf{x}$ to $\mathcal{W}$ is defined to be the length of the vector $\mathbf{x} - \mathbf{y}$, as shown in Fig. 7–11. Moreover, the vector $\mathbf{y}$ is characterized by the property that it is the *unique* vector in $\mathcal{W}$ such that $\mathbf{x} - \mathbf{y}$ is perpendicular to $\mathcal{W}$.*

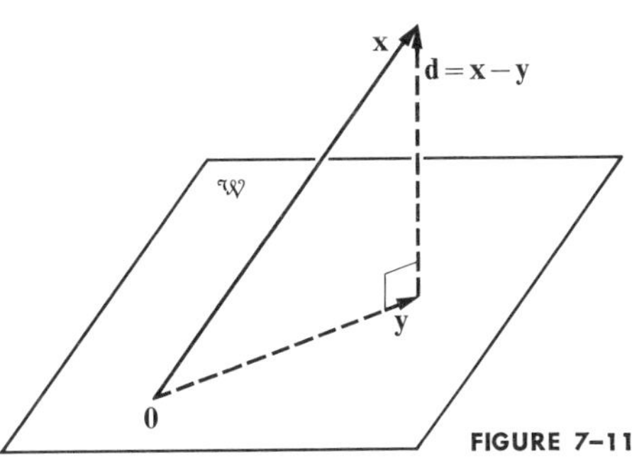

FIGURE 7–11

In this section we generalize these familiar concepts to arbitrary Euclidean spaces. To do so, however, it turns out that the subspace $\mathcal{W}$ must be restricted in some way in order to make things manageable. The most obvious and elementary restriction is the requirement that $\mathcal{W}$ be finite dimensional, for then a vector $\mathbf{d}$ will be orthogonal to $\mathcal{W}$ if and only if $\mathbf{d}$ is orthogonal to each of the vectors in a basis for $\mathcal{W}$ (Exercise 12(b), Section 7–4). Thus, throughout the following discussion we shall assume that $\mathcal{W}$ is a *finite dimensional* subspace of $\mathcal{V}$. Note, however, that we place no restriction on the dimension of $\mathcal{V}$ itself.

Our first objective is to establish the existence of perpendicular projections, which we do by proving the following theorem.

Theorem 7–5. *Let $\mathcal{W}$ be a finite dimensional subspace of a Euclidean space $\mathcal{V}$, and let $\mathbf{x}$ be an arbitrary vector in $\mathcal{V}$. Then $\mathbf{x}$ can be decomposed in* **precisely one** *way as*

$$\mathbf{x} = \mathbf{y} + \mathbf{d}, \tag{7–49}$$

where $\mathbf{y}$ is a vector in $\mathcal{W}$, and $\mathbf{d}$ is perpendicular to $\mathcal{W}$.

Proof. Since $\mathcal{W}$ is finite dimensional, we can apply Corollary 7–2 and find an orthonormal basis $\mathbf{e}_1, \ldots, \mathbf{e}_n$ for $\mathcal{W}$. Then if it exists at all, the vector $\mathbf{y}$ of Eq. (7–49) must be of the form

$$\mathbf{y} = \alpha_1\mathbf{e}_1 + \cdots + \alpha_n\mathbf{e}_n, \tag{7–50}$$

and it remains to show that the α_i can be so determined that the vector $\mathbf{d} = \mathbf{x} - \mathbf{y}$ is orthogonal to each of the basis vectors $\mathbf{e}_i$.† But if we substitute (7–50) into

---

* A vector $\mathbf{x}$ is said to be *perpendicular* or *orthogonal* to a *subspace* $\mathcal{W}$ of a Euclidean space if $\mathbf{x}$ is orthogonal to every vector in $\mathcal{W}$.

† Actually, we already know that this can be done, and that the answer is furnished by the $(n + 1)$st step in the Gram-Schmidt orthogonalization process applied to the linearly independent vectors $\mathbf{e}_1, \ldots, \mathbf{e}_n, \mathbf{x}$. The argument which follows is a repetition of that step.

(7–49) to obtain

$$\mathbf{x} = \alpha_1\mathbf{e}_1 + \cdots + \alpha_n\mathbf{e}_n + \mathbf{d},$$

and then apply the orthogonality condition

$$\mathbf{d} \cdot \mathbf{e}_i = 0, \qquad \mathbf{e}_i \cdot \mathbf{e}_j = \delta_{ij},$$

we find that

$$\alpha_1 = \mathbf{x} \cdot \mathbf{e}_1, \quad \ldots, \quad \alpha_n = \mathbf{x} \cdot \mathbf{e}_n. \tag{7–51}$$

This set of equations determines $\mathbf{y}$ *uniquely* in terms of $\mathbf{x}$ and the basis vectors $\mathbf{e}_1, \ldots, \mathbf{e}_n$ as

$$\mathbf{y} = (\mathbf{x} \cdot \mathbf{e}_1)\mathbf{e}_1 + \cdots + (\mathbf{x} \cdot \mathbf{e}_n)\mathbf{e}_n. \tag{7–52}$$

Finally, it is obvious that the vector $\mathbf{d} = \mathbf{x} - \mathbf{y}$ is perpendicular to $\mathcal{W}$, and the proof is complete. ▌

We now show that the vector $\mathbf{d}$, which is called the *component of* $\mathbf{x}$ *perpendicular to* $\mathcal{W}$, can be used to measure the distance from $\mathbf{x}$ to $\mathcal{W}$. To do so, let $\mathbf{z} \neq \mathbf{d}$ be any vector in $\mathcal{V}$ such that $\mathbf{x} - \mathbf{z}$ belongs to $\mathcal{W}$. (One can view $\mathbf{z}$ as a vector from $\mathcal{W}$ to the point $\mathbf{x}$; see Fig. 7–12.) Then since $\mathbf{z} - \mathbf{d}$ can be written as the difference of two vectors in $\mathcal{W}$, it too belongs to $\mathcal{W}$, and so is orthogonal to $\mathbf{d}$. It now follows from the Pythagorean theorem that

$$\|\mathbf{z}\|^2 = \|\mathbf{d}\|^2 + \|\mathbf{z} - \mathbf{d}\|^2,$$

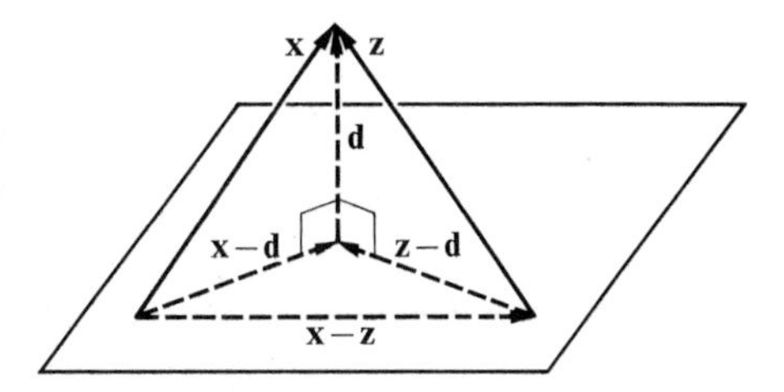

FIGURE 7–12

and hence, since $\|\mathbf{z} - \mathbf{d}\| > 0$, that $\|\mathbf{z}\| > \|\mathbf{d}\|$. In other words, *of all vectors* $\mathbf{z}$ *in* $\mathcal{V}$ *such that* $\mathbf{x} - \mathbf{z}$ *belongs to* $\mathcal{W}$, $\mathbf{d}$ *is the one whose length is smallest.* This serves to justify

Definition 7–6. If $\mathcal{W}$ is a finite dimensional subspace of a Euclidean space $\mathcal{V}$, and $\mathbf{x}$ is any vector in $\mathcal{V}$, then *the distance from* $\mathbf{x}$ *to* $\mathcal{W}$ is the length of the component of $\mathbf{x}$ perpendicular to $\mathcal{W}$.

In terms of this definition we can describe the perpendicular projection $\mathbf{x} - \mathbf{d}$ of $\mathbf{x}$ onto $\mathcal{W}$ as that point in $\mathcal{W}$ which is "closest to" $\mathbf{x}$, in the sense that if $\mathbf{w}$ is any other vector in $\mathcal{W}$, then $\|\mathbf{x} - \mathbf{w}\|$ is greater than $\|\mathbf{d}\|$.

Before we apply these results to a number of interesting special cases, we remark that the use of an orthogonal basis for $\mathcal{W}$ in the above computations was merely a matter of convenience, not necessity. In general, if $\mathbf{e}_1, \ldots, \mathbf{e}_n$ is an arbitrary basis for $\mathcal{W}$, and $\mathbf{d}$ is the component of $\mathbf{x}$ perpendicular to $\mathcal{W}$, then, because

$\mathbf{x} - \mathbf{d}$ belongs to $\mathcal{W}$, we can write

$$\mathbf{x} - \mathbf{d} = \alpha_1\mathbf{e}_1 + \cdots + \alpha_n\mathbf{e}_n.$$

This time the requirement that $\mathbf{d}$ be orthogonal to each of the $\mathbf{e}_i$ leads to the system of linear equations

$$\begin{aligned}
(\mathbf{e}_1 \cdot \mathbf{e}_1)\alpha_1 + (\mathbf{e}_1 \cdot \mathbf{e}_2)\alpha_2 + \cdots + (\mathbf{e}_1 \cdot \mathbf{e}_n)\alpha_n &= \mathbf{e}_1 \cdot \mathbf{x}, \\
(\mathbf{e}_2 \cdot \mathbf{e}_1)\alpha_1 + (\mathbf{e}_2 \cdot \mathbf{e}_2)\alpha_2 + \cdots + (\mathbf{e}_2 \cdot \mathbf{e}_n)\alpha_n &= \mathbf{e}_2 \cdot \mathbf{x}, \\
&\vdots \\
(\mathbf{e}_n \cdot \mathbf{e}_1)\alpha_1 + (\mathbf{e}_n \cdot \mathbf{e}_2)\alpha_2 + \cdots + (\mathbf{e}_n \cdot \mathbf{e}_n)\alpha_n &= \mathbf{e}_n \cdot \mathbf{x},
\end{aligned} \tag{7–53}$$

in the unknowns $\alpha_1, \ldots, \alpha_n$, which must be solved in order to find $\mathbf{d}$. Our earlier results guarantee that this system has a *unique* solution since we know that $\mathbf{d}$ is uniquely determined by $\mathbf{x}$ and $\mathcal{W}$.

Digressing for a moment, we recall that a system of n linear equations in n unknowns has a unique solution *if and only if* the determinant of its coefficients is different from zero.* Hence

$$\begin{vmatrix}
\mathbf{e}_1 \cdot \mathbf{e}_1 & \mathbf{e}_1 \cdot \mathbf{e}_2 & \cdots & \mathbf{e}_1 \cdot \mathbf{e}_n \\
\mathbf{e}_2 \cdot \mathbf{e}_1 & \mathbf{e}_2 \cdot \mathbf{e}_2 & \cdots & \mathbf{e}_2 \cdot \mathbf{e}_n \\
\vdots & & & \vdots \\
\mathbf{e}_n \cdot \mathbf{e}_1 & \mathbf{e}_n \cdot \mathbf{e}_2 & \cdots & \mathbf{e}_n \cdot \mathbf{e}_n
\end{vmatrix} \neq 0. \tag{7–54}$$

This determinant is known as the *Gram determinant* of the vectors $\mathbf{e}_1, \ldots, \mathbf{e}_n$, and the above argument shows that the Gram determinant of n linearly independent vectors in a Euclidean space is always different from zero. The converse of this statement is also true, and furnishes a method for testing a (finite) set of vectors in a Euclidean space for linear dependence. We shall leave the proof of this fact to the reader, and let matters rest with a formal statement.

Theorem 7–6. *The vectors $\mathbf{e}_1, \ldots, \mathbf{e}_n$ in a Euclidean space are linearly independent if and only if their Gram determinant is different from zero.*

One final comment is in order before we consider specific examples. In practice it is usually easier to find $\mathbf{d}$ by solving the system of equations (7–53) instead of trying to use Formula (7–52). The reason for this is to be found in the fact that (7–52) applies only when $\mathbf{e}_1, \ldots, \mathbf{e}_n$ is an orthonormal basis for $\mathcal{W}$, and the construction of such a basis is a tedious job at best. In essence, the issue here is that one ought to attack a simple problem directly, rather than try to adapt it to fit some general formula. Each of the following examples illustrates this point.

---

* The student who is unfamiliar with determinants should omit this paragraph, and resume reading after the statement of Theorem 7–6.

EXAMPLE 1. *The distance from a point to a line.* Let $\mathbf{x}$ be any vector in a Euclidean space, and let $\mathcal{L}$ be the *line* determined by the nonzero vector $\mathbf{y}$; i.e., $\mathcal{L}$ is the one-dimensional subspace spanned by $\mathbf{y}$ (Fig. 7–13). We propose to find the distance $\|\mathbf{d}\|$ from $\mathbf{x}$ to $\mathcal{L}$.

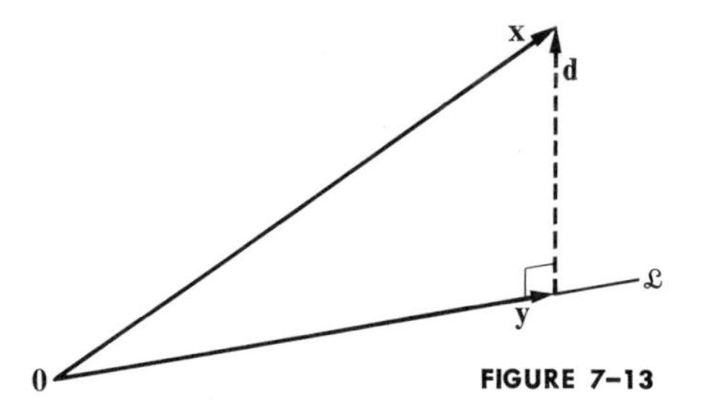

FIGURE 7–13

Since $\mathbf{y}$ spans $\mathcal{L}$, we can take $\mathbf{y}$ itself as a basis for $\mathcal{L}$. This done, we must determine α so that the vector $\mathbf{d} = \mathbf{x} - \alpha\mathbf{y}$ is orthogonal to $\mathbf{y}$. Thus

$$0 = \mathbf{d}\cdot\mathbf{y} = \mathbf{x}\cdot\mathbf{y} - \alpha(\mathbf{y}\cdot\mathbf{y}),$$

and

$$\alpha = \frac{\mathbf{x}\cdot\mathbf{y}}{\mathbf{y}\cdot\mathbf{y}}.$$

It follows that

$$\begin{aligned}\|\mathbf{d}\| &= (\mathbf{d}\cdot\mathbf{d})^{1/2}\\ &= [(\mathbf{x} - \alpha\mathbf{y})\cdot(\mathbf{x} - \alpha\mathbf{y})]^{1/2}\\ &= [\mathbf{x}\cdot\mathbf{x} - 2\alpha(\mathbf{x}\cdot\mathbf{y}) + \alpha^2(\mathbf{y}\cdot\mathbf{y})]^{1/2}\\ &= \left[\mathbf{x}\cdot\mathbf{x} - 2\frac{(\mathbf{x}\cdot\mathbf{y})^2}{\mathbf{y}\cdot\mathbf{y}} + \frac{(\mathbf{x}\cdot\mathbf{y})^2}{\mathbf{y}\cdot\mathbf{y}}\right]^{1/2},\end{aligned}$$

and we have the formula

$$\|\mathbf{d}\| = \left[\frac{(\mathbf{x}\cdot\mathbf{x})(\mathbf{y}\cdot\mathbf{y}) - (\mathbf{x}\cdot\mathbf{y})^2}{\mathbf{y}\cdot\mathbf{y}}\right]^{1/2} \tag{7–55}$$

for the distance from $\mathbf{x}$ to the line determined by the vector $\mathbf{y} \neq \mathbf{0}$ in any Euclidean space.

This formula assumes a particularly simple form in $\mathcal{R}^2$. For if $\mathbf{x} = (x_1, x_2)$ and $\mathbf{y} = (y_1, y_2)$, then

$$\begin{aligned}\|\mathbf{d}\| &= \left[\frac{(x_1^2 + x_2^2)(y_1^2 + y_2^2) - (x_1y_1 + x_2y_2)^2}{y_1^2 + y_2^2}\right]^{1/2}\\ &= \left[\frac{(x_1^2y_2^2 - 2x_1y_2x_2y_1 + x_2^2y_1^2}{y_1^2 + y_2^2}\right]^{1/2}\\ &= \left[\frac{(x_1y_2 - x_2y_1)^2}{y_1^2 + y_2^2}\right]^{1/2},\end{aligned}$$

which may be written

$$\|\mathbf{d}\| = \frac{|x_1y_2 - x_2y_1|}{\sqrt{y_1^2 + y_2^2}}. \tag{7–56}$$

For example, the distance from the point $(-1, 1)$ to the line (through the origin) determined by the point $(1, 1)$ is

$$\|\mathbf{d}\| = \frac{|-1 - 1|}{\sqrt{2}} = \sqrt{2}.$$

(See Fig. 7–14.)

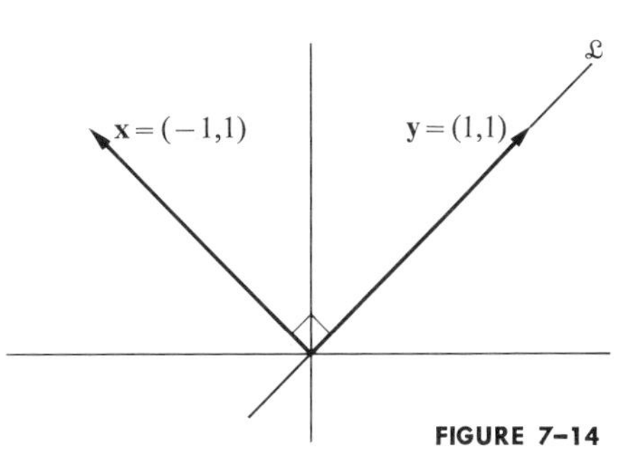

FIGURE 7–14

In $\mathcal{R}^3$ the formula for the distance from a point $\mathbf{x} = (x_1, x_2, x_3)$ to the line determined by the nonzero vector $\mathbf{y} = (y_1, y_2, y_3)$ is not nearly so simple as it is in $\mathcal{R}^2$. A straightforward calculation starting from (7–55) yields

$$\|\mathbf{d}\| = \left[\frac{(x_1y_2 - x_2y_1)^2 + (x_1y_3 - x_3y_1)^2 + (x_2y_3 - x_3y_2)^2}{y_1^2 + y_2^2 + y_3^2}\right]^{1/2}. \tag{7–57}$$

A similar formula can be established for $\mathcal{R}^n$, but it is obviously easier to use (7–55) directly.

EXAMPLE 2. Find the distance from the point $\mathbf{x} = (1, 3, 2)$ in $\mathcal{R}^3$ to the plane (through the origin) determined by the vectors $\mathbf{y}_1 = (1, 0, 0)$ and $\mathbf{y}_2 = (1, 1, 1)$.

We must first determine α and β so that the vector

$$\mathbf{d} = \mathbf{x} - \alpha\mathbf{y}_1 - \beta\mathbf{y}_2$$

is orthogonal to $\mathbf{y}_1$ and $\mathbf{y}_2$. This leads to the pair of equations

$$\begin{aligned} \mathbf{x}\cdot\mathbf{y}_1 - \alpha(\mathbf{y}_1\cdot\mathbf{y}_1) - \beta(\mathbf{y}_2\cdot\mathbf{y}_1) &= 0, \\ \mathbf{x}\cdot\mathbf{y}_2 - \alpha(\mathbf{y}_1\cdot\mathbf{y}_2) - \beta(\mathbf{y}_2\cdot\mathbf{y}_2) &= 0. \end{aligned}$$

Computing the values of the various inner products involved, we find that these equations become

$$\begin{aligned} 1 - \alpha - \beta &= 0 \\ 6 - \alpha - 3\beta &= 0, \end{aligned}$$

and hence $\alpha = -\frac{3}{2}$, $\beta = \frac{5}{2}$. It follows that $\mathbf{d} = (0, \frac{1}{2}, -\frac{1}{2})$, and

$$\|\mathbf{d}\| = \sqrt{\frac{1}{4} + \frac{1}{4}} = \frac{\sqrt{2}}{2}.$$

EXAMPLE 3. *The Fourier coefficients of $f(x)$.* Let $\mathcal{C}[-\pi, \pi]$ be the space of continuous functions on the interval $[-\pi, \pi]$ with the usual inner product

$$\mathbf{f}\cdot\mathbf{g} = \int_{-\pi}^{\pi} f(x)g(x)\,dx,$$

and let $\mathcal{T}_n$ be the $(2n + 1)$-dimensional subspace of $\mathcal{C}[-\pi, \pi]$ consisting of all trigonometric polynomials of degree $\leq 2n + 1$. (See Example 2, Section 7–3.)

We propose to compute the perpendicular projection onto $\mathfrak{J}_n$ of any function f in $\mathcal{C}[-\pi, \pi]$.

By (7–34) this projection is of the form

$$T(x) = \frac{a_0}{2} + a_1 \cos x + \cdots + a_n \cos nx + b_1 \sin x + \cdots + b_n \sin nx,$$

and we must determine values for the a_i and b_i so that

$$d(x) = f(x) - T(x)$$

is orthogonal to each of the functions $1, \cos x, \sin x, \ldots, \cos nx, \sin nx$. But, by (7–37), these functions are mutually orthogonal in $\mathcal{C}[-\pi, \pi]$, and so, by taking inner products, we have

$$\int_{-\pi}^{\pi} f(x)\,dx - \frac{a_0}{2}\int_{-\pi}^{\pi} dx = 0,$$

$$\int_{-\pi}^{\pi} f(x) \cos x\,dx - a_1 \int_{-\pi}^{\pi} \cos^2 x\,dx = 0,$$

$$\int_{-\pi}^{\pi} f(x) \sin x\,dx - b_n \int_{-\pi}^{\pi} \sin^2 x\,dx = 0,$$

$$\vdots$$

$$\int_{-\pi}^{\pi} f(x) \cos nx\,dx - a_n \int_{-\pi}^{\pi} \cos^2 nx\,dx = 0,$$

$$\int_{-\pi}^{\pi} f(x) \sin nx\,dx - b_1 \int_{-\pi}^{\pi} \sin^2 nx\,dx = 0.$$

Since

$$\int_{-\pi}^{\pi} dx = 2\pi,$$

and

$$\int_{-\pi}^{\pi} \sin^2 mx\,dx = \int_{-\pi}^{\pi} \cos^2 mx\,dx = \pi$$

if $m > 0$, it follows that

$$a_0 = \frac{1}{\pi}\int_{-\pi}^{\pi} f(x)\,dx,$$

$$a_1 = \frac{1}{\pi}\int_{-\pi}^{\pi} f(x)\cos x\,dx, \qquad b_1 = \frac{1}{\pi}\int_{-\pi}^{\pi} f(x)\sin x\,dx, \tag{7–58}$$

$$\vdots \qquad\qquad\qquad \vdots$$

$$a_n = \frac{1}{\pi}\int_{-\pi}^{\pi} f(x)\cos nx\,dx, \qquad b_n = \frac{1}{\pi}\int_{-\pi}^{\pi} f(x)\sin nx\,dx.$$

These coefficients are known as the *Fourier coefficients* of the function f. (The student should now appreciate the reason for associating the factor $\frac{1}{2}$ with the constant term of the trigonometric polynomials in $\mathfrak{T}_n$. It was done simply to assure that each of the Fourier coefficients has the same constant before the integral, for without the $\frac{1}{2}$ the formula for a_0 would have been $(1/2\pi)\int_{-\pi}^{\pi} f(x)\,dx$.)

We have now shown that the trigonometric polynomial T whose coefficients are the Fourier coefficients of f is the *best approximation* in $\mathfrak{T}_n$ to the function f, in the sense that of all the functions P belonging to $\mathfrak{T}_n$, T is *the* one which minimizes the integral

$$\int_{-\pi}^{\pi} [f(x) - P(x)]^2\,dx. \tag{7–59}$$

The value of this integral is often called the *mean deviation* (or *mean square deviation*) of P from f. In these terms, T is that trigonometric polynomial in $\mathfrak{T}_n$ with minimum mean deviation from f.

These considerations lead one naturally to the problem of determining whether the mean deviation of T from f tends to zero as $n \to \infty$. In other words, *can f be approximated arbitrarily closely by trigonometric polynomials*? This and related topics will be investigated in the chapters which follow.

EXERCISES

1. Find the distance from each of the following points $\mathbf{x}$ in $\mathfrak{R}^2$ to the line (through the origin) determined by the point $\mathbf{y}$.
   (a) $\mathbf{x} = (1, 0)$; $\mathbf{y} = (1, 1)$ (b) $\mathbf{x} = (-2, 1)$; $\mathbf{y} = (0, 1)$
   (c) $\mathbf{x} = (\frac{3}{2}, \frac{1}{8})$; $\mathbf{y} = (-3, 4)$ (d) $\mathbf{x} = (-2, 10)$; $\mathbf{y} = (1, \frac{5}{2})$
2. Find the distance from each of the following points $\mathbf{x}$ in $\mathfrak{R}^3$ to the line (through the origin) determined by the point $\mathbf{y}$.
   (a) $\mathbf{x} = (1, 1, 1)$; $\mathbf{y} = (1, 1, 0)$ (b) $\mathbf{x} = (0, 1, 1)$; $\mathbf{y} = (1, 1, 0)$
   (c) $\mathbf{x} = (2, 1, -3)$; $\mathbf{y} = (-1, 2, 2)$
   (d) $\mathbf{x} = (-\frac{1}{2}, 1, 3/\sqrt{3})$; $\mathbf{y} = (4, 3, 5\sqrt{3})$
3. Find the distance from each of the following points $\mathbf{x}$ in $\mathfrak{R}^4$ to the line (through the origin) determined by the point $\mathbf{y}$.
   (a) $\mathbf{x} = (0, 2, 1, 1)$; $\mathbf{y} = (1, 3, 1, 3)$ (b) $\mathbf{x} = (-1, 3, 2, 3)$; $\mathbf{y} = (1, 1, 1, 1)$
   (c) $\mathbf{x} = (16, \frac{3}{4}, 3, \frac{3}{2})$; $\mathbf{y} = (4, \frac{1}{2}, 2, 1)$ (d) $\mathbf{x} = (0, -1, 2, 2)$; $\mathbf{y} = (-1, 0, -2, 2)$
4. Let $\mathbf{x}$ and $\mathbf{x}'$ be linearly independent vectors in $\mathfrak{R}^2$ such that $\|\mathbf{x}\| = \|\mathbf{x}'\|$. Show that there are two lines through $(0, 0)$ for which the distances from $\mathbf{x}$ and $\mathbf{x}'$ are equal, and that these lines are orthogonal.
5. There are two lines through the origin in $\mathfrak{R}^2$ such that the distance from $(1, 7)$ is 5. Find them.
6. Find the locus of a point $\mathbf{x}$ in $\mathfrak{R}^2$ which is equidistant from each of two given distinct lines through the origin.

7. Find the distance from each of the following points $\mathbf{x}$ in $\mathcal{R}^3$ to the plane (through the origin) determined by the given points $\mathbf{y}$ and $\mathbf{z}$.
   (a) $\mathbf{x} = (1, 2, 0)$; $\mathbf{y} = (-2, 0, -2)$; $\mathbf{z} = (1, 1, 6)$
   (b) $\mathbf{x} = (2, -1, 1)$; $\mathbf{y} = (4, -3, -1)$; $\mathbf{z} = (-1, \frac{1}{2}, -2)$
   (c) $\mathbf{x} = (1, -2, -1)$; $\mathbf{y} = (0, 0, -\sqrt{21})$; $\mathbf{z} = (0, 2\sqrt{21}, -5\sqrt{21})$
8. Let $\mathcal{L}$ be the line through $(0, 0, 0)$ and $(10, 2, 11)$ in $\mathcal{R}^3$. Find the points on $\mathcal{L}$ which are at a distance 3 from $(2, -2, 1)$.
9. Find the coordinates of the point in $\mathcal{R}^3$ which lies on the plane determined by the origin, $(1, 2, 2)$, and $(2, -2, 1)$ and is at a minimum distance from the point $(3, 1, 1)$.
10. (a) Show that the area of the triangle in $\mathcal{R}^2$ with vertices at $(0, 0)$, (x_1, x_2), and (y_1, y_2) is $\frac{1}{2}|x_1y_2 - x_2y_1|$.
    (b) More generally, show that the area of the triangle in $\mathcal{R}^n$ with vertices at $\mathbf{0}$, $\mathbf{x}$, and $\mathbf{y}$ is $\frac{1}{2}[(\mathbf{x} \cdot \mathbf{x})(\mathbf{y} \cdot \mathbf{y}) - (\mathbf{x} \cdot \mathbf{y})^2]^{1/2}$.
11. Find the perpendicular projection of each of the following vectors in $\mathcal{C}[-\pi, \pi]$ onto the indicated subspace $\mathcal{W}$, and compute the distance from the vector to the subspace.
    (a) $f(x) = x$, $\mathcal{W} = \mathcal{S}(1, \cos x, \sin x)$
    (b) $f(x) = \cos^2 x$, $\mathcal{W} = \mathcal{S}(1, \cos 2x)$
    (c) $f(x) = x^2$, $\mathcal{W} = \mathcal{S}(1, \cos x, \cos 2x)$
12. Repeat Exercise 11 for the following vectors and subspaces of $\mathcal{C}[-\pi, \pi]$.
    (a) $f(x) = x^2$, $\mathcal{W} = \mathcal{S}(\sin x, \sin 2x)$
    (b) $f(x) = x^3$, $\mathcal{W} = \mathcal{S}(1, \cos x, \cos 2x)$
    (c) $f(x) = x^3$, $\mathcal{W} = \mathcal{S}(\sin x, \sin 2x)$

In Exercises 13–19 find the Fourier coefficients (for all values of n) for the given function in $\mathcal{C}[-\pi, \pi]$.

13. $f(x) = x$
14. $f(x) = x^2$
15. $f(x) = \sin^2 x$
16. $f(x) = x^m$, m a positive integer
17. $f(x) = |x|$
18. $f(x) = \cos^2 x$
19. $f(x) = e^x$

*20. Use the Gram determinant to test the following sets of vectors in $\mathcal{R}^3$ for linear independence.
    (a) $\mathbf{e}_1 = (2, -1, 1)$, $\mathbf{e}_2 = (1, 2, 1)$, $\mathbf{e}_3 = (-1, 2, 0)$
    (b) $\mathbf{e}_1 = (\frac{1}{2}, 2, -1)$, $\mathbf{e}_2 = (1, -1, 2)$, $\mathbf{e}_3 = (-2, 7, 4)$
    (c) $\mathbf{e}_1 = (\frac{1}{3}, \frac{2}{5}, 0)$, $\mathbf{e}_2 = (\frac{14}{3}, \frac{7}{3}, -1)$, $\mathbf{e}_3 = (3, -\frac{1}{6}, \frac{1}{2})$
21. Find the perpendicular projection of each of the following vectors in $\mathcal{C}[-1, 1]$ onto the subspace spanned by the polynomials $1, x, x^2 - \frac{1}{3}$, and compute the distance from these vectors to the subspace in question.
    (a) $f(x) = x^n$, n an integer
    (b) $f(x) = \sin x$
    (c) $f(x) = |x|$
    (d) $f(x) = 3x^2$
22. Let $\mathbf{e}_1, \ldots, \mathbf{e}_n$ be an orthonormal basis in a Euclidean space $\mathcal{V}$, and let $\mathcal{W}$ be the m-dimensional subspace of $\mathcal{V}$ spanned by $\mathbf{e}_1, \ldots, \mathbf{e}_m$, $m \leq n$.
    (a) Find the perpendicular projection of an arbitrary vector $\mathbf{x}$ in $\mathcal{V}$ onto $\mathcal{W}$.
    (b) Find the component of $\mathbf{x}$ perpendicular to $\mathcal{W}$.
    (c) Find the distance from $\mathbf{x}$ to $\mathcal{W}$.

23. Let $\mathcal{X}$ be an arbitrary nonempty subset of a Euclidean space $\mathcal{V}$, and let $\mathcal{X}^\perp$ (read "$\mathcal{X}$ perp") denote the set of all vectors in $\mathcal{V}$ which are orthogonal to every vector in $\mathcal{X}$. Show that $\mathcal{X}^\perp$ is a subspace of $\mathcal{V}$. What is $\mathcal{X}^\perp$ if $\mathcal{X} = \mathcal{V}$? If $\mathcal{X}$ contains only the zero vector?

24. With $\mathcal{X}$ and $\mathcal{X}^\perp$ as in Exercise 23, let $(\mathcal{X}^\perp)^\perp$ denote the set of all vectors in $\mathcal{V}$ which are orthogonal to every vector in $\mathcal{X}^\perp$.
(a) Show that $\mathcal{X}$ is a subset of $(\mathcal{X}^\perp)^\perp$, i.e., that every vector in $\mathcal{X}$ belongs to $(\mathcal{X}^\perp)^\perp$.
(b) Show that $(\mathcal{X}^\perp)^\perp = \mathcal{S}(\mathcal{X})$ if $\mathcal{V}$ is finite dimensional. [*Hint:* Choose an orthonormal basis for $\mathcal{S}(\mathcal{X})$, and extend it to a basis for $\mathcal{V}$.]

*25. (a) Let $\mathcal{V}$ consist of all infinite sequences

$$\mathbf{s} = \{a_0, a_1, \ldots, a_n, \ldots\}$$

of real numbers having only a finite number of nonzero entries, with addition and scalar multiplication defined termwise. If

$$\mathbf{s} = \{a_0, a_1, \ldots, a_n, \ldots\}$$

and

$$\mathbf{t} = \{b_0, b_1, \ldots, b_n, \ldots\},$$

define

$$\mathbf{s} \cdot \mathbf{t} = \sum_{n=0}^{\infty} a_n b_n.$$

Show that with this definition $\mathcal{V}$ becomes an infinite dimensional inner product space.
(b) Let $\mathcal{W}$ be the subspace of $\mathcal{V}$ consisting of all sequences of the form

$$\mathbf{s} = \{a, a, 0, 0, \ldots\},$$

where a is an arbitrary real number. Show that $\mathcal{W} = (\mathcal{W}^\perp)^\perp$.

26. Let $\mathcal{V}$ be a finite dimensional Euclidean space, and let $\mathcal{W}$ be a subspace of $\mathcal{V}$. Show that

$$\mathcal{V} = \mathcal{W} + \mathcal{W}^\perp,$$

and that $\mathcal{W} \cap \mathcal{W}^\perp$ contains only the zero vector (see Exercise 22, Section 1–4).

27. Find $\mathcal{W}^\perp$, where $\mathcal{W}$ is the subspace of $\mathbb{R}^3$ spanned by the following vector or vectors.
(a) $(1, 1, 0)$
(b) $(1, -1, 0)$, $(0, 0, 1)$
(c) $(1, 2, -1)$
(d) $(1, 1, 1)$, $(-1, -1, \frac{1}{2})$.

28. Let Π be an arbitrary plane and $\mathbf{x}$ an arbitrary vector. Show that the distance from $\mathbf{x}$ to Π is the perpendicular projection of $\mathbf{x}$ onto $\Pi^\perp$. (See Fig. 7–15.)

*29. Prove Theorem 7–6.

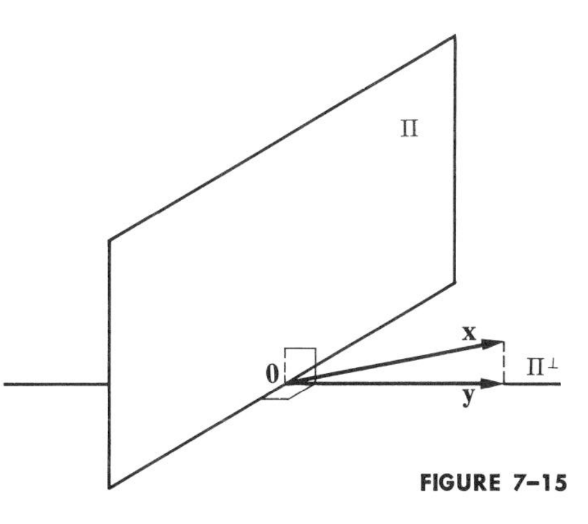

FIGURE 7–15

*7–6 THE METHOD OF LEAST SQUARES

One of the most important applications of the preceding material occurs in the theory of approximations. The problem here can be described most succinctly as the proper interpretation of experimental data, and a simple illustration is perhaps the best way of introducing the subject.

Suppose that we wish to determine the value of a certain physical constant c, such as the specific gravity of a given substance, and that an experimental method for measuring c is available. We then perform the experiment n times and obtain estimates $x_1, \ldots, x_n$ of c. In the absence of experimental errors each of the x_i would equal c, but in practice, of course, none of them will, and we are thus faced with the problem of finding the "best approximation" to c available from our experimental data.

To do so, we view the n experimental measurements as a vector $\mathbf{x} = (x_1, \ldots, x_n)$ in $\mathcal{R}^n$, and let $\mathbf{y}$ be the n-tuple $(1, \ldots, 1)$, so that

$$c\mathbf{y} = (c, \ldots, c).$$

Then, if we interpret the term "best approximation" as distance in $\mathcal{R}^n$ (which, after all, is the most reasonable interpretation conceivable), and if we let c' denote this approximation, we must choose c' so that the vector $c'\mathbf{y}$ is as close as possible to $\mathbf{x}$. In other words, c' is determined by the requirement that $c'\mathbf{y}$ be the perpendicular projection of $\mathbf{x}$ onto the one-dimensional subspace of $\mathcal{R}^n$ spanned by $\mathbf{y}$ (see Fig. 7–16). But, as we saw in the preceding section, this projection is

$$\frac{\mathbf{x} \cdot \mathbf{y}}{\mathbf{y} \cdot \mathbf{y}} \mathbf{y},$$

so that

$$c' = \frac{\mathbf{x} \cdot \mathbf{y}}{\mathbf{y} \cdot \mathbf{y}} = \frac{x_1 + \cdots + x_n}{n}.$$

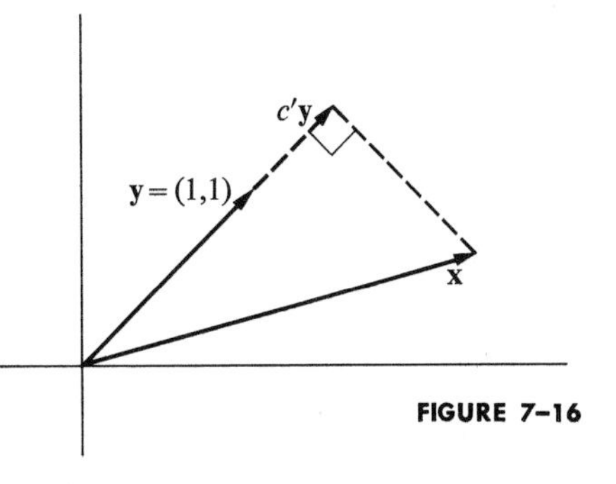

FIGURE 7–16

This, of course, is none other than the arithmetic average of the x_i, and we now have a theoretical interpretation of the popular practice of averaging separate (independent) measurements of the same quantity.

Appropriately generalized, the foregoing method will yield approximations to vectors as well as scalars. For simplicity we consider the case of a vector $\mathbf{c} = (c_1, c_2)$ in $\mathcal{R}^2$, and a set of measurements

$$\begin{aligned} \mathbf{x}_1 &= (x_{11}, x_{12}), \\ \mathbf{x}_2 &= (x_{21}, x_{22}), \\ &\vdots \\ \mathbf{x}_n &= (x_{n1}, x_{n2}) \end{aligned}$$

of $\mathbf{c}$. Again we wish to use the $\mathbf{x}_i$ to obtain the best possible approximation $\mathbf{c}' = (c'_1, c'_2)$ to $\mathbf{c}$.

This time we view the experimental results as a vector

$$\mathbf{x} = (x_{11}, x_{21}, \ldots, x_{n1}, x_{12}, \ldots, x_{n2})$$

in $\mathcal{R}^{2n}$, and let $\mathbf{y}_1$ and $\mathbf{y}_2$ be, respectively, the orthogonal vectors

$$(\underbrace{1, \ldots, 1}_{n}, \underbrace{0, \ldots, 0}_{n}) \quad \text{and} \quad (\underbrace{0, \ldots, 0}_{n}, \underbrace{1, \ldots, 1}_{n})$$

in $\mathcal{R}^{2n}$. Proceeding as before, we take as our approximation the scalars c'_1, c'_2 such that $c'_1\mathbf{y}_1 + c'_2\mathbf{y}_2$ is the perpendicular projection of $\mathbf{x}$ onto the subspace $\mathcal{S}(\mathbf{y}_1, \mathbf{y}_2)$. Thus the vector $\mathbf{x} - (c'_1\mathbf{y}_1 + c'_2\mathbf{y}_2)$ must be perpendicular to $\mathbf{y}_1$ and $\mathbf{y}_2$, whence

$$c'_1 = \frac{\mathbf{x} \cdot \mathbf{y}_1}{\mathbf{y}_1 \cdot \mathbf{y}_1} \quad \text{and} \quad c'_2 = \frac{\mathbf{x} \cdot \mathbf{y}_2}{\mathbf{y}_2 \cdot \mathbf{y}_2}.$$

This gives

$$c'_1 = \frac{x_{11} + x_{21} + \cdots + x_{n1}}{n}, \qquad c'_2 = \frac{x_{12} + x_{22} + \cdots + x_{n2}}{n},$$

and

$$\mathbf{c}' = \frac{1}{n}(\mathbf{x}_1 + \cdots + \mathbf{x}_n). \tag{7-60}$$

This vector may be familiar to some of our readers as the *centroid* of the vectors $\mathbf{x}_1, \ldots, \mathbf{x}_n$. We note that it may be characterized as *the* vector in $\mathcal{R}^2$ which minimizes the quantity

$$\sum_{i=1}^{n} \|\mathbf{x}_i - \mathbf{c}'\|^2 \tag{7-61}$$

(see Exercise 4 below).

In general, n experimental determinations $\mathbf{x}_1, \ldots, \mathbf{x}_n$ of a vector $\mathbf{c}$ in $\mathcal{R}^m$ can be handled in exactly the same way, and it is not difficult to show that Formulas (7-60) and (7-61) continue to describe the best approximation.

A related problem of this type occurs when one is given a scalar y which is known to depend linearly upon a scalar x, i.e.,

$$y = cx,$$

and one attempts to determine the value of c experimentally. (For instance, y might be the displacement of a spring under a weight x, in which case c would be the spring constant.) In this case our experiments yield a set of measured values $x_1, \ldots, x_n$ of x and corresponding values $y_1, \ldots, y_n$ for y, which can be displayed as a system of n linear equations

$$\begin{aligned} y_1 &= cx_1, \\ y_2 &= cx_2, \\ &\vdots \\ y_n &= cx_n, \end{aligned} \tag{7-62}$$

in the single unknown c. As a result of experimental errors, this system of equations will not be compatible (i.e., will not admit a unique solution), and our problem is to find the "best approximation" to c afforded by this data.

Again we pass to Euclidean n-space, $\mathcal{R}^n$, and consider the vectors

$$\mathbf{x} = (x_1, \ldots, x_n) \quad \text{and} \quad \mathbf{y} = (y_1, \ldots, y_n).$$

In this context our problem can be rephrased as follows: Find a scalar c' such that the vector $c'\mathbf{x}$ in the subspace $\mathcal{S}(\mathbf{x})$ is as close as possible to $\mathbf{y}$. This, of course, requires that c' be so chosen that the vector $c'\mathbf{x} - \mathbf{y}$ has the smallest possible length. Thus c' must minimize the quantity $\|c'\mathbf{x} - \mathbf{y}\|^2$. But

$$\begin{aligned} \|c'\mathbf{x} - \mathbf{y}\|^2 &= (c'\mathbf{x} - \mathbf{y}) \cdot (c'\mathbf{x} - \mathbf{y}) \\ &= \sum_{i=1}^{n} (c'x_i - y_i)^2, \end{aligned}$$

and we see that *the best approximation to c is the scalar which minimizes the sum*

$$\sum_{i=1}^{n} (c'x_i - y_i)^2. \tag{7-63}$$

For rather obvious reasons this method of approximation (and its generalization below) is called the *method of least squares.*

In the present example it is easy to compute the value of c' explicitly. Indeed, since $c'\mathbf{x} - \mathbf{y}$ must be perpendicular to $\mathbf{x}$ we have $(c'\mathbf{x} - \mathbf{y}) \cdot \mathbf{x} = 0$, and hence

$$c' = \frac{\mathbf{x} \cdot \mathbf{y}}{\mathbf{x} \cdot \mathbf{x}},$$

or

$$c' = \frac{x_1y_1 + \cdots + x_ny_n}{x_1^2 + \cdots + x_n^2}. \tag{7-64}$$

EXAMPLE 1. Use the method of least squares to find the best approximation to c available from the equations

$$\begin{aligned} 2 &= c, \\ 2 &= 3c, \\ 5 &= 4c, \\ 6 &= 6c. \end{aligned}$$

Here $\mathbf{x} = (1, 3, 4, 6)$ and $\mathbf{y} = (2, 2, 5, 6)$, so that

$$c' = \frac{2 + 6 + 20 + 36}{1 + 9 + 16 + 36} = \frac{32}{31}.$$

The type of approximation which has just been considered often arises as a problem in *curve fitting*. In this context one is asked to find a straight line $y = c'x$ through the origin (in $\mathcal{R}^2$) which best fits the points $(x_1, y_1), \ldots, (x_n, y_n)$. The accepted method of solution is to let Δ_i denote the difference $c'x_i - y_i$, as shown in Fig. 7–17, and then choose c' so that the *sum of the squares* of the Δ_i is a minimum.* But this sum is just the quantity given in (7–63), and the present problem is identical with the one solved above.

Finally, we consider the general situation in which a scalar y is an unknown linear combination of the scalars $x_1, \ldots, x_m$; i.e.,

$$y = c_1x_1 + \cdots + c_mx_m.$$

Again we imagine that n experiments have been performed (where $n > m$), and that the ith experiment has yielded the values $x_{i1}, \ldots, x_{im}$, and y_i, respectively, for $x_1, \ldots, x_m$ and y. Thus we have a system of linear equations

$$\begin{aligned} y_1 &= c_1x_{11} + c_2x_{12} + \cdots + c_mx_{1m}, \\ y_2 &= c_1x_{21} + c_2x_{22} + \cdots + c_mx_{2m}, \\ &\vdots \\ y_n &= c_1x_{n1} + c_2x_{n2} + \cdots + c_mx_{nm}, \end{aligned} \tag{7–65}$$

in which the c_i are unknowns. In general, of course, (7–65) cannot be solved, and our problem becomes that of finding values $c'_1, \ldots, c'_m$ for the c_i which make the expressions standing on the right of these equations approximate $y_1, \ldots, y_n$ as closely as possible.

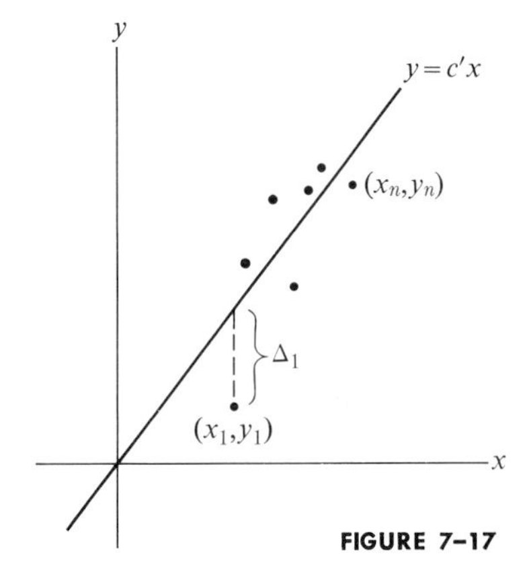

FIGURE 7–17

By now the method should be obvious. We consider the vectors

$$\begin{aligned} \mathbf{x}_1 &= (x_{11}, x_{21}, \ldots, x_{n1}), \\ \mathbf{x}_2 &= (x_{12}, x_{22}, \ldots, x_{n2}), \\ &\vdots \\ \mathbf{x}_m &= (x_{1m}, x_{2m}, \ldots, x_{nm}), \end{aligned}$$

and

$$\mathbf{y} = (y_1, \ldots, y_n)$$

* It makes excellent sense to use the quantities Δ_i^2 for this purpose rather than the Δ_i themselves, since in the latter case a large positive difference would obliterate the effect of several small negative differences, contrary to the requirement that each point be given equal weight in the fitting process.

formed from the columns of (7–65), and let $\mathcal{W}$ be the subspace of $\mathcal{R}^n$ spanned by $\mathbf{x}_1, \ldots, \mathbf{x}_m$ (recall that $m < n$). We now make the assumption that our experiments were so designed that $\mathbf{x}_1, \ldots, \mathbf{x}_m$ are linearly independent, so that they form a basis for $\mathcal{W}$. The $c'_1, \ldots, c'_m$ are then chosen so that the vector

$$c'_1\mathbf{x}_1 + \cdots + c'_m\mathbf{x}_m$$

is the perpendicular projection of $\mathbf{y}$ onto $\mathcal{W}$. Thus the c'_i must minimize the length of the vector

$$(c'_1\mathbf{x}_1 + \cdots + c'_m\mathbf{x}_m) - \mathbf{y},$$

or, equivalently, they must minimize the quantity

$$\sum_{i=1}^{n} [(c'_1x_{i1} + \cdots + c'_mx_{im}) - y_i]^2, \tag{7–66}$$

which is the square of the length of this vector.

In practice the c'_i are usually determined from the orthogonality relations

$$[(c'_1\mathbf{x}_1 + \cdots + c'_m\mathbf{x}_m) - \mathbf{y}] \cdot \mathbf{x}_i = 0,$$

$i = 1, \ldots, m$. They yield the system of m linear equations

$$\begin{aligned}
(\mathbf{x}_1 \cdot \mathbf{x}_1)c'_1 + (\mathbf{x}_1 \cdot \mathbf{x}_2)c'_2 + \cdots + (\mathbf{x}_1 \cdot \mathbf{x}_m)c'_m &= \mathbf{x}_1 \cdot \mathbf{y},\\
(\mathbf{x}_2 \cdot \mathbf{x}_1)c'_1 + (\mathbf{x}_2 \cdot \mathbf{x}_2)c'_2 + \cdots + (\mathbf{x}_2 \cdot \mathbf{x}_m)c'_m &= \mathbf{x}_2 \cdot \mathbf{y},\\
&\vdots\\
(\mathbf{x}_m \cdot \mathbf{x}_1)c'_1 + (\mathbf{x}_m \cdot \mathbf{x}_2)c'_2 + \cdots + (\mathbf{x}_m \cdot \mathbf{x}_m)c'_m &= \mathbf{x}_m \cdot \mathbf{y},
\end{aligned}$$

in the m unknowns $c'_1, \ldots, c'_m$, which, in slightly different notation, has already been discussed in the preceding section [see (7–53)]. These equations are called the *normal equations* for the approximation in question.

EXAMPLE 2. Let $y = c_1x_1 + c_2x_2$, and suppose that as a result of four separate experimental determinations we have found the set of equations

$$\begin{aligned}
15 &= c_1 + 2c_2,\\
12 &= 2c_1 + c_2,\\
10 &= c_1 + c_2,\\
0 &= c_1 - c_2.
\end{aligned}$$

Use the method of least squares to find the best approximation to c_1 and c_2.

In this case, $m = 2$, $n = 4$, and

$$\mathbf{x}_1 = (1, 2, 1, 1), \qquad \mathbf{x}_2 = (2, 1, 1, -1), \qquad \mathbf{y} = (15, 12, 10, 0).$$

(Note that $\mathbf{x}_1$ and $\mathbf{x}_2$ are linearly independent in $\mathbb{R}^4$, as required.) Thus the normal equations for this approximation are

$$7c_1' + 4c_2' = 49,$$
$$4c_1' + 7c_2' = 52,$$

and $c_1' = \frac{45}{11}$, $c_2' = \frac{56}{11}$. These then are the experimentally determined values of c_1 and c_2, and we have

$$11y = 45x_1 + 56x_2.$$

EXAMPLE 3. Find the equation of the parabola in the xy-plane which passes through the origin of coordinates and has vertical axis, and which best fits the points $(-1, 3)$, $(1, 1)$, $(2, 5)$ in the sense of the method of least squares.

The general equation of such a parabola is

$$y = c_1x^2 + c_2x,$$

and the above data give the set of equations

$$3 = c_1 - c_2,$$
$$1 = c_1 + c_2,$$
$$5 = 4c_1 + 2c_2.$$

Thus the least square approximation to c_1 and c_2 is obtained by solving the normal equations for the vectors

$$\mathbf{x}_1 = (1, 1, 4), \qquad \mathbf{x}_2 = (-1, 1, 2), \qquad \mathbf{y} = (3, 1, 5).$$

These equations are

$$18c_1' + 8c_2' = 24,$$
$$8c_1' + 6c_2' = 8,$$

and their solution is $c_1' = \frac{20}{11}$, $c_2' = -\frac{12}{11}$. Hence the desired parabola is $11y = 20x^2 - 12x$, and as can be seen in Fig. 7–18, this curve really does fit the given points extremely well.

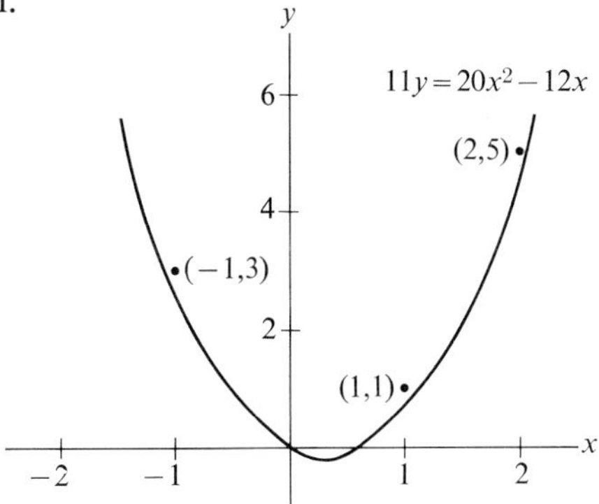

FIGURE 7–18

EXERCISES

1. When an experimental approximation is made by replacing a vector $\mathbf{x}$ in an n-dimensional Euclidean space $\mathcal{V}$ by its perpendicular projection $\mathbf{x}'$ on a subspace of $\mathcal{V}$ the quantity $\|\mathbf{x} - \mathbf{x}'\|^2/n$ is called the *variance* of the approximation. The magnitude of the variance of the approximation is taken as a measure of the consistency of the data used in obtaining it, and in judging between two experimental approximations the one which has the smaller variance is taken to be the more consistent.
   Suppose two students perform a series of five experiments to determine the valuc of a physical constant with the following results:

Student A:	1.402	1.420	1.395	1.418	1.406
Student B:	1.416	1.405	1.419	1.422	1.406

   (a) Find the best approximation to the constant which can be made by each student.
   (b) Which set of values is the more consistent?

2. Let $(x_1, \ldots, x_m)$ be the vector whose components represent the results of a series of m experimental determinations of a physical constant, and let $(x_{m+1}, \ldots, x_{m+n})$ represent the results of n experimental determinations of the same constant. Show that the variances for the two separate series of experiments do not each exceed the variance for the single series of experiments $(x_1, \ldots, x_{m+n})$. Under what conditions will equality hold?

3. (a) Repeat Exercise 1 for the following two series of measurements:

Student A:	16.02	15.99	16.12	16.07	16.10
Student B:	16.11	16.16	16.07	16.12	16.14

   (b) Consider the sequence obtained by combining the two sets of data in (a). Find the best approximation for this sequence, and compare its variance with the variances for the two approximations in (a).

4. Prove that the assertion made in the text concerning expression (7–61) is true.

5. The *center of population* of a geographical region is by definition the centroid of the residences of its inhabitants. Suppose that the population of a certain region is concentrated in four cities, with a negligible number of residents elsewhere, and suppose that the population and position (with respect to a Cartesian coordinate system in the plane) of each is as follows:

City A:	1,000,000 at (0, 0)	City B:	600,000 at (10, 5)
City C:	300,000 at (–5, 10)	City D:	100,000 at (0, –20)

   Find the center of population.

6. Repeat Exercise 5 for the following data:

City A:	2,000,000 at (0, 0)	City B:	500,000 at (6, 18)
City C:	300,000 at (–3, 15)	City D:	200,000 at (0, –12)

7. The *center of mass*, or *center of gravity*, of a system of k particles, each of mass m_i located at the point $\mathbf{x}_i$, $i = 1, 2, \ldots k$, in $\mathcal{R}^n$ is defined to be the centroid of the k vectors $m_i\mathbf{x}_i$. Find the center of mass of each of the following systems in $\mathcal{R}^3$.

(a) $m_1 = 2,\ \mathbf{x}_1 = (1, 3, 0)$
$m_2 = 3,\ \mathbf{x}_2 = (-2, 4, 1)$
$m_3 = 1,\ \mathbf{x}_3 = (0, 6, 2)$
$m_4 = 4,\ \mathbf{x}_4 = (-1, -5, 3)$

(b) $m_1 = 1,\ \mathbf{x}_1 = (1, 5, 0)$
$m_2 = 1,\ \mathbf{x}_2 = (2, -3, 3)$
$m_3 = 3,\ \mathbf{x}_3 = (7, 0, 1)$
$m_4 = 5,\ \mathbf{x}_4 = (2, -3, 6)$

(c) $m_1 = 5,\ \mathbf{x}_1 = (3, 1, -2)$
$m_2 = 1,\ \mathbf{x}_2 = (4, 0, 1)$
$m_3 = 4,\ \mathbf{x}_3 = (0, 0, 0)$
$m_4 = 3,\ \mathbf{x}_4 = (2, 2, 1)$
$m_5 = 5,\ \mathbf{x}_5 = (1, -1, 2)$
$m_6 = 2,\ \mathbf{x}_6 = (4, 1, 6)$

8. Repeat Exercise 7 for the following data:

(a) $m_1 = 1,\ \mathbf{x}_1 = (1, 2, -3)$
$m_2 = 2,\ \mathbf{x}_2 = (3, -4, 5)$
$m_3 = 3,\ \mathbf{x}_3 = (-1, 0, 0)$

(b) $m_1 = 1,\ \mathbf{x}_1 = (7, -5, 2)$
$m_2 = 2,\ \mathbf{x}_2 = (6, 3, 1)$
$m_3 = 5,\ \mathbf{x}_3 = (-1, -1, 2)$
$m_4 = 8,\ \mathbf{x}_4 = (2, 2, 0)$
$m_5 = 4,\ \mathbf{x}_5 = (3, 5, -1)$
$m_6 = 5,\ \mathbf{x}_6 = (7, 1, 0)$

(c) $m_1 = 2,\ \mathbf{x}_1 = (2, 3, -1)$
$m_2 = 5,\ \mathbf{x}_2 = (4, -2, 1)$
$m_3 = 4,\ \mathbf{x}_3 = (6, -3, 2)$
$m_4 = 3,\ \mathbf{x}_4 = (-1, -1, 0)$
$m_5 = 6,\ \mathbf{x}_5 = (0, 0, 0)$

9. (a) Show that the center of mass $\mathbf{x}$ (Exercise 7) of k particles, each of mass m_i located at the point $\mathbf{x}_i$, $i = 1, 2, \ldots, k$, minimizes the quantity $\|M\mathbf{x} - \sum_{i=1}^{k} m_i\mathbf{x}_i\|^2$, where $M = \sum_{i=1}^{k} m_i$.

(b) Let $\mathbf{x}_1 = (0, 0, 0)$, $\mathbf{x}_2 = (1, -1, 3)$, $\mathbf{x}_3 = (1, 1, 5)$, $\mathbf{x}_4 = (2, -8, -10)$. Find a set of masses m_1, m_2, m_3, m_4 such that $\|10\mathbf{x} - \sum_{i=1}^{4} m_i\mathbf{x}_i\|^2$ is minimized by the vector $\mathbf{x} = (1, -1, 2)$.

10. Let $m_1 = 4$, $\mathbf{x}_1 = (0, 0, 0)$; $m_2 = 3$, $\mathbf{x}_2 = (1, 1, 1)$; $m_3 = 2$, $\mathbf{x}_3 = (1, 1, 0)$; $m_4 = 1$. Determine $\mathbf{x}_4$ so that $\|10\mathbf{x} - \sum_{i=1}^{4} m_i\mathbf{x}_i\|^2$ is minimized by $\mathbf{x} = (1, 2, 1)$. (See Exercise 9(a).)

11. Let T be a linear transformation mapping $\mathcal{R}^n$ into itself, and let $\mathbf{x}$ be the centroid of k vectors $\mathbf{x}_1, \ldots, \mathbf{x}_k$ in $\mathcal{R}^n$. Show that $T(\mathbf{x})$ is the centroid of the vectors $T(\mathbf{x}_1), \ldots, T(\mathbf{x}_k)$.

12. Let T be the transformation which maps each vector in a Euclidean space $\mathcal{V}$ onto its best approximation (in the sense of least squares) in a given finite dimensional subspace of $\mathcal{V}$. Prove that T is a linear transformation. What are the null space and image of T?

13. Find the line $y = cx$ in $\mathcal{R}^2$ which best fits each of the following sets of points. Sketch the graph of each of these lines, and plot the given points relative to the standard coordinate system in $\mathcal{R}^2$.

(a) $(5, 9)$, $(10, 21)$, $(15, 19)$
(b) $(-4, -11)$, $(1, 3)$, $(5, 16)$, $(10, 29)$
(c) $(-6, 10)$, $(2, -2)$, $(8, -11)$, $(10, -16)$

14. Repeat Exercise 13 using the following data.

(a) $(-4, 9)$, $(4, -10)$, $(12, -25)$
(b) $(-5, -16)$, $(10, 31)$, $(15, 31)$, $(20, 40)$
(c) $(-4, -5)$, $(8, 9)$, $(12, 16)$, $(20, 24)$

In Exercises 15–18 suppose that $y = c_1x_1 + c_2x_2$, and use the method of least squares to obtain an approximate formula for y from the given data.

15. $x_{11} = 1, \quad x_{12} = 0, \quad y_1 = 2$
$x_{21} = 0, \quad x_{22} = 1, \quad y_2 = 3$
$x_{31} = 1, \quad x_{32} = 1, \quad y_3 = 2$
$x_{41} = 1, \quad x_{42} = -1, \quad y_4 = 0$

16. $x_{11} = 1, \quad x_{12} = 0, \quad y_1 = 1$
$x_{21} = 0, \quad x_{22} = 1, \quad y_2 = 1$
$x_{31} = -1, \quad x_{32} = 0, \quad y_3 = 3$

17. $x_{11} = 10, \quad x_{12} = 10, \quad y_1 = 0$
$x_{21} = 10, \quad x_{22} = -10, \quad y_2 = 19$
$x_{31} = -10, \quad x_{32} = 10, \quad y_3 = -21$
$x_{41} = -10, \quad x_{42} = -10, \quad y_4 = 0$

18. $x_{11} = 3, \quad x_{12} = 4, \quad y_1 = 0$
$x_{21} = 7, \quad x_{22} = 1, \quad y_2 = 10$
$x_{31} = 1, \quad x_{32} = 1, \quad y_3 = -5$
$x_{41} = 4, \quad x_{42} = 3, \quad y_4 = 0$

19. Find the best approximation to the formula $y = c_1x_1 + c_2x_2 + c_3x_3$ from the following data:

$$\begin{array}{llll} x_{11} = 0, & x_{12} = 0, & x_{13} = 0, & y_1 = 1 \\ x_{21} = 1, & x_{22} = 1, & x_{23} = -1, & y_2 = 2 \\ x_{31} = 1, & x_{32} = -1, & x_{33} = 1, & y_3 = -1 \\ x_{41} = 0, & x_{42} = 1, & x_{43} = 1, & y_4 = -2 \end{array}$$

20. Find the parabola through the origin in $\mathbb{R}^2$ with vertical axis which, in the sense of least squares, best fits each of the following sets of points. Sketch the graph of each parabola and plot the points it approximates.
(a) $(1, 2), (2, 5), (3, 9)$ (b) $(-1, -1), (1, 0), (3, -5)$
(c) $(-1, 4), (1, 2), (2, 10)$

21. Repeat Exercise 20 for the following sets of points.
(a) $(1, 4), (2, 5), (-1, -3)$ (b) $(1, 3), (2, 7), (-1, 2), (-2, 8)$
(c) $(1, -1), (2, -3), (-2, 1)$

22. Find a cubic curve through the origin in $\mathbb{R}^2$ which, in the sense of least squares, best fits each of the following sets of points. Sketch the graph of each curve obtained and plot the approximating points.
(a) $(1, -2), (-1, 1), (2, -5), (3, -20)$ (b) $(1, 2), (-1, 1), (2, 7), (-2, -4)$

23. Find the fourth-degree equation in the form

$$y = ax^4 + bx^2 + cx$$

which, in the sense of least squares, best fits the following points in $\mathbb{R}^2$:

$$(-2, 2), (-1, 1), (1, 2), (2, 1).$$

*7–7 AN APPLICATION TO LINEAR DIFFERENTIAL EQUATIONS

In Chapter 4 we learned how to obtain n distinct solutions of any nth-order homogeneous linear differential equation with constant coefficients from the roots of the auxiliary polynomial. At that time we also indicated how operator techniques could be used to prove the linear independence of these solutions in the real vector space $\mathcal{C}(I)$, where I is an arbitrary finite or infinite interval. We refrained from doing so, however, because such a proof is both tedious and uninteresting. But now that we have the notion of an inner product available, we shall give a particularly elegant proof of this result which has the added virtue of serving as an excellent example of the interplay between analysis and the theory of Euclidean spaces.

Specifically, we must prove that every set of functions of the form

$$x^m e^{ax} \sin bx \quad \text{and} \quad x^m e^{ax} \cos bx, \tag{7–67}$$

where a and b are real numbers ($b > 0$) and m is a non-negative integer, is linearly independent in the real vector space $\mathcal{C}(-\infty, \infty)$.* To this end suppose that F is a (finite) linear combination of such functions, and that $F(x) \equiv 0$. We must show that all of the coefficients in F are zero.

Our first step consists of rewriting F by grouping together those terms which contain the same exponential factor so that

$$F(x) = e^{a_1 x} P_1(x) + e^{a_2 x} P_2(x) + \cdots + e^{a_r x} P_r(x), \tag{7–68}$$

where $a_1 > a_2 > \cdots > a_r \geq 0$, and each $P_i(x)$ is a linear combination of functions of the form $x^m \sin bx$ and $x^m \cos bx$. Next we rewrite each $P_i(x)$ by grouping together those terms which contain the same power of x. This gives

$$P_i(x) = T_{i0}(x) + T_{i1}(x)x + \cdots + T_{is_i}(x)x^{s_i}, \tag{7–69}$$

in which the coefficients T_{ij} are expressions of the form

$$T_{ij}(x) = \sum_{k=1}^{n} (\alpha_k \cos a_k x + \beta_k \sin b_k x), \tag{7–70}$$

where the α_k, β_k are real numbers and the a_k, b_k are non-negative real numbers. In what follows we shall refer to expressions of this form as *trigonometric sums*, and we note that every such sum may be written as

$$\frac{\alpha_0}{2} + \sum_{k=1}^{n} (\alpha_k \cos a_k x + \beta_k \sin b_k x)$$

with a_k, b_k *positive.*

The essential step in proving that all of the coefficients of F are zero is furnished by the following lemma.

Lemma 7–2. *If T is a trigonometric sum with the property that* $\lim_{x\to\infty} T(x) = 0$, *then all of the coefficients of T are zero.*

---

* Note that it is sufficient to consider linear independence in $\mathcal{C}(-\infty, \infty)$ since by Theorem 3–2 (the uniqueness theorem) such a set of solutions is linearly independent in $\mathcal{C}(-\infty, \infty)$ if and only if it is linearly independent in $\mathcal{C}(I)$ for every subinterval I of $(-\infty, \infty)$. (See Exercise 22, Section 4–3).

Proof. Let $\mathfrak{I}$ be the real vector space consisting of all trigonometric sums, with the usual definitions of addition and scalar multiplication, and define an inner product on $\mathfrak{I}$ by

$$f \cdot g = \lim_{t \to \infty} \frac{1}{t} \int_0^t f(x)g(x)\,dx \tag{7–71}$$

for any pair of trigonometric sums f and g. It is easy to see that (7–71) does in fact define an inner product on $\mathfrak{I}$ provided that the limit in question always exists (see Exercise 1). To prove the existence of this limit we first observe that since integration is a linear operation (i.e., is performed term by term) it suffices to consider the case in which f and g are "monomials." In other words, we need only establish the existence of (7–71) when

Case 1. $f(x) = \cos a_1 x, \quad g(x) = \cos a_2 x, \quad a_1, a_2 \geq 0;$

Case 2. $f(x) = \cos a_1 x, \quad g(x) = \sin b_1 x, \quad a_1 \geq 0,\ b_1 > 0;$

Case 3. $f(x) = \sin b_1 x, \quad g(x) = \sin b_2 x, \quad b_1, b_2 > 0.$

Suppressing most of the details, we obtain the following results.

Case 1.

$$f \cdot g = \begin{cases} \lim\limits_{t \to \infty} \dfrac{1}{2t}\left[\dfrac{\sin (a_1 - a_2)t}{a_1 - a_2} + \dfrac{\sin (a_1 + a_2)t}{a_1 + a_2}\right] = 0, & \text{if } a_1 \neq a_2, \\[2ex] \lim\limits_{t \to \infty} \dfrac{1}{2t}\left[t + \dfrac{\sin 2a_1 t}{2a_1}\right] = \dfrac{1}{2}, & \text{if } a_1 = a_2 \neq 0, \\[2ex] \lim\limits_{t \to \infty} \dfrac{1}{t}\displaystyle\int_0^t dt = 1, & \text{if } a_1 = a_2 = 0. \end{cases}$$

Case 2.

$$f \cdot g = \begin{cases} \lim\limits_{t \to \infty} \dfrac{1}{2t}\left[-\dfrac{\cos (b_1 - a_1)t}{b_1 - a_1} - \dfrac{\cos (b_1 + a_1)t}{b_1 + a_1} + \dfrac{1}{b_1 - a_1} + \dfrac{1}{b_1 + a_1}\right] = 0, & \text{if } a_1 \neq b_1, \\[2ex] \lim\limits_{t \to \infty} \dfrac{1}{2t}\left[\dfrac{\sin^2 a_1 t}{a_1}\right] = 0, & \text{if } a_1 = b_1. \end{cases}$$

Case 3.

$$f \cdot g = \begin{cases} \lim\limits_{t \to \infty} \dfrac{1}{2t}\left[\dfrac{\sin (b_1 - b_2)t}{b_1 - b_2} - \dfrac{\sin (b_1 + b_2)t}{b_1 + b_2}\right] = 0, & \text{if } b_1 \neq b_2, \\[2ex] \lim\limits_{t \to \infty} \dfrac{1}{2t}\left[t - \dfrac{1}{2b_1}\sin 2b_1 t\right] = \dfrac{1}{2}, & \text{if } b_1 = b_2. \end{cases}$$

Hence the necessary limits exist, and (7–71) defines an inner product on $\mathfrak{I}$. Moreover, from the above computations we see that whenever a and b are positive real numbers the functions

$$1, \quad \cos ax, \quad \sin bx$$

are orthogonal in $\mathfrak{I}$, and

$$\|1\| = 1, \qquad \|\cos ax\| = \frac{1}{\sqrt{2}}, \qquad \|\sin bx\| = \frac{1}{\sqrt{2}}. \tag{7-72}$$

Now suppose that

$$T(x) = \frac{\alpha_0}{2} + \sum_{k=1}^{n} (\alpha_k \cos a_k x + \beta_k \sin b_k x)$$

is a trigonometric sum such that

$$\lim_{x \to \infty} T(x) = 0.$$

Making use of the orthogonality just established, together with the relations listed in (7–72), we have

$$\alpha_k = \frac{T(x) \cdot \cos a_k x}{\|\cos a_k x\|^2} = 2 \lim_{t \to \infty} \frac{1}{t} \int_0^t T(x) \cos a_k x \, dx, \quad k = 0, 1, \ldots, n,$$

$$\beta_k = \frac{T(x) \cdot \sin b_k x}{\|\sin b_k x\|^2} = 2 \lim_{t \to \infty} \frac{1}{t} \int_0^t T(x) \sin b_k x \, dx, \quad k = 1, 2, \ldots, n,$$

and to prove the lemma we must show that all of these coefficients are zero.

Let us examine the α_k. By assumption, $T(x) \to 0$ as $x \to \infty$. Hence, given any real number $\epsilon > 0$, there exists a real number $r > 0$ such that

$$|T(x)| < \epsilon \quad \text{whenever} \quad x > r.$$

Then

$$\begin{aligned}
|\alpha_k| &= \left| \lim_{t \to \infty} \frac{1}{t} \int_0^t T(x) \cos a_k x \, dx \right| \\
&= \lim_{t \to \infty} \left| \frac{1}{t} \int_0^r T(x) \cos a_k x \, dx + \frac{1}{t} \int_r^t T(x) \cos a_k x \, dx \right| \\
&\le \lim_{t \to \infty} \frac{1}{t} \left| \int_0^r T(x) \cos a_k x \, dx \right| + \lim_{t \to \infty} \frac{1}{t} \left| \int_r^t T(x) \cos a_k x \, dx \right| .
\end{aligned}$$

But

$$\lim_{t \to \infty} \frac{1}{t} \left| \int_0^r T(x) \cos a_k x \, dx \right| = 0,$$

while

$$\begin{aligned}
\lim_{t \to \infty} \frac{1}{t} \left| \int_r^t T(x) \cos a_k x \, dx \right| &\le \lim_{t \to \infty} \frac{1}{t} \int_r^t |T(x) \cos a_k x| \, dx \\
&< \lim_{t \to \infty} \frac{1}{t} \int_r^t \epsilon \, dx \\
&= \lim_{t \to \infty} \frac{\epsilon(t - r)}{t} = \epsilon.
\end{aligned}$$

Hence $|\alpha_k| < \epsilon$ for any positive number ϵ, and it follows that $\alpha_k = 0$. A similar argument shows that $\beta_k = 0$ for all k, and the proposition is proved. ▮

It is now relatively easy to establish our main result, which we state formally as

Theorem 7–7. *The set of all functions of the form* (7–67) *is linearly independent in* $\mathcal{C}(-\infty, \infty)$.

Proof. Let F be a finite linear combination of these functions such that $F(x) \equiv 0$. We rewrite F in the manner described above as

$$F(x) = e^{a_1 x}P_1(x) + \cdots + e^{a_r x}P_r(x), \tag{7–73}$$

where $a_1 > a_2 > \cdots > a_r$, and

$$P_i(x) = T_{i0}(x) + \cdots + T_{is_i}(x)x^{s_i}. \tag{7–74}$$

We shall assume that F has at least one nonzero coefficient, and deduce a contradiction.

Indeed, if such is the case there exists at least one integer i such that P_i has a nonzero coefficient. With i chosen as small as possible, we have

$$F(x) = e^{a_i x}P_i(x) + e^{a_{i+1}x}P_{i+1}(x) + \cdots + e^{a_r x}P_r(x),$$

and hence

$$P_i(x) = e^{-a_i x}F(x) - e^{(a_{i+1}-a_i)x}P_{i+1}(x) - \cdots - e^{(a_r - a_i)x}P_r(x).$$

Since $F(x) \equiv 0$,

$$\lim_{x\to\infty} e^{-a_i x}F(x) = 0,$$

and since

$$\lim_{x\to\infty} e^{(a_j - a_i)x}P_j(x) = 0$$

whenever $j > i$, it follows that

$$\lim_{x\to\infty} P_i(x) = 0.$$

But P_i has at least one nonzero coefficient. Hence there exists at least one index j in (7–74) such that the trigonometric sum T_{ij} has a nonzero coefficient. If we choose j as large as possible, then

$$P_i(x) = T_{i0}(x) + \cdots + T_{ij}(x)x^j,$$

whence

$$T_{ij}(x) = \frac{P_i(x)}{x^j} - \frac{T_{i0}(x)}{x^j} - \frac{T_{i1}(x)}{x^{j-1}} - \cdots - \frac{T_{i,j-1}(x)}{x}.$$

Since it is clear that each of the terms on the right-hand side of this equation tends to zero as $x \to \infty$, we conclude that

$$\lim_{x\to\infty} T_{ij}(x) = 0.$$

This, however, contradicts Lemma 7–2 since T_{ij} has nonzero coefficients. It follows that all of the coefficients of F must be zero, and the theorem is proved. ▌

EXERCISES

1. Verify that (7–71) defines an inner product on the vector space $\mathfrak{T}$ of all trigonometric sums. In particular, if

$$f(x) = \frac{\alpha_0}{2} + \sum_{k=1}^{n} (\alpha_k \cos a_k x + \beta_k \sin b_k x),$$

show that

$$\mathbf{f}\cdot\mathbf{f} = \frac{\alpha_0^2}{4} + \frac{1}{2}\sum_{k=1}^{n} (\alpha_k^2 + \beta_k^2).$$

2. Suppose that the roots of the auxiliary polynomial for a constant coefficient linear differential equation are $2, 2, 2 \pm 2i, 2 \pm 2i, 2 \pm 3i, 2 \pm 3i, 2 \pm 3i, 1 \pm i, 1 \pm i$. Write out F as in the text, and group the terms in the way described.

3. (a) In the proof of Theorem 7–7 we asserted that

$$\lim_{x\to\infty} e^{(a_j - a_i)x} P_j(x) = 0$$

whenever $j > i$. Prove the assertion.

(b) Prove that

$$\lim_{x\to\infty} \frac{T(x)}{x^n} = 0$$

whenever T is a trigonometric sum and n is a positive integer.

8
convergence in euclidean spaces*

8–1 SEQUENTIAL CONVERGENCE

In this chapter we prepare the way for the study of such topics as Fourier series and boundary value problems by introducing the notion of convergence in Euclidean spaces. As we shall see, the really interesting applications of this concept occur in infinite dimensional spaces, but for the sake of completeness the basic definition will be given without reference to the dimension of the underlying space.

Definition 8–1. A sequence $\{\mathbf{x}_k\} = \{\mathbf{x}_1, \mathbf{x}_2, \ldots\}$ of vectors in a Euclidean space $\mathcal{V}$ is said to *converge to the vector* $\mathbf{x}$ in $\mathcal{V}$ if and only if

$$\lim_{k\to\infty} \|\mathbf{x}_k - \mathbf{x}\| = 0. \tag{8–1}$$

In this case $\mathbf{x}$ is said to be the *limit* of $\{\mathbf{x}_k\}$, and we denote the fact that the sequence $\{\mathbf{x}_k\}$ converges to $\mathbf{x}$ by writing

$$\lim_{k\to\infty} \{\mathbf{x}_k\} = \mathbf{x}.$$

We recall that the above definition is an abbreviation for the statement: $\{\mathbf{x}_k\}$ *converges to* $\mathbf{x}$ *if and only if for each real number* $\epsilon > 0$, *an integer* K *can be found such that*

$$\|\mathbf{x}_k - \mathbf{x}\| < \epsilon \tag{8–2}$$

for all $k > K$. In general, of course, the integer K depends upon ϵ, and increases as ϵ approaches 0. Moreover, since the quantity $\|\mathbf{x}_k - \mathbf{x}\|$ is just the distance from $\mathbf{x}_k$ to $\mathbf{x}$, (8–2) asserts that $\{\mathbf{x}_k\}$ converges to $\mathbf{x}$ if and only if the distance from $\mathbf{x}_k$ to $\mathbf{x}$ approaches 0 as k becomes large. And this is precisely what intuition demands of sequential convergence.

---

* Although this chapter is logically self-contained, we assume that the reader is familiar with the notions of sequential convergence and infinite series as studied in elementary calculus. A review of this material can be found in Appendix I.

Before Definition 8–1 can be accepted as a reasonable description of convergence we must prove that the limit of a convergent sequence of vectors is *unique*, in the sense that a given sequence can never converge to more than one vector. This we do by establishing

Lemma 8–1. *If*

$$\lim_{k\to\infty} \{\mathbf{x}_k\} = \mathbf{x} \quad \textit{and} \quad \lim_{k\to\infty} \{\mathbf{x}_k\} = \mathbf{y},$$

then $\mathbf{x} = \mathbf{y}$.

Proof. Let $\epsilon > 0$ be given. Then, by definition, there exists an integer K such that

$$\|\mathbf{x}_k - \mathbf{x}\| < \epsilon/2 \quad \text{and} \quad \|\mathbf{x}_k - \mathbf{y}\| < \epsilon/2$$

whenever $k > K$. Thus, by the triangle inequality,

$$\begin{aligned} \|\mathbf{x} - \mathbf{y}\| &\le \|\mathbf{x} - \mathbf{x}_k\| + \|\mathbf{x}_k - \mathbf{y}\| \\ &< \frac{\epsilon}{2} + \frac{\epsilon}{2} \\ &= \epsilon. \end{aligned}$$

Since $\epsilon > 0$ was arbitrary, this inequality implies that $\|\mathbf{x} - \mathbf{y}\| = 0$. Hence $\mathbf{x} = \mathbf{y}$, as asserted. ▮

Example 1. Let $\{a_k\}$ be a sequence of real numbers, viewed as vectors in $\mathcal{R}^1$. Then $\{a_k\}$ converges to the real number (i.e., vector) a if and only if

$$\lim_{k\to\infty} \|a_k - a\| = 0.$$

Recalling that the inner product in $\mathcal{R}^1$ is just ordinary multiplication of real numbers, we find that

$$\begin{aligned} \|a_k - a\| &= [(a_k - a)\cdot(a_k - a)]^{1/2} \\ &= [(a_k - a)^2]^{1/2} \\ &= |a_k - a|. \end{aligned}$$

Hence $\{a_k\}$ converges to a if and only if

$$\lim_{k\to\infty} |a_k - a| = 0.$$

But this equation is none other than the definition of sequential convergence given in elementary calculus, and it therefore follows that *sequential convergence in $\mathcal{R}^1$ is identical with the usual convergence of sequences of real numbers.*

EXAMPLE 2. Let $\mathbf{e}_1, \ldots, \mathbf{e}_n$ be an orthonormal basis in $\mathcal{R}^n$, and let $\{\mathbf{x}_k\}$ be a sequence of vectors in $\mathcal{R}^n$. Then if

$$\mathbf{x} = \alpha_1\mathbf{e}_1 + \cdots + \alpha_n\mathbf{e}_n$$

is an arbitrary vector in $\mathcal{R}^n$, and if, for each integer k,

$$\mathbf{x}_k = \alpha_{1k}\mathbf{e}_1 + \cdots + \alpha_{nk}\mathbf{e}_n,$$

we have

$$\|\mathbf{x}_k - \mathbf{x}\|^2 = (\alpha_{1k} - \alpha_1)^2 + \cdots + (\alpha_{nk} - \alpha_n)^2.$$

Hence

$$\lim_{k\to\infty} \|\mathbf{x}_k - \mathbf{x}\| = 0$$

if and only if the sequences

$$\{\alpha_{1k}\} = \{\alpha_{11}, \alpha_{12}, \ldots\},$$
$$\{\alpha_{2k}\} = \{\alpha_{21}, \alpha_{22}, \ldots\},$$
$$\vdots$$
$$\{\alpha_{nk}\} = \{\alpha_{n1}, \alpha_{n2}, \ldots\},$$

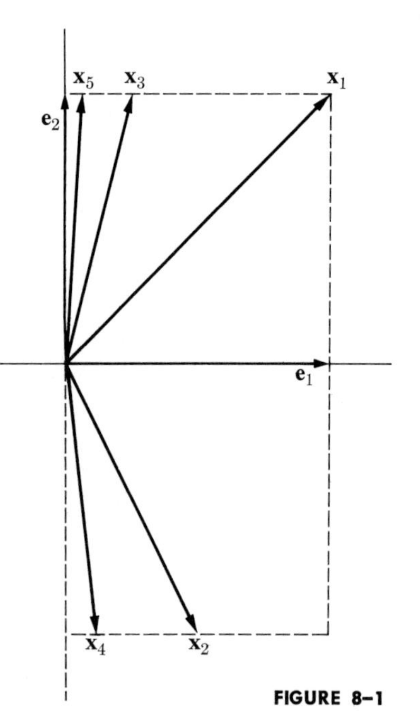

FIGURE 8–1

converge (as ordinary sequences of real numbers) to $\alpha_1, \ldots, \alpha_n$, respectively.

For instance, if $\mathbf{e}_1$ and $\mathbf{e}_2$ are the standard basis vectors in $\mathcal{R}^2$, then

$$\{\mathbf{x}_k\} = \left\{\frac{1}{2^{k-1}}\mathbf{e}_1 + (-1)^{k+1}\mathbf{e}_2\right\}$$
$$= \{\mathbf{e}_1 + \mathbf{e}_2, \tfrac{1}{2}\mathbf{e}_1 - \mathbf{e}_2, \tfrac{1}{4}\mathbf{e}_1 + \mathbf{e}_2, \ldots\}$$

does *not* converge because the sequence $\{1, -1, 1, \ldots\}$ formed from the components of $\mathbf{e}_2$ is not a convergent sequence of real numbers (see Fig. 8–1). On the other hand, the sequence

$$\{(1/2^{k-1})\mathbf{e}_1 + \mathbf{e}_2\}$$

converges to the vector $\mathbf{e}_2$, since $\{1, \frac{1}{2}, \frac{1}{4}, \ldots\}$ converges to 0, and $\{1, 1, 1, \ldots\}$ converges to 1 (Fig. 8–2).

The preceding examples show that the study of convergence in *finite* dimensional Euclidean spaces is essentially the same as the study of convergence of sequences of real numbers. However, in infinite dimensional spaces such as $\mathcal{C}[a, b]$ the situation becomes much more complex, and correspondingly more interesting. For then the type of convergence defined above is radically different from that studied in calculus under the name of pointwise convergence.* Indeed, we shall see momentarily that in a function space with an integral inner product the assertion that

$$\lim_{k\to\infty} \|\mathbf{f}_k - \mathbf{f}\| = \lim_{k\to\infty} \left(\int_a^b [f_k(x) - f(x)]^2\, dx\right)^{1/2} = 0$$

is not at all the same as saying that the sequence $\{\mathbf{f}_k\}$ converges to the function $\mathbf{f}$ at every point of $[a, b]$. In analysis such convergence is known as *mean convergence* to emphasize that it is computed by integration, which, in a sense, is a generalized averaging process.

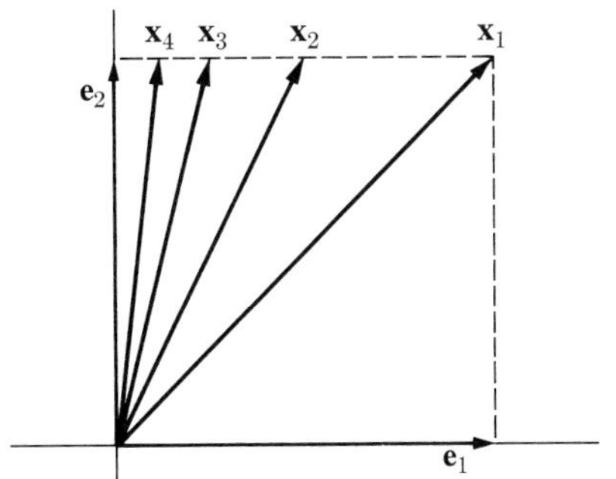

FIGURE 8–2

EXAMPLE 3. The sequence of functions $\{x, x^2, x^3, \ldots\}$ converges in the mean in $\mathcal{C}[-1, 1]$ to the zero function since

$$\begin{aligned}\lim_{k\to\infty} \|x^k - 0\| &= \lim_{k\to\infty} \left(\int_{-1}^{1} x^{2k}\, dx\right)^{1/2} \\ &= \lim_{k\to\infty} \left(\frac{2}{2k+1}\right)^{1/2} \\ &= 0.\end{aligned}$$

This not withstanding, $\{x, x^2, x^3, \ldots\}$ does *not* converge to zero at *each point* in the interval $[-1, 1]$. In fact, at $x = 1$ the sequence converges to 1, while at $x = -1$ it does not converge at all. (See Fig. 8–3.)

The example just given shows that mean convergence is different from pointwise convergence. The one which follows shows how different it really is.

EXAMPLE 4. Let $\mathcal{PC}[a, b]$ denote the set of all *piecewise continuous* functions on $[a, b]$; that is, the set of all functions which are continuous everywhere on $[a, b]$ except (possibly) at a finite number of points where they have jump discontinuities (see Section 5–1). In Section 9–2 we will prove that $\mathcal{PC}[a, b]$ can be regarded as a Euclidean space under the usual definitions of addition, scalar multiplication,

* Recall that a sequence of functions $\{f_k\}$ defined on $[a, b]$ is said to converge *pointwise* to the function f if $\lim_{k\to\infty} f_k(x) = f(x)$ for each point x in $[a, b]$. (See Appendix I.)

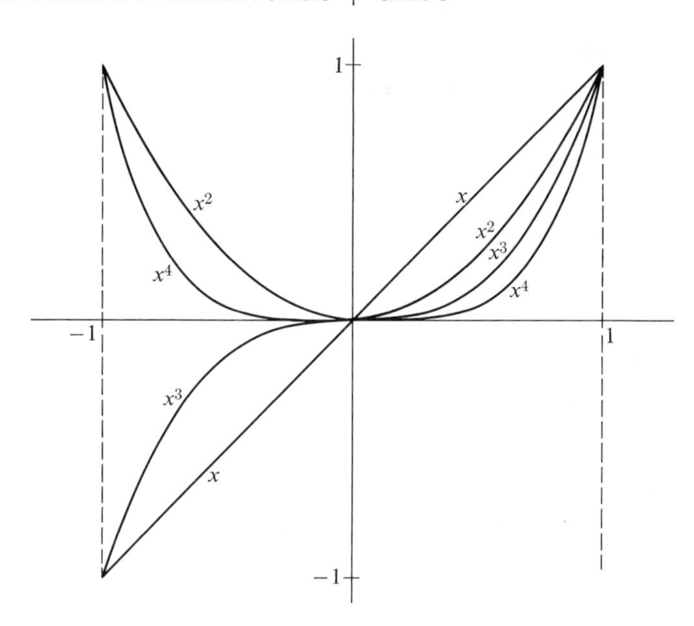

FIGURE 8–3

and inner product, the last being defined by the formula

$$\mathbf{f} \cdot \mathbf{g} = \int_a^b f(x)g(x)\,dx.$$

Accepting the truth of this assertion, let $[a, b]$ be the unit interval $[0, 1]$, and let $I_1, I_2, \ldots$ be the sequence of subintervals

$$\begin{aligned} I_1 &= [0, 1], & & \\ I_2 &= [0, \tfrac{1}{2}], & I_3 &= [\tfrac{1}{2}, 1], \\ I_4 &= [0, \tfrac{1}{4}], & I_5 &= [\tfrac{1}{4}, \tfrac{1}{2}], \\ I_6 &= [\tfrac{1}{2}, \tfrac{3}{4}], & I_7 &= [\tfrac{3}{4}, 1], \\ &\vdots & & \end{aligned}$$

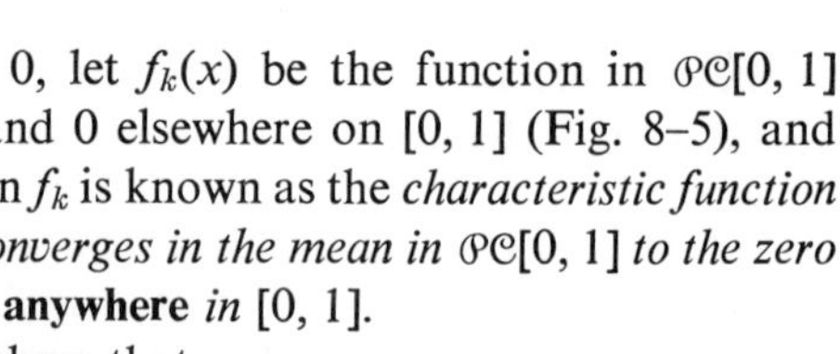

FIGURE 8–4

(See Fig. 8–4.) For each integer $k > 0$, let $f_k(x)$ be the function in $\mathcal{PC}[0, 1]$ which has the value 1 when x is in I_k and 0 elsewhere on $[0, 1]$ (Fig. 8–5), and consider the sequence $\{f_k\}$. (The function f_k is known as the *characteristic function* of the interval I_k.) We assert that *$\{f_k\}$ converges in the mean in $\mathcal{PC}[0, 1]$ to the zero function, but does* **not** *converge pointwise* **anywhere** *in* $[0, 1]$.

To prove mean convergence we must show that

$$\lim_{k\to\infty} \|f_k - 0\| = 0.$$

But

$$\|f_k - 0\| = \left(\int_0^1 [f_k(x)]^2\,dx\right)^{1/2},$$

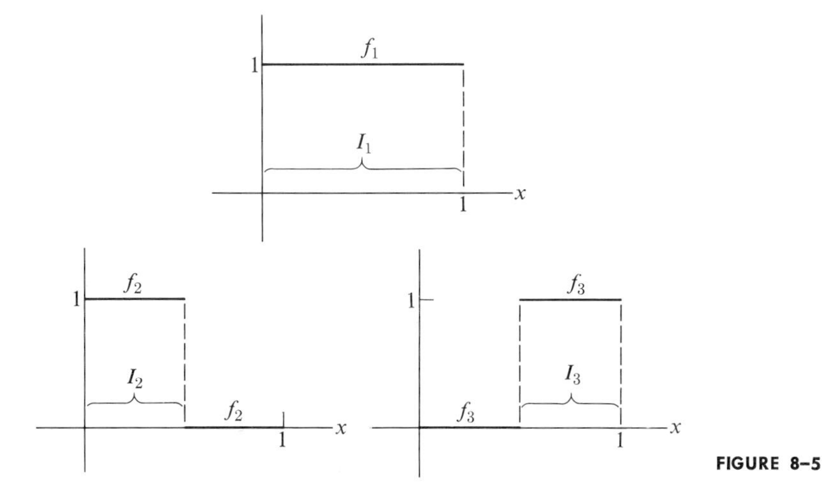

FIGURE 8-5

and since f_k is identically 1 on I_k and is zero elsewhere, the value of this integral is just the square root of the length of I_k. However, the length of I_k tends to zero with increasing k, and it follows that

$$\lim_{k\to\infty}\left(\int_0^1 [f_k(x)]^2\,dx\right)^{1/2} = 0,$$

as asserted.

Now let x_0 be a fixed point in [0, 1], and consider the sequence of real numbers $\{f_k(x_0)\}$. We contend that this sequence does *not* converge. For, by definition, $f_k(x_0)$ is either 1 or 0 depending on whether x_0 is or is not in I_k. But the I_k were constructed in such a way that there exist arbitrarily large values of k for which $f_k(x_0) = 0$ *and* arbitrarily large values of k for which $f_k(x_0) = 1$. Thus $\{f_k(x_0)\}$ contains *both* zeros and ones no matter how far out in the sequence we go, and hence does not converge.

EXERCISES

Determine which of the following sequences $\{a_k\}$ converge in $\mathcal{R}^1$, and find the limit of each convergent sequence.

1. $a_k = \dfrac{(-1)^k}{k}$

2. $a_k = \dfrac{k}{2k+1}$

3. $a_k = 1 + (-1)^k$

4. $a_k = \dfrac{k^2}{k+1}$

5. $a_k = 1 + e^{-k}$

6. $a_k = \dfrac{k^2+1}{(k+1)(k-1)}$

7. $a_k = \dfrac{2^{k+1} + (-1)^k}{2^k}$

8. $a_k = \dfrac{\ln k}{k}$

Determine which of the following sequences $\{\mathbf{x}_k\}$ converge in $\mathcal{R}^3$, and find the limit of each convergent sequence. (In each case the vectors $\mathbf{e}_1, \mathbf{e}_2, \mathbf{e}_3$ are an orthonormal basis for $\mathcal{R}^3$.)

9. $\mathbf{x}_k = \left(1 + \frac{1}{k}\right)\mathbf{e}_1 + \mathbf{e}_2 + \frac{k+1}{k+2}\mathbf{e}_3$

10. $\mathbf{x}_k = \frac{1}{k^2}\mathbf{e}_1 + k^2\mathbf{e}_2 + (-1)^k\mathbf{e}_3$

11. $\mathbf{x}_k = (1 - 2^{-k})\mathbf{e}_1 + \frac{\sin k}{k}\mathbf{e}_2 + (\ln k)\mathbf{e}_3$

12. $\mathbf{x}_k = \frac{k}{k^2+1}\mathbf{e}_1 + \frac{\ln k}{k}\mathbf{e}_2 + \frac{1}{k \ln k}\mathbf{e}_3$

13. Let $\{\mathbf{x}_k\}$ be a convergent sequence of vectors in a Euclidean space. Prove that for every real number $\epsilon > 0$ there exists an integer K such that

$$\|\mathbf{x}_m - \mathbf{x}_n\| < \epsilon$$

whenever $m, n > K$. [*Hint:* Use the triangle inequality.]

14. In Chapter 10 it will be shown that

$$\lim_{k\to\infty} \int_{-\pi}^{\pi} f(x) \sin kx \, dx = 0$$

for any function f in $\mathcal{C}[-\pi, \pi]$. Use this fact to prove that the sequence $\{\sin kx\}$, $k = 1, 2, \ldots$, does *not* converge in the mean in $\mathcal{C}[-\pi, \pi]$.

15. Prove that the sequence $\{(\sin kx)/k\}$ converges in the mean in $\mathcal{C}[-\pi, \pi]$ to the zero function. Prove that this sequence also converges to zero for each x in the interval $[-\pi, \pi]$.

*16. (For students familiar with uniform convergence.) Let $\{f_k\}$ be a sequence of functions in $\mathcal{C}[a, b]$, and suppose that $\{f_k\}$ converges uniformly on $[a, b]$ to a function f in $\mathcal{C}[a, b]$; i.e., given any $\epsilon > 0$ there exists an integer k such that $|f_k(x) - f(x)| < \epsilon$ for all $k > K$ and *all* x in $[a, b]$. Prove that $\{f_k\}$ also converges *in the mean* to f.

8–2 SEQUENCES AND SERIES

In this section we establish a number of elementary facts concerning convergence in Euclidean spaces. The first asserts that limits of sums and scalar products behave entirely as expected, in the sense of

Lemma 8–2. *Let $\{\mathbf{x}_k\}$ and $\{\mathbf{y}_k\}$ be convergent sequences in a Euclidean space with limits $\mathbf{x}$ and $\mathbf{y}$ respectively. Then the sequence $\{\alpha\mathbf{x}_k + \beta\mathbf{y}_k\}$ is convergent for every pair of real numbers α, β, and*

$$\lim_{k\to\infty} \{\alpha\mathbf{x}_k + \beta\mathbf{y}_k\} = \alpha\mathbf{x} + \beta\mathbf{y}.$$

The proof is an immediate consequence of Definition 8–1, and has been left to the student as an exercise (see Exercise 1 below).

For our purposes, a much more important property of convergence is that given by

Theorem 8–1. *Let* $\{\mathbf{x}_k\}$ *and* $\{\mathbf{y}_k\}$ *be convergent sequences in a Euclidean space with limits* $\mathbf{x}$ *and* $\mathbf{y}$, *respectively. Then* $\{\mathbf{x}_k \cdot \mathbf{y}_k\}$ *is a convergent sequence of real numbers, and*

$$\lim_{k \to \infty} \{\mathbf{x}_k \cdot \mathbf{y}_k\} = \mathbf{x} \cdot \mathbf{y}.$$

In short, the inner product is a continuous function on a Euclidean space.

Proof. We must show that

$$\lim_{k \to \infty} |(\mathbf{x}_k \cdot \mathbf{y}_k) - (\mathbf{x} \cdot \mathbf{y})| = 0.$$

To this end we set $\mathbf{u}_k = \mathbf{x}_k - \mathbf{x}$, $\mathbf{v}_k = \mathbf{y}_k - \mathbf{y}$, and write

$$\begin{aligned} |(\mathbf{x}_k \cdot \mathbf{y}_k) - (\mathbf{x} \cdot \mathbf{y})| &= |(\mathbf{u}_k + \mathbf{x}) \cdot (\mathbf{v}_k + \mathbf{y}) - (\mathbf{x} \cdot \mathbf{y})| \\ &= |(\mathbf{u}_k \cdot \mathbf{v}_k) + (\mathbf{x} \cdot \mathbf{v}_k) + (\mathbf{y} \cdot \mathbf{u}_k)| \\ &\leq |\mathbf{u}_k \cdot \mathbf{v}_k| + |\mathbf{x} \cdot \mathbf{v}_k| + |\mathbf{y} \cdot \mathbf{u}_k|, \end{aligned}$$

the last step following from Eq. (7–27), applied to $\mathcal{R}^1$. We now use the Schwarz inequality to deduce that

$$\begin{aligned} |\mathbf{u}_k \cdot \mathbf{v}_k| &\leq \|\mathbf{u}_k\| \, \|\mathbf{v}_k\|, \\ |\mathbf{x} \cdot \mathbf{v}_k| &\leq \|\mathbf{x}\| \, \|\mathbf{v}_k\|, \\ |\mathbf{y} \cdot \mathbf{u}_k| &\leq \|\mathbf{y}\| \, \|\mathbf{u}_k\|. \end{aligned}$$

Hence

$$|(\mathbf{x}_k \cdot \mathbf{y}_k) - (\mathbf{x} \cdot \mathbf{y})| \leq \|\mathbf{u}_k\| \, \|\mathbf{v}_k\| + \|\mathbf{x}\| \, \|\mathbf{v}_k\| + \|\mathbf{y}\| \, \|\mathbf{u}_k\|,$$

and since $\|\mathbf{u}_k\| \to 0$ and $\|\mathbf{v}_k\| \to 0$ as $k \to \infty$, we have

$$\lim_{k \to \infty} |(\mathbf{x}_k \cdot \mathbf{y}_k) - (\mathbf{x} \cdot \mathbf{y})| = 0. \blacksquare$$

Now that we have introduced sequential convergence in Euclidean spaces, we can also discuss convergence of *infinite series* in the same context. Thus, with each series $\sum_{k=1}^{\infty} \mathbf{x}_k$ in a Euclidean space we associate its sequence of partial sums

$$\{\mathbf{x}_1, \ \mathbf{x}_1 + \mathbf{x}_2, \ \mathbf{x}_1 + \mathbf{x}_2 + \mathbf{x}_3, \ \ldots\},$$

and then take the convergence of this sequence as the criterion for the convergence of the series in question. This is the content of

Definition 8–2. An infinite series $\sum_{k=1}^{\infty} \mathbf{x}_k$ of vectors in a Euclidean space $\mathcal{V}$ is said to *converge to the vector* $\mathbf{x}$ in $\mathcal{V}$ if and only if the associated sequence of partial sums converges to $\mathbf{x}$ in the sense of Definition 8–1. If this is the case we write

$$\mathbf{x} = \sum_{k=1}^{\infty} \mathbf{x}_k, \tag{8–3}$$

and say that $\mathbf{x}$ has been expanded as an infinite series. In greater detail, $\sum_{k=1}^{\infty} \mathbf{x}_k$ *converges to* $\mathbf{x}$ *if and only if for each real number* $\epsilon > 0$, *there exists an integer* K *such that*

$$\left\| \sum_{k=1}^{n} \mathbf{x}_k - \mathbf{x} \right\| < \epsilon \tag{8–4}$$

whenever $n > K$.

EXAMPLE. Let $\mathbf{e}_1, \ldots, \mathbf{e}_n$ be an orthonormal basis for $\mathcal{R}^n$, and suppose $\sum_{k=1}^{\infty} \mathbf{x}_k$ is an infinite series of vectors in $\mathcal{R}^n$. Then

$$\sum_{k=1}^{\infty} \mathbf{x}_k = \sum_{k=1}^{\infty} \alpha_{1k}\mathbf{e}_1 + \cdots + \sum_{k=1}^{\infty} \alpha_{nk}\mathbf{e}_k,$$

where the α_{ik} are the coordinates of $\mathbf{x}_k$ with respect to the given basis, and it follows from Example 2 of the preceding section that $\sum_{k=1}^{\infty} \mathbf{x}_k$ converges to the vector $\mathbf{x} = \alpha_1\mathbf{e}_1 + \cdots + \alpha_n\mathbf{e}_n$ if and only if

$$\alpha_1 = \sum_{k=1}^{\infty} \alpha_{1k}, \quad \ldots \quad, \alpha_n = \sum_{k=1}^{\infty} \alpha_{nk}.$$

In particular, this result implies that the usual tests for the convergence of a series of real numbers such as the comparison test, the ratio test, etc., can be applied to determine whether or not an infinite series in $\mathcal{R}^n$ converges. The only difference is that the tests must be applied n times, component by component.

EXERCISES

1. (a) Prove Lemma 8–2.

   (b) Let $\{\alpha_k\}$ and $\{\beta_k\}$ be convergent sequences of real numbers with limits α and β, respectively, and let $\{\mathbf{x}_k\}$ and $\{\mathbf{y}_k\}$ be as in Lemma 8–2. Prove that $\{\alpha_k\mathbf{x}_k + \beta_k\mathbf{y}_k\}$ converges to $\alpha\mathbf{x} + \beta\mathbf{y}$.

2. (a) Let $\{\mathbf{x}_k\}$ be a convergent sequence of vectors in a Euclidean space $\mathcal{V}$, and suppose that $\lim_{k\to\infty} \{\mathbf{x}_k\} = \mathbf{x}$. Prove that

$$\lim_{k\to\infty} [(\mathbf{x}_k - \mathbf{x}) \cdot \mathbf{y}] = 0$$

   for every vector $\mathbf{y}$ in $\mathcal{V}$.

(b) Let $\mathcal{V}$ be a finite dimensional Euclidean space, and let $\{\mathbf{x}_k\}$ be a sequence of vectors in $\mathcal{V}$. Suppose that there exists a vector $\mathbf{x}$ in $\mathcal{V}$ such that

$$\lim_{k\to\infty} [(\mathbf{x}_k - \mathbf{x}) \cdot \mathbf{y}] = 0$$

for *all* $\mathbf{y}$ in $\mathcal{V}$. Prove that $\{\mathbf{x}_k\}$ converges to $\mathbf{x}$ in the sense of Definition 8-1. [*Hint:* Choose an orthonormal basis in $\mathcal{V}$.]

(c) Give an example to show that the result in (b) fails when $\mathcal{V}$ is infinite dimensional. [*Hint:* Let $\mathcal{V}$ be the space of all continuously differentiable functions on $[-\pi, \pi]$ with the standard inner product, let $\mathbf{x}_k = \cos kx$, $k = 1, 2, \ldots$, and let $\mathbf{x} = 0$.]

8-3 BASES IN INFINITE DIMENSIONAL EUCLIDEAN SPACES

In Section 7-4 we used the Gram-Schmidt orthogonalization process to prove that every *finite* dimensional Euclidean space has an orthonormal basis $\mathbf{e}_1, \ldots, \mathbf{e}_n$, and that every vector in such a space can be written *uniquely* in the form

$$\mathbf{x} = (\mathbf{x} \cdot \mathbf{e}_1)\mathbf{e}_1 + \cdots + (\mathbf{x} \cdot \mathbf{e}_n)\mathbf{e}_n. \tag{8-5}$$

But the Gram-Schmidt process can be applied equally well in an infinite dimensional Euclidean space where it can be used to produce an *infinite* orthonormal set $\mathbf{e}_1, \mathbf{e}_2, \ldots$. This fact suggests that an extended version of our earlier results may be within our grasp, and the remaining sections of the present chapter are devoted to realizing this suggestion.

Before we begin, we emphasize that the ideas we are about to develop are natural and inevitable consequences of the study of finite dimensional Euclidean spaces, and should be understood as such. This, however, is not to deny their importance or to suggest that they are being presented as an academic exercise in generalization. Quite the contrary; for, as we shall see, these ideas lie at the heart of much of analysis and its applications to mathematical physics.

The most obvious way in which to attempt to generalize (8-5) in the presence of an orthonormal set $\mathbf{e}_1, \mathbf{e}_2, \ldots$ in an infinite dimensional space $\mathcal{V}$ is to replace its right-hand side by the infinite series

$$\sum_{k=1}^{\infty} (\mathbf{x} \cdot \mathbf{e}_k)\mathbf{e}_k. \tag{8-6}$$

However, in the absence of any further information there is clearly no *a priori* reason for supposing that this series converges, much less that it converges to $\mathbf{x}$.* Nevertheless, it is convenient to have a notation which expresses the fact that

---

* Such a series is sometimes called a *formal series*, to emphasize the fact that on the face of things it is nothing more than an expression which actually may be devoid of meaning.

(8–6) is deduced from **x**, albeit in a purely formal way. The one most commonly used is

$$\mathbf{x} \sim \sum_{k=1}^{\infty} (\mathbf{x} \cdot \mathbf{e}_k)\mathbf{e}_k, \tag{8–7}$$

the symbol ~ (which, by the way, is no relative of the one used earlier for equivalence relations) being used to emphasize that the series in question may not converge to **x**. Of course, if it does, we write

$$\mathbf{x} = \sum_{k=1}^{\infty} (\mathbf{x} \cdot \mathbf{e}_k)\mathbf{e}_k, \tag{8–8}$$

and say that the series *converges* **in the mean** *to* **x**. In either case, the inner products $\mathbf{x} \cdot \mathbf{e}_k$ are called the *coordinates* or (generalized) *Fourier coefficients* of **x** with respect to the orthonormal set $\mathbf{e}_1, \mathbf{e}_2, \ldots$.

It is clear that the Fourier coefficients of **x** depend upon the orthonormal set with respect to which they are computed. Not quite so clear, but equally true, is that (8–6) may converge in the mean to **x** for one orthonormal set but not for another. The following examples illustrate both of these points.

EXAMPLE 1. Compute the generalized Fourier coefficients of the function

$$f(x) = x, \quad -\pi \leq x \leq \pi,$$

in $\mathcal{C}[-\pi, \pi]$ with respect to the orthonormal set

$$\frac{\sin x}{\sqrt{\pi}}, \frac{\sin 2x}{\sqrt{\pi}}, \frac{\sin 3x}{\sqrt{\pi}}, \ldots \tag{8–9}$$

(see Example 2, Section 7–3). In this case the coefficients are

$$\mathbf{x} \cdot \mathbf{e}_k = \frac{1}{\sqrt{\pi}} \int_{-\pi}^{\pi} x \sin kx \, dx, \quad k = 1, 2, \ldots,$$

and, using integration by parts, we obtain

$$\begin{aligned}
\frac{1}{\sqrt{\pi}} \int_{-\pi}^{\pi} x \sin kx \, dx &= \frac{1}{\sqrt{\pi}} \left[- \frac{x \cos kx}{k} \Big|_{-\pi}^{\pi} + \frac{1}{k} \int_{-\pi}^{\pi} \cos kx \, dx \right] \\
&= - \frac{2\sqrt{\pi}}{k} \cos k\pi \\
&= \begin{cases} \dfrac{2\sqrt{\pi}}{k}, & k = 1, 3, 5, \ldots, \\[2ex] -\dfrac{2\sqrt{\pi}}{k}, & k = 2, 4, 6, \ldots . \end{cases}
\end{aligned}$$

Hence the Fourier coefficients of x with respect to (8–9) are $(-1)^{k-1}(2\sqrt{\pi}/k)$, and (8–7) becomes

$$x \sim 2\left(\sin x - \frac{\sin 2x}{2} + \frac{\sin 3x}{3} - \cdots\right).$$

In the next chapter we shall see that this series actually converges in the mean to x, so that we are justified in writing

$$x = 2\sum_{k=1}^{\infty}(-1)^{k-1}\frac{\sin kx}{k}.$$

EXAMPLE 2. If we replace (8–9) with the orthonormal set

$$\frac{1}{\sqrt{2\pi}}, \frac{\cos x}{\sqrt{\pi}}, \frac{\cos 2x}{\sqrt{\pi}}, \ldots, \tag{8–10}$$

the generalized Fourier coefficients of x become

$$\frac{1}{\sqrt{2\pi}}\int_{-\pi}^{\pi} x\,dx, \quad \text{and} \quad \frac{1}{\sqrt{\pi}}\int_{-\pi}^{\pi} x\cos kx\,dx, \quad k = 1, 2, \ldots.$$

But all of these integrals are zero (see Exercise 1), and hence (8–7) becomes

$$x \sim 0.$$

This time it is clear that we do not have an equality since

$$\|x\| = \left(\int_{-\pi}^{\pi} x^2\,dx\right)^{1/2} = \left(\frac{2\pi^3}{3}\right)^{1/2} \neq 0.$$

We now return to the general situation in which $\mathbf{e}_1, \mathbf{e}_2, \ldots$ is an (infinite) orthonormal set in $\mathcal{V}$, and examine (8–6) a little more closely. The sum of the first n terms of this series is

$$(\mathbf{x} \cdot \mathbf{e}_1)\mathbf{e}_1 + \cdots + (\mathbf{x} \cdot \mathbf{e}_n)\mathbf{e}_n,$$

which we recognize from Section 7–5 as the perpendicular projection of $\mathbf{x}$ onto the subspace $\mathcal{V}_n$ of $\mathcal{V}$ spanned by the orthonormal vectors $\mathbf{e}_1, \ldots, \mathbf{e}_n$. Hence, if $\mathbf{w}$ is *any* vector in $\mathcal{V}_n$,

$$\left\|\mathbf{x} - \sum_{k=1}^{n}(\mathbf{x} \cdot \mathbf{e}_k)\mathbf{e}_k\right\| \leq \|\mathbf{x} - \mathbf{w}\|. \tag{8–11}$$

This simple observation furnishes the key to the proof of the following theorem.

Theorem 8–2. *Let* $\sum_{k=1}^{\infty} a_k \mathbf{e}_k$ *be* **any** *infinite series which converges in the mean to* $\mathbf{x}$; *i.e.,*

$$\mathbf{x} = \sum_{k=1}^{\infty} a_k \mathbf{e}_k. \tag{8–12}$$

Then $a_k = \mathbf{x} \cdot \mathbf{e}_k$ *for each integer* k.

This theorem asserts that whenever $\mathbf{x}$ can be expressed as an infinite series the coefficients of the series are *uniquely determined* and *must* be the Fourier coefficients of $\mathbf{x}$ with respect to the orthonormal set $\mathbf{e}_1, \mathbf{e}_2, \ldots$. This, of course, is the infinite dimensional analog of the uniqueness of (8–5) in the finite dimensional case.

Proof. Since

$$\mathbf{x} = \sum_{k=1}^{\infty} a_k \mathbf{e}_k,$$

we have

$$\lim_{n\to\infty} \left\| \mathbf{x} - \sum_{k=1}^{n} a_k \mathbf{e}_k \right\| = 0.$$

But $\sum_{k=1}^{n} a_k \mathbf{e}_k$ is a vector in $\mathcal{V}_n$, and hence, by (8–11),

$$\left\| \mathbf{x} - \sum_{k=1}^{n} (\mathbf{x} \cdot \mathbf{e}_k)\mathbf{e}_k \right\| \leq \left\| \mathbf{x} - \sum_{k=1}^{n} a_k \mathbf{e}_k \right\|$$

for all n. Thus

$$\lim_{n\to\infty} \left\| \mathbf{x} - \sum_{k=1}^{n} (\mathbf{x} \cdot \mathbf{e}_k)\mathbf{e}_k \right\| = 0,$$

from which it follows that

$$\mathbf{x} = \sum_{k=1}^{\infty} (\mathbf{x} \cdot \mathbf{e}_k)\mathbf{e}_k.$$

Subtracting this series from (8–12) gives

$$\sum_{k=1}^{\infty} [a_k - (\mathbf{x} \cdot \mathbf{e}_k)]\mathbf{e}_k = \mathbf{0}, \tag{8–13}$$

and we must now show that this equation implies $a_k - (\mathbf{x} \cdot \mathbf{e}_k) = 0$ for all k. But since the series in (8–13) converges in the mean to $\mathbf{0}$, we have

$$\lim_{n\to\infty} \left\| \sum_{k=1}^{n} [a_k - (\mathbf{x} \cdot \mathbf{e}_k)]\mathbf{e}_k - \mathbf{0} \right\| = \lim_{n\to\infty} \left(\sum_{k=1}^{n} [a_k - (\mathbf{x} \cdot \mathbf{e}_k)]^2 \right)^{1/2} = 0.$$

The desired conclusion now follows from the fact that each of the terms $[a_k - (\mathbf{x} \cdot \mathbf{e}_k)]^2$ in this sum is non-negative. ▮

In view of this result it seems reasonable to extend the meaning of the term "basis" to infinite dimensional Euclidean spaces as follows.

Definition 8-3. An orthogonal set of vectors $\mathbf{e}_1, \mathbf{e}_2, \ldots$ in a Euclidean space $\mathcal{V}$ is said to be a *basis* for $\mathcal{V}$ if and only if each vector $\mathbf{x}$ in $\mathcal{V}$ can be written *uniquely* in the form

$$\mathbf{x} = \sum_{k=1}^{\infty} a_k \mathbf{e}_k. \tag{8-14}$$

(Remember that such an equality must be interpreted as asserting that the series in question converges *in the mean* to $\mathbf{x}$.) The coefficients a_k in this series are called the *generalized Fourier coefficients* of $\mathbf{x}$ with respect to the given basis, and the series itself is called the *generalized Fourier series expansion of* $\mathbf{x}$.

The reader should note that the vectors in a basis for $\mathcal{V}$ are only required to be orthogonal, not orthonormal. Of course, any basis for $\mathcal{V}$ can be normalized in the standard fashion, but it would be a hinderance to restrict the definition in this way. It is also possible to define a basis for an infinite dimensional Euclidean space by using a linearly independent set of vectors rather than an orthogonal one. This would be somewhat closer to the spirit of the definition in Chapter 1, but since any such set can be orthogonalized by the Gram-Schmidt process, it would not result in any significant gain in generality.*

For all we know at the moment, we may never encounter an infinite dimensional Euclidean space with a basis. This notwithstanding, we shall assemble a number of results about such spaces pending the time when we finally meet one. However, in the absence of an immediate example it is only fair to state that these results are more than speculations in a vacuum since *all* of the Euclidean spaces which appear in analysis do, in fact, have bases.

We begin by making a number of simple observations, the first of which we state formally as follows:

Lemma 8-3. *The zero vector is the only vector which is orthogonal to* **every** *vector in a basis for a Euclidean space* $\mathcal{V}$.

This is an immediate consequence of the postulated uniqueness of series expansions relative to a basis, and the fact that $\sum_{k=1}^{\infty} a_k \mathbf{e}_k$ converges in the mean to zero whenever $a_k = 0$ for all k. Despite its simplicity, Lemma 8-3 is often useful, particularly in showing that an orthogonal set is *not* a basis for a given Euclidean

* A word of warning. Definition 8-3 is not the only meaning assigned to the term "basis" for infinite dimensional spaces. There is another in common use which is not restricted to Euclidean spaces, but applies to arbitrary vector spaces as well. We mention this point only to alert the student to the fact that when he encounters this term he must check the definition to avoid serious errors.

space. For instance, when this result is combined with Example 2 above, it allows us to assert that the set of functions in (8–10) is not a basis for $\mathcal{C}[-\pi, \pi]$.

An equally simple observation is that *an orthonormal set* $\mathbf{e}_1, \mathbf{e}_2, \ldots$ *is a basis for a Euclidean space* $\mathcal{V}$ *if and only if every vector* $\mathbf{x}$ *in* $\mathcal{V}$ *can be written in the form*

$$\mathbf{x} = \sum_{k=1}^{\infty} (\mathbf{x} \cdot \mathbf{e}_k)\mathbf{e}_k.$$

For then, by Theorem 8–2, this expression must be unique, and Definition 8–3 is therefore satisfied.

And while on the subject of orthonormal bases, there is another point it would do well to settle. Recall that if $\mathbf{e}_1, \ldots, \mathbf{e}_n$ is such a basis in a *finite* dimensional space, and if

$$\mathbf{x} = \sum_{k=1}^{n} (\mathbf{x} \cdot \mathbf{e}_k)\mathbf{e}_k \quad \text{and} \quad \mathbf{y} = \sum_{k=1}^{n} (\mathbf{y} \cdot \mathbf{e}_k)\mathbf{e}_k$$

are any two vectors in this space, then

$$\mathbf{x} \cdot \mathbf{y} = \sum_{k=1}^{n} (\mathbf{x} \cdot \mathbf{e}_k)(\mathbf{y} \cdot \mathbf{e}_k).$$

That is, the inner product of $\mathbf{x}$ and $\mathbf{y}$ is the sum of the products of their corresponding components when computed with respect to an orthonormal basis. We assert that *this result is also valid for an orthonormal basis in an infinite dimensional space* $\mathcal{V}$.

To prove this assertion, let

$$\mathbf{x} = \sum_{k=1}^{\infty} (\mathbf{x} \cdot \mathbf{e}_k)\mathbf{e}_k \quad \text{and} \quad \mathbf{y} = \sum_{k=1}^{\infty} (\mathbf{y} \cdot \mathbf{e}_k)\mathbf{e}_k$$

be any pair of vectors in $\mathcal{V}$, and for each positive integer n, set

$$\mathbf{x}_n = \sum_{k=1}^{n} (\mathbf{x} \cdot \mathbf{e}_k)\mathbf{e}_k \quad \text{and} \quad \mathbf{y}_n = \sum_{k=1}^{n} (\mathbf{y} \cdot \mathbf{e}_k)\mathbf{e}_k.$$

Then $\{\mathbf{x}_n\} \to \mathbf{x}$ and $\{\mathbf{y}_n\} \to \mathbf{y}$, and hence by Theorem 8–1,

$$\{\mathbf{x}_n \cdot \mathbf{y}_n\} \to \mathbf{x} \cdot \mathbf{y}. \tag{8–15}$$

But the fact that $\mathbf{e}_1, \mathbf{e}_2, \ldots$ is an ortho*normal* set implies that

$$\mathbf{x}_n \cdot \mathbf{y}_n = \sum_{k=1}^{n} (\mathbf{x} \cdot \mathbf{e}_k)(\mathbf{y} \cdot \mathbf{e}_k),$$

and it follows at once from (8–15) that

$$\mathbf{x} \cdot \mathbf{y} = \sum_{k=1}^{\infty} (\mathbf{x} \cdot \mathbf{e}_k)(\mathbf{y} \cdot \mathbf{e}_k). \tag{8–16}$$

EXERCISES

1. Verify that

$$\frac{1}{\sqrt{2\pi}}\int_{-\pi}^{\pi} x\,dx = 0 \qquad \text{and} \qquad \frac{1}{\sqrt{\pi}}\int_{-\pi}^{\pi} x \cos kx\,dx = 0, \quad k = 1, 2, \ldots.$$

2. Compute the Fourier coefficients of the function

$$f(x) = x^2, \quad -\pi \leq x \leq \pi$$

in $\mathcal{C}[-\pi, \pi]$ with respect to the orthonormal set

$$\frac{1}{\sqrt{2\pi}}, \frac{\cos x}{\sqrt{\pi}}, \frac{\cos 2x}{\sqrt{\pi}}, \ldots,$$

and also with respect to the orthonormal set

$$\frac{\sin x}{\sqrt{\pi}}, \frac{\sin 2x}{\sqrt{\pi}}, \ldots.$$

3. Repeat Exercise 2 for the function e^x, $-\pi \leq x \leq \pi$.
4. Let $\mathbf{e}_1, \mathbf{e}_2, \ldots$ be an infinite orthonormal set in a Euclidean space $\mathcal{V}$, and suppose that

$$\sum_{k=1}^{\infty} a_k \mathbf{e}_k = 0.$$

Prove that $a_k = 0$ for all k.

8-4 BESSEL'S INEQUALITY: PARSEVAL'S EQUALITY

This section bears the title of two of the most important results in the theory of infinite dimensional Euclidean spaces. As we shall see, both of them can be interpreted as natural extensions of familiar statements about finite dimensional spaces. This fact, however, should not be allowed to obscure the importance of the generalizations themselves, and the reader is advised that they will ultimately furnish the justification for most of our work with orthogonal sequences in function spaces.

Theorem 8-3. *Let $\mathbf{e}_1, \mathbf{e}_2, \ldots$ be an orthonormal set of vectors in an infinite dimensional Euclidean space $\mathcal{V}$, and let $\mathbf{x}$ be an arbitrary vector in $\mathcal{V}$. Then*

$$\sum_{k=1}^{\infty} (\mathbf{x} \cdot \mathbf{e}_k)^2 \leq \|\mathbf{x}\|^2 \quad (\textit{Bessel's inequality}). \tag{8-17}$$

Moreover, $\mathbf{e}_1, \mathbf{e}_2, \ldots$ is a basis for $\mathcal{V}$ if and only if

$$\sum_{k=1}^{\infty} (\mathbf{x} \cdot \mathbf{e}_k)^2 = \|\mathbf{x}\|^2 \quad (\textit{Parseval's equality}). \tag{8-18}$$

Proof. The proof rests upon computing the value of

$$\left\| \mathbf{x} - \sum_{k=1}^{n} (\mathbf{x} \cdot \mathbf{e}_k)\mathbf{e}_k \right\|^2$$

for any $\mathbf{x}$ in $\mathcal{V}$ and any integer n. Now

$$\begin{aligned}\left\| \mathbf{x} - \sum_{k=1}^{n} (\mathbf{x} \cdot \mathbf{e}_k)\mathbf{e}_k \right\|^2 &= \left(\mathbf{x} - \sum_{k=1}^{n} (\mathbf{x} \cdot \mathbf{e}_k)\mathbf{e}_k\right) \cdot \left(\mathbf{x} - \sum_{k=1}^{n} (\mathbf{x} \cdot \mathbf{e}_k)\mathbf{e}_k\right) \\ &= \mathbf{x} \cdot \mathbf{x} - 2\sum_{k=1}^{n} (\mathbf{x} \cdot \mathbf{e}_k)(\mathbf{x} \cdot \mathbf{e}_k) + \left(\sum_{j=1}^{n} (\mathbf{x} \cdot \mathbf{e}_j)\mathbf{e}_j\right) \cdot \left(\sum_{k=1}^{n} (\mathbf{x} \cdot \mathbf{e}_k)\mathbf{e}_k\right) \cdot\end{aligned}$$

(We have changed the index of summation on the first factor of the last term for convenience in computation.) But since the $\mathbf{e}_k$ are orthonormal, $\mathbf{e}_j \cdot \mathbf{e}_k = \delta_{jk}$, and it follows that

$$\begin{aligned}\left(\sum_{j=1}^{n} (\mathbf{x} \cdot \mathbf{e}_j)\mathbf{e}_j\right) \cdot \left(\sum_{k=1}^{n} (\mathbf{x} \cdot \mathbf{e}_k)\mathbf{e}_k\right) &= \sum_{j=1}^{n} \sum_{k=1}^{n} (\mathbf{x} \cdot \mathbf{e}_j)(\mathbf{x} \cdot \mathbf{e}_k)(\mathbf{e}_j \cdot \mathbf{e}_k) \\ &= \sum_{k=1}^{n} (\mathbf{x} \cdot \mathbf{e}_k)^2.\end{aligned}$$

Thus

$$\left\| \mathbf{x} - \sum_{k=1}^{n} (\mathbf{x} \cdot \mathbf{e}_k)\mathbf{e}_k \right\|^2 = \|\mathbf{x}\|^2 - \sum_{k=1}^{n} (\mathbf{x} \cdot \mathbf{e}_k)^2. \tag{8–19}$$

To prove Bessel's inequality, it suffices to note that

$$0 \le \left\| \mathbf{x} - \sum_{k=1}^{n} (\mathbf{x} \cdot \mathbf{e}_k)\mathbf{e}_k \right\|^2 = \|\mathbf{x}\|^2 - \sum_{k=1}^{n} (\mathbf{x} \cdot \mathbf{e}_k)^2$$

for all $\mathbf{x}$ in $\mathcal{V}$ and all n. Hence

$$\sum_{k=1}^{n} (\mathbf{x} \cdot \mathbf{e}_k)^2 \le \|\mathbf{x}\|^2$$

for all n, and the partial sums of the series $\sum_{k=1}^{\infty} (\mathbf{x} \cdot \mathbf{e}_k)^2$ form a bounded non-decreasing sequence of non-negative real numbers. By a well-known theorem from calculus (see Appendix I) we conclude that this series converges and that

$$\sum_{k=1}^{\infty} (\mathbf{x} \cdot \mathbf{e}_k)^2 \le \|\mathbf{x}\|^2.$$

Thus (8–17) holds.

To finish the proof we now suppose that $\mathbf{e}_1, \mathbf{e}_2, \ldots$ is an orthonormal basis for $\mathcal{V}$. Then by Theorem 8–2 we know that

$$\mathbf{x} = \sum_{k=1}^{\infty} (\mathbf{x} \cdot \mathbf{e}_k)\mathbf{e}_k.$$

Hence

$$\lim_{n\to\infty} \left\| \mathbf{x} - \sum_{k=1}^{n} (\mathbf{x} \cdot \mathbf{e}_k)\mathbf{e}_k \right\|^2 = 0,$$

and it follows from (8–19) that

$$\lim_{n\to\infty} \left[\|\mathbf{x}\|^2 - \sum_{k=1}^{n} (\mathbf{x} \cdot \mathbf{e}_k)^2 \right] = 0.$$

Thus

$$\sum_{k=1}^{\infty} (\mathbf{x} \cdot \mathbf{e}_k)^2 = \|\mathbf{x}\|^2,$$

which is Parseval's equality. Finally, since the steps in this chain of reasoning are reversible, we conclude that $\mathbf{e}_1, \mathbf{e}_2, \ldots$ is a basis whenever Parseval's equality is satisfied, and the proof is complete. ▮

In a finite dimensional Euclidean space the square of the length of any vector is equal to the sum of the squares of the lengths of its components relative to an orthonormal basis. Conversely, if this is true the set in question is an orthonormal basis for the space. Parseval's equality asserts that this result also holds for infinite dimensional spaces, and thus can be viewed as *an infinite dimensional version of the Pythagorean theorem.* Its value lies in the fact that it provides an analytic tool for determining whether an orthonormal set is a basis for a Euclidean space, and as such is much easier to apply than the definition itself.

Bessel's inequality—whose geometric interpretation is now obvious—can be used to obtain estimates of the magnitude of the Fourier coefficients of $\mathbf{x}$ with respect to $\mathbf{e}_1, \mathbf{e}_2, \ldots$. One of the most important of the many results of this sort is given in the following corollary.

Corollary 8–1. *If* $\mathbf{e}_1, \mathbf{e}_2, \ldots$ *is an orthonormal set in an infinite dimensional Euclidean space* $\mathcal{V}$, *then*

$$\lim_{k\to\infty} (\mathbf{x} \cdot \mathbf{e}_k) = 0 \tag{8–20}$$

for every $\mathbf{x}$ *in* $\mathcal{V}$. *That is, the Fourier coefficients of* $\mathbf{x}$ *tend to zero as* $k \to \infty$ *for any orthonormal set in* $\mathcal{V}$.

Indeed, Bessel's inequality implies that $\sum_{k=1}^{\infty} (\mathbf{x} \cdot \mathbf{e}_k)^2$ is a convergent series of real numbers. Hence the individual terms in this series approach zero with increasing k, which, in turn, implies (8–20).

In applications one frequently wants to work with an *orthogonal* set in $\mathcal{V}$ rather than an orthonormal one. To do so we must know which of the results proved for orthonormal sets remain true for orthogonal sets. The fact of the matter is that, save for minor notational changes, they all do. This statement finds its justification in the following assertion, whose proof we leave as an exercise. *An orthogonal set* $\mathbf{e}_1, \mathbf{e}_2, \ldots$ *is a basis in a Euclidean space* $\mathcal{V}$ *if and only if its associated orthonormal set* $\mathbf{e}_1/\|\mathbf{e}_1\|, \mathbf{e}_2/\|\mathbf{e}_2\|, \ldots$ *is a basis.*

From this it follows that we can rephrase all of our earlier results in terms of orthogonal sets. In particular, if $\mathbf{e}_1, \mathbf{e}_2, \ldots$ is an arbitrary *basis* for $\mathcal{V}$, Parseval's equality becomes

$$\|\mathbf{x}\|^2 = \sum_{k=1}^{\infty} \frac{(\mathbf{x} \cdot \mathbf{e}_k)^2}{\mathbf{e}_k \cdot \mathbf{e}_k} = \sum_{k=1}^{\infty} \frac{(\mathbf{x} \cdot \mathbf{e}_k)^2}{\|\mathbf{e}_k\|^2}, \tag{8–21}$$

and, conversely, if (8–21) is satisfied for all $\mathbf{x}$ in $\mathcal{V}$, $\mathbf{e}_1, \mathbf{e}_2, \ldots$ is a basis for $\mathcal{V}$. Finally, the series expansion of a vector in terms of an arbitrary basis is

$$\mathbf{x} = \sum_{k=1}^{\infty} \frac{\mathbf{x} \cdot \mathbf{e}_k}{\|\mathbf{e}_k\|^2} \mathbf{e}_k, \tag{8–22}$$

where the

$$\frac{\mathbf{x} \cdot \mathbf{e}_k}{\|\mathbf{e}_k\|^2}$$

are the generalized Fourier coefficients of $\mathbf{x}$ with respect to $\mathbf{e}_1, \mathbf{e}_2, \ldots$. This formula will be used repeatedly throughout the following chapters.

EXERCISES

1. Let $\mathbf{e}_1, \mathbf{e}_2, \ldots$ be an orthogonal set in a Euclidean space $\mathcal{V}$. Prove that this set is a basis for $\mathcal{V}$ if and only if its associated orthonormal set is a basis.
2. Prove Formula (8–21).
3. Assume that the functions $1, \sin x, \cos x, \ldots, \sin kx, \cos kx, \ldots$ are a basis for the Euclidean space $\mathcal{C}[-\pi, \pi]$. (They are.) Find the Fourier coefficients of f in $\mathcal{C}[-\pi, \pi]$ in terms of this basis. What is the series expansion of f?

*8–5 CLOSED SUBSPACES

Roughly speaking, a Euclidean space is an object in which the notions of linearity and convergence are studied simultaneously. A particularly simple yet important example of the way in which these ideas enrich and support one another is furnished by the study of subspaces of Euclidean spaces. In this instance the notion of convergence intervenes to produce what is known as a "closed" subspace, defined as follows.

Definition 8–4. A subspace $\mathcal{W}$ of a Euclidean space $\mathcal{V}$ is said to be *closed in* $\mathcal{V}$ if the limit of every convergent sequence of vectors in $\mathcal{W}$ belongs to $\mathcal{W}$.

(In those cases where the context is clear we shall omit the reference to $\mathcal{V}$, and simply say that $\mathcal{W}$ is a closed subspace.)

As with every new definition, one's first thought is to produce examples. In this case we can furnish a plentiful supply by examining finite dimensional spaces, for there we can prove

Lemma 8–4. *Every subspace $\mathcal{W}$ of a finite dimensional Euclidean space $\mathcal{V}$ is closed.*

Proof. If $\mathcal{W}$ is the trivial subspace of $\mathcal{V}$ it is obviously closed, and we are done. Otherwise we can find an orthonormal basis $\mathbf{e}_1, \ldots, \mathbf{e}_n$ for $\mathcal{V}$ whose first m vectors $\mathbf{e}_1, \ldots, \mathbf{e}_m$ are a basis for $\mathcal{W}$.* It follows that an arbitrary vector $\mathbf{x}$ in $\mathcal{V}$ belongs to $\mathcal{W}$ if and only if its last $n - m$ components with respect to this basis are zero.

Now suppose that $\{\mathbf{x}_k\}$ is a sequence of vectors in $\mathcal{W}$, and that $\{\mathbf{x}_k\} \to \mathbf{x}$. Then, by Example 2 of Section 8–1, we know that the sequences of real numbers formed from the components of the $\mathbf{x}_k$ converge to the components of $\mathbf{x}$. It follows that the last $n - m$ components of $\mathbf{x}$ are zero, and hence $\mathbf{x}$ belongs to $\mathcal{W}$, as asserted. ▮

More generally, it can be shown that every *finite* dimensional subspace of an arbitrary Euclidean space is closed. The proof is almost identical with the one just given, and has been left as an exercise.

This lemma tells us that the only interesting illustrations of Definition 8–4 are to be found in infinite dimensional spaces, if any exist at all. They do, as the following example shows.

EXAMPLE. Let ℓ_2 (read "little ℓ two") denote the set of all infinite sequences of real numbers

$$\mathbf{x} = (x_1, x_2, x_3, \ldots)$$

with the property that

$$\sum_{k=1}^{\infty} x_k^2 < \infty.$$

Then ℓ_2 becomes a Euclidean space if, with $\mathbf{x}$ as above, α a real number, and $\mathbf{y} = (y_1, y_2, y_3, \ldots)$, we define

$$\begin{aligned} \alpha\mathbf{x} &= (\alpha x_1, \alpha x_2, \alpha x_3, \ldots) \\ \mathbf{x} + \mathbf{y} &= (x_1 + y_1, x_2 + y_2, x_3 + y_3, \ldots) \\ \mathbf{x} \cdot \mathbf{y} &= x_1 y_1 + x_2 y_2 + x_3 y_3 + \cdots \end{aligned}$$

---

* This assertion can be proved at once by applying the Gram-Schmidt process to the basis described in Theorem 1–7.

(see Exercise 2). Moreover, since ℓ_2 contains the infinite orthornormal sequence

$$\begin{aligned} \mathbf{e}_1 &= (1, 0, 0, \ldots), \\ \mathbf{e}_2 &= (0, 1, 0, \ldots), \\ \mathbf{e}_3 &= (0, 0, 1, \ldots), \\ &\vdots \end{aligned}$$

it is infinite dimensional.

Now let $\mathcal{W}$ be the set of all vectors in ℓ_2 which have only a *finite* number of nonzero components. Then $\mathcal{W}$ is a subspace of ℓ_2. (Proof?) However, $\mathcal{W}$ is not closed in ℓ_2 since the sequence

$$\begin{aligned} \mathbf{x}_1 &= (\tfrac{1}{2}, 0, 0, \ldots), \\ \mathbf{x}_2 &= (\tfrac{1}{2}, \tfrac{1}{4}, 0, \ldots), \\ \mathbf{x}_3 &= (\tfrac{1}{2}, \tfrac{1}{4}, \tfrac{1}{8}, \ldots), \\ &\vdots \end{aligned}$$

each of whose vectors belongs to $\mathcal{W}$, converges to the vector $(\frac{1}{2}, \frac{1}{4}, \frac{1}{8}, \ldots, 1/2^k, \ldots)$ which is not in $\mathcal{W}$.

(The Euclidean space ℓ_2 is an example of what mathematicians call a *Hilbert space.*)

Having shown that there exist subspaces which are not closed, and hence that there is good reason for studying closed subspaces as entities in themselves, we begin with the usual elementary observations. First, every Euclidean space $\mathcal{V}$, viewed as a subspace of itself, is closed, and hence every subspace of $\mathcal{V}$ (closed or not) is contained in at least one closed subspace of $\mathcal{V}$. Next, the intersection of any collection of closed subspaces of $\mathcal{V}$ is itself a closed subspace. (Recall that this intersection consists of those vectors which belong to *every one* of the subspaces in question. Again the proof is deferred to the exercises.) Conclusion: *If $\mathcal{W}$ is an arbitrary subspace of a Euclidean space $\mathcal{V}$, there exists a smallest closed subspace of $\mathcal{V}$ containing $\mathcal{W}$*; namely the intersection of *all* the closed subspaces of $\mathcal{V}$ which contain $\mathcal{W}$. This subspace is denoted $\overline{\mathcal{W}}$, and, for obvious reasons, is called the *closure of $\mathcal{W}$ in $\mathcal{V}$*. Finally, $\overline{\mathcal{W}}$ can be described as follows.

Lemma 8–5. *If $\mathcal{W}$ is a subspace of a Euclidean space $\mathcal{V}$, its closure (in $\mathcal{V}$) is the set of all vectors in $\mathcal{V}$ which are limits of sequences of vectors in $\mathcal{W}$.*

In other words, to get $\overline{\mathcal{W}}$ we simply adjoin to $\mathcal{W}$ all limits of sequences of vectors in $\mathcal{W}$. The proof of this statement is an obvious consequence of our definitions.

In specific cases this lemma is far too general to be of much use in finding $\overline{\mathcal{W}}$. In fact, many of the deeper investigations of the theory of Euclidean spaces center around the problem of determining $\overline{\mathcal{W}}$ for a given $\mathcal{W}$. Unfortunately, most of this work is highly technical, and beyond the scope of this book.

The sequence of observations leading to Lemma 8–5 admits a useful and natural generalization. Suppose that instead of starting with a subspace of $\mathcal{V}$ we begin with an arbitrary nonempty sub*set* $\mathcal{X}$ in $\mathcal{V}$. Then we can form the subspace $\mathcal{S}(\mathcal{X})$ spanned by $\mathcal{X}$, and follow this with the closure $\overline{\mathcal{S}(\mathcal{X})}$ of $\mathcal{S}(\mathcal{X})$ in $\mathcal{V}$, in which case we say that $\overline{\mathcal{S}(\mathcal{X})}$ is the closed subspace of $\mathcal{V}$ *generated* by $\mathcal{X}$. Note that the subspace *spanned* by $\mathcal{X}$ coincides with the subspace *generated* by $\mathcal{X}$ if and only if $\mathcal{S}(\mathcal{X}) = \overline{\mathcal{S}(\mathcal{X})}$. This will always happen, of course, if $\mathcal{V}$ is finite dimensional (Lemma 8–4), or if $\mathcal{X}$ is a finite subset of $\mathcal{V}$ (Exercise 10). However, the above example shows that when $\mathcal{X}$ is an infinite subset of an infinite dimensional space, $\mathcal{S}(\mathcal{X})$ will in general be different from $\overline{\mathcal{S}(\mathcal{X})}$. In Chapter 1 we gave a description of the vectors belonging to $\mathcal{S}(\mathcal{X})$ in terms of the vectors belonging to $\mathcal{X}$ (see Theorem 1–1). When combined with Lemma 8–5, that description allows us to assert that $\overline{\mathcal{S}(\mathcal{X})}$ consists of all vectors in $\mathcal{V}$ which are limits of linear combinations of vectors in $\mathcal{X}$. Stated more formally, we have

Lemma 8–6. *Let $\mathcal{X}$ be a set of generators of a (necessarily closed) subspace $\mathcal{W}$ of a Euclidean space $\mathcal{V}$. Then, if ϵ is any positive real number, and $\mathbf{x}$ any vector in $\mathcal{W}$, there exists a finite linear combination $\sum_{k=1}^{N} a_k\mathbf{x}_k$ of vectors $\mathbf{x}_k$ in $\mathcal{S}$ such that*

$$\left\| \mathbf{x} - \sum_{k=1}^{N} a_k\mathbf{x}_k \right\| < \epsilon.$$

Finally, suppose that $\mathcal{X}$ is a set of generators for $\mathcal{V}$ itself, and that the vectors in $\mathcal{X}$ are also linearly independent.* On the strength of our experience with finite dimensional spaces the reader could hardly be blamed for assuming that every vector $\mathbf{x}$ in $\mathcal{V}$ could then be written in the form

$$\mathbf{x} = \sum_{k=1}^{\infty} a_k\mathbf{x}_k,$$

where the a_k are scalars, and the $\mathbf{x}_k$ belong to $\mathcal{X}$. Surprisingly, this is *not* the case, and an example of this apparently paradoxical situation can be found in Exercise 13 below. However, things improve remarkably when $\mathcal{X}$ is also an *orthonormal* (or orthogonal) set, for then we can prove

Lemma 8–7. *If $\mathbf{e}_1, \mathbf{e}_2, \ldots$ is an orthonormal set of generators of a Euclidean space $\mathcal{V}$, then $\mathbf{e}_1, \mathbf{e}_2, \ldots$ is a basis for $\mathcal{V}$.*

Proof. We must show that every vector $\mathbf{x}$ in $\mathcal{V}$ can be written in the form

$$\mathbf{x} = \sum_{k=1}^{\infty} (\mathbf{x} \cdot \mathbf{e}_k)\mathbf{e}_k.$$

---

* It can be shown that such a set always exists in any Euclidean space $\mathcal{V}$. The proof, however, is not easy.

To do so, let $\epsilon > 0$ be given. Then, since the $\mathbf{e}_k$ generate $\mathcal{V}$ we can find a linear combination $\sum_{k=1}^{N} a_k\mathbf{e}_k$ such that

$$\left\| \mathbf{x} - \sum_{k=1}^{N} a_k\mathbf{e}_k \right\| < \epsilon.$$

If $n > N$, it follows by two applications of (8–11) that

$$\begin{aligned} \left\| \mathbf{x} - \sum_{k=1}^{n} (\mathbf{x} \cdot \mathbf{e}_k)\mathbf{e}_k \right\| &\leq \left\| \mathbf{x} - \sum_{k=1}^{N} (\mathbf{x} \cdot \mathbf{e}_k)\mathbf{e}_k \right\| \\ &\leq \left\| \mathbf{x} - \sum_{k=1}^{N} a_k\mathbf{e}_k \right\| < \epsilon. \end{aligned}$$

Hence, by the definition of convergence,

$$\mathbf{x} = \sum_{k=1}^{\infty} (\mathbf{x} \cdot \mathbf{e}_k)\mathbf{e}_k,$$

and the lemma is proved. ▮

This result explains why orthonormal sets enjoy a special status in the theory of infinite dimensional Euclidean spaces beyond what they have in finite dimensional spaces. In the latter their use is simply a matter of convenience, in the former it is often a matter of necessity. Since we shall be dealing almost exclusively with orthonormal (and orthogonal) sets in the following chapters, it may be well to summarize what we now know about them. Our results may be stated in the following form.

Theorem 8–4. *If* $\mathbf{e}_1, \mathbf{e}_2, \ldots$ *is an orthonormal set in an infinite dimensional Euclidean space* $\mathcal{V}$, *then the following three statements are equivalent* (*i.e., any one implies the other two*):

(i) $\mathbf{e}_1, \mathbf{e}_2, \ldots$ *is a basis for* $\mathcal{V}$;
(ii) $\mathbf{e}_1, \mathbf{e}_2, \ldots$ *is a linearly independent set of generators for* $\mathcal{V}$;
(iii) *If* $\mathbf{x}$ *is any vector in* $\mathcal{V}$, *then* $\|\mathbf{x}\|^2 = \sum_{k=1}^{\infty} (\mathbf{x} \cdot \mathbf{e}_k)^2$.

There is, of course, a corresponding theorem for orthogonal sets.

EXERCISES

1. Prove that every finite dimensional subspace $\mathcal{W}$ of a Euclidean space $\mathcal{V}$ is closed. [*Hint:* If $\mathbf{x}$ is the limit of a sequence of vectors in $\mathcal{W}$, apply Lemma 8–4 to the subspace of $\mathcal{V}$ spanned by $\mathcal{W}$ and $\mathbf{x}$.]

2. Let ℓ_2 be as defined in the example in the text, and let

$$\mathbf{x} = (x_1, x_2, \ldots, x_k, \ldots), \qquad \mathbf{y} = (y_1, y_2, \ldots, y_k, \ldots)$$

be arbitrary vectors in ℓ_2.

(a) Prove that $\mathbf{x} + \mathbf{y}$ belongs to ℓ_2. [*Hint*: Use Eq. (7–27) to deduce that

$$\left(\sum_{k=1}^{n} (x_k + y_k)^2\right)^{1/2} \leq \left(\sum_{k=1}^{n} x_k^2\right)^{1/2} + \left(\sum_{k=1}^{n} y_k^2\right)^{1/2},$$

and then use the fact that $\mathbf{x}$ and $\mathbf{y}$ belong to ℓ_2 to obtain an upper bound, independent of n, for the right-hand side of this inequality.]

(b) Prove that $\mathbf{x} \cdot \mathbf{y}$ is defined. [*Hint:* For any two real numbers x, y, we have $xy = \frac{1}{2}[(x + y)^2 - x^2 - y^2]$. Hence

$$\sum_{k=1}^{n} x_k y_k = \tfrac{1}{2}\left(\sum_{k=1}^{n} (x_k + y_k)^2 - \sum_{k=1}^{n} x_k^2 - \sum_{k=1}^{n} y_k^2\right).\Bigg]$$

(c) Now complete the verification that ℓ_2 is a Euclidean space.

3. (a) Prove that the subset $\mathcal{W}$ defined in the text is a subspace of ℓ_2.

(b) Find the closure of $\mathcal{W}$ in ℓ_2.

4. In the example above it was asserted that a certain sequence in ℓ_2 converged to the vector $(\frac{1}{2}, \frac{1}{4}, \frac{1}{8}, \ldots)$. Prove this assertion.

5. Prove that the intersection of any (nonempty) collection of closed subspaces of a Euclidean space $\mathcal{V}$ is closed.

6. Give an example of a Euclidean space $\mathcal{W}$ which is a subspace of the Euclidean spaces $\mathcal{V}_1$ and $\mathcal{V}_2$, and which has the property that it is closed in $\mathcal{V}_1$ but not in $\mathcal{V}_2$. (Thus in the definition of a closed subspace reference *must* be made to the space in which the closure is taking place.)

7. Prove that $\overline{\overline{\mathcal{W}}} = \overline{\mathcal{W}}$ for any subspace $\mathcal{W}$ of a Euclidean space $\mathcal{V}$. (This result is sometimes expressed by saying that "the closure of the closure is the closure.")

8. Prove Lemma 8–5.

9. Prove that $\mathcal{W}$ is a closed subspace of $\mathcal{V}$ if and only if $\overline{\mathcal{W}} = \mathcal{W}$. [*Hint:* See Exercise 7.]

10. Prove that $\mathcal{S}(\mathcal{X})$ is closed in $\mathcal{V}$ whenever $\mathcal{X}$ is a finite subset of $\mathcal{V}$.

11. Suppose $\mathbf{x}$ is orthogonal to a subspace $\mathcal{W}$ of a Euclidean space $\mathcal{V}$ (i.e., $\mathbf{x} \cdot \mathbf{y} = 0$ for every $\mathbf{y}$ in $\mathcal{W}$). Prove that $\mathbf{x}$ is orthogonal to $\overline{\mathcal{W}}$. [*Hint:* Use Theorem 8–1.]

*12. Let $\mathcal{W}$ be an arbitrary subspace of a Euclidean space $\mathcal{V}$, and let $\mathcal{W}^{\perp}$ denote the set of all vectors in $\mathcal{V}$ which are orthogonal to $\mathcal{W}$ (see Exercise 23, Section 7–5). Prove that $\mathcal{W}^{\perp}$ is a closed subspace of $\mathcal{V}$. [*Hint:* Let $\{\mathbf{x}_k\}$ be a sequence in $\mathcal{W}^{\perp}$ which converges to $\mathbf{x}$, and let $\mathbf{y}$ be any vector in $\mathcal{W}$. To prove that $\mathbf{x} \cdot \mathbf{y} = 0$, use the identity

$$\mathbf{x} \cdot \mathbf{y} = \mathbf{y} \cdot (\mathbf{x} - \mathbf{x}_k)$$

and the Schwarz inequality.]

13. Let ℓ_2 be the Euclidean space defined above, and for each integer $k \geq 2$ let $\mathbf{x}_k$ be the vector in ℓ_2 defined by

$$\mathbf{x}_k = (1, 0, \ldots, 0, 1, 0, \ldots),$$

where the ones occur in the first and kth positions. Let $\mathfrak{X}$ denote the set consisting of the vectors $\mathbf{x}_2, \mathbf{x}_3, \ldots$

(a) Show that $\mathfrak{X}$ is a linearly independent set in ℓ_2.

(b) Describe the subspace $\mathcal{S}(\mathfrak{X})$. That is, give a rule which can be applied to test whether or not a given vector in ℓ_2 belongs to $\mathcal{S}(\mathfrak{X})$.

(c) Show that the vector $\mathbf{e}_1 = (1, 0, 0, \ldots)$ belongs to $\overline{\mathcal{S}(\mathfrak{X})}$. [*Hint:* Compute the value of

$$\left\| \mathbf{e}_1 - \frac{1}{n} \sum_{k=2}^{\infty} \mathbf{x}_k \right\|,$$

and show that

$$\lim_{n \to \infty} \left\| \mathbf{e}_1 - \frac{1}{n} \sum_{k=2}^{\infty} \mathbf{x}_k \right\| = 0.]$$

(d) Prove that $\overline{\mathcal{S}(\mathfrak{X})} = \ell_2$. [*Hint:* Begin by showing that each of the vectors $\mathbf{e}_1$, $\mathbf{e}_2, \ldots$ as defined in the text belong to $\overline{\mathcal{S}(\mathfrak{X})}$, and then show that these vectors are a basis for ℓ_2.]

(e) Show that $\mathbf{e}_1$ *cannot* be written in the form $\sum_{k=2}^{\infty} a_k \mathbf{x}_k$, thereby furnishing the example mentioned just before Lemma 8–7.

14. Let $\{a_k\}$, $k = 1, 2, \ldots$, be a sequence of real numbers, and suppose that $\sum_{k=1}^{\infty} a_k^2$ converges. Prove that $\sum_{k=1}^{\infty} a_k/k$ also converges. Give an example to show that the converse of this statement is false.

9 fourier series

9–1 INTRODUCTION

Although we now know a great deal about the general theory of orthogonal series in infinite dimensional Euclidean spaces we have yet to produce a single concrete example of such a series. To fill this gap in our knowledge we propose to devote the next three chapters to a detailed study of several types of series expansions in infinite dimensional function spaces. In each case we shall begin, as of course we must, by exhibiting a basis for the particular space in which the series is to be constructed. Once this has been done the series in question will appear as a special version of our earlier results, and for this reason the following discussion can be viewed as a collection of elaborate examples in illustration of material that is already known.

At the same time, however, it is only fair to warn the reader that each of the series which we shall construct is of significant importance in physics and applied mathematics. Indeed, since 1822 when Jean Baptiste Fourier first solved the problem of heat flow in solid bodies by means of those series which now bear his name, this subject has grown until, at present, it is an entire branch of mathematics and mathematical physics. Later in this book we shall examine some of the applications of this theory, but first we discuss the series themselves.

9–2 THE SPACE OF PIECEWISE CONTINUOUS FUNCTIONS

Given the importance of the various series which we are about to consider, it is clearly of some interest to phrase the following discussion so as to encompass as wide a class of functions as possible. Unfortunately, most of the generalizations which go beyond the space of continuous functions are highly technical and inaccessible in a course at this level. Still, it is possible to extend our results to include piecewise continuous functions, and since this modest generalization is well worth making, we shall devote the present section to showing how the set of all such functions on a fixed interval can be made into a Euclidean space. For convenience we begin by recalling the definition of piecewise continuity.

Definition 9–1. A real valued function f is said to be *piecewise continuous* on an interval $[a, b]$ if

(i) f is defined and continuous at all but a finite number of points of $[a, b]$ and

(ii) the limits

$$\begin{aligned} f(x_0^+) &= \lim_{h\to 0^+} f(x_0 + h), \\ f(x_0^-) &= \lim_{h\to 0^+} f(x_0 - h) \end{aligned} \tag{9–1}$$

exist at each point x_0 in $[a, b]$. (Note that only one of these limits is relevant if x_0 is an end point of $[a, b]$.)

We remind the reader that the notation $h \to 0^+$ means h approaches zero through positive values only, and that the two limits appearing in (9–1) are called, respectively, the *right-* and *left-hand limits* of f at x_0. When x_0 is a point of continuity of f each of these limits is equal to the value of f at x_0, and we then have

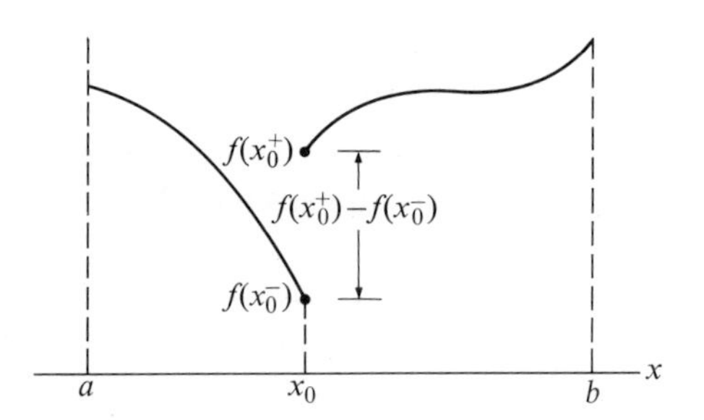

FIGURE 9–1

$$f(x_0^+) = f(x_0^-) = f(x_0).$$

More generally, the requirement that both of these limits be finite everywhere in $[a, b]$ implies that the only discontinuities of f are "jump discontinuities" of the type shown in Fig. 9–1. Moreover, the difference

$$f(x_0^+) - f(x_0^-)$$

measures the magnitude of the jump of the function f at x_0.

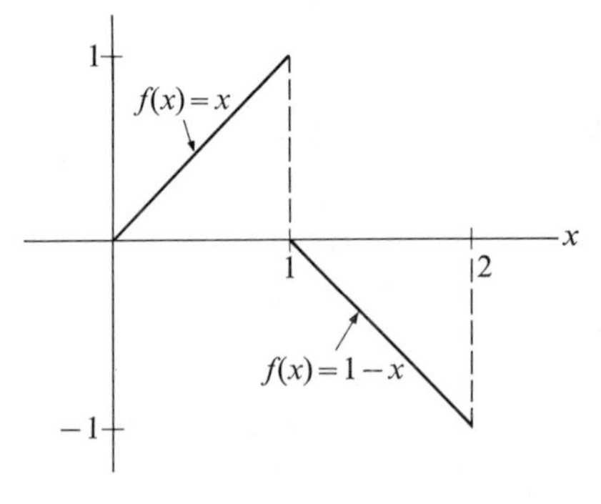

FIGURE 9–2

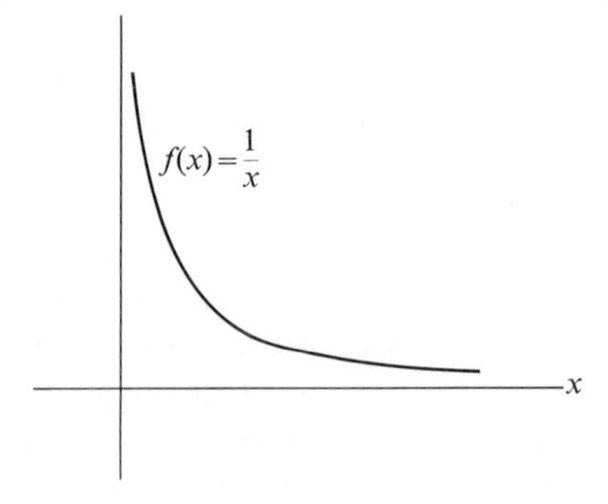

FIGURE 9–3

Thus, for example, the function

$$f(x) = \begin{cases} x, & 0 < x < 1, \\ 1 - x, & 1 < x < 2, \end{cases}$$

(see Fig. 9–2) is piecewise continuous on the interval [0, 2] and has a jump discontinuity of magnitude -1 at $x = 1$, since $f(1^+) = 0$ while $f(1^-) = 1$. On the other hand, the functions $1/x$ and $\sin 1/x$ fail to be piecewise continuous on any interval containing the origin because of their behavior as $x \to 0$. (See Figs. 9–3 and 9–4.)

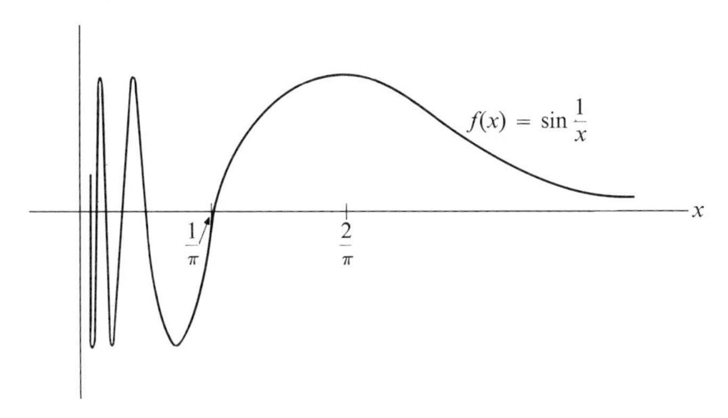

FIGURE 9–4

The following facts concerning piecewise continuous functions are of particular importance, and will be used repeatedly hereafter.

1. If f is piecewise continuous on $[a, b]$, then

$$\int_a^b f(x)\,dx$$

exists, and is independent of whatever values (if any) f assumes at its points of discontinuity. In particular, if f and g are identical everywhere in $[a, b]$ save at their points of discontinuity, then

$$\int_a^b f(x)\,dx = \int_a^b g(x)\,dx.$$

2. If f and g are piecewise continuous on $[a, b]$, then so is their product fg. (See Exercise 11.) This, together with (1) implies that the integral of the product of two piecewise continuous functions always exists.

3. Every continuous function on $[a, b]$ is piecewise continuous.

This said, we now turn our attention to the problem of converting the set of piecewise continuous functions on $[a, b]$ into a Euclidean space. In view of the fact that this set includes the continuous functions on $[a, b]$ it is only reasonable to require our construction to be so conceived that the resulting Euclidean space has $\mathcal{C}[a, b]$ as a subspace. This, in turn, suggests that we define $f \cdot g$ by the formula

$$f \cdot g = \int_a^b f(x)g(x)\,dx. \tag{9-2}$$

But does (9–2) actually yield an inner product on the set of piecewise continuous functions on $[a, b]$? The unpleasant answer is, No. To see what goes wrong, let $n(x)$ be a function which is zero everywhere in $[a, b]$ except at a finite number of points (see Fig. 9–5). Such a function is said to be a *null function*, and has the annoying property that

$$\int_a^b n(x)^2\,dx = 0$$

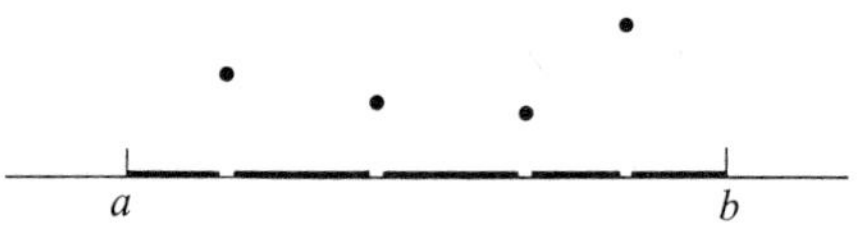

FIGURE 9–5

in spite of the fact that n is not the zero function. This, of course, vitiates using (9–2) as an inner product, since, by definition, the inner product of a nonzero vector with itself cannot be zero.

It is perfectly clear, however, that the above difficulty will disappear if we overlook the fact that a null function is not identically zero, and treat it as if it were. But then, to be consistent, *we must also regard any two piecewise continuous functions as* **identical** *whenever they differ at only a finite number of points.* And this is just what needs to be done to make (9–2) yield an inner product.

As usual, there are a number of facts which have to be verified before such an assertion can be accepted. Among them we call attention to the need to ascertain that (9–2) respects our identification of functions. In other words, we must show that whenever f_1 is identified with f, and g_1 with g, then

$$\int_a^b f_1(x)g_1(x)\,dx = \int_a^b f(x)g(x)\,dx.$$

For only then will (9–2) unambiguously define an inner product on the set of piecewise continuous functions identified as above. This, however, is easy to prove, and has been left to the reader as an exercise (see Exercise 5).

Finally, to complete the argument, we observe that whenever f_1 is identified with f, and g_1 with g, then $f_1 + g_1$ is identified with $f + g$, and αf_1 with αf for all real numbers α. Thus functional addition and multiplication by real numbers also respect our identification of functions, and it follows that the set of piecewise continuous functions on a fixed interval $[a, b]$, identified as above, is a Euclidean space. We shall denote this space by $\mathcal{PC}[a, b]$, and assert without proof that it contains $\mathcal{C}[a, b]$ as a subspace (Exercise 2).

Now that we have rigorously constructed the space $\mathcal{PC}[a, b]$ and insisted that its vectors are collections of functions identified with one another, it is perfectly clear that we can ignore this fact and treat these vectors as though they were ordinary functions. This is precisely what we shall do henceforth, but always with the tacit understanding that the facts of the matter are as outlined above, and that all of our arguments can be rigorously restated, if necessary. As with all such abuses of terminology, this involves no real danger of error, and has the positive result of simplifying language and notation.

EXERCISES

1. For each of the following functions evaluate the right- and left-hand limits at all points of discontinuity, and state whether or not the function is piecewise continuous on $[0, 2]$.

(a) $f(x) = \begin{cases} x, & 0 \le x < 1 \\ x - 2, & 1 < x \le 2 \end{cases}$ (b) $f(x) = \begin{cases} 0, & x = 0 \\ \dfrac{1}{x}, & 0 < x \le 2 \end{cases}$

(c) $f(x) = \begin{cases} \dfrac{x^2 - 2x + 1}{x - 1}, & 0 \le x < 1 \\ 1, & 1 < x \le 2 \end{cases}$ (d) $f(x) = \begin{cases} 1 - x, & 0 \le x < 1 \\ x - 1, & 1 < x \le 2 \end{cases}$

(e) $f(x) = \begin{cases} 1, & 0 \le x \le \frac{1}{2} \\ 0, & \frac{1}{2} < x \le \frac{3}{2} \\ \frac{1}{2}, & \frac{3}{2} < x \le 2 \end{cases}$ (f) $f(x) = \begin{cases} \dfrac{x}{x - 1}, & 0 \le x < 1 \\ 2x, & 1 < x \le 2 \end{cases}$

2. (a) Suppose a piecewise continuous function f is identified with a continuous function g in $\mathcal{PC}[a, b]$. What can be said about the magnitudes of the jump discontinuities of f?

   (b) Use the result in (a) to prove that $\mathcal{C}[a, b]$ is a subspace of $\mathcal{PC}[a, b]$.

3. Determine whether or not the following functions are piecewise continuous on $[0, 1]$.

(a) $f(x) = \begin{cases} x \sin \dfrac{\pi}{x}, & 0 < x \le 1 \\ 0, & x = 0 \end{cases}$

(b) $f(x) = \begin{cases} 1, & \text{if } x \text{ is irrational} \\ 0, & \text{if } x \text{ is rational} \end{cases}$

(c) $f(x) = \dfrac{1}{n + 1}, \quad \dfrac{1}{2^{n+1}} < x \le \dfrac{1}{2^n}, \quad n = 0, 1, 2, \ldots$

4. Prove that f_1 and f_2 are identified in $\mathcal{PC}[a, b]$ if and only if $f_1 - f_2$ is a null function.

5. Let f and g be piecewise continuous on $[a, b]$, and suppose that f_1 is identified with f, and g_1 with g. Prove that

$$\int_a^b f_1(x) g_1(x)\, dx = \int_a^b f(x) g(x)\, dx.$$

   [*Hint:* Note that $f_1 g_1 - fg = f_1 g_1 - f_1 g + f_1 g - fg$.]

6. Verify the assertion that $\mathcal{PC}[a, b]$ is a Euclidean space.

7. Are the functions $1, x, x^2, \ldots$ linearly independent in $\mathcal{PC}[a, b]$? Why?

8. Prove that the functions

$$1,\ \cos\frac{\pi x}{a},\ \sin\frac{\pi x}{a},\ \ldots,\ \cos\frac{n\pi x}{a},\ \sin\frac{n\pi x}{a},\ \ldots$$

are linearly independent in $\mathcal{PC}[-a, a]$.

9. Prove that $\|f - g\| = 0$ in $\mathcal{PC}[a, b]$ if and only if $f - g$ is a null function.

*10. Let f_1 and g_1 be piecewise continuous on $[a, b]$, and write $f_1 \sim f_2$ if and only if $f_1 - f_2$ is a null function. Prove that this defines an equivalence relation on the set of all piecewise continuous functions on $[a, b]$, and that the equivalence classes it defines are the elements of $\mathcal{PC}[a, b]$. (See Section 1–9.)

11. Let f and g be piecewise continuous on the interval $[a, b]$, and let fg denote their product; that is, fg is the function defined by

$$fg(x) = f(x)g(x)$$

for all x in $[a, b]$. Prove that fg is piecewise continuous on $[a, b]$, and thus deduce that the integral in (9–2) exists. [*Hint:* Use the fact that the product of continuous functions is continuous.]

9–3 EVEN AND ODD FUNCTIONS

The task of evaluating the integrals which arise in the study of orthogonal series can often be simplified by exploiting the symmetry of the functions involved. This technique is usually formalized by introducing the notions of even and odd functions, as follows.

Definition 9–2. A function f, defined on an interval centered at the origin, is said to be *even* if

$$f(-x) = f(x) \tag{9–3}$$

for all x in the domain of f, and *odd* if

$$f(-x) = -f(x). \tag{9–4}$$

This, of course, is just another way of saying that a function is even if its graph is symmetric about the vertical axis (Fig. 9–6), and odd if its graph is symmetric

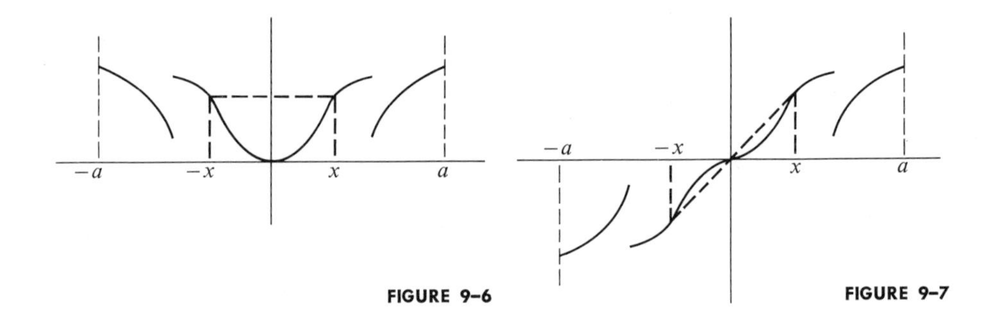

FIGURE 9–6

FIGURE 9–7

about the origin (Fig. 9–7). Thus, for integral values of n, x^n is even if n is even, odd if n is odd, a fact which helps explain the particular terminology used in this context.

The importance of even and odd functions for our work stems from the equalities

$$\int_{-a}^{a} f(x)\,dx = 2\int_{0}^{a} f(x)\,dx \tag{9–5}$$

whenever f is even and integrable, and

$$\int_{-a}^{a} f(x)\,dx = 0 \tag{9–6}$$

whenever f is odd and integrable. Both of these assertions are easy consequences of the above definition (see Exercise 2), and are also evident from the geometric interpretation of the definite integral as area.

An equally elementary observation is that the product of two functions is even whenever both of the functions are even or both are odd, and is odd whenever one of the functions is even and one odd. In short, the multiplication of even and odd functions obeys the rules

$$\begin{aligned}(\text{Even})(\text{Even}) &= (\text{Odd})(\text{Odd}) = \text{Even},\\ (\text{Even})(\text{Odd}) &= (\text{Odd})(\text{Even}) = \text{Odd}.\end{aligned}$$

From this, and (9–6), we deduce that

$$\int_{-a}^{a} f(x)g(x)\,dx = 0$$

whenever f and g have the opposite parity. Hence *even and odd functions in* $\mathcal{PC}[-a, a]$ *are mutually orthogonal.*

EXAMPLE 1. The functions

$$1,\ \cos x,\ \cos 2x, \ldots$$

are even in $\mathcal{PC}[-a, a]$, and

$$\sin x,\ \sin 2x, \ldots$$

are odd. Thus

$$\int_{-a}^{a} f(x) \cos kx\,dx = 0$$

if f is odd, and

$$\int_{-a}^{a} f(x) \sin kx\,dx = 0$$

if f is even. The value of these results for computing Fourier coefficients is too obvious to need comment.

A somewhat less obvious property of even and odd functions is established in the following lemma.

Lemma 9–1. *Every function on the interval* $[-a, a]$ *can be written in exactly one way as the sum of an even function and an odd function.*

Proof. Let f be an arbitrary function on $[-a, a]$, and set

$$f_E(x) = \frac{f(x) + f(-x)}{2}, \qquad f_O(x) = \frac{f(x) - f(-x)}{2}. \tag{9–7}$$

It is trivial to verify that f_E is even, f_O odd, and that $f = f_E + f_O$. Thus f has *at least one* decomposition of the desired form, and it remains to show that this is the only such. To this end, suppose that we also had $f = g_E + g_O$, with g_E even, and g_O odd. Then $f_E + f_O = g_E + g_O$, and

$$f_E - g_E = g_O - f_O.$$

But the difference of two even functions is even, and the difference of two odd functions is odd. Thus the function defined by the above identity is simultaneously even and odd, and so must be the zero function. In other words, $f_E - g_E = g_O - f_O = 0$, and it follows that $g_E = f_E$, $g_O = f_O$, as desired. ▮

The functions f_E and f_O defined in (9–7) are known respectively as the *even* and *odd parts* of f.

EXAMPLE 2. If $f(x) = e^x$, then

$$f_E(x) = \frac{e^x + e^{-x}}{2} \quad \text{and} \quad f_O(x) = \frac{e^x - e^{-x}}{2}.$$

Thus the even and odd parts of the exponential function are the hyperbolic cosine and hyperbolic sine, respectively.

EXERCISES

1. Classify each of the following functions as even, odd, or neither even nor odd.

(a) $\tan x$ (b) e^{x^2} (c) $\dfrac{x+1}{x-1}$ (d) $\ln |x|$

(e) $\dfrac{x}{(x+1)(x-1)}$ (f) $\sin^{-1} x$ (g) $\cos^{-1} x$ (h) $\dfrac{x^2}{(x+1)(x-1)}$

(i) $\dfrac{x(x+1)}{x-1}$ (j) $f(|x|)$, f defined in $[0, a]$

2. Prove Eqs. (9–5) and (9–6) of the text.
3. Establish the multiplicative rules for even and odd functions given in the text.

4. Show that

$$\int_0^{\pi}\int_{-\pi}^{\pi} f(t)\cos u(t-x)\,dt\,du = \tfrac{1}{2}\int_{-\pi}^{\pi}\int_{-\pi}^{\pi} f(t)\cos u(t-x)\,dt\,du$$

for any piecewise continuous function f on the interval $[-\pi, \pi]$.

5. Let f be differentiable in the interval $[-a, a]$. Prove that f' is odd whenever f is even, and even whenever f is odd.

6. Let f be an integrable function defined in the interval $[-a, a]$, and let

$$F(x) = \int_0^x f(t)\,dt, \quad -a \leq x \leq a.$$

Prove that F is even if f is odd, and odd if f is even.

7. Decompose each of the following functions into its even and odd parts.

(a) $\dfrac{x+1}{x-1}$ (b) $\dfrac{1}{x+1}$ (c) $x\cos x - \cos 2x$ (d) $\dfrac{x^2+1}{x-1}$

(e) $\dfrac{1}{ax^2+bx+c}$, a, b, c constants, not all zero

8. Find the even and odd parts of the function

$$f(x) = \sum_{k=0}^{\infty} a_k x^k, \quad a_k \text{ constants.}$$

9. Find the even and odd parts of the function

$$f(x) = \frac{a_0}{2} + \sum_{k=1}^{\infty}(a_k\cos kx + b_k \sin kx), \quad a_k, b_k \text{ constants.}$$

10. Let $\mathcal{E}[-a, a]$ denote the set of all even piecewise continuous functions on the interval $[-a, a]$, and let $\mathcal{O}[-a, a]$ denote the set of all odd piecewise continuous functions on $[-a, a]$. Prove that $\mathcal{E}[-a, a]$ and $\mathcal{O}[-a, a]$ are subspaces of $\mathcal{PC}[-a, a]$.

11. Prove that the zero function is the only function which is simultaneously even and odd on $[-a, a]$.

12. Let $p_0(x), p_1(x), \ldots$ be the sequence of polynomials obtained by applying the Gram-Schmidt orthogonalization process to $1, x, x^2, \ldots$ in $[-1, 1]$ (see Example 3, Section 7–4). Show that $p_{2k}(x)$ is an even function, and that $p_{2k+1}(x)$ is an odd function for all values of k. [*Hint:* Use mathematical induction.]

13. Let f be piecewise continuous on $[-a, a]$, and suppose that f is orthogonal to *every* even function in $\mathcal{PC}[-a, a]$. Prove that f is odd. [*Hint:* Use Lemma 9–1.]

9–4 FOURIER SERIES

In this section we begin our study of orthogonal series by considering series expansions relative to the functions

$$1, \cos x, \sin x, \cos 2x, \sin 2x, \ldots. \tag{9–8}$$

We have already seen that these functions are mutually orthogonal in $\mathcal{PC}[-\pi, \pi]$, and we shall prove shortly that they are a *basis* as well. Granting the truth of this fact, we can then use Formula (8–22) to write any piecewise continuous function f on the interval $[-\pi, \pi]$ in the form

$$f(x) = \frac{f \cdot 1}{\|1\|^2} + \sum_{k=1}^{\infty} \left[\frac{f \cdot \cos kx}{\|\cos kx\|^2} \cos kx + \frac{f \cdot \sin kx}{\|\sin kx\|^2} \sin kx \right] \quad \text{(mean)} \tag{9–9}$$

where the notation "(mean)" indicates that the series in question converges in the mean to f. But since

$$\|1\|^2 = \int_{-\pi}^{\pi} dx = 2\pi,$$

$$\|\cos kx\|^2 = \int_{-\pi}^{\pi} \cos^2 kx \, dx = \pi,$$

$$\|\sin kx\|^2 = \int_{-\pi}^{\pi} \sin^2 kx \, dx = \pi,$$

(9–9) may be rewritten

$$f(x) = \frac{a_0}{2} + \sum_{k=1}^{\infty} (a_k \cos kx + b_k \sin kx) \quad \text{(mean)}, \tag{9–10}$$

where

$$a_k = \frac{1}{\pi} \int_{-\pi}^{\pi} f(x) \cos kx \, dx,$$
$$b_k = \frac{1}{\pi} \int_{-\pi}^{\pi} f(x) \sin kx \, dx, \tag{9–11}$$

for all k. This particular representation of f is known as its *Fourier series expansion* on the interval $[-\pi, \pi]$, and the a_k and b_k are called the *Fourier* or *Euler-Fourier coefficients* of f.

Once again we emphasize that (9–10) must be read as asserting that the series in question converges *in the mean* to f, *not* that it converges *pointwise* in the sense that

$$f(x_0) = \frac{a_0}{2} + \sum_{k=1}^{\infty} (a_k \cos kx_0 + b_k \sin kx_0)$$

for all x_0 in $[-\pi, \pi]$. Indeed, since the value of f at x_0 can be changed arbitrarily without changing the value of its Fourier coefficients, this would be entirely too much to expect. Moreover, in view of Example 4 in Section 8–1, there is no *a priori* reason to expect that the Fourier series for f will converge to the value of f at so much as a single point in $[-\pi, \pi]$. But, surprisingly, whenever f is reasonably well-behaved, it converges to $f(x)$ for *all* x. We shall have more to say on this point as soon as we have considered an example.

EXAMPLE 1. Find the Fourier series expansion of the function

$$f(x) = \begin{cases} -1, & -\pi < x < 0, \\ 1, & 0 < x < \pi, \end{cases}$$

(see Fig. 9–8).

In this case f is an odd function on $[-\pi, \pi]$. Hence so is $f(x) \cos kx$, and (9–6) implies that $a_k = 0$ for all k. On the other hand, $f(x) \sin kx$ is even, and we therefore have

$$\begin{aligned} b_k &= \frac{2}{\pi} \int_0^\pi \sin kx \, dx \\ &= \frac{2}{k\pi}(1 - \cos k\pi) \\ &= \begin{cases} \dfrac{4}{k\pi}, & k = 1, 3, 5, \ldots, \\ 0, & k = 2, 4, 6, \ldots. \end{cases} \end{aligned}$$

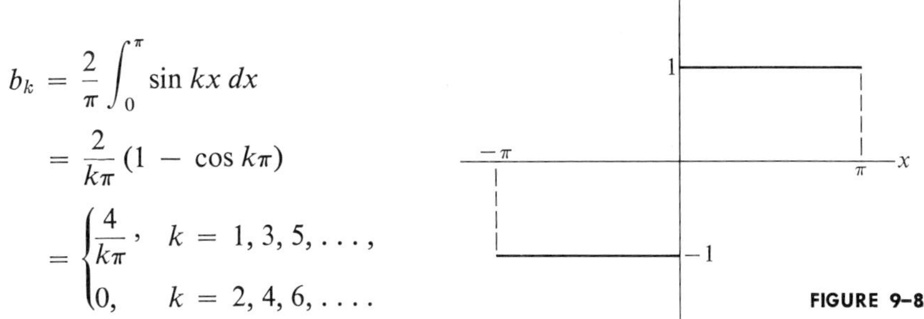

FIGURE 9–8

Hence the Fourier series expansion of f is

$$\begin{aligned} f(x) &= \frac{4}{\pi}\left[\sin x + \frac{\sin 3x}{3} + \frac{\sin 5x}{5} + \cdots\right] \\ &= \frac{4}{\pi} \sum_{k=1}^{\infty} \frac{\sin (2k-1)x}{2k-1} \quad \text{(mean)}. \end{aligned} \tag{9–12}$$

In Fig. 9–9 we have sketched the graph of f and the sum of the first two terms of its Fourier series, which, as can be seen, already furnishes a fairly good approximation to f throughout the interval $[-\pi, \pi]$. This approximation improves considerably if we use the first four terms in the series (Fig. 9–10), and it is not difficult to see that it continues to improve as additional terms are considered. In so doing one quickly becomes convinced that the series actually converges *pointwise* to f everywhere in $[-\pi, \pi]$ where f is defined. Moreover, when $x = 0$ and

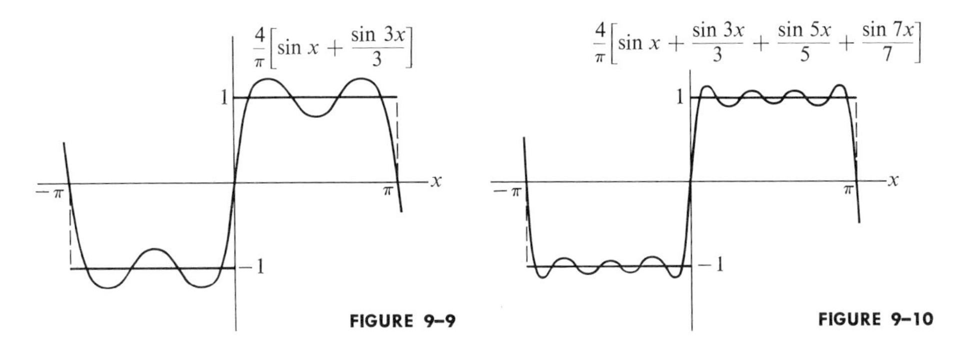

FIGURE 9–9 FIGURE 9–10

$\pm\pi$, the series obviously converges to zero even though f is not defined at those points. Here then, in our very first example, we have come upon a Fourier series which converges pointwise in the interval $[-\pi, \pi]$, and which represents the function from which it was derived at each point in the domain of that function.

It would be easy to multiply the number of such examples indefinitely until the reader became convinced (erroneously, as it turns out) that all Fourier series converge at each point in the interval $[-\pi, \pi]$. Instead, however, we prefer to cite a theorem which will account for this phenomenon within the framework of the general theory of Fourier series. For the present we content ourselves with a statement of the result in question, leaving any attempts at a proof to the next chapter.

Theorem 9–1. *Let f be a* **piecewise smooth** *function in $\mathcal{PC}[-\pi, \pi]$, by which we mean that f has a piecewise continuous first derivative on $[-\pi, \pi]$. Then the Fourier series expansion for f converges pointwise everywhere in $[-\pi, \pi]$, and has the value*

$$\frac{f(x_0^+) + f(x_0^-)}{2} \tag{9–13}$$

at each point x_0 in the interior of the interval, and

$$\frac{f(-\pi^+) + f(\pi^-)}{2} \tag{9–14}$$

at $\pm\pi$.

This theorem is one of the most important in the entire theory of Fourier series, and the student should make certain that he understands it thoroughly before going on. In particular, he should note that the expression

$$\frac{f(x_0^+) + f(x_0^-)}{2}$$

is none other than the *average* of the right- and left-hand limits of f at x_0, and is equal to $f(x_0)$ whenever x_0 is a point of continuity of f. Hence *the Fourier series expansion of a piecewise smooth function f in $\mathcal{PC}[-\pi, \pi]$ converges to $f(x_0)$ whenever x_0 is a point of continuity of f.* On the other hand, if f has a jump discontinuity at x_0, (9–13) implies that the Fourier series for f converges at x_0 to the value located at the *midpoint* of the jump, as shown in Fig. 9–11.

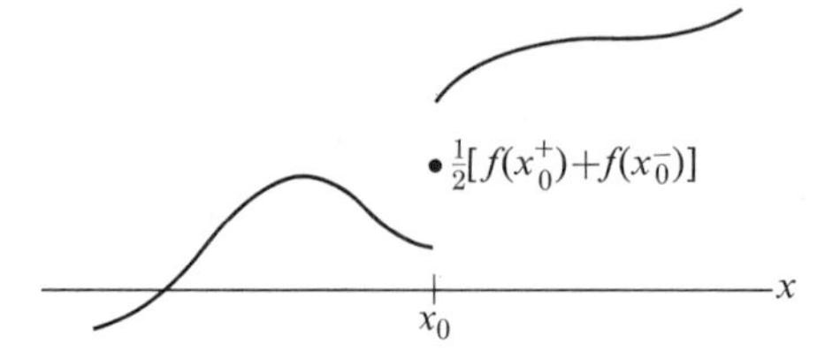

FIGURE 9–11

When these results are applied to Example 1 above they allow us to assert that the series

$$\frac{4}{\pi}\left[\sin x + \frac{\sin 3x}{3} + \frac{\sin 5x}{5} + \cdots\right] \tag{9–15}$$

converges pointwise in the interval $[-\pi, \pi]$ to

$$\begin{aligned} -1 &\text{ if } \; -\pi < x < 0, \\ 0 &\text{ if } \; x = -\pi, 0, \pi, \\ 1 &\text{ if } \; 0 < x < \pi. \end{aligned}$$

Thus, for example, when $x = \pi/2$, the value of this series is 1, and we conclude that

$$1 = \frac{4}{\pi}\left[1 - \frac{1}{3} + \frac{1}{5} - \frac{1}{7} + \cdots\right],$$

or

$$\frac{\pi}{4} = 1 - \frac{1}{3} + \frac{1}{5} - \frac{1}{7} + \cdots.$$

Similarly, when $x = \pi/4$, (9–15) yields the numerical series

$$1 = \frac{4}{\pi}\left[\frac{\sqrt{2}}{2} + \frac{\sqrt{2}}{2\cdot 3} - \frac{\sqrt{2}}{2\cdot 5} - \frac{\sqrt{2}}{2\cdot 7} + \cdots\right],$$

and we have a second representation of $\pi/4$ as

$$\frac{\pi}{4} = \frac{\sqrt{2}}{2}\left(1 + \frac{1}{3} - \frac{1}{5} - \frac{1}{7} + \cdots\right).$$

Before continuing, it may be appropriate to remark that there exist *continuous* functions whose Fourier series *diverge* at finitely many points in $[-\pi, \pi]$. Thus the requirement that f have a piecewise continuous first derivative is imposed in order to guarantee the pointwise convergence of its Fourier series. Incidentally, the problem of determining whether there exists a continuous function whose Fourier series diverges *everywhere* in $[-\pi, \pi]$ is still unsolved. Needless to say, such a function, if it exists, will be rather bizarre, and is not likely to arise in applications.

By now the reader has undoubtedly observed that our description of the pointwise behavior of a Fourier series is incomplete. For if a trigonometric series

$$\frac{a_0}{2} + \sum_{k=1}^{\infty} (a_k \cos kx + b_k \sin kx) \tag{9–16}$$

converges to the value K_0 when $x = x_0$, it will also converge to K_0 at all points of the form $x_0 + 2\pi n$, n an arbitrary integer. This, of course, is an immediate

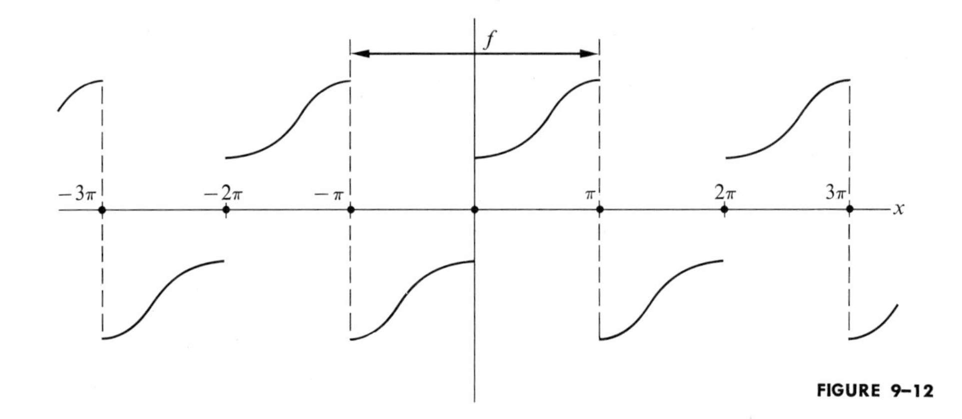

FIGURE 9–12

consequence of the fact that the functions $\sin kx$ and $\cos kx$ are periodic, with 2π as a period, and allows us to make the following important observation: *If* (9–16) *converges pointwise to a function f everywhere in* $[-\pi, \pi)$, *then it actually converges on the entire x-axis to the function F obtained by repeating f successively along the x-axis in intervals of length* 2π (see Fig. 9–12). It is obvious that the function F obtained in this way is periodic with 2π as a period, in the sense that

$$F(x + 2\pi) = F(x)$$

for all x. It is known as the *periodic extension* of f.

When these remarks are combined with Theorem 9–1, they yield

Theorem 9–2. *The Fourier series expansion of a piecewise smooth function f in* $\mathcal{PC}[-\pi, \pi]$ *converges pointwise on the entire real line. Moreover, if F denotes the periodic extension of f, then the value of the series is* $F(x_0)$ *when* x_0 *is a point of continuity of F, and*

$$\frac{F(x_0^+) + F(x_0^-)}{2}$$

when x_0 *is a jump discontinuity of F.*

In particular, we note that the Fourier series for f will converge pointwise to a *continuous* function on the entire x-axis if and only if f is continuous on $[-\pi, \pi]$ *and* $f(-\pi) = f(\pi)$. For only then will F be free of jump discontinuities. We shall have occasion to use this fact later in the chapter.

At the risk of belaboring the obvious, we point out that Theorem 9–2 can be used to sketch the graph of the Fourier series of any piecewise smooth function f in $\mathcal{PC}[-\pi, \pi]$. The procedure is as follows: First sketch the graph of the periodic extension F of f; then plot the midpoint of each jump discontinuity of F. The resulting picture, *with these isolated points included*, will be the graph of the series

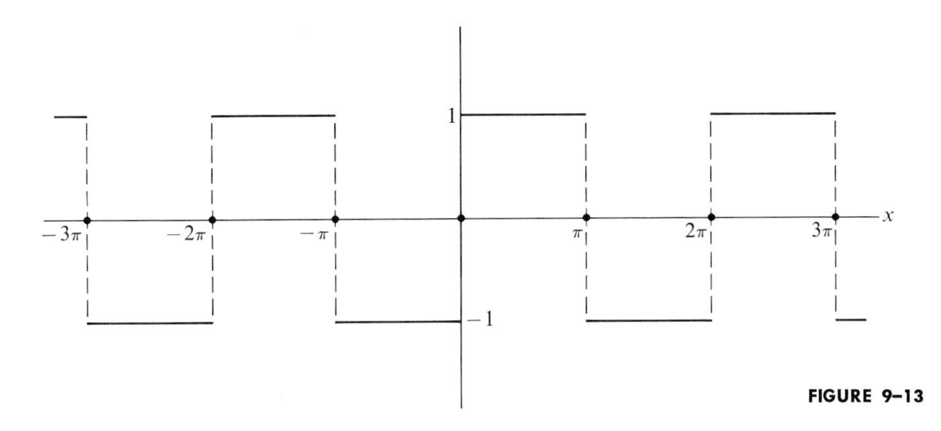

FIGURE 9–13

in question. Thus, for instance, the graph of the series

$$\frac{4}{\pi}\left[\sin x + \frac{\sin 3x}{3} + \cdots\right]$$

found in Example 1 appears as shown in Fig. 9–13.

EXAMPLE 2. Find the Fourier series expansion of the function

$$f(x) = |x|, \quad -\pi < x < \pi,$$

and sketch the graph of the series.

In this case f is an even function on $[-\pi, \pi]$. Hence $b_k = 0$ for all k, while, for $k \neq 0$,

$$\begin{aligned} a_k &= \frac{2}{\pi}\int_0^{\pi} x \cos kx \, dx \\ &= \frac{2}{\pi}\left[\frac{x \sin kx}{k}\bigg|_0^{\pi} - \frac{1}{k}\int_0^{\pi} \sin kx \, dx\right] \\ &= \frac{2}{\pi k^2} \cos kx \bigg|_0^{\pi} \\ &= \frac{2}{\pi k^2}(\cos k\pi - 1) \\ &= \begin{cases} -\dfrac{4}{\pi k^2}, & k = 1, 3, 5, \ldots, \\ 0, & k = 2, 4, 6, \ldots. \end{cases} \end{aligned}$$

Finally, when $k = 0$, we have

$$a_0 = \frac{2}{\pi}\int_0^{\pi} x \, dx = \pi,$$

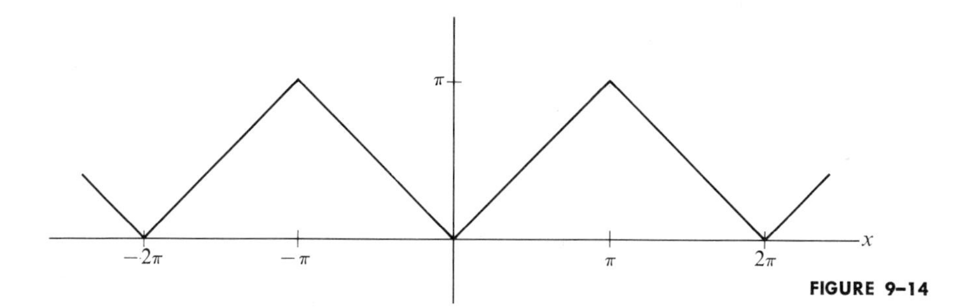

FIGURE 9–14

and it follows that

$$|x| = \frac{\pi}{2} - \frac{4}{\pi}\left(\cos x + \frac{\cos 3x}{3^2} + \frac{\cos 5x}{5^2} + \cdots\right) \quad \text{(mean)}$$

in $\mathcal{PC}[-\pi, \pi]$. The graph of this series is shown in Fig. 9–14, and, for comparison, we have sketched the graph of the sum of the first three terms of the series in Fig. 9–15.

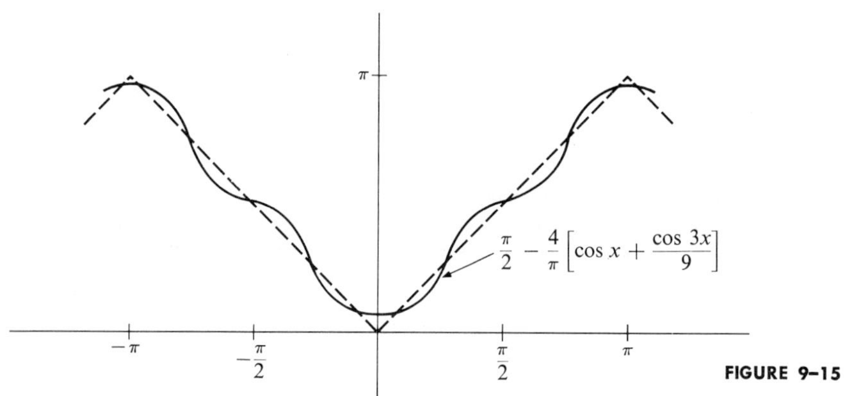

FIGURE 9–15

EXAMPLE 3. Let g be the function in $\mathcal{PC}[-\pi, \pi]$ defined by

$$g(x) = \begin{cases} 0, & -\pi < x < 0, \\ 1, & 0 < x < \pi. \end{cases}$$

Then, with f as in Example 1,

$$g = \tfrac{1}{2}[1 + f] = \tfrac{1}{2} + \tfrac{1}{2}f,$$

and we conclude that the Fourier series expansion of g is

$$\frac{1}{2} + \frac{2}{\pi}\left(\sin x + \frac{\sin 3x}{3} + \frac{\sin 5x}{5} + \cdots\right).$$

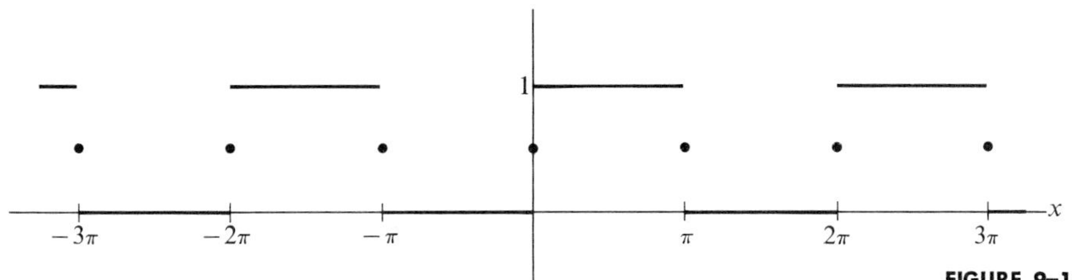

FIGURE 9–16

The moral of this example is that a Fourier series can sometimes be found without recourse to integration. We refer the reader to Exercise 21 for a discussion of this technique, and to Fig. 9–16 for the graph of the series.

EXERCISES

In Exercises 1–10 find the Fourier series expansion of the given function. Sketch the graph of the series obtained, paying particular attention to its values at any points of discontinuity.

1. $f(x) = x, \quad -\pi < x < \pi$

2. $f(x) = \begin{cases} 1, & -\pi < x < 0 \\ \frac{1}{2}, & 0 < x < \pi \end{cases}$

3. $f(x) = e^x, \quad -\pi < x < \pi$

4. $f(x) = |\sin x|$

5. $f(x) = \begin{cases} -|x - \frac{1}{2}|, & -\pi \le x \le -\frac{1}{2} \\ 0, & -\frac{1}{2} \le x \le \frac{1}{2} \\ x - \frac{1}{2}, & \frac{1}{2} \le x \le \pi \end{cases}$

6. $f(x) = \begin{cases} \frac{1}{2}, & -\pi \le x \le -\frac{1}{2} \\ |x|, & -\frac{1}{2} \le x \le \frac{1}{2} \\ \frac{1}{2}, & \frac{1}{2} \le x \le \pi \end{cases}$

7. $f(x) = (\pi - x)(\pi + x),$
   $-\pi \le x \le \pi$

8. $f(x) = e^{|x|}, \quad -\pi \le x \le \pi$

9. $f(x) = \begin{cases} x + \pi, & -\pi < x < 0 \\ x - \pi, & 0 < x < \pi \end{cases}$

10. $f(x) = x^2, \quad -\pi < x < \pi$

11. (a) Sketch the graph of the Fourier series expansion of the function

$$f(x) = \begin{cases} -1, & -\pi < x < -\dfrac{\pi}{2}, \\ 1, & -\dfrac{\pi}{2} < x < \dfrac{\pi}{2}, \\ -1, & \dfrac{\pi}{2} < x < \pi, \end{cases}$$

over the intervals $[2\pi, 3\pi]$ and $[-2\pi, 0]$.

(b) What is the value of the Fourier series for f when $x = k\pi$, k an integer? When

$$x = (2k + 1)\frac{\pi}{2},$$

k an integer?

12. (a) Sketch the graph of the Fourier series expansion of the function

$$f(x) = \begin{cases} -1, & -\pi < x < -1, \\ \dfrac{x}{2}, & -1 < x < 1, \\ 1, & 1 < x < \pi, \end{cases}$$

over the intervals $[2\pi, 5\pi]$ and $[-3\pi, -\pi]$.

(b) Find the value of the Fourier series for this function at the following points:

$$x = 1, \quad x = \pi, \quad x = -6, \quad x = 3, \quad x = 7, \quad x = -5\pi.$$

13. (a) Sketch the graph of the Fourier series expansion of the function

$$f(x) = \begin{cases} x + \pi, & -\pi < x < -\dfrac{2\pi}{3}, \\ x + \dfrac{2\pi}{3}, & -\dfrac{2\pi}{3} < x < 0, \\ x, & 0 < x < \dfrac{\pi}{3}, \\ x - \dfrac{\pi}{3}, & \dfrac{\pi}{3} < x < \pi. \end{cases}$$

(b) What is the value of the Fourier series for this function when $x = k\pi$, k an integer? When $x = (4k + 1)(\pi/3)$, k an integer?

14. (a) Find the Fourier series expansion of the function

$$f(x) = \begin{cases} 0, & -\pi < x \leq 0, \\ x^2, & 0 \leq x < \pi, \end{cases}$$

and sketch the graph of the series obtained.

(b) Use this series to show that

$$\frac{\pi^2}{6} = 1 + \frac{1}{2^2} + \frac{1}{3^2} + \frac{1}{4^2} + \cdots.$$

15. (a) Find the Fourier series expansion of the function

$$f(x) = \begin{cases} 0, & -\pi < x \leq 0, \\ x, & 0 \leq x < \pi, \end{cases}$$

and sketch the graph of the series obtained.

(b) Use this series to show that

$$\frac{\pi^2}{8} = 1 + \frac{1}{3^2} + \frac{1}{5^2} + \frac{1}{7^2} + \cdots.$$

16. Find the Fourier series expansion of the function

$$f(x) = \begin{cases} 0, & -\pi < x < 0, \\ \cos x, & 0 < x < \pi, \end{cases}$$

and sketch the graph of the series obtained.

17. (a) Find the Fourier series expansion of the function

$$f(x) = \cos \alpha x, \quad -\pi < x < \pi,$$

for any real number α.

(b) Use this series to prove that

$$\cot \alpha\pi = \frac{1}{\pi}\left[\frac{1}{\alpha} - \sum_{k=1}^{\infty} \frac{2\alpha}{k^2 - \alpha^2}\right]$$

whenever α is not an integer. Justify the validity of your computations.

18. A function f, defined and continuous at all but a finite number of points in any closed interval of the x-axis, is said to be *periodic* if there exists a real number $p > 0$ such that

$$f(x + p) = f(x)$$

for all x in the domain of f. The smallest positive real number p with this property (if such exists) is called the *fundamental period* of f; otherwise p is simply called a *period*.

(a) Prove that if f is periodic with period p, then

$$f(x + kp) = f(x)$$

for all integral values of k.

(b) Give an example of a periodic function which does not have a fundamental period.

(c) Let f_1 and f_2 be periodic functions with fundamental periods p_1 and p_2, respectively. Prove that $\alpha_1 f_1 + \alpha_2 f_2$ is periodic for every pair of real numbers α_1 and α_2 if and only if p_1/p_2 is a rational number.

(d) Generalize the result in (c) to linear combinations of n periodic functions with fundamental periods $p_1, \ldots, p_n$.

19. Determine which of the following functions are periodic, and find the fundamental period of each if it has one. (See Exercise 18.)

(a) $\sin \dfrac{\pi x}{4}$ (b) $2 \sin 3x - \cos 2x$ (c) $\tan x^2$

(d) $(\sin 2x)(\cos x)$ (e) $\sin \dfrac{1}{x}$ (f) $\tan 2x + 3 \sin \pi x$

(g) $f(x) = \begin{cases} 0 & \text{if } x \text{ is rational} \\ 1 & \text{if } x \text{ is irrational} \end{cases}$

*20. Let

$$f(x) = \frac{P(x)}{Q(x)},$$

where $P(x)$ and $Q(x)$ are polynomials and $Q(0) \neq 0$. Prove that if f is periodic it is constant. [*Hint:* Let $f(0) = \alpha$, and consider the equation $P(x) - \alpha Q(x) = 0$.]

21. Let

$$\frac{a_0}{2} + \sum_{k=1}^{\infty} (a_k \cos kx + b_k \sin kx) \quad \text{and} \quad \frac{A_0}{2} + \sum_{k=1}^{\infty} (A_k \cos kx + B_k \sin kx)$$

be, respectively, the Fourier series expansions of the functions f and g in $\mathcal{PC}[-\pi, \pi]$. Prove that the series

$$\frac{\alpha a_0 + \beta A_0}{2} + \sum_{k=1}^{\infty} [(\alpha a_k + \beta A_k) \cos kx + (\alpha b_k + \beta B_k) \sin kx]$$

is the Fourier series expansion of the function $\alpha f + \beta g$ for any pair of real numbers α, β.

22. Use Exercise 21 and the series found in Examples 1 and 2 and Exercise 1 above to obtain the Fourier series expansions of the functions shown in Fig. 9–17.

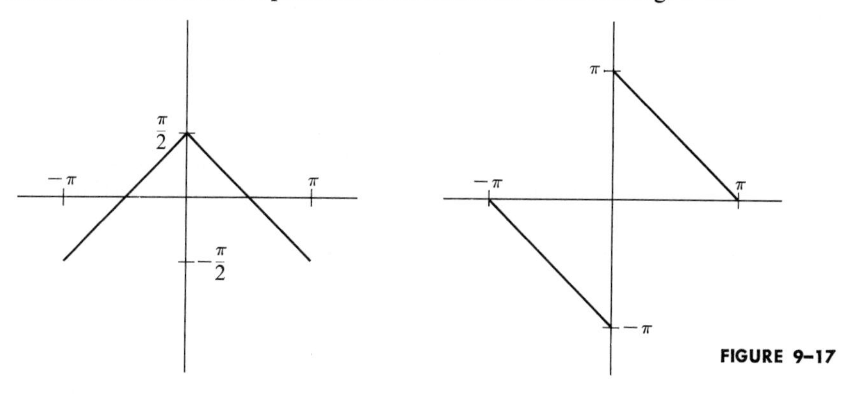

FIGURE 9–17

23. It can be shown that the series

$$\frac{1}{2} + \frac{2}{\pi} \sum_{k=1}^{\infty} \frac{\sin (2k - 1)x}{2k - 1}$$

and

$$\frac{2}{\pi} \sum_{k=1}^{\infty} \frac{(-1)^{k+1}}{k} \sin kx$$

are, respectively, the Fourier series expansions of the functions

$$f_1(x) = \begin{cases} 0, & -\pi < x < 0, \\ 1, & 0 < x < \pi, \end{cases} \quad \text{and} \quad f_2(x) = \frac{x}{\pi}, \quad -\pi < x < \pi.$$

Use these series, and Exercise 21, to find the Fourier series expansion of each of the following functions.

(a) $f(x) = \begin{cases} \frac{1}{2}, & -\pi < x < 0 \\ -\frac{1}{2}, & 0 < x < \pi \end{cases}$

(b) $f(x) = \begin{cases} x + \pi, & -\pi < x < 0 \\ x, & 0 < x < \pi \end{cases}$

(c) $f(x) = \begin{cases} -x, & -\pi < x < 0 \\ -x + 2\pi, & 0 < x < \pi \end{cases}$

24. What is the Fourier series expansion of $2 + 7 \cos 3x - 4 \sin 2x$, considered as a function in $\mathcal{PC}[-\pi, \pi]$?

25. Without resorting to integration, find the Fourier series expansion of each of the following functions in $\mathcal{PC}[-\pi, \pi]$.

(a) $\sin^2 x$ (b) $\sin x \cos x$

(c) $\sin^3 x$ (d) $\sin x \left(\cos^2 \frac{x}{2}\right)$

(e) $\cos 3x \cos 2x$ (f) $\cos^4 x$

26. Show that Parseval's equality assumes the form

$$\frac{1}{\pi}\int_{-\pi}^{\pi} f(x)^2\, dx = \frac{a_0^2}{2} + \sum_{k=1}^{\infty} (a_k^2 + b_k^2),$$

with respect to the basis $1, \cos x, \sin x, \ldots$ in $\mathcal{PC}[-\pi, \pi]$, where the a_k and b_k are the Fourier coefficients of f.

27. (a) Find the Fourier series expansion of the function $f(-x)$ in terms of the expansion of $f(x)$.

(b) Use the series found above, together with the appropriate result from Section 9–3 to find the Fourier series expansions of f_E and f_O, the even and odd parts of f.

(c) Under the assumption that the functions $1, \cos x, \sin x, \ldots$ form a basis for $\mathcal{PC}[-\pi, \pi]$, use the results in (a) and (b) to find bases for the subspaces $\mathcal{E}[-\pi, \pi]$ and $\mathcal{O}[-\pi, \pi]$ of $\mathcal{PC}[-\pi, \pi]$.

28. Use the results in Exercise 27, together with Exercise 21 and the Fourier series expansion of e^x, $-\pi < x < \pi$, to find the Fourier series expansions of $\sinh x$ and $\cosh x$.

*29. (a) Prove that the functions $1, \cos x, \sin x, \ldots, \cos kx, \sin kx, \ldots$ are orthogonal in $\mathcal{PC}[a, a + 2\pi]$ for any real number a.

(b) Under the assumption that the functions in (a) are a basis for $\mathcal{PC}[-\pi, \pi]$, prove that they also form a basis for $\mathcal{PC}[a, a + 2\pi]$ for any real number a. [*Hint:* Use Parseval's equality.]

9–5 SINE AND COSINE SERIES

In the examples of the last section we took advantage of the fact that the functions considered were even or odd to simplify the task of finding their Fourier series expansions. This technique can be exploited more often than one might expect, and is of sufficient importance to be brought out into the open.

Specifically, if f is an *even* function in $\mathcal{PC}[-\pi, \pi]$, then, for all values of k, $f(x) \cos kx$ is even, and $f(x) \sin kx$ is odd. Thus, by (9–5) and (9–6),

$$\int_{-\pi}^{\pi} f(x) \cos kx\, dx = 2\int_{0}^{\pi} f(x) \cos kx\, dx,$$

$$\int_{-\pi}^{\pi} f(x) \sin kx\, dx = 0,$$

and it follows that *the Fourier series expansion of an even function in* $\mathcal{PC}[-\pi, \pi]$

involves only cosine terms and may be computed according to the formula

$$f(x) = \frac{a_0}{2} + \sum_{k=1}^{\infty} a_k \cos kx \quad \text{(mean)}, \tag{9–17}$$

where

$$a_k = \frac{2}{\pi} \int_0^{\pi} f(x) \cos kx \, dx. \tag{9–18}$$

A similar argument shows that the *Fourier series expansion of an odd function in* $\mathcal{PC}[-\pi, \pi]$ *involves only sine terms, and may be computed according to the formula*

$$f(x) = \sum_{k=1}^{\infty} b_k \sin kx \quad \text{(mean)}, \tag{9–19}$$

where

$$b_k = \frac{2}{\pi} \int_0^{\pi} f(x) \sin kx \, dx. \tag{9–20}$$

Actually these results are more than mere formulas. For if we combine them with the fact that the functions

$$1, \cos x, \sin x, \cos 2x, \sin 2x, \ldots$$

are a basis for $\mathcal{PC}[-\pi, \pi]$, they imply, in turn, that

$$1, \cos x, \cos 2x, \ldots$$

is a basis for the space of piecewise continuous *even* functions on $[-\pi, \pi]$, and that

$$\sin x, \sin 2x, \ldots$$

is a basis for the space of piecewise continuous *odd* functions on $[-\pi, \pi]$. Indeed, this is the import of the assertion that the series in (9–17) and (9–19) converge in the mean.

In applications of the theory of Fourier series one frequently needs to obtain a series expansion for a piecewise continuous function f which is defined only on the interval $[0, \pi]$. One way of doing this is to *extend* f to the entire interval $[-\pi, \pi]$ (where by this we mean that a function F is defined on $[-\pi, \pi]$ in such a way that F coincides with f on $[0, \pi]$), and then expand F as a Fourier series. In those cases where f is reasonably well-behaved, Theorem 9–1 guarantees that the Fourier series expansion of F will be a good approximation to f on $[0, \pi]$.

The crux of this method concerns the manner in which f is extended to $[-\pi, \pi]$. This, of course, can be done in any way whatsoever (so long as the resulting function belongs to $\mathcal{PC}[-\pi, \pi]$), but the following two extensions are the most convenient and important. The first is the so-called *even extension* of f, denoted E_f, and defined by

$$E_f(x) = \begin{cases} f(x), & 0 \le x \le \pi, \\ f(-x), & -\pi \le x < 0, \end{cases} \tag{9–21}$$

while the second is the *odd extension* of f, denoted O_f, and defined by

$$O_f(x) = \begin{cases} f(x), & 0 < x \leq \pi, \\ -f(-x), & -\pi \leq x < 0. \end{cases} \tag{9–22}$$

Figure 9–18 illustrates the even and odd extensions of a particular f, and furnishes visual evidence of the easily proved assertion that E_f is even and O_f is odd for any f. This being the case, we can use Formulas (9–17) through (9–20) to obtain the Fourier series expansions of E_f and O_f. They are

$$\begin{aligned} E_f(x) &= \frac{a_0}{2} + \sum_{k=1}^{\infty} a_k \cos kx \quad \text{(mean)}, \\ a_k &= \frac{2}{\pi} \int_0^{\pi} f(x) \cos kx \, dx, \end{aligned} \tag{9–23}$$

and

$$\begin{aligned} O_f(x) &= \sum_{k=1}^{\infty} b_k \sin kx \quad \text{(mean)}, \\ b_k &= \frac{2}{\pi} \int_0^{\pi} f(x) \sin kx \, dx. \end{aligned} \tag{9–24}$$

These series are called, respectively, the *Fourier cosine* and *Fourier sine series expansions of f*, a mild misuse of terminology which rarely causes any misunderstanding. Unfortunately, the somewhat benighted term "half-range expansion" is also used in this context.

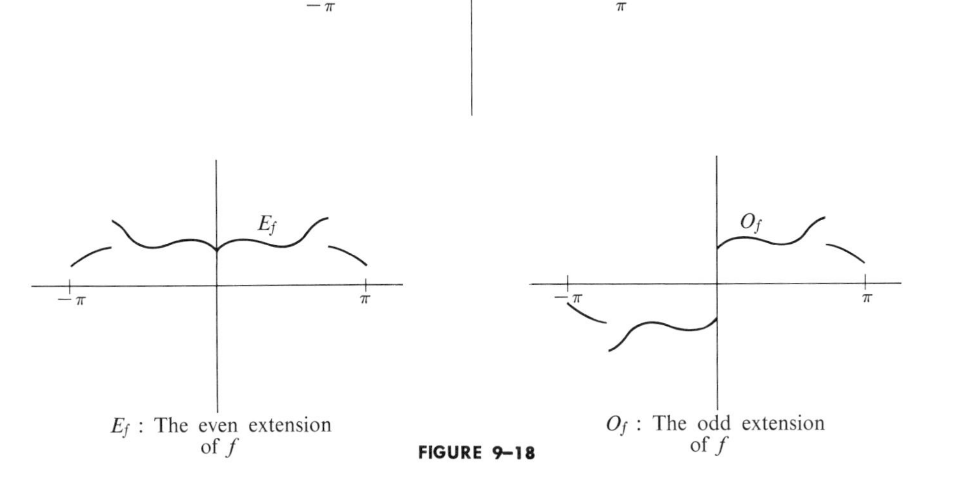

FIGURE 9–18

As an example, we compute the Fourier cosine and sine series for the function

$$f(x) = x^2, \quad 0 < x < \pi.$$

Here

$$E_f(x) = x^2, \quad -\pi < x < \pi,$$

and integration by parts yields

$$\begin{aligned} a_k &= \frac{2}{\pi}\int_0^\pi x^2 \cos kx\, dx \\ &= -\frac{4}{\pi k}\int_0^\pi x \sin kx\, dx \\ &= -\frac{4}{\pi k}\left[-\frac{x\cos kx}{k}\Big|_0^\pi + \frac{1}{k}\int_0^\pi \cos kx\, dx\right] \\ &= \frac{4}{k^2}\cos k\pi \\ &= (-1)^k \frac{4}{k^2}, \quad k \neq 0. \end{aligned}$$

Finally,

$$a_0 = \frac{2}{\pi}\int_0^\pi x^2\, dx = \frac{2\pi^2}{3},$$

and we have

$$E_f(x) = \frac{\pi^2}{3} - 4\left(\cos x - \frac{\cos 2x}{2^2} + \frac{\cos 3x}{3^2} - \frac{\cos 4x}{4^2} + \cdots\right) \quad \text{(mean)}.$$

To compute the series expansion of O_f we use (9–24) with $f(x) = x^2$. This gives

$$\begin{aligned} b_k &= \frac{2}{\pi}\int_0^\pi x^2 \sin kx\, dx \\ &= \frac{2}{\pi}\left[-\frac{x^2\cos kx}{k}\Big|_0^\pi + \frac{2}{k}\int_0^\pi x\cos kx\, dx\right] \\ &= -\frac{2\pi\cos k\pi}{k} + \frac{4}{k\pi}\left[\frac{x\sin kx}{k}\Big|_0^\pi - \frac{1}{k}\int_0^\pi \sin kx\, dx\right] \\ &= -\frac{2\pi\cos k\pi}{k} + \frac{4}{k^3\pi}\cos kx\Big|_0^\pi \\ &= -\frac{2\pi\cos k\pi}{k} + \frac{4}{k^3\pi}(\cos k\pi - 1) \\ &= \begin{cases} \dfrac{2\pi}{k} - \dfrac{8}{k^3\pi}, & k = 1, 3, 5, \ldots, \\[2ex] -\dfrac{2\pi}{k}, & k = 2, 4, 6, \ldots. \end{cases} \end{aligned}$$

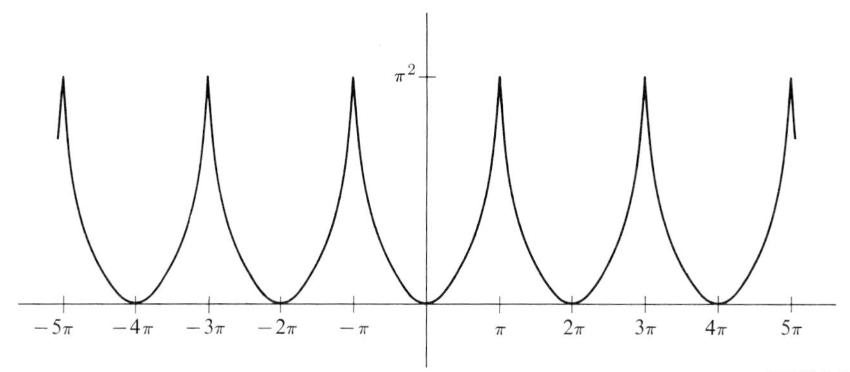

FIGURE 9–19

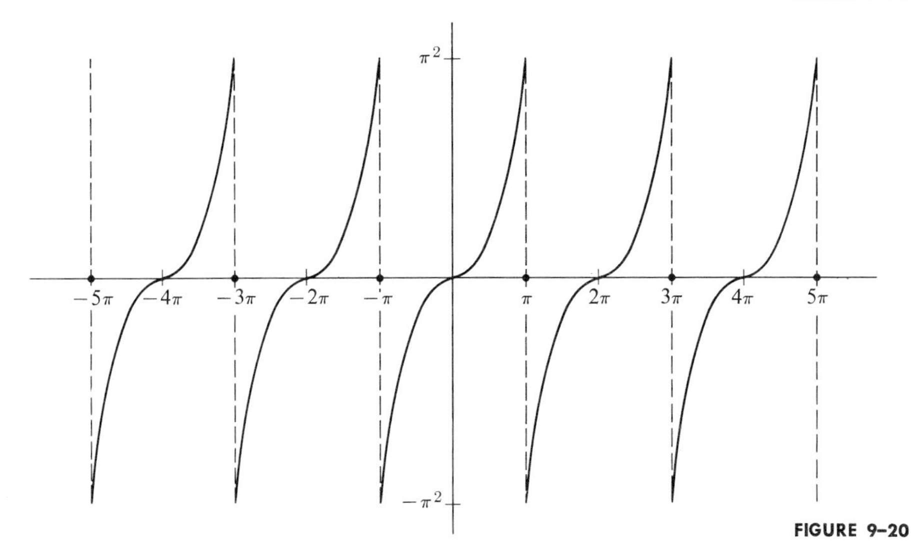

FIGURE 9–20

Thus

$$O_f(x) = 2\pi\left(\sin x - \frac{\sin 2x}{2} + \frac{\sin 3x}{3} - \cdots\right)$$
$$- \frac{8}{\pi}\left(\sin x + \frac{\sin 3x}{3^3} + \frac{\sin 5x}{5^3} + \cdots\right) \quad \text{(mean)}.$$

The graphs of these two series are shown in Figs. 9–19 and 9–20, respectively.

EXERCISES

1. (a) Find the even and odd extensions of each of the following functions in $\mathcal{PC}[0, \pi]$, and sketch their graphs on the interval $[-\pi, \pi]$.

   (i) $f(x) = 1$ (ii) $f(x) = x^2$ (iii) $f(x) = \pi - x$ (iv) $f(x) = e^x$

   (b) Sketch the graphs of the Fourier sine and cosine series for each of the functions in (a).

2. Let f be an arbitrary function in $\mathcal{PC}[0, \pi]$. Prove that E_f is an even function in $\mathcal{PC}[-\pi, \pi]$, and that O_f is odd.
3. Find the Fourier sine series expansion of the function

$$f(x) = \cos x, \quad 0 < x < \pi,$$

and use this result to deduce that

$$\frac{\sqrt{2}\,\pi}{16} = \frac{1}{2^2 - 1} - \frac{3}{6^2 - 1} + \frac{5}{10^2 - 1} - \frac{7}{14^2 - 1} + \cdots.$$

4. Find the Fourier cosine series expansion of the function

$$f(x) = \sin x, \quad 0 < x < \pi,$$

and sketch the graph of this series.
5. Find the Fourier sine series expansion of e^x, $0 < x < \pi$.
6. Find the Fourier cosine series expansion of e^x, $0 < x < \pi$.
7. Find the Fourier sine series expansion of the function

$$f(x) = \pi - x, \quad 0 < x < \pi.$$

8. Let f be a function in $\mathcal{PC}[0, \pi]$ which is symmetric about the line $x = \pi/2$. Show that the only nonzero terms in the Fourier sine series expansion of f are of the form

$$B_k \sin kx, \quad k \text{ odd},$$

and find a formula for B_k.
9. Let f be the function in $\mathcal{PC}[0, \pi]$ which is obtained by reflecting the function x^2, $0 < x < \pi/2$, across the line $x = \pi/2$. Use the results of Exercise 8 to find the Fourier sine series expansion of f.
10. Find the Fourier sine series expansion of the function

$$f(x) = x^2 - \pi x, \quad 0 < x < \pi.$$

11. Let f be a function in $\mathcal{PC}[0, \pi]$ with the property that

$$f\left(\frac{\pi}{2} + x\right) = -f\left(\frac{\pi}{2} - x\right), \quad 0 < x < \pi/2.$$

Show that the only nonzero terms in the Fourier cosine series expansion of f are of the form

$$A_k \cos kx, \quad k \text{ odd},$$

and find a formula for A_k.
12. Use the results of Exercise 11 to find the Fourier cosine series expansion of the function

$$f(x) = \begin{cases} 1, & 0 < x < \pi/2, \\ -1, & \pi/2 < x < \pi. \end{cases}$$

13. Use the results of Exercise 11 to find the Fourier cosine series expansion of the function

$$f(x) = \sin\left(x - \frac{\pi}{2}\right), \quad 0 < x < \pi.$$

*14. (a) Prove that the functions

$$\sin x, \sin 2x, \ldots, \sin kx, \ldots$$

form an orthogonal set in $\mathcal{PC}[0, \pi]$.

(b) Use the fact that the functions $1, \cos x, \sin x, \ldots$ form a basis for $\mathcal{PC}[-\pi, \pi]$ to prove that the set of functions in (a) is a basis for $\mathcal{PC}[0, \pi]$.

(c) Prove that the functions

$$1, \cos x, \cos 2x, \ldots$$

also form a basis for $\mathcal{PC}[0, \pi]$.

9–6 CHANGE OF INTERVAL

Up to this point we have dealt exclusively with functions on the intervals $[-\pi, \pi]$ and $[0, \pi]$. For many purposes, however, this setting is too restrictive, and we now propose to generalize our results to an arbitrary interval $[a, b]$. But rather than begin at once with the most general case, it will prove simpler if we first consider intervals of the form $[-p, p]$ and their associated Euclidean spaces $\mathcal{PC}[-p, p]$. For here the situation can be handled with dispatch.

Indeed, it is all but obvious that the functions

$$1, \cos\frac{\pi x}{p}, \sin\frac{\pi x}{p}, \cos\frac{2\pi x}{p}, \sin\frac{2\pi x}{p}, \ldots \tag{9–25}$$

are mutually orthogonal in $\mathcal{PC}[-p, p]$ (Exercise 1 below).* Moreover, just as in the case where $p = \pi$, it can be shown that these functions are a *basis* for this space, and hence that their associated orthogonal series (which, by the way, are still called Fourier series) converge in the mean. And finally, with due allowance being made for the length of the interval, all of our earlier remarks concerning pointwise convergence are valid in this setting.

To obtain formulas for the Fourier coefficients of a function in $\mathcal{PC}[-p, p]$ we note that

$$\int_{-p}^{p} dx = 2p,$$

and

$$\int_{-p}^{p} \cos^2\frac{k\pi x}{p}\,dx = \int_{-p}^{p} \sin^2\frac{k\pi x}{p}\,dx = p.$$

Thus, by Formula (8–22),

$$f(x) = \frac{a_0}{2} + \sum_{k=1}^{\infty}\left(a_k\cos\frac{k\pi x}{p} + b_k\sin\frac{k\pi x}{p}\right) \quad \text{(mean)}, \tag{9–26}$$

* Actually, the entire issue comes down to making a change in scale on the x-axis by substituting $\pi x/p$ for x in the functions used earlier.

where

$$a_k = \frac{1}{p}\int_{-p}^{p} f(x)\cos\frac{k\pi x}{p}\,dx,$$
$$b_k = \frac{1}{p}\int_{-p}^{p} f(x)\sin\frac{k\pi x}{p}\,dx \tag{9–27}$$

for all k. And with this we are done.

The above discussion can easily be adapted to handle the Euclidean space $\mathcal{PC}[a, b]$. Indeed, if we set $2p = b - a$, so that $[a, b] = [a, a + 2p]$, the functions in (9–25) are also a basis for $\mathcal{PC}[a, a + 2p]$. This leads at once to the following formulas for computing the Fourier series expansion of a function f in $\mathcal{PC}[a, b]$:

$$f(x) = \frac{a_0}{2} + \sum_{k=1}^{\infty}\left(a_k\cos\frac{2k\pi x}{b-a} + b_k\sin\frac{2k\pi x}{b-a}\right) \quad \text{(mean)}, \tag{9–28}$$

where

$$a_k = \frac{2}{b-a}\int_a^b f(x)\cos\frac{2k\pi x}{b-a}\,dx,$$
$$b_k = \frac{2}{b-a}\int_a^b f(x)\sin\frac{2k\pi x}{b-a}\,dx, \tag{9–29}$$

for all k.

EXAMPLE 1. Find the Fourier series expansion in $\mathcal{PC}[0, 1]$ of the function $f(x) = x$.

Here $b - a = 1$, and (9–29) becomes

$$a_k = 2\int_0^1 x\cos(2k\pi x)\,dx,$$
$$b_k = 2\int_0^1 x\sin(2k\pi x)\,dx.$$

Integration by parts now yields

$$a_0 = 1, \quad a_k = 0, \quad k \neq 0, \quad b_k = -\frac{1}{k\pi}\cdot$$

Hence

$$f(x) = \frac{1}{2} - \frac{1}{\pi}\left(\sin 2\pi x + \frac{\sin 4\pi x}{2} + \frac{\sin 6\pi x}{3} + \cdots\right) \quad \text{(mean)}.$$

The graph of this series is given in Fig. 9–21.

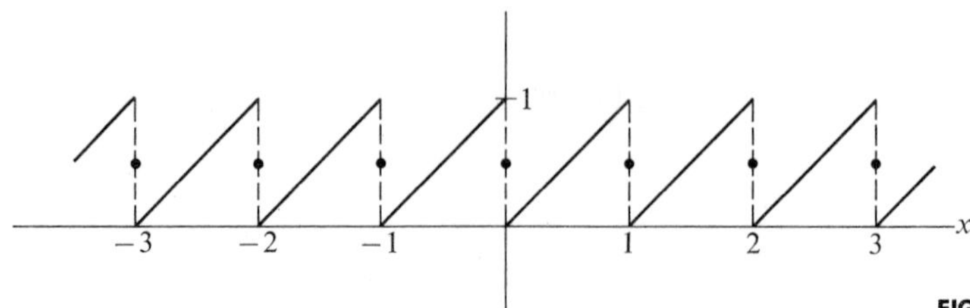

FIGURE 9–21

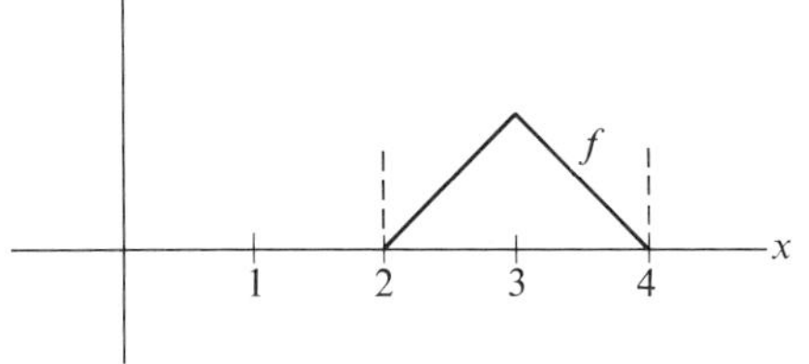

FIGURE 9–22

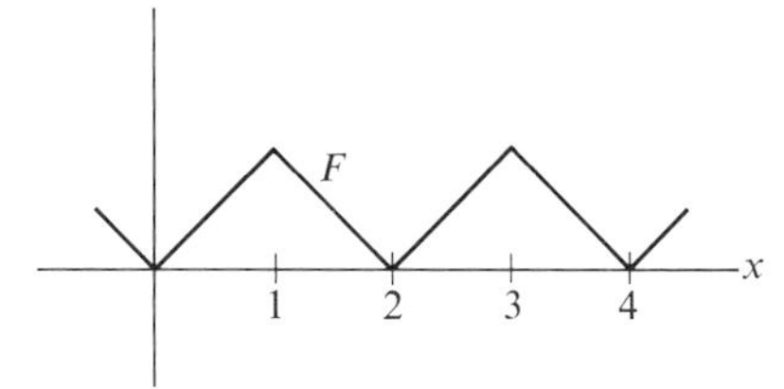

FIGURE 9–23

EXAMPLE 2. Find the Fourier series expansion of the function f shown in Fig. 9–22.

In this case,

$$f(x) = \begin{cases} x - 2, & 2 \le x \le 3, \\ 4 - x, & 3 \le x \le 4, \end{cases}$$

and Formulas (9–29) yield

$$\begin{aligned} a_k &= \int_2^4 f(x) \cos k\pi x \, dx \\ &= \int_2^3 (x - 2) \cos k\pi x \, dx + \int_3^4 (4 - x) \cos k\pi x \, dx, \\ b_k &= \int_2^4 f(x) \sin k\pi x \, dx \\ &= \int_2^3 (x - 2) \sin k\pi x \, dx + \int_3^4 (4 - x) \sin k\pi x \, dx. \end{aligned}$$

Although these integrals can be evaluated directly, the computations can be considerably simplified by invoking the following argument.

Let F denote the periodic extension of f to the entire x-axis (Fig. 9–23). Then the functions $F(x) \cos k\pi x$ and $F(x) \sin k\pi x$ are periodic with period 2, and we have

$$\begin{aligned} \int_a^{a+2} F(x) \cos k\pi x \, dx &= \int_2^4 f(x) \cos k\pi x \, dx, \\ \int_a^{a+2} F(x) \sin k\pi x \, dx &= \int_2^4 f(x) \sin k\pi x \, dx, \end{aligned} \tag{9–30}$$

for any real number a. [At this point we are using the obvious fact that if g is piecewise continuous on $(-\infty, \infty)$ with period $2p$, then

$$\int_a^{a+2p} g(x) \, dx = \int_b^{b+2p} g(x) \, dx$$

for any pair of real numbers a, b.] We now set $a = -1$ in (9–30) to obtain

$$\begin{aligned} a_k &= \int_{-1}^1 F(x) \cos k\pi x \, dx, \\ b_k &= \int_{-1}^1 F(x) \sin k\pi x \, dx. \end{aligned}$$

But on the interval $[-1, 1]$, F coincides with the even function $|x|$. Hence $b_k = 0$ for all k, and $a_k = 2\int_0^1 x \cos k\pi x \, dx$. Thus

$$a_0 = 1, \qquad a_k = \begin{cases} -\dfrac{4}{k^2\pi^2}, & k \text{ odd}, \\ 0, & k \text{ even}, \quad k \neq 0, \end{cases}$$

and the Fourier series expansion of f is

$$f(x) = \frac{1}{2} - \frac{4}{\pi^2}\left(\cos \pi x + \frac{\cos 3\pi x}{3^2} + \frac{\cos 5\pi x}{5^2} + \cdots\right) \quad \text{(mean)}.$$

The student will find this technique useful in some of the following exercises.

EXERCISES

1. Prove that the functions

$$1, \cos\frac{\pi x}{p}, \sin\frac{\pi x}{p}, \ldots, \cos\frac{k\pi x}{p}, \sin\frac{k\pi x}{p}, \ldots$$

are mutually orthogonal in $\mathcal{PC}[-p, p]$.

2. Let f be piecewise continuous and periodic on the entire x-axis, and suppose that f has period $2p$. Prove that

$$\int_a^{a+2p} f(x)\, dx = \int_b^{b+2p} f(x)\, dx$$

for any pair of real numbers a, b.

3. Find the Fourier series expansion of the function in $\mathcal{PC}[-2, 2]$ defined by

$$f(x) = \begin{cases} 0, & -2 < x < -1, \\ |x|, & -1 < x < 1, \\ 0, & 1 < x < 2. \end{cases}$$

Sketch the graph of the series.

4. Find the Fourier series expansion of the function in $\mathcal{PC}[1, 3]$ defined by

$$f(x) = \begin{cases} 2 - x, & 1 < x < 2, \\ x - 2, & 2 < x < 3. \end{cases}$$

5. Find the Fourier series expansion of $\sin x$ as a function in $\mathcal{PC}[0, \pi/2]$, and sketch the graph of the series.
6. Find the Fourier series expansion of $\cos x$ as a function in $\mathcal{PC}[\pi/2, 3\pi/2]$, and sketch the graph of the series.
7. Find the Fourier series expansion of the function

$$f(x) = \begin{cases} 1, & 8 < x < 9, \\ 10 - x, & 9 < x < 10. \end{cases}$$

8. Find the Fourier series expansion of the function

$$f(x) = \begin{cases} 1, & 2 < x < 3, \\ 4 - x, & 3 < x < 4, \\ x - 4, & 4 < x < 5, \\ 1, & 5 < x < 6. \end{cases}$$

9. Find a Fourier series which contains only sine terms and which converges pointwise to the function $x - 1$ for $1 < x < 2$.

10. Find a Fourier series which contains only cosine terms and which converges pointwise to the function $x - 1$ for $1 < x < 2$.

11. Find a Fourier series which contains only sine terms of odd "degree" and which converges pointwise to the function $x^2 - 4$ when $2 < x < 3$. [*Hint*: See Exercise 8, Section 9–5.]

12. Let f be a piecewise continuous function of period 2π defined on the entire x-axis, and suppose that the Fourier series expansion of f is

$$\frac{a_0}{2} + \sum_{k=1}^{\infty} (a_k \cos kx + b_k \sin kx).$$

Let the Fourier series expansion of $f(x + \pi)$ be

$$\frac{A_0}{2} + \sum_{k=1}^{\infty} (A_k \cos kx + B_k \sin kx).$$

Show that

$$A_k = (-1)^k a_k$$

and

$$B_k = (-1)^k b_k.$$

13. Given that the Fourier series expansion of x, $-\pi < x < \pi$, is

$$2\left[\sin x - \frac{\sin 2x}{2} + \frac{\sin 3x}{3} - \cdots\right],$$

use Exercise 12 to find the Fourier series expansion of the function

$$f(x) = \begin{cases} x + \pi, & -\pi < x < 0, \\ x - \pi, & 0 < x < \pi. \end{cases}$$

14. Let f be piecewise continuous on $(-\infty, \infty)$, and assume that $f(x + \pi) = -f(x)$ for all x where f is defined.

(a) Prove that f is periodic with period 2π.

(b) Show that the Fourier series for f has only odd terms.

(c) Prove, conversely, that $f(x) = -f(x + \pi)$ whenever the Fourier series for f has only terms of odd degree.

(d) What can be said about a function whose Fourier series contains only even terms? Why is this situation not particularly interesting?

Remark: Functions satisfying $f(x) = -f(x + \pi)$ on $(-\infty, \infty)$ are said to possess *half-wave symmetry* and are of interest in electrical engineering.

*9–7 THE BASIS THEOREM†

We have already had occasion to remark that the entire theory of Fourier series ultimately rests upon the fact that the set of functions

$$1, \cos x, \sin x, \ldots, \cos kx, \sin kx, \ldots$$

is a basis for the Euclidean space $\mathcal{PC}[-\pi, \pi]$. As we shall see, this result can be made to depend upon one of the really fundamental theorems in analysis, the famous Weierstrass Approximation Theorem. This theorem, whose proof will be given in Section 10–8, is usually stated in one of two equivalent forms, the first of which involves trigonometric polynomials, the second ordinary polynomials. For our present purposes the statement involving trigonometric polynomials is the more appropriate.

Theorem 9–3. (Weierstrass Approximation Theorem.) *Let f be a continuous function on the interval $[-\pi, \pi]$, and suppose that $f(-\pi) = f(\pi)$. Then, given any real number $\epsilon > 0$, there exists a trigonometric polynomial*

$$T(x) = A_0 + \sum_{k=1}^{N(\epsilon)} (A_k \cos kx + B_k \sin kx)$$

such that

$$|T(x) - f(x)| < \epsilon$$

for every x in $[-\pi, \pi]$.

Descriptively, this theorem asserts that the graph of $T(x)$ lies between the graphs of $f + \epsilon$ and $f - \epsilon$ *throughout the entire interval* $[-\pi, \pi]$, as illustrated in Fig. 9–24. The notation $N(\epsilon)$ in the formula for $T(x)$ is a reflection of the fact that in general the number of terms in this trigonometric polynomial will depend upon ϵ, increasing as ϵ becomes small.

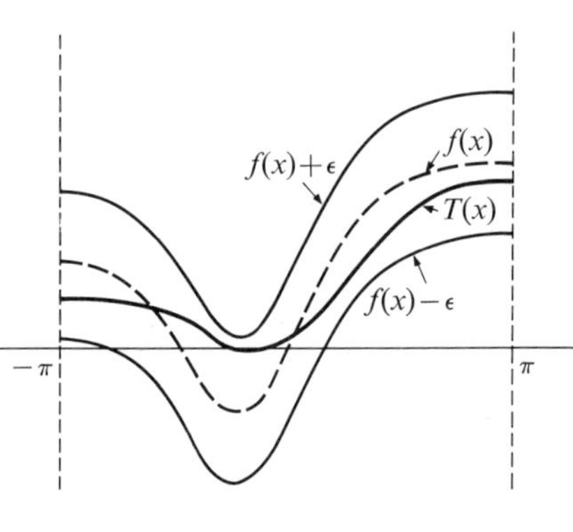

FIGURE 9–24

Theorem 9–3 is often stated in the following terms: "*Any* **continuous** *periodic function can be* **uniformly** *approximated by trigonometric polynomials*," phraseology which requires some explanation. In the first place, we have already observed that any function f in $\mathcal{PC}[-\pi, \pi]$ can be extended to the entire real line in such a way that the extended function is periodic. Now if f happens to be continuous *and* if $f(-\pi) = f(\pi)$, it is clear that the

† The discussion in this section uses the notion of a closed subspace defined in Section 8–5.

periodic extension F of f will be continuous on the whole real line (Fig. 9–25). It follows that a trigonometric polynomial which approximates such a function everywhere in $[-\pi, \pi]$ will actually approximate the periodic extension of f. This accounts for the term "periodic" in the above statement.

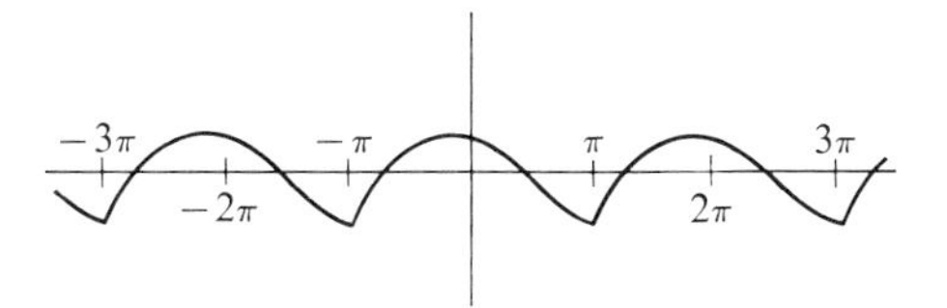

FIGURE 9–25

As far as the word "uniform" is concerned, it will suffice for now to remark that it is used to call attention to the fact that the approximation of f by T holds for *all* x in the interval $[-\pi, \pi]$.

We are now ready to prove that the set 1, $\cos x$, $\sin x, \ldots$, which we shall denote by $\mathcal{B}$, is a basis for $\mathcal{PC}[-\pi, \pi]$. To do so we introduce the subspace $\mathcal{P}[-\pi, \pi]$ consisting of all *continuous* functions in $\mathcal{PC}[-\pi, \pi]$ which satisfy the equation $f(-\pi) = f(\pi)$, and establish two preliminary results concerning this subspace. The first is

Theorem 9–4. *Every function in $\mathcal{P}[-\pi, \pi]$ belongs to $\overline{\mathcal{S}}(\mathcal{B})$, the closure of $\mathcal{B}$ in $\mathcal{PC}[-\pi, \pi]$.*

Proof. Recall that $\overline{\mathcal{S}}(\mathcal{B})$ is, by definition, the subspace of $\mathcal{PC}[-\pi, \pi]$ consisting of those functions which belong to $\mathcal{S}(\mathcal{B})$ or are limits *in the mean* of sequences of functions in $\mathcal{S}(\mathcal{B})$, or both.* Thus to prove the theorem we must show that every f in $\mathcal{P}[-\pi, \pi]$ can be approximated arbitrarily closely *in the mean* by a trigonometric polynomial.

To accomplish this we apply the approximation theorem and, for each integer $k > 0$, find a trigonometric polynomial T_k such that

$$|T_k(x) - f(x)| < \frac{1}{k\sqrt{2\pi}}$$

for *all* x in $[-\pi, \pi]$. Then

$$\begin{aligned} \|T_k - f\| &= \left(\int_{-\pi}^{\pi} [T_k(x) - f(x)]^2\,dx\right)^{1/2} \\ &< \left(\int_{-\pi}^{\pi} \frac{dx}{2\pi k^2}\right)^{1/2} \\ &= \frac{1}{k}\cdot \end{aligned}$$

---

* $\mathcal{S}(\mathcal{B})$ is, of course, the set of *all* trigonometric polynomials.

It follows at once that the sequence $\{T_k\}$, $k = 1, 2, \ldots$, so obtained, converges in the mean to f, and the theorem is proved. ▌

The second result is as follows:

Theorem 9–5. *Given any real number $\epsilon > 0$, and any f in $\mathcal{PC}[-\pi, \pi]$, there exists a function g in $\mathcal{P}[-\pi, \pi]$ such that*

$$\|f - g\| < \epsilon.$$

In other words, any function in $\mathcal{PC}[-\pi, \pi]$ can be approximated arbitrarily closely in the mean by a function in $\mathcal{P}[-\pi, \pi]$.

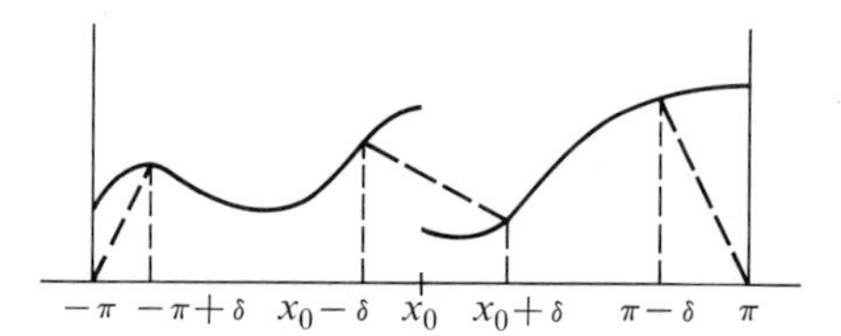

FIGURE 9–26

Proof. The function g is constructed from f by "mending the discontinuities" of f as suggested in Fig. 9–26. Specifically, suppose that f is a function in $\mathcal{PC}[-\pi, \pi]$ which has a single jump discontinuity at x_0. In this case, g is the *continuous* function obtained by redefining f near $-\pi$, x_0, and π as shown by the broken lines in Fig. 9–26. (Note that this has been done in such a way that $g(-\pi) = g(\pi)$, so that g does indeed belong to $\mathcal{P}[-\pi, \pi]$.)

Calculating $\|f - g\|^2$ we find that

$$\begin{aligned}\|f - g\|^2 &= \int_{-\pi}^{\pi} [f(x) - g(x)]^2\, dx \\ &= \int_{-\pi}^{-\pi+\delta} [f(x) - g(x)]^2\, dx + \int_{x_0-\delta}^{x_0+\delta} [f(x) - g(x)]^2\, dx \\ &\qquad + \int_{\pi-\delta}^{\pi} [f(x) - g(x)]^2\, dx,\end{aligned}$$

since $f(x) = g(x)$ whenever x belongs to the intervals $[-\pi + \delta, x_0 - \delta]$ and $[x_0 + \delta, \pi - \delta]$. Now let M be an upper bound for $|f(x)|$ on $[-\pi, \pi]$; i.e., $|f(x)| \leq M$ for all x in $[-\pi, \pi]$. Then $|f(x) - g(x)|^2 \leq (2M)^2$ everywhere on $[-\pi, \pi]$, and it follows that

$$\|f - g\|^2 \leq 4M^2(\delta + 2\delta + \delta) = 16M^2\delta.$$

To make this quantity less than any preassigned $\epsilon > 0$, it obviously suffices to choose $\delta < \epsilon/16M^2$. Thus the theorem holds for functions with a single discontinuity. To give the proof for *any* f in $\mathcal{PC}[-\pi, \pi]$ just repeat this process at each point of discontinuity. ▌

Before stating the basis theorem, let us pause a moment to take stock of the situation. The preceding result tells us that any function in $\mathcal{PC}[-\pi, \pi]$ can be approximated arbitrarily closely in the mean by a function in $\mathcal{P}[-\pi, \pi]$. But Theorem 9–4 asserts that every function in $\mathcal{P}[-\pi, \pi]$ can in turn be approximated arbitrarily closely in the mean by a trigonometric polynomial. Taken together these two statements ought to yield a proof of the fact that $\mathcal{B}$ is a basis for $\mathcal{PC}[-\pi, \pi]$. They do, in the following manner.

Theorem 9–6. *$\mathcal{B}$ is a basis for $\mathcal{PC}[-\pi, \pi]$.*

Proof. We must show that every f in $\mathcal{PC}[-\pi, \pi]$ is the limit of a sequence of trigonometric polynomials. But, by Theorem 9–5 we can find functions f_k in $\mathcal{P}[-\pi, \pi]$ such that

$$\|f - f_k\| < \frac{1}{2k}, \quad k = 1, 2, \ldots,$$

while by Theorem 9–4 we can find trigonometric polynomials T_k such that

$$\|f_k - T_k\| < \frac{1}{2k}.$$

Hence, by the triangle inequality,

$$\begin{aligned} \|f - T_k\| &= \|(f - f_k) + (f_k - T_k)\| \\ &\leq \|f - f_k\| + \|f_k - T_k\| \\ &< \frac{1}{2k} + \frac{1}{2k} = \frac{1}{k}, \end{aligned}$$

and it follows that the sequence $\{T_k\}$ converges in the mean to f. ▌

For later purposes we also record the following easy consequence of the basis theorem:

Corollary 9–1. (Parseval's Equality.) *If f is any function in $\mathcal{PC}[-\pi, \pi]$, then*

$$\frac{1}{\pi}\int_{-\pi}^{\pi} f(x)^2\, dx = \frac{a_0^2}{2} + \sum_{k=1}^{\infty} (a_k^2 + b_k^2), \tag{9–31}$$

where the a_k and b_k are the Fourier coefficients of f.

Proof. Since the functions $1, \cos x, \sin x, \ldots$ are a basis for $\mathcal{PC}[-\pi, \pi]$, Parseval's equality is satisfied, and we have

$$\|f\|^2 = \frac{(f \cdot 1)^2}{\|1\|^2} + \sum_{k=1}^{\infty} \left[\frac{(f \cdot \cos kx)^2}{\|\cos kx\|^2} + \frac{(f \cdot \sin kx)^2}{\|\sin kx\|^2}\right] \tag{9–32}$$

(Formula 8–21). But

$$\frac{(f \cdot 1)^2}{\|1\|^2} = \frac{(\int_{-\pi}^{\pi} f(x)\,dx)^2}{2\pi} = \frac{\pi a_0^2}{2},$$

$$\frac{(f \cdot \cos kx)^2}{\|\cos kx\|^2} = \frac{(\int_{-\pi}^{\pi} f(x)\cos kx\,dx)^2}{\pi} = \pi a_k^2,$$

$$\frac{(f \cdot \sin kx)^2}{\|\sin kx\|^2} = \frac{(\int_{-\pi}^{\pi} f(x)\sin kx\,dx)^2}{\pi} = \pi b_k^2,$$

and (9–31) now follows at once from (9–32). ▌

9–8 ORTHOGONAL SERIES IN TWO VARIABLES

With an eye to later applications we now sketch the theory of series expansions for functions of two variables. A corresponding theory also exists for functions of any finite number of variables, but since we shall have no occasion to use these results their formulation is left to the reader.

Our first task, of course, is to define a function space in which to conduct the discussion. This is accomplished by generalizing the notion of piecewise continuity to accommodate functions of two variables, and then using these functions to construct the two-dimensional analog of the space $\mathcal{PC}[a, b]$, as follows.

A function f is said to be *piecewise continuous* on a rectangle R in the plane if

(i) f is continuous everywhere in and on the boundary of R, with the possible exception of a finite number of points, or along a finite number of simple differentiable arcs, or both;* and

(ii) $\lim_{(x,y)\to(x_0,y_0)} f(x, y)$ exists whenever (x_0, y_0) is a point of discontinuity of f and (x, y) approaches (x_0, y_0) from the interior of any *one* of the regions into which R is separated by the arcs of discontinuity.

Any function which is continuous in and on the boundary of R is piecewise continuous, so that, in particular, such functions as $\sin(mx)\sin(ny)$, $\sin(mx)\cos(ny)$, etc., m and n integers, are piecewise continuous on any rectangle in the plane. However, the set of piecewise continuous functions is clearly larger than the set of continuous functions for any rectangle R. In Fig. 9–27 we have illustrated a rather general piecewise continuous function, in order to dispel any doubts about the nature of such functions. We also remark that it is quite legitimate to consider piecewise continuous functions in regions other than rectangles. The basic definition remains unchanged, except that we replace R by a region whose boundary

---

* A plane curve

$$\left.\begin{aligned} x &= x(t) \\ y &= y(t) \end{aligned}\right\} \quad 0 \le t \le 1,$$

is said to be a *differentiable arc* if the functions $x(t)$, $y(t)$ have continuous derivatives with respect to t. A differentiable arc which does not intersect itself is said to be *simple*.

consists of a finite number of simple differentiable arcs. This not withstanding, we shall restrict our attention to rectangular regions throughout this section. The reasons behind this prejudice in favor of rectangles will become apparent as soon as we state Theorem 9–7.

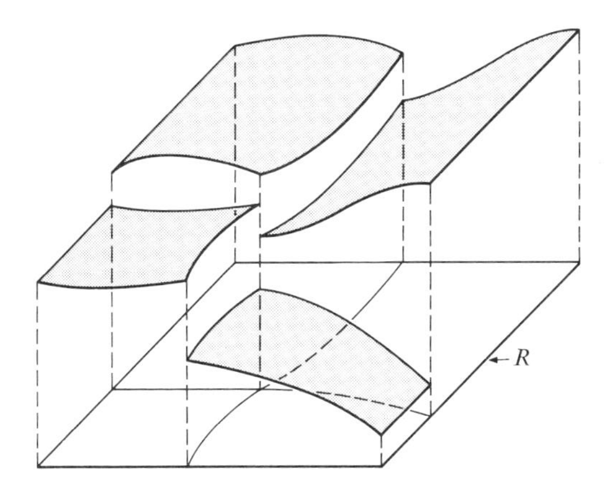

FIGURE 9–27

As with functions of a single variable, we agree to *identify* two piecewise continuous functions whenever they agree everywhere in R except at their points of discontinuity. This done, the totality of piecewise continuous functions on R becomes a Euclidean space under the usual operations of addition and scalar multiplication, and the integral inner product

$$f \cdot g = \iint\limits_R f(x, y)g(x, y)\, dR. \tag{9–33}$$

(The student will recall that when R is the rectangle $a \leq x \leq b$, $c \leq y \leq d$, this double integral may be evaluated by computing either one of the iterated integrals

$$\int_a^b \int_c^d f(x, y)g(x, y)\, dy\, dx \qquad \text{or} \qquad \int_c^d \int_a^b f(x, y)g(x, y)\, dx\, dy.)$$

We shall denote the resulting Euclidean space by $\mathcal{PC}(R)$.

Our next task is to find a basis for $\mathcal{PC}(R)$. And here the fact that R is a rectangle becomes important, for we can then reduce this problem to that of finding a basis for a Euclidean space of the type $\mathcal{PC}[a, b]$. This is the content of

Theorem 9–7. *Let $\{f_i(x)\}$ and $\{g_j(y)\}$ be orthogonal bases for the Euclidean spaces $\mathcal{PC}[a, b]$ and $\mathcal{PC}[c, d]$, respectively. Then the set of all products*

$$\{f_i(x)g_j(y)\}, \quad i = 1, 2, \ldots, \quad j = 1, 2, \ldots, \tag{9–34}$$

is a basis for $\mathcal{PC}(R)$, where R is the rectangle $a \leq x \leq b$, $c \leq y \leq d$.

The proof of this theorem is given below, following the examples. First, however, we observe that the generalized Fourier coefficients of any function F in $\mathcal{PC}(R)$, computed with respect to the functions in (9–34), are

$$\alpha_{ij} = \frac{F \cdot (f_i g_j)}{(f_i g_j) \cdot (f_i g_j)},$$

or, in greater detail,

$$\alpha_{ij} = \frac{\iint\limits_R F(x, y) f_i(x) g_j(y)\, dR}{\iint\limits_R f_i(x)^2 g_j(y)^2\, dR}. \tag{9–35}$$

Thus the series expansion of F can be written as a double series

$$\sum_{i,j=1}^{\infty} \alpha_{ij} f_i(x) g_j(y), \tag{9–36}$$

and Theorem 9–7 allows us to assert that this series converges *in the mean* to F. (The order of summation in such a series is a matter of indifference since the assertion that the functions $f_i g_j$ are a basis for $\mathcal{PC}(R)$ is not affected by the order in which they are displayed.)

EXAMPLE 1. Let R be the rectangle $-\pi \le x \le \pi$, $-\pi \le y \le \pi$. Then the set of functions

$$\begin{array}{ll} \sin mx \, \sin ny, & \sin mx \, \cos qy, \\ \cos px \, \sin ny, & \cos px \, \cos qy, \end{array} \tag{9–37}$$

where m and n range independently over the integers $1, 2, \ldots$, and p and q over the integers $0, 1, 2, \ldots$, is a basis for $\mathcal{PC}(R)$. More generally, the set of functions

$$\begin{array}{ll} \sin \dfrac{m\pi x}{a} \sin \dfrac{n\pi y}{b}, & \sin \dfrac{m\pi x}{a} \cos \dfrac{q\pi y}{b}, \\ \cos \dfrac{p\pi x}{a} \sin \dfrac{n\pi y}{b}, & \cos \dfrac{p\pi x}{a} \cos \dfrac{q\pi y}{b} \end{array} \tag{9–38}$$

is a basis for the Euclidean space of piecewise continuous functions on the rectangle $-a \le x \le a$, $-b \le y \le b$.

EXAMPLE 2. Find the series expansion of the function

$$F(x, y) = xy$$

in the rectangle $-\pi \le x \le \pi$, $-\pi \le y \le \pi$ relative to the basis given in the preceding example.

Here we must evaluate the coefficients

$$\alpha_{mn}, \; \alpha_{mq}, \; \alpha_{pn}, \; \alpha_{pq}$$

of Formula (9–35) for the various functions in (9–37) and the given function F. But since $x \cos px$ and $y \cos qy$ are odd functions of x and y,

$$\int_{-\pi}^{\pi}\int_{-\pi}^{\pi} (x \cos px)(y \sin ny)\, dx\, dy = 0,$$

$$\int_{-\pi}^{\pi}\int_{-\pi}^{\pi} (x \cos px)(y \cos qy)\, dx\, dy = 0,$$

$$\int_{-\pi}^{\pi}\int_{-\pi}^{\pi} (x \sin nx)(y \cos qy)\, dx\, dy = 0.$$

Thus all of the Fourier coefficients of F except the α_{mn} are zero. To evaluate them we note that

$$\int_{-\pi}^{\pi} \sin^2 mx\, dx = \int_{-\pi}^{\pi} \cos^2 ny\, dy = \pi$$

for all positive values of m and n. Hence

$$\begin{aligned}\alpha_{mn} &= \frac{\int_{-\pi}^{\pi}\int_{-\pi}^{\pi} xy \sin mx \sin ny\, dx\, dy}{\int_{-\pi}^{\pi}\int_{-\pi}^{\pi} \sin^2 mx \cos^2 ny\, dx\, dy} \\ &= \frac{1}{\pi^2}\int_{-\pi}^{\pi} x \sin mx\, dx \int_{-\pi}^{\pi} y \sin ny\, dy \\ &= \frac{4}{\pi^2}\int_{0}^{\pi} x \sin mx\, dx \int_{0}^{\pi} y \sin ny\, dy.\end{aligned}$$

But

$$\int_0^{\pi} t \sin kt\, dt = (-1)^{k+1}\frac{\pi}{k},$$

and it follows that

$$\begin{aligned}\alpha_{mn} &= \frac{4}{\pi^2}\left[(-1)^{m+1}\frac{\pi}{m}\right]\left[(-1)^{n+1}\frac{\pi}{n}\right] \\ &= (-1)^{m+n}\frac{4}{mn}\,.\end{aligned}$$

Thus

$$\begin{aligned}xy &= 4\left[\sin x \sin y - \frac{\sin x \sin 2y}{1 \cdot 2} - \frac{\sin 2x \sin y}{2 \cdot 1}\right. \\ &\quad \left. + \frac{\sin x \sin 3y}{1 \cdot 3} + \frac{\sin 2x \sin 2y}{2 \cdot 2} + \frac{\sin 3x \sin y}{3 \cdot 1} - \cdots\right] \\ &= 4 \sum_{m,n=1}^{\infty} (-1)^{m+n} \frac{\sin mx \sin ny}{mn}\,.\end{aligned}$$

Series of this sort are called *double Fourier series*, and, as we shall see, arise in the study of boundary value problems involving partial differential equations.

We now return to the proof of Theorem 9–7. For the sake of simplicity we shall assume that $\{f_i(x)\}$ and $\{g_j(y)\}$ are ortho*normal* bases for $\mathcal{PC}[a, b]$ and $\mathcal{PC}[c, d]$. As we know, this involves no loss of generality since an orthogonal set can always be normalized in the usual fashion. In this case the set of products $\{f_i(x)g_j(y)\}$ is also orthonormal, and hence, of course, orthogonal, as asserted in the theorem. We leave the task of establishing this fact as an easy exercise, and turn to the problem of proving that these functions are a basis for $\mathcal{PC}(R)$.

By Theorem 8–3 we know that the set $\{f_i(x)g_j(y)\}$ will be a basis if and only if it satisfies Parseval's equality. Thus it suffices to prove that

$$\iint_R F(x, y)^2 \, dR = \sum_{i,j=1}^{\infty} \alpha_{ij}^2 \tag{9–39}$$

for any function F in $\mathcal{PC}(R)$.

We first consider the case in which F is continuous everywhere in R. Then, for each value of y in the interval $[c, d]$, $F(x, y)$ is a continuous function of x, and as such can be viewed as a member of $\mathcal{PC}[a, b]$. But then we can apply Parseval's equality to it and the basis $\{f_i(x)\}$ to obtain

$$\begin{aligned}\int_a^b F(x, y)^2 \, dx &= \sum_{i=1}^{\infty} (F \cdot f_i)^2 \\ &= \sum_{i=1}^{\infty} \left[\int_a^b F(x, y) f_i(x) \, dx\right]^2.\end{aligned}$$

Moreover, each of the integrals appearing in this equality is a *continuous* function of y. For convenience of notation we now set

$$h_i(y) = \int_a^b F(x, y) f_i(x) \, dx,$$

and rewrite the above equality as

$$\int_a^b F(x, y)^2 \, dx = \sum_{i=1}^{\infty} h_i(y)^2. \tag{9–40}$$

We now call upon the result from the theory of infinite series which says that a series of *positive* continuous functions which converges pointwise on a closed interval to a continuous function may be integrated term-by-term.* Thus, integrating (9–40), we obtain

$$\int_c^d \int_a^b F(x, y)^2 \, dx \, dy = \sum_{i=1}^{\infty} \left[\int_c^d h_i(y)^2 \, dy\right]. \tag{9–41}$$

---

* This result is a consequence of Dini's Theorem (see R. Courant, *Calculus II*, Interscience, New York) which guarantees that such a series is uniformly convergent and can be integrated term-by-term. (See Appendix I.)

But since $\{g_j(y)\}$ is a basis for $\mathcal{PC}[c, d]$, Parseval's equality implies that

$$\int_c^d h_i(y)^2\,dy = \sum_{j=1}^{\infty} (h_i \cdot g_j)^2$$
$$= \sum_{j=1}^{\infty} \left[\int_c^d h_i(y)g_j(y)\,dy\right]^2,$$

for each integer i. If we substitute these values in (9–41), and recall the definition of $h_i(y)$, we find that

$$\iint_R F(x, y)^2\,dR = \sum_{i=1}^{\infty}\sum_{j=1}^{\infty}\left[\int_c^d h_i(y)g_j(y)\,dy\right]^2$$
$$= \sum_{i,j=1}^{\infty}\left[\int_c^d \int_a^b F(x, y)f_i(x)g_j(y)\,dx\,dy\right]^2$$
$$= \sum_{i,j=1}^{\infty} \alpha_{ij}^2.$$

Thus (9–39) holds, and the proof is complete in the case where F is continuous.

The proof when F is piecewise continuous, but not continuous, requires a more sophisticated version of the "mending of discontinuities" theorem of Section 9–7 to show that F can be approximated arbitrarily closely in the mean by a continuous function. Although the procedure necessary to prove this result is conceptually clear, its details are both complicated and unenlightening, and we therefore omit the argument. But once this fact has been accepted, the general conclusion follows from the continuous case proved above. ▮

As might be expected, questions concerning the pointwise convergence of double series are rather difficult, and particular orthogonal systems must be considered individually. However, we can state one theorem which, though somewhat restricted in scope, is sufficient to answer all such questions that arise in this book.

Theorem 9–8. *Let R be the rectangle $-\pi \le x \le \pi$, $-\pi \le y \le \pi$, and suppose that F is continuous on R, and that*

$$\frac{\partial F}{\partial x}, \quad \frac{\partial F}{\partial y}, \quad \text{and} \quad \frac{\partial^2 F}{\partial x\,\partial y}$$

exist and are bounded everywhere in R. Then the double Fourier series for F converges **pointwise** *to F everywhere in R.**

With obvious modifications this theorem remains true for any rectangle.

---

* See E. W. Hobson, *The Theory of Functions of a Real Variable*, Second Ed., Cambridge University Press, 1921, 1926.

EXERCISES

1. (a) Verify that the set of functions $\{f_i(x)g_j(y)\}$ of Theorem 9–7 is orthogonal in $\mathcal{PC}(R)$, and that this set is orthonormal whenever $\{f_i(x)\}$ and $\{g_j(y)\}$ are orthonormal.

   (b) Find the norms of these functions in terms of the norms of the functions $f_i(x)$ and $g_j(y)$; i.e., find a formula for $\|f_i(x)g_j(y)\|$ in terms of $\|f_i(x)\|$ and $\|g_j(y)\|$.
2. What are the norms of the functions in (9–37)? In (9–38)?
3. What is the form of the double Fourier series expansion of a function F if $F(-x, y) = F(x, y)$ and $F(x, -y) = F(x, y)$? If $F(-x, y) = -F(x, y)$ and $F(x, -y) = -F(x, y)$?
4. (a) Repeat Exercise 3 for a function F such that

$$F(-x, y) = F(x, y) \quad \text{and} \quad F(x, -y) = -F(x, y).$$

   (b) Repeat Exercise 3 for a function F such that

$$F(-x, y) = -F(x, y) \quad \text{and} \quad F(x, -y) = F(x, y).$$

In each of the following exercises find the double Fourier series expansion of the given function in $\mathcal{PC}(R)$; R the rectangle $-\pi \leq x \leq \pi$, $-\pi \leq y \leq \pi$.

5. $F(x, y) = x$
6. $F(x, y)$ the function which is 1 when x and y are both positive or both negative, and -1 otherwise.
7. $F(x, y) = \sin^2 (x + y)$
8. $F(x, y) = e^{xy}$
9. $F(x, y) = xy^2$
10. $F(x, y) = |xy|$

10
convergence of fourier series*

10–1 INTRODUCTION

In this chapter we shall investigate some of the convergence problems which arise in the study of Fourier series. Our first efforts will be devoted to proving the theorem cited in Section 9–4 describing the pointwise behavior of the Fourier series for a piecewise smooth periodic function. Once this has been done, we shall consider the more delicate (and interesting) problem of uniform convergence, and the related questions of term-by-term differentiability and integrability of Fourier series. Finally, we shall introduce the important notion of "summability" for infinite series, and use it to extend these results to arbitrary functions in $\mathcal{PC}[-\pi, \pi]$. Throughout this discussion we shall assume that the reader is familiar with the notion of uniform convergence, and the results contained in Sections I-3 and I-4 of Appendix I. They will be essential in all that follows.

10–2 THE RIEMANN-LEBESGUE LEMMA

We begin the formal work of this chapter by establishing a result which, in addition to being essential for the proof of our first convergence theorem, is also of considerable interest in itself.

Lemma 10–1. (The Riemann-Lebesgue lemma.) *If g is piecewise continuous on $[a, b]$, then*

$$\lim_{\lambda\to\infty} \int_a^b g(x) \sin(\lambda x)\, dx = \lim_{\lambda\to\infty} \int_a^b g(x) \cos(\lambda x)\, dx = 0. \qquad (10\text{–}1)$$

The reader should note that (10–1) has already been established when $\lambda \to \infty$ through the values $2\pi k/(b - a)$, $k = 1, 2, \ldots$ (see Corollary 8–1). The burden of the present assertion is that this result is still valid as λ tends *continuously* to infinity. Intuitively, of course, this is reasonable, since the positive and negative portions of the area under each of the curves $g(x) \sin \lambda x$ and $g(x) \cos \lambda x$ tend to cancel one another as $\lambda \to \infty$.

---

* This chapter may be omitted in its entirety without loss of continuity.

Proof. Since the argument is similar for both functions we shall give the proof only for $g(x) \sin \lambda x$. Here if

$$I(\lambda) = \int_a^b g(x) \sin (\lambda x)\, dx, \tag{10–2}$$

and if $\epsilon > 0$ is given, we must show that there exists a constant λ_0 such that $|I(\lambda)| < \epsilon$ for all $\lambda > \lambda_0$. To this end we assume for the moment that g is continuous, and make the substitution $x = t + (\pi/\lambda)$ in (10–2) to obtain

$$I(\lambda) = -\int_{a-\pi/\lambda}^{b-\pi/\lambda} g\left(t + \frac{\pi}{\lambda}\right) \sin (\lambda t)\, dt,$$

or, reverting to the variable x,

$$I(\lambda) = -\int_{a-\pi/\lambda}^{b-\pi/\lambda} g\left(x + \frac{\pi}{\lambda}\right) \sin (\lambda x)\, dx. \tag{10–3}$$

Adding (10–2) and (10–3), we find that

$$\begin{aligned} 2I(\lambda) = &-\int_{a-\pi/\lambda}^{a} g\left(x + \frac{\pi}{\lambda}\right) \sin (\lambda x)\, dx + \int_{b-\pi/\lambda}^{b} g(x) \sin (\lambda x)\, dx \\ &+ \int_a^{b-\pi/\lambda} \left[g(x) - g\left(x + \frac{\pi}{\lambda}\right)\right] \sin (\lambda x)\, dx. \end{aligned}$$

Hence if M denotes the maximum value of the function $|g|$ on $[a, b]$, and if $\pi/\lambda \leq b - a$ (which, of course, we may assume), then

$$\begin{aligned} 2|I(\lambda)| &\leq M \int_{a-\pi/\lambda}^{a} |\sin \lambda x|\, dx + M \int_{b-\pi/\lambda}^{b} |\sin \lambda x|\, dx \\ &\quad + \int_a^{b-\pi/\lambda} \left|g(x) - g\left(x + \frac{\pi}{\lambda}\right)\right| |\sin \lambda x|\, dx \\ &\leq \frac{2M\pi}{\lambda} + \int_a^{b-\pi/\lambda} \left|g(x) - g\left(x + \frac{\pi}{\lambda}\right)\right| dx \end{aligned}$$

(recall that $|\sin \lambda x| \leq 1$ for all x). Thus

$$|I(\lambda)| \leq \frac{M\pi}{\lambda} + \frac{1}{2} \int_a^{b-\pi/\lambda} \left|g(x) - g\left(x + \frac{\pi}{\lambda}\right)\right| dx.$$

To complete the proof in the case under consideration we now use the fact that g is *uniformly continuous* on $[a, b]$ to find a constant λ_0 such that

$$\left|g(x) - g\left(x + \frac{\pi}{\lambda}\right)\right| < \frac{\epsilon}{b - a}$$

for all $\lambda > \lambda_0$, and *all* x in $[a, b]$.* In addition, we suppose that λ_0 is chosen so that, at the same time, $M\pi/\lambda < \epsilon/2$ whenever $\lambda > \lambda_0$. Then

$$|I(\lambda)| < \frac{\epsilon}{2} + \frac{\epsilon}{2} = \epsilon$$

for all $\lambda > \lambda_0$, as required. Finally, to establish the assertion for an arbitrary function in $\mathcal{PC}[a, b]$, we merely apply the above argument to each of its continuous pieces. ▮

10-3 POINTWISE CONVERGENCE OF FOURIER SERIES

In this section we use the Riemann-Lebesgue lemma to establish our first convergence theorem. The main step in the proof consists of deriving a formula for the partial sums of an arbitrary Fourier series

$$\frac{a_0}{2} + \sum_{k=1}^{\infty} (a_k \cos kx + b_k \sin kx),$$

where

$$a_k = \frac{1}{\pi}\int_{-\pi}^{\pi} f(x) \cos kx\, dx, \qquad b_k = \frac{1}{\pi}\int_{-\pi}^{\pi} f(x) \sin kx\, dx,$$

and f is any function in $\mathcal{PC}[-\pi, \pi]$. The derivation goes as follows.

Let

$$S_n(x) = \frac{a_0}{2} + \sum_{k=1}^{n} (a_k \cos kx + b_k \sin kx).$$

Then

$$\begin{aligned} S_n(x) &= \frac{1}{2\pi}\int_{-\pi}^{\pi} f(t)\, dt \\ &\quad + \frac{1}{\pi}\sum_{k=1}^{n}\left[\cos (kx)\int_{-\pi}^{\pi} f(t) \cos kt\, dt + \sin (kx)\int_{-\pi}^{\pi} f(t) \sin kt\, dt\right] \\ &= \frac{1}{\pi}\int_{-\pi}^{\pi} f(t)\left[\frac{1}{2} + \sum_{k=1}^{n} (\cos kx \cos kt + \sin kx \sin kt)\right] dt \\ &= \frac{1}{\pi}\int_{-\pi}^{\pi} f(t)\left[\frac{1}{2} + \sum_{k=1}^{n} \cos k(t - x)\right] dt. \end{aligned}$$

But by summing the trigonometric identity

$$\sin (k + \tfrac{1}{2})s - \sin (k - \tfrac{1}{2})s = 2 \sin \frac{s}{2} \cos ks$$

* Recall that a continuous function on a *closed* interval is uniformly continuous. (Theorem I-13, Appendix I.)

as k runs from 1 to n, we find that

$$\sin\,(n + \tfrac{1}{2})s - \sin\frac{s}{2} = 2\sin\frac{s}{2}\sum_{k=1}^{n}\cos ks,$$

or

$$\frac{1}{2} + \sum_{k=1}^{n}\cos ks = \frac{\sin\,(n + \frac{1}{2})s}{2\sin\,(s/2)}. \tag{10–4}$$

Hence

$$S_n(x) = \frac{1}{\pi}\int_{-\pi}^{\pi} f(t)\,\frac{\sin\,(n + \frac{1}{2})(t - x)}{2\sin\,(t - x)/2}\,dt.$$

We now regard x as fixed, and make the change of variable $s = t - x$, to obtain

$$S_n(x) = \frac{1}{\pi}\int_{-\pi-x}^{\pi-x} f(x + s)\,\frac{\sin\,(n + \frac{1}{2})s}{2\sin\,(s/2)}\,ds.$$

Finally, if we now assume that f is periodic on $(-\infty, \infty)$ with period 2π (i.e., if we replace f by its periodic extension to the entire real line), then $S_n(x)$ is also periodic with period 2π (Exercise 9), and we can write

$$S_n(x) = \frac{1}{\pi}\int_{-\pi}^{\pi} f(x + s)\,\frac{\sin\,(n + \frac{1}{2})s}{2\sin\,(s/2)}\,ds. \tag{10–5}$$

This is the desired result, which is known as *Dirichlet's formula* for S_n. Moreover, for future reference, we also note that when (10–4) is integrated from $-\pi$ to π, we have

$$\int_{-\pi}^{\pi}\frac{\sin\,(n + \frac{1}{2})s}{2\sin\,(s/2)}\,ds = \pi. \tag{10–6}$$

Now that these facts have been established we can easily prove

Theorem 10–1. *Let f be piecewise continuous on $(-\infty, \infty)$, with period 2π, and suppose that $f(x) = \frac{1}{2}[f(x^+) + f(x^-)]$ for all x. Then the Fourier series expansion for f converges to $f(x_0)$ at each point x_0 where f has a right- and left-hand derivative. In particular, if f is piecewise smooth its Fourier series converges to $f(x)$ for all x.**

---

* To accommodate points x_0 where f has a jump discontinuity, we define the right- and left-hand derivatives of f to be, respectively,

$$\lim_{h\to 0^+}\frac{f(x_0 + h) - f(x_0^+)}{h} \quad\text{and}\quad \lim_{h\to 0^+}\frac{f(x_0 - h) - f(x_0^-)}{h}$$

provided these limits exist. The reader should note that these definitions reduce to the usual ones at those points where f is continuous. Moreover, if both of these limits exist and are equal and if x_0 is a point of continuity of f, then f is differentiable at x_0, and the value of its derivative is the common value of the above limits. Finally, we recall that a function is said to be piecewise smooth if it has a piecewise continuous first derivative.

Proof. We begin by considering the case where x_0 is a point of continuity of f. Then $f(x_0^+) = f(x_0^-) = f(x_0)$, and the assumption that f has a right- and left-hand derivative at x_0 requires that both of the limits

$$\lim_{h\to 0^+} \frac{f(x_0 + h) - f(x_0)}{h} \quad \text{and} \quad \lim_{h\to 0^+} \frac{f(x_0 - h) - f(x_0)}{h}$$

exist. We must show that the difference $S_n(x_0) - f(x_0)$ tends to zero with increasing n. (Here, as always, S_n denotes the nth partial sum of the Fourier series for f).

Now by (10–5) and (10–6),

$$\begin{aligned} S_n(x_0) - f(x_0) &= \frac{1}{\pi}\int_{-\pi}^{\pi} f(x_0 + s)\frac{\sin (n + \frac{1}{2})s}{2\sin (s/2)}\,ds \\ &\quad - \frac{1}{\pi}\int_{-\pi}^{\pi} f(x_0)\frac{\sin (n + \frac{1}{2})s}{2\sin (s/2)}\,ds \\ &= \frac{1}{\pi}\int_{-\pi}^{\pi} [f(x_0 + s) - f(x_0)]\frac{\sin (n + \frac{1}{2})s}{2\sin (s/2)}\,ds \\ &= \frac{1}{\pi}\int_{-\pi}^{\pi} \left[\frac{f(x_0 + s) - f(x_0)}{s}\,\frac{\frac{1}{2}s}{\sin \frac{1}{2}s}\right]\sin (n + \tfrac{1}{2})s\,ds. \end{aligned}$$

Moreover, since f has both a right- and left-hand derivative at x_0, the function

$$g(s) = \frac{f(x_0 + s) - f(x_0)}{s}\,\frac{\frac{1}{2}s}{\sin \frac{1}{2}s} \tag{10–7}$$

is piecewise continuous on $[-\pi, \pi]$ (see Exercise 1). Hence by the Riemann-Lebesgue lemma, $\lim_{n\to\infty}[S_n(x_0) - f(x_0)] = 0$, as required.

To complete the proof it remains to treat the case where x_0 is a point of discontinuity of f. Here we must show that the series

$$\frac{a_0}{2} + \sum_{k=1}^{\infty} (a_k \cos kx_0 + b_k \sin kx_0),$$

where a_k and b_k are the Fourier coefficients of f, converges to $f(x_0) = \frac{1}{2}[f(x_0^+) + f(x_0^-)]$. But this is equivalent to showing that the Fourier series expansion of the function

$$G(x) = f(x + x_0) - f(x_0)$$

converges to zero at $x = 0$ (see Exercise 4). To do so we decompose G into its even and odd parts as

$$G = G_E + G_O,$$

and observe that the Fourier series for G_O converges to zero at $x = 0$. Hence it is sufficient to prove that the same is true of the Fourier series for G_E. However, it follows at once from the definition of G that G_E is continuous at $x = 0$, and has a right- and left-hand derivative there (see Exercise 4). Thus, by the continuous case treated above, the Fourier series for G_E converges to $\frac{1}{2}[G_E(0^+) + G_E(0^-)] = G_E(0) = 0$. From this we conclude that the Fourier series for G, which is just the sum of the corresponding series for G_E and G_O, converges to zero at $x = 0$, and the proof is complete. ▮

In Section 10–7 we will prove a far reaching generalization of this result. In the meantime, however, we conclude our discussion of pointwise convergence by evaluating a certain improper integral which will be encountered shortly.

Lemma 10–2. $\displaystyle\int_0^\infty \frac{\sin x}{x}\,dx = \frac{\pi}{2}.$

Proof. Let f be the function in $\mathcal{PC}[-\pi, \pi]$ defined by

$$f(x) = \begin{cases} \dfrac{1}{x}\sin\dfrac{x}{2}, & -\pi < x < \pi, \quad x \neq 0, \\ \frac{1}{2}, & x = 0, \end{cases}$$

and let S_n denote the nth partial sum of the Fourier series for f. Then since f is piecewise smooth on $[-\pi, \pi]$, Theorem 10–1 implies that $S_n(x_0) \to f(x_0)$ as $n \to \infty$ for all x_0 in $[-\pi, \pi]$. Setting $x_0 = 0$, and using Formula (10–5), we therefore have

$$\frac{1}{\pi}\int_{-\pi}^{\pi} \frac{\sin(s/2)}{s}\,\frac{\sin(n + \frac{1}{2})s}{2\sin(s/2)}\,ds \to \frac{1}{2}$$

as $n \to \infty$. We now use the fact that the integrand appearing in this expression is even to deduce that

$$\int_0^{\pi} \frac{\sin(n + \frac{1}{2})s}{s}\,ds \to \frac{\pi}{2}$$

as $n \to \infty$. Hence

$$\int_0^{(n+1/2)\pi} \frac{\sin x}{x}\,dx \to \frac{\pi}{2}$$

as $n \to \infty$, and we conclude that

$$\int_0^\infty \frac{\sin x}{x}\,dx = \frac{\pi}{2},$$

provided the integral exists.

To settle this last point, we set

$$A_k = \left|\int_{(k-1)\pi}^{k\pi} \frac{\sin x}{x}\,dx\right|,$$

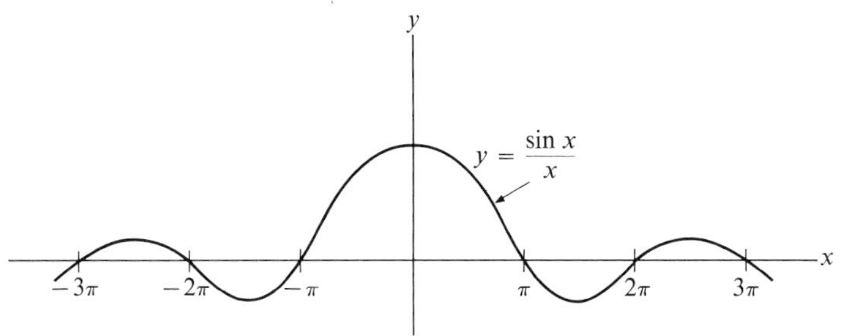

FIGURE 10–1

$k = 1, 2, \ldots$. Then, referring to Fig. 10–1, we see that the A_k measure the area between the x-axis and successive oscillations of the curve $\sin x/x$. Thus $A_1 > A_2 > \cdots$, and $A_k \to 0$ as $k \to \infty$ (see Exercise 2). From this it follows that the alternating series

$$A_1 - A_2 + A_3 - \cdots$$

converges, and that

$$\left| \sum_{k=N+1}^{\infty} (-1)^{k+1} A_k \right| \leq |A_{N+1}|.$$

Hence, if T is chosen so that $(k - 1)\pi < T < k\pi$, then

$$\left| \int_T^{\infty} \frac{\sin x}{x}\, dx \right| \leq \left| \int_T^{k\pi} \frac{\sin x}{x}\, dx \right| + \left| \sum_{j=k+1}^{\infty} (-1)^{j+1} A_j \right|$$

$$\leq |A_k| + |A_{k+1}|,$$

and since $A_k \to 0$, the integral in question must converge. ▮

EXERCISES

1. Prove that the function $g(s)$ defined by Formula (10–7) is piecewise continuous on $[-\pi, \pi]$.
2. Let

$$A_k = \left| \int_{(k-1)\pi}^{k\pi} \frac{\sin x}{x}\, dx \right|, \quad k = 1, 2, \ldots.$$

   Prove that $A_1 > A_2 > \cdots$, and that $A_k \to 0$ as $k \to \infty$.
3. Show that

$$\int_0^x \frac{\sin t}{t}\, dt \leq \int_0^{\pi} \frac{\sin t}{t}\, dt$$

   for all x.

4. (a) Let a_k, b_k denote the Fourier coefficients of f. Prove that the series

$$\frac{a_0}{2} + \sum_{k=1}^{\infty} (a_k \cos kx_0 + b_k \sin kx_0)$$

converges to $f(x_0)$ if and only if

$$\frac{a_0 - 2f(x_0)}{2} + \sum_{k=1}^{\infty} [(a_k \cos kx_0 + b_k \sin kx_0) \cos kx + (b_k \cos kx_0 - a_k \sin kx_0) \sin kx]$$

converges to zero at $x = 0$.
(b) Show that the second of the above series is the Fourier series expansion of the function $G(x) = f(x + x_0) - f(x_0)$. [*Hint:* Make the change of variable $t = x + x_0$ in the formulas for the Fourier coefficients of G on $[-\pi, \pi]$ and use the periodicity of f.]
(c) Show that

$$\lim_{x \to 0^+} G_E(x) = \lim_{x \to 0^+} G_E(-x) = G_E(0) = 0,$$

where G_E is as above, thereby establishing the continuity of G_E at $x = 0$.
(d) Show that the right- and left-hand derivatives of G_E exist at $x = 0$. [*Hint:* Note that

$$\begin{aligned} \frac{G_E(h) - G_E(0)}{h} &= \frac{1}{2}\left[\frac{G(h) - G(0)}{h} + \frac{G(-h) - G(0)}{h}\right] \\ &= \frac{1}{2}\left[\frac{f(x_0 + h) - f(x_0)}{h} + \frac{f(x_0 - h) - f(x_0)}{h}\right].\end{aligned}$$]

5. A continuous function f is said to satisfy a *Lipschitz condition of order* α if there exist positive constants M and α such that

$$|f(x_1) - f(x_2)| < M|x_1 - x_2|^\alpha$$

for all x_1 and x_2 in the domain of f.
(a) Show that each of the following functions satisfies a Lipschitz condition on the entire x-axis.
   (i) $f(x) = c$, c a constant
   (ii) $f(x) = \sin x$
   (iii) $f(x) = \sin^2 x$
   (iv) f the periodic extension to $(-\infty, \infty)$ of the function x^2 in $\mathcal{PC}[-\pi, \pi]$
   (v) f the periodic extension to $(-\infty, \infty)$ of the function $|x^3|$ in $\mathcal{PC}[-\pi, \pi]$
(b) Let f be a piecewise continuous function of period 2π, and suppose that f satisfies a Lipschitz condition of order α on the entire x-axis. Prove that the Fourier coefficients of f then satisfy the inequalities

$$|a_k| < \frac{M\pi^\alpha}{k^\alpha}, \qquad |b_k| < \frac{M\pi^\alpha}{k^\alpha}$$

for all $k > 0$. [*Hint:* Make the change of variable $x = t + (\pi/k)$ in the formulas for a_k and b_k, and deduce that

$$a_k = \frac{1}{2\pi}\int_{-\pi}^{\pi}\left[f(x) - f\left(x + \frac{\pi}{k}\right)\right]\cos kx\,dx,$$

$$b_k = \frac{1}{2\pi}\int_{-\pi}^{\pi}\left[f(x) - f\left(x + \frac{\pi}{k}\right)\right]\sin kx\,dx.]$$

(c) Why is it uninteresting to permit $\alpha > 1$?

6. (a) Prove that every continuously differentiable function f on a closed interval $[a, b]$ satisfies a Lipschitz condition of order one on that interval. (See Exercise 5 above.) [*Hint:* Use the mean value theorem and set $M = \max|f'(x)|$ on $[a, b]$.]

   (b) Prove the following generalization of Theorem 10–1. *Let f be piecewise continuous on $(-\infty, \infty)$ with period 2π, and suppose that*

   (i) *f is continuous at the point x_0, and*

   (ii) *f satisfies a Lipschitz condition of order α for all x in an interval about x_0.*

   Then the Fourier series for f converges to $f(x_0)$ when $x = x_0$. [*Hint:* Follow the proof of Theorem 10–1.]

7. Let f be piecewise smooth on $(-\infty, \infty)$ with period 2π, and suppose that $f(x) = \frac{1}{2}[f(x^+) + f(x^-)]$ for all x. Prove that

$$f(x) = \frac{1}{\pi}\lim_{\lambda\to\infty}\int_{-\pi}^{\pi} f(x+s)\frac{\sin\lambda s}{s}\,ds.$$

[*Hint:* Show that

$$\int_{-\pi}^{\pi} f(x+s)\frac{\sin\lambda s}{s}\,ds$$
$$= \int_0^{\pi}\frac{f(x+s) - f(x^+)}{s}\sin(\lambda s)\,ds + f(x^+)\int_0^{\pi}\frac{\sin\lambda s}{s}\,ds$$
$$+ \int_{-\pi}^{0}\frac{f(x+s) - f(x^-)}{s}\sin(\lambda s)\,ds + f(x^-)\int_{-\pi}^{0}\frac{\sin\lambda s}{s}\,ds,$$

and then apply the Riemann-Lebesgue lemma and Lemma 10–2.]

8. Prove that

$$\frac{1}{2} + \sum_{k=1}^{n}\cos kx = \frac{\sin(n + \frac{1}{2})x}{2\sin(x/2)},$$

as follows.
Use Euler's formula

$$e^{ix} = \cos x + i\sin x$$

to deduce that

$$\frac{1}{2} + \sum_{k=1}^{n}\cos kx$$

is the real part of

$$\frac{1}{2} + \sum_{k=1}^{n} e^{ikx}.$$

Evaluate this expression by using the formula for the sum of the first n terms of a geometric series, and then find its real part.

9. Let f be piecewise continuous and periodic on $(-\infty, \infty)$, with period 2π. Prove that the same is true of the function

$$f(x+s)\frac{\sin(n+\frac{1}{2})s}{2\sin(s/2)}.$$

10–4 UNIFORM CONVERGENCE OF FOURIER SERIES

Now that we have settled the question of pointwise convergence for the Fourier series expansions of piecewise smooth functions, we propose to determine conditions under which this convergence is uniform on a closed interval $[a, b]$. Here, of course, we must impose additional hypotheses on the functions considered, and at first sight one might expect them to be rather stringent. Surprisingly, however, we need only demand that the functions be *continuous* in order to guarantee both uniform and absolute convergence; an assertion which we prove as

Theorem 10–2. *Let f be a continuous function on $(-\infty, \infty)$ with period 2π, and suppose that f has a piecewise continuous first derivative. Then the Fourier series for f converges uniformly and absolutely to f on every closed interval of the x-axis.*

Proof. Let

$$\frac{a_0}{2} + \sum_{k=1}^{\infty} (a_k \cos kx + b_k \sin kx) \qquad \text{and} \qquad \frac{a_0'}{2} + \sum_{k=1}^{\infty} (a_k' \cos kx + b_k' \sin kx)$$

be the Fourier series for f and f', respectively. Then, since f is periodic with period 2π,

$$a_0' = \frac{1}{\pi}\int_{-\pi}^{\pi} f'(x)\,dx = \frac{1}{\pi}[f(\pi) - f(-\pi)] = 0,$$

while, for $k > 0$,

$$\begin{aligned} a_k' &= \frac{1}{\pi}\int_{-\pi}^{\pi} f'(x)\cos kx\,dx \\ &= \frac{1}{\pi}\left[f(x)\cos kx\Big|_{-\pi}^{\pi} + k\int_{-\pi}^{\pi} f(x)\sin kx\,dx\right] \\ &= \frac{k}{\pi}\int_{-\pi}^{\pi} f(x)\sin kx\,dx \\ &= kb_k, \end{aligned}$$

and

$$\begin{aligned} b'_k &= \frac{1}{\pi}\int_{-\pi}^{\pi} f'(x)\sin kx\,dx \\ &= \frac{1}{\pi}\left[f(x)\sin kx\Big|_{-\pi}^{\pi} - k\int_{-\pi}^{\pi} f(x)\cos kx\,dx\right] \\ &= -ka_k. \end{aligned}$$

Moreover, by Bessel's inequality (Theorem 8–3), we have

$$\sum_{k=1}^{\infty} (a_k'^2 + b_k'^2) \le \frac{1}{\pi}\int_{-\pi}^{\pi} f'(x)^2\,dx < \infty.$$

Thus

$$\sum_{k=1}^{\infty} [k\sqrt{a_k^2 + b_k^2}]^2 < \infty,$$

and we conclude that the sequence $\{k\sqrt{a_k^2 + b_k^2}\}$, $k = 1, 2, \ldots$, belongs to the Euclidean space ℓ_2 of all "square summable" sequences of real numbers introduced in Section 8–5. But the sequence $\{1/k\}$, $k = 1, 2, \ldots$, also belongs to ℓ_2. Hence the inner product of these two sequences exists, and it follows that the series

$$\sum_{k=1}^{\infty}\left(\frac{1}{k}\cdot k\sqrt{a_k^2 + b_k^2}\right) = \sum_{k=1}^{\infty}\sqrt{a_k^2 + b_k^2}$$

must converge.

Now, given any pair of real numbers a and b, the Cauchy-Schwarz inequality in $\mathcal{R}^2$ applied to the vectors $a\mathbf{e}_1 + b\mathbf{e}_2$, and $(\cos kx)\mathbf{e}_1 + (\sin kx)\mathbf{e}_2$, $\mathbf{e}_1$, $\mathbf{e}_2$ the standard basis, implies that

$$\begin{aligned} |a\cos kx + b\sin kx| &\le \sqrt{a^2 + b^2}\sqrt{\sin^2 kx + \cos^2 kx} \\ &= \sqrt{a^2 + b^2} \end{aligned}$$

for all x. This allows us to compare the series

$$\left|\frac{a_0}{2}\right| + \sum_{k=1}^{\infty} |a_k\cos kx + b_k\sin kx|$$

with the convergent series of positive constants

$$\left|\frac{a_0}{2}\right| + \sum_{k=1}^{\infty}\sqrt{a_k^2 + b_k^2},$$

and the Weierstrass M-test (Appendix I) implies that

$$\frac{a_0}{2} + \sum_{k=1}^{\infty} (a_k\cos kx + b_k\sin kx)$$

is uniformly and absolutely convergent on any closed interval of the x-axis. Finally, by Theorem 10–1, we know that this series converges pointwise to f, and the proof is complete. ▌

The reader should note that the hypotheses of Theorem 10–2 merely require f' to be *piecewise* continuous, and hence allow f' to be undefined at isolated points of the x-axis. Thus the Fourier series expansion of a function such as the one shown in Fig. 10–2 converges uniformly and absolutely in every closed interval of the x-axis in spite of the fact that f' does not exist at the points $2\pi n$, n an integer. On the other hand, the theorem is certain to fail in any interval where f itself is discontinuous, since it is well known that the limit of a *uniformly convergent* sequence of continuous functions (in this case the $S_k(x)$) is continuous. Thus the above result is the best that can be expected if we demand uniform convergence in *every* closed interval. By relaxing this requirement, however, Theorem 10–2 can be generalized to include functions with jump discontinuities. In this case the result in question reads as follows.

Theorem 10–3. *Let f be piecewise smooth and periodic on $(-\infty, \infty)$ with period 2π. Then the Fourier series for f converges uniformly to f in any closed interval of the x-axis which does not contain a point of discontinuity of f.*

The proof is an easy consequence of the following lemma, which itself is a special case of the theorem.

Lemma 10–3. *Let φ be the piecewise smooth, periodic function on $(-\infty, \infty)$ whose definition in the interval $[-\pi, \pi]$ is*

$$\varphi(x) = \begin{cases} -\dfrac{1}{2}\left(1 + \dfrac{x}{\pi}\right), & -\pi \le x < 0, \\ 0, & x = 0, \\ \dfrac{1}{2}\left(1 - \dfrac{x}{\pi}\right), & 0 < x \le \pi, \end{cases} \tag{10–8}$$

(see Fig. 10–3). Then the Fourier series for φ converges pointwise to φ for all x, and the convergence is uniform on any closed interval which does not contain a point of the form $2\pi n$, n an integer.

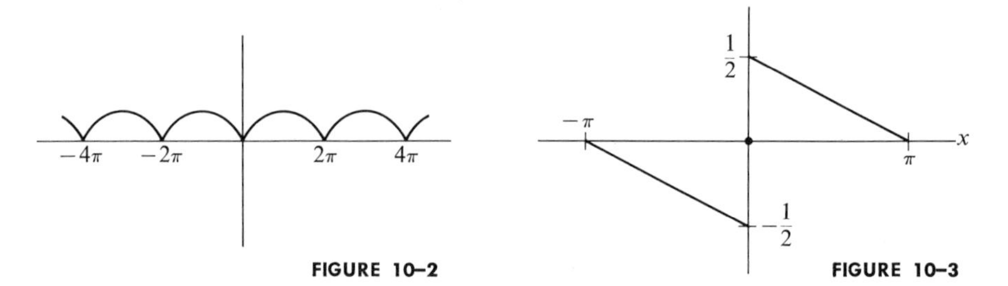

FIGURE 10–2

FIGURE 10–3

Granting the truth of this assertion, Theorem 10–3 is proved in the following manner.

Let $x_1, x_2, \ldots, x_m$ be the points in $(-\pi, \pi)$ where f is discontinuous, and for each i, $i = 1, \ldots, m$, let ω_i denote the magnitude of the jump discontinuity at x_i [that is, $\omega_i = f(x_i^+) - f(x_i^-)$]. Then the function $\omega_i\varphi(x - x_i)$ also has jump discontinuities of magnitude ω_i at the points $x_i + 2\pi n$, but is continuous for all other values of x. Hence

$$f(x) - \omega_i\varphi(x - x_i)$$

is continuous both at the points where f is continuous, *and* at the points $x_i + 2\pi n$. In short, by subtracting $\omega_i\varphi(x - x_i)$ from f we have removed the discontinuities at the points $x_i + 2\pi n$ *without introducing any new points of discontinuity.* We now repeat this process for each index i, to obtain the function

$$f(x) - \sum_{i=1}^{m} \omega_i\varphi(x - x_i),$$

which is piecewise smooth on $(-\infty, \infty)$, periodic with period 2π, and continuous everywhere except possibly at the points $\pm\pi, \pm 3\pi, \ldots$. [Such discontinuities will occur whenever $f(-\pi) \neq f(\pi)$.] To remove these last discontinuities we set $\omega_\pi = f(\pi^+) - f(\pi^-)$, and construct the function

$$F(x) = f(x) - \sum_{i=1}^{m} \omega_i\varphi(x - x_i) - \omega_\pi\varphi(x - \pi).$$

Then F satisfies the hypotheses of Theorem 10–2, and its Fourier series therefore converges uniformly to F in every closed interval of the x-axis. Moreover, Lemma 10–3 allows us to assert that the Fourier series for the function

$$\Phi(x) = \sum_{i=1}^{m} \omega_i\varphi(x - x_i) + \omega_\pi\varphi(x - \pi)$$

converges uniformly to Φ in any closed interval not containing a point of discontinuity of f. Hence the Fourier series for f, being the sum of the series for F and Φ, must also converge uniformly in any such interval, and the theorem is proved. ▌

To complete the argument, we now establish Lemma 10–3. Here we reason as follows.

A routine calculation reveals that the Fourier series for φ is

$$\frac{1}{\pi}\sum_{k=1}^{\infty} \frac{\sin kx}{k}, \tag{10–9}$$

and hence to prove the lemma it suffices to show that this series converges uniformly on every closed subinterval of $[-\pi, \pi]$ not containing the origin.

To this end we set

$$S_n(x) = \sin x + \frac{\sin 2x}{2} + \cdots + \frac{\sin nx}{n},$$

and let

$$T_n(x) = \sin x + \sin 2x + \cdots + \sin nx.$$

(Note that S_n is the nth partial sum of the Fourier series for the function $\pi\varphi$.) Then, since $\sin kx = T_k(x) - T_{k-1}(x)$ for all $k > 1$, we have

$$S_n(x) = T_1(x) + \frac{T_2(x) - T_1(x)}{2} + \cdots + \frac{T_n(x) - T_{n-1}(x)}{n}$$

or

$$S_n(x) = \frac{T_1(x)}{1 \cdot 2} + \frac{T_2(x)}{2 \cdot 3} + \cdots + \frac{T_{n-1}(x)}{(n-1)n} + \frac{T_n(x)}{n}.$$

Moreover, by summing the trigonometric identity

$$2 \sin \frac{x}{2} \sin kx = \cos (k - \tfrac{1}{2})x - \cos (k + \tfrac{1}{2})x$$

as k runs from 1 to n, we find that

$$T_n(x) = \frac{\cos (x/2) - \cos (n + \frac{1}{2})x}{2 \sin (x/2)}, \quad x \neq 0, \tag{10–10}$$

from which it easily follows that

$$|T_n(x)| \leq \frac{1}{|\sin \frac{1}{2}x|}$$

for *all* n and *all* $x \neq 0$ in $[-\pi, \pi]$. Thus, for $n > m$, and $x \neq 0$ in $[-\pi, \pi]$,

$$\begin{aligned} |S_n(x) - S_m(x)| &= \left| \frac{T_m(x)}{m(m+1)} + \cdots + \frac{T_{n-1}(x)}{(n-1)n} + \frac{T_n(x)}{n} - \frac{T_m(x)}{m} \right| \\ &\leq \frac{1}{|\sin (x/2)|} \left\{ \left[\frac{1}{m(m+1)} + \cdots + \frac{1}{(n-1)n} \right] + \frac{1}{n} + \frac{1}{m} \right\}. \end{aligned}$$

Now let x be restricted by the inequalities $0 < \delta \leq |x| \leq \pi$. Then

$$\left| \sin \frac{x}{2} \right| \geq \sin \frac{\delta}{2},$$

and we have

$$|S_n(x) - S_m(x)| \leq \frac{1}{\sin (\delta/2)} \left\{ \left[\frac{1}{m(m+1)} + \cdots + \frac{1}{(n-1)n} \right] + \frac{1}{n} + \frac{1}{m} \right\}.$$

But since the series

$$\sum_{k=1}^{\infty} \frac{1}{k(k+1)}$$

is convergent, the quantity

$$\left[\frac{1}{m(m+1)} + \cdots + \frac{1}{(n-1)n}\right] + \frac{1}{n} + \frac{1}{m}$$

can be made arbitrarily small by taking m and n sufficiently large. Hence, given any $\epsilon > 0$, there exists an integer N such that

$$|S_n(x) - S_m(x)| < \epsilon$$

for all $m, n > N$, and all x with $\delta \leq |x| \leq \pi$. This, of course, implies that

$$\sum_{k=1}^{\infty} \frac{\sin kx}{k}$$

is uniformly convergent whenever $0 < \delta \leq |x| \leq \pi$. The same is therefore true of the series for φ, and since the choice of δ in $(0, \pi)$ was arbitrary, we are done. ▌

We have already observed that the Fourier series for a piecewise smooth function cannot converge uniformly on any closed interval $[a, b]$ containing a point of discontinuity of the function in its interior. To complete our discussion of uniform convergence, it remains to consider the behavior of Fourier series

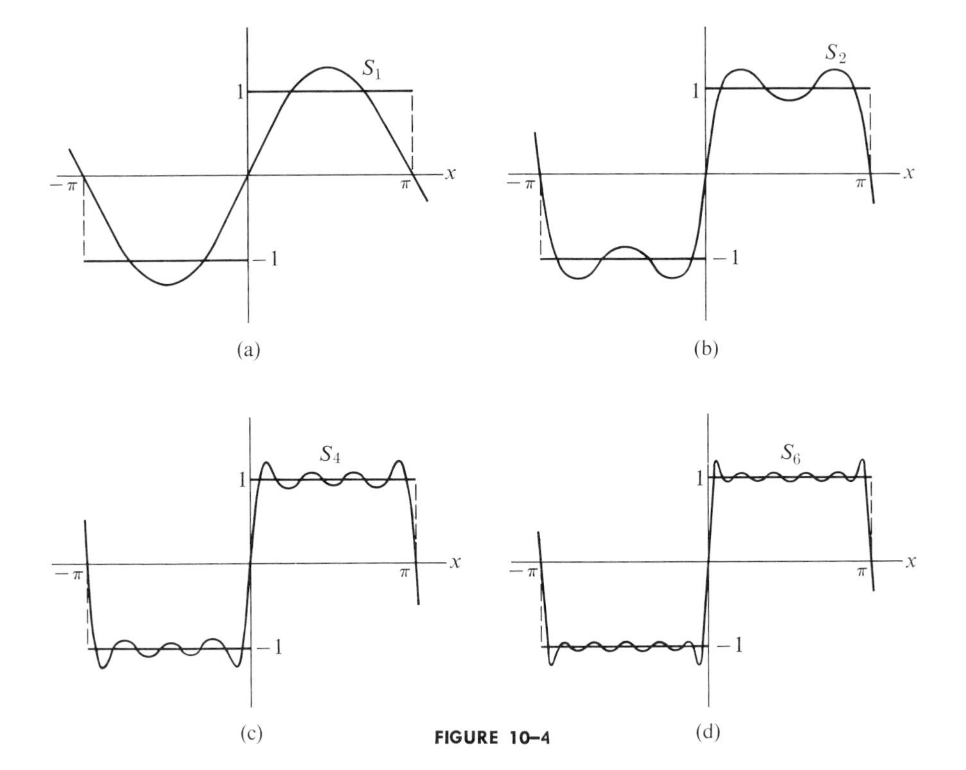

FIGURE 10–4

in the vicinity of such discontinuities. The diagrams in Fig. 10–4, depicting the function

$$f(x) = \begin{cases} -1, & -\pi < x < 0, \\ 0, & x = \pi, 0, \pi, \\ 1, & 0 < x < \pi, \end{cases}$$

and several of the partial sums of its Fourier series, are typical of the situation which obtains. From these diagrams it is apparent that the oscillations in the partial sums, S_n, of the Fourier series for f do *not* decrease at a uniform rate on the interval $(0, \pi)$ as $n \to \infty$. On the contrary, the oscillations toward either end of this interval remain rather large for *all* values of n, and though these exceptional oscillations move toward the ends of the interval as n increases, they do not die out in the process. This peculiar behavior of the partial sums of a Fourier series, wherein these sums seem to gather momentum before plunging across a jump discontinuity, is known as the *Gibbs phenomenon*, after the American mathematician and physicist, J. W. Gibbs, who first discovered it. In the next section we shall analyze this phenomenon in some detail, and obtain a limiting value for the amplitude of the oscillations involved.

EXERCISES

Each of the following exercises refers to an arbitrary trigonometric series of the form

$$\frac{a_0}{2} + \sum_{k=1}^{\infty} (a_k \cos kx + b_k \sin kx) \tag{10–11}$$

which, initially, is not assumed to be the Fourier series expansion of a function in $\mathcal{PC}[-\pi, \pi]$.

1. Prove that whenever (10–11) converges in the mean in $\mathcal{PC}[-\pi, \pi]$ to a function f, it is the Fourier series for f.
2. Suppose that (10–11) converges uniformly on every closed interval of the x-axis. Prove that (10–11) then converges in the mean in $\mathcal{PC}[-\pi, \pi]$, and is the Fourier series expansion of its limit. (See Exercise 1.)
3. Suppose that

$$\sum_{k=1}^{\infty} (|a_k| + |b_k|)$$

converges. Prove that (10–11) converges uniformly and absolutely on every closed interval of the x-axis, and is the Fourier series expansion of its limit.

4. Prove that the conclusion of Exercise 3 holds whenever

$$|a_k| \le \frac{1}{k^2}, \qquad |b_k| \le \frac{1}{k^2}$$

for all $k > 0$.

*10-5 THE GIBBS PHENOMENON

We begin our analysis of the Gibbs phenomenon by examining the behavior of the partial sums of the Fourier series for the periodic function φ defined in the interval $[-\pi, \pi]$ by

$$\varphi(x) = \begin{cases} -\dfrac{\pi - x}{2}, & -\pi \leq x < 0, \\ 0, & x = 0, \\ \dfrac{\pi - x}{2}, & 0 < x \leq \pi. \end{cases} \qquad (10\text{–}12)$$

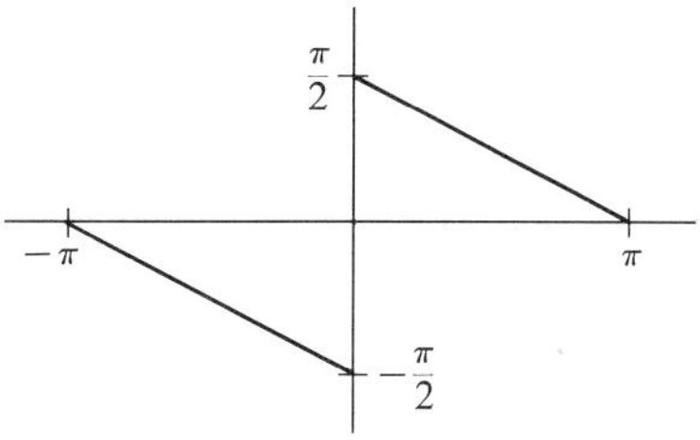

FIGURE 10-5

(See Fig. 10–5.) The series in question is

$$\sum_{k=1}^{\infty} \frac{\sin kx}{k}, \qquad (10\text{–}13)$$

and by our earlier results we know that this series converges to $\varphi(x)$ for all x, the convergence being uniform on any closed interval not containing a point of the form $2\pi n$, n an integer. Figure 10–6 shows the graph of φ on the interval $[-\pi, \pi]$, together with the graph of the sum of the first six terms of its Fourier series, and again furnishes visual evidence of the fact that the partial sums of the series tend to "overshoot" the values of the function near a point of discontinuity.

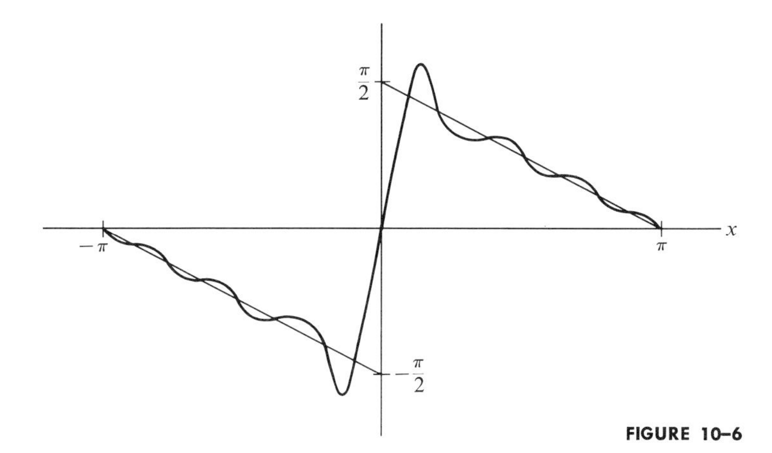

FIGURE 10-6

Thus, the values of φ between any two jump discontinuities range through the interval $(-\pi/2, \pi/2)$, while those of S_n, the nth partial sum of (10–13), range through a somewhat larger interval $[-\alpha_n, \alpha_n]$. The limiting value of α_n as $n \to \infty$ determines what is known as the *Gibbs interval* for φ, and our first objective is to obtain a precise description of this interval.

To this end we observe that a real number y will belong to the Gibbs interval for φ if and only if there exists an increasing sequence of positive integers $\{n_k\}$,

$k = 1, 2, \ldots$, and a sequence of real numbers $\{x_k\}$ converging to zero, such that

$$\lim_{k\to\infty} S_{n_k}(x_k) = y.^* \tag{10–14}$$

Informally, (10–14) asserts that we can approach y as closely as we wish by points which lie on the graphs of a certain subsequence $\{S_{n_k}\}$ of the sequence of partial sums $\{S_n\}$. (See Fig. 10–7.) To find all points y which satisfy this condition we consider the expression

$$\frac{x}{2} + S_n(x) = \frac{x}{2} + \sum_{k=1}^{n} \frac{\sin kx}{k} = \int_0^x \left[\frac{1}{2} + \sum_{k=1}^{n} \cos kt\right] dt.$$

Then, by Formula (10–4), we have

$$\begin{aligned}\int_0^x \left[\frac{1}{2} + \sum_{k=1}^{n} \cos kt\right] dt &= \int_0^x \frac{\sin (n + \frac{1}{2})t}{2 \sin (t/2)}\, dt \\ &= \frac{1}{2}\int_0^x \frac{\sin (n + \frac{1}{2})t}{t/2}\, dt \\ &\quad + \frac{1}{2}\int_0^x \left[\frac{1}{\sin (t/2)} - \frac{1}{t/2}\right] \sin (n + \tfrac{1}{2})t\, dt.\end{aligned}$$

Thus

$$\begin{aligned}S_n(x) = -\frac{x}{2} &+ \frac{1}{2}\int_0^x \frac{\sin (n + \frac{1}{2})t}{t/2}\, dt \\ &+ \frac{1}{2}\int_0^x \left[\frac{1}{\sin (t/2)} - \frac{1}{t/2}\right] \sin (n + \tfrac{1}{2})t\, dt.\end{aligned} \tag{10–15}$$

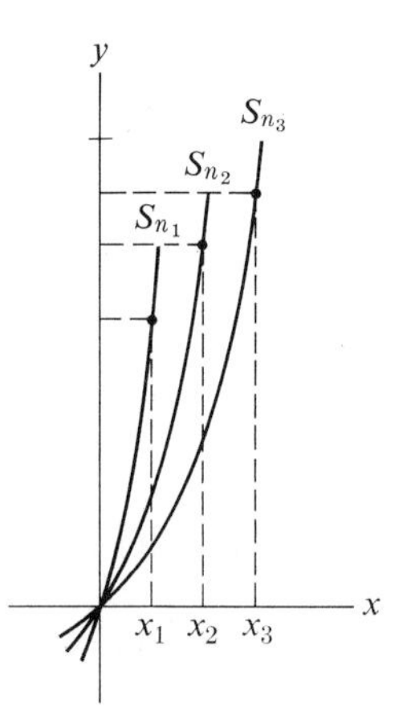

FIGURE 10–7

We now make the substitution $u = (n + \frac{1}{2})t$ in the first of these integrals to obtain

$$\frac{1}{2}\int_0^x \frac{\sin (n + \frac{1}{2})t}{t/2}\, dt = \int_0^{(n+1/2)x} \frac{\sin u}{u}\, du.$$

Next, an easy application of l'Hôpital's rule shows that

$$\lim_{t\to 0}\left[\frac{1}{\sin (t/2)} - \frac{1}{t/2}\right] = 0,$$

and hence that the function

$$\frac{1}{\sin (t/2)} - \frac{1}{t/2}$$

---

* The reader who is so inclined may take this statement as a formal definition of the Gibbs interval for φ.

is continuous on the closed interval $[0, x]$, provided it is assigned the value 0 when $t = 0$. Moreover, if

$$F(n; x) = \frac{1}{2} \int_0^x \left[\frac{1}{\sin(t/2)} - \frac{1}{t/2} \right] \sin(n + \tfrac{1}{2})t \, dt,$$

it is not difficult to show that

$$\lim_{x \to 0} F(n; x) = 0$$

uniformly in n; i.e., given any $\epsilon > 0$, a $\delta > 0$ can be found such that *for all* n, $|F(n; x)| < \epsilon$ whenever $|x| < \delta$. (See Exercise 1 below.) Thus (10–15) may be rewritten

$$S_n(x) = -\frac{x}{2} + \int_0^{(n+1/2)x} \frac{\sin t}{t} \, dt + F(n; x),$$

and it follows that if $\{x_k\}$ is any sequence of real numbers which converges to zero, and if $\{n_k\}$ is an increasing sequence of positive integers, both chosen so that $\{x_k n_k\}$ has a limit, h, as $k \to \infty$, then

$$\lim_{k \to \infty} S_{n_k}(x_k) = \int_0^h \frac{\sin t}{t} \, dt,$$

(Note that *every* real number h can be expressed as a limit of this form, and that in certain cases h may assume the values $\pm\infty$.) Hence every point of the form

$$\int_0^h \frac{\sin t}{t} \, dt, \quad -\infty \le h \le \infty,$$

belongs to the Gibbs interval for φ.

Conversely, if y_0 belongs to the Gibbs interval for φ, there exist sequences $\{x_k\}$ and $\{n_k\}$ with the property that $S_{n_k}(x_k) \to y_0$ as $k \to \infty$. But then there exists a subsequence $\{n_{k'} x_{k'}\}$ of $\{n_k x_k\}$ which either converges or else diverges to $\pm\infty$.* If h denotes the limit of this subsequence, then (10–15) (in its amended form) implies that

$$y_0 = \lim_{k \to \infty} S_{n_k}(x_k) = \lim_{k \to \infty} S_{n_{k'}}(x_{k'}) = \int_0^h \frac{\sin t}{t} \, dt,$$

and we have shown that y_0 can be expressed in the form

$$\int_0^h \frac{\sin t}{t} \, dt$$

---

* This is a consequence of the famous Bolzano-Weierstrass theorem which asserts that every bounded infinite set of real numbers has a limit point. For a proof see Buck, *Advanced Calculus*, 2nd Ed., McGraw-Hill, New York, 1965.

for a suitable value of h. Hence we have proved the following theorem.

Theorem 10–4. *The Gibbs interval for the function φ defined above is the set of all real numbers y of the form*

$$y = \int_0^h \frac{\sin t}{t}\, dt \tag{10–16}$$

with $-\infty \leq h \leq \infty$.

Actually, we can say even more than this. For, referring to Fig. 10–8, and using the interpretation of the integral as area under the curve, we see that (10–16) assumes its maximum value when $h = \pi$. Thus the Gibbs interval for φ is

$$\left[-\int_0^\pi \frac{\sin t}{t}\, dt,\ \int_0^\pi \frac{\sin t}{t}\, dt\right].$$

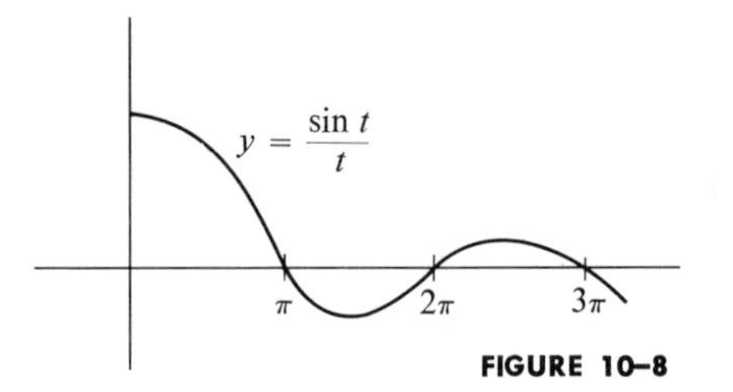

FIGURE 10–8

Moreover, since the value of

$$\int_0^\pi \frac{\sin t}{t}\, dt$$

is known to be $(\pi/2)\,1.089490\ldots$, it follows that the length of the Gibbs interval for φ is appreciably greater than the magnitude of the jump discontinuity in φ at $x = 0$. Once again this implies that the Fourier series for φ is *not* uniformly convergent in any open interval having a point of discontinuity of φ as one of its end points.

It is now a simple matter to pass to the case of an arbitrary piecewise smooth function. Indeed, if f is piecewise smooth on $[-\pi, \pi]$ with a jump discontinuity at x_0, then, arguing as in the proof of Theorem 10–3, the function

$$g(x) = f(x) - \frac{f(x_0^+) - f(x_0^-)}{\pi}\varphi(x - x_0)$$

is continuous in some closed interval of the form $[x_0 - \delta, x_0 + \delta]$ about x_0, and has a uniformly convergent Fourier series on that interval. But since

$$f(x) = g(x) + \frac{f(x_0^+) - f(x_0^-)}{\pi}\varphi(x - x_0),$$

we conclude that the Gibbs interval for f at x_0 must be the same as the Gibbs interval for the function

$$\frac{f(x_0^+) - f(x_0^-)}{\pi}\varphi(x - x_0).$$

Thus, on the vertical line $x = x_0$ the Gibbs interval for f consists of all points y such that

$$\begin{aligned}\left| y - \frac{f(x_0^+) + f(x_0^-)}{2} \right| &\le \frac{|f(x_0^+) - f(x_0^-)|}{\pi} \int_0^\pi \frac{\sin t}{t}\, dt \\ &= \frac{|f(x_0^+) - f(x_0^-)|}{2} \cdot \frac{2}{\pi} \cdot \int_0^\pi \frac{\sin t}{t}\, dt \\ &= \frac{|f(x_0^+) - f(x_0^-)|}{2}\, 1.089490\ldots.\end{aligned}$$

In short, the length of the Gibbs interval for f at x_0 exceeds the magnitude of the jump discontinuity in f at that point by the factor $1.089490\ldots$.

EXERCISES

1. Let

$$F(n; x) = \frac{1}{2} \int_0^x \left[\frac{1}{\sin(t/2)} - \frac{1}{t/2} \right] \sin\left(n + \tfrac{1}{2}\right)t\, dt.$$

Prove that $\lim_{x \to 0} F(n; x) = 0$ uniformly in n.

10–6 DIFFERENTIATION AND INTEGRATION OF FOURIER SERIES

With the results we now have available it is an easy matter to settle the basic questions concerning the termwise differentiability and integrability of Fourier series. The relevant theorems in this connection are as follows.

Theorem 10–5. (The differentiation theorem.) *Let f be a continuous function on $(-\infty, \infty)$, with period 2π, and suppose that f has a piecewise continuous first derivative, f'. Then the Fourier series for f' can be obtained by differentiating the series for f term-by-term, and the differentiated series converges pointwise to $f'(x)$ wherever f'' exists.*

Proof. Let

$$\frac{a_0}{2} + \sum_{k=1}^{\infty} (a_k \cos kx + b_k \sin kx) \tag{10–17}$$

and

$$\frac{a_0'}{2} + \sum_{k=1}^{\infty} (a_k' \cos kx + b_k' \sin kx) \tag{10–18}$$

be, respectively, the Fourier series for f and f'. Then, as was shown in the proof of Theorem 10–2, $a_0' = 0$, and, for $k > 0$,

$$a_k' = kb_k, \qquad b_k' = -ka_k.$$

Thus, (10–18) is

$$\sum_{k=1}^{\infty} k[b_k \cos kx - a_k \sin kx],$$

which is precisely the result obtained by differentiating (10–17) term-by-term. This proves the first statement in the theorem, while the second is a consequence of Theorem 10–1. ▮

From this, and Theorem 10–2, we immediately deduce

Corollary 10–1. *Let f be continuous and periodic on $(-\infty, \infty)$, with period 2π, and suppose that f has a continuous first derivative and piecewise continuous second derivative. Then the Fourier series for f' is uniformly and absolutely convergent on every closed interval of the x-axis, and can be obtained by differentiating the Fourier series for f term-by-term. More generally, if $f, f', \ldots, f^{(n-1)}$ are continuous, while $f^{(n)}$ is piecewise continuous, then the Fourier series for $f^{(j)}, j = 1, \ldots, n-1$ converges uniformly and absolutely to $f^{(j)}$ on every closed interval of the x-axis, and can be obtained by differentiating the Fourier series for f term-by-term j times.*

Turning to the integration of Fourier series we now prove

Theorem 10–6. *Let f be a piecewise continuous function on $(-\infty, \infty)$ with period 2π, and let*

$$\frac{a_0}{2} + \sum_{k=1}^{\infty} (a_k \cos kx + b_k \sin kx)$$

be the Fourier series for f. Then

$$\int_a^b f(x)\,dx = \frac{a_0}{2}(b-a) + \sum_{k=1}^{\infty} \frac{a_k(\sin kb - \sin ka) - b_k(\cos kb - \cos ka)}{k}. \tag{10–19}$$

In other words, the definite integral of f from a to b can be evaluated by integrating the Fourier series for f term-by-term.

Proof. Set

$$F(x) = \int_0^x \left[f(t) - \frac{a_0}{2}\right] dt. \tag{10–20}$$

Then F is continuous, has a piecewise continuous first derivative, and is periodic with period 2π (Exercise 1). Thus F can be expanded in an everywhere convergent

Fourier series as

$$F(x) = \frac{A_0}{2} + \sum_{k=1}^{\infty} (A_k \cos kx + B_k \sin kx),$$

and an easy computation reveals that

$$\begin{aligned} A_k &= -\frac{b_k}{k}, \\ B_k &= \frac{a_k}{k}, \quad k \geq 1. \end{aligned} \tag{10–21}$$

Hence

$$F(x) = \frac{A_0}{2} + \sum_{k=1}^{\infty} \frac{a_k \sin kx - b_k \cos kx}{k},$$

and it follows that

$$\int_0^x f(t)\, dt = \frac{a_0}{2} x + \frac{A_0}{2} + \sum_{k=1}^{\infty} \frac{a_k \sin kx - b_k \cos kx}{k}. \tag{10–22}$$

We now use the fact that

$$\int_a^b f(t)\, dt = \int_0^b f(t)\, dt - \int_0^a f(t)\, dt$$

to deduce that

$$\begin{aligned} \int_a^b f(x)\, dx = \frac{a_0}{2}(b - a) &+ \sum_{k=1}^{\infty} \frac{a_k \sin kb - b_k \cos kb}{k} \\ &- \sum_{k=1}^{\infty} \frac{a_k \sin ka - b_k \cos ka}{k}. \end{aligned}$$

Finally, since both of these series are absolutely convergent, the necessary rearrangement of terms leading to (10–19) can be effected, and we are done. ▮

In much the same way we can also prove the following theorem.

Theorem 10–7. (The integration theorem.) *Let f be an arbitrary function in $\mathcal{PC}[-\pi, \pi]$ with Fourier series*

$$\frac{a_0}{2} + \sum_{k=1}^{\infty} (a_k \cos kx + b_k \sin kx). \tag{10–23}$$

Then the function

$$\int_0^x f(t)\, dt, \quad -\pi < x < \pi,$$

has a Fourier series which converges pointwise for all x in the interval $(-\pi, \pi)$, *and*

$$\int_0^x f(t)\,dt = \sum_{k=1}^{\infty} \frac{b_k}{k} + \sum_{k=1}^{\infty} \frac{-b_k \cos kx + [a_k + (-1)^{k+1} a_0] \sin kx}{k}. \tag{10–24}$$

Remark. The reader should note that since f is arbitrary in $\mathcal{PC}[-\pi, \pi]$, the series in (10–23) need not converge to $f(x)$ everywhere in $[-\pi, \pi]$. This not withstanding, the series for $\int_0^x f(t)\,dt$ does converge pointwise as asserted in the theorem.

Proof. Reasoning as in the proof of Theorem 10–6, we have

$$\int_0^x f(t)\,dt = \frac{a_0}{2} x + \frac{A_0}{2} + \sum_{k=1}^{\infty} \frac{a_k \sin kx - b_k \cos kx}{k},$$

$-\pi < x < \pi$. (See Formula 10–22.) We now set $x = 0$ and find that

$$\frac{A_0}{2} = \sum_{k=1}^{\infty} \frac{b_k}{k}, \tag{10–25}$$

and hence that

$$\int_0^x f(t)\,dt = \frac{a_0}{2} x + \sum_{k=1}^{\infty} \frac{b_k}{k} + \sum_{k=1}^{\infty} \frac{a_k \sin kx - b_k \cos kx}{k}.$$

But when $-\pi < x < \pi$,

$$\frac{x}{2} = \sum_{k=1}^{\infty} (-1)^{k+1} \frac{\sin kx}{k}.$$

Thus

$$\int_0^x f(t)\,dt = a_0 \sum_{k=1}^{\infty} (-1)^{k+1} \frac{\sin kx}{k} + \sum_{k=1}^{\infty} \frac{b_k}{k} + \sum_{k=1}^{\infty} \frac{a_k \sin kx - b_k \cos kx}{k},$$

an expression which is clearly equivalent to (10–24). ▮

EXERCISES

1. Verify that the function defined by Eq. (10–20) is periodic with period 2π, and compute the coefficients A_k and B_k $(k > 0)$ of its Fourier series.
2. Let f be a function in $\mathcal{PC}[-\pi, \pi]$ with Fourier series

$$\frac{a_0}{2} + \sum_{k=1}^{\infty} (a_k \cos kx + b_k \sin kx).$$

Prove that

$$\int_{-\pi}^{x} f(t)\,dt = \frac{a_0(x+\pi)}{2} + \sum_{k=1}^{\infty} \frac{a_k \sin kx - b_k(\cos kx - \cos k\pi)}{k}$$

for $-\pi < x < \pi$.

3. Suppose that the function f considered in Theorem 10–7 is piecewise continuous on $(-\infty, \infty)$ with period 2π. Is Formula (10–24) then true for *all* x? Why?

4. Let f be a piecewise continuous *odd* function on $(-\infty, \infty)$ with period 2π. Prove that

$$\int_{0}^{x} f(t)\,dt = \sum_{k=1}^{\infty} \frac{b_k}{k} - \sum_{k=1}^{\infty} \frac{b_k}{k} \cos kx$$

for *all* x.

5. Starting with the series

$$\frac{x}{2} = \sum_{k=1}^{\infty} (-1)^{k+1} \frac{\sin kx}{k}, \qquad -\pi < x < \pi,$$

use Theorem 10–7 to prove that

$$\frac{x^2}{4} = \frac{\pi^2}{12} + \sum_{k=1}^{\infty} (-1)^k \frac{\cos kx}{k^2},$$

and that

$$\frac{x^3}{12} - \frac{\pi^2 x}{12} = \sum_{k=1}^{\infty} (-1)^k \frac{\sin kx}{k^3}$$

for $-\pi < x < \pi$. See Exercises 14(b) and 15(b) of Section 9–4.

6. Show that the trigonometric series $\sum_{k=2}^{\infty} \sin kx/\ln k$ is *not* the Fourier series of any function $\mathcal{PC}[-\pi, \pi]$. [*Hint:* See (10–25).]

7. (a) Let $\{f_k\}$, $k = 1, 2, \ldots$, be a sequence of functions in $\mathcal{PC}[-\pi, \pi]$ which converges in the mean to f. Prove that

$$\lim_{k \to \infty} \int_{-\pi}^{\pi} f_k(x)g(x)\,dx = \int_{-\pi}^{\pi} f(x)g(x)\,dx$$

for all g in $\mathcal{PC}[-\pi, \pi]$. [*Hint:* Apply the Cauchy-Schwarz inequality to the functions $f - f_k$ and g in $\mathcal{PC}[-\pi, \pi]$.]

(b) Use the result in (a) to deduce that if $\sum_{k=1}^{\infty} f_k$ converges in the mean to f in $\mathcal{PC}[-\pi, \pi]$, then

$$\int_{-\pi}^{\pi} f(x)g(x)\,dx = \sum_{k=1}^{\infty} \left[\int_{-\pi}^{\pi} f_k(x)g(x)\,dx\right]$$

for all g in $\mathcal{PC}[-\pi, \pi]$.

(c) Let f and g be piecewise continuous on $(-\infty, \infty)$ with period 2π, and let a_k, b_k and α_k. β_k be, respectively, the Fourier coefficients for f and g. Prove that

$$\frac{1}{\pi} \int_{-\pi}^{\pi} f(x)g(x)\,dx = \frac{a_0\alpha_0}{2} + \sum_{k=1}^{\infty} (a_k\alpha_k + b_k\beta_k).$$

10–7 SUMMABILITY OF FOURIER SERIES; FEJER'S THEOREM

We have already had occasion to remark that the Fourier series for an arbitrary function f in $\mathcal{PC}[-\pi, \pi]$ may be divergent for certain values of x. In such a case one would naturally be inclined to feel that the series in question was a rather poor approximation to f, and, in particular, that it would be impossible to determine f from its series in the absence of any additional information. One of the really remarkable properties of Fourier series is that this impression is entirely false, and that the value of $f(x)$ can be found at *all* x where f is continuous even though the Fourier series for f is divergent in the usual sense. Needless to say, the method of "summation" which is used to accomplish this feat must be quite different from the standard one involving the partial sums of the series, and yet, at the same time, must yield the value which would be obtained by the method of partial sums whenever the latter do converge. In this section we shall introduce this new technique of summation, and then use it to prove Fejer's theorem—one of the most important results in the entire theory of Fourier series.

Let

$$a_0 + a_1 + a_2 + \cdots \tag{10–26}$$

be an infinite series of constants with partial sums

$$s_k = a_0 + a_1 + \cdots + a_k,$$

$k = 0, 1, 2, \ldots$. Then the sequence of *arithmetic means* associated with (10–26) is, by definition, the sequence $\{\sigma_k\}$, $k = 1, 2, \ldots$, with

$$\sigma_k = \frac{s_0 + s_1 + \cdots + s_{k-1}}{k} \tag{10–27}$$

for all $k > 0$. If this sequence converges to a limit σ as $k \to \infty$, that is, if

$$\lim_{k\to\infty} \sigma_k = \sigma,$$

we then say that (10–26) is *summable by the method of arithmetic means*, or *Cesàro summable*, and σ is called the *Cesàro sum* of the series.

EXAMPLE 1. Consider the series

$$1 - 1 + 1 - 1 + \cdots \tag{10–28}$$

with partial sums

$$s_0 = 1, \quad s_1 = 0, \quad s_2 = 1, \quad s_3 = 0, \ldots.$$

In this case the sequence of arithmetic means is

$$\{1, \tfrac{1}{2}, \tfrac{2}{3}, \tfrac{2}{4}, \tfrac{3}{5}, \tfrac{3}{6}, \ldots\}$$

with general terms

$$\sigma_{2k} = \frac{1}{2}, \qquad \sigma_{2k+1} = \frac{k+1}{2k+1},$$

and has $\frac{1}{2}$ as a limit. Thus (10–28) is Cesàro summable, with sum $\frac{1}{2}$, even though the series diverges in the usual sense of the term.

EXAMPLE 2. The geometric series

$$1 + \tfrac{1}{2} + \tfrac{1}{4} + \cdots = \sum_{k=0}^{\infty} \frac{1}{2^k},$$

with partial sums

$$s_0 = 1, \quad s_1 = 1 + \tfrac{1}{2}, \quad \ldots, \quad s_k = 2 - \frac{1}{2^k}, \quad \ldots,$$

converges to the value 2. In addition, it is Cesàro summable *to the same value*, since, by (10–27),

$$\begin{aligned} \sigma_k &= \frac{1}{k}(s_0 + s_1 + \cdots + s_{k-1}) \\ &= \frac{1}{k}\sum_{j=0}^{k-1}\left(2 - \frac{1}{2^j}\right) \\ &= 2 - \frac{1}{k}\sum_{j=0}^{k-1}\frac{1}{2^j} \\ &= 2 - \frac{1}{k}\left(2 - \frac{1}{2^{k-1}}\right), \end{aligned}$$

and

$$\lim_{k\to\infty} \sigma_k = 2.$$

These examples suggest that the notion of Cesàro summability is a generalization of the notion of ordinary convergence of the type stipulated above, in that the Cesàro sum of a convergent series always exists and is equal to the ordinary sum of the series. We now prove that this is the case by establishing

Theorem 10–8. *If an infinite series*

$$a_0 + a_1 + a_2 + \cdots \tag{10–29}$$

converges to the value σ, then the series is Cesàro summable, and its Cesàro sum is σ.

Proof. Let $\epsilon > 0$ be given. Then since (10–29) converges to σ, there exists an integer N such that

$$|s_n - \sigma| < \frac{\epsilon}{2}$$

for all $n > N$. We now consider the quantity

$$\begin{aligned}\sigma_n - \sigma &= \frac{(s_0 + s_1 + \cdots + s_{n-1}) - n\sigma}{n} \\ &= \frac{1}{n}\sum_{k=0}^{n-1}(s_k - \sigma).\end{aligned}$$

Then for $n > N$, we have

$$\begin{aligned}|\sigma_n - \sigma| &= \frac{1}{n}\left|\sum_{k=0}^{N}(s_k - \sigma) + \sum_{k=N+1}^{n}(s_k - \sigma)\right| \\ &\leq \frac{1}{n}\sum_{k=0}^{N}|s_k - \sigma| + \frac{1}{n}\sum_{k=N+1}^{n}|s_k - \sigma|.\end{aligned}$$

But by assumption, $|s_k - \sigma| < \epsilon/2$ for $k > N$. Hence

$$\begin{aligned}\frac{1}{n}\sum_{k=N+1}^{n}|s_k - \sigma| &< \frac{n-N}{n}\left(\frac{\epsilon}{2}\right) \\ &= \left(1 - \frac{N}{n}\right)\left(\frac{\epsilon}{2}\right) \\ &\leq \frac{\epsilon}{2}\cdot\end{aligned}$$

Moreover, since N is fixed, the quantity

$$\frac{1}{n}\sum_{k=0}^{N}|s_k - \sigma|$$

can also be made less than $\epsilon/2$ by choosing n sufficiently large, say $n > N'$, and when this is done we have

$$|\sigma_n - \sigma| < \frac{\epsilon}{2} + \frac{\epsilon}{2} = \epsilon$$

whenever $n > \max[N, N']$. ▮

The method of Cesàro summability can, of course, be applied to an infinite series of functions

$$\sum_{k=0}^{\infty} a_k(x). \tag{10–30}$$

In this case we say that the series is *Cesàro summable at the point* x_0 if the numerical series

$$\sum_{k=0}^{\infty} a_k(x_0)$$

is Cesàro summable in the sense of the above definition, and that (10–30) is *uniformly* (Cesàro) *summable* on an interval $[a, b]$ if the sequence $\{\sigma_k(x)\}$ of arithmetic means associated with (10–30) converges uniformly on $[a, b]$ to a function $\sigma(x)$; i.e., if, given any $\epsilon > 0$, an integer N can be found such that $|\sigma_n(x) - \sigma(x)| < \epsilon$ for *all* x in $[a, b]$, and all $n > N$. In these terms the proof of Theorem 10–8 also serves to establish

Theorem 10–9. *If the series*

$$\sum_{k=0}^{\infty} a_k(x)$$

converges uniformly to $\sigma(x)$ *on an interval* $[a, b]$, *where the* $a_k(x)$ *belong to* $\mathcal{PC}[a, b]$, *then the series is uniformly summable on* $[a, b]$ *to the same function* $\sigma(x)$.

With these preliminaries out of the way, we now state the theorem which justifies introducing the notion of Cesàro summability.

Theorem 10–10. (Fejer's theorem.) *If f is a continuous function on* $(-\infty, \infty)$ *with period* 2π, *then its Fourier series is* **uniformly summable** *to f on every closed interval of the x-axis.*

We have already stated that this theorem is one of the most important in the theory of Fourier series, and before giving a proof we point out some of its far-reaching consequences. In the first place, the conditions imposed here are *not* sufficient even to guarantee that the Fourier series for f converges pointwise, much less that the convergence be uniform. In spite of this, Fejer's theorem asserts that we can "sum" the series in question, and thereby recover the function f. Moreover this summability is sufficiently well behaved so as to proceed *uniformly* on closed intervals—a truly remarkable fact. Hence one can legitimately say that the Fourier series for a continuous periodic function in $\mathcal{PC}[-\pi, \pi]$ *does* serve to determine the function from which it was derived. (At the end of this section we shall generalize these results in the usual way to piecewise continuous functions as well.)

Finally, we note that Fejer's theorem also implies the following fact.

Theorem 10–11. *If a trigonometric series*

$$\frac{a_0}{2} + \sum_{k=1}^{\infty} (a_k \cos kx + b_k \sin kx) \tag{10–31}$$

is known to be the Fourier series of a continuous function f on $(-\infty, \infty)$, *and if this series converges in the usual sense when* $x = x_0$, *then the series* **must** *converge to* $f(x_0)$.

Proof. By Fejer's theorem, (10–31) is Cesàro summable to $f(x)$ for all x. If, in addition, (10–31) converges pointwise when $x = x_0$, then, by Theorem 10–8, the value of the series must be the same as its Cesàro sum, namely $f(x_0)$. ▌

At first sight Theorem 10–11 may seem to be stating the obvious, since the reader has probably long since assumed that a Fourier series must converge to the function from which it was obtained if it converges at all. Until now, however, we have proved this fact only for piecewise *smooth* functions, and there is nothing in our earlier results to prevent the Fourier series for a continuous function from converging pointwise to an entirely different function. This, as we now see, is impossible.

Turning to the proof of Fejer's theorem we begin by establishing an elementary lemma on approximating continuous functions, which, though obvious, does stand in need of proof. The result in question asserts that every continuous function on a closed interval can be *uniformly approximated* by a "broken line function," meaning a continuous function whose graph is made up of a finite number of line segments as shown in Fig. 10–9. Since we shall have occasion to refer to this result in our later work, we state it formally as follows.

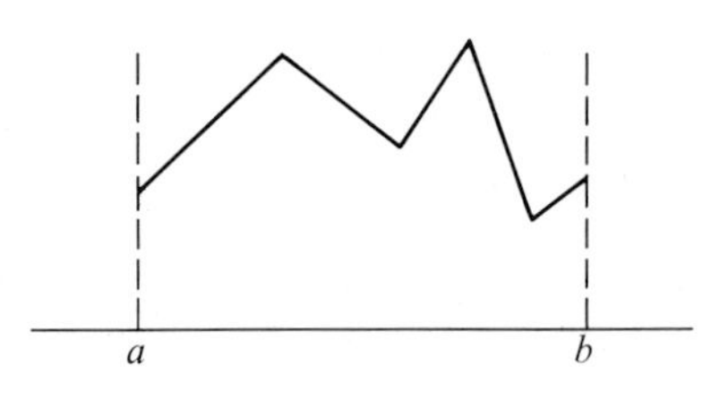

FIGURE 10–9

Lemma 10–4. *Let f be a continuous function on a closed interval $[a, b]$. Then, given any $\epsilon > 0$, there exists a broken line function B on $[a, b]$ such that*

$$|f(x) - B(x)| < \epsilon$$

for all x in $[a, b]$.

Proof. The proof follows the construction illustrated in Fig. 10–10, and goes as follows.

Since f is uniformly continuous on $[a, b]$ there exists a $\delta > 0$ such that

$$|f(x_1) - f(x_2)| < \frac{\epsilon}{2} \tag{10–32}$$

whenever $|x_1 - x_2| < \delta$. (Theorem I–13, Appendix I.) Moreover, since $[a, b]$ is of finite length we can find points

$$a = x_0 < x_1 < \cdots < x_n = b$$

in $[a, b]$ such that each of the intervals $I_k = [x_{k-1}, x_k]$ has length less than δ. The function B is now constructed by successively joining the points $(x_0, f(x_0))$, $(x_1, f(x_1)), \ldots, (x_n, f(x_n))$ on the graph of f by straight line segments. Thus if x belongs to the interval I_k,

$$B(x) = f(x_{k-1}) + \frac{f(x_k) - f(x_{k-1})}{x_k - x_{k-1}}(x - x_{k-1})$$

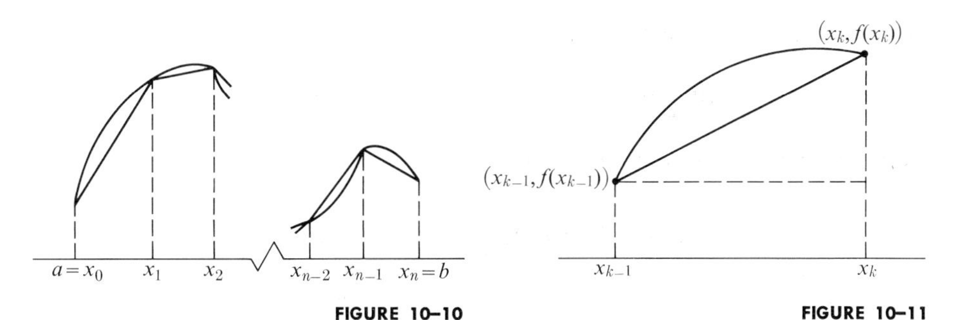

FIGURE 10–10

FIGURE 10–11

(Fig. 10–11), and we have

$$\begin{aligned} |f(x) - B(x)| &= \left| f(x) - f(x_{k-1}) - \frac{f(x_k) - f(x_{k-1})}{x_k - x_{k-1}} (x - x_{k-1}) \right| \\ &\leq |f(x) - f(x_{k-1})| + \frac{|f(x_k) - f(x_{k-1})|}{|x_k - x_{k-1}|} |x - x_{k-1}|. \end{aligned}$$

But by (10–32),

$$|f(x) - f(x_{k-1})| < \frac{\epsilon}{2}, \qquad |f(x_k) - f(x_{k-1})| < \frac{\epsilon}{2},$$

and since

$$\frac{|x - x_{k-1}|}{|x_k - x_{k-1}|} \leq 1,$$

we conclude that

$$|f(x) - B(x)| < \frac{\epsilon}{2} + \frac{\epsilon}{2} = \epsilon,$$

as required. ▌

Our next result is the key to the proof of Fejer's theorem, and is of sufficient independent interest to be stated separately.

Lemma 10–5. *Let f be piecewise continuous on $(-\infty, \infty)$ with period 2π, and let M denote the maximum value of $|f(x)|$ for all x. Then if σ_n denotes the* nth *arithmetic mean of the Fourier series for f,*

$$|\sigma_n(x)| \leq M$$

for **all** *n, and* **all** *x.*

Proof. If $s_n(x)$ denotes the nth partial sum of the Fourier series for f, then by (10–5),

$$s_n(x) = \frac{1}{\pi} \int_{-\pi}^{\pi} f(x + s) \frac{\sin (n + \frac{1}{2})s}{2 \sin (s/2)} \, ds.$$

Hence

$$\sigma_n(x) = \frac{1}{n\pi} \int_{-\pi}^{\pi} \frac{f(x + s)}{2 \sin (s/2)} \left[\sum_{k=0}^{n-1} \sin (k + \tfrac{1}{2})s \right] ds.$$

But by summing the identity

$$2 \sin \frac{s}{2} \sin (k + \tfrac{1}{2})s = \cos ks - \cos (k + 1)s$$

as k runs from 0 to $n - 1$, we obtain

$$\begin{aligned} 2 \sin \frac{s}{2} \sum_{k=0}^{n-1} \sin (k + \tfrac{1}{2})s &= 1 - \cos ns \\ &= 2 \sin^2 \frac{n}{2} s, \end{aligned}$$

and it follows that

$$\sigma_n(x) = \frac{1}{n\pi} \int_{-\pi}^{\pi} f(x + s) \frac{\sin^2 (n/2)s}{2 \sin^2 (s/2)} \, ds. \tag{10–33}$$

Now when $f(x) \equiv 1$, it is clear that $s_k(x) \equiv 1$ for all k (consider the Fourier series for f). Thus, in this case, $\sigma_k(x) \equiv 1$ for all k, and (10–33) yields

$$\frac{1}{n\pi} \int_{-\pi}^{\pi} \frac{\sin^2 (n/2)s}{2 \sin^2 (s/2)} \, ds = 1. \tag{10–34}$$

Hence if $|f(x)| \leq M$ for all x,

$$|\sigma_n(x)| \leq \frac{M}{n\pi} \int_{-\pi}^{\pi} \frac{\sin^2 (n/2)s}{2 \sin^2 (s/2)} \, ds = M,$$

as asserted. ▮

With these facts in hand, the proof of Fejer's theorem is all but obvious. Indeed, let f be continuous and periodic on $(-\infty, \infty)$ with period 2π, and let $\epsilon > 0$ be given. We must show that if n is sufficiently large,

$$|f(x) - \sigma_n(x)| < \epsilon$$

for *all* x, where σ_n is the nth arithmetic mean of the Fourier series for f.

But by Lemma 10–4, we can find a continuous broken line function B, with period 2π, such that

$$|f(x) - B(x)| < \frac{\epsilon}{3} \tag{10–35}$$

for all x. Setting

$$g(x) = f(x) - B(x),$$

so that $|g(x)| < \epsilon/3$ for all x, we let σ'_n and σ''_n denote, respectively, the nth arithmetic means of the Fourier series for g and B. Then since

$$f(x) = g(x) + B(x),$$

we have

$$\sigma_n(x) = \sigma'_n(x) + \sigma''_n(x)$$

for all x, and it follows that

$$\begin{aligned}|f(x) - \sigma_n(x)| &= |g(x) - \sigma_n'(x) + B(x) - \sigma_n''(x)| \\ &\leq |g(x)| + |\sigma_n'(x)| + |B(x) - \sigma_n''(x)|.\end{aligned}$$

But since $|g(x)| < \epsilon/3$ for all x, Lemma 10–5 implies that *the same is true of* $|\sigma_n'(x)|$. Finally, since the Fourier series for B converges uniformly to B on every closed interval of the x-axis, Theorem 10–9 allows us to conclude that its associated sequence of arithmetic means also converges uniformly to B. Hence there exists an integer N such that

$$|B(x) - \sigma_n''(x)| < \frac{\epsilon}{3}$$

for all $n > N$ and all x, and we have

$$|f(x) - \sigma_n(x)| < \frac{\epsilon}{3} + \frac{\epsilon}{3} + \frac{\epsilon}{3} = \epsilon,$$

which is precisely what had to be shown. ▌

Functions with jump discontinuities can be handled in much the same way, in which case we have the following generalized version of Fejer's theorem.

Theorem 10–12. *If f is piecewise continuous on $(-\infty, \infty)$, with period 2π, then the Fourier series for f is Cesàro summable for all x, with sum*

$$\frac{f(x^+) + f(x^-)}{2}.$$

Furthermore, this summability is uniform on any closed interval of the x-axis not containing a point of discontinuity of f.

The proof has been left to the reader as an exercise (Exercise 8).

At this point we can truthfully say that the theory of Fourier series for piecewise continuous functions is complete. Indeed, we now know that every such series is "summable," either in the standard fashion when the function involved is piecewise smooth, or by the method of arithmetic means. Moreover, the series will *always* converge pointwise in either the standard or Cesàro sense to the function from which it was derived (so long as the obvious proviso is made for points of discontinuity), and this convergence will be uniform on any closed interval in which the function is continuous. Truly, then, there is little more to be said.

At the same time, however, the reader should not be misled into thinking that we have uttered the last word on the subject of Fourier series. This is far from being the case. Nevertheless, at this point the direction of inquiry changes abruptly, and addresses itself to the task of generalizing the above results to wider classes of functions. Such generalizations do exist, but are based upon an entirely different type of integral than the one we have been using, and are beyond the reach of an introductory text. Thus, as soon as we have settled the one point which still remains outstanding (i.e., the Weierstrass approximation theorem) our discussion will be complete, and we can turn our attention to other types of orthogonal series, and their applications to physical problems.

EXERCISES

1. Let $\sum_{k=0}^{\infty} a_k$ and $\sum_{k=0}^{\infty} b_k$ be Cesàro summable, with sums σ and τ, respectively. Prove that

$$\sum_{k=0}^{\infty} (\alpha a_k + \beta b_k)$$

is Cesàro summable with sum $\alpha\sigma + \beta\tau$ for all real numbers α, β. What does this imply about the set of all Cesàro summable infinite series?

2. Let $\sum_{k=0}^{\infty} a_k$ be Cesàro summable, with sum σ. Prove that $\sum_{k=n}^{\infty} a_k$ is also Cesàro summable for all $n \geq 0$, and find the sum of the series.

3. Find the Cesàro sum of each of the following series.
   (a) $1 + 0 - 1 + 0 + 1 + 0 - 1 + \cdots$
   (b) $1 + 0 + 0 - 1 + 0 + 0 + 1 + 0 + 0 - 1 + \cdots$

4. (a) Show that the series

$$\sin x + \sin 2x + \sin 3x + \cdots$$

is Cesàro summable to zero in the interval $(0, 2\pi)$. [*Hint:* Use Formula (10–10).]
   (b) Show that the series

$$\tfrac{1}{2} + \cos x + \cos 2x + \cos 3x + \cdots$$

is Cesàro summable to zero in the interval $(0, 2\pi)$. [*Hint:* Use Formula (10–4).]

5. Let $(a_0/2) + \sum_{k=1}^{\infty} (a_k \cos kx + b_k \sin kx)$ be the Fourier series expansion of a function f in $\mathcal{PC}[-\pi, \pi]$, and let $\{\sigma_k\}$ be the sequence of arithmetic means associated with this series, that is, $\sigma_n = (s_0 + \cdots + s_{n-1})/n$. Prove that

$$\frac{1}{\pi}\int_{-\pi}^{\pi} [\sigma_n(x) - f(x)]^2\, dx = \sum_{k=1}^{n} \left(\frac{k}{n}\right)^2 (a_k^2 + b_k^2) + \sum_{k=n+1}^{\infty} (a_k^2 + b_k^2).$$

6. Let $a_0 + a_1 + a_2 + \cdots$ be an infinite series with arithmetic means σ_k, $k = 1, 2, \ldots,$ and, for each k, set

$$\tau_k = \frac{\sigma_1 + \sigma_2 + \cdots + \sigma_k}{k}.$$

The sequence $\{\tau_k\}$ obtained in this way is known as the sequence of second arithmetic means associated with the given series, and the series is said to be summable to the value τ by the method of second arithmetic means if $\lim_{k\to\infty} \tau_k = \tau$. (Higher orders of summability by arithmetic means can also be defined.)
   (a) Prove that the series

$$1 - 2 + 3 - 4 + \cdots$$

is not Cesàro summable, but is summable by the method of second arithmetic means.

(b) Prove that every series which is Cesàro summable is also summable to the same value by the method of second arithmetic means.

7. Let s_k and σ_k denote, respectively, the kth partial sum and arithmetic mean of the series

$$a_0 + a_1 + a_2 + \cdots.$$

(a) Show that

$$s_{k+1} = (k+1)\sigma_{k+1} - k\sigma_k, \quad k > 0,$$

and use this result to prove that $\lim_{k\to\infty} s_k/k = 0$ whenever $a_0 + a_1 + a_2 + \cdots$ is Cesàro summable.

(b) Use the result in (a) to deduce that the series

$$1^2 - 2^2 + 3^2 - 4^2 + \cdots$$

is *not* Cesàro summable.

8. Prove Theorem 10–12. [*Hint:* First show that it is sufficient to prove

$$\lim_{n\to\infty} \frac{1}{n\pi} \int_0^{\pi} [f(x+s) - f(x^+)] \frac{\sin^2 (n/2)s}{2\sin^2 (s/2)}\, ds = 0$$

and

$$\lim_{n\to\infty} \frac{1}{n\pi} \int_{-\pi}^{0} [f(x+s) - f(x^-)] \frac{\sin^2 (n/2)s}{2\sin^2 (s/2)}\, ds = 0$$

for all x. Let $\epsilon > 0$ be given, and divide the integral appearing in this expression into two parts as

$$\frac{1}{n\pi} \int_0^{\delta} [f(x+s) - f(x^+)] \frac{\sin^2 (n/2)s}{2\sin^2 (s/2)}\, ds + \frac{1}{n\pi} \int_{\delta}^{\pi} [f(x+s) - f(x^+)] \frac{\sin^2 (n/2)s}{2\sin^2 (s/2)}\, ds$$

where δ is chosen so that $|f(x+s) - f(x)| < \epsilon$ for all s in the interval $(0, \delta]$. Treat the second limit similarly.]

*10–8 THE WEIERSTRASS APPROXIMATION THEOREM

To establish the version of the Weierstrass approximation theorem cited in the preceding chapter we must first show that every continuous periodic function on $(-\infty, \infty)$ can be uniformly approximated by a "smooth" function. This is the content of

Theorem 10–13. *Let f be continuous and periodic on $(-\infty, \infty)$ with period 2π, and let $\epsilon > 0$ be given. Then there exists a function g which is continuously differentiable and periodic on $(-\infty, \infty)$ with period 2π, such that $|f(x) - g(x)| < \epsilon$ for all x.*

Proof. For each $\delta > 0$ let

$$F_\delta(x) = \frac{1}{2\delta} \int_{-\delta}^{\delta} f(x+t)\, dt.$$

(Note that $F_\delta(x)$ is the average value of $f(x)$ in the interval $x - \delta \leq t \leq x + \delta$.) Then F_δ is continuous and periodic on $(-\infty, \infty)$ with period 2π, and if we set $u = x + t$, so that

$$F_\delta(x) = \frac{1}{2\delta} \int_{x-\delta}^{x+\delta} f(u)\, du,$$

we find that

$$F_\delta'(x) = \frac{1}{2\delta} [f(x + \delta) - f(x - \delta)]$$

(see Appendix I). Thus F_δ' is also continuous and periodic on $(-\infty, \infty)$.

Next, we observe that

$$\begin{aligned} |F_\delta(x) - f(x)| &= \left| \frac{1}{2\delta} \int_{x-\delta}^{x+\delta} f(u)\, du - \frac{1}{2\delta} \int_{x-\delta}^{x+\delta} f(x)\, du \right| \\ &= \frac{1}{2\delta} \left| \int_{x-\delta}^{x+\delta} [f(u) - f(x)]\, du \right| \\ &\leq \frac{1}{2\delta} \int_{x-\delta}^{x+\delta} |f(u) - f(x)|\, du. \end{aligned}$$

But since f is continuous and periodic on $(-\infty, \infty)$, it is *uniformly continuous*, and hence there exists a number $\delta = \delta(\epsilon)$ such that $|f(u) - f(x)| < \epsilon$ whenever $|x - u| < \delta(\epsilon)$. We now set $g = F_{\delta(\epsilon)}$, and use the above inequality to deduce that

$$|g(x) - f(x)| \leq \frac{1}{2\,\delta(\epsilon)} \int_{x-\delta(\epsilon)}^{x+\delta(\epsilon)} |f(u) - f(x)|\, du < \epsilon. \blacksquare$$

Theorem 10–14. (The Weierstrass approximation theorem for trigonometric polynomials.) *If f is continuous and periodic on $(-\infty, \infty)$ with period 2π, then, given any $\epsilon > 0$, there exists a trigonometric polynomial*

$$T_N(x) = A_0 + \sum_{k=1}^{N} (A_k \cos kx + B_k \sin kx)$$

such that

$$|f(x) - T_N(x)| < \epsilon$$

for all x.

Proof. According to the preceding theorem we can find a function g such that g and g' are continuous and periodic on $(-\infty, \infty)$ with period 2π, and $|f(x) - g(x)| < \epsilon/2$ for all x. Thus if T_n denotes the trigonometric polynomial consisting of the nth partial sum of the Fourier series for g, Theorem 10–2 implies that there exists an integer N such that $|g(x) - T_N(x)| < \epsilon/2$ for all x. Hence

$$|f(x) - T_N(x)| \leq |f(x) - g(x)| + |g(x) - T_N(x)| < \frac{\epsilon}{2} + \frac{\epsilon}{2} = \epsilon,$$

and we are done. $\blacksquare$

In the next chapter we shall have occasion to refer to the version of the Weierstrass approximation theorem which uses ordinary polynomials, and hence we now prove this result as well. Our argument is based on Theorem 10–14, and begins with the following lemma.

Lemma 10–6. *If*

$$T(x) = a_0 + \sum_{k=1}^{n} a_n \cos kx \tag{10–36}$$

is a trigonometric polynomial involving only cosine terms, then there exist constants $A_0, A_1, \ldots, A_n$ such that

$$T(x) = A_0 + A_1 \cos x + \cdots + A_n \cos^n x.$$

Proof. When $n = 0$ or $n = 1$ there is nothing to prove. Thus we can proceed by induction, assuming the validity of the lemma for all trigonometric polynomials of degree $n - 1$ or less ($n \geq 1$). Let T be given as in (10–36). Then

$$T(x) = a_0 + \left(\sum_{k=1}^{n-1} a_k \cos kx\right) + a_n \cos nx,$$

and, by assumption, we can find constants $A_0, \ldots, A_{n-1}$ such that

$$T(x) = A_0 + A_1 \cos x + \cdots + A_{n-1} \cos^{n-1} x + a_n \cos nx.$$

But

$$\cos nx = 2 \cos [(n - 1)x] \cos x - \cos (n - 2)x,$$

and, applying the induction assumption once more, we can write $\cos (n - 1)x$ and $\cos (n - 2)x$ as polynomials involving powers of $\cos x$. This, of course, implies that T can be written in the desired form. ▮

Theorem 10–15. *Let f be continuous on the interval $[-1, 1]$, and let $\epsilon > 0$ be given. Then there exists a polynomial P such that*

$$|f(x) - P(x)| < \epsilon$$

for all x in $[-1, 1]$.

Proof. Consider the function

$$F(t) \equiv f[\cos t]$$

for $-\pi \leq t \leq \pi$. Then F is continuous, periodic, and even. Hence the F_δ approximations of Theorem 10–13 are also *even*, and it follows that their Fourier series contain only cosine terms. Thus, by the preceding lemma a partial sum $T(t) = A_0 + A_1 \cos t + \cdots + A_n \cos^n t$ of the Fourier series for some F_δ will satisfy

$$|F(t) - T(t)| < \epsilon$$

for all t in $[-\pi, \pi]$. Setting $x = \cos t$ yields $|f(x) - P(x)| < \epsilon$ for all x in $[-1, 1]$, with

$$P(x) = A_0 + A_1 x + \cdots + A_n x^n. \blacksquare$$

From here it is an easy step to our final result.

Theorem 10–16. (The Weierstrass approximation theorem for ordinary polynomials.) *Let f be a continuous function on a closed interval* $[a, b]$. *Then, given any* $\epsilon > 0$, *there exists a polynomial P such that*

$$|f(x) - P(x)| < \epsilon$$

for all x in $[a, b]$. *In other words, a continuous function on a closed interval can be uniformly approximated by polynomials.*

The asserted result follows immediately from Theorem 10–15 and the fact that the mapping

$$F(x) \to F\left(\frac{b+a}{2} + \frac{b-a}{2}x\right)$$

establishes a one-to-one correspondence between $\mathcal{C}[-1, 1]$ and $\mathcal{C}[a, b]$.

11 orthogonal series of polynomials

11–1 INTRODUCTION

In this chapter we continue the study of series expansions in Euclidean spaces by introducing three classic orthogonal series expressed in terms of polynomials. At the moment, of course, it is not at all clear that there is anything to be gained by using polynomials in place of trigonometric functions, and we shall have to ask the reader to reserve judgment on this point until we come to the study of boundary value problems. In fact, this entire chapter can be omitted without serious prejudice until Chapter 13 has been read, and even then it will be possible to continue with no more than a knowledge of Legendre polynomials and series (Sections 11–2 through 11–4). Nevertheless, the reader who pursues this discussion to its conclusion will enhance his appreciation for the scope and subtility of the theory of orthogonal series, and will be that much better prepared for the material which follows.

Before we begin, it may be appropriate to remark, once more, that our present investigations are a natural outgrowth of the ideas developed in the study of Euclidean spaces. Thus, although certain portions of the following discussion are technically involved, the issues at stake are simple and familiar. The student who keeps this point firmly in mind as he reads on should then be able to see the forest while among the trees.

11–2 LEGENDRE POLYNOMIALS

In Section 7–4 we applied the Gram-Schmidt orthogonalization process to the linearly independent set $1, x, x^2, \ldots$ in $\mathcal{PC}[-1, 1]$ to obtain an orthogonal sequence of *polynomials*

$$p_0(x),\ p_1(x),\ p_2(x), \ldots. \tag{11–1}$$

As we shall see, these polynomials actually form a basis for $\mathcal{PC}[-1, 1]$, and were it not for the complications involved in applying the orthogonalization process we could proceed directly to the study of series expansions relative to (11–1). But these complications *do* exist, and are serious enough to make it easier to start

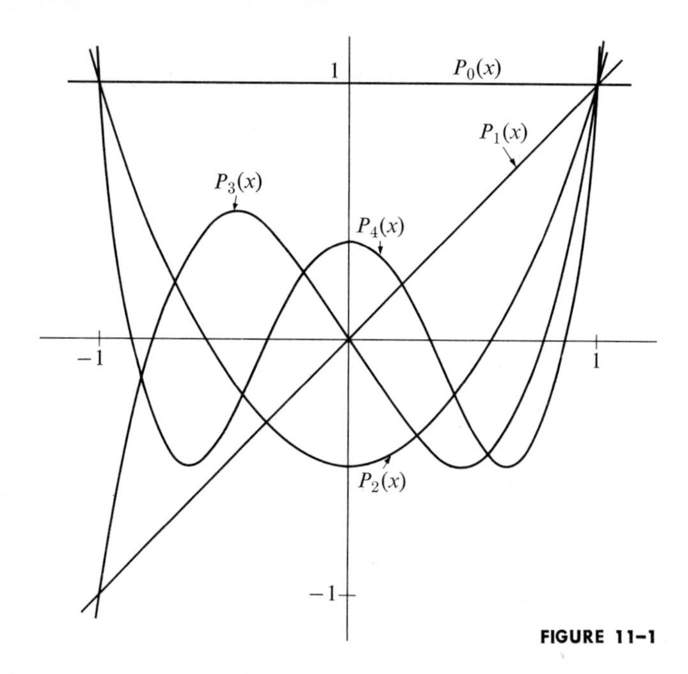

FIGURE 11–1

afresh with a slightly different polynomial basis. This, then, is the reason for introducing the so-called Legendre polynomials—they are a basis for $\mathcal{PC}[-1, 1]$, and are reasonably amenable to computations.

Definition 11–1. Let $\{P_n(x)\}$, $n = 0, 1, 2, \ldots$, be the sequence of polynomials defined as follows:

$$P_0(x) = 1, \tag{11–2}$$

and

$$P_n(x) = \frac{1}{2^n n!} \frac{d^n}{dx^n} (x^2 - 1)^n, \tag{11–3}$$

for $n > 0$. Then $P_n(x)$ is called the *Legendre polynomial* of degree n, and (11–3) is known as *Rodrigues' formula* for these polynomials.*

It is clear from (11–3) that P_n is a polynomial of degree n, and by direct computation we see that

$$P_0(x) = 1, \qquad P_1(x) = x,$$
$$P_2(x) = \tfrac{3}{2}x^2 - \tfrac{1}{2}, \qquad P_3(x) = \tfrac{5}{2}x^3 - \tfrac{3}{2}x,$$
$$P_4(x) = \tfrac{35}{8}x^4 - \tfrac{15}{4}x^2 + \tfrac{3}{8}, \qquad P_5(x) = \tfrac{63}{8}x^5 - \tfrac{35}{4}x^3 + \tfrac{15}{8}x.$$

---

* This formula is also valid when $n = 0$, provided D^0 is interpreted as the identity operator.

(See Fig. 11-1.) Moreover, since all of the powers of x in $(x^2 - 1)^n$ are even, P_{2n} contains only even powers of x, while P_{2n+1} contains only odd powers of x. This phenomenon is already apparent in the above list, and will prove useful as we continue.

Before going on to establish some of the basic properties of the Legendre polynomials, we prove a general theorem concerning orthogonal sequences of polynomials in $\mathcal{PC}[a, b]$ which will serve to relate the Legendre polynomials to the polynomials in (11-1). The theorem we have in mind is an immediate consequence of the following lemma.

Lemma 11-1. *Let $\{R_n(x)\}$, $n = 0, 1, 2, \ldots$, be an orthogonal sequence of polynomials in $\mathcal{PC}[a, b]$ indexed by degree (i.e., R_n is of degree n). Then for each n, R_n is orthogonal in $\mathcal{PC}[a, b]$ to every polynomial of degree $<n$.*

Proof. Let $\mathcal{P}_m$ denote the m-dimensional subspace of $\mathcal{PC}[a, b]$ consisting of all polynomials of degree $<m$, together with the zero polynomial. Then, R_0, $R_1, \ldots, R_{m-1}$ is an orthogonal basis for $\mathcal{P}_m$, and every polynomial Q of degree $<m$ can be written in the form

$$Q = \alpha_0 R_0 + \cdots + \alpha_{m-1} R_{m-1},$$

where

$$\alpha_k = \frac{Q \cdot R_k}{R_k \cdot R_k}, \quad k = 0, \ldots, m - 1.$$

Hence

$$Q \cdot R_n = \alpha_0 (R_0 \cdot R_n) + \cdots + \alpha_{m-1}(R_{m-1} \cdot R_n),$$

and since $R_k \cdot R_n = 0$ if $k \neq n$, it follows that $Q \cdot R_n = 0$, as asserted. ▮

Theorem 11-1. *Let $\{Q_n(x)\}$ and $\{R_n(x)\}$, $n = 0, 1, 2, \ldots$, be orthogonal sequences of polynomials in $\mathcal{PC}[a, b]$ indexed by degree. Then, for each n, Q_n and R_n are scalar multiples of one another.*

Proof. Since $Q_0, \ldots, Q_n$ is an orthogonal basis for $\mathcal{P}_{n+1}$,

$$R_n(x) = \frac{R_n \cdot Q_0}{Q_0 \cdot Q_0} Q_0(x) + \frac{R_n \cdot Q_1}{Q_1 \cdot Q_1} Q_1(x) + \cdots + \frac{R_n \cdot Q_n}{Q_n \cdot Q_n} Q_n(x).$$

However, by the preceding lemma, R_n is orthogonal to $Q_0, \ldots, Q_{n-1}$. Hence this equation reduces to

$$R_n(x) = \frac{R_n \cdot Q_n}{Q_n \cdot Q_n} Q_n(x),$$

and we are done. ▮

Among other things, Theorem 11–1 implies that, up to scalar multiples, $\mathcal{PC}[-1, 1]$ contains only one orthogonal sequence of polynomials $\{R_n(x)\}$ indexed by degree. Thus if each of the R_n is chosen so that its leading coefficient is one, the sequence in question is *uniquely* determined. But these two properties are enjoyed by the polynomials in the orthogonal sequence (11–1) (see Exercise 16, Section 7–4), and we therefore have

Corollary 11–1. *The only orthogonal sequence of polynomials* $\{p_n(x)\}$, $n = 0, 1, 2, \ldots$, *in* $\mathcal{PC}[-1, 1]$ *with the property that* p_n *has leading coefficient one and is of degree n for each n is the sequence obtained by orthogonalizing* $1, x, x^2, \ldots$.

Finally, if we accept the assertion that the Legendre polynomials are mutually orthogonal in $\mathcal{PC}[-1, 1]$, Theorem 11–1 implies that each P_n is a constant multiple of the corresponding polynomial in (11–1). This is, in fact, the case, and we shall see later that

$$P_n(x) = \frac{(2n)!}{2^n(n!)^2}\, p_n(x), \tag{11–4}$$

a formula which gives a second possible definition of the Legendre polynomials.

EXERCISES

1. Use Rodrigues' formula to compute the values of $P_1, \ldots, P_5$.
2. Prove that P_{2m} and P_{2n+1} are orthogonal in $\mathcal{PC}[-1, 1]$ for all m and n.
3. Use Rodrigues' formula to prove that P_1 is orthogonal to P_n in $\mathcal{PC}[-1, 1]$ for all $n \neq 1$.
4. Prove that if $\mathcal{PC}[-1, 1]$ has a basis $\mathcal{B}$ consisting of polynomials, then $\mathcal{B}$ must contain exactly one polynomial of each degree. Hence deduce that up to scalar multiples the sequence of polynomials given in (11–1) is the only polynomial basis for $\mathcal{PC}[-1, 1]$.

11–3 ORTHOGONALITY: THE RECURRENCE RELATION

Our first task is to prove that the Legendre polynomials are mutually orthogonal in $\mathcal{PC}[-1, 1]$. To this end we begin by showing that P_n is a solution of a certain linear differential equation, a fact which in itself is of considerable importance.* Thus let $(x^2 - 1)^n = w$, and let $w^{(k)}$ denote the kth derivative of w. Then

$$w^{(1)} = 2nx(x^2 - 1)^{n-1},$$

* A direct proof of orthogonality can also be given along the lines suggested in Exercise 5 below.

and, multiplying by $x^2 - 1$, we have

$$(x^2 - 1)w^{(1)} - 2nxw = 0. \tag{11-5}$$

Repeated differentiation of (11-5) yields

$$\begin{aligned}
&(x^2 - 1)w^{(2)} - 2x(n - 1)w^{(1)} - 2nw = 0,\\
&(x^2 - 1)w^{(3)} - 2x(n - 2)w^{(2)} - 2[n + (n - 1)]w^{(1)} = 0,\\
&\quad\vdots\\
&(x^2 - 1)w^{(k+2)} - 2x[n - (k + 1)]w^{(k+1)}\\
&\qquad\qquad -2[n + (n - 1) + \cdots + (n - k)]w^{(k)} = 0.
\end{aligned}$$

But since

$$n + (n - 1) + \cdots + (n - k) = \frac{(2n - k)(k + 1)}{2}$$

(Exercise 7), this last equation reduces to

$$(x^2 - 1)w^{(k+2)} - 2x[n - (k + 1)]w^{(k+1)} - (2n - k)(k + 1)w^{(k)} = 0.$$

We now set $k = n$, and observe that, by definition,

$$P_n = \frac{w^{(n)}}{2^n n!}.$$

Thus, if the above equation is multiplied by $-1/2^n n!$, it can be rewritten

$$(1 - x^2)P_n'' - 2xP_n' + n(n + 1)P_n = 0,$$

and we have proved

Theorem 11-2. *The* nth *Legendre polynomial* P_n *is a solution of the second-order linear differential equation*

$$(1 - x^2)y'' - 2xy' + n(n + 1)y = 0. \tag{11-6}$$

It is not difficult to show that every polynomial solution of Equation (11-6)—which, by the way, is known as *Legendre's equation of order* n—is a constant multiple of P_n (see Example 2, Section 6-5). Thus (11-6) characterizes P_n up to a constant multiple. Many treatments of the theory of Legendre polynomials start at this point, and *define* P_n as the polynomial solution of Legendre's equation of order n which assumes the value one when $x = 1$ (cf., Theorem 11-5 below).

Using (11-6) it is easy to establish the orthogonality of the Legendre polynomials in $\mathcal{PC}[-1, 1]$. Indeed, starting with the pair of equations

$$\begin{aligned}
(1 - x^2)P_n'' - 2xP_n' + n(n + 1)P_n &= 0,\\
(1 - x^2)P_m'' - 2xP_m' + m(m + 1)P_m &= 0,
\end{aligned}$$

we multiply the first by P_m, the second by P_n, and subtract to get

$$(1 - x^2)[P_m P_n'' - P_m'' P_n] - 2x[P_m P_n' - P_m' P_n] = P_m P_n[m(m + 1) - n(n + 1)]. \tag{11-7}$$

But the left-hand side of this equation is just the derivative of

$$(1 - x^2)[P_m P_n' - P_m' P_n],$$

and hence, integrating (11–7) from -1 to 1, we obtain

$$(1 - x^2)[P_m P_n' - P_m' P_n]\Big|_{-1}^{1} = [m(m + 1) - n(n + 1)]\int_{-1}^{1} P_m(x)P_n(x)\,dx.$$

Finally, since $1 - x^2$ vanishes at the upper and lower limits of integration, we have

$$[m(m + 1) - n(n + 1)]\int_{-1}^{1} P_m(x)P_n(x)\,dx = 0,$$

and thus, if $m \neq n$,

$$\int_{-1}^{1} P_m(x)P_n(x)\,dx = 0.$$

This completes the proof of

Theorem 11–3. *The Legendre polynomials are mutually orthogonal in* $\mathcal{PC}[-1, 1]$.

We now anticipate the construction of series expansions relative to the Legendre polynomials by computing $\|P_n\|$, $n = 0, 1, 2, \ldots$. Rather than attack this problem directly, we first derive an important formula known as the *recurrence relation* for the Legendre polynomials from which these values can be deduced without difficulty.

Here we begin by considering the function $xP_n(x)$, which is obviously a polynomial of degree $n + 1$. Thus, since $P_0, P_1, \ldots, P_{n+1}$, is an orthogonal basis for the subspace $\mathcal{P}_{n+2}$ of $\mathcal{PC}[-1, 1]$, we have

$$xP_n(x) = \sum_{k=0}^{n+1} \frac{(xP_n)\cdot P_k}{P_k \cdot P_k} P_k(x).$$

But, by Lemma 11–1, P_n is orthogonal to every polynomial of degree $<n$, so that

$$(xP_n)\cdot P_k = P_n \cdot (xP_k) = 0$$

whenever $k < n - 1$. Hence

$$xP_n(x) = \frac{(xP_n)\cdot P_{n-1}}{P_{n-1}\cdot P_{n-1}} P_{n-1}(x) + \frac{(xP_n)\cdot P_n}{P_n \cdot P_n} P_n(x) + \frac{(xP_n)\cdot P_{n+1}}{P_{n+1}\cdot P_{n+1}} P_{n+1}(x).$$

This equation can be simplified still further if we note that $xP_n(x)^2$ is an odd function (it contains only odd powers of x), for it then follows that

$$(xP_n) \cdot P_n = \int_{-1}^{1} xP_n(x)^2 \, dx = 0.$$

This allows us to write

$$xP_n = \alpha P_{n+1} + \beta P_{n-1}, \tag{11–8}$$

where α and β are real numbers which we now propose to determine.

In the first place, since

$$P_k(x) = \frac{1}{2^k k!} \frac{d^k}{dx^k} (x^2 - 1)^k,$$

the coefficient of x^k in P_k is

$$\frac{1}{2^k k!} 2k(2k - 1) \cdots [2k - (k - 1)] = \frac{(2k)!}{2^k (k!)^2}. \tag{11–9}$$

Secondly, since

$$(x^2 - 1)^k = x^{2k} - kx^{2k-2} + \frac{k(k - 1)}{2!} x^{2k-4} - \cdots,$$

the coefficient of x^{k-2} in P_k is

$$\frac{1}{2^k k!} (-k)(2k - 2) \cdots [2k - (k + 1)] = - \frac{(2k - 2)!}{2^k (k - 1)!(k - 2)!}. \tag{11–10}$$

We now use (11–9) to compute the coefficients of x^{n+1} on both sides of (11–8) and equate the results, obtaining

$$\frac{(2n)!}{2^n (n!)^2} = \alpha \frac{[2(n + 1)]!}{2^{n+1}[(n + 1)!]^2}.$$

This implies that

$$\alpha = \frac{n + 1}{2n + 1},$$

and (11–8) becomes

$$xP_n = \frac{n + 1}{2n + 1} P_{n+1} + \beta P_{n-1}. \tag{11–11}$$

To find β we use (11–9) and (11–10) to compute the coefficients of x^{n-1} on both sides of (11–11), again equating results. This gives

$$- \frac{(2n - 2)!}{2^n (n - 1)!(n - 2)!} = - \frac{n + 1}{2n + 1} \cdot \frac{[2(n + 1) - 2]!}{2^{n+1} n!(n - 1)!} + \beta \frac{[2(n - 1)]!}{2^{n-1}[(n - 1)!]^2},$$

which, after a little arithmetic, yields

$$\beta = \frac{n}{2n+1}.$$

We now substitute this value in (11–11) and solve the resulting expression for P_{n+1} to obtain

Theorem 11–4. *The $(n + 1)$st Legendre polynomial satisfies the identity*

$$P_{n+1} = \frac{2n+1}{n+1} xP_n - \frac{n}{n+1} P_{n-1} \tag{11–12}$$

for all $n \geq 1$, and all x.

Remark. This result is also valid when $n = 0$ if we agree to set $P_{-1}(x) \equiv 0$.

Equation (11–12) is known as the *recurrence relation* for the Legendre polynomials and can be used to deduce properties of P_{n+1} from those of the two immediately preceding polynomials. For example, given that $P_0(x) = 1$ and that $P_1(x) = x$, we can compute P_2 from (11–12) by setting $n = 1$. This gives

$$P_2(x) = \tfrac{3}{2}x^2 - \tfrac{1}{2}.$$

From this, and the known value of P_1, we find that

$$\begin{aligned} P_3(x) &= \tfrac{5}{3}x(\tfrac{3}{2}x^2 - \tfrac{1}{2}) - \tfrac{2}{3}x \\ &= \tfrac{5}{2}x^3 - \tfrac{3}{2}x \\ &\ \vdots \end{aligned}$$

As a somewhat more substantial application of the recurrence relation we prove

Theorem 11–5. $P_n(1) = 1$ *for all n.*

Proof. The assertion obviously holds for P_0 and P_1. Moreover, if we assume that $P_0(1) = \cdots = P_n(1) = 1$, $n \geq 1$, the recurrence relation gives

$$P_{n+1}(1) = \frac{2n+1}{n+1} - \frac{n}{n+1} = 1.$$

The desired result now follows by mathematical induction. ▌

In much the same way it can be shown that $P_n(-1) = (-1)^n$.

We now use the recurrence relation to compute the value of $\|P_n\|$. Again a trick is needed, and this time it is furnished by the polynomial

$$p(x) = P_n(x) - \frac{2n-1}{n} xP_{n-1}(x). \tag{11–13}$$

A simple calculation using (11–9) reveals that this polynomial is of degree *less* than n (see Exercise 8), and hence, by Lemma 11–1, is orthogonal to P_n in $\mathcal{PC}[-1, 1]$. Thus, if we multiply (11–13) by P_n and integrate from -1 to 1 we get

$$\int_{-1}^{1} P_n(x)^2\,dx = \frac{2n-1}{n}\int_{-1}^{1} xP_{n-1}(x)P_n(x)\,dx. \tag{11–14}$$

This is half of what we need. The other half is found by multiplying the recurrence relation by P_{n-1} and then integrating. This gives

$$\frac{n}{n+1}\int_{-1}^{1} P_{n-1}(x)^2\,dx = \frac{2n+1}{n+1}\int_{-1}^{1} xP_{n-1}(x)P_n(x)\,dx.$$

Thus, by (11–14),

$$\int_{-1}^{1} P_n(x)^2\,dx = \frac{2n-1}{2n+1}\int_{-1}^{1} P_{n-1}(x)^2\,dx,$$

or

$$\|P_n\|^2 = \frac{2n-1}{2n+1}\|P_{n-1}\|^2, \quad n = 1, 2, \ldots$$

We now use the fact that $\|P_0\|^2 = 2$ to deduce that

$$\begin{aligned}
\|P_1\|^2 &= \tfrac{1}{3}\cdot 2\\
\|P_2\|^2 &= \tfrac{3}{5}\cdot\tfrac{1}{3}\cdot 2\\
&\vdots\\
\|P_n\|^2 &= \frac{2n-1}{2n+1}\cdot\frac{2n-3}{2n-1}\cdot\frac{2n-5}{2n-3}\cdots\frac{1}{3}\cdot 2,
\end{aligned}$$

and we have proved

Theorem 11–6. $\|P_n\|^2 = 2/(2n+1)$ *for all n.*

By now the reader may suspect that the list of formulas involving Legendre polynomials is almost endless. It is! But rather than continue the task of collecting them—interesting as that may be—we let matters rest where they are, and bring this discussion to a close with a theorem which will be of some importance in the following chapters.

Theorem 11–7. *The Legendre polynomial of degree n has n distinct (real) roots between* -1 *and* 1.

Proof. The assertion is obviously true when $n = 0$. Moreover, for $n \geq 1$,

$$\int_{-1}^{1} P_n(x)\,dx = \int_{-1}^{1} P_n(x)P_0(x)\,dx = 0,$$

by orthogonality, and it follows that P_n must change sign *at least once* in the interval $-1 < x < 1$, and hence has a root in this interval. Now let $x_1, x_2, \ldots, x_m$ be the roots of P_n in $(-1, 1)$, and consider the polynomial

$$Q(x) = (x - x_1)(x - x_2)\cdots(x - x_m).$$

(Note that Q is of degree m, with $1 \leq m \leq n$.) Then since P_n has no repeated roots (see Exercise 19), and since P_n and Q change sign at the same points in $(-1, 1)$, their product is either positive or negative throughout the entire interval, and it follows that

$$\int_{-1}^{1} P_n(x)Q(x)\,dx \neq 0.$$

However, by Lemma 11–1, P_n is orthogonal to every polynomial of degree less than n, and hence would be orthogonal to Q were $m < n$. The preceding inequality excludes that possibility, and forces us to conclude that $m = n$, as asserted. ▌

[*Note:* For convenience of reference we have summarized the essential information relating to the various polynomials studied in this chapter in tabular form at the end of the chapter.]

EXERCISES

1. Use the recurrence relation to compute P_3 and P_4.
2. Prove that

$$P_n(-x) = (-1)^n P_n(x)$$

for all n, and use this result together with Theorem 11–5 to deduce that $P_n(-1) = (-1)^n$.

3. Compute P_6.
4. Prove that

$$P_n(x) = [(2n)!/2^n(n!)^2]p_n(x),$$

where p_n is the nth polynomial in the sequence (11–1).

5. (a) Let $(x^2 - 1)^n = w$, and denote the kth derivative of w by $w^{(k)}$. Prove that $w, w^{(1)}, \ldots, w^{(n-1)}$ all vanish when $x = \pm 1$.

(b) Prove that

$$\int_{-1}^{1} P_n(x)x^m\,dx = 0$$

for every non-negative integer $m < n$. [*Hint:* Use integration by parts, and the result in (a).]

(c) Use the result in (b) to deduce that the Legendre polynomials are mutually orthogonal in $\mathcal{PC}[-1, 1]$.

6. (a) With w and $w^{(n)}$ as in Exercise 5, use integration by parts to prove that

$$\int_{-1}^{1} w^{(n)}w^{(n)}\,dx = (2n)!\int_{-1}^{1} (1-x)^n(1+x)^n\,dx.$$

(b) Prove that

$$\int_{-1}^{1} (1-x)^n(1+x)^n\,dx = \frac{(n!)^2}{(2n)!(2n+1)}\,2^{2n+1}.$$

(c) Use the results in (a) and (b) to prove that

$$\int_{-1}^{1} P_n(x)^2\,dx = \frac{2}{2n+1}.$$

7. Let k and n be non-negative integers with $k \leq n$. Prove that

$$n + (n-1) + \cdots + (n-k) = \frac{(2n-k)(k+1)}{2}.$$

[*Hint:* Use the formula $1 + 2 + \cdots + n = n(n+1)/2$.]

8. Prove that $P_n(x) - [(2n-1)/n]xP_{n-1}(x)$ is a polynomial of degree $< n$.

9. Use mathematical induction to prove that the nth derivative of the product of two functions is given by

$$(uv)^{(n)} = \sum_{k=0}^{n} \binom{n}{k} u^{(k)}v^{(n-k)},$$

where

$$\binom{n}{k} = \frac{n!}{k!(n-k)!},$$

and $u^{(0)} = u$, $v^{(0)} = v$.

10. (a) Write

$$P_{n+1}(x) = \frac{1}{2(n+1)}\frac{1}{2^n n!}[(x^2-1)^n(x^2-1)]^{(n+1)},$$

apply Exercise 9, and then differentiate the result to obtain

$$P'_{n+1} = \frac{1}{2(n+1)}(x^2-1)P''_n + \frac{n+2}{n+1}xP'_n + \frac{n+2}{2}P_n.$$

(b) Use the result just obtained and the differential equation for P_n to prove that

$$P'_{n+1} = xP'_n + (n+1)P_n.$$

11. Differentiate the recurrence relation for P_{n+1}, and use the result obtained together with that of Exercise 10(b) to prove that

$$xP'_n - P'_{n-1} = nP_n.$$

12. Prove that

(a) $P'_{n+1} - P'_{n-1} = (2n+1)P_n$; (b) $(1-x^2)P'_n = nP_{n-1} - nxP_n$.

13. Establish each of the following results.

(a) $$\int_{-1}^{1} xP_n(x)P_{n-1}(x)\,dx = \frac{2n}{4n^2 - 1}, \quad n \geq 1$$

(b) $$\int_{-1}^{1} P_n(x)P'_{n+1}(x)\,dx = 2, \quad n \geq 0$$

(c) $$\int_{-1}^{1} xP'_n(x)P_n(x)\,dx = \frac{2n}{2n+1}, \quad n \geq 0$$

14. (a) Use Exercise 12(b) and some other suitable identity to prove that

$$P_n = xP_{n-1} + \frac{x^2 - 1}{n} P'_{n-1} \cdot$$

(b) Replace n by $n - 1$ in the equation of Exercise 10(b), and square the result. Square the identity in 14(a), and use these two relations to establish

$$\frac{1 - x^2}{n^2} [P'_n]^2 + [P_n]^2 = \frac{1 - x^2}{n^2} [P'_{n-1}]^2 + [P_{n-1}]^2$$

for $n \geq 1$.

15. Use the result of Exercise 14(b) to show that

$$\frac{1 - x^2}{n^2} [P'_n]^2 + [P_n]^2 \leq 1$$

whenever $|x| \leq 1$, and $n > 0$. Now prove that

$$|P_n(x)| \leq 1$$

for all n if $|x| \leq 1$.

16. (a) Let $x_0, x_1, \ldots, x_m$ be $m + 1$ distinct numbers between -1 and 1. Find a polynomial $Q_i(x)$ of degree m which has roots at $x_0, \ldots, x_{i-1}, x_{i+1}, \ldots, x_m$ and takes the value a_i at x_i.

(b) Let $Q(x)$ denote the sum of the $m + 1$ polynomials $Q_i(x)$ obtained in (a). Show that $Q(x_i) = a_i$ for $0 \leq i \leq m$, and then prove that no other polynomial of degree less than or equal to m has this property.

17. Let $F(x)$ be *any* polynomial of degree $\leq 2m + 1$, and let $x_0, \ldots, x_m$ be the roots of P_{m+1}. Divide $F(x)$ by $P_{m+1}(x)$ to obtain

$$F(x) = P_{m+1}(x)P(x) + R(x),$$

where $P(x)$ is of degree $\leq m$, and $R(x)$ is either zero or of degree $\leq m$.

(a) Prove that

$$\int_{-1}^{1} F(x)\,dx = \int_{-1}^{1} R(x)\,dx.$$

(b) Prove that $R(x)$ is the polynomial $Q(x)$ constructed in Exercise 16(b) if a_i is taken as $F(x_i)$, $i = 0, 1, \ldots, m$.

(c) Prove that

$$\int_{-1}^{1} F(x)\,dx = \sum_{i=0}^{m} I_i F(x_i),$$

where the constants $I_0, I_1, \ldots, I_m$ depend *only* on $x_0, x_1, \ldots, x_m$, respectively, and not on any particular properties of $F(x)$.

18. Prove that there exists a positive constant M such that

$$\int_{-1}^{1} |P_n(x)|\,dx < \frac{M}{\sqrt{n}}$$

for all $n \geq 1$. [*Hint:* Use the Cauchy-Schwarz inequality.]

19. Prove that none of the Legendre polynomials has a repeated root. [*Hint:* Observe that any repeated root of P_n is also a root of P_n', and then use the uniqueness theorem for initial-value problems for second-order linear differential equations.]

11-4 LEGENDRE SERIES

Now that we have established the orthogonality of the Legendre polynomials, it is only natural to ask if they form a basis for $\mathcal{PC}[-1, 1]$. The answer is that they do, and a proof of this fact will be given toward the end of the present section. For the moment, however, let us accept the truth of this assertion, and consider the series expansion of an arbitrary function f in $\mathcal{PC}[-1, 1]$ computed with respect to the Legendre polynomials. By Formula (8-22), this series assumes the form

$$\frac{f \cdot P_0}{\|P_0\|^2} P_0(x) + \frac{f \cdot P_1}{\|P_1\|^2} P_1(x) + \cdots,$$

and converges *in the mean* to f. Thus we are entitled to write

$$f(x) = \sum_{n=0}^{\infty} \frac{f \cdot P_n}{\|P_n\|^2} P_n(x) \quad \text{(mean)}, \tag{11-15}$$

or, by Theorem 11-6,

$$f(x) = \sum_{n=0}^{\infty} a_n P_n(x) \quad \text{(mean)}, \tag{11-16}$$

where

$$a_n = \frac{2n+1}{2} \int_{-1}^{1} f(x) P_n(x)\,dx, \quad n = 0, 1, 2, \ldots. \tag{11-17}$$

A series of this type is known as the *Legendre series expansion* of the function f, and the a_n are called the *Legendre* or *Fourier-Legendre coefficients* of f.

EXAMPLE. Find the Legendre series expansion of the function

$$f(x) = \begin{cases} -1, & -1 < x < 0, \\ 1, & 0 < x < 1. \end{cases}$$

Since f is an odd function on $[-1, 1]$, so is $f(x)P_{2n}(x)$ for all n, and we have

$$a_{2n} = 0, \quad n = 0, 1, 2, \ldots.$$

Moreover,

$$\int_{-1}^{1} f(x)P_{2n+1}(x)\,dx = 2\int_{0}^{1} P_{2n+1}(x)\,dx,$$

and (11–17) yields

$$a_{2n+1} = (4n + 3)\int_{0}^{1} P_{2n+1}(x)\,dx.$$

To evaluate this integral we write the differential equation for P_n in the form

$$[(1 - x^2)P_n']' + n(n + 1)P_n = 0,$$

and integrate from 0 to 1 to obtain

$$\begin{aligned} n(n+1)\int_{0}^{1} P_n(x)\,dx &= -[(1 - x^2)P_n'(x)]\Big|_{0}^{1} \\ &= P_n'(0). \end{aligned}$$

But, by Exercise 12(b) of the last section,

$$P_n'(0) = nP_{n-1}(0), \quad n > 0,$$

and we therefore have

$$\int_{0}^{1} P_n(x)\,dx = \frac{1}{n+1}P_{n-1}(0), \quad n > 0.$$

Hence

$$a_{2n+1} = \frac{4n + 3}{2(n + 1)}P_{2n}(0), \quad n = 0, 1, 2, \ldots.$$

Finally, since

$$P_{2n}(0) = (-1)^n \frac{1 \cdot 3 \cdot 5 \cdots (2n - 1)}{2 \cdot 4 \cdot 6 \cdots (2n)}, \quad n = 1, 2, \ldots,$$

(Exercise 2 below),

$$a_1 = \tfrac{3}{2},$$

$$a_{2n+1} = (-1)^n \frac{4n + 3}{2(n + 1)} \cdot \frac{1 \cdot 3 \cdot 5 \cdots (2n - 1)}{2 \cdot 4 \cdot 6 \cdots (2n)}, \quad n = 1, 2, \ldots,$$

and we have

$$f(x) = \tfrac{3}{2}P_1(x) - \tfrac{7}{8}P_3(x) + \tfrac{11}{16}P_5(x) - \tfrac{75}{128}P_7(x) + \cdots$$
$$= \tfrac{3}{2}x + \sum_{n=1}^{\infty} (-1)^n \frac{4n+3}{2(n+1)} \cdot \frac{1 \cdot 3 \cdot 5 \cdots (2n-1)}{2 \cdot 4 \cdot 6 \cdots (2n)} P_{2n+1}(x).$$

In general, the theorems describing the convergence of Legendre series are similar to the corresponding results for Fourier series, and are proved in much the same way.* Without going into the details here (see Section 11–5), the basic result in this connection reads as follows.

Theorem 11–8. *The Legendre series for a piecewise smooth function f in $\mathcal{PC}[-1, 1]$ converges pointwise everywhere in the (open) interval $(-1, 1)$, and has the value*

$$\frac{f(x^+) + f(x^-)}{2}$$

at each point in the interval. Moreover, the convergence is uniform on any closed subinterval of $(-1, 1)$ in which f is continuous.

This said, we now turn to the one major item of unfinished business—a proof of the fact that the Legendre polynomials form a basis for $\mathcal{PC}[-1, 1]$.† Again the most important step in the proof is furnished by the Weierstrass approximation theorem, phrased this time for ordinary polynomials.

Theorem 11–9. *Let f be a continuous function on a closed interval $[a, b]$, and let ϵ be a positive real number. Then there exists a* **polynomial** *p such that*

$$|f(x) - p(x)| < \epsilon$$

for all x in $[a, b]$.

More succinctly, Theorem 11–9 asserts that *any continuous function on a closed interval can be* **uniformly** *approximated by polynomials.* Properly interpreted, Fig. 9–24 furnishes a suitable illustration of this result, and a proof can be found in Section 10–8.

For simplicity we now let $\mathcal{L}$ denote the set of all Legendre polynomials. One of the implications of Theorem 11–1 is that $\mathcal{L}$ spans the same subspace of $\mathcal{PC}[-1, 1]$ as the set $1, x, x^2, \ldots,$ from which it follows that $\mathcal{S}(\mathcal{L})$ is the subspace of $\mathcal{PC}[-1, 1]$ consisting of *all* polynomials. Thus $\bar{\mathcal{S}}(\mathcal{L})$, the closure of $\mathcal{S}(\mathcal{L})$ in

---

* In fact, it can be shown that the Legendre series for a function converges pointwise in the interval $(-1, 1)$ if and only if its Fourier series does.

† The following argument makes use of the results in Section 9–7, and should be omitted by those who are unfamiliar with that material.

$\mathcal{PC}[-1, 1]$, must also contain all polynomials. This fact, combined with the Weierstrass approximation theorem and Theorem 9–5 will allow us to show that every function in $\mathcal{PC}[-1, 1]$ is the limit in the mean of a sequence of functions in $\mathcal{S}(\mathcal{L})$, which, of course, will prove the basis theorem.

Indeed, if f is a continuous function in $\mathcal{PC}[-1, 1]$, Theorem 11–9 provides a sequence of polynomials $\{Q_k\}$, $k = 1, 2, \ldots$, (the subscript does not indicate the degree of the polynomial this time) such that

$$|f(x) - Q_k(x)| < \frac{1}{\sqrt{8}\,k}$$

for all x in $[-1, 1]$. But then

$$\begin{aligned}\|f - Q_k\| &= \left(\int_{-1}^{1} [f(x) - Q_k(x)]^2\,dx\right)^{1/2}\\ &< \left(\frac{2}{8k^2}\right)^{1/2} = \frac{1}{2k},\end{aligned}$$

and we conclude that $\{Q_k\}$ converges *in the mean* to f. Hence $\bar{\mathcal{S}}(\mathcal{L})$ contains all *continuous* functions on $[-1, 1]$.

Finally, let g be *any* function in $\mathcal{PC}[-1, 1]$. If g is not continuous we apply Theorem 9–5 (modified in the obvious way) to obtain a sequence of continuous functions $\{g_k\}$, $k = 1, 2, \ldots$, such that

$$\|g - g_k\| < \frac{1}{2k}.$$

Since each g_k is continuous we can apply the preceding argument to find a *polynomial* Q_k such that

$$\|g_k - Q_k\| < \frac{1}{2k}.$$

Hence

$$\begin{aligned}\|g - Q_k\| &= \|(g - g_k) + (g_k - Q_k)\|\\ &\le \|g - g_k\| + \|g_k - Q_k\|\\ &< \frac{1}{2k} + \frac{1}{2k}\\ &= \frac{1}{k},\end{aligned}$$

and the sequence of polynomials $\{Q_k\}$ converges *in the mean* to g. Thus g also belongs to $\bar{\mathcal{S}}(\mathcal{L})$, and we have proved

Theorem 11–10. *The Legendre polynomials form a basis for the Euclidean space* $\mathcal{PC}[-1, 1]$.

EXERCISES

1. What is the formula for the Legendre series expansion of an even function in $\mathcal{PC}[-1, 1]$? Of an odd function?
2. Prove that $P_{2n+1}(0) = 0, \quad n = 0, 1, 2, \ldots,$ and that

$$P_{2n}(0) = (-1)^n \frac{1 \cdot 3 \cdot 5 \cdots (2n-1)}{2 \cdot 4 \cdot 6 \cdots (2n)}, \quad n = 1, 2, \ldots.$$

3. Find the Legendre series expansion of the function $|x|$ in $\mathcal{PC}[-1, 1]$. [*Hint:* Use integration by parts and Exercise 12(a) of the preceding section to evaluate the coefficients.]
4. Prove that $P_n' = (2n - 1)P_{n-1} + (2n - 5)P_{n-3} + (2n - 9)P_{n-5} + \cdots$, for all $n \geq 1$, by expanding P_n' in a Legendre series.
5. Find the Legendre series expansion of each of the following functions in $\mathcal{PC}[-1, 1]$.

   (a) x^3 (b) $x^5 - x^3 + 2$ (c) $4x^4 + 2x^2 - x$.

6. Find the Legendre series expansion of the function

$$f(x) = \begin{cases} 1, & -1 \leq x < 0, \\ 0, & 0 < x \leq 1. \end{cases}$$

*11-5 CONVERGENCE OF LEGENDRE SERIES

We conclude our discussion of Legendre series by proving the convergence theorem cited in the preceding section. As with the proof of the corresponding result for Fourier series, we begin the argument by deriving a formula for the partial sums of the series in question.

Thus let f be an arbitrary function in $\mathcal{PC}[-1, 1]$, and let

$$S_n(x) = \sum_{k=0}^{n} a_k P_k(x),$$

where

$$a_k = \frac{2k+1}{2} \int_{-1}^{1} f(t) P_k(t)\, dt.$$

Then

$$\begin{aligned} S_n(x) &= \sum_{k=0}^{n} \left[\frac{2k+1}{2} \int_{-1}^{1} f(t) P_k(t)\, dt \right] P_k(x) \\ &= \int_{-1}^{1} \left[\sum_{k=0}^{n} \frac{2k+1}{2} P_k(t) P_k(x) \right] f(t)\, dt \\ &= \int_{-1}^{1} K_n(t, x) f(t)\, dt, \end{aligned}$$

where

$$K_n(t, x) = \sum_{k=0}^{n} \frac{2k + 1}{2} P_k(t)P_k(x).$$

To put this expression in more manageable form, we rewrite the recurrence relation for the Legendre polynomials as

$$(2k + 1)xP_k(x) = (k + 1)P_{k+1}(x) + kP_{k-1}(x),$$

and multiply by $P_k(t)$. This gives

$$(2k + 1)xP_k(t)P_k(x) = (k + 1)P_k(t)P_{k+1}(x) + kP_k(t)P_{k-1}(x). \tag{11–18}$$

We now interchange the roles of x and t in this expression and subtract (11–18) from the result to obtain

$$\begin{aligned}(2k + 1)(t - x)P_k(t)P_k(x) = (k + 1)&[P_{k+1}(t)P_k(x) - P_k(t)P_{k+1}(x)] \\ &-k[P_k(t)P_{k-1}(x) - P_{k-1}(t)P_k(x)].\end{aligned}$$

Finally, summing this identity as k runs from 0 to n, we find that

$$(t - x)\sum_{k=0}^{n} (2k + 1)P_k(t)P_k(x) = (n + 1)[P_{n+1}(t)P_n(x) - P_n(t)P_{n+1}(x)],$$

or

$$K_n(t, x) = \frac{n + 1}{2}\left[\frac{P_{n+1}(t)P_n(x) - P_n(t)P_{n+1}(x)}{t - x}\right]; \tag{11–19}$$

a formula which is known as *Christoffel's identity*.

In particular, we now have

$$S_n(x) = \frac{n + 1}{2}\int_{-1}^{1} \frac{P_{n+1}(t)P_n(x) - P_n(t)P_{n+1}(x)}{t - x} f(t)\, dt. \tag{11–20}$$

Moreover, since $S_n(x) = 1$ for all n when $f(x) \equiv 1$, (11–20) also implies that

$$\frac{n + 1}{2}\int_{-1}^{1} \frac{P_{n+1}(t)P_n(x) - P_n(t)P_{n+1}(x)}{t - x}\, dt = 1. \tag{11–21}$$

The two formulas just derived allow us to express the difference between $S_n(x)$ and $f(x)$ in integral form. The following lemmas will enable us to show that this difference approaches zero with increasing n whenever f is reasonably well-behaved, and therefore play the same role in the present argument that the Riemann-Lebesgue lemma played in the study of Fourier series.

Lemma 11–2. *If g is an arbitrary function in* $\mathcal{PC}[-1, 1]$, *then*

$$\lim_{n\to\infty} n^{1/2}\left|\int_{-1}^{1} g(x)P_n(x)\,dx\right| = 0. \tag{11–22}$$

Proof. Applying Parseval's equality to the function g, we have

$$\begin{aligned}\int_{-1}^{1} g(x)^2\,dx &= \sum_{n=0}^{\infty} \frac{\left(\int_{-1}^{1} g(x)P_n(x)\,dx\right)^2}{\int_{-1}^{1} P_n(x)^2\,dx}\\ &= \sum_{n=0}^{\infty}\left[\int_{-1}^{1} P_n(x)^2\,dx\right]a_n^2\\ &= \sum_{n=0}^{\infty}\left(\frac{2}{2n+1}\right)a_n^2.\end{aligned}$$

Thus the series

$$\sum_{n=0}^{\infty}\left(\frac{2}{2n+1}\right)a_n^2$$

is convergent, and it follows that

$$\lim_{n\to\infty}\left(\frac{2}{2n+1}\right)^{1/2}|a_n| = 0.$$

Hence

$$\lim_{n\to\infty}\left(\frac{2}{2n+1}\right)^{1/2}|a_n| = \lim_{n\to\infty}\left(\frac{2n+1}{2}\right)^{1/2}\left|\int_{-1}^{1} g(x)P_n(x)\,dx\right| = 0,$$

which, in turn, implies the desired result. ▮

Lemma 11–3. *There exists a positive constant M such that*

$$|P_n(x)| < \frac{M}{n^{1/2}(1-x^2)^{1/2}} \tag{11–23}$$

for all $n > 0$, *and all x in the interval* $(-1, 1)$.

The proof of this inequality is long and somewhat involved. Hence, rather than run the risk of obscuring our present argument, we have deferred the proof to the exercises where sufficient directions have been given to enable the student to work the details through for himself.

Now let f be an arbitrary function in $\mathcal{PC}[-1, 1]$, and let x_0 be a point in the open interval $(-1, 1)$ at which f is continuous and has a right- and left-hand derivative, i.e., a point where

$$\lim_{t\to x_0^+}\frac{f(t)-f(x_0)}{t-x_0} \quad\text{and}\quad \lim_{t\to x_0^-}\frac{f(t)-f(x_0)}{t-x_0}$$

exist. Then (11–20) and (11–21) imply that

$$S_n(x_0) - f(x_0) = \frac{n+1}{2}\int_{-1}^{1} \frac{f(t) - f(x_0)}{t - x_0}[P_{n+1}(t)P_n(x_0) - P_n(t)P_{n+1}(x_0)]\,dt$$

$$= \frac{n+1}{2}P_n(x_0)\int_{-1}^{1}\frac{f(t) - f(x_0)}{t - x_0}P_{n+1}(t)\,dt$$

$$- \frac{n+1}{2}P_{n+1}(x_0)\int_{-1}^{1}\frac{f(t) - f(x_0)}{t - x_0}P_n(t)\,dt.$$

But, by Lemma 11–3, there exists a constant M such that, for all $n \geq 1$,

$$\frac{n+1}{2}P_n(x_0) < \frac{M}{2(1 - x_0^2)^{1/2}} \cdot \frac{n+1}{n^{1/2}}$$

$$\leq \frac{M}{2(1 - x_0^2)^{1/2}} \cdot \frac{2n}{n^{1/2}}$$

$$= Kn^{1/2},$$

where

$$K = M/(1 - x_0^2)^{1/2}.$$

Moreover (and this is the critical point in the argument), the assumption that f has a right- and left-hand derivative at x_0 implies that the function

$$\frac{f(t) - f(x_0)}{t - x_0}, \quad -1 \leq t \leq 1,$$

belongs to $\mathcal{PC}[-1, 1]$. Hence, by Lemma 11–2,

$$\lim_{n\to\infty}\left|\frac{n+1}{2}P_n(x_0)\int_{-1}^{1}\frac{f(t) - f(x_0)}{t - x_0}P_{n+1}(t)\,dt\right|$$

$$< K \lim_{n\to\infty} n^{1/2}\left|\int_{-1}^{1}\frac{f(t) - f(x_0)}{t - x_0}P_{n+1}(t)\,dt\right|$$

$$= 0.$$

Similarly,

$$\lim_{n\to\infty}\left|\frac{n+1}{2}P_{n+1}(x_0)\int_{-1}^{1}\frac{f(t) - f(x_0)}{t - x_0}P_n(t)\,dt\right| = 0,$$

and we have therefore shown that

$$\lim_{n\to\infty}|S_n(x_0) - f(x_0)| = 0,$$

completing the proof of

Theorem 11–11. *The Legendre series for a function f in $\mathcal{PC}[-1, 1]$ converges to the value $f(x_0)$ at each point x_0 in the open interval $(-1, 1)$ where f is continuous and has a right- and left-hand derivative.*

In addition, it can also be shown that this convergence is uniform whenever f is continuous. Specifically, we have the following theorem.

Theorem 11–12. *The Legendre series for a function f in $\mathcal{PC}[-1, 1]$ converges uniformly to f on any closed subinterval of $(-1, 1)$ in which f is continuous and piecewise smooth.*

We omit the proof.

To complete the argument leading to Theorem 11–8 we must now determine the behavior of the Legendre series at a point of discontinuity of f. To this end we first prove the following rather surprising result.

Lemma 11–4. *Let x_0 be an arbitrary point in the open interval $(-1, 1)$. Then*

$$\lim_{n\to\infty} \frac{n+1}{2} \int_{-1}^{x_0} \frac{P_{n+1}(t)P_n(x_0) - P_n(t)P_{n+1}(x_0)}{t - x_0}\, dt = \tfrac{1}{2},$$

and

$$\lim_{n\to\infty} \frac{n+1}{2} \int_{x_0}^{1} \frac{P_{n+1}(t)P_n(x_0) - P_n(t)P_{n+1}(x_0)}{t - x_0}\, dt = \tfrac{1}{2}.$$

Proof. We begin by computing the Legendre series expansion of the function

$$f(x) = \begin{cases} 1, & -1 \le x < x_0, \\ 0, & x_0 < x \le 1. \end{cases}$$

Since

$$P_n(x) = \frac{1}{2n+1}[P'_{n+1}(x) - P'_{n-1}(x)]$$

for all $n \geq 1$ (Exercise 12(a), Section 11–3), the general coefficient in this series is

$$\begin{aligned} a_n &= \frac{2n+1}{2} \int_{-1}^{x_0} P_n(x)\, dx \\ &= \frac{1}{2} \int_{-1}^{x_0} [P'_{n+1}(x) - P'_{n-1}(x)]\, dx \\ &= \tfrac{1}{2}[P_{n+1}(x_0) - P_{n-1}(x_0)] - \tfrac{1}{2}[P_{n+1}(-1) - P_{n-1}(-1)] \\ &= \tfrac{1}{2}[P_{n+1}(x_0) - P_{n-1}(x_0)]; \end{aligned}$$

the last step following from the fact that $P_n(-1) = (-1)^n$.

Moreover,

$$a_0 = \frac{1}{2}\int_{-1}^{x_0} dx = \frac{x_0 + 1}{2} = \frac{1}{2} + \frac{P_1(x_0)}{2}.$$

Hence the Legendre series for f is

$$\frac{1}{2} + \frac{P_1(x_0)}{2} + \frac{1}{2}\sum_{n=1}^{\infty} [P_{n+1}(x_0) - P_{n-1}(x_0)]P_n(x),$$

and it follows that the value of the nth partial sum of this series at the point x_0 is

$$\begin{aligned} S_n(x_0) &= \frac{1}{2} + \frac{P_1(x_0)}{2} + \frac{1}{2}\sum_{k=1}^{n} [P_{k+1}(x_0) - P_{k-1}(x_0)]P_k(x_0) \\ &= \tfrac{1}{2} + \tfrac{1}{2}P_{n+1}(x_0)P_n(x_0). \end{aligned}$$

But by (11–20),

$$S_n(x_0) = \frac{n+1}{2}\int_{-1}^{x_0} \frac{P_{n+1}(t)P_n(x_0) - P_n(t)P_{n+1}(x_0)}{t - x_0}\, dt.$$

Hence, by (11–23),

$$\begin{aligned} \lim_{n\to\infty} \frac{n+1}{2}\int_{-1}^{x_0} &\frac{P_{n+1}(t)P_n(x_0) - P_n(t)P_{n+1}(x_0)}{t - x_0}\, dt \\ &= \tfrac{1}{2} + \tfrac{1}{2}\lim_{n\to\infty} P_{n+1}(x_0)P_n(x_0) \\ &= \tfrac{1}{2}. \end{aligned}$$

This proves the first statement in the lemma, and the second follows by subtracting this result from (11–21). ▮

Now let f be any function in $\mathcal{PC}[-1, 1]$ with the property that

$$f(x) = \frac{f(x^+) + f(x^-)}{2}$$

for all x in $(-1, 1)$, and let x_0 be any point at which f has a right- and left-hand derivative. Then, if $S_n(x_0)$ denotes the value of the nth partial sum of the Legendre series for f, (11–20) implies that

$$S_n(x_0) = I_1 + I_2,$$

where

$$I_1 = \frac{n+1}{2}\int_{-1}^{x_0} \frac{P_{n+1}(t)P_n(x_0) - P_n(t)P_{n+1}(x_0)}{t - x_0} f(t)\, dt,$$

$$I_2 = \frac{n+1}{2}\int_{x_0}^{1} \frac{P_{n+1}(t)P_n(x_0) - P_n(t)P_{n+1}(x_0)}{t - x_0} f(t)\, dt.$$

Thus

$$S_n(x_0) - f(x_0) = \left[I_1 - \frac{f(x_0^-)}{2}\right] + \left[I_2 - \frac{f(x_0^+)}{2}\right],$$

and to prove that

$$\lim_{n\to\infty} [S_n(x_0) - f(x_0)] = 0,$$

it suffices to show separately that

$$\lim_{n\to\infty} \left[I_1 - \frac{f(x_0^-)}{2}\right] = \lim_{n\to\infty} \left[I_2 - \frac{f(x_0^+)}{2}\right] = 0.$$

By the preceding lemma,

$$\frac{f(x_0^-)}{2} = \lim_{n\to\infty} \frac{n+1}{2} \int_{-1}^{x_0} \frac{P_{n+1}(t)P_n(x_0) - P_n(t)P_{n+1}(x_0)}{t - x_0} f(x_0^-)\, dt,$$

whence

$$\lim_{n\to\infty} \left[I_1 - \frac{f(x_0^-)}{2}\right]$$
$$= \lim_{n\to\infty} \frac{n+1}{2} \int_{-1}^{x_0} \frac{f(t) - f(x_0^-)}{t - x_0} [P_{n+1}(t)P_n(x_0) - P_n(t)P_{n+1}(x_0)]\, dt.$$

Now let

$$g(t) = \begin{cases} \dfrac{f(t) - f(x_0^-)}{t - x_0}, & -1 \le t < x_0, \\ 0, & x_0 < t \le 1. \end{cases}$$

Then, since f has a left-hand derivative at x_0, g is piecewise continuous on $[-1, 1]$, and we have

$$\lim_{n\to\infty} \left[I_1 - \frac{f(x_0^-)}{2}\right] = \lim_{n\to\infty} \frac{n+1}{2} \int_{-1}^{1} g(t)[P_{n+1}(t)P_n(x_0) - P_n(t)P_{n+1}(x_0)]\, dt.$$

We now repeat the argument leading to Theorem 11–11 to deduce that

$$\lim_{n\to\infty} \left[I_1 - \frac{f(x_0^-)}{2}\right] = 0.$$

Finally, by introducing the function

$$h(t) = \begin{cases} 0, & -1 \le t < x_0, \\ \dfrac{f(t) - f(x_0^+)}{t - x_0}, & x_0 < t \le 1, \end{cases}$$

and reasoning as above, we also find that

$$\lim_{n\to\infty}\left[I_2 - \frac{f(x_0^+)}{2}\right] = 0,$$

thereby proving

Theorem 11–13. *The Legendre series for a function f in $\mathcal{PC}[-1, 1]$ converges to the value*

$$\frac{f(x_0^+) + f(x_0^-)}{2}$$

at each point x_0 in the open interval $(-1, 1)$ where f has a right- and left-hand derivative.

EXERCISES

1. During the proof of Lemma 11–2 it was asserted that if

$$\lim_{n\to\infty}\left(\frac{2n+1}{2}\right)^{1/2}\left|\int_{-1}^{1} g(x)P_n(x)\,dx\right| = 0,$$

then

$$\lim_{n\to\infty} n^{1/2}\left|\int_{-1}^{1} g(x)P_n(x)\,dx\right| = 0.$$

Why is this true?

2. At what point does the proof of Theorem 11–11 break down if f has a jump discontinuity at x_0?

The following exercises furnish a proof of Lemma 11–3.

3. For each integer $n \geq 0$, let

$$f_n(x) = \frac{1}{\pi}\int_0^{\pi} [x + (x^2 - 1)^{1/2}\cos\varphi]^n\,d\varphi.$$

(a) Show that $f_0(x) = 1$ and $f_1(x) = x$.

(b) Show that

$$(n+1)f_{n+1}(x) - (2n+1)xf_n(x) + nf_{n-1}(x)$$
$$= \frac{1}{\pi}\int_0^{\pi}\{-n(x^2-1)\sin^2\varphi + (x^2-1)^{1/2}\cos\varphi[x + (x^2-1)^{1/2}\cos\varphi]\}$$
$$\times [x + (x^2-1)^{1/2}\cos\varphi]^{n-1}\,d\varphi.$$

(c) Use integration by parts with

$$u = [x + (x^2-1)^{1/2}\cos\varphi]^n, \qquad dv = \cos\varphi\,d\varphi$$

to prove that

$$\int_0^\pi (x^2 - 1)^{1/2} \cos\varphi[x + (x^2 - 1)^{1/2}\cos\varphi]^n \, d\varphi$$
$$= \int_0^\pi n(x^2 - 1)\sin^2\varphi[x + (x^2 - 1)^{1/2}\cos\varphi]^{n-1} \, d\varphi.$$

(d) Use the results of (a), (b), and (c) to deduce that

$$P_n(x) = \frac{1}{\pi}\int_0^\pi [x + (x^2 - 1)^{1/2}\cos\varphi]^n \, d\varphi.$$

4. (a) Rewrite the formula in Exercise 3(d) as

$$P_n(x) = \frac{1}{\pi}\int_0^\pi [x + i(1 - x^2)^{1/2}\cos\varphi]^n \, d\varphi,$$

where $i = \sqrt{-1}$, and show that

$$|P_n(x)| \le \frac{1}{\pi}\int_0^\pi (\cos^2\varphi + x^2\sin^2\varphi)^{n/2} \, d\varphi.$$

[Recall that if $z = a + bi$, then $|z| = (a^2 + b^2)^{1/2}$.]

(b) Use the result of (a) to prove that

$$|P_n(x)| < 1$$

for all n and all x in $(-1, 1)$.

(c) Show that the inequality in (a) can be rewritten

$$|P_n(x)| \le \frac{2}{\pi}\int_0^{\pi/2} [1 - (1 - x^2)\sin^2\varphi]^{n/2} \, d\varphi.$$

[*Hint:* The graph of $\sin\varphi$ is symmetric about the line $\varphi = \pi/2$.]

(d) Prove that $\sin\varphi \ge 2\varphi/\pi$ for $0 \le \varphi \le \pi/2$.

(e) Set $u = (2/\pi)(1 - x^2)^{1/2}$, and use the inequality in (d) to deduce that

$$1 - (1 - x^2)\sin^2\varphi \le 1 - u^2\varphi^2, \quad 0 \le \varphi \le \pi/2.$$

Then use (c) to prove that

$$|P_n(x)| \le \frac{2}{\pi}\int_0^{\pi/2} e^{-nu^2\varphi^2/2} d\varphi < \frac{2}{\pi n^{1/2} u}\int_0^\infty e^{-t^2/2} \, dt.$$

(f) Now use the fact that

$$\int_0^\infty e^{-t^2/2} \, dt < \infty$$

to conclude that there exists a positive constant M such that

$$|P_n(x)| < \frac{M}{n^{1/2}(1 - x^2)^{1/2}}$$

for all $n \ge 1$ and all x in $(-1, 1)$.

11-6 HERMITE POLYNOMIALS

We now know a great deal about the behavior of polynomials in the Euclidean space $\mathcal{PC}[-1, 1]$, all the way from some very special properties of the Legendre polynomials to the central result which states that the orthogonalized sequence obtained from $1, x, x^2, \ldots$ is a basis for the space. In point of fact, however, the essential portions of our earlier discussion remain valid over any *finite* interval $[a, b]$, and thus, for theoretical purposes at least, we need not consider the apparently more general space $\mathcal{PC}[a, b]$. But when we come to study piecewise continuous functions on the entire real line, $(-\infty, \infty)$, the situation becomes much different, and requires special consideration.*

In the first place, if f and g are arbitrary piecewise continuous functions on $(-\infty, \infty)$, there is no guarantee that the improper integral

$$\int_{-\infty}^{\infty} f(x)g(x)\,dx = \lim_{\substack{a\to\infty \\ b\to\infty}} \int_{-a}^{b} f(x)g(x)\,dx \tag{11–24}$$

converges. Indeed, (11–24) is undefined even when f and g are polynomials. Thus our usual definition of an inner product is not valid, and this, in turn, implies that if we wish to use the general theory of Euclidean spaces in this context we must either consider another set of functions, or another inner product, or both.

As our point of departure, let us insist that any function space which we consider contain all polynomials, and that its inner product be defined by means of an improper integral over the entire real line. This being the case, it is clear that we must introduce a *weight function* w into our integral in order to guarantee that

$$\int_{-\infty}^{\infty} w(x)p(x)q(x)\,dx$$

exists for every pair of polynomials p, q. (See Example 3, Section 7–1). This requirement suggests that w be piecewise continuous on the entire real line, and that it tend to zero as $|x| \to \infty$. Moreover, it is convenient to require that $\lim_{|x|\to\infty} w(x)x^n = 0$ for any positive integer n. (Why?) Our experience from calculus suggests that we try an exponential function (with negative exponent), and since we allow x to assume negative as well as positive values, the exponent should involve only even powers of x. Thus we are led to try the weight function

$$w(x) = e^{-x^2/2}. \tag{11–25}$$

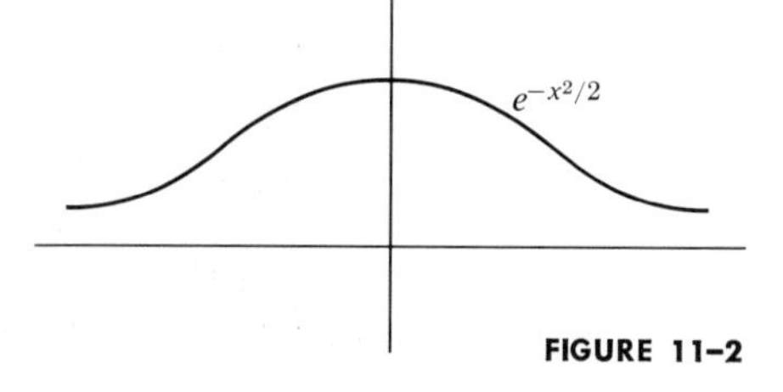

FIGURE 11–2

* A function is said to be piecewise continuous on $(-\infty, \infty)$ whenever it is piecewise continuous on *every* finite interval $[a, b]$.

(See Fig. 11–2.)* The following lemma guarantees that this choice will be successful.

Lemma 11–5. *Let* $I_n = \int_{-\infty}^{\infty} e^{-x^2/2}x^n\,dx$, $n = 0, 1, 2, \ldots$ *Then*

$$I_{2n+1} = 0,$$

$$I_{2n} = \frac{(2n)!}{2^n n!}\sqrt{2\pi}$$

for all n.

Proof. We begin by considering

$$I_{2n+1} = \lim_{\substack{a\to\infty \\ b\to\infty}} \int_{-a}^{b} e^{-x^2/2}x^{2n+1}\,dx.$$

In this case the integrand is an odd function of x, and hence

$$\begin{aligned}\int_{-a}^{b} e^{-x^2/2}x^{2n+1}\,dx &= \int_{-a}^{a} e^{-x^2/2}x^{2n+1}\,dx + \int_{a}^{b} e^{-x^2/2}x^{2n+1}\,dx \\ &= \int_{a}^{b} e^{-x^2/2}x^{2n+1}\,dx.\end{aligned}$$

But

$$\begin{aligned}\int_{a}^{b} x^{2n+1}e^{-x^2/2}\,dx &= \int_{a}^{b} x^{2n}(xe^{-x^2/2})\,dx \\ &= -x^{2n}e^{-x^2/2}\Big|_{a}^{b} + 2n\int_{a}^{b} x^{2n-1}e^{-x^2/2}\,dx.\end{aligned}$$

Thus

$$\begin{aligned}I_{2n+1} &= \lim_{\substack{a\to\infty \\ b\to\infty}} \left[-x^{2n}e^{-x^2/2}\Big|_{a}^{b} + 2n\int_{a}^{b} x^{2n-1}e^{-x^2/2}\,dx\right] \\ &= 2n \lim_{\substack{a\to\infty \\ b\to\infty}} \int_{a}^{b} x^{2n-1}e^{-x^2/2}\,dx \\ &= 2n \lim_{\substack{a\to\infty \\ b\to\infty}} \left[\int_{-a}^{a} x^{2n-1}e^{-x^2/2}\,dx + \int_{a}^{b} x^{2n-1}e^{-x^2/2}\,dx\right] \\ &= 2n \lim_{\substack{a\to\infty \\ b\to\infty}} \int_{-a}^{b} x^{2n-1}e^{-x^2/2}\,dx,\end{aligned}$$

---

* Some authors set $w(x) = e^{-x^2}$. Except for the obvious modifications this necessitates, all of the results in this section still remain valid. However, the Hermite polynomials defined below then do not have 1 as their leading coefficient.

and it follows that $I_{2n+1} = 2nI_{2n-1}$ for all $n > 0$. Moreover, by direct computation, we find that $I_1 = 0$. Hence $I_{2n+1} = 0$ for all $n \geq 0$, as asserted.

To evaluate I_{2n} we must use the fact (proved in Exercise 3 below) that

$$I_0 = \int_{-\infty}^{\infty} e^{-x^2/2}\, dx = \sqrt{2\pi}.$$

Then if $n > 0$,

$$\begin{aligned} I_{2n} &= \int_{-\infty}^{\infty} x^{2n-1}(xe^{-x^2/2})\, dx \\ &= \lim_{\substack{a\to\infty \\ b\to\infty}} \left[- x^{2n-1}e^{-x^2/2}\Big|_{-a}^{b} + (2n-1)\int_{-a}^{b} x^{2n-2}e^{-x^2/2}\, dx \right] \\ &= (2n-1)\, I_{2n-2}. \end{aligned}$$

Hence

$$\begin{aligned} I_2 &= (2-1)\sqrt{2\pi}, \\ I_4 &= (4-1)(2-1)\sqrt{2\pi}, \\ &\vdots \\ I_{2n} &= (2n-1)(2n-3)\cdots(2-1)\sqrt{2\pi}. \end{aligned}$$

But

$$(2n-1)(2n-3)\cdots(2-1) = \frac{(2n)!}{2^n n!} \quad \text{if } n > 0,$$

and the lemma is proved. ▮

From the way things are developing, it is obvious that we stand in imminent prospect of having to consider various integrals of the form

$$\int_{-\infty}^{\infty} e^{-x^2/2} f(x)g(x)\, dx.$$

This being so, it will be well to dispose of convergence questions once and for all, which we do by proving

Lemma 11–6. *Let f and g be piecewise continuous on $(-\infty, \infty)$, and suppose that*

$$\int_{-\infty}^{\infty} e^{-x^2/2} f(x)^2\, dx < \infty \quad \textit{and} \quad \int_{-\infty}^{\infty} e^{-x^2/2} g(x)^2\, dx < \infty.$$

Then

(i) $\int_{-\infty}^{\infty} e^{-x^2/2}[\alpha f(x)]^2\, dx < \infty$ *for all real* α;

(ii) $\int_{-\infty}^{\infty} e^{-x^2/2}[f(x) + g(x)]^2\, dx < \infty$;

(iii) $\int_{-\infty}^{\infty} e^{-x^2/2} f(x)g(x)\, dx < \infty$.

Proof. The first result is obvious, since

$$\int_{-\infty}^{\infty} e^{-x^2/2}[\alpha f(x)]^2\,dx = \alpha^2 \int_{-\infty}^{\infty} e^{-x^2/2} f(x)^2\,dx.$$

To prove (ii), we introduce the function max (f, g) which, for each x, is defined to be the larger of the two numbers $f(x)$, $g(x)$. It is clear that max (f, g) is piecewise continuous on $(-\infty, \infty)$ (see Fig. 11–3) and, in addition, that

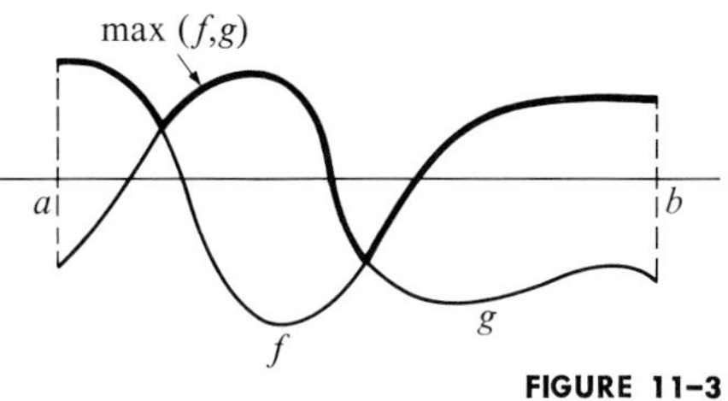

FIGURE 11–3

$$\begin{aligned}[f + g]^2 &\le [|f| + |g|]^2 \\ &\le [2 \max(|f|, |g|)]^2 \\ &= 4 \max(f^2, g^2) \\ &\le 4[f^2 + g^2].\end{aligned}$$

Thus

$$\int_{-\infty}^{\infty} e^{-x^2/2}[f(x) + g(x)]^2\,dx \le 4\int_{-\infty}^{\infty} e^{-x^2/2} f(x)^2\,dx + 4\int_{-\infty}^{\infty} e^{-x^2/2} g(x)^2\,dx.$$

By assumption these last two integrals are finite. Hence

$$\int_{-\infty}^{\infty} e^{-x^2/2}[f(x) + g(x)]^2\,dx < \infty.$$

Finally, since

$$f(x)g(x) = \frac{[f(x) + g(x)]^2}{2} - \frac{f(x)^2 + g(x)^2}{2},$$

$$\int_{-\infty}^{\infty} e^{-x^2/2} f(x)g(x)\,dx = \tfrac{1}{2}\int_{-\infty}^{\infty} e^{-x^2/2}[f(x) + g(x)]^2\,dx - \tfrac{1}{2}\int_{-\infty}^{\infty} e^{-x^2/2} f(x)^2\,dx - \tfrac{1}{2}\int_{-\infty}^{\infty} e^{-x^2/2} g(x)^2\,dx.$$

But we now know that all of these integrals are finite, and we are done. ▌

These technical details out of the way, we consider the set of all piecewise continuous functions on $(-\infty, \infty)$, identified in accordance with the convention in Section 9–2, which have the property that

$$\int_{-\infty}^{\infty} e^{-x^2/2} f(x)^2\,dx < \infty. \tag{11–26}$$

This set will be denoted by $\mathcal{I}_2^w$, the subscript serving as a reminder that f^2, rather than f, appears in (11–26), the superscript recalling that a weight function is involved, and $\mathcal{I}$ standing for "integrable." The functions in $\mathcal{I}_2^w$ are sometimes referred to as being "square integrable with respect to the weight function $e^{-x^2/2}$."

Theorem 11–14. *$\mathcal{G}_2^w$ is a Euclidean space under the usual definitions of addition and scalar multiplication of piecewise continuous functions, and inner product*

$$f \cdot g = \int_{-\infty}^{\infty} e^{-x^2/2} f(x)g(x)\, dx. \tag{11–27}$$

Proof. From Lemma 11–6 we know that $f + g$ and αf belong to $\mathcal{G}_2^w$ whenever f and g do, and also that the integral defining $f \cdot g$ is finite. This disposes of all but the straightforward details in the proof, and these we leave to the reader. ▮

With this, we have realized our goal of constructing a function space on $(-\infty, \infty)$ which contains all polynomials, and also, by the way, all *bounded* piecewise continuous functions. [For if f is such a function, then

$$\int_{-\infty}^{\infty} e^{-x^2/2} f(x)^2\, dx \leq M^2 \int_{-\infty}^{\infty} e^{-x^2/2}\, dx = M^2\sqrt{2\pi} < \infty.]$$

Moreover, our patient insistence on keeping polynomials in view is now rewarded in the form of the following basic result.

Theorem 11–15. *$\mathcal{G}_2^w$ is the set of all piecewise continuous functions on $(-\infty, \infty)$ which can be expressed as limits, in the mean, of sequences of polynomials. Hence, if f is an arbitrary function in $\mathcal{G}_2^w$, there exists a sequence of polynomials $\{p_n\}$, $n = 0, 1, 2, \ldots$, such that*

$$\lim_{n\to\infty} \|f - p_n\| = \lim_{n\to\infty} \left(\int_{-\infty}^{\infty} e^{-x^2/2}[f(x) - p_n(x)]^2\, dx\right)^{1/2} = 0.$$

The proof is difficult, and will be omitted.

From this point the discussion proceeds very much as with Legendre polynomials. Indeed, the sequence $1, x, x^2, \ldots$ is linearly independent in $\mathcal{G}_2^w$ (Exercise 6), and can be orthogonalized by the Gram-Schmidt process to produce a *basis*

$$p_0(x),\ p_1(x),\ \ldots,\ p_n(x),\ \ldots$$

in which p_n is a polynomial of degree n. (This is one of the consequences of Theorem 11–15.) But again this particular basis leads to computational difficulties, and since Theorem 11–1 holds, *mutatis mutandis*, in this situation, we are free to seek another basis composed of polynomials, one of each degree. This time the most convenient basis consists of the so-called *Hermite polynomials*, which are defined by the formula

$$H_n(x) = (-1)^n e^{x^2/2} \frac{d^n}{dx^n} e^{-x^2/2}, \quad n = 0, 1, 2, \ldots \tag{11–28}$$

By direct computation we see that

$$\begin{aligned} H_0(x) &= 1, & H_1(x) &= x, \\ H_2(x) &= x^2 - 1, & H_3(x) &= x^3 - 3x, \\ H_4(x) &= x^4 - 6x^2 + 3, & H_5(x) &= x^5 - 10x^3 + 15x. \end{aligned}$$

The fact that H_n is a polynomial of degree n with leading coefficient 1 can be determined by inspection of (11–28), or by the technique suggested in Exercise 5. An even more obvious remark is that H_{2n} is even, and H_{2n+1} odd, for all n.

Theorem 11–16. *The Hermite polynomials are an orthogonal sequence in $\mathcal{G}_2^w$ with*

$$\|H_n\|^2 = n!\sqrt{2\pi}. \tag{11–29}$$

Proof. Let

$$\begin{aligned} I &= \int_{-\infty}^{\infty} e^{-x^2/2} H_m(x) H_n(x)\, dx \\ &= (-1)^n \int_{-\infty}^{\infty} H_m(x) \frac{d^n}{dx^n} (e^{-x^2/2})\, dx, \end{aligned}$$

and assume that $m \leq n$. Integrating by parts, we obtain

$$I = (-1)^n \left[H_m(x) \frac{d^{n-1}}{dx^{n-1}} (e^{-x^2/2}) \Big|_{-\infty}^{\infty} - \int_{-\infty}^{\infty} H_m'(x) \frac{d^{n-1}}{dx^{n-1}} (e^{-x^2/2})\, dx \right] \cdot$$

But the first term in this quantity is the product of a polynomial and $e^{-x^2/2}$, and hence vanishes as $|x| \to \infty$. Thus

$$I = (-1)^{n+1} \int_{-\infty}^{\infty} H_m'(x) \frac{d^{n-1}}{dx^{n-1}} (e^{-x^2/2})\, dx.$$

We repeat this procedure a total of m times, and get

$$I = (-1)^{m+n} \int_{-\infty}^{\infty} H_m^{(m)}(x) \frac{d^{n-m}}{dx^{n-m}} (e^{-x^2/2})\, dx,$$

where $H_m^{(m)}$ denotes the m-fold derivative of H_m. But $H_m^{(m)} = m!$, since H_m is a polynomial of degree m and leading coefficient 1. Hence

$$I = (-1)^{m+n} m! \int_{-\infty}^{\infty} \frac{d^{n-m}}{dx^{n-m}} (e^{-x^2/2})\, dx.$$

Now suppose $m < n$. Then

$$\begin{aligned} I &= (-1)^{m+n} m! \frac{d^{n-m-1}}{dx^{n-m-1}} (e^{-x^2/2}) \Big|_{-\infty}^{\infty} \\ &= 0, \end{aligned}$$

since this quantity is also the product of a polynomial and $e^{-x^2/2}$. Thus H_m and H_n are orthogonal in $\mathcal{G}_2^w$ whenever $m \neq n$.

Finally, if $m = n$,

$$\begin{aligned} I &= (-1)^{2n} n! \int_{-\infty}^{\infty} e^{-x^2/2}\, dx \\ &= n!\sqrt{2\pi}, \end{aligned}$$

and the theorem is proved. ▮

One of the principal reasons for the importance of Hermite polynomials is that they appear as solutions of a certain linear differential equation. To derive this equation we first establish the following *recurrence relation.*

Theorem 11–17. *The Hermite polynomials satisfy the identity*

$$H_{n+1} = xH_n - nH_{n-1}, \quad n > 0. \tag{11–30}$$

(This identity is also valid for $n = 0$ if we agree that $H_{-1} \equiv 0$.)

Proof. The proof is similar to the one given in Section 11–3 for the Legendre polynomials. Since $xH_n(x)$ is a polynomial of degree $n + 1$, its *Hermite series* is of the form

$$xH_n(x) = \sum_{k=0}^{n+1} \frac{xH_n \cdot H_k}{H_k \cdot H_k} H_k(x). \tag{11–31}$$

By Lemma 11–1, which is also valid in the Euclidean space $\mathcal{I}_2^w$, we have $xH_n \cdot H_k = H_n \cdot xH_k = 0$ if $k < n - 1$. Moreover,

$$\begin{aligned} xH_n \cdot H_n &= \int_{-\infty}^{\infty} xe^{-x^2/2} H_n(x)^2 \, dx \\ &= 0 \end{aligned}$$

since the integrand is an odd function (see Lemma 11–5). Thus (11–31) is of the form

$$xH_n(x) = \alpha_n H_{n+1}(x) + \beta_n H_{n-1}(x).$$

If we equate the coefficients of x^{n+1} in this identity, and recall that the leading coefficient of H_k is 1, we see at once that $\alpha_n = 1$.

Rather than equate coefficients to obtain β_n, we reason as follows. Since

$$\alpha_n = \frac{xH_n \cdot H_{n+1}}{H_{n+1} \cdot H_{n+1}} = 1$$

for all n, it follows that

$$xH_n \cdot H_{n+1} = \|H_{n+1}\|^2 = \sqrt{2\pi}\,(n + 1)!.$$

But then

$$\begin{aligned} \beta_n &= \frac{xH_n \cdot H_{n-1}}{H_{n-1} \cdot H_{n-1}} \\ &= \frac{\sqrt{2\pi}\, n!}{\sqrt{2\pi}\,(n - 1)!} = n. \end{aligned}$$

Thus

$$xH_n = H_{n+1} + nH_{n-1},$$

as asserted. ▮

To derive the differential equation we start with the defining formula for H_n:

$$H_n(x) = (-1)^n e^{x^2/2} \frac{d^n}{dx^n} e^{-x^2/2}.$$

This may be rewritten

$$\frac{d^n}{dx^n} e^{-x^2/2} = (-1)^n e^{-x^2/2} H_n(x),$$

and, when differentiated, gives

$$\frac{d^{n+1}}{dx^{n+1}} e^{-x^2/2} = (-1)^n [-x H_n(x) + H_n'(x)] e^{-x^2/2}.$$

On the other hand,

$$\frac{d^{n+1}}{dx^{n+1}} e^{-x^2/2} = (-1)^{n+1} e^{-x^2/2} H_{n+1}(x).$$

Thus

$$H_{n+1} = x H_n - H_n', \tag{11–32}$$

a relation which is sometimes useful in itself.

The recurrence relation and (11–32) together give

$$H_n' - n H_{n-1} = 0,$$

or

$$H_{n+1}' - (n+1) H_n = 0.$$

We now differentiate (11–32) to get

$$H_{n+1}' = x H_n' + H_n - H_n'',$$

and it follows that

$$H_n'' - x H_n' + n H_n = 0.$$

Thus we have proved

Theorem 11–18. *The Hermite polynomial of degree n is a solution of the second-order linear differential equation*

$$y'' - xy' + ny = 0. \tag{11–33}$$

As with the Legendre polynomials and their differential equation, it can be shown that H_n is the only *polynomial* solution of (11–33) with leading coefficient 1. We shall encounter this equation again in Chapter 13 when we discuss the Schrödinger wave equation from quantum mechanics.

EXERCISES

1. Determine whether or not the following functions, defined on $(-\infty, \infty)$, belong to $\mathcal{G}_2^w$.

   (a) $e^{x^2/4}$ (b) $x \sin x$ (c) $|x|$ (d) $f(x) = n,\ n \leq x < n + 1$

   (e) $xe^{x^2/8}$ (f) $f(x) = \begin{cases} \ln |x|, & |x| \geq 1 \\ 0, & |x| \leq 1 \end{cases}$ (g) e^x

2. Prove that $\lim_{|x| \to \infty} p(x)e^{-x^2/2} = 0$ for any polynomial p.

3. Let $I = \int_{-\infty}^{\infty} e^{-x^2/2}\, dx$. Then

$$I^2 = \left(\int_{-\infty}^{\infty} e^{-x^2/2}\, dx\right)\left(\int_{-\infty}^{\infty} e^{-y^2/2}\, dy\right)$$
$$= \int_{-\infty}^{\infty}\int_{-\infty}^{\infty} e^{-(x^2+y^2)/2}\, dx\, dy.$$

   Change to polar coordinates, evaluate this integral, and thus show that $I = \sqrt{2\pi}$.

4. Prove that

$$(2n - 1)(2n - 3)\cdots(2 - 1) = \frac{(2n)!}{2^n n!}, \quad n > 0.$$

   [*Hint:* Use mathematical induction.]

5. Given that $H_0(x) = 1$, use (11–32) and mathematical induction to prove that $H_n(x)$ is a polynomial of degree n with leading coefficient 1.

6. Prove that the functions $1, x, x^2, \ldots$ are linearly independent in $\mathcal{G}_2^w$.

7. (a) Use the recurrence relation to verify that the polynomials listed in the text as $H_2, \ldots, H_5$ are correct.

   (b) Compute H_6 and H_7.

8. Prove that

$$H_{2n+1}(0) = 0, \quad \text{and} \quad H_{2n}(0) = (-1)^n 1 \cdot 3 \cdot 5 \cdots (2n - 1)$$

   for all n.

In Exercises 9–11 we outline an alternate approach to the materials of this section.

9. Let $\mathcal{G}_2$ denote the set of all piecewise continuous functions f on $(-\infty, \infty)$, identified in the usual way, with the property that

$$\int_{-\infty}^{\infty} f(x)^2\, dx < \infty.$$

   (Such functions are said to be "square integrable" on $(-\infty, \infty)$.)

   (a) If f and g belong to $\mathcal{G}_2$, set

$$f \cdot g = \int_{-\infty}^{\infty} f(x)g(x)\, dx.$$

   Prove that this defines an inner product on $\mathcal{G}_2$, and that with the usual definitions of addition and scalar multiplication $\mathcal{G}_2$ is a Euclidean space.

(b) Prove that $\mathcal{G}_2$ contains all functions of the form $e^{-x^2/4}p(x)$, where p is a polynomial.

10. With $\mathcal{G}_2$ as above, define the *Hermite functions* h_n, $n = 0, 1, 2, \ldots$, by the formula

$$h_n(x) = (-1)^n \left[\frac{d}{dx^n} e^{-x^2/2}\right] e^{x^2/4}.$$

(a) Prove that $h_n(x) = e^{-x^2/4}H_n(x)$, where H_n is the nth Hermite polynomial.

(b) Prove that $h_m \cdot h_n = 0$ if $m \neq n$, and $h_n \cdot h_n = n!\sqrt{2\pi}$ with respect to the inner product in $\mathcal{G}_2$. (The sequence $\{h_n\}$, $n = 0, 1, 2, \ldots$, of Hermite functions actually forms a basis for $\mathcal{G}_2$.)

11. Show that the Hermite functions satisfy the same recurrence relation that the Hermite polynomials do.

12. Show that the functions $h_n(x) = e^{-x^2/4}H_n(x)$ satisfy the linear differential equation

$$y'' + \left(n + \frac{1}{2} - \frac{x^2}{4}\right) y = 0.$$

13. Let $\{e^{-x^2/4}p_n(x)\}$ and $\{e^{-x^2/4}q_n(x)\}$, $n = 0, 1, 2, \ldots$, be orthogonal sequences in $\mathcal{G}_2$ (see Exercise 9), and suppose that p_n and q_n are polynomials of degree n. Prove that for each n, p_n is a scalar multiple of q_n.

14. Prove that

$$(-1)^n e^{\alpha^2 x^2/4} \left[\frac{d^n}{dt^n}(e^{-t^2/2})\right]_{t=\alpha x}, \qquad n = 0, 1, 2, \ldots,$$

α a nonzero real number, is an orthogonal sequence in $\mathcal{G}_2$ (see Exercise 9). (Actually, this sequence is a basis for $\mathcal{G}_2$.)

11-7 LAGUERRE POLYNOMIALS

The last of the various sets of orthogonal polynomials which we shall formally discuss—the so-called *Laguerre polynomials*—arises in the study of functions on the semi-infinite interval $[0, \infty)$. In this setting we use the weight function e^{-x}, and let $\mathcal{G}_2^w[0, \infty)$ denote the set of all piecewise continuous functions f on $[0, \infty)$ with the property that

$$\int_0^\infty e^{-x} f(x)^2 \, dx$$

converges. It is then easy to show that $\mathcal{G}_2^w$ is a Euclidean space under the usual addition, scalar multiplication, and inner product, the last being defined by the formula

$$f \cdot g = \int_0^\infty e^{-x} f(x)g(x) \, dx.$$

Moreover, since

$$\int_0^\infty e^{-x}x^n dx = n!, \quad n = 0, 1, 2, \ldots, \tag{11-34}$$

(Exercise 1), $\mathcal{G}_2^w$ contains all polynomials, and hence, as before, we have the following theorem.

Theorem 11–19. *$\mathcal{G}_2^w$ is the set of all piecewise continuous functions on $[0, \infty)$ which can be expressed as limits in the mean of sequences of polynomials.*

From this it follows that any mutually orthogonal set of polynomials in $\mathcal{G}_2^w$, one of each degree, will be a basis for this space. One such set is the sequence $\{L_n(x)\}$, $n = 0, 1, 2, \ldots$, of *Laguerre polynomials*, where

$$L_n(x) = (-1)^n e^x \frac{d^n}{dx^n}(x^n e^{-x}). \tag{11–35}$$

Indeed, when the argument used to establish Theorem 11–16 is adapted to this situation, we find that

$$\begin{aligned} L_m \cdot L_n &= 0, \quad m \neq n, \\ L_n \cdot L_n &= (n!)^2. \end{aligned} \tag{11–36}$$

Furthermore, we have

Lemma 11–7. *For each positive integer n, L_n is a polynomial of degree n with 1 as its leading coefficient and $-n^2$ as the coefficient of x^{n-1}.*

Proof. By repeated differentiation we find that

$$\frac{d^n}{dx^n}(x^n e^{-x}) = (-1)^n[x^n - n^2 x^{n-1} + \cdots]e^{-x} \tag{11–37}$$

(see Exercise 9, Section 11–3) where the terms omitted are of degree $< n - 1$ in x. The lemma now follows by substituting this expression in (11–35). ▌

To derive the recurrence relation for the Laguerre polynomials we reason as follows. The polynomial $xL_n(x)$ is of degree $n + 1$, and hence can be written in the form

$$xL_n(x) = \sum_{k=0}^{n+1} a_k L_k(x), \tag{11–38}$$

where

$$a_k = \frac{(xL_n) \cdot L_k}{\|L_k\|^2}.$$

But, by Lemma 11–1, $(xL_n) \cdot L_k = 0$ for all $k < n - 1$. Hence (11–38) reduces to

$$xL_n = a_{n-1}L_{n-1} + a_n L_n + a_{n+1} L_{n+1}. \tag{11–39}$$

We now equate the coefficients of x^{n+1} on both sides of this equation to deduce that $a_{n+1} = 1$. In other words,

$$\frac{(xL_n) \cdot L_{n+1}}{L_{n+1} \cdot L_{n+1}} = 1,$$

and we have

$$(xL_n) \cdot L_{n+1} = [(n+1)!]^2$$

see (11–36). Hence

$$a_{n-1} = \frac{(xL_n) \cdot L_{n-1}}{[(n-1)!]^2} = \frac{(xL_{n-1}) \cdot L_n}{[(n-1)!]^2}$$
$$= \left[\frac{n!}{(n-1)!}\right]^2 = n^2,$$

and (11–39) can be rewritten

$$xL_n = n^2 L_{n-1} + a_n L_n + L_{n+1}.$$

Finally, to find a_n we equate the coefficients of x^n on both sides of this expression and use Lemma 11–7. This gives $a_n = 2n + 1$, and we have proved

Theorem 11–20. *The Laguerre polynomials satisfy the recurrence relation*

$$L_{n+1} + (2n + 1 - x)L_n + n^2 L_{n-1} = 0 \qquad (11\text{–}40)$$

for all $n \geq 1$, and all x.

(Again this relation is valid when $n = 0$ if we set $L_{-1} \equiv 0$.)

Starting with the value $L_0(x) = 1$ derived from (11–35), the above formula yields

$$L_0(x) = 1,$$
$$L_1(x) = x - 1,$$
$$L_2(x) = x^2 - 4x + 2,$$
$$L_3(x) = x^3 - 9x^2 + 18x - 6,$$
$$L_4(x) = x^4 - 16x^3 + 72x^2 - 96x + 24,$$
$$L_5(x) = x^5 - 25x^4 + 200x^3 - 600x^2 + 600x - 120.$$

To complete our discussion it remains to derive the differential equation for L_n. Here we start by differentiating (11–35) to obtain

$$\frac{d}{dx} L_n(x) = (-1)^n e^x \frac{d^{n+1}}{dx^{n+1}} (x^n e^{-x}) + L_n(x)$$
$$= (-1)^n e^x \frac{d^n}{dx^n} (nx^{n-1} e^{-x} - x^n e^{-x}) + L_n(x)$$
$$= (-1)^n n e^x \frac{d^n}{dx^n} (x^{n-1} e^{-x}).$$

Similarly,

$$\frac{d}{dx} L_{n-1}(x) = (-1)^{n-1} e^x \frac{d^n}{dx^n} (x^{n-1} e^{-x}) + L_{n-1}(x),$$

and we therefore have

$$L_n' = nL_{n-1} - nL_{n-1}', \quad n \geq 1. \tag{11–41}$$

But, by (11–40),

$$L_{n+1}' + (2n + 1 - x)L_n' - L_n + n^2 L_{n-1}' = 0,$$

and since (11–41) implies that

$$L_{n+1}' = (n + 1)L_n - (n + 1)L_n' \qquad \text{and} \qquad n^2 L_{n-1}' = n^2 L_{n-1} - nL_n',$$

we find that

$$xL_n' = nL_n + n^2 L_{n-1}. \tag{11–42}$$

We now use (11–41) again to deduce that $n^2 L_{n-1}' + nL_n' = n^2 L_{n-1}$ and hence, by (11–42), that

$$n^2 L_{n-1}' + nL_n' = xL_n' - nL_n. \tag{11–43}$$

Finally, differentiating (11–42), we obtain

$$xL_n'' + L_n' = nL_n' + n^2 L_{n-1}',$$

or, by (11–43),

$$xL_n'' + L_n' = xL_n' - nL_n,$$

thereby proving

Theorem 11–21. *The Laguerre polynomial of degree n is a solution of the second-order linear differential equation*

$$xy'' + (1 - x)y' + ny = 0. \tag{11–44}$$

Equation (11–44) is known as *Laguerre's equation* of order n, and, as we shall see (Exercise 11, Section 15–3), L_n is the only polynomial solution of this equation with leading coefficient 1.

EXERCISES

1. Prove that

$$\int_0^\infty e^{-x} x^n \, dx = n!, \quad n = 0, 1, 2, \ldots.$$

2. Verify the formulas for L_2, L_3, and L_4 given above.

3. Prove that $\mathcal{G}_2^w[0, \infty)$ is a Euclidean space.

4. Prove that

$$L_m \cdot L_n = \begin{cases} 0 & \text{if } m \neq n, \\ (n!)^2 & \text{if } m = n. \end{cases}$$

5. Use the differential equation for the L_n to prove the orthogonality of these polynomials in $\mathcal{G}_2^w[0, \infty)$. [*Hint:* Mimic the argument used in the case of the Legendre polynomials.]

In the following exercises we give the student the opportunity to explore for himself an interesting variation of the material in this section.

6. Let $\alpha > -1$ be a real number, and let

$$L_n^{(\alpha)}(x) = (-1)^n x^{-\alpha} e^x \frac{d^n}{dx^n} (x^{n+\alpha} e^{-x}),$$

$n = 0, 1, 2, \ldots$. Show that $L_n^{(\alpha)}$ is a polynomial in x of degree n with leading coefficient 1 for all α and all n.

7. Prove that

$$\int_0^\infty x^\alpha e^{-x} L_m^{(\alpha)}(x) L_n^{(\alpha)}(x)\, dx = 0$$

for $m \neq n$, and hence deduce that $\{L_n^{(\alpha)}(x)\}$, $n = 0, 1, 2, \ldots$, is a mutually orthogonal sequence of polynomials in the space $\mathcal{G}_2^w[0, \infty)$ with weight function $x^\alpha e^{-x}$.

8. Prove that

$$\|L_n^{(\alpha)}\|^2 = n!\, \Gamma(n + \alpha + 1),$$

where $\Gamma(\alpha)$ denotes the *gamma function.* (See Exercises 25 and 26, Section 5-7.)

9. (a) Prove that the Laguerre polynomials $\{L_n^{(\alpha)}\}$ satisfy the recurrence relation

$$L_{n+1}^{(\alpha)} + (2n + \alpha + 1 - x)L_n^{(\alpha)} + n(n + \alpha)L_{n-1}^{(\alpha)} = 0.$$

(b) Compute the first four Laguerre polynomials $L_n^{(2)}$.

*10. Prove that $L_n^{(\alpha)}$ is a solution of the differential equation

$$xy'' + (\alpha + 1 - x)y' + ny = 0.$$

*11-8 GENERATING FUNCTIONS

We have seen that each of the various types of orthogonal polynomials encountered in this chapter can be characterized in several different ways; by means of a Rodrigues formula valid for all n, by a recurrence relation, or as polynomial solutions of certain linear differential equations. Each of these definitions has something to recommend it, and the particular one chosen is as much a matter of taste as anything else. In this section we shall introduce still another way of obtaining these polynomials—namely, by means of their so-called *generating functions.* Here the motivating idea is that the polynomials in question can be made to appear as the coefficients in the power series expansions of certain functions of two variables, and in this sense are "generated" by these functions. Although this method may appear highly artificial at first—particularly since we make scant effort to motivate the choice of generating function—it is not so at all. In fact, in many physical problems the orthogonal functions we have been considering arise out of their generating functions in a perfectly natural way, and for this reason alone the following discussion has much to recommend it.

I. *The Legendre polynomials.* Here we start with the function

$$G(x, r) = \frac{1}{(1 - 2xr + r^2)^{1/2}}, \tag{11–45}$$

which we expand as a power series in r under the assumption that x and r are chosen so that the resulting series converges. This allows us to write

$$\begin{aligned} G(x, r) &= P_0(x) + P_1(x)r + P_2(x)r^2 + \cdots \\ &= \sum_{n=0}^{\infty} P_n(x)r^n, \end{aligned} \tag{11–46}$$

where $P_0, P_1, \ldots$ are functions of x alone. These coefficients can be computed directly from (11–45) and (11–46) in the usual fashion, and it is not difficult to show that P_n must be a polynomial in x of degree n. For instance, setting $r = 0$ in (11–46) and noting that $G(x, 0) = 1$, we find that

$$P_0(x) = 1.$$

For our present purposes, however, it is sufficient to observe that this series is uniformly and absolutely convergent whenever $|2xr - r^2| < 1$, and hence, under these conditions, can be differentiated term-by-term with respect to r to yield

$$\frac{\partial G}{\partial r} = \sum_{n=1}^{\infty} nP_n(x)r^{n-1}. \tag{11–47}$$

But, by (11–45),

$$\frac{\partial G}{\partial r} = -\frac{r - x}{(1 - 2xr + r^2)^{3/2}},$$

or

$$(1 - 2xr + r^2)\frac{\partial G}{\partial r} + (r - x)G = 0.$$

Substituting (11–46) and (11–47) into this expression yields

$$(1 - 2xr + r^2)\sum_{n=1}^{\infty} nP_n(x)r^{n-1} + (r - x)\sum_{n=0}^{\infty} P_n(x)r^n = 0,$$

and it follows that

$$\sum_{n=1}^{\infty} nP_n(x)r^{n-1} - xP_0(x) - \sum_{n=1}^{\infty} (2n+1)xP_n(x)r^n + P_0(x)r + \sum_{n=1}^{\infty} (n+1)P_n(x)r^{n+1} = 0,$$

or

$$-xP_0(x) + P_0(x)r + \sum_{n=0}^{\infty} (n+1)P_{n+1}(x)r^n - \sum_{n=1}^{\infty} (2n+1)xP_n(x)r^n + \sum_{n=2}^{\infty} nP_{n-1}(x)r^n = 0.$$

Thus

$$[P_1(x) - xP_0(x)] + [2P_2(x) - 3xP_1(x) + P_0(x)]r + \sum_{n=2}^{\infty} [(n+1)P_{n+1}(x) - (2n+1)xP_n(x) + nP_{n-1}(x)]r^n = 0,$$

and we have

$$P_1 - xP_0 = 0,$$
$$(n+1)P_{n+1} - (2n+1)xP_n + nP_{n-1} = 0, \quad n \geq 1.$$

But the second of these expressions is none other than the recurrence relation for the Legendre polynomials, and since $P_0(x) = 1$ and $P_1(x) = x$, we conclude that the coefficients in (11–46) are, in fact, the Legendre polynomials. Thus the function

$$G(x, r) = \frac{1}{(1 - 2xr + r^2)^{1/2}}$$

is a generating function for the Legendre polynomials.

In physics, the function $G(x, r)$ is encountered in the study of planetary motion and electrostatic potential, among other places. In fact, Legendre's original memoir on this subject, published in 1785, was devoted to the study of the gravitational attraction of "spheroids," and it is only fitting that we conclude our discussion of the Legendre polynomials with a brief indication of how this comes to pass. For various reasons we shall consider the case involving electrostatic potential, which, theoretically, is the same as the one treated by Legendre.

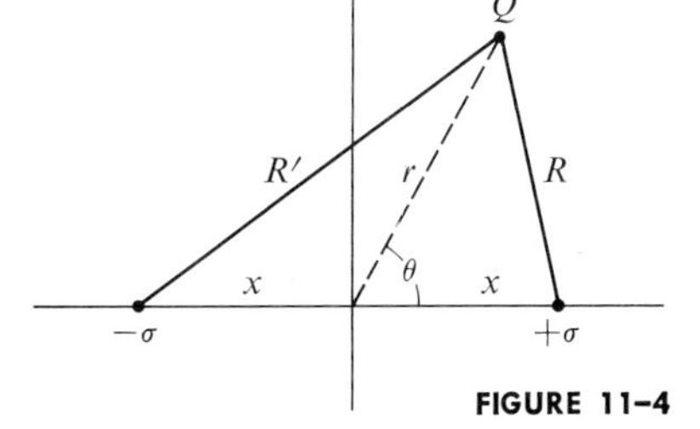

FIGURE 11–4

Specifically, we pose the problem of determining the potential at an arbitrary point Q of the plane arising from a pair of point charges of magnitudes $+\sigma$ and $-\sigma$ located as shown in Fig. 11–4. Since the potential at Q due to a single point charge σ located R units away is σ/R, the potential at Q in this case is

$$V = \sigma\left(\frac{1}{R} - \frac{1}{R'}\right).$$

But, referring to Fig. 11–4 we see that

$$\begin{aligned} R^2 &= r^2 + x^2 - 2rx\cos\theta, \\ R'^2 &= r^2 + x^2 - 2rx\cos(\pi - \theta), \end{aligned}$$

from which it follows that

$$\frac{1}{R} = \frac{1}{r}\left[1 - \left(2\frac{x}{r}\cos\theta - \frac{x^2}{r^2}\right)\right]^{-1/2}$$

and

$$\frac{1}{R'} = \frac{1}{r}\left[1 - \left(2\frac{x}{r}\cos(\pi - \theta) - \frac{x^2}{r^2}\right)\right]^{-1/2}.$$

Save for notation, each of these expressions is the generating function for the Legendre polynomials, and we therefore have

$$\frac{1}{R} = \frac{1}{r}\sum_{n=0}^{\infty} P_n(\cos\theta)\left(\frac{x}{r}\right)^n$$

and

$$\frac{1}{R'} = \frac{1}{r}\sum_{n=0}^{\infty} P_n(\cos(\pi - \theta))\left(\frac{x}{r}\right)^n = \frac{1}{r}\sum_{n=0}^{\infty} P_n(-\cos\theta)\left(\frac{x}{r}\right)^n,$$

provided that $|2(x/r)\cos\theta - (x/r)^2| < 1$. Thus

$$V = \frac{\sigma}{r}\left\{\sum_{n=0}^{\infty}\left[P_n(\cos\theta) - P_n(-\cos\theta)\right]\left(\frac{x}{r}\right)^n\right\}.$$

Finally, recalling that P_n is even if n is even, and odd if n is odd, the above expression reduces to

$$V = \frac{2\sigma}{r}\sum_{n=0}^{\infty} P_{2n+1}(\cos\theta)\left(\frac{x}{r}\right)^{2n+1}.$$

In elementary applications it is customary to approximate V by the first term in this series, in which case we have

$$V \approx \frac{\sigma}{r^2}(2x)\cos\theta = \frac{\sigma}{r^2}d\cos\theta,$$

where d is the distance between the charges.

II. *The Hermite polynomials.* Here a generating function is

$$G(x, r) = e^{xr - r^2/2}. \tag{11-48}$$

Indeed, since the series

$$e^z = 1 + z + \frac{z^2}{2!} + \cdots$$

is uniformly and absolutely convergent for all z,

$$\begin{aligned} G(x, r) &= 1 + \left(xr - \frac{r^2}{2}\right) + \frac{1}{2!}\left(xr - \frac{r^2}{2}\right)^2 + \cdots \\ &= 1 + xr + \frac{1}{2!}(x^2 - 1)r^2 + \cdots \\ &= H_0(x) + H_1(x)r + \frac{1}{2!}H_2(x)r^2 + \cdots \end{aligned}$$

for all x and all r. Differentiating this series term-by-term we find that

$$\frac{\partial G}{\partial r} = \sum_{n=0}^{\infty} \frac{H_{n+1}(x)}{n!} r^n.$$

But (11-48) implies that

$$\frac{\partial G}{\partial r} = (x - r)G,$$

and hence that

$$\begin{aligned} \frac{\partial G}{\partial r} &= (x - r)\sum_{n=0}^{\infty} \frac{H_n(x)}{n!} r^n \\ &= xH_0(x) + \sum_{n=1}^{\infty} \frac{xH_n(x) - nH_{n-1}(x)}{n!} r^n. \end{aligned}$$

Thus

$$H_1(x) + \sum_{n=1}^{\infty} \frac{H_{n+1}(x)}{n!} r^n = xH_0(x) + \sum_{n=1}^{\infty} \frac{xH_n(x) - nH_{n-1}(x)}{n!} r^n,$$

and it follows that

$$\begin{aligned} H_1 - xH_0 &= 0, \\ H_{n+1} - xH_n + nH_{n-1} &= 0, \quad n \geq 0, \end{aligned}$$

as required.

III. *The Laguerre polynomials.* In this case a generating function is

$$G(x, r) = \frac{1}{1 - r} e^{-rx/(1-r)}; \tag{11-49}$$

an assertion which is proved as follows.

The Taylor series expansion of G in powers of r can be written in the form

$$G(x, r) = \sum_{n=0}^{\infty} (-1)^n L_n(x) \frac{r^n}{n!}, \tag{11-50}$$

where $L_n(x)$ is a function of x alone. Moreover, this series converges uniformly and absolutely for all x and all r with $|r| < 1$ since $G(x, r)$ is the product of two functions whose series are thus convergent. Hence

$$\frac{\partial G}{\partial r} = \sum_{n=1}^{\infty} (-1)^n L_n(x) \frac{r^{n-1}}{(n-1)!}.$$

Moreover, by (11–49),

$$(1 - r)^2 \frac{\partial G}{\partial r} = (1 - x - r)G,$$

and we therefore have

$$(1 - r)^2 \sum_{n=1}^{\infty} (-1)^n L_n(x) \frac{r^{n-1}}{(n-1)!} + (x - 1 + r) \sum_{n=0}^{\infty} (-1)^n L_n(x) \frac{r^n}{n!} = 0. \tag{11-51}$$

An easy computation now reveals that

$$L_{n+1} + (2n + 1 - x)L_n + n^2 L_{n-1} = 0, \quad n \geq 1,$$

and since $L_0(x) = 1$ and $L_1(x) = x - 1$, we are done.

EXERCISES

1. Let

$$G(x, r) = \frac{1}{(1 - 2xr + r^2)^{1/2}}.$$

Prove that

(a) $(1 - 2xr + r^2) \dfrac{\partial G}{\partial x} - rG \equiv 0;$

(b) $(x - r) \dfrac{\partial G}{\partial x} - r \dfrac{\partial G}{\partial r} \equiv 0;$

(c) $r \dfrac{\partial}{\partial r} (rG) - (1 - rx) \dfrac{\partial G}{\partial x} \equiv 0.$

2. (a) Substitute the series

$$G(x, r) = \sum_{n=0}^{\infty} P_n(x) r^n$$

into the identities in 1(b) and 1(c), and deduce in turn that

$$nP_n(x) - xP_n'(x) + P_{n-1}'(x) \equiv 0, \quad n \geq 1,$$

and

$$(n + 1)P_n(x) - P_{n+1}'(x) + xP_n'(x) \equiv 0.$$

(b) Use these results to prove that

$$P_{n+1}' - P_{n-1}' = (2n + 1)P_n$$

for all $n \geq 1$.

3. Use the identities in the preceding exercise to derive the differential equation for P_n. [*Hint:* Replace n by $n - 1$ in the second identity in 2(a) and then substitute the value of P_{n-1}' found in 2(b). Differentiate, and again eliminate P_{n-1}'.]

4. Starting with the series

$$G(x, r) = \sum_{n=0}^{\infty} \frac{H_n(x)}{n!} r^n,$$

prove that

$$H_n(x) = (-1)^n e^{x^2/2} \frac{d^n}{dx^n} e^{-x^2/2}.$$

[*Hint:* $e^{xr - r^2/2} = e^{x^2/2} e^{-(x-r)^2/2}$.]

5. (a) With G as in Exercise 4, compute $\partial G/\partial r$, and show that $H_n'(x) = nH_{n-1}(x)$.
(b) Use the result in (a) and the recurrence formula for the H_n to obtain the differential equation for H_n.

6. Show that the recurrence relation for the Laguerre polynomials follows from (11–51).

7. Let the Laguerre polynomials be defined by their generating function

$$G(x, r) = \frac{1}{1 - r} e^{-rx/(1-r)}.$$

(a) Prove that

$$(1 - r)\frac{\partial G}{\partial x} + rG \equiv 0,$$

and

$$r(1 - r)\frac{\partial G}{\partial r} - (x - 1 + r)\frac{\partial G}{\partial x} \equiv 0.$$

(b) Use these identities and the series expansion

$$G(x, r) = \sum_{n=0}^{\infty} (-1)^n L_n(x) \frac{r^n}{n!}$$

to prove that L_n is a solution of the differential equation

$$xy'' + (1 - x)y' + ny = 0.$$

8. (a) Starting with the formula

$$L_n(x) = (-1)^n e^x \frac{d^n}{dx^n}(x^n e^{-x}),$$

show that

$$L_n(x) = \sum_{k=0}^{n} (-1)^k \left[\frac{n!}{(n-k)!}\right]^2 \frac{x^{n-k}}{k!}.$$

*(b) Use the result in (a) to sum the series

$$\sum_{n=0}^{\infty} (-1)^n L_n(x) \frac{r^n}{n!},$$

and show that

$$\frac{1}{1-r} e^{-rx/(1-r)} = \sum_{n=0}^{\infty} (-1)^n L_n(x) \frac{r^n}{n!}.$$

[*Hint:* Interchange the order of summation.]

*9. Prove that the function

$$G(r, x) = \frac{1}{(1-r)^{\alpha+1}} e^{-rx/(1-r)}$$

is a generating function for the Laguerre polynomials $L_n^{(\alpha)}$. (See Exercises 6 through 10 of the preceding section.)

A Table of Orthogonal Functions

The following table summarizes the results obtained in this chapter. For convenience of reference we have also included the corresponding information for Bessel functions (see Chapter 15).

LEGENDRE POLYNOMIALS

Definition	$P_n(x) = \frac{1}{2^n n!} \frac{d^n}{dx^n}(x^2 - 1)^n$
*Recurrence relation	$P_{-1} \equiv 0, \quad P_0 \equiv 1$ $(n+1)P_{n+1} - (2n+1)xP_n + nP_{n-1} \equiv 0, \quad n \geq 0$
Differential equation	$(1 - x^2)y'' - 2xy' + n(n+1)y = 0$
Generating function	$\frac{1}{(1 - 2rx + r^2)^{1/2}}$
Orthogonality	Orthogonal in $\mathcal{PC}[-1, 1]$ $P_m \cdot P_n = \begin{cases} 0, & m \neq n \\ \frac{2}{2n+1}, & m = n \end{cases}$

* The index -1 is used as a subscript *only* when working with recurrence relations.

HERMITE POLYNOMIALS

Definition	$H_n(x) = (-1)^n e^{x^2/2} \dfrac{d^n}{dx^n} e^{-x^2/2}$
*Recurrence relation	$H_{-1} \equiv 0, \quad H_0 \equiv 1$ $H_{n+1} - xH_n + nH_{n-1} \equiv 0, \quad n \geq 0$
Differential equation	$y'' - xy' + ny = 0$
Generating function	$e^{xr-r^2/2}$
Orthogonality	Orthogonal in $\mathcal{G}_2^w(-\infty, \infty)$; weight function $e^{-x^2/2}$ $H_m \cdot H_n = \begin{cases} 0, & m \neq n \\ \sqrt{2\pi}\, n!, & m = n \end{cases}$

LAGUERRE POLYNOMIALS L_n

Definition	$L_n(x) = (-1)^n e^x \dfrac{d^n}{dx^n} (x^n e^{-x})$
*Recurrence relation	$L_{-1} \equiv 0, \quad L_0 \equiv 1$ $L_{n+1} + (2n + 1 - x)L_n + n^2 L_{n-1} \equiv 0, \quad n \geq 0$
Differential equation	$xy'' + (1 - x)y' + ny = 0$
Generating function	$\dfrac{1}{1 - r} e^{-rx/(1-r)}$
Orthogonality	Orthogonal in $\mathcal{G}_2^w[0, \infty)$; weight function e^{-x} $L_m \cdot L_n = \begin{cases} 0, & m \neq n \\ (n!)^2, & m = n \end{cases}$

* The index -1 is used as a subscript *only* when working with recurrence relations.

LAGUERRE POLYNOMIALS $L_n^{(\alpha)}$

Definition	$L_n^{(\alpha)}(x) = (-1)^n x^{-\alpha} e^x \dfrac{d^n}{dx^n}(x^{n+\alpha}e^{-x}), \quad \alpha > -1$
*Recurrence relation	$L_{-1}^{(\alpha)} \equiv 0, \quad L_0^{(\alpha)} \equiv 1$ $L_{n+1}^{(\alpha)} + (2n + \alpha + 1 - x)L_n^{(\alpha)} + n(n + \alpha)L_{n-1}^{(\alpha)} \equiv 0, \quad n \geq 0$
Differential equation	$xy'' + (\alpha + 1 - x)y' + ny = 0$
Generating function	$\dfrac{1}{(1 - r)^{\alpha+1}} e^{-rx/(1-r)}$
Orthogonality	Orthogonal in $\mathcal{G}_2^w[0, \infty)$; weight function $x^\alpha e^{-x}$ $L_m^{(\alpha)} \cdot L_n^{(\alpha)} = \begin{cases} 0, & m \neq n \\ n!\Gamma(n + \alpha + 1), & m = n \end{cases}$

BESSEL FUNCTIONS

Definition	Defined by the differential equation
Recurrence relation	$xJ_{p+1} - 2pJ_p + xJ_{p-1} \equiv 0$, p an arbitrary real number
Differential equation	$x^2y'' + xy' + (x^2 - p^2)y = 0$
Generating function	$e^{(x/2)(t-1/t)}$
Orthogonality	Too lengthy to be summarized here. See Section 15–9.

* The index -1 is used as a subscript *only* when working with recurrence relations.

12 boundary-value problems for ordinary differential equations

12–1 DEFINITIONS AND EXAMPLES

By far the most important application of the ideas we have been considering in the past few chapters occurs in the study of *boundary-value problems* for second-order linear differential equations. Formally, such a problem consists of

(i) an equation of the type

$$Ly = h, \tag{12–1}$$

in which L is a second-order linear differential operator defined on a (finite) interval $[a, b]$, and h a function in $\mathcal{C}[a, b]$; and

(ii) a pair of *boundary* or *endpoint conditions* of the form

$$\begin{aligned} \alpha_1 y(a) + \alpha_2 y(b) + \alpha_3 y'(a) + \alpha_4 y'(b) &= \gamma_1, \\ \beta_1 y(a) + \beta_2 y(b) + \beta_3 y'(a) + \beta_4 y'(b) &= \gamma_2, \end{aligned} \tag{12–2}$$

where the α_i, β_i, and γ_i are constants. The problem, of course, is to find all functions y in $\mathcal{C}^2[a, b]$ which simultaneously satisfy (12–1) and (12–2).*

For instance, the equation

$$y'' + y = 0 \tag{12–3}$$

with boundary conditions

$$y(0) = 0, \qquad y(\pi) = 0 \tag{12–4}$$

is a problem of this type on the interval $[0, \pi]$. To solve it we simply apply the boundary conditions to the general solution $c_1 \sin x + c_2 \cos x$ of (12–3) to

* Boundary conditions of a different sort must be imposed if the interval is infinite. An example of such a problem will be found in Section 13–8.

deduce that $c_2 = 0$ and that c_1 is arbitrary. Thus

$$y = c \sin x,$$

c an arbitrary constant, is the general solution of this particular problem.

As we shall see, boundary-value problems abound in physics and applied mathematics, and the solution of all but the simplest of them involves some type of orthogonal series. Indeed, the need to solve boundary-value problems was the stimulus which originally led to the invention of orthogonal series, and as we proceed it will become apparent that their study is a mathematical discipline in its own right; not, as one might expect, an adjunct to the general theory of differential equations.

In order to exclude certain trivial cases from the following discussion, we demand that at least one of the α_i *and* one of the β_i appearing in (12–2) be different from zero, *and* that the left-hand sides of these equations be linearly independent in the sense that they are not constant multiples of one another. Furthermore, to ensure that we are actually looking at a boundary-value problem, and not an initial-value problem, we also require that (12–2) contain nonzero terms involving each of the endpoints of the interval. Finally, we shall say that the given boundary conditions are *homogeneous* whenever $\gamma_1 = \gamma_2 = 0$. In this case the set of twice continuously differentiable functions on $[a, b]$ which satisfy (12–2) is a subspace $\mathcal{S}$ of $\mathcal{C}^2[a, b]$, and by viewing L as a linear transformation from $\mathcal{S}$ to $\mathcal{C}[a, b]$ the problem in question becomes one of solving a certain operator equation; to wit, the equation $Ly = h$, where h is a known function in $\mathcal{C}[a, b]$, and $L: \mathcal{S} \to \mathcal{C}[a, b]$. Despite its simplicity this observation is important because it shows that the boundary conditions influence the problem only to the extent of determining the *domain space* for L. Strictly speaking, some symbol other than "L" ought to be used to represent the operator from $\mathcal{S}$ to $\mathcal{C}[a, b]$, since heretofore we have considered L as acting on *all* of $\mathcal{C}^2[a, b]$. (Recall that operators with different domains are different, even though the operators themselves are defined by the same formula.) However, such changes in notation would be as confusing as they are correct, and will therefore be avoided. But by the same token, it then becomes mandatory to specify the domain of the operator being considered whenever there is any possibility of confusion.

At this point we invoke a familiar argument to reduce the study of boundary-value problems with nonhomogeneous boundary conditions to the homogeneous case (see Section 2–9 and Exercise 8 below). Hence, unless otherwise stated, we shall assume from now on that all boundary conditions imposed are homogeneous.

The solutions of a boundary-value problem involving a linear differential operator $L: \mathcal{S} \to \mathcal{C}[a, b]$ are intimately related to the solutions of the equation

$$Ly = \lambda y, \tag{12–5}$$

where λ is an unknown parameter. In this setting we are required to find all values of λ for which (12–5) admits *nontrivial* solutions in $\mathcal{S}$, and then find the solutions corresponding to these λ. The reader should note that (12-5) may be rewritten

$$(L - \lambda I)y = 0, \tag{12-6}$$

where I denotes the identity transformation which sends each function in $\mathcal{C}[a, b]$ onto itself, or as

$$a_2(x)y'' + a_1(x)y' + [a_0(x) - \lambda]y = 0$$

if $L = a_2(x)D^2 + a_1(x)D + a_0(x)$. Thus, for each value of λ, (12-5) is a homogeneous second-order linear differential equation, and the solution set of any (homogeneous) boundary-value problem involving this equation is the null space, in $\mathcal{S}$, of the operator $L - \lambda I$.

This having been said, we observe that the above problem can be rephrased in the language of linear algebra without specific reference to the nature of the operator L, as follows:

Given a linear transformation $L: \mathcal{S} \to \mathcal{V}$, where $\mathcal{S}$ is a subspace of $\mathcal{V}$, find all values of λ for which the equation

$$L\mathbf{x} = \lambda\mathbf{x}$$

has nontrivial solutions; then find all solutions corresponding to these values of λ.

And this is the problem with which we shall begin our investigations. But first, an example.

EXAMPLE. Solve the boundary-value problem

$$\begin{gathered} y'' + \lambda y = 0, \\ y(0) = 0, \qquad y(\pi) = 0. \end{gathered} \tag{12-7}$$

Here $\mathcal{S}$ is the space of all twice continuously differentiable functions on $[0, \pi]$ which vanish at the endpoints of the interval, and L is the second-order linear differential operator $-D^2$. (The minus sign has been introduced merely to simplify the final results. Without it the relevant values of λ would be negative.) We distinguish three cases, according as $\lambda = 0, \lambda < 0, \lambda > 0$.*

Case 1: $\lambda = 0$. Here the differential equation has $c_1 + c_2x$ as its general solution, and the boundary conditions imply that $c_1 = c_2 = 0$. Thus (12-7) has no nontrivial solutions when $\lambda = 0$.

Case 2: $\lambda < 0$. In this case $y = c_1e^{\sqrt{-\lambda}x} + c_2e^{-\sqrt{-\lambda}x}$, and the boundary conditions again yield $y \equiv 0$.

---

* At this point we are tacitly assuming that λ must be real. This assumption will be justified later.

Case 3: $\lambda > 0$. Here the general solution of $y'' + \lambda y = 0$ is

$$y = c_1 \sin \sqrt{\lambda}\, x + c_2 \cos \sqrt{\lambda}\, x, \tag{12–8}$$

and the boundary conditions now yield the pair of equations

$$c_2 = 0, \qquad c_1 \sin \sqrt{\lambda}\, \pi = 0.$$

Thus (12–7) admits nontrivial solutions if and only if

$$\sin \sqrt{\lambda}\, \pi = 0;$$

that is, if and only if λ assumes one of the values $\lambda_n = n^2, n = 1, 2, \ldots$. Furthermore, for each of these values of λ the constant c_1 in (12–8) remains arbitrary, and it follows that the solution space corresponding to λ_n is the one-dimensional subspace of $\mathcal{C}[0, \pi]$ spanned by the function $\sin nx$.

The numbers $\lambda_n = n^2$ which determine the cases in which (12–7) has nontrivial solutions are called the *eigenvalues* for this problem, and each nontrivial solution corresponding to the eigenvalue λ_n is called an *eigenvector* or *eigenfunction belonging to* λ_n. In the next section this terminology will be generalized to include a much wider class of problems, and for the moment we merely ask the reader to note that any set of eigenfunctions, such as $\sin x, \sin 2x, \sin 3x, \ldots$, one for each eigenvalue, is *orthogonal* in $\mathcal{C}[0, \pi]$. This, as we shall see, is no accident, and when properly generalized will be of fundamental importance in the study of boundary-value problems.

EXERCISES

Find all solutions of each of the following boundary-value problems.

1. $y'' + y = 0;\ y(0) = 1,\ y(\pi) = -1$
2. $y'' + y = 0;\ y(0) = 0,\ y'(\pi) = 0$
3. $y'' + 4y = \sin 2x;\ y(0) = 0,\ y(\pi) = 0$
4. $y'' + 5y = e^x;\ y(0) = 0,\ y(\pi) = 0$
5. $y'' + 9y = 0;\ y'(0) = 0,\ y'(\pi) = 0$
6. $y'' + 9y = \cos 2x;\ y'(0) = 0,\ y'(\pi) = 0$
7. $y'' + 9y = x;\ y'(0) = 0,\ y'(\pi) = 0$
8. Let

$$Ly = h;$$
$$B_1 = \gamma_1, \qquad B_2 = \gamma_2,$$

be a boundary-value problem of the type described by Eqs. (12–1) and (12–2) above. Prove that the solution set of this problem consists of all functions in $\mathcal{C}^2[a, b]$ of the form $y_p + y_h$, where y_p is a fixed solution of the problem, and y_h a solution of the associated homogeneous problem

$$Ly = 0,$$
$$B_1 = B_2 = 0.$$

9. Find all eigenvalues and eigenfunctions for the boundary-value problem

$$y'' + \lambda y = 0,$$
$$y(0) = y(2\pi), \qquad y'(0) = y'(2\pi).$$

(Assume that λ is real.)

10. Consider the boundary-value problem

$$y'' + 4y' + (4 + 9\lambda)y = 0,$$
$$y(0) = 0, \qquad y(a) = 0,$$

with λ real.

(a) Show that this problem has no nontrivial solutions for $\lambda \leq 0$. [*Hint:* Consider the cases $\lambda < 0$ and $\lambda = 0$ separately.]

(b) Show that the only positive values of λ for which this problem has nontrivial solutions are

$$\lambda_n = \frac{\pi^2 n^2}{9a^2}, \quad n = 1, 2, \ldots,$$

and find the corresponding solutions.

12-2 EIGENVALUES AND EIGENVECTORS

In this section we consider the general problem of solving an operator equation of the form

$$L\mathbf{x} = \lambda\mathbf{x}, \tag{12-9}$$

where L is a linear transformation from $\mathcal{S}$ to $\mathcal{V}$, $\mathcal{S}$ a subspace of $\mathcal{V}$, and λ an unknown parameter which can assume real or complex values. Technically, this problem is known as the *eigenvalue problem* for the operator L, and requires that we find all λ for which (12-9) has nontrivial solutions, and all solutions corresponding to these λ. Our primary interest, of course, is in the case where $\mathcal{S}$ and $\mathcal{V}$ are Euclidean spaces, and L a second-order linear differential operator. Nevertheless, we shall refrain from imposing these restrictions until Section 12-4 in order that our preliminary results be valid for general linear transformations and arbitrary vector spaces.

As usual, we begin by introducing some terminology.

Definition 12-1. The values of λ for which Eq. (12-9) has nonzero solutions are called the *eigenvalues* (or *characteristic values*) of L, and for each eigenvalue λ_0 the *nonzero* vectors in $\mathcal{S}$ which satisfy the equation $L\mathbf{x} = \lambda_0\mathbf{x}$ are called the *eigenvectors* (or *characteristic vectors*) of L *belonging to* λ_0.

The reader should observe that, by definition, the zero vector is *never* an eigenvector for L. Furthermore, zero is an eigenvalue for L if and only if the equation $L\mathbf{x} = \mathbf{0}$ has nonzero solutions, i.e., if and only if L is *not* one-to-one. Failure to appreciate these simple distinctions is a frequent source of confusion.

If λ_0 is an eigenvalue for L, and $\mathbf{x}_0$ an eigenvector belonging to λ_0, then

$$L(\alpha\mathbf{x}_0) = \alpha L(\mathbf{x}_0) = \alpha(\lambda_0\mathbf{x}_0) = \lambda_0(\alpha\mathbf{x}_0)$$

for all real numbers α. Thus $\alpha\mathbf{x}_0$ is also an eigenvector belonging to λ_0 whenever α is different from zero. This, combined with the obvious fact that the sum of two eigenvectors belonging to λ_0 is again an eigenvector belonging to λ_0, yields

> **Lemma 12–1.** *The solution set of the equation $L\mathbf{x} = \lambda_0\mathbf{x}$ is a nontrivial subspace of $\mathcal{S}$ for each eigenvalue λ_0 of L.*

In other words, the zero vector together with the eigenvectors for L which belong to λ_0 constitute a subspace of $\mathcal{S}$ (and hence, by implication, of $\mathcal{V}$ as well). We shall denote this subspace by $\mathcal{S}_{\lambda_0}$, and observe in passing that $\dim \mathcal{S}_{\lambda_0} \geq 1$ for all λ_0. Geometrically, L acts on $\mathcal{S}_{\lambda_0}$ by "stretching" each of its vectors by the scalar factor λ_0 as indicated in Fig. 12–1.

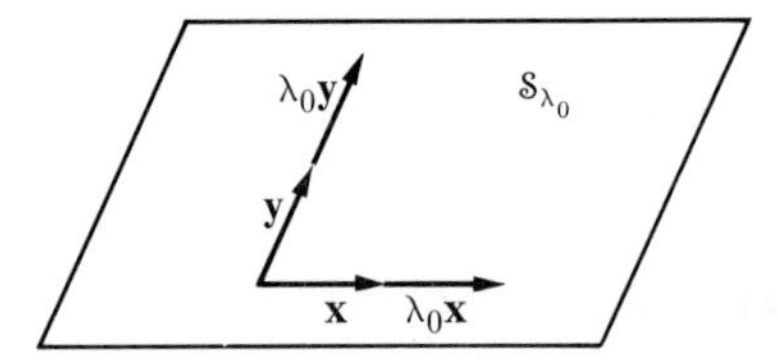

FIGURE 12–1

We have just seen that $L\mathbf{x}_0$ belongs to $\mathcal{S}_{\lambda_0}$ for all $\mathbf{x}_0$ in $\mathcal{S}_{\lambda_0}$. This fact is sometimes expressed by saying that $\mathcal{S}_{\lambda_0}$ is "invariant" under L in accordance with the following definition.

> **Definition 12–2.** Let $L: \mathcal{S} \to \mathcal{V}$ be a linear transformation, and suppose that $\mathcal{S}$ is a subspace of $\mathcal{V}$. Then a subspace $\mathcal{W}$ of $\mathcal{S}$ is said to be *invariant* under L if and only if $L\mathbf{w}$ belongs to $\mathcal{W}$ for all $\mathbf{w}$ in $\mathcal{W}$.

We hasten to point out that there is nothing in this definition to imply that the nonzero vectors in an invariant subspace for L need be eigenvectors for L. Indeed, as we shall see momentarily, such a conclusion is false. Rather, the implication goes the other way: the $\mathcal{S}_{\lambda_0}$ are invariant subspaces for L consisting of vectors with the special property that $L\mathbf{x} = \lambda_0\mathbf{x}$.

Having introduced the notion of invariant subspace, we can now rephrase the definition of an eigenvector for a linear transformation $L: \mathcal{S} \to \mathcal{V}$ to read as follows: *A nonzero vector $\mathbf{x}$ in $\mathcal{S}$ is an eigenvector for L if and only if the* one-*dimensional subspace of $\mathcal{S}$ spanned by $\mathbf{x}$ is invariant under L.* This observation is frequently useful in the search for eigenvectors.

EXAMPLE 1. Let $\mathbf{e}_1$, $\mathbf{e}_2$ be the standard basis vectors in $\mathcal{R}^2$, and let $L: \mathcal{R}^2 \to \mathcal{R}^2$ be reflection across the $\mathbf{e}_1$-axis; that is,

$$L\mathbf{e}_1 = \mathbf{e}_1, \qquad L\mathbf{e}_2 = -\mathbf{e}_2.$$

Then, from geometric considerations alone, it is clear that the only subspaces of $\mathcal{R}^2$ which are invariant under L are (i) the trivial subspace, (ii) $\mathcal{R}^2$ itself, and (iii) the two one-dimensional subspaces spanned by $\mathbf{e}_1$ and $\mathbf{e}_2$. By the remark made a moment ago, the last two must be *eigenspaces* for L, and are the only such. Further, it is obvious from the definition of L that these subspaces are associated with the eigenvalues 1 and -1, respectively.

EXAMPLE 2. If $L: \mathcal{R}^2 \to \mathcal{R}^2$ is reflection across the origin, then *every* subspace of $\mathcal{R}^2$ is invariant under L. In this case $L\mathbf{x} = -\mathbf{x}$ for all $\mathbf{x}$, and it follows that -1 is the only eigenvalue for L, and that $\mathcal{S}_{-1} = \mathcal{R}^2$. The reader should note that here the invariant subspace associated with the eigenvalue is *two*-dimensional.

EXAMPLE 3. Let L be a rotation of $\mathcal{R}^2$ about the origin through an angle θ. Then, if θ is not an integral multiple of π, there are no one-dimensional invariant subspaces, and L has no eigenvectors at all.

EXAMPLE 4. Let $\mathcal{S}$ be the subspace of $\mathcal{C}[0, \pi]$ consisting of all twice continuously differentiable functions y such that $y(0) = y(\pi) = 0$, and let $L: \mathcal{S} \to \mathcal{C}[0, \pi]$ be the operator $-D^2$. Then, by the example in the previous section, L has an infinite sequence of eigenvalues

$$1, 4, 9, \ldots, n^2, \ldots$$

with associated eigenvectors $c_n \sin nx$, $\quad c_n \neq 0$.

All this is simple enough, but hardly explains why these notions were introduced in the first place. The following theorem furnishes a partial answer to this question, and gives an indication of the importance of eigenvectors in the study of linear transformations.

Theorem 12–1. *Any set of eigenvectors belonging to* **distinct** *eigenvalues for a linear transformation $L: \mathcal{S} \to \mathcal{V}$ is linearly independent in $\mathcal{S}$.*

(Note that this result would fail if $\mathbf{0}$ were allowed to be an eigenvector. Thus the prejudice against the zero vector found in Definition 12–1.)

Proof. The theorem is obviously true when applied to a single eigenvector. Beyond this we reason by induction, as follows.

Assume that the theorem has been proved for every set of $n - 1$ eigenvectors for L, $n > 1$, let $\mathbf{x}_1, \ldots, \mathbf{x}_n$ be n eigenvectors belonging, respectively, to distinct eigenvalues $\lambda_1, \ldots, \lambda_n$, and let

$$\alpha_1\mathbf{x}_1 + \cdots + \alpha_{n-1}\mathbf{x}_{n-1} + \alpha_n\mathbf{x}_n = \mathbf{0}. \tag{12–10}$$

Then, applying L to both sides of this equation, we have

$$\alpha_1 L\mathbf{x}_1 + \cdots + \alpha_{n-1} L\mathbf{x}_{n-1} + \alpha_n L\mathbf{x}_n = \mathbf{0},$$

or

$$\alpha_1(\lambda_1 \mathbf{x}_1) + \cdots + \alpha_{n-1}(\lambda_{n-1}\mathbf{x}_{n-1}) + \alpha_n(\lambda_n \mathbf{x}_n) = \mathbf{0}. \qquad (12\text{–}11)$$

We now multiply (12–10) by λ_n and subtract the resulting equation from (12–11) to obtain

$$\alpha_1(\lambda_1 - \lambda_n)\mathbf{x}_1 + \cdots + \alpha_{n-1}(\lambda_{n-1} - \lambda_n)\mathbf{x}_{n-1} = \mathbf{0}.$$

But, by assumption, the vectors $\mathbf{x}_1, \ldots, \mathbf{x}_{n-1}$ are linearly independent. Hence each of the coefficients in this expression vanishes, and since $\lambda_i - \lambda_n \neq 0$ whenever $i \neq n$, we conclude that $\alpha_i = 0$ for $i = 1, \ldots, n-1$. This, together with (12–10), implies that α_n is also zero, and we are done. ▌

EXERCISES

1. Every vector space $\mathcal{V}$ has at least two subspaces which are invariant under a linear transformation $L: \mathcal{V} \to \mathcal{V}$. What are they?

2. Let L be a linear transformation on a finite dimensional vector space $\mathcal{V}$ (i.e., $L: \mathcal{V} \to \mathcal{V}$), and let $\mathcal{S}_1$ be invariant under L.

   (a) Prove that there exists a subspace $\mathcal{S}_2$ of $\mathcal{V}$ such that every vector $\mathbf{x}$ in $\mathcal{V}$ can be written in one and only one way as

   $$\mathbf{x} = \mathbf{x}_1 + \mathbf{x}_2$$

   with $\mathbf{x}_1$ in $\mathcal{S}_1$, $\mathbf{x}_2$ in $\mathcal{S}_2$. [*Hint:* Choose an appropriate basis for $\mathcal{V}$.]

   (b) Is the subspace $\mathcal{S}_2$ found in (a) necessarily unique? Why?

   (c) Prove, by example, that it may be impossible to choose the subspace $\mathcal{S}_2$ in (a) so that $\mathcal{S}_2$ is invariant under L. [*Hint:* Let $\mathcal{V} = \mathcal{P}_n$, the space of polynomials of degree $< n$, and let L be the differentiation operator.]

3. Let $\mathcal{S}_1$ and $\mathcal{S}_2$ be invariant under a linear transformation $L: \mathcal{S} \to \mathcal{V}$, $\mathcal{S}$ a subspace of $\mathcal{V}$. Prove that the subspace of $\mathcal{S}$ spanned by $\mathcal{S}_1$ and $\mathcal{S}_2$ is also invariant under L.

4. (a) Show that the null space of a linear transformation $L: \mathcal{V} \to \mathcal{V}$ is invariant under L.

   (b) Let $L: \mathcal{V} \to \mathcal{V}$ be a linear transformation with the property that $L^2 = L$ (i.e., L is idempotent on $\mathcal{V}$). Prove that the image of L is an invariant subspace of $\mathcal{V}$.

5. Let $\mathcal{S}$ denote the subspace of $\mathcal{C}^2[a, b]$ determined by a pair of homogeneous boundary conditions

   $$\alpha_1 y(a) + \alpha_2 y'(a) = 0,$$
   $$\beta_1 y(b) + \beta_2 y'(b) = 0,$$

   and let $L: \mathcal{S} \to \mathcal{C}[a, b]$ be the second-order linear differential operator defined by

   $$Ly = (py')' + qy,$$

where p is a function in $\mathcal{C}^1[a, b]$ with $p(x) > 0$ for all x in $[a, b]$, and q is an arbitrary function in $\mathcal{C}[a, b]$. Let y_1 and y_2 be a basis for the null space of L, *viewed as an operator from* $\mathcal{C}^2[a, b]$ *to* $\mathcal{C}[a, b]$. Prove that zero is an eigenvalue for $L: \mathcal{S} \to \mathcal{C}[a, b]$ if and only if the determinant

$$\begin{vmatrix} \alpha_1 y_1(a) + \alpha_2 y_1'(a) & \alpha_1 y_2(a) + \alpha_2 y_2'(a) \\ \beta_1 y_1(b) + \beta_2 y_1'(b) & \beta_1 y_2(b) + \beta_2 y_2'(b) \end{vmatrix}$$

vanishes.

12-3 EIGENVECTORS IN FINITE DIMENSIONAL SPACES

We have seen that eigenvectors belonging to distinct eigenvalues for a linear transformation are linearly independent. In the finite dimensional case this fact immediately yields

Theorem 12-2. *A linear transformation L mapping an n-dimensional vector space $\mathcal{V}$ into itself has at most n distinct eigenvalues. Moreover, when the number of distinct eigenvalues is equal to n, any complete set of eigenvectors, one for each eigenvalue, is a* **basis** *for $\mathcal{V}$, and the matrix of L with respect to such a basis is*

$$\begin{bmatrix} \lambda_1 & & & 0 \\ & \lambda_2 & & \\ & & \ddots & \\ 0 & & & \lambda_n \end{bmatrix},$$

*with the eigenvalues on the main diagonal and zeros elsewhere.**

Of course, such bases need not exist for a given $L: \mathcal{V} \to \mathcal{V}$ (see Examples 2 and 3 above). When they do, however, a number of pleasant things happen. For one, we can then solve operator equations involving L; and rather efficiently too. The following example will illustrate the technique.

EXAMPLE 1. Let L be a linear transformation mapping $\mathcal{R}^3$ into itself, and suppose that L has distinct eigenvalues $\lambda_1, \lambda_2, \lambda_3$. Let $\mathbf{e}_1, \mathbf{e}_2, \mathbf{e}_3$ be eigenvectors belonging to these eigenvalues, and consider the equation

$$L\mathbf{x} = \mathbf{y}, \tag{12–12}$$

$\mathbf{y}$ known, $\mathbf{x}$ unknown. Then, since the vectors $\mathbf{e}_1, \mathbf{e}_2, \mathbf{e}_3$ are a basis for $\mathcal{R}^3$, we have

$$\mathbf{x} = x_1\mathbf{e}_1 + x_2\mathbf{e}_2 + x_3\mathbf{e}_3,$$
$$\mathbf{y} = y_1\mathbf{e}_1 + y_2\mathbf{e}_2 + y_3\mathbf{e}_3,$$

---

* Such a matrix is said to be in *diagonal form*.

and (12–12) can be written

$$L(x_1\mathbf{e}_1 + x_2\mathbf{e}_2 + x_3\mathbf{e}_3) = y_1\mathbf{e}_1 + y_2\mathbf{e}_2 + y_3\mathbf{e}_3.$$

Hence

$$(x_1\lambda_1)\mathbf{e}_1 + (x_2\lambda_2)\mathbf{e}_2 + (x_3\lambda_3)\mathbf{e}_3 = y_1\mathbf{e}_1 + y_2\mathbf{e}_2 + y_3\mathbf{e}_3,$$

and it follows that x_1, x_2, x_3 must be chosen so that

$$x_1\lambda_1 = y_1, \qquad x_2\lambda_2 = y_2, \qquad x_3\lambda_3 = y_3.$$

In particular, we see that (12–12) has a *unique* solution

$$\mathbf{x} = \frac{y_1}{\lambda_1}\mathbf{e}_1 + \frac{y_2}{\lambda_2}\mathbf{e}_2 + \frac{y_3}{\lambda_3}\mathbf{e}_3$$

whenever the λ_i are different from zero. If, on the other hand, one of the λ_i, say λ_1, is zero, (12–12) has no solutions at all *unless* $y_1 = 0$. In the latter case the equation $x_1\lambda_1 = y_1$ is satisfied for *all* values of x_1, and the solution set of (12–12) then consists of all vectors of the form

$$\mathbf{x} = x_1\mathbf{e}_1 + \frac{y_2}{\lambda_2}\mathbf{e}_2 + \frac{y_3}{\lambda_3}\mathbf{e}_3,$$

with x_1 arbitrary.

The generalization of these results to n-dimensional spaces is obvious, and has been left to the reader.

The technique introduced in the above example is known as the *eigenvalue method* for solving an operator equation. Its success depends upon the existence of enough eigenvectors for L to span $\mathcal{V}$, and upon our ability to find them. But both of these questions can be settled by computing the eigenvalues for L, and thus we now address ourselves to this problem.

In Section 2–9 we saw that an equation of the form $L\mathbf{x} = \mathbf{y}$ involving a linear transformation mapping an n-dimensional vector space $\mathcal{V}$ into itself can be written in terms of a basis for $\mathcal{V}$ as a system of linear equations

$$\begin{aligned}
\alpha_{11}x_1 + \alpha_{12}x_2 + \cdots + \alpha_{1n}x_n &= y_1,\\
\alpha_{21}x_1 + \alpha_{22}x_2 + \cdots + \alpha_{2n}x_n &= y_2,\\
&\ \ \vdots\\
\alpha_{n1}x_1 + \alpha_{n2}x_2 + \cdots + \alpha_{nn}x_n &= y_n,
\end{aligned}$$

where

$$\begin{bmatrix}
\alpha_{11} & \alpha_{12} & \cdots & \alpha_{1n}\\
\alpha_{21} & \alpha_{22} & \cdots & \alpha_{2n}\\
\vdots & & & \vdots\\
\alpha_{n1} & \alpha_{n2} & \cdots & \alpha_{nn}
\end{bmatrix}$$

is the matrix of L, and $x_1, \ldots, x_n$ and $y_1, \ldots, y_n$ are the components of $\mathbf{x}$ and $\mathbf{y}$, all with respect to the chosen basis. In particular, this is true of the equation

$$L\mathbf{x} = \lambda\mathbf{x},$$

which, as we have already observed, is equivalent to

$$(L - \lambda I)\mathbf{x} = \mathbf{0},$$

where I is the identity transformation on $\mathcal{V}$. Noting that $L - \lambda I$ can be represented by the matrix

$$\begin{bmatrix} \alpha_{11} - \lambda & \alpha_{12} & \cdots & \alpha_{1n} \\ \alpha_{21} & \alpha_{22} - \lambda & \cdots & \alpha_{2n} \\ \vdots & \vdots & & \vdots \\ \alpha_{n1} & \alpha_{n2} & \cdots & \alpha_{nn} - \lambda \end{bmatrix},$$

we therefore conclude that *the eigenvalues for L are simply the values of* λ *for which the system of homogeneous equations*

$$\begin{aligned} (\alpha_{11} - \lambda)x_1 + \alpha_{12}x_2 + \cdots + \alpha_{1n}x_n &= 0 \\ \alpha_{21}x_1 + (\alpha_{22} - \lambda)x_2 + \cdots + \alpha_{2n}x_n &= 0 \\ &\ \ \vdots \\ \alpha_{n1}x_1 + \alpha_{n2}x_2 + \cdots + (\alpha_{nn} - \lambda)x_n &= 0 \end{aligned} \tag{12–13}$$

has **nontrivial** *solutions.* But this will occur if and only if the determinant of the coefficients of (12–13) vanishes (see Appendix III), i.e., if and only if

$$\begin{vmatrix} \alpha_{11} - \lambda & \alpha_{12} & \cdots & \alpha_{1n} \\ \alpha_{21} & \alpha_{22} - \lambda & \cdots & \alpha_{2n} \\ \vdots & \vdots & & \vdots \\ \alpha_{n1} & \alpha_{n2} & \cdots & \alpha_{nn} - \lambda \end{vmatrix} = 0. \tag{12–14}$$

Thus the eigenvalues for L can be computed by solving (12–14) for λ, and since the left-hand side of this equation is an nth degree polynomial in λ, this can even be done by the methods of elementary algebra (at least for small values of n). The polynomial appearing in (12–14) is known as the *characteristic polynomial* of the linear transformation L, and the equation itself is called the *characteristic equation* of L. As its name suggests, the characteristic polynomial is independent of the particular basis used to compute it; a fact which is also proved in Appendix III.

EXAMPLE 2. Find the eigenvalues and eigenvectors for the linear transformation $L: \mathcal{R}^2 \to \mathcal{R}^2$, given that the matrix of L with respect to the standard basis $\mathbf{e}_1, \mathbf{e}_2$ is

$$\begin{bmatrix} 1 & 0 \\ 2 & 1 \end{bmatrix}.$$

In this case the characteristic equation of L is

$$\begin{vmatrix} 1-\lambda & 0 \\ 2 & 1-\lambda \end{vmatrix} = 0,$$

or

$$(1-\lambda)^2 = 0,$$

and it follows that $\lambda = 1$ is the only eigenvalue for L. Hence a nonzero vector $\mathbf{x} = x_1\mathbf{e}_1 + x_2\mathbf{e}_2$ will be an eigenvector for L if and only if $L\mathbf{x} = \mathbf{x}$. Rewriting this equation in matrix form as

$$\begin{bmatrix} 1 & 0 \\ 2 & 1 \end{bmatrix}\begin{bmatrix} x_1 \\ x_2 \end{bmatrix} = \begin{bmatrix} x_1 \\ x_2 \end{bmatrix},$$

we find that x_1 and x_2 must satisfy the equations

$$\begin{aligned} x_1 &= x_1, \\ 2x_1 + x_2 &= x_2. \end{aligned}$$

Thus x_1 must be zero, while x_2 is arbitrary, and the eigenvectors for L are of the form $x_2\mathbf{e}_2$, $x_2 \neq 0$. Finally, the eigenspace is the one-dimensional subspace of $\mathcal{R}^2$ spanned by $\mathbf{e}_2$.

The reader is encouraged to interpret these results geometrically by viewing L as a "shear" of the xy-plane parallel to the y-axis.

EXAMPLE 3. Find the eigenvalues and eigenvectors of the linear transformation L on $\mathcal{R}^3$ whose matrix with respect to the standard basis $\mathbf{e}_1, \mathbf{e}_2, \mathbf{e}_3$ is

$$\begin{bmatrix} 0 & 0 & 1 \\ 0 & -1 & 0 \\ 2 & 2 & 1 \end{bmatrix}.$$

Since the characteristic polynomial of L is

$$\begin{vmatrix} -\lambda & 0 & 1 \\ 0 & -(1+\lambda) & 0 \\ 2 & 2 & 1-\lambda \end{vmatrix} = -(\lambda - 2)(\lambda + 1)^2,$$

the eigenvalues for L are $\lambda = 2$ and $\lambda = -1$. To find the associated eigenvectors we set $\mathbf{x} = x_1\mathbf{e}_1 + x_2\mathbf{e}_2 + x_3\mathbf{e}_3$, and solve the equation

$$L\mathbf{x} = \lambda\mathbf{x} \tag{12–15}$$

when $\lambda = 2$ and $\lambda = -1$.

In the first case, (12–15) becomes

$$\begin{bmatrix} 0 & 0 & 1 \\ 0 & -1 & 0 \\ 2 & 2 & 1 \end{bmatrix} \begin{bmatrix} x_1 \\ x_2 \\ x_3 \end{bmatrix} = \begin{bmatrix} 2x_1 \\ 2x_2 \\ 2x_3 \end{bmatrix},$$

and we have

$$\begin{aligned} x_3 &= 2x_1, \\ -x_2 &= 2x_2, \\ 2x_1 + 2x_2 + x_3 &= 2x_3. \end{aligned}$$

Thus $x_2 = 0$, $2x_1 = x_3$, and the relevant eigenvectors are $x_1\mathbf{e}_1 + 2x_1\mathbf{e}_3$, x_1 an arbitrary nonzero constant. Here the associated invariant subspace, $\mathcal{S}_2$, is the one-dimensional subspace of $\mathcal{R}^3$ spanned by $\mathbf{e}_1 + 2\mathbf{e}_3$.

Finally, a similar computation reveals that the eigenvectors belonging to the eigenvalue -1 are of the form $x_1\mathbf{e}_1 - x_1\mathbf{e}_3$, x_1 an arbitrary nonzero constant. In this case the associated invariant subspace is the one-dimensional subspace of $\mathcal{R}^3$ spanned by the vector $\mathbf{e}_1 - \mathbf{e}_3$.

EXAMPLE 4. In the preceding section we invoked a geometric argument to prove that any rotation of $\mathcal{R}^2$ through an angle $\theta \neq n\pi$ has no (real) eigenvalues or eigenvectors. We now establish this fact algebraically, as follows.

The matrix of a rotation L of $\mathcal{R}^2$ with respect to the standard basis is

$$\begin{bmatrix} \cos\theta & -\sin\theta \\ \sin\theta & \cos\theta \end{bmatrix},$$

where θ denotes the angle of rotation. Hence the characteristic equation of L is

$$\begin{vmatrix} \cos\theta - \lambda & -\sin\theta \\ \sin\theta & \cos\theta - \lambda \end{vmatrix} = 0,$$

and we have

$$\lambda^2 - 2(\cos\theta)\lambda + 1 = 0.$$

Thus

$$\lambda = \cos\theta \pm i\sin\theta,$$

and it follows that λ is real if and only if $\theta = n\pi$. Moreover, when this is the case, λ assumes one of the values ± 1, and has all of $\mathcal{R}^2$ as its invariant subspace. Otherwise, λ is complex, and L has no eigenvectors.

EXERCISES

1. Find all eigenvalues and eigenvectors for the linear transformations on $\mathcal{R}^2$ defined by the following matrices.

(a) $\begin{bmatrix} 1 & 1 \\ 0 & 2 \end{bmatrix}$ (b) $\begin{bmatrix} 1 & 0 \\ 0 & 0 \end{bmatrix}$ (c) $\begin{bmatrix} 1 & 1 \\ 1 & 1 \end{bmatrix}$ (d) $\begin{bmatrix} 1 & 4 \\ 1 & 1 \end{bmatrix}$

2. Find all eigenvalues for the linear transformations on $\mathcal{R}^3$ defined by the following matrices, and in each case find the eigenvectors belonging to the real eigenvalues.

(a) $\begin{bmatrix} 2 & 0 & 1 \\ -1 & 2 & 3 \\ 1 & 0 & 2 \end{bmatrix}$ (b) $\begin{bmatrix} 1 & 2 & 0 \\ 0 & 1 & 1 \\ 0 & 1 & 1 \end{bmatrix}$

(c) $\begin{bmatrix} -1 & 1 & 0 \\ 1 & 2 & 2 \\ 0 & -1 & 0 \end{bmatrix}$ (d) $\begin{bmatrix} 2 & 0 & 0 \\ 0 & -1 & 0 \\ 1 & 0 & 1 \end{bmatrix}$

3. Repeat Exercise 2 for the following matrices.

(a) $\begin{bmatrix} 1 & 0 & 1 \\ 0 & 1 & 0 \\ 1 & 0 & 1 \end{bmatrix}$ (b) $\begin{bmatrix} 1 & 0 & 1 \\ 0 & 2 & 0 \\ 1 & 0 & -1 \end{bmatrix}$

(c) $\begin{bmatrix} 5 & -6 & -6 \\ -1 & 4 & 2 \\ 3 & -6 & -4 \end{bmatrix}$ (d) $\begin{bmatrix} 1 & 2 & 1 \\ 1 & 2 & 1 \\ 0 & 1 & 2 \end{bmatrix}$

4. Find the eigenvalues and eigenvectors for the linear transformations on $\mathcal{R}^4$ defined by the matrices

(a) $\begin{bmatrix} 1 & 0 & -1 & 0 \\ 0 & 1 & 1 & 0 \\ -1 & 1 & 2 & 1 \\ 0 & 0 & 1 & -1 \end{bmatrix}$ (b) $\begin{bmatrix} -1 & 2 & 1 & 3 \\ 0 & 0 & -2 & 1 \\ 0 & 0 & 2 & -3 \\ 0 & 0 & 0 & 4 \end{bmatrix}$

5. Let $L: \mathcal{R}^3 \to \mathcal{R}^3$ be the linear transformation whose matrix with respect to the standard basis $\mathbf{e}_1, \mathbf{e}_2, \mathbf{e}_3$ is

$$\begin{bmatrix} 1 & 0 & 2 \\ 0 & 1 & 0 \\ 2 & 0 & 1 \end{bmatrix}.$$

Use the eigenvalue method to solve the equation $L\mathbf{x} = \mathbf{y}$ for $\mathbf{x}$, given that

(a) $\mathbf{y} = 2\mathbf{e}_1 + \mathbf{e}_2$; (b) $\mathbf{y} = \mathbf{e}_1 + \mathbf{e}_3$; (c) $\mathbf{y} = 4\mathbf{e}_1 - 2\mathbf{e}_2 - 2\mathbf{e}_3$.

6. Repeat Exercise 5 when the matrix of L is

$$\begin{bmatrix} 0 & 2 & 1 \\ 0 & 1 & 3 \\ 0 & 0 & 2 \end{bmatrix},$$

and

(a) $\mathbf{y} = -2\mathbf{e}_1 + \mathbf{e}_2$;
(b) $\mathbf{y} = 4\mathbf{e}_1 + 4\mathbf{e}_2 + 2\mathbf{e}_3$;
(c) $\mathbf{y} = \mathbf{e}_1 + 9\mathbf{e}_2 + 2\mathbf{e}_3$.

7. Prove that every linear transformation on an *odd*-dimensional vector space has at least one real eigenvalue. Give a geometric interpretation of this result for linear transformations on $\mathcal{R}^3$.

8. An $(n \times n)$-matrix is said to be *triangular* if all of the entries above (or below) the main diagonal are zero. Prove that each of the diagonal entries in such a matrix is an eigenvalue for the linear transformation on $\mathcal{R}^n$ defined by the matrix.

*9. (a) Let L be a linear transformation on a finite dimensional vector space $\mathcal{V}$, and let λ_0 be a real eigenvalue for L of multiplicity m, by which we mean that $(\lambda - \lambda_0)^m$ is a factor of the characteristic polynomial for L. Prove that the dimension of the invariant subspace associated with λ_0 is *at most m*. [*Hint:* Consider the characteristic polynomial of the linear transformation obtained by restricting L to the subspace $\mathcal{S}_{\lambda_0}$.]

(b) Give an example to show that this dimension can, in certain cases, be less than m. [*Hint:* Consider the operator $-D$ on the space of polynomials $\mathcal{P}_n$.]

12–4 SYMMETRIC LINEAR TRANSFORMATIONS

In this section we begin the process of extending the eigenvalue method introduced above to include operator equations defined on infinite dimensional Euclidean spaces. Our objective is to isolate a class of linear operators whose eigenvectors can be used to construct bases, and which, at the same time, includes the operators arising in the study of boundary-value problems. Clearly, the first step in such a program is to select a criterion which will guarantee that eigenvectors belonging to distinct eigenvalues are *orthogonal*, and to this end we now introduce the notion of a symmetric linear transformation, as follows.

Definition 12–3. Let L be a linear transformation from $\mathcal{S}$ to $\mathcal{V}$, where $\mathcal{S}$ and $\mathcal{V}$ are *Euclidean* spaces, $\mathcal{S}$ a subspace of $\mathcal{V}$. Then L is said to be *symmetric* with respect to the inner product on $\mathcal{V}$ if and only if

$$(L\mathbf{x}) \cdot \mathbf{y} = \mathbf{x} \cdot (L\mathbf{y}) \tag{12–16}$$

for all $\mathbf{x}$ and $\mathbf{y}$ in $\mathcal{S}$.

Before giving any examples, we prove that this definition accomplishes our objective by establishing

Theorem 12–3. *Every pair of eigenvectors belonging to distinct eigenvalues for a symmetric linear transformation $L: \mathcal{S} \to \mathcal{V}$ are orthogonal in $\mathcal{V}$.*

Proof. Let $\mathbf{x}_1$ and $\mathbf{x}_2$ be nonzero vectors in $\mathcal{S}$, and suppose that $L\mathbf{x}_1 = \lambda_1\mathbf{x}_1$, $L\mathbf{x}_2 = \lambda_2\mathbf{x}_2$, with $\lambda_1 \neq \lambda_2$. Then

$$\begin{aligned}(L\mathbf{x}_1)\cdot\mathbf{x}_2 &= \lambda_1(\mathbf{x}_1\cdot\mathbf{x}_2),\\ \mathbf{x}_1\cdot(L\mathbf{x}_2) &= \lambda_2(\mathbf{x}_1\cdot\mathbf{x}_2),\end{aligned}$$

and since $(L\mathbf{x}_1)\cdot\mathbf{x}_2 = \mathbf{x}_1\cdot(L\mathbf{x}_2)$, it follows that

$$(\lambda_1 - \lambda_2)(\mathbf{x}_1\cdot\mathbf{x}_2) = 0.$$

But, by assumption, $\lambda_1 - \lambda_2 \neq 0$. Hence $\mathbf{x}_1\cdot\mathbf{x}_2 = 0$, as asserted. ▌

EXAMPLE 1. Let L be a symmetric linear transformation mapping a finite dimensional Euclidean space $\mathcal{V}$ into itself, and let $\mathbf{e}_1, \ldots, \mathbf{e}_n$ be an *orthonormal* basis in $\mathcal{V}$. Then if

$$L\mathbf{e}_j = \alpha_{1j}\mathbf{e}_1 + \cdots + \alpha_{nj}\mathbf{e}_n, \quad j = 1, \ldots, n,$$

we have

$$\begin{aligned}\mathbf{e}_i\cdot(L\mathbf{e}_j) &= \mathbf{e}_i\cdot(\alpha_{1j}\mathbf{e}_1 + \cdots + \alpha_{nj}\mathbf{e}_n)\\ &= \alpha_{1j}(\mathbf{e}_i\cdot\mathbf{e}_1) + \cdots + \alpha_{ij}(\mathbf{e}_i\cdot\mathbf{e}_i) + \cdots + \alpha_{nj}(\mathbf{e}_i\cdot\mathbf{e}_n)\\ &= \alpha_{ij},\end{aligned}$$

the last step following from the fact that

$$\mathbf{e}_i\cdot\mathbf{e}_j = \delta_{ij} = \begin{cases}0, & i \neq j,\\ 1, & i = j.\end{cases}$$

On the other hand,

$$\begin{aligned}(L\mathbf{e}_i)\cdot\mathbf{e}_j &= (\alpha_{1i}\mathbf{e}_1 + \cdots + \alpha_{ni}\mathbf{e}_n)\cdot\mathbf{e}_j\\ &= \alpha_{1i}(\mathbf{e}_1\cdot\mathbf{e}_j) + \cdots + \alpha_{ji}(\mathbf{e}_j\cdot\mathbf{e}_j) + \cdots + \alpha_{ni}(\mathbf{e}_n\cdot\mathbf{e}_j)\\ &= \alpha_{ji},\end{aligned}$$

and the equality $(L\mathbf{e}_i)\cdot\mathbf{e}_j = \mathbf{e}_i\cdot(L\mathbf{e}_j)$ implies that $\alpha_{ij} = \alpha_{ji}$ for all i and all j. Thus the matrix of L with respect to the basis $\mathbf{e}_1, \ldots, \mathbf{e}_n$ is a *symmetric matrix* in the sense that its rows and columns may be interchanged without changing the matrix itself.

Conversely, suppose that the matrix of L is symmetric with respect to an orthonormal basis $\mathbf{e}_1, \ldots, \mathbf{e}_n$ in $\mathcal{V}$, and let

$$L\mathbf{e}_j = \alpha_{1j}\mathbf{e}_1 + \cdots + \alpha_{nj}\mathbf{e}_n, \quad j = 1, \ldots, n.$$

Then

$$(L\mathbf{e}_i) \cdot \mathbf{e}_j = \alpha_{ji} = \alpha_{ij} = \mathbf{e}_i \cdot (L\mathbf{e}_j)$$

for all i and j, and hence if $\mathbf{x}$ and $\mathbf{y}$ are arbitrary vectors in $\mathcal{V}$ with

$$\mathbf{x} = x_1\mathbf{e}_1 + \cdots + x_n\mathbf{e}_n,$$
$$\mathbf{y} = y_1\mathbf{e}_1 + \cdots + y_n\mathbf{e}_n,$$

we have

$$\begin{aligned}(L\mathbf{x}) \cdot \mathbf{y} &= \left(\sum_{i=1}^{n} x_i L\mathbf{e}_i\right) \cdot \left(\sum_{j=1}^{n} y_j \mathbf{e}_j\right) \\ &= \sum_{i=1}^{n}\sum_{j=1}^{n} x_i y_j (L\mathbf{e}_i) \cdot \mathbf{e}_j \\ &= \sum_{i=1}^{n}\sum_{j=1}^{n} x_i y_j \mathbf{e}_i \cdot (L\mathbf{e}_j) \\ &= \left(\sum_{i=1}^{n} x_i \mathbf{e}_i\right) \cdot \left(\sum_{j=1}^{n} y_j L\mathbf{e}_j\right) \\ &= \mathbf{x} \cdot (L\mathbf{y}).\end{aligned}$$

Thus L is symmetric on $\mathcal{V}$, and we have proved

Theorem 12–4. *A linear transformation on a finite dimensional Euclidean space $\mathcal{V}$ is symmetric if and only if the matrix of the transformation with respect to any orthonormal basis in $\mathcal{V}$ is a symmetric matrix.*

EXAMPLE 2. Let $L: \mathcal{R}^3 \to \mathcal{R}^3$ be defined by the matrix

$$\begin{bmatrix} 1 & 0 & 1 \\ 0 & 1 & 0 \\ 1 & 0 & 1 \end{bmatrix}$$

with respect to the standard basis $\mathbf{e}_1, \mathbf{e}_2, \mathbf{e}_3$. Then, by the above theorem, L is a symmetric linear transformation, and an easy computation reveals that its characteristic equation is

$$\lambda(\lambda - 1)(\lambda - 2) = 0.$$

Hence the eigenvalues for L are 0, 1, 2, and Theorems 12–2 and 12–3 imply that any complete set of eigenvectors for L will be an orthogonal basis for $\mathcal{R}^3$; a fact which can be readily verified by direct computation. Finally, we note that the matrix of L with respect to such a basis assumes the diagonal form

$$\begin{bmatrix} 0 & 0 & 0 \\ 0 & 1 & 0 \\ 0 & 0 & 2 \end{bmatrix}.$$

EXAMPLE 3. Let $\mathcal{S}$ denote the subspace of $\mathcal{C}[0, \pi]$ consisting of all twice continuously differentiable functions y such that $y(0) = y(\pi) = 0$, and let $L: \mathcal{S} \to \mathcal{C}[0, \pi]$ be the operator $-D^2$. Then, using integration by parts, we find that if y_1 and y_2 belong to $\mathcal{S}$,

$$\begin{aligned} (Ly_1) \cdot y_2 &= -\int_0^\pi y_1''(x)y_2(x)\, dx \\ &= -y_1'(x)y_2(x)\Big|_0^\pi + \int_0^\pi y_1'(x)y_2'(x)\, dx \end{aligned}$$

and

$$\begin{aligned} y_1 \cdot (Ly_2) &= -\int_0^\pi y_1(x)y_2''(x)\, dx \\ &= -y_1(x)y_2'(x)\Big|_0^\pi + \int_0^\pi y_1'(x)y_2'(x)\, dx. \end{aligned}$$

But, by assumption, $y_1(0) = y_1(\pi) = 0$, $y_2(0) = y_2(\pi) = 0$. Thus

$$(Ly_1) \cdot y_2 = \int_0^\pi y_1'(x)y_2'(x)\, dx = y_1 \cdot (Ly_2),$$

and L is symmetric on $\mathcal{S}$. Here again Theorem 12–3 applies, and we can assert that any complete set of eigenvectors for L is an orthogonal set in $\mathcal{C}[0, \pi]$. This agrees with the results obtained in Section 12–1 where we found that the eigenvalues for L are the integers $\lambda_n = n^2$, $n = 1, 2, \ldots$, and that the corresponding eigenvectors (or *eigenfunctions* as they are called in this case) are $c_n \sin nx$, c_n an arbitrary nonzero constant.

The reader may have noticed that all of the eigenvalues for the linear transformations considered in the last two examples were real. This, as it turns out, is always true of symmetric linear transformations; a fact which we state formally as

Theorem 12–5. *All of the eigenvalues for a symmetric linear transformation are real.*

The proof of this result, though not difficult, would necessitate introducing complex inner product spaces, and is therefore omitted. The interested reader can find the argument in any standard text on linear algebra.

EXERCISES

1. Let L be the symmetric linear transformation on $\mathcal{R}^2$ defined by the matrix

$$\begin{bmatrix} a & b \\ b & c \end{bmatrix}$$

with respect to the standard basis.

(a) Show that L has two distinct real eigenvalues except in the trivial case where $b = 0$ and $a = c$.

(b) Find a basis for $\mathcal{R}^2$ composed of eigenvectors for L in the nontrivial case.

2. Let L_1 and L_2 be symmetric linear transformations mapping $\mathcal{V}$ into itself. Prove that the transformation L_1L_2 is symmetric if and only if $L_1L_2 = L_2L_1$.

3. Let $\mathcal{S}$ denote the subspace of $\mathcal{C}[0, \pi]$ consisting of all twice continuously differentiable functions y such that $y'(0) = y'(\pi) = 0$, and let $L: \mathcal{S} \to \mathcal{C}[0, \pi]$ be the operator $-D^2$. Prove that L is a symmetric linear transformation, and find its eigenvalues.

4. Determine whether or not the following linear transformations are symmetric on the Euclidean space of polynomials with inner product $p \cdot q = \int_{-1}^{1} p(x)q(x)\, dx$.

(a) $Lp(x) = xp(x)$;

(b) $Lp(x) = p'(x)$;

(c) $Lp(x) = p(x + 1) - p(x)$;

(d) $Lp(x) = \dfrac{p(x) - p(-x)}{2}$.

5. Let $\mathfrak{F}$ denote the set of all real-valued functions of an integer variable k, $-\infty < k < \infty$, and let $\Delta^2: \mathfrak{F} \to \mathfrak{F}$ be defined by

$$(\Delta^2 F)(k) = F(k + 1) - 2F(k) + F(k - 1)$$

for all F in $\mathfrak{F}$ and all k.

(a) Show that $\mathfrak{F}$ is a real vector space under the usual definitions of addition and scalar multiplication, and that Δ^2 is a linear transformation on $\mathfrak{F}$.

(b) Find the null space of Δ^2.

*6. Let $\mathfrak{F}_0$ denote the space of all real-valued functions defined on the finite set of integers $0, 1, \ldots, N, N + 1$, and for each pair of functions F, G in $\mathfrak{F}_0$, set

$$F \cdot G = \sum_{k=0}^{N+1} F(k)G(k).$$

(a) Prove that $\mathfrak{F}_0$ is a Euclidean space.

(b) Let $\mathfrak{F}_1$ be the subspace of $\mathfrak{F}_0$ consisting of those functions F for which $F(0) = F(N + 1) = 0$, and let Δ^2 be defined as in Exercise 5 above. Prove that Δ^2 is a symmetric linear transformation on $\mathfrak{F}_1$. [The symbols $F(-1)$, $F(N + 2)$, $G(-1)$, and $G(N + 2)$ will appear in the computations, but no matter what values they are given, the asserted result holds.]

(c) Find the eigenvalues and eigenvectors for the operator Δ^2 on $\mathfrak{F}_1$, and check by direct computation that the eigenvectors belonging to distinct eigenvalues are orthogonal.

12–5 SELF-ADJOINT DIFFERENTIAL OPERATORS; STURM-LIOUVILLE PROBLEMS

We have seen that the operator $-D^2$ becomes a symmetric linear transformation when restricted to the space of twice continuously differentiable functions on $[0, \pi]$ which vanish at the endpoints of the interval. Such behavior is typical of a large number of differential operators, and when properly generalized furnishes the key to the study of boundary-value problems. To effect this generalization we now introduce the class of self-adjoint linear differential operators.

Definition 12–4. A second-order linear differential operator L defined on an interval $[a, b]$ is said to be in *self-adjoint form* if

$$L = D\big(p(x)D\big) + q(x), \tag{12–17}$$

where p is any function in $\mathcal{C}^1[a, b]$ such that $p(x) > 0$, or $p(x) < 0$, for all x in the open interval (a, b), and q is an arbitrary function in $\mathcal{C}[a, b]$.

Despite its appearance, (12–17) is sufficiently general to include all *normal* second-order linear differential operators on $[a, b]$. For if $L = a_2(x)D^2 + a_1(x)D + a_0(x)$ is such an operator, then L can be written in self-adjoint form by setting

$$p(x) = e^{\int [a_1(x)/a_2(x)]dx}$$

and

$$q(x) = \frac{a_0(x)}{a_2(x)} e^{\int [a_1(x)/a_2(x)]dx}$$

(see Section 6–3 and Exercise 1 below). Thus, without any real loss of generality, we can (and shall) restrict ourselves to the study of self-adjoint operators, and to differential equations of the form

$$\frac{d}{dx}\left(p(x)\frac{dy}{dx}\right) + q(x)y = h(x). \tag{12–18}$$

Finally, we note that the function p appearing as the leading coefficient in this equation is allowed to vanish at the endpoints of $[a, b]$. This fact will be of some importance later.

Our immediate objective is to determine conditions under which a self-adjoint operator will be symmetric when viewed as a linear transformation from $\mathcal{S}$ to $\mathcal{C}[a, b]$, $\mathcal{S}$ a subspace of $\mathcal{C}^2[a, b]$ determined by a pair of homogeneous boundary conditions

$$\begin{aligned} \alpha_1 y(a) + \alpha_2 y(b) + \alpha_3 y'(a) + \alpha_4 y'(b) &= 0, \\ \beta_1 y(a) + \beta_2 y(b) + \beta_3 y'(a) + \beta_4 y'(b) &= 0. \end{aligned} \tag{12–19}$$

To this end we first prove the following lemma.

Lemma 12–2. (The Lagrange Identity.) *If*

$$L = D\big(p(x)D\big) + q(x)$$

is any self-adjoint linear differential operator on $[a, b]$, *and if* y_1 *and* y_2 *are twice differentiable on* $[a, b]$, *then*

$$y_1(Ly_2) - (Ly_1)y_2 = [p(y_1y_2' - y_2y_1')]'. \tag{12–20}$$

(As usual, the primes denote differentiation.)

Proof. We simply apply the definition of L, and rearrange terms, as follows:

$$\begin{aligned} y_1(Ly_2) - (Ly_1)y_2 &= y_1[(py_2')' + qy_2] - y_2[(py_1')' + qy_1] \\ &= y_1(py_2')' - y_2(py_1')' \\ &= y_1[py_2'' + p'y_2'] - y_2[py_1'' + p'y_1'] \\ &= p'[y_1y_2' - y_2y_1'] + p[y_1y_2'' - y_2y_1''] \\ &= [p(y_1y_2' - y_2y_1')]'. \blacksquare \end{aligned}$$

Formula (12–20) can be written in much more suggestive form by integrating from a to b. For then its left-hand side becomes $y_1 \cdot (Ly_2) - (Ly_1) \cdot y_2$, and we therefore have

$$y_1 \cdot (Ly_2) - (Ly_1) \cdot y_2 = p(y_1y_2' - y_2y_1')\Big|_a^b, \tag{12–21}$$

from which we immediately deduce

Theorem 12–6. *Let* $\mathcal{S}$ *be a subspace of* $\mathcal{C}^2[a, b]$ *determined by boundary conditions of the type given in* (12–19), *and let* L *be any self-adjoint linear differential operator mapping* $\mathcal{S}$ *into* $\mathcal{C}[a, b]$. *Then* L *will be symmetric with respect to the standard inner product on* $\mathcal{C}[a, b]$ *if and only if*

$$p(y_1y_2' - y_2y_1')\Big|_a^b = 0 \tag{12–22}$$

for all y_1 *and* y_2 *in* $\mathcal{S}$; i.e., *if and only if*

$$p(b)[y_1(b)y_2'(b) - y_2(b)y_1'(b)] - p(a)[y_1(a)y_2'(a) - y_2(a)y_1'(a)] = 0.$$

Before giving any examples, we use Theorem 12–6 to determine several of the more obvious and important boundary conditions which lead to symmetric operators.

Case 1. $p(a) = p(b) = 0$. Here (12–22) is satisfied without restriction, and we can set $\mathcal{S} = \mathcal{C}^2[a, b]$.

Case 2. Let $\mathcal{S}$ be the set of all y in $\mathcal{C}^2[a, b]$ such that

$$\begin{aligned} \alpha_1 y(a) + \alpha_2 y'(a) &= 0, \\ \beta_1 y(b) + \beta_2 y'(b) &= 0, \end{aligned} \tag{12–23}$$

with $|\alpha_1| + |\alpha_2| \neq 0$ and $|\beta_1| + |\beta_2| \neq 0$. (These last conditions are imposed to force at least one of the α's and one of the β's to be different from zero.) Then, if y_1 and y_2 are any two functions in $\mathcal{S}$,

$$y_1(a)y_2'(a) - y_2(a)y_1'(a) = 0$$

and

$$y_1(b)y_2'(b) - y_2(b)y_1'(b) = 0.$$

Hence (12–22) vanishes on $\mathcal{S}$, as desired.

A boundary condition of the form

$$\alpha_1 y(a) + \alpha_2 y'(a) = 0$$

which involves the values of y and y' at only a single point is said to be *unmixed.* In these terms the above argument asserts that *a self-adjoint linear differential operator is symmetric on every subspace of* $\mathcal{C}^2[a, b]$ *described by a pair of unmixed boundary conditions.* (Note that only one such condition need be given if p vanishes at a or at b.)

Case 3. Assume that $p(a) = p(b)$, and let $\mathcal{S}$ be the subspace of $\mathcal{C}^2[a, b]$ consisting of all y such that

$$\begin{aligned} y(a) &= y(b), \\ y'(a) &= y'(b). \end{aligned} \tag{12–24}$$

Then (12–22) is obviously satisfied for all y_1 and y_2 in $\mathcal{S}$, and L is again symmetric. This is known as the case of *periodic boundary conditions.*

EXAMPLE 1. Let $\mathcal{S}$ be the subspace of $\mathcal{C}^2[0, \pi]$ consisting of all functions y satisfying the pair of unmixed boundary conditions

$$y(0) = y(\pi) = 0,$$

and let $L = -D^2$. Then, by Case 2, L is symmetric on $\mathcal{S}$,

EXAMPLE 2. By Case 3, the operator $-D^2$ is symmetric on the subspace of $\mathcal{C}^2[0, 2\pi]$ described by the periodic boundary conditions

$$\begin{aligned} y(0) &= y(2\pi), \\ y'(0) &= y'(2\pi). \end{aligned} \tag{12–25}$$

To find its eigenvalues and eigenfunctions we again apply the given boundary conditions to the general solution of

$$y'' + \lambda y = 0. \tag{12–26}$$

The argument proceeds by cases, as with the example in Section 12–1.*

---

* Note that by Theorem 12–5 we need only consider real values of λ.

When $\lambda < 0$,

$$y = c_1 e^{\sqrt{-\lambda}x} + c_2 e^{-\sqrt{-\lambda}x},$$

and (12–25) implies that $c_1 = c_2 = 0$. Hence there are no negative eigenvalues.

When $\lambda = 0$, the general solution of (12–26) is $c_1 + c_2 x$, and the boundary conditions can be satisfied by setting $c_2 = 0$ and leaving c_1 arbitrary. Thus $\lambda = 0$ is an eigenvalue, and its associated eigenfunctions are the constant functions on $[0, 2\pi]$.

Finally, when $\lambda > 0$,

$$y = c_1 \sin \sqrt{\lambda}x + c_2 \cos \sqrt{\lambda}x,$$

and (12–25) leads to the pair of equations

$$c_1[1 - \cos(2\pi\sqrt{\lambda})] = c_1 \sin(2\pi\sqrt{\lambda}),$$
$$c_2[1 - \cos(2\pi\sqrt{\lambda})] = -c_2 \sin(2\pi\sqrt{\lambda}),$$

which can be satisfied with $y \not\equiv 0$ by setting $\sqrt{\lambda} = 1, 2, 3, \ldots$. Thus the integers

$$1^2, 2^2, 3^2, \ldots$$

are eigenvalues, and the invariant subspace associated with n^2 is the *two*-dimensional subspace of $\mathcal{C}[0, 2\pi]$ spanned by the functions $\sin nx$ and $\cos nx$.

Boundary-value problems involving self-adjoint linear differential operators with mutually orthogonal eigenfunctions are usually referred to as *Sturm-Liouville problems* after the mathematicians J. C. F. Sturm and J. Liouville who first investigated them. Thus, each of the problems considered above was a Sturm-Liouville problem. More generally, a *Sturm-Liouville problem*, or *system*, is, by definition, a second-order homogeneous linear differential equation of the form

$$\frac{d}{dx}\left(p(x)\frac{dy}{dx}\right) + [q(x) - \lambda]y = 0,$$

p and q as above, together with a pair of (homogeneous) boundary conditions chosen in such a way that eigenfunctions belonging to distinct eigenvalues for the operator

$$D(p(x)D) + q(x)$$

are orthogonal.* In the next chapter we shall see that such systems arise naturally in the study of boundary-value problems involving *partial* differential equations, and for this reason are extremely important in physics and applied mathematics.

* This terminology will be generalized in Section 12–8 to a somewhat wider class of problems.

EXERCISES

1. Prove that every normal second-order linear differential operator $a_2(x)D^2 + a_1(x)D + a_0(x)$ can be put in self-adjoint form by setting

$$p(x) = e^{\int [a_1(x)/a_2(x)]dx},$$

$$q(x) = \frac{a_0(x)}{a_2(x)} e^{\int [a_1(x)/a_2(x)]dx}.$$

2. Rewrite each of the following linear differential operators in self-adjoint form.

   (a) $D^2 + \frac{1}{x} D + 1, \quad x > 0$

   (b) $(\cos x)D^2 + (\sin x)D - 1, \quad -\pi/2 < x < \pi/2$

   (c) $x^2D^2 + xD + (x^2 - p^2), \quad x > 0, \quad p$ a real number

   (d) $(1 - x^2)D^2 - 2xD + n(n + 1), \quad -1 < x < 1$, n a non-negative integer

3. Find all eigenvalues and eigenvectors for the Sturm-Liouville system

$$y'' + \lambda y = 0,$$
$$y'(-\pi) = 0, \qquad y'(\pi) = 0.$$

4. Repeat Exercise 3 for the Sturm-Liouville system

$$y'' + \lambda y = 0,$$
$$y(0) = 0, \qquad y'(\pi) = 0.$$

12–6 FURTHER EXAMPLES

In this section we consider a number of Sturm-Liouville problems which will be encountered repeatedly in our later work, and which, for convenience of reference, we solve here, once and for all. In each case we shall limit ourselves to finding eigenvalues and eigenfunctions, and postpone any discussion of how this information is used to solve specific boundary-value problems involving the given operator.

EXAMPLE 1. Solve the Sturm-Liouville system

$$y'' + \lambda y = 0,$$
$$y(0) = 0, \qquad y(L) = 0.$$

This is a variant of a problem we have already considered many times over, and this time we find that the constants

$$\lambda_n = \frac{n^2\pi^2}{L^2}, \quad n = 1, 2, \ldots,$$

and functions

$$y_n(x) = \sin \frac{n\pi x}{L}, \quad n = 1, 2, \ldots,$$

are complete sets of eigenvalues and eigenfunctions.

EXAMPLE 2. Solve the Sturm-Liouville system

$$y'' + \lambda y = 0,$$
$$y'(0) = 0, \qquad y'(L) = 0.$$

A computation similar in all respects to the one used to solve Example 1 reveals that the eigenvalues for this problem are the non-negative constants

$$\lambda_n = \frac{n^2\pi^2}{L^2}, \quad n = 0, 1, 2, \ldots,$$

and that

$$y_n(x) = \cos \frac{n\pi x}{L}, \quad n = 0, 1, 2, \ldots,$$

is a complete set of eigenfunctions. The details are left to the reader.

EXAMPLE 3. Solve the Sturm-Liouville system

$$\begin{aligned} y'' + \lambda y &= 0, \\ y(0) &= 0, \\ hy(L) + y'(L) &= 0, \end{aligned}$$

given that h and L are positive constants. (Note that the boundary conditions are unmixed, and that the problem falls under Case 2 above.)

As usual, we argue by cases, depending upon the algebraic sign of λ, and again find that there are no eigenvalues ≤ 0. On the other hand, when $\lambda > 0$,

$$y = c_1 \sin \sqrt{\lambda}\, x + c_2 \cos \sqrt{\lambda}\, x,$$

and the first boundary condition implies that $c_2 = 0$. Thus it remains (if possible) to choose λ so that the function

$$y = c_1 \sin \sqrt{\lambda}\, x,$$

with $c_1 \neq 0$, satisfies the equation $hy(L) + y'(L) = 0$. This, in turn, implies that λ must be chosen so that

$$\sin(\sqrt{\lambda}\, L) = -\frac{\sqrt{\lambda}\, L}{hL} \cos(\sqrt{\lambda}\, L); \tag{12–27}$$

an equation which we rewrite as

$$\tan \mu = -\frac{1}{hL}\, \mu \tag{12–28}$$

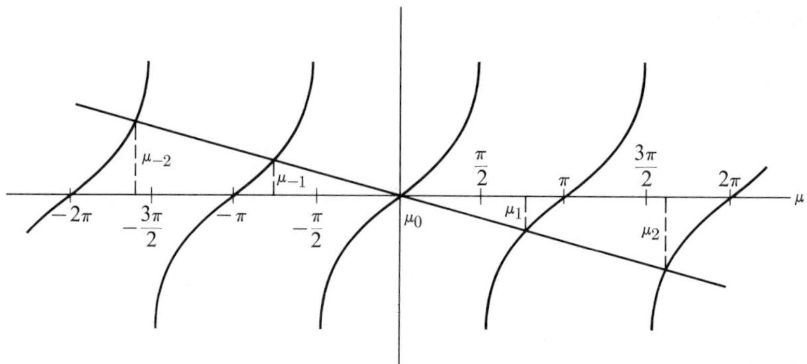

FIGURE 12–2

by setting $\mu = \sqrt{\lambda}L$. Although it is impossible to solve (12–27) explicitly for λ, its solutions can be visualized as arising, via (12–28), from the points of intersection of the graphs of the functions $\tan \mu$ and $-\mu/hL$. As indicated in Fig. 12–2, there are infinitely many such points

$$\ldots, \mu_{-2}, \mu_{-1}, \mu_0 = 0, \mu_1, \mu_2, \ldots$$

located symmetrically across the origin. Thus the given problem has an infinite number of positive eigenvalues

$$\lambda_n = \frac{\mu_n^2}{L^2}, \quad n = 1, 2, \ldots,$$

with $\lambda_n < \lambda_{n+1}$ for all n, and $\lim_{n\to\infty} \lambda_n = \infty$. [Also note that from the geometry of the situation we have $\lim_{n\to\infty} (\lambda_{n+1} - \lambda_n) = \pi$.] Finally, the functions

$$\begin{aligned} y_n(x) &= \sin \frac{\mu_n x}{L} \\ &= \sin \sqrt{\lambda_n}\, x, \quad n = 1, 2, \ldots, \end{aligned}$$

constitute a complete set of eigenfunctions for this problem.

Example 4.* Solve the Sturm-Liouville problem

$$\frac{d}{dx}\left[(1 - x^2)\frac{dy}{dx}\right] + \lambda y = 0 \tag{12–29}$$

on the interval $[-1, 1]$.

The leading coefficient of this equation vanishes at the endpoints of the interval, and hence, by Case 1 above, the operator $D[(1 - x^2)D]$ is symmetric on all of $\mathcal{C}^2[-1, 1]$. Moreover, since (12–29) is the self-adjoint form of Legendre's equation (of order λ), our earlier results imply that the integers $n(n + 1)$, $n = 0, 1, 2, \ldots,$

* This example should be omitted by anyone who is not familiar with the material in Sections 11–2 through 11–4.

are eigenvalues, and that the Legendre polynomials

$$P_0(x),\ P_1(x),\ P_2(x), \ldots$$

are eigenfunctions for the problem.* To complete the discussion it remains to show that these polynomials are a *complete* set of eigenfunctions for (12–29). Here we argue as follows.

Were $\lambda_0 \neq n(n+1)$ an eigenvalue of (12–29), and y_0 an eigenfunction belonging to λ_0, then y_0 would be orthogonal in $\mathcal{C}[-1, 1]$ to *all* of the P_n. But, as we know, the Legendre polynomials are a *basis* for $\mathcal{C}[-1, 1]$, and hence, by Lemma 8–3, $y_0 \equiv 0$. Since this cannot be, no such eigenvalue exists.

EXERCISES

1. Verify that the eigenvalues and eigenvectors listed in Example 2 above are correct.

2. (a) Show that the boundary-value problem consisting of the fourth-order differential equation

$$\frac{d^4y}{dx^4} - \omega^2 y = 0$$

and boundary conditions

$$y(0) = y(1) = 0, \qquad y'(0) = y'(1) = 0$$

has nontrivial solutions if and only if

$$\cos\sqrt{\omega} = \frac{1}{\cosh\sqrt{\omega}}.$$

(b) Use the technique introduced in Example 3 above to prove that the boundary-value problem in (a) has infinitely many non-negative eigenvalues ω_n, $n = 0, 1, 2, \ldots$. How do these eigenvalues behave as $n \to \infty$?

(c) What is the general solution of the boundary-value problem in (a) corresponding to the eigenvalue ω_n?

3. Let L denote the fourth-order linear differential operator D^4, and let $\mathcal{S}$ denote the subspace of $\mathcal{C}^4[a, b]$ consisting of all functions y such that

$$y(a) = y''(a) = y(b) = y''(b) = 0.$$

(a) Prove that

$$y_1(Ly_2) - y_2(Ly_1) = [y_1y_2''' - y_2y_1''' - y_1'y_2'' + y_2'y_1'']'$$

for all y_1 and y_2 in $\mathcal{S}$.

(b) Use the result in (a) to prove that eigenfunctions belonging to distinct eigenvalues for the boundary-value problem $L\colon \mathcal{S} \to \mathcal{C}[a, b]$ are orthogonal.

---

* Note that at this point Theorem 12–6 allows us to assert, without further proof, that the Legendre polynomials are mutually orthogonal in $\mathcal{PC}\,[-1, 1]$.

12-7 BOUNDARY-VALUE PROBLEMS AND SERIES EXPANSIONS

We began this chapter by defining a boundary-value problem to be an operator equation of the form

$$Ly = h \tag{12-30}$$

in which h is a known function in $\mathcal{C}[a, b]$, and L a second-order linear differential operator acting on a subspace $\mathcal{S}$ of $\mathcal{C}^2[a, b]$ described by a pair of boundary conditions of the form (12-2). Since then we have said very little about such problems and have, instead, been off chasing eigenvalues and eigenfunctions. It is now time to justify this apparent digression by returning to our original problem and applying what we have learned to solve it.

The method we propose to use is a straightforward generalization to infinite dimensional Euclidean spaces of the eigenvalue method introduced in Section 12-3, and thus depends upon the existence of an eigenfunction *basis* for $\mathcal{C}[a, b]$. As yet, of course, we have no assurance that such a basis will exist, even when L is symmetric, but if it does we can argue as follows.

Let

$$\lambda_0, \lambda_1, \lambda_2, \ldots$$

be the eigenvalues for L, and let

$$\varphi_0(x), \varphi_1(x), \varphi_2(x), \ldots$$

be a complete set of eigenfunctions belonging to the λ_n. Then, since the φ_i are a *basis* for $\mathcal{C}[a, b]$, we have

$$h(x) = \sum_{n=0}^{\infty} c_n \varphi_n(x),$$

where

$$c_n = \frac{h \cdot \varphi_n}{\|\varphi_n\|^2} = \frac{\int_a^b h(x)\varphi_n(x)\,dx}{\int_a^b [\varphi_n(x)]^2\,dx},$$

and the series converges *in the mean* to h. We now set

$$y(x) = \sum_{n=0}^{\infty} \alpha_n \varphi_n(x), \tag{12-31}$$

with the α_n unknown, and substitute in (12-30) to obtain

$$L\left[\sum_{n=0}^{\infty} \alpha_n \varphi_n(x)\right] = \sum_{n=0}^{\infty} c_n \varphi_n(x).$$

Thus if L can be applied to (12-31) term-by-term, we have

$$\sum_{n=0}^{\infty} \alpha_n \lambda_n \varphi_n(x) = \sum_{n=0}^{\infty} c_n \varphi_n(x)$$

(recall that $L\varphi_n = \lambda_n\varphi_n$), and it follows that (12–31) will be a solution of the given equation whenever

(i) the α_n can be chosen so that

$$\alpha_n\lambda_n = c_n$$

for all n, and

(ii) with these as the values of α_n, the series

$$\sum_{n=0}^{\infty} \alpha_n\varphi_n(x)$$

defines a function in $\mathcal{C}^2[a, b]$ whose first two derivatives can be computed by termwise differentiation.

It is clear that the first of these requirements can be met by setting $\alpha_n = c_n/\lambda_n$ so long as $\lambda_n \neq 0$ for all n (i.e., so long as L is one-to-one). Furthermore, the resulting solution is then unique. If, on the other hand, one of the λ_n, say λ_0, is zero, the problem has no solution at all when $c_0 \neq 0$, and an infinite number of solutions when $c_0 = 0$.

Unfortunately, no such simple analysis can be used to dispose of (ii), since here we must investigate the convergence of the series

$$\sum_{n=0}^{\infty} \frac{c_n}{\lambda_n} \varphi_n(x).$$

As we have already seen, this is a delicate problem whose solution depends upon the properties of the function h from which the c_n are derived, and upon the particular orthogonal system φ_n appearing in the series. Thus different orthogonal systems must be examined individually, and the best that can be said in general is that the desired termwise differentiability will be possible whenever h is "sufficiently smooth." (However, see Theorem 12–7 below.) In the absence of specific information as to what degree of smoothness is "sufficient" in any given instance, it is standard practice to proceed formally, as above, and then attempt to verify that the resulting series has the required properties. This method will be illustrated in some detail in the next chapter.

EXAMPLE 1. Let $\mathcal{S}$ be the subspace of $\mathcal{C}^2[0, \pi]$ described by the boundary conditions $y(0) = y(\pi) = 0$, and let $L = -D^2$. Then

$$\lambda_n = n^2, \qquad \varphi_n(x) = \sin nx,$$

$n = 1, 2, \ldots$, and the φ_n are a *basis* for $\mathcal{C}[0, \pi]$. (See Section 9–5.) Hence the boundary-value problem

$$\begin{aligned} -y'' &= h(x), \\ y(0) &= y(\pi) = 0 \end{aligned} \tag{12–32}$$

has the *formal* solution

$$y(x) = \sum_{n=1}^{\infty} \frac{c_n}{n^2} \sin nx, \tag{12–33}$$

with

$$c_n = \frac{2}{\pi} \int_0^{\pi} h(x) \sin nx \, dx.$$

In this case the validity of (12–33) can be guaranteed by demanding that h be continuous and have a piecewise continuous first derivative on $[0, \pi]$. For then the Fourier sine series for h will converge uniformly and absolutely on every closed subinterval of $(0, \pi)$, and the same will therefore be true of the series obtained by twice differentiating (12–33) term-by-term.*

As a concrete illustration, let $h(x) = x$. Then (12–32) becomes

$$\begin{aligned} -y'' &= x, \\ y(0) = y(\pi) &= 0, \end{aligned} \tag{12–34}$$

and we have

$$y(x) = \sum_{n=1}^{\infty} \frac{c_n}{n^2} \sin nx,$$

where

$$c_n = \frac{2}{\pi} \int_0^{\pi} x \sin nx \, dx.$$

A routine calculation gives

$$c_n = (-1)^{n+1} \frac{2}{n};$$

whence

$$y(x) = 2 \sum_{n=1}^{\infty} (-1)^{n+1} \frac{\sin nx}{n^3}.$$

Of course, (12–34) can also be solved in closed form by applying the given boundary conditions to the general solution of $y'' = -x$. Lest the reader feel that we have been somewhat dishonest in using Fourier series when this easier method was at hand we point out that frequently no such option exists, and the only available solutions are those expressed as series in terms of eigenfunctions for the problem.

---

* The reader should note that (12–33) will satisfy the given boundary conditions *and* reduce to the value of h at 0 and π only if $h(0) = h(\pi) = 0$. Thus, in general, we can neither demand nor expect the solution of (12–32) to satisfy the differential equation on the *closed* interval $[0, \pi]$.

The technique used in the above example was successful precisely because the operator $-D^2$ had a sufficient number of mutually orthogonal eigenfunctions when restricted to $\mathcal{S}$ to allow us to construct an eigenfunction *basis* for $\mathcal{C}[0, \pi]$. This immediately raises the problem of determining conditions which are sufficient to guarantee the existence of such a basis for an arbitrary self-adjoint linear differential operator acting on a given subspace of $\mathcal{C}^2[a, b]$. As might be expected, this is a very difficult problem, and any attempt to answer it, even for the symmetric operators introduced in Section 12–5, would carry us too far afield. Thus we let matters rest with the statement of the following theorem which, as it turns out, is adequate to handle most of the problems we shall discuss.

Theorem 12–7. *Let L be a* **normal** *second-order linear differential operator defined on a closed interval* $[a, b]$*, and let* $\mathcal{S}$ *be a subspace of* $\mathcal{C}^2[a, b]$ *described by a pair of unmixed boundary conditions. Then L has an infinite sequence of real eigenvalues* $\{\lambda_n\}$, $n = 0, 1, 2, \ldots$, *such that*

$$|\lambda_0| < |\lambda_1| < |\lambda_2| < \cdots,$$

and

$$\lim_{n\to\infty} |\lambda_n| = \infty.$$

Moreover, the invariant subspaces of $\mathcal{C}[a, b]$ *associated with the* λ_n *are all one-dimensional; any complete set of eigenfunctions for L, one for each eigenvalue, is a basis for* $\mathcal{C}[a, b]$*; and the series expansion of any piecewise smooth function y on* $[a, b]$ *relative to such a basis converges uniformly and absolutely to y on any closed subinterval in which y is continuous.**

EXERCISES

Find the formal series expansion of the solution of the boundary-value problems in Exercises 1–6 in terms of the eigenfunctions for the associated Sturm-Liouville system.

1. $y'' = x(x - 2\pi), \quad y(0) = 0, \quad y'(\pi) = 0$
2. $y'' = x^2 - \pi^2, \quad y'(0) = 0, \quad y(\pi) = 0$
3. $y'' = \sin\dfrac{\pi x}{L}, \quad y'(0) = 0, \quad y'(L) = 0$
4. $y'' = \sin\dfrac{\pi x}{L}, \quad y(0) = 0, \quad y'(L) = 0$
5. $y'' = \begin{cases} -x, & 0 \le x < \pi/2, \\ x - \pi/2, & \pi/2 < x \le \pi, \end{cases} \quad y(0) = 0, \quad y(\pi) = 0.$
6. $y'' = -\sin^2 x, \quad y(0) = y(\pi), \quad y'(0) = y'(\pi).$

*7. Use the technique introduced in this section to discuss the boundary-value problem

$$y'' = -h(x),$$
$$y(0) = y(2\pi), \quad y'(0) = y'(2\pi),$$

---

* For a proof see E. L. Ince, *Ordinary Differential Equations*, Dover, New York, 1956.

given that h is a function in $\mathcal{C}[0, 2\pi]$. [*Hint:* Consider the cases

$$\int_0^{2\pi} h(x)\,dx = 0 \quad \text{and} \quad \int_0^{2\pi} h(x)\,dx \neq 0$$

separately.]

*8. By citing appropriate theorems in the text, verify the assertion made above concerning the convergence of the series given in (12–33).

12–8 ORTHOGONALITY AND WEIGHT FUNCTIONS

The considerations of the preceding section admit an easy and important generalization to boundary-value problems involving a Sturm-Liouville system consisting of

(i) a second-order homogeneous linear differential equation of the form

$$\frac{d}{dx}\left(p(x)\frac{dy}{dx}\right) + [q(x) - \lambda r(x)]y = 0 \tag{12–35}$$

defined on an interval $[a, b]$, and

(ii) a pair of homogeneous boundary conditions which serve to determine the domain space for the operator

$$L = D\big(p(x)D\big) + q(x).$$

As before, we assume that p and q belong, respectively, to $\mathcal{C}^1[a, b]$ and $\mathcal{C}[a, b]$, and that $p(x)$ does not vanish in (a, b). In addition, we demand that r be a *continuous*, *non-negative* function on $[a, b]$ which vanishes at most finitely many times in the interval.

The values of λ for which such a problem admits nontrivial solutions are again called eigenvalues and the associated nontrivial solutions of (12–35) are called eigenfunctions. Our task, of course, is to find all eigenvalues and eigenfunctions once L and its domain space have been given, and, more generally, to investigate the possibility of extending our earlier results to this setting.

We begin by noting that if λ_1 and λ_2 are distinct eigenvalues for (12–35), and if y_1 and y_2 are eigenfunctions belonging to these eigenvalues, then

$$Ly_1 = \lambda_1 r(x)y_1(x),$$
$$Ly_2 = \lambda_2 r(x)y_2(x),$$

and Lagrange's identity implies that

$$\begin{aligned}(\lambda_1 - \lambda_2)r(x)y_1(x)y_2(x) &= y_2(x)[Ly_1(x)] - y_1(x)[Ly_2(x)]\\ &= \frac{d}{dx}\{p(x)[y_1(x)y_2'(x) - y_2(x)y_1'(x)]\}.\end{aligned}$$

Hence

$$(\lambda_1 - \lambda_2)\int_a^b r(x)y_1(x)y_2(x)\,dx = p(x)[y_1(x)y_2'(x) - y_2(x)y_1'(x)]\Big|_a^b, \tag{12–36}$$

and it follows that

$$\int_a^b r(x)y_1(x)y_2(x)\,dx = 0 \tag{12-37}$$

whenever the boundary conditions are such that the expression on the right-hand side of (12–36) vanishes. Assuming this to be the case, (12–37) allows us to assert that the functions $\sqrt{r}y_1$ and $\sqrt{r}y_2$ are orthogonal in $\mathcal{C}[a, b]$, or, equivalently, that y_1 and y_2 are orthogonal in $\mathcal{C}[a, b]$ *with respect to the weight function* r. (See Example 3, Section 7–1.) This latter terminology has the advantage of banishing the cumbersome factor $\sqrt{r}$ from the discussion of orthogonality, and amounts to redefining the inner product on $\mathcal{C}[a, b]$ to be

$$f \cdot g = \int_a^b f(x)g(x)r(x)\,dx, \tag{12-38}$$

a definition we know to be valid whenever r satisfies the conditions imposed above. And with this we have proved the following generalization of Theorem 12–6.

Theorem 12–8. *Let L be a self-adjoint linear differential operator on an interval $[a, b]$, let r be any weight function on $[a, b]$, and let $\mathcal{S}$ be a subspace of $\mathcal{C}^2[a, b]$ such that*

$$p(x)[y_1(x)y_2'(x) - y_2(x)y_1'(x)]\Big|_a^b = 0$$

for every pair of functions y_1 and y_2 in $\mathcal{S}$. Then any set of eigenfunctions belonging to distinct eigenvalues for the Sturm-Liouville problem

$$Ly = \lambda r y$$

is orthogonal in $\mathcal{C}[a, b]$ when the inner product is computed with respect to the weight function r.

We call the reader's attention to the fact that in general the operator L will *not* be symmetric on $\mathcal{S}$ with respect to the weighted inner product defined by (12–38). Nevertheless, Theorem 12–8 asserts that eigenfunctions belonging to distinct eigenvalues are still orthogonal, and this is really what is needed to construct eigenfunction bases. Moreover, since the conditions required to ensure orthogonality here are the same as those imposed in Section 12–5, we see that the conclusion of Theorem 12–8 is assured whenever $\mathcal{S}$ is described by boundary conditions of Type 1, 2, or 3 of that section. And finally if $L\colon \mathcal{S} \to \mathcal{C}[a, b]$ is both one-to-one and normal, and if $r(x) > 0$ for all x in $[a, b]$, it can be shown that $\mathcal{C}[a, b]$ has a *basis* composed of eigenfunctions for L. Again we omit the proof.

Example. Find the eigenvalues and eigenfunctions for the Sturm-Liouville problem

$$\begin{gathered} y'' + 4y' + (4 - 9\lambda)y = 0, \\ y(0) = y(a) = 0. \end{gathered} \tag{12-39}$$

In the first place, (12–39) can be rewritten in self-adjoint form as

$$\frac{d}{dx}\left(e^{4x}\frac{dy}{dx}\right) + 4e^{4x}y = \lambda(9e^{4x})y,$$
$$y(0) = 0, \qquad y(a) = 0,$$

and therefore satisfies the hypotheses of Theorem 12–8. Thus eigenfunctions belonging to distinct eigenvalues for this problem will be mutually orthogonal in the Euclidean space $\mathcal{C}[0, a]$ with inner product computed relative to the weight function $r(x) = 9e^{4x}$. Furthermore, it is not difficult to show that the operator $L = D(e^{4x}D) + 4e^{4x}$ is a one-to-one linear transformation from the subspace described by the given boundary conditions to $\mathcal{C}[0, a]$. (See Exercise 5, Section 12–2, and Lemma 12–3 below). Hence, since L is normal and $r(x) > 0$ for all x in $[0, a]$, the result cited a moment ago guarantees the existence of an eigenfunction basis for $\mathcal{C}[0, a]$. To compute such a basis we argue as follows.

Case 1. $\lambda > 0$. Here the general solution of $y'' + 4y' + (4 - 9\lambda)y = 0$ is

$$y = c_1e^{(-2+3\sqrt{\lambda})x} + c_2e^{(-2-3\sqrt{\lambda})x},$$

and the boundary conditions imply that

$$\begin{aligned} c_1 \qquad\qquad &+ c_2 &= 0, \\ c_1e^{(-2+3\sqrt{\lambda})a} &+ c_2e^{(-2-3\sqrt{\lambda})a} &= 0. \end{aligned}$$

Thus $c_1 = c_2 = 0$, and (12–39) has no positive eigenvalues.

Case 2. $\lambda = 0$. This time the general solution of the equation is $(c_1 + c_2x)e^{-2x}$, and we again find that $c_1 = c_2 = 0$.

Case 3. $\lambda < 0$. Here

$$y = e^{-2x}(c_1 \sin 3\sqrt{-\lambda}\, x + c_2 \cos 3\sqrt{-\lambda}\, x), \tag{12–40}$$

and the requirement that $y(0) = y(a) = 0$ yields

$$c_2 = 0,$$
$$c_1 \sin 3\sqrt{-\lambda}\, a = 0.$$

Hence (12–39) has nontrivial solutions if and only if λ satisfies the equation $\sin 3\sqrt{-\lambda}\, a = 0$, and it follows that the eigenvalues for this problem are

$$\lambda_n = -\frac{\pi^2n^2}{9a^2}, \quad n = 1, 2, \ldots.$$

To find a corresponding set of eigenfunctions we now set $c_1 = 1$, $c_2 = 0$, and $\lambda = \lambda_n$ in (12–40), thereby obtaining the functions

$$\varphi_n(x) = e^{-2x} \sin \frac{n\pi x}{a}, \quad n = 1, 2, \ldots.$$

EXERCISES

Compute the eigenvalues and eigenfunctions for the boundary-value problems in Exercises 1–8, and in each case determine a Euclidean space in which a complete set of eigenfunctions for the given problem is an orthogonal set.

1. $y'' + (1 + \lambda)y = 0; \quad y(0) = 0, \quad y(\pi) = 0$
2. $y'' + (1 - \lambda)y = 0; \quad y'(0) = 0, \quad y'(1) = 0$
3. $y'' + 2y' + (1 - \lambda)y = 0; \quad y(0) = 0, \quad y(1) = 0$
4. $y'' - 4y' + (4 - \lambda)y = 0; \quad y(0) = 0, \quad y(\pi) = 0$
5. $4y'' - 4y' + (1 + \lambda)y = 0; \quad y(-1) = 0, \quad y(1) = 0$
6. $y'' + (1 - \lambda)y = 0; \quad y(0) + y'(0) = 0, \quad y(1) + y'(1) = 0$
7. $y'' + 2y' + (1 - \lambda)y = 0; \quad y'(0) = 0, \quad y'(\pi) = 0$
8. $y'' - 3y' + 2(1 + \lambda)y = 0; \quad y(0) = 0, \quad y(1) = 0$

*9. Find all eigenvalues and eigenfunctions for the "singular" Sturm-Liouville problem

$$x^2y'' - xy' + (1 + \lambda)y = 0,$$

given that $y(1) = 0$, and $\lim_{x\to 0^+} |y(x)| < \infty$. How does the set of eigenvalues of this problem differ from those encountered earlier in this chapter? [*Hint:* Note that the given equation is an Euler equation, and recall that when its indicial polynomial has complex roots $\alpha \pm \beta i$, the solution space on $(0, \infty)$ is spanned by the functions $x^\alpha \sin(\beta \ln x)$, $x^\alpha \cos(\beta \ln x)$.]

10. Repeat Exercise 9 for the boundary-value problem

$$x^2y'' + xy' - 9\lambda y = 0,$$

$$y(1) = 0, \quad \lim_{x\to 0^+} |y(x)| < \infty.$$

*12–9 GREEN'S FUNCTIONS FOR BOUNDARY-VALUE PROBLEMS: AN EXAMPLE

In an earlier chapter we saw that the equation

$$L\mathbf{x} = \mathbf{y}, \quad L: \mathcal{S} \to \mathcal{V}, \tag{12–41}$$

can be solved for $\mathbf{x}$ by "inverting the operator" whenever L is a *one-to-one* linear transformation mapping $\mathcal{S}$ *onto* $\mathcal{V}$. For, under these conditions, there exists a linear transformation L^{-1} from $\mathcal{V}$ to $\mathcal{S}$ such that $LL^{-1} = I$, I the identity map on $\mathcal{S}$, and hence $\mathbf{x} = L^{-1}\mathbf{y}$. In the sections which follow we shall apply this technique to the case where (12–41) is a boundary-value problem involving a normal self-adjoint linear differential operator and unmixed boundary conditions, thereby obtaining a method for solving such problems which is independent of the elaborate theory of series expansions in Euclidean spaces.

The best place to begin, perhaps, is at the point where we left the discussion of initial-value problems in Chapter 4. As the reader will recall, we saw there that if

$$L: \mathcal{C}^2[a, b] \to \mathcal{C}[a, b]$$

is a *normal* second-order linear differential operator, and if $\mathcal{S}$ is the subspace of $\mathcal{C}^2[a, b]$ consisting of all functions y which satisfy the initial conditions

$$y(x_0) = y_0, \qquad y'(x_0) = y_1,$$

then L, *restricted to* $\mathcal{S}$, has an inverse which can be expressed as an integral operator of the form

$$L^{-1}h(x) = \int_a^b K(x, t)h(t)\, dt.$$

Moreover, the function $K(x, t)$, known as the Green's function for L for initial-value problems, can be constructed in a perfectly definite way from the coefficients of L and a basis for the null space of L in $\mathcal{C}^2[a, b]$. Our present objective is to prove that a similar construction is possible whenever $\mathcal{S}$ is a subspace of $\mathcal{C}^2[a, b]$ determined by a pair of unmixed boundary conditions

$$\begin{aligned} \alpha_1 y(a) + \alpha_2 y'(a) &= 0, \\ \beta_1 y(b) + \beta_2 y'(b) &= 0, \end{aligned} \tag{12–42}$$

and L is one-to-one when restricted to $\mathcal{S}$.

Obviously then, our first task is to devise a criterion which will guarantee the one-to-oneness of L. In other words, we must impose restrictions on the boundary conditions appearing in (12–42) which will ensure that the only solution of the equation

$$Ly = 0 \tag{12–43}$$

which belongs to $\mathcal{S}$ is the trivial solution $y \equiv 0$. This can be done as follows. Let y_1 and y_2 be a basis for the null space of L in $\mathcal{C}^2[a, b]$, and let

$$y(x) = c_1 y_1(x) + c_2 y_2(x)$$

be the general solution of (12–43). Then $y(x)$ will be identically zero if and only if $c_1 = c_2 = 0$, *and* $y(x)$ will belong to $\mathcal{S}$ if and only if

$$\begin{aligned} \alpha_1[c_1 y_1(a) + c_2 y_2(a)] + \alpha_2[c_1 y_1'(a) + c_2 y_2'(a)] &= 0, \\ \beta_1[c_1 y_1(b) + c_2 y_2(b)] + \beta_2[c_1 y_1'(b) + c_2 y_2'(b)] &= 0, \end{aligned}$$

that is, if and only if

$$\begin{aligned} c_1[\alpha_1 y_1(a) + \alpha_2 y_1'(a)] + c_2[\alpha_1 y_2(a) + \alpha_2 y_2'(a)] &= 0, \\ c_1[\beta_1 y_1(b) + \beta_2 y_1'(b)] + c_2[\beta_1 y_2(b) + \beta_2 y_2'(b)] &= 0. \end{aligned} \tag{12–44}$$

Viewing (12–44) as a pair of equations in the unknowns c_1 and c_2, it follows that $y(x) \equiv 0$ if and only if (12–44) has the *unique* solution $c_1 = c_2 = 0$. From this, and the elementary theory of systems of linear equations, we immediately deduce

Lemma 12–3. *Let L be a normal second-order linear differential operator defined on an interval $[a, b]$, let y_1 and y_2 be a basis for the null space of L in $\mathcal{C}^2[a, b]$, and let $\mathcal{S}$ be the subspace of $\mathcal{C}^2[a, b]$ determined by the unmixed boundary conditions in (12–42). Then L will be one-to-one when restricted to $\mathcal{S}$ if and only if the determinant*

$$\begin{vmatrix} \alpha_1 y_1(a) + \alpha_2 y_1'(a) & \alpha_1 y_2(a) + \alpha_2 y_2'(a) \\ \beta_1 y_1(b) + \beta_2 y_1'(b) & \beta_1 y_2(b) + \beta_2 y_2'(b) \end{vmatrix} \tag{12–45}$$

is different from zero.

Example. Let $L = D^2$, and let $\mathcal{S}$ be the subspace of $\mathcal{C}^2[a, b]$ defined by

$$y(a) = 0, \qquad y(b) = 0. \tag{12–46}$$

Then using the functions $1, x$ as a basis for the null space of L, the above determinant becomes

$$\begin{vmatrix} 1 & a \\ 1 & b \end{vmatrix} = b - a,$$

and it follows that L is one-to-one on $\mathcal{S}$.

In this case we can also prove that L maps $\mathcal{S}$ *onto* $\mathcal{C}[a, b]$ by the simple expedient of constructing L^{-1} directly from the formula for L. Indeed, since the equation $Ly = h$ is simply $y''(x) = h(x)$, two integrations yield

$$y'(s) = \int_a^s h(t)\, dt + c,$$

and

$$y(x) = \int_a^x \left[\int_a^s h(t)\, dt \right] ds + c(x - a) + d, \tag{12–47}$$

where c and d are arbitrary constants. Applying the given boundary conditions we find that $d = 0$ and that

$$\int_a^b \left[\int_a^s h(t)\, dt \right] ds + c(b - a) = 0.$$

Thus

$$c = -\frac{1}{b - a} \int_a^b \left[\int_a^s h(t)\, dt \right] ds,$$

and (12–47) becomes

$$y(x) = \int_a^x \left[\int_a^s h(t)\, dt \right] ds - \frac{x-a}{b-a} \int_a^b \left[\int_a^s h(t)\, dt \right] ds.$$

We now use the unit step function

$$u_0(s) = \begin{cases} 0, & s \leq 0, \\ 1, & s > 0, \end{cases}$$

to rewrite this expression as

$$\begin{aligned} y(x) &= \int_a^b u_0(x-s) \left[\int_a^s h(t)\, dt \right] ds - \frac{x-a}{b-a} \int_a^b \left[\int_a^s h(t)\, dt \right] ds \\ &= \int_a^b \left\{ \int_a^s h(t) \left[u_0(x-s) - \frac{x-a}{b-a} \right] dt \right\} ds. \end{aligned}$$

Using the unit step function again, we have

$$\begin{aligned} y(x) &= \int_a^b \left\{ \int_a^b u_0(s-t) \left[u_0(x-s) - \frac{x-a}{b-a} \right] h(t)\, dt \right\} ds \\ &= \int_a^b \left\{ \int_a^b u_0(s-t) \left[u_0(x-s) - \frac{x-a}{b-a} \right] ds \right\} h(t)\, dt, \end{aligned}$$

and it follows that

$$y(x) = \int_a^b K(x, t) h(t)\, dt, \tag{12–48}$$

where

$$\begin{aligned} K(x, t) &= \int_a^b u_0(s-t) \left[u_0(x-s) - \frac{x-a}{b-a} \right] ds \\ &= \begin{cases} \dfrac{(x-a)(t-b)}{b-a}, & x \leq t, \\ \dfrac{(x-b)(t-a)}{b-a}, & x \geq t. \end{cases} \end{aligned} \tag{12–49}$$

(See Exercise 3.) The function $K(x, t)$ defined by this formula is known as the *Green's function* for the operator $L = D^2$ for the given boundary-value problem, and Eq. (12–48) can be read as the definition of $L^{-1}\colon \mathcal{C}[a, b] \to \mathcal{S}$ with

$$L^{-1}h(x) = \int_a^b K(x, t) h(t)\, dt. \tag{12–50}$$

In the next section we shall prove that an analogous result is valid for *any* normal second-order linear differential operator L acting on $\mathcal{S}$ so long as $\mathcal{S}$ is determined by a pair of unmixed boundary conditions, and L is one-to-one when restricted to $\mathcal{S}$.

EXERCISES

1. Determine which of the following operators are one-to-one when restricted to the given subspace $\mathcal{S}$ of $\mathcal{C}^2[a, b]$.

   (a) $L = D^2 + 4D + 4$
   $\mathcal{S}: y(0) = 0,\ y(a) = 0$

   (b) $L = D^2 + 1$
   $\mathcal{S}: y(-\pi) = 0,\ y(\pi) = 0$

   (c) $L = D^2 - 1$
   $\mathcal{S}: y(0) + y'(0) = 0,\ y(a) = 0$

   (d) $L = D^2 - 1$
   $\mathcal{S}: y(0) - y'(0) = 0,\ y(a) = 0$

   (e) $L = x^2D^2 + xD + 1$
   $\mathcal{S}: y(1) = 0,\ y(e^\pi) = 0$

2. Show that if the condition in Lemma 12–3 is satisfied for a pair of functions y_1 and y_2, it will also be satisfied for every other pair of linearly independent functions of the form

$$Y_1(x) = \gamma_1 y_1(x) + \gamma_2 y_2(x),$$
$$Y_2(x) = \gamma_3 y_1(x) + \gamma_4 y_2(x).$$

3. Prove that

$$\int_a^b u_0(s - t)\left[u_0(x - s) - \frac{x - a}{b - a}\right] ds = \begin{cases} \dfrac{(x - a)(t - b)}{b - a}, & x \le t, \\[2ex] \dfrac{(x - b)(t - a)}{b - a}, & x \ge t, \end{cases}$$

   where u_0 is the unit step function, and $a \le x \le b, a \le t \le b$.

4. Prove that the first derivative *with respect to* x of the Green's function $K(x, t)$ constructed above has a jump discontinuity of magnitude 1 along the line $x = t$, but is continuous at all other points in the region $a \le x \le b,\ a \le t \le b$.

5. Let t_0 be a fixed point in the interval $[a, b]$, and let $K(x, t)$ be the Green's function constructed above. Prove that the function $K(x, t_0)$ is a solution of the boundary-value problem

$$D^2y = 0,$$
$$y(a) = y(b) = 0,$$

   for all $x \ne t_0$.

6. Use the technique introduced above to find the Green's function for the operator D^2 on the subspace of $\mathcal{C}^2[0, 1]$ defined by the boundary conditions $y(0) = y'(1) = 0$.

7. Repeat Exercise 6 for the boundary conditions

$$y(0) - y'(0) = 0,$$
$$y(1) + y'(1) = 0.$$

8. Let $\mathcal{S}$ denote the subspace of $\mathcal{C}^2[a, b]$ determined by the periodic boundary conditions

$$y(a) = y(b),$$
$$y'(a) = y'(b),$$

and let $L = D(p(x)D) + q(x)$ be normal on $[a, b]$. Prove that L is one-to-one when restricted to $\mathcal{S}$ if and only if

$$\begin{vmatrix} y_1(a) - y_1(b) & y_2(a) - y_2(b) \\ y_1'(a) - y_1'(b) & y_2'(a) - y_2'(b) \end{vmatrix} \neq 0$$

whenever y_1 and y_2 are linearly independent solutions of the equation $Ly = 0$.

*12–10 GREEN'S FUNCTIONS FOR BOUNDARY-VALUE PROBLEMS: UNMIXED BOUNDARY CONDITIONS

Throughout this section we shall assume that $L = D[p(x)D] + q(x)$ is a *normal* second-order linear differential operator defined on an interval $[a, b]$, that $\mathcal{S}$ is a subspace of $\mathcal{C}^2[a, b]$ determined by a pair of unmixed boundary conditions, and that L is one-to-one when restricted to $\mathcal{S}$. As was stated above, we propose to show that such a transformation necessarily maps $\mathcal{S}$ *onto* $\mathcal{C}[a, b]$, and admits an inverse which can be expressed as an integral operator of the form

$$L^{-1}h(x) = \int_a^b K(x, t)h(t)\,dt$$

for all h in $\mathcal{C}[a, b]$. In order to motivate the axiomatic definition of the function $K(x, t)$ given below, we begin by presenting a heuristic argument which will simultaneously suggest the existence of L^{-1}, the validity of the above formula, and, most important of all, the correct determination of $K(x, t)$.*

Assume that the equation

$$Ly = h \tag{12–51}$$

describes the behavior of a physical system under the influence of a given input function h, and assume, in addition, that the response $y(x)$ of the system at the point x due to the unit input

$$\varphi_t(x) = \begin{cases} 1 & \text{if } x = t, \\ 0 & \text{if } x \neq t, \end{cases}$$

is $K(x, t)$.† Then, in view of the linearity of L, it is reasonable to expect that the

* The following argument is essentially the one given by R. Courant and D. Hilbert in *Methods of Mathematical Physics*, Vol. I, Interscience Publishers, Inc., New York, 1953.

† For instance, (12–51) might be the equation of equilibrium of an elastic string stretched along the interval $[a, b]$ and subjected to a continuously distributed force $h = h(x)$. In that case, $\varphi_t(x)$ represents a unit force applied at the point t, and $K(x, t)$ the deflection of the string at x due to this force.

response of the system at x to a unit input applied continuously throughout the entire interval should be obtained by "summing" the responses due to unit inputs applied at each point of the interval, and hence should be of the form

$$y(x) = \int_a^b K(x, t)\, dt.$$

In the case of a general input function h, this reasoning leads to the formula

$$y(x) = \int_a^b K(x, t)h(t)\, dt, \tag{12–52}$$

where the integrand is now viewed as the response at x to that portion of the input applied at the point t.

Granting the validity of these considerations, we can easily deduce a number of properties of the function $K(x, t)$. In the first place, it is clear that $K(x, t)$ must be defined and continuous for $a \leq x \leq b$, $a \leq t \leq b$, and, for each value of t, must satisfy whatever boundary conditions have been imposed on the problem. Moreover, by its very definition, $K(x, t)$ is a solution of the equation

$$Ly = \varphi_t(x),$$

and hence, for each fixed t_0 in $[a, b]$, *the function* $K(x, t_0)$ *satisfies the homogeneous equation*

$$Ly = 0$$

for all $x \neq t_0$. Finally, to determine the behavior of $K(x, t_0)$ when $x = t_0$, let f_{t_0} denote the function which vanishes outside the interval $|x - t_0| > \epsilon$, and which contributes a total input of one in the interval $|x - t_0| < \epsilon$; that is,

$$\int_{t_0-\epsilon}^{t_0+\epsilon} f_{t_0}(x)\, dx = 1.$$

(We assume that ϵ has been chosen sufficiently small so that the interval $[t_0 - \epsilon, t_0 + \epsilon]$ is contained in $[a, b]$.) Then, if $\bar{K}(x, t_0)$ denotes the response of the system at x to f_{t_0}, we have

$$L[\bar{K}(x, t_0)] = f_{t_0}(x),$$

and it follows that

$$\int_{t_0-\epsilon}^{t_0+\epsilon} L[\bar{K}(x, t_0)]\, dx = 1. \tag{12–53}$$

But, by assumption, $L = D[p(x)D] + q(x)$, whence (12–53) becomes

$$\int_{t_0-\epsilon}^{t_0+\epsilon} \frac{d}{dx}[p(x)\bar{K}'(x, t_0)]\, dx + \int_{t_0-\epsilon}^{t_0+\epsilon} q(x)\bar{K}(x, t_0)\, dx = 1,$$

or

$$p(x)\bar{K}'(x, t_0)\Big|_{t_0-\epsilon}^{t_0+\epsilon} + \int_{t_0-\epsilon}^{t_0+\epsilon} q(x)\bar{K}(x, t_0)\, dt = 1. \tag{12–54}$$

We now make the not unreasonable assumption that as $\epsilon \to 0$, $\bar{K}(x, t_0) \to K(x, t_0)$ for all x, and that $\bar{K}'(x, t_0) \to K'(x, t_0)$ for all x different from t_0. Then the continuity of q and $\bar{K}$ imply that the second term in (12–54) vanishes as $\epsilon \to 0$, while the first reduces to

$$p(t_0)\left[\frac{d}{dx} K(x, t_0)\right]_{t_0^-}^{t_0^+}.$$

Thus

$$\left.\frac{d}{dx} K(x, t_0)\right|_{t_0^-}^{t_0^+} = \frac{1}{p(t_0)}, \tag{12–55}$$

an equation which asserts that at the point t_0 the derivative of $K(x, t_0)$ has a jump discontinuity of magnitude $1/p(t_0)$.

Although these considerations are admittedly nonrigorous, they do agree with the results obtained in the preceding section (see Exercises 4 and 5, Section 12–9), and serve to motivate the following definition.

Definition 12–5. A *Green's function* for the boundary-value problem $L: \mathcal{S} \to \mathcal{C}[a, b]$ described above is a function $K(x, t)$ of two variables satisfying the following three conditions:

(1) $K(s, t)$ is defined and continuous for $a \leq x \leq b$, $a \leq t \leq b$, and, as a function of x, is twice continuously differentiable except when $x = t$;

(2) For each fixed t_0 in $[a, b]$, $K(x, t_0)$ belongs to the subspace $\mathcal{S}$ (i.e., satisfies the boundary conditions imposed on the problem), and, in addition, is a solution of the equation $Ly = 0$, except at the point $x = t_0$;

(3) $\left.\dfrac{d}{dx} K(x, t_0)\right|_{t_0^-}^{t_0^+} = \dfrac{1}{p(t_0)}.$

With this as our definition, we now state the following basic theorem.

Theorem 12–9. *If $L: \mathcal{S} \to \mathcal{C}[a, b]$ is a boundary-value problem of the type described at the beginning of this section, and if L is a one-to-one mapping of $\mathcal{S}$ into $\mathcal{C}[a, b]$, then L maps $\mathcal{S}$* **onto** *$\mathcal{C}[a, b]$, and for each h in $\mathcal{C}[a, b]$ the (necessarily unique) solution in $\mathcal{S}$ of the equation*

$$Ly = h$$

is given by the formula

$$y(x) = \int_a^b K(x, t)h(t)\, dt,$$

where $K(x, t)$ is a Green's function for L. Moreover, $K(x, t)$ is **uniquely** *determined by the operator L and the boundary conditions which define the subspace $\mathcal{S}$.*

We defer the proof of this result to the next section in favor of showing how the Green's function for L can be explicitly computed once L and $\mathcal{S}$ are known. Here we argue as follows.

Let y_1 and y_2 be solutions of the homogeneous equation $Ly = 0$ chosen so that y_1 satisfies the boundary condition imposed at $x = a$, and y_2 the boundary condition imposed at $x = b$; that is,

$$\alpha_1 y_1(a) + \alpha_2 y_1'(a) = 0,$$
$$\beta_1 y_2(b) + \beta_2 y_2'(b) = 0.$$

(See Exercise 1 for a proof that such solutions do, in fact, exist.) Then y_1 and y_2 are linearly independent in $\mathcal{C}^2[a, b]$. For otherwise, there would exist a constant $c \neq 0$ such that $y_2(x) = cy_1(x)$, and the function $cy_1(x)$ would be a *nontrivial* solution of $Ly = 0$ satisfying *both* of the boundary conditions imposed on $\mathcal{S}$. This, however, contradicts the assumption that L is one-to-one when restricted to $\mathcal{S}$, and is therefore impossible.

We now use the fact that the Wronskian of y_1 and y_2 never vanishes on the interval $[a, b]$ to find functions $A_1(t)$ and $A_2(t)$ such that

$$\begin{aligned} A_2(t)y_2(t) - A_1(t)y_1(t) &\equiv 0, \\ A_2(t)y_2'(t) - A_1(t)y_1'(t) &= \frac{1}{p(t)} \end{aligned} \tag{12–56}$$

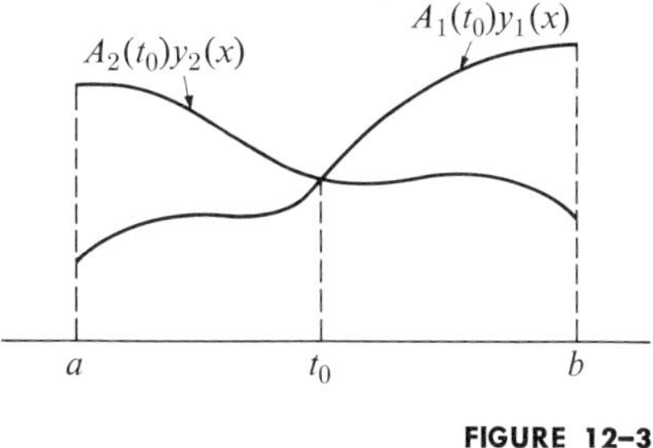

FIGURE 12–3

for all t in $[a, b]$. The first of these equations guarantees that for each t_0 in the interval (a, b) the curves $A_1(t_0)y_1(x)$ and $A_2(t_0)y_2(x)$ intersect at the point $x = t_0$ (see Fig. 12–3), while the second guarantees that the slope of $A_2(t_0)y_2(x)$ differs from the slope of $A_1(t_0)y_1(x)$ by $1/p(t_0)$ at $x = t_0$. Thus the function

$$K(x, t) = \begin{cases} A_1(t)y_1(x), & x \le t, \\ A_2(t)y_2(x), & x \ge t, \end{cases} \tag{12–57}$$

satisfies the various conditions imposed in Definition 12–5, and is *the* Green's function for L. Finally, by solving (12–56) for A_1 and A_2 we obtain the formula

$$K(x, t) = \begin{cases} \dfrac{y_1(x)y_2(t)}{p(t)[y_1(t)y_2'(t) - y_1'(t)y_2(t)]}, & x \le t, \\[2ex] \dfrac{y_1(t)y_2(x)}{p(t)[y_1(t)y_2'(t) - y_1'(t)y_2(t)]}, & x \ge t. \end{cases} \tag{12–58}$$

Remark: The argument just given provides a rigorous proof of the *existence* of a Green's function for the problem under consideration. Uniqueness will be established in the next section.

EXAMPLE. Let $L = D^2$, and let $\mathcal{S}$ be the subspace of $\mathcal{C}^2[a, b]$ for which

$$y(a) = y(b) = 0.$$

Then, as we know, L is one-to-one when restricted to $\mathcal{S}$, and Theorem 12–9 applies. To find the Green's function for L on $\mathcal{S}$ we set

$$y_1(x) = x - a, \qquad y_2(x) = x - b$$

in (12–58), to obtain

$$K(x, t) = \begin{cases} \dfrac{(x - a)(t - b)}{b - a}, & x \leq t, \\[2ex] \dfrac{(t - a)(x - b)}{b - a}, & x \geq t, \end{cases}$$

in agreement with the formula found in Section 12–9.

EXERCISES

1. Prove that the equation $Ly = 0$, L as above, has a pair of solutions y_1 and y_2 such that

$$\alpha_1 y_1(a) + \alpha_2 y_1'(a) = 0,$$
$$\beta_1 y_2(b) + \beta_2 y_2'(b) = 0.$$

[*Hint:* Use the existence theorem for solutions of initial-value problems.]

2. Use the method of this section to obtain the Green's functions for the boundary-value problems of Exercises 6 and 7 of the preceding section.

3. (a) Find the Green's function for the boundary-value problem

$$y'' + k^2 y = 0, \qquad y(0) = y(1) = 0.$$

(b) Use the result in (a) to find the solution of

$$y'' + k^2 y = h(x), \qquad y(0) = y(1) = 0,$$

given that h belongs to $\mathcal{C}[0, 1]$.

*12–11 GREEN'S FUNCTIONS: A PROOF OF THE MAIN THEOREM

Continuing with the notation introduced above, we now show that Definition 12–5 does, in fact, uniquely characterize the Green's function for L. This is the content of

Lemma 12–4. *Let $L = D[p(x)D] + q(x)$ be a normal second-order linear differential operator on $[a, b]$, and let $H(x, t)$ and $K(x, t)$ be two functions satisfying Definition 12–5. Then $H(x, t) \equiv K(x, t)$ for all x and t in $[a, b]$.*

Proof. Let t_0 be any point in $[a, b]$, and set $F(x) = H(x, t_0) - K(x, t_0)$. Then F and F' are continuous for all x in $[a, b]$, and

$$L(F) = L(H - K) = L(H) - L(K) = 0 \tag{12–59}$$

for all $x \neq t_0$.* Furthermore, by applying L to F and solving the identity

$$p(x)F'' + p'(x)F' + q(x)F \equiv 0$$

for F'', we conclude that F'' exists and is continuous on $[a, b]$, and hence that F belongs to $\mathcal{C}^2[a, b]$. Finally, since $H(x, t_0)$ and $K(x, t_0)$ belong to the subspace $\mathcal{S}$, the same is true of F, and (12–59) together with the fact that L is one-to-one when restricted to $\mathcal{S}$ now implies that $F(x) \equiv 0$. Hence $H(x, t_0) \equiv K(x, t_0)$ for all x, and since t_0 was arbitrary in $[a, b]$, the proof is complete. ▮

At this point we are in a position to assert that every boundary-value problem $L: \mathcal{S} \to \mathcal{C}[a, b]$ involving a normal second-order linear differential operator which is one-to-one on the subspace $\mathcal{S}$ has *a unique* Green's function $K(x, t)$ and that this function is given by Formula (12–58). Thus to complete the proof of Theorem 12–9 we need but show that the function

$$y(x) = \int_a^b K(x, t)h(t)\,dt$$

belongs to $\mathcal{S}$ and satisfies the equation $Ly = h$, for every h in $\mathcal{C}[a, b]$.

To this end we write

$$y(x) = \int_a^x K(x, t)h(t)\,dt + \int_x^b K(x, t)h(t)\,dt,$$

and differentiate to obtain

$$\begin{aligned} y'(x) &= K(x, x)h(x) + \int_a^x \frac{\partial}{\partial x} K(x, t)h(t)\,dt \\ &\quad - K(x, x)h(x) + \int_x^b \frac{\partial}{\partial x} K(x, t)h(t)\,dt \\ &= \int_a^b \frac{\partial}{\partial x} K(x, t)h(t)\,dt. \end{aligned}$$

(See Appendix I.) Thus

$$\begin{aligned} \alpha_1 y(a) + \alpha_2 y'(a) &= \int_a^b \left[\alpha_1 K(a, t) + \alpha_2 \frac{\partial}{\partial x} K(a, t)\right] h(t)\,dt \\ &= 0, \end{aligned}$$

* Strictly speaking, F' has a *removable* discontinuity at $x = t_0$, which we can ignore because the discontinuities in H and K cancel by subtraction.

the last step following from the fact that, as a function of x, $K(x, t)$ belongs to $\mathcal{S}$ for each t in $[a, b]$. Similarly

$$\beta_1 y(b) + \beta_2 y'(b) = 0,$$

and $y(x)$ does belong to $\mathcal{S}$, as required.

Finally,

$$\begin{aligned} y''(x) &= \frac{d}{dx}\left[\int_a^x \frac{\partial}{\partial x} K(x, t)h(t)\,dt + \int_x^b \frac{\partial}{\partial x} K(x, t)h(t)\,dt\right] \\ &= \int_a^x \frac{\partial^2}{\partial x^2} K(x, t)h(t)\,dt + \frac{\partial K}{\partial x}(x, x^-)h(x) \\ &\quad + \int_x^b \frac{\partial^2}{\partial x^2} K(x, t)h(t)\,dt - \frac{\partial K}{\partial x}(x, x^+)h(x) \\ &= \int_a^b \frac{\partial^2}{\partial x^2} K(x, t)h(t)\,dt + h(x)\left[\frac{\partial K}{\partial x}(x, x^-) - \frac{\partial K}{\partial x}(x, x^+)\right]. \end{aligned}$$

But, referring to Fig. 12–4, we observe that the continuity of $\partial K/\partial x$ in triangle ABC implies

$$\frac{\partial K}{\partial x}(x, x^+) = \frac{\partial K}{\partial x}(x^-, x).$$

Similarly, using triangle ABD, we obtain

$$\frac{\partial K}{\partial x}(x, x^-) = \frac{\partial K}{\partial x}(x^+, x),$$

and it follows that

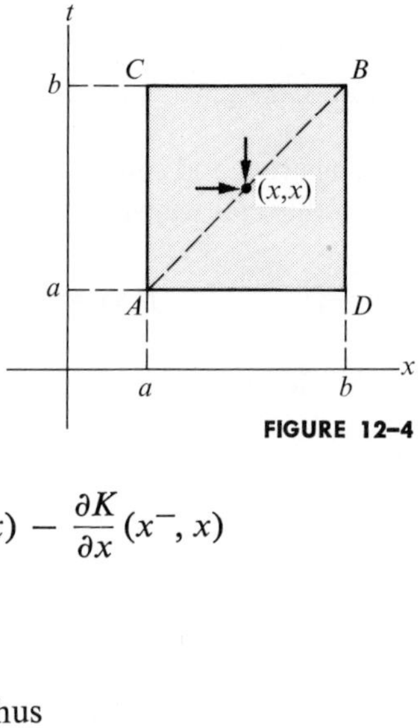

FIGURE 12–4

$$\begin{aligned} \frac{\partial K}{\partial x}(x, x^-) - \frac{\partial K}{\partial x}(x, x^+) &= \frac{\partial K}{\partial x}(x^+, x) - \frac{\partial K}{\partial x}(x^-, x) \\ &= \frac{1}{p(x)}, \end{aligned}$$

the prescribed jump in $\partial K/\partial x$ at the point (x, x). Thus

$$\begin{aligned} L[y(x)] &= p(x)y''(x) + p'(x)y'(x) + q(x)y(x) \\ &= h(x) + \int_a^b \left[p(x)\frac{\partial^2}{\partial x^2} K(x, t) + p'(x)\frac{\partial}{\partial x} K(x, t) + q(x)K(x, t)\right] h(t)\,dt \\ &= h(x) + \int_a^b L[K(x, t)]h(t)\,dt, \end{aligned}$$

and since $L[K(x, t)] = 0$ for each t in $[a, b]$,

$$L[y(x)] = h(x),$$

and we are done. ∎

As our final result on Green's functions, we now prove

Theorem 12–10. *If $K(x, t)$ is the Green's function for the boundary-value problem $L: \mathcal{S} \to \mathcal{C}[a, b]$ described above, then $K(x, t) \equiv K(t, x)$ for all x and all t.*

Proof. Let s_0 and t_0 be fixed points in $[a, b]$ (with $t_0 \leq s_0$), and set $u = K(x, s_0)$, $v = K(x, t_0)$. Then $Lu = Lv = 0$ for all x in $[a, b]$ different from s_0 and t_0, and the Lagrange identity applied to u and v yields

$$\frac{d}{dx}[p(uv' - u'v)] = 0.$$

We now integrate this expression from a to b, taking account of the discontinuities of u' and v' at s_0 and t_0, to obtain

$$\begin{aligned} p(x)&\left[\frac{\partial K}{\partial x}(x, t_0) \cdot K(x, s_0) - \frac{\partial K}{\partial x}(x, s_0) \cdot K(x, t_0)\right]\Bigg|_a^{t_0^-} \\ &+ p(x)\left[\frac{\partial K}{\partial x}(x, t_0) \cdot K(x, s_0) - \frac{\partial K}{\partial x}(x, s_0) \cdot K(x, t_0)\right]\Bigg|_{t_0^+}^{s_0^-} \\ &+ p(x)\left[\frac{\partial K}{\partial x}(x, t_0) \cdot K(x, s_0) - \frac{\partial K}{\partial x}(x, s_0) \cdot K(x, t_0)\right]\Bigg|_{s_0^+}^{b} = 0. \end{aligned}$$

Thus

$$\begin{aligned} p(t_0)&\left[\frac{\partial K}{\partial x}(t_0^-, t_0) \cdot K(t_0, s_0) - \frac{\partial K}{\partial x}(t_0^+, t_0) \cdot K(t_0, s_0)\right] \\ &+ p(s_0)\left[\frac{\partial K}{\partial x}(s_0^+, s_0) \cdot K(s_0, t_0) - \frac{\partial K}{\partial x}(s_0^-, s_0) \cdot K(s_0, t_0)\right] \\ &+ p(b)\left[\frac{\partial K}{\partial x}(b, t_0) \cdot K(b, s_0) - \frac{\partial K}{\partial x}(b, s_0) \cdot K(b, t_0)\right] \\ &- p(a)\left[\frac{\partial K}{\partial x}(a, t_0) \cdot K(a, s_0) - \frac{\partial K}{\partial x}(a, s_0) \cdot K(a, t_0)\right] = 0. \end{aligned}$$

But, using the known jumps in $\partial K/\partial x$, this expression can be rewritten

$$\begin{aligned} -K(t_0, s_0) + K(s_0, t_0) + p(b)&\left[\frac{\partial K}{\partial x}(b, t_0) \cdot K(b, s_0) - \frac{\partial K}{\partial x}(b, s_0) \cdot K(b, t_0)\right] \\ - p(a)&\left[\frac{\partial K}{\partial x}(a, t_0) \cdot K(a, s_0) - \frac{\partial K}{\partial x}(a, s_0) \cdot K(a, t_0)\right] = 0. \end{aligned}$$

Finally, using the boundary conditions

$$\begin{cases} \alpha_1 K(a, t_0) = -\alpha_2 \dfrac{\partial K}{\partial x}(a, t_0), \\[2ex] \alpha_1 K(a, s_0) = -\alpha_2 \dfrac{\partial K}{\partial x}(a, s_0), \end{cases}$$

and

$$\begin{cases} \beta_1 K(b, t_0) = -\beta_2 \dfrac{\partial K}{\partial x}(b, t_0), \\[2ex] \beta_1 K(b, s_0) = -\beta_2 \dfrac{\partial K}{\partial x}(b, t_0), \end{cases}$$

we see that the bracketed terms vanish. Thus $K(s_0, t_0) = K(t_0, s_0)$, as asserted. ▮

13

boundary-value problems for partial differential equations: the wave and heat equations

13–1 INTRODUCTION

Historically the theory of boundary-value problems grew out of the study of certain partial differential equations encountered in classical physics, and many of the ideas treated in this book originated in attempts to solve these problems. For this reason, if for no other, any introduction to the subject of boundary-value problems would be incomplete without a discussion of partial differential equations. But there are, in fact, compelling reasons for pursuing this subject which are quite unrelated to any feelings of historical nicety. And though these reasons are bound to become obvious as the chapter unfolds, it may not be out of place to mention some of them before we begin.

For one thing, this discussion will serve to unite the various results on eigenfunctions and orthogonal series expansions that have been obtained in the preceding chapters, bringing them into sharper focus, and reenforcing the point of view which sees them as a unified body of mathematical thought. For another, we will at last be in a position to consider nontrivial physical problems, and the fact that they can be solved with comparative ease should increase the student's appreciation for the power of the techniques we now have at hand. And finally, this material will in its turn suggest further problems leading to new results in the subject.

13–2 PARTIAL DIFFERENTIAL EQUATIONS

The definition of a linear differential operator given in Chapter 3 can easily be extended to include operators which involve partial differentiation. Such operators act on vector spaces whose members are functions of several variables, and their associated operator equations are known as *linear partial differential equations.* Thus if $\mathcal{C}^1(R)$ is the space of all continuously differentiable functions defined in a region R of the xy-plane, the most general first-order linear differential operator defined on $\mathcal{C}^1(R)$ has the form

$$L = a(x, y)D_x + b(x, y)D_y + c(x, y) = a(x, y)\frac{\partial}{\partial x} + b(x, y)\frac{\partial}{\partial y} + c(x, y),$$

where $a(x, y)$, $b(x, y)$, $c(x, y)$ are continuous everywhere in R. Here L may be viewed as a linear transformation from $\mathcal{C}^1(R)$ to $\mathcal{C}(R)$, the space of all functions continuous in R, and if h is any preassigned function in $\mathcal{C}(R)$, the equation $Lu = h$, u unknown, is a first-order (linear) partial differential equation.

It is clear that analogous, but more cumbersome formulas can be given for linear differential operators (and equations) of higher order involving any number of variables. Moreover, it is equally clear that all of the standard facts pertaining to linearity continue to hold in this more general setting. This not withstanding, the general theory of linear partial differential equations has very little in common with that of ordinary equations for the simple reason that the solution space of *every* (homogeneous) linear partial differential equation is *infinite* dimensional. For instance, it is not difficult to show that the general solution of the first-order equation

$$\frac{\partial u}{\partial x} + \frac{\partial u}{\partial y} = 0 \tag{13–1}$$

is $u(x - y)$, where u is an everywhere differentiable but otherwise arbitrary function of a single variable. From this it follows that each of the functions

$$\sin (x - y),\ \cos (x - y),\ e^{(x-y)^2},\ (x - y)^\alpha,\quad \alpha \geq 1$$

is a solution of (13–1), and it is clear that these functions are linearly independent in $\mathcal{C}(R)$ for any R. The fact that even so simple an equation as this has such a wealth of linearly independent solutions gives some indication of the difficulties which must be surmounted in the study of partial differential equations.

These remarks go far to explain our insistence upon treating partial differential equations strictly within the context of boundary-value problems. For there the difficulties just mentioned vanish, and all of the problems we shall consider *do* have unique solutions. But before we go on to describe these problems in detail we must define what is meant by a solution of a boundary-value problem involving a partial differential equation. Surprisingly, this is not so easy as it sounds, and thus, for the sake of simplicity, we shall give the definition only in the two-dimensional case with boundary conditions involving a single function. Once this has been done the reader should have no difficulty in extending the definition to higher-dimensional regions and more complicated boundary conditions.

In the particular case just mentioned the ingredients of a boundary-value problem are

(i) a two-dimensional plane region R with boundary B,
(ii) a partial differential equation defined everywhere in R, and
(iii) a function f defined on B.

As used here the word "region" is a technical term reserved to describe a connected subset of the plane each point of which can be surrounded by a circle lying entirely within the set in question.* Thus the upper half plane, an infinite or semi-infinite vertical strip of the plane, the interior of a rectangle, or the annulus between two concentric circles are all regions in this sense, and as such are typical of the two-dimensional regions in which boundary-value problems involving partial differential equations are defined. In addition, we shall assume henceforth that the boundary of each of the regions we consider is made up of a finite number of simple differentiable arcs in the sense of the definition given in Section 9–8.†

Whenever (i), (ii), (iii) are given, the problem, of course, is to find all functions $u = u(x, y)$ which satisfy the differential equation in R and reduce to f on B. (Note that we do *not* require u to satisfy the differential equation on B.) But were this all that was required we could immediately solve the problem by letting u be any solution whatever of the differential equation in R and then redefining u on B to coincide with f. However, this clearly violates the spirit of the problem, and to ensure that it violates the letter as well we must impose some restriction which will guarantee that the values of u near B are related to the values of f on B. Although this can be done in a variety of ways, the constraining condition is usually taken to be one of the following:

a. For each point b_0 on B and each p in R, the limiting value of $u(p)$ as p approaches b_0 along *any* smooth curve in R is $f(b_0)$. (See Fig. 13–1.)

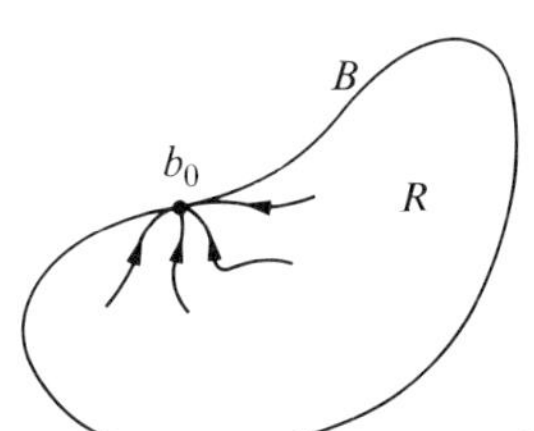

FIGURE 13–1

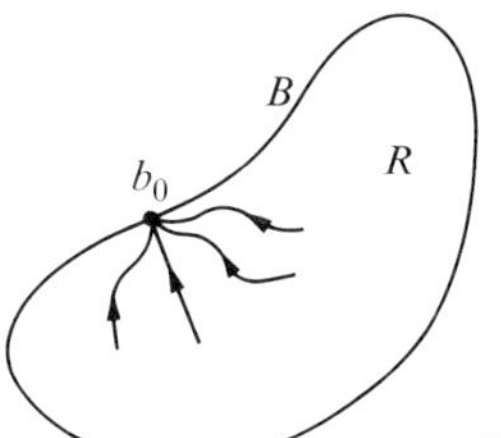

FIGURE 13–2

b. For each point b_0 on B and each p in R the limiting value of $u(p)$ as p approaches b_0 along any smooth curve in R which is *normal* to B at b_0 is $f(b_0)$. (See Fig. 13–2.)

In practice the choice between (a) and (b) is reflected in the hypotheses which must be imposed on f to guarantee the existence of solutions of the desired type. Since the second of these conditions is less restrictive than the first, it allows theorems to be proved in greater generality and is therefore in wider use. It is the condition which we shall adopt in the following chapters.

---

* A subset R of the plane is said to be *connected* or *pathwise connected* if every pair of points in R can be joined by a smooth curve lying entirely in R.

† The *boundary* of a region R is, by definition, the set B of all points in the plane with the property that *every* circle centered at a point of B contains points in R and points not in R.

EXERCISES

1. (a) Give the formula for the most general linear differential operator $L: \mathcal{C}^1(R) \to \mathcal{C}(R)$ when R is a region of xyz-space.

   (b) Give the formula for the most general linear differential operator $L: \mathcal{C}^2(R) \to \mathcal{C}(R)$ when R is a region of the xy-plane.

2. Determine which of the following partial differential equations are linear.

   (a) $\dfrac{\partial^2 u}{\partial x^2} + \dfrac{\partial^2 u}{\partial y^2} + \dfrac{\partial^2 u}{\partial z^2} = 0$ (b) $xy \dfrac{\partial^2 u}{\partial x^2} + x \dfrac{\partial u}{\partial x} \dfrac{\partial u}{\partial y} + y \dfrac{\partial^2 u}{\partial y^2} = \sin(xy)$

   (c) $\dfrac{\partial^2 u}{\partial x\, \partial y} + 2 \dfrac{\partial^2 u}{\partial y\, \partial z} + xyz = u$ (d) $\left(\dfrac{\partial^2 u}{\partial x^2} + \dfrac{\partial^2 u}{\partial y^2}\right) u - \dfrac{\partial u}{\partial x} - \dfrac{\partial u}{\partial y} = e^{xy}$

   (e) $\dfrac{\partial^2 u}{\partial x^2} + \dfrac{\partial^2 u}{\partial y^2} + (xyu)^2 = 0$

   (f) $x \dfrac{\partial^2 u}{\partial x^2} + y \dfrac{\partial^2 u}{\partial y^2} + 2xy \left(\dfrac{\partial u}{\partial x} + \dfrac{\partial u}{\partial y}\right) + u = (xy)^2 u$

3. (a) Show that $u(x - y)$ is a solution of the partial differential equation $u_x + u_y = 0$ whenever u is a differentiable function of a single variable.
   (b) Let $F(x, y)$ be a solution of $u_x + u_y = 0$. Set $p = x + y$, $q = x - y$, and write $F(x, y)$ as

$$F(x, y) = F\left(\frac{p + q}{2}, \frac{p - q}{2}\right) \equiv G(p, q).$$

   Show that G is actually a function of q alone by computing $\partial G/\partial p$, and then deduce that every solution of $u_x + u_y = 0$ can be written in the form $u(x - y)$, where u is a differentiable function of a single variable.

13–3 THE CLASSICAL PARTIAL DIFFERENTIAL EQUATIONS

Throughout the next three chapters, we shall, with but one exception, be exclusively concerned with boundary-value problems involving various forms of the following second-order linear partial differential equations:

$$\frac{\partial^2 u}{\partial x^2} + \frac{\partial^2 u}{\partial y^2} + \frac{\partial^2 u}{\partial z^2} = \frac{1}{a^2} \frac{\partial^2 u}{\partial t^2}, \quad a > 0, \tag{13–2}$$

$$\frac{\partial^2 u}{\partial x^2} + \frac{\partial^2 u}{\partial y^2} + \frac{\partial^2 u}{\partial z^2} = a^2 \frac{\partial u}{\partial t}, \quad a > 0, \tag{13–3}$$

$$\frac{\partial^2 u}{\partial x^2} + \frac{\partial^2 u}{\partial y^2} + \frac{\partial^2 u}{\partial z^2} = 0. \tag{13–4}$$

Each of these equations first arose in classical (i.e., Newtonian) physics as the mathematical description of a particular type of physical system, and the first two are still known by the names of the simplest systems they describe. Thus (13–2) is called the *wave equation*, and (13–3) the *heat equation*. Equation (13–4), on the other hand, is known as *Laplace's equation* in honor of one of the mathematicians who first studied it. For the present we shall content ourselves with the observation that Laplace's equation can be viewed as the time independent version of the heat equation, and shall turn our attention to (13–2) and (13–3).

FIGURE 13–3

As its name indicates, the wave equation furnishes a satisfactory mathematical description of certain vibrating physical systems. In particular, the so-called *one-dimensional wave equation*

$$\frac{\partial^2 u}{\partial x^2} = \frac{1}{a^2}\frac{\partial^2 u}{\partial t^2}, \tag{13–5}$$

in which a is a positive constant, is the differential equation governing the motion of a vibrating string of constant density. To see why this is so, consider a stretched elastic string of arbitrary length which is vibrating vertically in the xu-plane and whose position of rest lies along the x-axis (Fig. 13–3). Throughout this discussion we shall make the following simplifying assumptions:

(a) The amplitude of vibration of the string is small, and each point on the string moves only in a vertical direction;

(b) All frictional forces (both internal and external) may be neglected;

(c) The mass of the string per unit length is sufficiently small in comparison with the tension in the string that gravitational forces may be neglected.*

In Fig. 13–4 we have isolated a small segment of the string and have indicated by T and T' the forces of tension acting on its endpoints. Since the string moves only in the vertical direction, the horizontal components of T and T' must cancel, and we have

$$T\cos\alpha = T'\cos\alpha' = k, \tag{13–6}$$

where k denotes the constant horizontal tension in the string. On the other hand, the total force acting on this element of the string in the vertical direction is $T'\sin\alpha' - T\sin\alpha$. Hence, by Newton's second law,

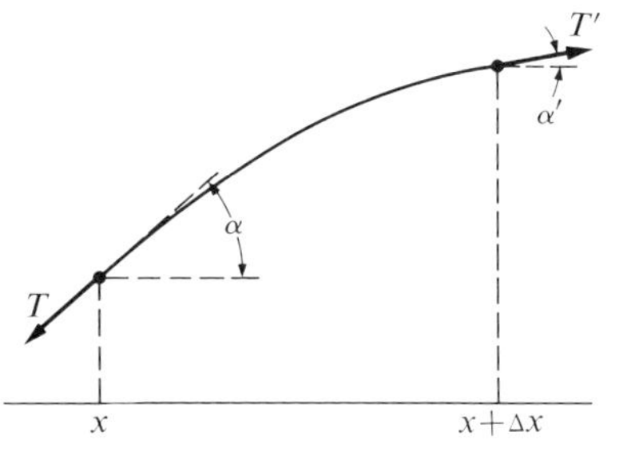

FIGURE 13–4

$$T'\sin\alpha' - T\sin\alpha = \rho\,\Delta x\,\frac{\partial^2 u}{\partial t^2}, \tag{13–7}$$

---

* The question as to whether these assumptions are permissible from a physical point of view and do not prejudice the solution obtained is one which must be settled by the physicist in his laboratory. In most situations they are, in fact, physically acceptable.

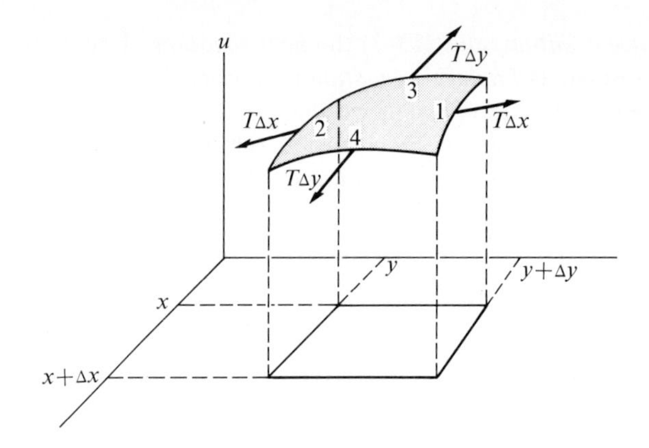

FIGURE 13–5

where ρ denotes the mass per unit length of the string, Δx the length of the segment in question, and $\partial^2 u/\partial t^2$ the acceleration of the segment at an appropriate point between x and $x + \Delta x$. Using (13–6), this equation may be rewritten

$$\frac{T' \sin \alpha'}{T' \cos \alpha'} - \frac{T \sin \alpha}{T \cos \alpha} = \frac{\rho \, \Delta x}{k} \frac{\partial^2 u}{\partial t^2},$$

or

$$\tan \alpha' - \tan \alpha = \frac{\rho \, \Delta x}{k} \frac{\partial^2 u}{\partial t^2}.$$

But $\tan \alpha' = \partial u/\partial x$ evaluated at $x + \Delta x$, and $\tan \alpha = \partial u/\partial x$ evaluated at x. Hence

$$\frac{1}{\Delta x}\left[\left.\frac{\partial u}{\partial x}\right|_{x+\Delta x} - \left.\frac{\partial u}{\partial x}\right|_{x}\right] = \frac{\rho}{k}\frac{\partial^2 u}{\partial t^2},$$

and passing to the limit as $\Delta x \to 0$, we obtain

$$\frac{\partial^2 u}{\partial x^2} = \frac{1}{a^2}\frac{\partial^2 u}{\partial t^2},$$

where $a = \sqrt{k/\rho}$.

The *two-dimensional wave equation*

$$\frac{\partial^2 u}{\partial x^2} + \frac{\partial^2 u}{\partial y^2} = \frac{1}{a^2}\frac{\partial^2 u}{\partial t^2} \tag{13–8}$$

arises in physics as the differential equation governing the motion of a thin flexible membrane of constant density which is tightly stretched and then fixed along its boundary, and which vibrates in the u-direction from its position of rest in the xy-plane. In this case the simplifying physical assumptions under which the equation is derived are as follows:

(a) the amplitude of vibration is small, and every point of the membrane moves only in the u-direction;

(b) all frictional and gravitational forces may be neglected;

(c) the tension per unit length in any direction is constant throughout the membrane.

To obtain the equation of motion under these assumptions we analyze the forces acting on the portion of the membrane shown in Fig. 13–5. If T denotes tension per unit length, then the vertical components of the forces acting along edges 1 and 2 of the indicated portion of the membrane are $T\,\Delta x \sin\alpha_1$ and $T\,\Delta x \sin\alpha_2$ for appropriate angles α_1 and α_2. But since the amplitude of deflection is small, we can replace $\sin\alpha_1$ and $\sin\alpha_2$ by $\tan\alpha_1$ and $\tan\alpha_2$, respectively.*

Thus the total vertical force contributed by edges 1 and 2 is

$$T\,\Delta x(\tan\alpha_1 - \tan\alpha_2).$$

Similarly, edges 3 and 4 contribute a vertical force

$$T\,\Delta y(\tan\alpha_3 - \tan\alpha_4)$$

for appropriate angles α_3 and α_4. Thus, by Newton's second law,

$$T\,\Delta x(\tan\alpha_1 - \tan\alpha_2) + T\,\Delta y(\tan\alpha_3 - \tan\alpha_4) = \rho\,\Delta x\,\Delta y\,\frac{\partial^2 u}{\partial t^2}, \tag{13–9}$$

where ρ is the mass of the membrane per unit area, and $\partial^2 u/\partial t^2$ is computed at some point in the region under consideration. But

$$\tan\alpha_1 = \left.\frac{\partial u}{\partial y}\right|_{(x_1,y+\Delta y)}, \qquad \tan\alpha_2 = \left.\frac{\partial u}{\partial y}\right|_{(x_2,y)},$$

$$\tan\alpha_3 = \left.\frac{\partial u}{\partial x}\right|_{(x,y_1)}, \qquad \tan\alpha_4 = \left.\frac{\partial u}{\partial x}\right|_{(x+\Delta x,y_2)},$$

where x_1 and x_2 lie between x and $x + \Delta x$, y_1 and y_2 between y and $y + \Delta y$. Thus (13–9) may be rewritten

$$\frac{1}{\Delta y}\left[\left.\frac{\partial u}{\partial y}\right|_{(x_1,y+\Delta y)} - \left.\frac{\partial u}{\partial y}\right|_{(x_2,y)}\right] + \frac{1}{\Delta x}\left[\left.\frac{\partial u}{\partial x}\right|_{(x,y_1)} - \left.\frac{\partial u}{\partial x}\right|_{(x+\Delta x,y_2)}\right] = \frac{\rho}{T}\frac{\partial^2 u}{\partial t^2},$$

---

* Compare

$$\sin x = x - \frac{x^3}{3!} + \frac{x^5}{5!} - \cdots$$

with

$$\tan x = x + \frac{x^3}{3} + \frac{2x^5}{15} + \cdots.$$

Their difference is of the order of magnitude of $x^3/2$, which is small if x is near zero.

and passing to the limit as Δx and Δy tend to zero, we obtain

$$\frac{\partial^2 u}{\partial x^2} + \frac{\partial^2 u}{\partial y^2} = \frac{1}{a^2}\frac{\partial^2 u}{\partial t^2},$$

where $a = \sqrt{T/\rho}$.

Finally, the three-dimensional wave equation arises, among other places, in that branch of physics which deals with electric and magnetic fields in space. In fact, by using Maxwell's equations from electromagnetic field theory it can be shown that each of the components of both the electric and magnetic field strengths in a region of space are governed by Eq. (13–2).

Next we consider the heat equation, and show that under appropriate assumptions it serves to describe the temperature distribution in material bodies as a function of position and time. To obtain the one-dimensional version of this equation we consider a slender homogeneous rod, lying along the x-axis, and insulated so that no heat can escape across its longitudinal surface. In addition, we make the simplifying assumption that the temperature in the rod is constant on each cross section perpendicular to the x-axis, and thus that the flow of heat in the rod takes place only in the x-direction.

Now, for empirical reasons it is assumed that the quantity of heat, ΔH, which flows across any cross section of the rod is proportional to the rate of change of temperature u on that cross section. In other words,

$$\Delta H = -k\frac{\partial u}{\partial x}, \quad k > 0, \tag{13–10}$$

where the minus sign is introduced because heat flows in the direction opposite to the positive direction of $\partial u/\partial x$. But it is also known that the amount of heat which accumulates in any portion of the rod is proportional to the product of its mass and the (average) time rate of change of temperature in that mass. Hence

$$\Delta H = cm\frac{\partial u}{\partial t} \tag{13–11}$$

for an appropriate positive constant c, known as the *specific heat* of the material in question.

To obtain the heat equation we now focus our attention on the portion of the rod between the points x and $x + \Delta x$. If ρ denotes the mass of the rod per unit length, then by (13–11) the amount of heat accumulating in this portion of the rod per unit time is

$$\Delta H = c\rho\,\Delta x\frac{\partial u}{\partial t},$$

where $\partial u/\partial t$ is computed at some point between x and $x + \Delta x$. But by (13–10) the amount of heat flowing across the two faces of this portion of the rod is

$$\Delta H = -k\left[-\left.\frac{\partial u}{\partial x}\right|_{x+\Delta x} + \left.\frac{\partial u}{\partial x}\right|_{x}\right],$$

and since these two expressions must be equal, we have

$$\frac{1}{\Delta x}\left[\left.\frac{\partial u}{\partial x}\right|_{x+\Delta x} - \left.\frac{\partial u}{\partial x}\right|_{x}\right] = \frac{c\rho}{k}\frac{\partial u}{\partial t}.$$

Taking the limit as $\Delta x \to 0$, we obtain

$$\frac{\partial^2 u}{\partial x^2} = a^2 \frac{\partial u}{\partial t},$$

where the constant $c\rho/k$ has been replaced by a^2 to emphasize that it is positive.*

An argument similar in almost all respects to the one just given can be used to show that the temperature distribution in a thin rectangular plate, insulated so that no heat flows across its faces, is governed by the *two-dimensional heat equation*

$$\frac{\partial^2 u}{\partial x^2} + \frac{\partial^2 u}{\partial y^2} = a^2 \frac{\partial u}{\partial t}.$$

We leave this derivation as an exercise.

Each of the physical problems described above must also be subjected to certain boundary and initial conditions before the future behavior of the system can be determined. In the case of the one-dimensional wave equation one is interested in finding solutions on a finite interval $[0, L]$ under the assumption that at time $t = 0$ both $u(x, 0)$ and $u_x(x, 0)$ are known functions of x. Physically this corresponds to finding the equation of motion for a vibrating string of length L, given its initial displacement and initial velocity. Furthermore, if A denotes the end points $x = 0$ and $x = L$, then the boundary conditions imposed on the problem are always chosen from among the following:

1. $u(A, t) = 0$;
2. $u_x(A, t) = 0$;
3. $u(A, t) = (1/h)u_x(A, t)$, h a constant.

The first of these conditions simply means that the string is held fixed at the endpoint A, while the second is known as a *free-end* condition. Here the string can move in a vertical direction at A, but is constrained to do so in such a way that it always remains horizontal there (see Fig. 13–6). The third condition says that at $x = A$ the displacement of the string is proportional to its slope. But for small

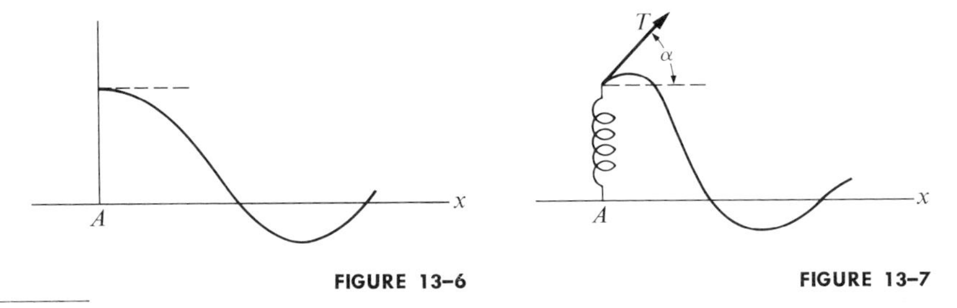

FIGURE 13–6 FIGURE 13–7

* This equation is also known as the *diffusion equation*.

vibrations we know that $u_x(A, t) = \tan \alpha \approx \sin \alpha$, where α is the angle which the string makes with the horizontal at A. Thus $u(A, t) \approx (1/h) \sin \alpha$, and the string behaves as though it were attached to the end of a spring as shown in Fig. 13–7. Indeed, the requirement that such a system be in equilibrium at a given displacement $u(A, t)$ is $T \sin \alpha = ku(A, t)$, where k denotes the spring constant, and (3) now follows by setting $h = k/T$. Finally, if the coupling constant h is very small we approach the free-end condition given in (2).

When solving the one-dimensional heat equation it is customary to start with a given initial temperature distribution $u(x, 0) = f(x)$ along the conducting rod, and then choose the boundary conditions from among the following:

1. $u(A, t) = k$, k a constant;
2. $u_x(A, t) = 0$;
3. $u_x(A, t) = hu(A, t)$, h a constant.

The first of these conditions means that the end of the rod at A is maintained at the constant temperature k, the second that the rod is insulated at A and neither gains nor loses heat at that end. This time the third condition may be read as asserting that the rate at which heat passes through the end at A is proportional to the temperature at A.

EXERCISES

1. Derive the two-dimensional heat equation

$$\frac{\partial^2 u}{\partial x^2} + \frac{\partial^2 u}{\partial y^2} = a^2 \frac{\partial u}{\partial t}$$

under the assumptions given in the text.

2. Show that the equation governing the temperature distribution in a thin homogeneous rod is

$$a^2 \frac{\partial u}{\partial t} = \frac{\partial^2 u}{\partial x^2} + bg(x, t), \quad b \text{ a constant},$$

when heat is being generated in the rod (say by an electric current) at the rate $g(x, t)$ per unit length.

3. Suppose that the assumptions under which the equation of motion of the vibrating string was derived are modified to include a retarding force due to air resistance which is proportional to the velocity of the string. Show that the equation governing the motion then becomes

$$a^2 \frac{\partial^2 u}{\partial x^2} = k \frac{\partial u}{\partial t} + \frac{\partial^2 u}{\partial t^2}.$$

4. Suppose that an external force of magnitude $G(x, t)$ per unit length acts on a vibrating string. (This is the case of "forced vibrations.") Show that the equation governing

the motion is

$$\frac{\partial^2 u}{\partial t^2} = a^2 \frac{\partial^2 u}{\partial x^2} + \frac{1}{\rho} G(x, t),$$

where ρ is the (constant) density of the string.

In the following exercises we sketch a method for solving the one-dimensional wave equation discovered by the 18th-century French mathematician and philosopher Jean d'Alembert and called after him *d'Alembert's solution* of the wave equation.

5. Prove that the function

$$u(x, t) = f(x + at) + g(x - at)$$

is a solution of the one-dimensional wave equation whenever f and g are twice differentiable functions of a single variable.

6. (a) Let $G(p, q)$ and its derivatives G_p, G_q, G_{pp}, G_{pq}, and G_{qq} be continuous throughout the pq-plane. Prove that there exist twice continuously differentiable functions G_1 and G_2 of a single variable such that

$$G(p, q) = G_1(p) + G_2(q)$$

if and only if $G_{pq} \equiv 0$.

(b) Let $F(x, t)$ be a solution of the one-dimensional wave equation, and suppose that F_{xx}, F_{xt}, and F_{tt} are continuous. Set $p = x + at$, $q = x - at$, and, as in Exercise 3(b) of Section 13–2, rewrite $F(s, t)$ in the form $G(p, q)$. Prove that $G_{pq} \equiv 0$, and then use the result in (a) to conclude that *every* twice continuously differentiable solution of the one-dimensional wave equation has the form

$$f(x + at) + g(x - at).$$

7. Let $u(x, t) = f(x + at) + g(x - at)$ be any twice continuously differentiable solution of the one-dimensional wave equation, and note that

$$u(x, 0) = f(x) + g(x),$$
$$u_t(x, 0) = a[f'(x) - g'(x)].$$

Set

$$r(x) = f(x) + g(x),$$
$$s(x) = a[f'(x) - g'(x)],$$

and show that

$$u(x, t) = \frac{1}{2}[r(x + at) + r(x - at)] + \frac{1}{2a}\int_{x-at}^{x+at} s(\varphi)\, d\varphi.$$

(This is d'Alembert's solution of the wave equation.)

8. (a) Show that the physical units of the constant appearing in the equation

$$\frac{\partial^2 u}{\partial x^2} = \frac{1}{a^2}\frac{\partial^2 u}{\partial t^2}$$

are those of velocity; i.e., length/time. [*Hint:* Recall that $a = \sqrt{T/\rho}$, where the units of T and ρ are, respectively, force/length and mass/length, and then use Newton's second law.]

(b) Use the above result together with Exercise 6(b) to show that every twice continuously differentiable solution of the one-dimensional wave equation can be interpreted as the superposition of two standing waves, one of which moves to the right, and one to the left, both with velocity a.

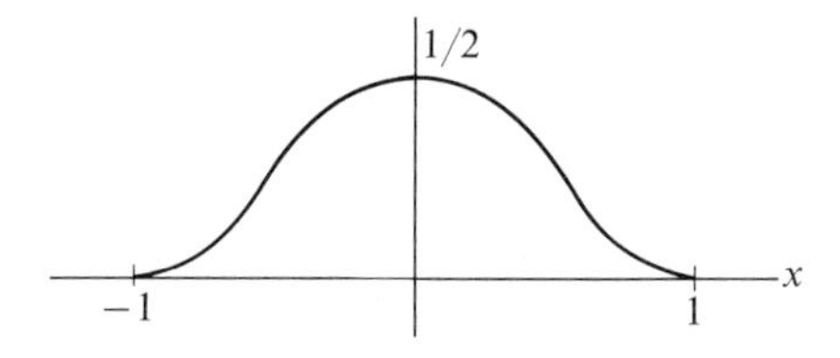

FIGURE 13–8

9. Suppose that at time $t = 0$ a homogeneous string of infinite length is displaced as shown in Fig. 13–8 and then released from rest. Sketch the displacement of the string at time $t = \frac{1}{2}$, $t = 1$, and $t = 2$, under the assumption that $a = 1$. [*Hint:* See Exercise 8(b).]

13–4 SEPARATION OF VARIABLES: THE ONE-DIMENSIONAL WAVE EQUATION

In this section we introduce the method of *separation of variables* for solving boundary-value problems involving partial differential equations. In essence this method consists of finding solutions of the equation which are products of functions *of a single variable*, and then combining these solutions in such a way that the given boundary conditions are satisfied. To illustrate how this is done we consider the one-dimensional wave equation

$$\frac{\partial^2 u}{\partial x^2} = \frac{1}{a^2}\frac{\partial^2 u}{\partial t^2}, \tag{13–12}$$

subject to the boundary conditions

$$\begin{aligned} u(0, t) &= u(\pi, t) = 0, \\ u(x, 0) &= f(x), \\ u_t(x, 0) &= g(x), \end{aligned} \tag{13–13}$$

where f and g are assumed known. Physically this problem consists of finding the equation of motion of an elastic string which is stretched along the x-axis from 0 to π, clamped at its endpoints, given initial position $f(x)$, initial velocity $g(x)$, and then allowed to vibrate freely.

We begin by seeking solutions of (13–12) of the form

$$u(x, t) = X(x)T(t), \tag{13–14}$$

where X and T are, respectively, functions of x and t alone. Furthermore, we demand that these solutions be such that $u(0, t) = u(\pi, t) = 0$ for all $t > 0$ (see 13-13). This, in turn, implies that X must satisfy the endpoint conditions $X(0) = X(\pi) = 0$, else (13-14) would yield $T(t) \equiv 0$, and $u(x, t)$ would then be the trivial solution of (13-12). If we now assume (as we must) that X and T are twice differentiable, then

$$\frac{\partial^2 u}{\partial x^2} = X''T, \qquad \frac{\partial^2 u}{\partial t^2} = XT''.$$

Substituting in (13-12), we obtain

$$X''T = \frac{1}{a^2} XT'',$$

whence

$$\frac{X''}{X} = \frac{1}{a^2}\frac{T''}{T} \tag{13-15}$$

whenever $XT \neq 0$. At this point we make the crucial observation that the left-hand side of this expression is a function of x alone, while the right-hand side involves only t. Thus each is a constant, λ, and (13-15) is equivalent to the pair of *ordinary* linear differential equations

$$\begin{aligned} X'' - \lambda X &= 0, \\ T'' - \lambda a^2 T &= 0, \end{aligned} \tag{13-16}$$

the first of which must satisfy the endpoint conditions $X(0) = X(\pi) = 0$.

To solve these equations we first note that the boundary-value problem

$$X'' - \lambda X = 0, \qquad X(0) = X(\pi) = 0$$

is essentially the one solved as Example 1 in Section 12-6. Thus we know that up to multiplicative constants the only nontrivial solutions of this problem are

$$X_n(x) = \sin nx, \quad n = 1, 2, \ldots, \tag{13-17}$$

corresponding to the eigenvalues $\lambda_n = -n^2$. Moreover, when $\lambda = -n^2$ the general solution of $T'' - \lambda a^2 T = 0$ is

$$T_n(t) = A_n \sin nat + B_n \cos nat, \tag{13-18}$$

where A_n and B_n are arbitrary constants. Hence, forming the product of the functions in (13-17) and (13-18), we see that *each of the functions*

$$u_n(x, t) = \sin nx(A_n \sin nat + B_n \cos nat) \tag{13-19}$$

is a solution of the one-dimensional wave equation which vanishes when $x = 0$ *and* $x = \pi$.

This done, we now attempt to use these functions to construct a solution $u(x, t)$ of (13–12) such that

$$u(x, 0) = f(x), \qquad u_t(x, 0) = g(x). \tag{13–20}$$

In general, of course, no one of the $u_n(x, t)$ by itself will satisfy these conditions. Neither, for that matter, will any *finite* sum of them, and it therefore appears that the only possible choice is an infinite series of the form

$$\begin{aligned} u(x, t) &= \sum_{n=1}^{\infty} u_n(x, t) \\ &= \sum_{n=1}^{\infty} \sin nx(A_n \sin nat + B_n \cos nat) \end{aligned} \tag{13–21}$$

for suitable values of A_n and B_n. Now when $t = 0$, this series reduces to

$$u(x, 0) = \sum_{n=1}^{\infty} B_n \sin nx, \tag{13–22}$$

and we see that the first condition in (13–20) will be satisfied if the B_n are chosen in such a way that (13–22) converges (pointwise) to the function f in the interval $[0, \pi]$. But this is a familiar problem which, as we know, can be solved in either of the following equivalent ways.

I. Let O_f denote the *odd* extension of f to the interval $[-\pi, \pi]$ (see Section 9–5), and let B_n be the nth Fourier coefficient of O_f. Then

$$B_n = \frac{2}{\pi} \int_0^{\pi} f(x) \sin nx \, dx, \tag{13–23}$$

and whenever f is sufficiently well behaved, the series obtained from (13–22) using these values for the coefficients will converge to f everywhere in $[0, \pi]$.

II. The functions $\sin nx$, $n = 1, 2, \ldots$, form a complete set of eigenfunctions for the Sturm-Liouville problem

$$X'' - \lambda x = 0, \qquad X(0) = X(\pi) = 0,$$

and are an orthogonal basis for the space $\mathcal{PC}[0, \pi]$ (Theorem 12–7). Thus the B_n in (13–22) may also be computed as the coefficients of the expansion of f in terms of the eigenfunctions $\sin nx$. The student should realize that the same series is obtained in either case.

The technique for determining the A_n is much the same. We differentiate (13–21) term-by-term with respect to t (under the assumption, of course, that this can be done), and then set $t = 0$. Using the fact that $u_t(x, 0) = g(x)$, we obtain

$$g(x) = \sum_{n=1}^{\infty} naA_n \sin nx, \tag{13–24}$$

and it follows that naA_n must be the nth coefficient of the eigenfunction expansion of g. Thus

$$A_n = \frac{2}{n\pi a} \int_0^\pi g(x) \sin nx \, dx, \tag{13–25}$$

and we are done.

An excellent picture of the physical significance of this solution can be given in the special case where $g(x) \equiv 0$. Then

$$u(x, t) = \sum_{n=1}^{\infty} B_n \sin nx \cos nat,$$

and the components of the motion assume the simple form

$$u_n(x, t) = B_n \sin nx \cos \text{nat}, \quad n = 1, 2, \ldots .$$

In this case the frequency of vibration of each of the components of u is an integral multiple of the *fundamental frequency* $\nu_1 = a/2\pi$ of u_1.* This frequency determines what is known as the *fundamental tone* of the vibration, and its multiples determine the *overtones* or *harmonics*.

The way in which the eigenvalues determine the frequencies of the various modes of vibration of the string becomes strikingly clear if the $u_n(x, t)$ are graphed for various values of t, as in Fig. 13–9. For a rapidly moving string these figures present a reasonably accurate picture of the fundamental vibration and its first three harmonics. The reader should note that the different overtones are characterized by the appearance of *nodes* or stationary points which may be regarded as visual manifestations of the eigenvalues for this problem. In the general case where $g \not\equiv 0$ the situation is much the same, but does not lend itself to such an easy graphical realization.

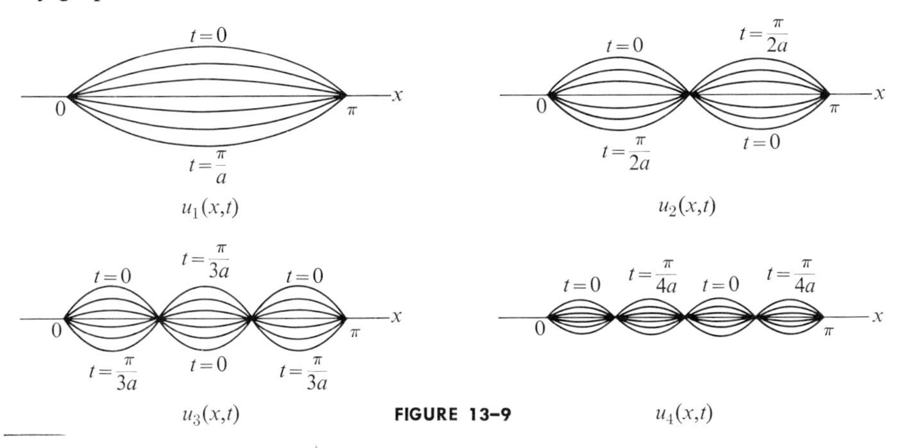

FIGURE 13–9

---

* The *frequency* ν of the wave described by the function $\cos \alpha t$ is, by definition, $\alpha/2\pi$. Physically, ν may be interpreted as the number of waves which pass a given point per unit time.

EXERCISES

In Exercises 1 through 5 find the solution $u(x, t)$ of the one-dimensional wave equation on the interval $[0, L]$ subject to the endpoint conditions $u(0, t) = u(L, t) = 0$ and initial conditions as given.

1. $u(x, 0) = \sin \frac{n\pi x}{L}, \quad u_t(x, 0) \equiv 0$

2. $u(x, 0) \equiv 0, \quad u_t(x, 0) = \sin \frac{n\pi x}{L}$

3. $u(x, 0) = x(L - x), \quad u_t(x, 0) = 0$

4. $u(x, 0) \equiv 0, \quad u_t(x, 0) = \begin{cases} 0 & \text{for } 0 \le x \le L/3 \text{ and } 2L/3 \le x \le L \\ 1 & \text{for } L/3 \le x \le 2L/3 \end{cases}$

5. $u(x, 0) = \sum_{n=1}^{N} A_n \sin \frac{n\pi x}{L}, \quad u_t(x, 0) = \sum_{n=1}^{N} B_n \sin \frac{n\pi x}{L}$

6. If gravitational acceleration is taken into account the motion of a vibrating string is governed by the equation

$$\frac{\partial^2 u}{\partial t^2} = a^2 \frac{\partial^2 u}{\partial x^2} - g.$$

(See Exercise 4 of the preceding section.) Find the time independent solution of this equation, and show how to use it to treat the general case.

7. Find the equation of motion of a vibrating string with fixed ends if it is released from rest in the plucked position shown in Fig. 13–10. Sketch the position of the string at times $L/2a$, L/a, $3L/2a$, $2L/a$.

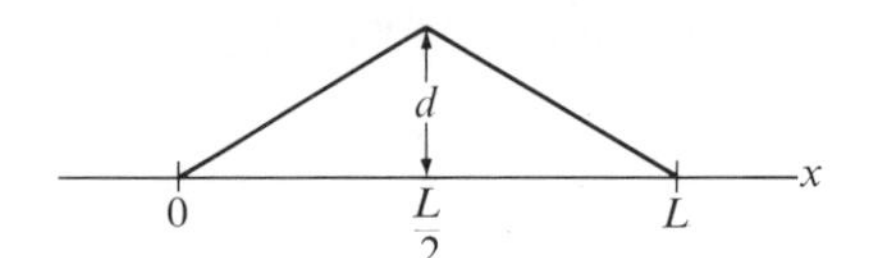

FIGURE 13–10

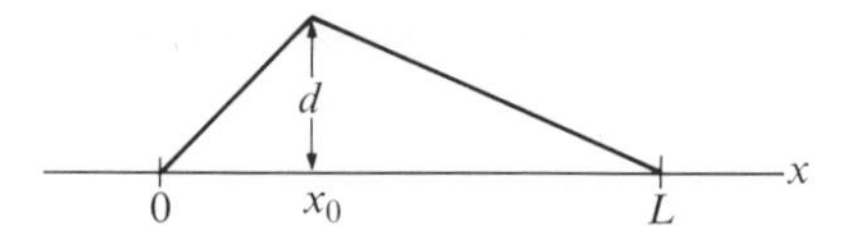

FIGURE 13–11

8. Generalize the results of the preceding exercise to the case where the string is plucked as shown in Fig. 13–11.

9. (a) Solve the one-dimensional wave equation subject to the conditions

$$u(0, t) = u(L, t) \equiv 0,$$
$$u(x, 0) \equiv 0,$$
$$u_t(x, 0) = \begin{cases} 0 & \text{for } \delta < |x - L/2| < L/2, \\ 1/(2\rho\delta) & \text{for } |x - L/2| \le \delta, \end{cases}$$

where ρ is the linear density of the string.

(b) Take the limit of the solution found in (a) as $\delta \to 0$, and interpret the result physically.

10. (a) Show that the solution of the boundary-value problem

$$a^2 \frac{\partial^2 u}{\partial x^2} - \frac{\partial^2 u}{\partial t^2} = F(x, t),$$

$$u(0, t) = u(L, t) = 0,$$

$$u(x, 0) = f(x),$$

$$u_t(x, 0) = g(x),$$

is $v(x, t) + w(x, t)$, where $v(x, t)$ is a solution of

$$a^2 \frac{\partial^2 u}{\partial x^2} - \frac{\partial^2 u}{\partial t^2} = F(x, t)$$

such that $v(0, t) = v(L, t) = 0$, and $w(x, t)$ is the solution of

$$a^2 \frac{\partial^2 u}{\partial x^2} - \frac{\partial^2 u}{\partial t^2} = 0$$

such that $w(0, t) = w(L, t) = 0$, $w(x, 0) = f(x) - v(x, 0)$, $w_t(x, 0) = g(x) - v_t(x, 0)$.

(b) Use the technique suggested in (a) to solve the boundary-value problem

$$\frac{\partial^2 u}{\partial x^2} - \frac{\partial^2 u}{\partial t^2} = \sin x \cos t,$$

$$u(0, t) = u(L, t) = 0,$$

$$u(x, 0) = f(x),$$

$$u_t(x, 0) = g(x).$$

11. Under suitable assumptions it can be shown that the torsional vibrations of a homogeneous metallic shaft of uniform circular cross section are governed by the partial differential equation

$$\frac{\partial^2 \varphi}{\partial t^2} = a^2 \frac{\partial^2 \varphi}{\partial x^2},$$

where a is a positive constant, and $\varphi(x, t)$ is the angular displacement from equilibrium at time t of the cross section of the shaft at x. (See Fig. 13–12.) Assume a shaft of length L with $\varphi(x, 0) = f(x)$, $\varphi_t(x, 0) = g(x)$.

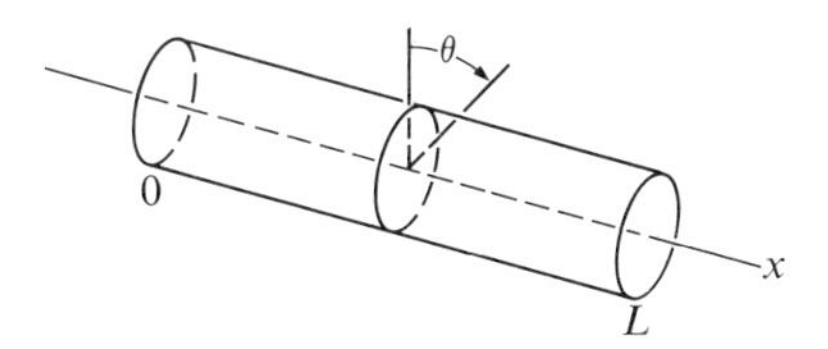

FIGURE 13–12

(a) Find $\varphi(x, t)$ if the ends of the shaft are clamped, i.e., if $\varphi(0, t) = \varphi(L, t) = 0$.

(b) Find $\varphi(x, t)$ if $\varphi_x(0, t) = \varphi_x(L, t) = 0$. (These conditions obtain when the ends of the shaft are free to twist and no torque is transmitted across them.)

(c) Find $\varphi(x, t)$ if the shaft is clamped at $x = 0$ and free at $x = L$.

*13–5 THE WAVE EQUATION; VALIDITY OF THE SOLUTION

In the preceding section we saw that the series

$$u(x, t) = \sum_{n=1}^{\infty} \sin nx(A_n \sin nat + B_n \cos nat) \tag{13–26}$$

provides a *formal* solution of the boundary-value problem

$$\frac{\partial^2 u}{\partial x^2} = \frac{1}{a^2}\frac{\partial^2 u}{\partial t^2},$$
$$u(0, t) = u(\pi, t) = 0,$$

and that the initial conditions $u(x, 0) = f(x)$ and $u_t(x, 0) = g(x)$ can be formally satisfied by choosing A_n and B_n as the coefficients in the orthogonal series expansions

$$\sum_{n=1}^{\infty} B_n \sin nx = f(x) \quad \text{and} \quad \sum_{n=1}^{\infty} naA_n \sin nx = g(x) \tag{13–27}$$

on $[0, \pi]$. To complete the discussion of this problem we now impose conditions on f and g which are sufficient to guarantee the validity of these results. Our choice in this respect will be guided by the various convergence theorems for Fourier series proved in Chapter 10, and we shall assume that the reader is familiar with these results.

The first step in our argument consists of the simple observation that with A_n and B_n as above

$$u_1(x, t) = \sum_{n=1}^{\infty} B_n \sin nx \cos nat \tag{13–28}$$

is the formal solution of the boundary-value probem

$$\frac{\partial^2 u}{\partial x^2} = \frac{1}{a^2}\frac{\partial^2 u}{\partial t^2},$$
$$u(0, t) = u(\pi, t) = 0, \tag{13–29}$$
$$u(x, 0) = f(x), \qquad u_t(x, 0) \equiv 0,$$

and

$$u_2(x, t) = \sum_{n=1}^{\infty} A_n \sin nx \sin nat \tag{13–30}$$

is the formal solution of

$$\frac{\partial^2 u}{\partial x^2} = \frac{1}{a^2}\frac{\partial^2 u}{\partial t^2},$$
$$u(0, t) = u(\pi, t) = 0 \tag{13–31}$$
$$u(x, 0) \equiv 0, \qquad u_t(x, 0) = g(x).$$

Thus (13–26) can be viewed as the sum of the solutions of two simpler problems, and it suffices to direct our attention to them.

Beginning with the first, that is with (13–28) and (13–29), we recall that the series for f given in (13–27) will converge (pointwise) to $f(x)$ for each x in $[0, \pi]$ whenever

(i) f is continuous and f' piecewise continuous on $[0, \pi]$, and

(ii) $f(0) = f(\pi) = 0$.

Moreover, under these hypotheses, this series is actually *uniformly* convergent on $(-\infty, \infty)$ where it represents the *odd* extension of f on $[-\pi, \pi]$ repeated periodically along the entire x-axis. Denoting this extension by F, so that

$$F(x) = \sum_{n=1}^{\infty} B_n \sin nx \tag{13–32}$$

for all x, we now use the identity

$$\sin nx \cos nat = \tfrac{1}{2} \sin n(x - at) + \tfrac{1}{2} \sin n(x + at)$$

to rewrite (13–28) as

$$u_1(x, t) = \tfrac{1}{2}\sum_{n=1}^{\infty} B_n \sin n(x - at) + \tfrac{1}{2}\sum_{n=1}^{\infty} B_n \sin n(x + at). \tag{13–33}$$

(The rearrangement of terms here is justified by the fact that under the hypotheses in effect this series is absolutely convergent.) Thus

$$u_1(x, t) = \tfrac{1}{2}[F(x - at) + F(x + at)],$$

and it now follows that

$$\begin{aligned} u_1(0, t) &= \tfrac{1}{2}[F(-at) + F(at)] \\ &= \tfrac{1}{2}[-F(at) + F(at)] \\ &\equiv 0, \end{aligned}$$

$$\begin{aligned} u_1(\pi, t) &= \tfrac{1}{2}[F(\pi - at) + F(\pi + at)] \\ &= \tfrac{1}{2}[F(-\pi - at) + F(\pi + at)] \\ &= \tfrac{1}{2}[-F(\pi + at) + F(\pi + at)] \\ &\equiv 0, \end{aligned}$$

and

$$\begin{aligned} u_1(x, 0) &= \tfrac{1}{2}[F(x) + F(x)] \\ &= F(x) \\ &\equiv f(x) \end{aligned}$$

on $[0, \pi]$. Thus u_1 satisfies the first three boundary conditions.

To ensure that it satisfies the last as well, we now demand that

(iii) f' be continuous on $[0, \pi]$.*

Under this assumption it is easy to show that F' exists and is continuous for *all* x (see Exercise 3), whence

$$\frac{\partial u_1}{\partial t} = \tfrac{1}{2}[-aF'(x - at) + aF'(x + at)]$$

and

$$\left.\frac{\partial u_1}{\partial t}\right|_{t=0} = \tfrac{1}{2}[-aF'(x) + aF'(x)] \equiv 0,$$

as required.

Finally, to complete the argument and prove that u_1 is also a solution of the one-dimensional wave equation we impose the following additional restrictions on f:

(iv) f'' is continuous on $[0, \pi]$, and

(v) $f''(0) = f''(\pi) = 0$.

For then, arguing as above (Exercise 3 again) we find that F'' is everywhere continuous, and that

$$\frac{\partial^2 u_1}{\partial t^2} = \frac{a^2}{2}[F''(x - at) + F(x + at)],$$

$$\frac{\partial^2 u_1}{\partial x^2} = \tfrac{1}{2}[F''(x - at) + F''(x + at)].$$

Thus

$$\frac{\partial^2 u_1}{\partial x^2} = \frac{1}{a^2}\frac{\partial^2 u_1}{\partial t^2},$$

and the proof is complete.

This done, we turn our attention to the formal solution u_2 of the boundary-value problem described in (13–31), and begin by assuming that

(i) g is continuous and g' piecewise continuous on $[0, \pi]$, and

(ii) $g(0) = g(\pi) = 0$.

Under these conditions we know that the Fourier sine series

$$g(x) = \sum_{n=1}^{\infty} C_n \sin nx$$

---

* In particular, this means that f has a right-hand derivative f'_R at zero, a left-hand derivative f'_L at π, and that

$$\lim_{x\to 0^+} f'(x) = f'_R(0), \qquad \lim_{x\to\pi^-} f'(x) = f'_L(\pi).$$

converges uniformly and absolutely to $g(x)$ for all x in $[0, \pi]$. Since $A_n = C_n/na$, we have

$$u_2(x, t) = \frac{1}{a} \sum_{n=1}^{\infty} \frac{C_n}{n} \sin nx \sin nat. \tag{13-34}$$

Thus if we tentatively allow the necessary term-by-term differentiation, we find that

$$\frac{\partial u_2}{\partial t} = \sum_{n=1}^{\infty} C_n \sin nx \cos nat$$

or, using the trigonometric identity introduced earlier in this section,

$$\frac{\partial u_2}{\partial t} = \tfrac{1}{2} \sum_{n=1}^{\infty} C_n \sin n(x - at) + \tfrac{1}{2} \sum_{n=1}^{\infty} C_n \sin n(x + at). \tag{13-35}$$

But the assumptions imposed on g imply that these two series are uniformly and absolutely convergent on $[0, \pi]$. Hence so too are

$$\sum_{n=1}^{\infty} C_n \sin nx \cos nat \qquad \text{and} \qquad \frac{1}{a} \sum_{n=1}^{\infty} \frac{C_n}{n} \sin nx \sin nat,$$

and the term-by-term differentiation of (13-34) is therefore legitimate. Moreover, if G denotes the *odd periodic* extension of g to the whole real line, then

$$\frac{\partial u_2}{\partial t} = \tfrac{1}{2}[G(x - at) + G(x + at)],$$

and

$$u_2(x, t) = \tfrac{1}{2} \int_0^t G(x - at)\, dt + \tfrac{1}{2} \int_0^t G(x + at)\, dt,$$

or

$$u_2(x, t) = \frac{1}{2a} \int_{x-at}^{x+at} G(s)\, ds. \tag{13-36}$$

From this it follows at once that $u_2(x, 0) \equiv 0$ and that

$$\left.\frac{\partial u_2}{\partial t}\right|_{t=0} \equiv g(x)$$

on $[0, \pi]$. Thus the function defined by (13-36) satisfies the initial conditions prescribed in (13-31). Furthermore, since G is odd and periodic with period 2π, it satisfies the end conditions $u_2(0, t) = u_2(\pi, t) = 0$ as well. Hence, to complete the argument we need only show that this function is also a solution of the one-dimensional wave equation, *and* that its series expansion is

$$\frac{1}{a} \sum_{n=1}^{\infty} \frac{C_n}{n} \sin nx \sin nat,$$

as required by (13-34).

To this end we recall that

$$\begin{aligned}\frac{\partial u_2}{\partial t} &= \tfrac{1}{2}[G(x - at) + G(x + at)] \\ &\equiv \sum_{n=1}^{\infty} C_n \sin nx \cos nat,\end{aligned}$$

and that this series converges uniformly and absolutely to

$$\tfrac{1}{2}[G(x - at) + G(x + at)]$$

on $[0, \pi]$. Hence we can integrate term-by-term to obtain

$$\begin{aligned}u_2(x, t) &= \frac{1}{2a}\int_{x-at}^{x+at} G(s)\, ds \\ &= \sum_{n=1}^{\infty} C_n \sin nx \int_0^t \cos nas\, ds \\ &= \frac{1}{a}\sum_{n=1}^{\infty} \frac{C_n}{n} \sin nx \sin nat,\end{aligned}$$

as required. In addition, this series can be differentiated term-by-term with respect to either variable, from which it follows that

$$\begin{aligned}\frac{\partial u_2}{\partial x} &= \frac{1}{2a}[G(x + at) - G(x - at)], \\ \frac{\partial u_2}{\partial t} &= \tfrac{1}{2}[G(x + at) + G(x - at)].\end{aligned}$$

Finally, to assure the existence of the second partials of u_2 we now assume that

(iii) g' is continuous on $[0, \pi]$.

This implies that G' exists and is continuous on $(-\infty, \infty)$, and that

$$\begin{aligned}\frac{\partial^2 u_2}{\partial x^2} &= \frac{1}{2a}[G'(x + at) - G'(x - at)], \\ \frac{\partial^2 u_2}{\partial t^2} &= \frac{a}{2}[G'(x + at) - G'(x - at)].\end{aligned}$$

Thus

$$\frac{\partial^2 u_2}{\partial x^2} = \frac{1}{a^2}\frac{\partial^2 u_2}{\partial t^2},$$

and we are done.

When taken together these two arguments furnish a proof of the following theorem.

Theorem 13–1. *The series*

$$u(x, t) = \sum_{n=1}^{\infty} \sin nx(A_n \sin nat + B_n \cos nat)$$

with

$$A_n = \frac{2}{n\pi a} \int_0^{\pi} g(x) \sin nx \, dx,$$

$$B_n = \frac{2}{\pi} \int_0^{\pi} f(x) \sin nx \, dx,$$

converges uniformly and absolutely to a solution of the boundary-value problem

$$\frac{\partial^2 u}{\partial x^2} = \frac{1}{a^2} \frac{\partial^2 u}{\partial t^2},$$

$$u(0, t) = u(\pi, t) = 0,$$

$$u(x, 0) = f(x), \qquad u_t(x, 0) = g(x),$$

whenever

(1) *f, f' and f'' are continuous on $[0, \pi]$ with*

$$f(0) = f''(0) = f(\pi) = f''(\pi) = 0,$$

and

(2) *g and g' are continuous on $[0, \pi]$ with*

$$g(0) = g(\pi) = 0.$$

Although Theorem 13–1 is sufficient for most purposes, it is clearly not strong enough to handle every boundary-value problem involving the one-dimensional wave equation that arises in practice. Perhaps the best known example which falls outside the scope of this theorem is that of the "plucked string," and involves finding the equation of motion of a string which is released from rest in the position shown in Fig. 13–13. Here too a formal solution can be obtained by the method of separation of variables (see Exercise 8, Section 13–4), but a much more careful analysis than the one given above is required to treat those points where f' and f'' have discontinuities. We shall omit this discussion since it would carry us further afield than we care to go, and content ourselves with the remark that in most cases the formal solution can in fact be proved valid.*

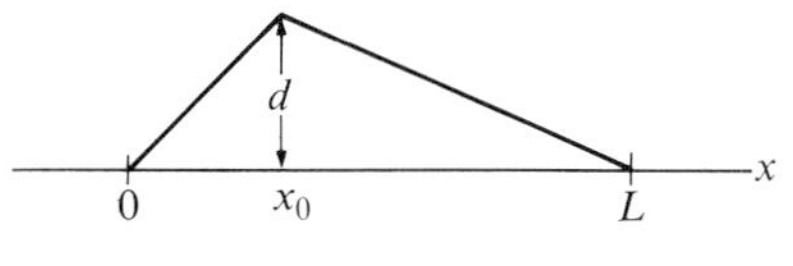

FIGURE 13–13

* For instance, see Chapter 4 of H. Sagan, *Boundary and Eigenvalue Problems in Mathematical Physics*, John Wiley & Sons, Inc., New York, 1961.

EXERCISES

1. Give a rigorous discussion modeled on the one appearing above to establish the validity of the formal solution of the boundary-value problem

$$a^2 \frac{\partial^2 u}{\partial x^2} = \frac{\partial^2 u}{\partial t^2},$$

$$u(0, t) = 0, \qquad u_x(\pi, t) = 0$$

when

(a) $u(x, 0) = f(x), \quad u_t(x, 0) = 0;$ (b) $u(x, 0) = 0, \quad u_t(x, 0) = g(x).$

(Physically this is the problem of a torsionally vibrating bar with one fixed end and one free end. See Exercise 11, Section 13–4.)

2. Repeat Exercise 1 for

$$a^2 \frac{\partial^2 u}{\partial x^2} = \frac{\partial^2 u}{\partial t^2},$$

$$u_x(0, t) = 0, \qquad u_x(\pi, t) = 0$$

when

(a) $u(x, 0) = f(x), \quad u_t(x, 0) = 0;$ (b) $u(x, 0) = 0, \quad u_t(x, 0) = g(x).$

(Physically this is the problem of a torsionally vibrating bar with both ends free.)

3. Let f be continuously differentiable on the interval $[0, a)$, and suppose that

$$\lim_{x \to 0^+} f'(x) = f'_R(0),$$

where $f'_R(0)$ denotes the right-hand derivative of f at zero; i.e.,

$$f'_R(0) = \lim_{h \to 0^+} \frac{f(h) - f(0)}{h}.$$

Assume that $f(0^+) = f(0) = 0$.

(a) Prove that O_f, the *odd* extension of f to $(-a, a)$ is continuously differentiable on $(-a, a)$.

(b) Show that O'_f need not have a continuous derivative at $x = 0$ even when f'' is continuous on $[0, a)$ and $f''_R(0)$ exists. [*Hint:* Consider the function $x + x^2$.]

(c) Prove that O''_f is continuous on $(-a, a)$ whenever f is twice continuously differentiable on $[0, a)$ and $f''_R(0) = 0$.

13–6 THE ONE-DIMENSIONAL HEAT EQUATION

In this section we shall use the method of separation of variables to solve boundary-value problems involving the one-dimensional heat equation

$$\frac{\partial^2 u}{\partial x^2} = a^2 \frac{\partial u}{\partial t}. \tag{13–37}$$

As our first example we consider (13–37) in conjunction with the boundary conditions

$$\begin{aligned} u(0, t) &= 0, \\ u_x(L, t) &= -hu(L, t), \quad h \text{ a constant}, \\ u(x, 0) &= f(x). \end{aligned} \tag{13–38}$$

Physically these equations describe a slender insulated rod of length L, with initial temperature distribution $f(x)$, whose left-hand end is kept at 0°, and which loses (or gains) heat through its right end at a rate proportional to the temperature at that end. The problem is to find the temperature $u(x, t)$ in the rod as a function of position and time.

We begin by seeking solutions of (13–37) of the form

$$u(x, t) = X(x)T(t) \tag{13–39}$$

in which X is twice differentiable, T once differentiable, and each is a function of a single variable. In addition we demand that these solutions satisfy the given boundary conditions $u(0, t) = 0$, $u_x(L, t) = -hu(L, t)$. Substituting (13–39) into (13–37) and dividing by XT we obtain

$$\frac{X''}{X} = a^2 \frac{T'}{T}$$

whenever $XT \neq 0$, and it follows that

$$\begin{aligned} X'' - \lambda X &= 0, \\ T' - \frac{\lambda}{a^2} T &= 0, \end{aligned} \tag{13–40}$$

λ a constant. Moreover, it is easy to see that the only way in which XT can be nontrivial and yet satisfy the required boundary conditions is for X itself to satisfy those conditions. Hence we must begin by solving the Sturm-Liouville system

$$\begin{aligned} X'' - \lambda X &= 0; \\ X(0) &= 0, \\ hX(L) + X'(L) &= 0. \end{aligned}$$

But this problem has already been discussed in Section 12–6, and we know that its eigenfunctions are

$$X_n(x) = \sin \lambda_n x, \quad n = 1, 2, \ldots,$$

where X_n belongs to the eigenvalue $-\lambda_n^2$ obtained by solving a certain transcendental equation (see Eq. 12–27). Furthermore, for each of these values of λ the general solution of $T' + (\lambda_n^2/a^2)T = 0$ is

$$T_n(t) = A_n e^{-(\lambda_n^2/a^2)t}.$$

Hence the only nontrivial solutions of (13–37) which are of the form XT and which satisfy the first two boundary conditions in (13–38) are

$$u_n(x, t) = A_n \sin(\lambda_n x) e^{-(\lambda_n^2/a^2)t}, \quad n = 1, 2, \ldots, \tag{13–41}$$

A_n an arbitrary constant.

It now remains to use these functions to construct a solution $u(x, t)$ of (13–37) which also satisfies the boundary condition $u(x, 0) = f(x)$. To this end we form the series

$$u(x, t) = \sum_{n=1}^{\infty} A_n e^{-(\lambda_n^2/a^2)t} \sin(\lambda_n x)$$

and set $t = 0$ to obtain

$$u(x, 0) = \sum_{n=1}^{\infty} A_n \sin \lambda_n x.$$

From this it follows that the A_n must be chosen as the coefficients of the series expansion of f in terms of the eigenfunctions $\sin \lambda_n x$.* Thus

$$A_n = \frac{\int_0^L f(x) \sin(\lambda_n x)\, dx}{\int_0^L \sin^2(\lambda_n x)\, dx},$$

and the series

$$u(x, t) = \sum_{n=1}^{\infty} \frac{\int_0^L f(x) \sin(\lambda_n x)\, dx}{\int_0^L \sin^2(\lambda_n x)\, dx} e^{-(\lambda_n^2/a^2)t} \sin(\lambda_n x), \tag{13–42}$$

which converges in the mean in $\mathcal{PC}[0, L]$, is a formal solution of the given problem.

As our second example we solve the one-dimensional heat equation given that

$$\begin{aligned} u(0, t) &= 100, \\ u_x(L, t) &= -hu(L, t), \quad h \text{ a constant}, \\ u(x, 0) &= f(x). \end{aligned} \tag{13–43}$$

Here the first boundary condition is *nonhomogeneous*, and we must therefore begin by finding a *particular* solution $u(x, t) = X(x)T(t)$ of (13–37) such that $u(0, t) = 100$, $u_x(L, t) = -hu(L, t)$. As before, X and T will then be solutions of

$$X'' - \lambda X = 0,$$

$$T' - \frac{\lambda}{a^2} T = 0,$$

* The reader should note that there is no possibility of using a Fourier series here. The expansion *must* be given in terms of the $\sin \lambda_n x$.

and by setting $\lambda = 0$ in these equations we immediately obtain $XT = Ax + B$, A and B constants. The boundary conditions in effect imply that $A = -100h/(1 + hL)$, $B = 100$, and it follows that

$$u_0(x, t) = Kx + 100, \qquad K = -\frac{100h}{1 + hL}$$

is a solution with the desired properties.

This done, we now observe that the sum of u_0 and any solution of (13–37) which satisfies the *homogeneous* boundary conditions $u(0, t) = 0$, $u_x(L, t) = -hu(L, t)$ will in turn satisfy the first two boundary conditions in (13–43). In particular, our earlier results imply that

$$u(x, t) = (Kx + 100) + \sum_{n=1}^{\infty} A_n\, e^{-(\lambda_n^2/a^2)t} \sin(\lambda_n x) \qquad (13\text{–}44)$$

is such a solution, and the problem will be solved as soon as the A_n are chosen so that $u(x, 0) = f(x)$. But when $t = 0$, (13–44) becomes

$$u(x, 0) = (Kx + 100) + \sum_{n=1}^{\infty} A_n \sin \lambda_n x.$$

Thus

$$A_n = \frac{\int_0^L [f(x) - Kx - 100] \sin(\lambda_n x)\, dx}{\int_0^L \sin^2(\lambda_n x)\, dx},$$

and we are done.

In Section 13–9 we shall prove that these series are solutions of the boundary-value problems in question whenever f and f' are piecewise continuous on $[0, L]$, and $f(x) = \frac{1}{2}[f(x^+) + f(x^-)]$ everywhere in the interval. (See Theorem 13–3.)

EXERCISES

In each of the following exercises find the solution $u(x, t)$ of the one-dimensional heat equation which satisfies the given boundary conditions.

1. $u(0, t) = u(L, t) = 0;\ u(x, 0) = \begin{cases} x, & 0 < x \le L/2 \\ L - x, & L/2 \le x < L \end{cases}$
2. $u(0, t) = u(L, t) = 0;\ u(x, 0) = k \sin \dfrac{\pi x}{L}$, k a constant
3. $u(0, t) = u(L, t) = 0;\ u(x, 0) = x(L - x)$
4. $u(0, t) = u(L, t) = 0;\ u(x, 0) = f(x)$
5. $u_x(0, t) = u_x(L, t) = 0;\ u(x, 0) = k \sin \dfrac{\pi x}{L}$, k a constant

6. $u_x(0, t) = u_x(L, t) = 0;\ u(x, 0) = x(L - x^2)$
7. $u_x(0, t) = u_x(L, t) = 0;\ u(x, 0) = f(x)$
8. $u(0, t) = u_x(L, t) = 0;\ u(x, 0) = f(x)$
9. $u(0, t) = 100,\ u(L, t) = 0;\ u(x, 0) = 100 \cos \dfrac{\pi x}{2L}$
10. $u_x(0, t) = u_x(L, t) = k,\quad k$ a constant; $u(x, 0) \equiv 0$
11. The temperature in a slender insulated rod of length L satisfies the endpoint conditions $u(0, t) = 0$, $u(L, t) = 1$, and initial condition $u(x, 0) = \sin \pi x/L$.
    (a) Find the temperature in the rod as a function of position and time.
    (b) What is the *steady-state* temperature in the rod (i.e., the temperature as $t \to \infty$)?
12. Find the steady-state temperature in a slender insulated rod of length L given that

$$u(0, t) = k_0, \qquad u(L, t) = k_1, \quad k_0 \text{ and } k_1 \text{ constants,}$$
$$u(x, 0) = f(x).$$

    (See Exercise 11b.)
13. Suppose that the temperature in the rod described in Exercise 12 is allowed to reach equilibrium and that the temperature at the endpoints 0 and L is then suddenly changed to 0 and 100, respectively. Find the new temperature distribution in the rod as a function of x and t.
14. The equation governing the temperature distribution in a slender insulated rod in which heat is being generated at a constant rate per unit length is

$$a^2 \frac{\partial u}{\partial t} = \frac{\partial^2 u}{\partial x^2} + k,$$

    k a constant. (See Exercise 2, Section 13–3.) Solve this equation given that

$$u(0, t) = u(L, t) = 0; \qquad u(x, 0) = f(x).$$

13–7 THE TWO-DIMENSIONAL HEAT EQUATION; BIORTHOGONAL SERIES

In this section we shall study the flow of heat in a thin rectangular plate R of length L and width M, situated in the xy-plane as shown in Fig. 13–14. Under the assumption that heat is neither gained nor lost across the faces of the plate, the flow is two-dimensional, and is described by the equation

$$\frac{\partial^2 u}{\partial x^2} + \frac{\partial^2 u}{\partial y^2} = a^2 \frac{\partial u}{\partial t}. \tag{13–45}$$

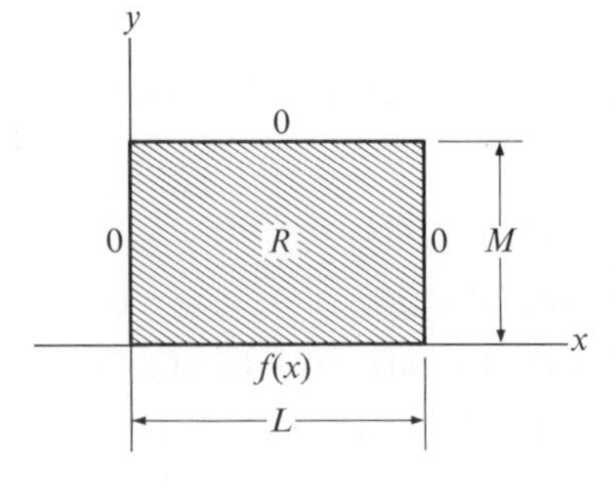

FIGURE 13–14

We now propose to solve this equation in the presence of the following boundary conditions:

$$\begin{cases} u(x, 0, t) = 0, \\ u(x, M, t) = 0; \end{cases} \quad \text{the temperature on the horizontal sides of the plate is held at 0;} \tag{13-46}$$

$$\begin{cases} u_x(0, y, t) = 0, \\ u_x(L, y, t) = 0; \end{cases} \quad \text{the vertical sides are insulated and no heat flows across them;} \tag{13-47}$$

$$u(x, y, 0) = f(x, y); \quad \text{the initial temperature distribution is known.} \tag{13-48}$$

As usual we seek product solutions of the form

$$u(x, y, t) = X(x)Y(y)T(t)$$

which are nonzero in the region under consideration. Substituting this expression in (13–45) and dividing by XYT, we obtain

$$\frac{X''}{X} + \frac{Y''}{Y} = a^2 \frac{T'}{T}. \tag{13-49}$$

Thus

$$\frac{Y''}{Y} = a^2 \frac{T'}{T} - \frac{X''}{X} = \lambda, \quad \lambda \text{ a constant,}$$

and we have

$$Y'' - \lambda Y = 0.$$

The boundary conditions (13–46) imply that $Y(0) = Y(M) = 0$, and lead to the familiar set of eigenvalues $-n^2\pi^2/M^2$, and eigenfunctions

$$Y_n(y) = A_n \sin \frac{n\pi y}{M}, \quad n = 1, 2, \ldots. \tag{13-50}$$

Next we substitute these eigenvalues in (13–49), obtaining

$$\frac{X''}{X} = a^2 \frac{T'}{T} + \frac{n^2\pi^2}{M^2}, \tag{13-51}$$

from which it follows that X''/X is also a constant, μ. Taking (13–47) into consideration we find that we must now solve the Sturm-Liouville system

$$X'' - \mu X = 0; \quad X'(0) = X'(L) = 0.$$

Leaving the details to the reader, we assert that in this case the eigenvalues are

$$\mu_0 = 0, \quad \mu_1 = -\frac{\pi^2}{L^2}, \ldots, \quad \mu_m = -\frac{m^2\pi^2}{L^2}, \ldots,$$

and that the eigenfunctions belonging to the μ_m are

$$X_m(x) = B_m \cos \frac{m\pi x}{L}, \quad m = 0, 1, 2, \ldots \tag{13-52}$$

For these values of X, (13–51) becomes

$$a^2 \frac{T'}{T} = -\pi^2 \left(\frac{m^2}{L^2} + \frac{n^2}{M^2}\right)$$

and

$$T = e^{-(\pi/a)^2[(m/L)^2+(n/M)^2]t}. \tag{13-53}$$

(The constant of integration may be omitted without prejudice at this point.) We now combine (13–50), (13–52), and (13–53) to conclude that each of the functions

$$u_{mn}(x, y, t) = A_{mn} \cos\left(\frac{m\pi x}{L}\right) \sin\left(\frac{n\pi y}{M}\right) e^{-(\pi/a)^2[(m/L)^2+(n/M)^2]t},$$
$$m = 0, 1, 2, \ldots, \quad n = 1, 2, \ldots, \quad A_{mn} \text{ arbitrary}, \tag{13-54}$$

is a solution of the two-dimensional heat equation which also satisfies the boundary conditions given in (13–46) and (13–47).

To complete the solution of the given problem we must now choose the A_{mn} so that the (double) series

$$u(x, y, t) = \sum_{\substack{m=0 \\ n=1}}^{\infty} u_{mn}(x, y, t)$$

satisfies the boundary condition $u(x, y, 0) = f(x, y)$. This means that the A_{mn} must be chosen so that

$$f(x, y) = \sum_{\substack{m=0 \\ n=1}}^{\infty} A_{mn} \cos \frac{m\pi x}{L} \sin \frac{n\pi y}{M},$$

and hence must be the coefficients of the double Fourier series expansion of f in the rectangular region R under consideration (see Section 9–8). Thus

$$A_{0n} = \frac{2}{LM} \int_0^M \int_0^L f(x, y) \sin\left(\frac{n\pi y}{M}\right) dx\, dy,$$

and, when $m \neq 0$,

$$A_{mn} = \frac{4}{LM} \int_0^M \int_0^L f(x, y) \cos\left(\frac{m\pi x}{L}\right) \sin\left(\frac{n\pi y}{M}\right) dx\, dy,$$

and $u(x, y, t)$ has been completely determined.

In this case we shall omit the argument needed to establish the validity of these computations. Suffice it to say that $u(x, y, t)$ as determined above will in fact be a solution of the given problem whenever the function f is "sufficiently" smooth in R. Moreover, it can also be shown that this problem admits only one solution, and hence our discussion is complete.

This example is but one of an almost endless number of three-dimensional boundary-value problems encountered in mathematical physics.* In this case the solution was expressed as a double Fourier series, but from our earlier experience it is clear that one can describe physically meaningful boundary conditions which lead to Sturm-Liouville systems with eigenfunctions other than the sine and cosine functions encountered here. (This would have happened, for instance, had we imposed the boundary conditions

$$u_x(0, y, t) = -hu(0, y, t) \quad \text{and} \quad u_x(L, y, t) = -hu(L, y, t)$$

on the vertical edges of the plate.) Such sets of eigenfunctions lead to the notions of *biorthogonal* sets of functions and *biorthogonal* series expansions. The basic theorem concerning biorthogonal sets in rectangular regions was given in Section 9–8, where it was shown that whenever $\{f_m\}$ and $\{g_n\}$ are orthogonal bases in $\mathcal{PC}[a, b]$ and $\mathcal{PC}[c, d]$, respectively, $\{f_m g_n\}$ is a (bi)orthogonal basis in $\mathcal{PC}(R)$, R the rectangle $a \leq x \leq b$, $c \leq y \leq d$. The import of this theorem is now obvious: it allows us to apply the theory of eigenfunction expansions in rectangular regions, and use them to solve three-dimensional boundary-value problems such as the one discussed above.

EXERCISES

1. Verify that the eigenvalues and eigenfunctions for the Sturm-Liouville problem

$$X'' - \mu X = 0, \qquad X'(0) = X'(L) = 0$$

are as given in the text.

2. Solve the boundary-value problem discussed above when $f(x, y) = \sin^2 \pi y/M$.

3. Repeat Exercise 2 with $f(x, y) = y(M - y) \cos \pi x/L$.

4. Solve the two-dimensional heat equation in the rectangular region $0 < x < \pi$, $0 < y < \pi$ given that

$$u(0, y, t) = u(\pi, y, t) = u(x, 0, t) = u(x, \pi, t) = 0,$$
$$u(x, y, 0) = f(x, y).$$

5. (a) Solve the partial differential equation

$$\frac{\partial^2 u}{\partial x^2} + \frac{\partial^2 u}{\partial y^2} + \frac{\partial^2 u}{\partial z^2} = 0$$

---

* The three-dimensional region in question is the semi-infinite slab $0 < x < L$, $0 < y < M$, $0 < t$.

in the region $0 < x < \pi$, $0 < y < \pi$, $0 < z < \pi$ under the boundary conditions $u(x, y, 0) = f(x, y)$, $u \equiv 0$ on all the other sides of the region.

(b) Generalize the result in (a) to the case where u assumes arbitrary boundary values on all six sides of the region.

6. Solve the boundary-value problem

$$\frac{\partial^2 u}{\partial t^2} = a^2\left(\frac{\partial^2 u}{\partial x^2} + \frac{\partial^2 u}{\partial y^2}\right),$$

$$u(0, y, t) = u(\pi, y, t) = 0,$$

$$u(x, 0, t) = u(x, \pi, t) = 0,$$

$$u(x, y, 0) = f(x, y), \; u_t(x, y, 0) = 0.$$

7. Solve the boundary-value problem in Exercise 6 when the last boundary condition is replaced by $u_t(x, y, 0) = g(x, y)$.

*13–8 THE SCHRÖDINGER WAVE EQUATION

The second-order partial differential equation

$$\frac{\partial^2 \psi}{\partial x^2} + \frac{\partial^2 \psi}{\partial y^2} + \frac{\partial^2 \psi}{\partial z^2} - V(x, y, z)\psi = k\frac{\partial \psi}{\partial t}, \tag{13–55}$$

k a constant, is encountered in quantum mechanics as the *Schrödinger wave equation* for a single particle. Its eigenfunctions are among the most interesting in applied mathematics, and their study has stimulated a great deal of modern research in this field. Unfortunately most of these results are beyond the scope of this book, and any attempt to solve (13–55) as it stands would be quite out of place here. Instead we shall limit ourselves to the one-dimensional version of this equation where $V(x, y, z) = c^2x^2$, c a positive constant. Artificial as these restrictions may seem, the problem is actually of considerable physical interest.†

Thus we consider the simplified equation

$$\frac{\partial^2 \psi}{\partial x^2} - c^2x^2\psi = k\frac{\partial \psi}{\partial t}, \tag{13–56}$$

which we propose to solve in the region $-\infty < x < \infty$, $t > 0$, subject to the restriction that the solutions tend to zero as $|x| \to \infty$. Applying the method of separation of variables with $\psi(x, t) = \psi(x)\varphi(t)$, we find that φ and ψ satisfy the equations

$$\frac{d\varphi}{dt} + \frac{\lambda}{k}\varphi = 0, \tag{13–57}$$

and

$$\frac{d^2\psi}{dx^2} + (\lambda - c^2x^2)\psi = 0, \tag{13–58}$$

† For a discussion of the physical significance of (13–55) and the problem we are about to consider, the reader is referred to any standard text on quantum mechanics.

λ a constant, the second of which is subject to the "boundary" condition $\psi(x) \to 0$ as $|x| \to \infty$. (This equation is known as the *amplitude equation* for the particle.)

The very fact that we are now dealing with functions on $(-\infty, \infty)$ suggests that we attempt to solve this problem by using the Hermite polynomials $H_n(x)$ defined in Section 11–6. With this in mind we recall that for each non-negative integer n the function $e^{-x^2/4}H_n(x)$ is a solution of the differential equation

$$y'' + \left(n + \frac{1}{2} - \frac{x^2}{4}\right) y = 0 \tag{13–59}$$

(see Exercise 12, Section 11–6). The strong similarity between this equation and (13–58) leads us to seek a solution of the latter in the form

$$S(x) = e^{-(ax)^2/4}H_n(ax) \tag{13–60}$$

for a suitable constant a. Differentiating (13–60) we obtain

$$S'(x) = e^{-(a^2x^2)/4}[aH_n'(ax) - \tfrac{1}{2}a^2xH_n(ax)],$$

$$S''(x) = a^2e^{-(a^2x^2)/4}\left[H_n''(ax) - (ax)H_n'(ax) - \frac{1}{2}\left(1 - \frac{a^2x^2}{2}\right)H_n(ax)\right].$$

But since H_n is a solution of

$$y'' - xy' + ny = 0,$$

we have

$$H_n''(ax) - (ax)H_n'(ax) + nH_n(ax) \equiv 0,$$

and it follows that

$$\begin{aligned} S''(x) &= a^2e^{-(a^2x^2)/4}\left[-n - \frac{1}{2} + \frac{a^2x^2}{4}\right]H_n(ax) \\ &= a^2\left[-n - \frac{1}{2} + \frac{a^2x^2}{4}\right]S(x). \end{aligned}$$

Thus

$$S''(x) + a^2\left(n + \frac{1}{2} - \frac{a^2x^2}{4}\right)S(x) \equiv 0,$$

and we conclude that $S(x)$ will be a solution of (13–58) if $a = (4c^2)^{1/4} = \sqrt{2c}$. This, in turn, implies that the eigenvalues and eigenfunctions for the problem under consideration are

$$\lambda_n = (2n + 1)c, \quad n = 0, 1, 2, \ldots, \tag{13–61}$$

and

$$\psi_n(x) = e^{-(cx^2)/2}H_n(\sqrt{2c}\,x). \tag{13–62}$$

(It can be shown that up to multiplicative constants these are the *only* eigenfunctions for this problem.) Moreover, with λ_n as above,

$$\varphi_n(t) = e^{-[(2n+1)ct]/k},$$

and we therefore conclude that the functions

$$\psi_n(x, t) = A_n e^{-(cx^2/2)} H_n(\sqrt{2c}\, x) e^{-[(2n+1)ct]/k},$$

$n = 0, 1, 2, \ldots, A_n$ arbitrary, are solutions of (13–56). Finally, as with all of the other problems discussed in this chapter, these functions can be used to construct series solutions of (13–56) which satisfy initial conditions of the type $\psi(x, 0) = f(x)$ provided f is a reasonably well behaved function. We leave the details to the reader.

*13–9 THE HEAT EQUATION; VALIDITY OF THE SOLUTION

In order to establish the validity of the formal computations in Section 13–6 we must introduce the notion of a *uniformly bounded, monotonic nonincreasing* sequence of functions $\{F_n(t)\}$ on an interval I. The relevant definitions are as follows: $\{G_n(t)\}$ is said to be

(i) *monotonic nonincreasing* if $G_n(t) \geq G_{n+1}(t)$ for all t in I and all n;

(ii) *uniformly bounded* if there exists a real number M such that $|G_n(t)| \leq M$ for all t in I and all n. (The word "uniform" is used here because M simultaneously bounds *all* of the G_n.)

Perhaps the simplest nontrivial example of a sequence with these properties is $\{t^n\}$, $n = 1, 2, \ldots,$ on the unit interval $[0, 1]$. In this case monotonicity is obvious, and uniform boundedness follows from the inequality $|t^n| \leq 1$ whenever $|t| \leq 1$. This last inequality actually shows that $\{t^n\}$ is uniformly bounded on the larger interval $[-1, 1]$, though it is no longer monotonic there since the t^n alternate in sign to the left of the origin. For our purposes a more pertinent example is furnished by the sequence with

$$G_n(t) = e^{-(\lambda_n^2/a^2)t},$$

λ_n as in Section 13–6. Indeed, since $\lambda_1 < \lambda_2 < \cdots$, $G_n(t) \geq G_{n+1}(t)$ for all $t > 0$, and $\{G_n(t)\}$ is monotonic nonincreasing. In addition, $|G_n(t)| \leq e^0 = 1$ for all $t \geq 0$, and hence the sequence is uniformly bounded on the semi-infinite interval $[0, \infty)$.

With this terminology in force, we now prove

Theorem 13–2. *Let $\sum_{n=1}^{\infty} F_n(x)$ be a uniformly convergent series of continuous functions on the interval $a \leq x \leq b$, and let $\{G_n(t)\}$ be a uniformly bounded monotonic nonincreasing sequence of continuous functions on*

$c \leq t \leq d$. Then the series

$$\sum_{n=1}^{\infty} F_n(x)G_n(t) \tag{13-63}$$

is uniformly convergent in the rectangle $a \leq x \leq b$, $c \leq t \leq d$.

Proof. The proof is based upon the fact that an infinite series is uniformly convergent *if and only if* its associated sequence of partial sums is uniformly a Cauchy sequence (see Appendix I). Thus if

$$s_k(x, t) = \sum_{n=1}^{k} F_n(x)G_n(t)$$

denotes the kth partial sum of (13–63), we will be done if we can show that for each $\epsilon > 0$ there exists an integer K such that whenever $k > K$, $p > 0$,

$$|s_{k+p}(x, t) - s_k(x, t)| < \epsilon \tag{13-64}$$

for all x and t in the prescribed rectangle.

To this end let

$$\sigma_k(x) = \sum_{n=1}^{k} F_n(x)$$

be the kth partial sum of the uniformly convergent series $\sum_{n=1}^{\infty} F_n(x)$. Then $F_k(x) = \sigma_k(x) - \sigma_{k-1}(x)$ for all $k \geq 1$, and we have

$$\begin{aligned}
s_{k+p} - s_k &= F_{k+p}G_{k+p} + \cdots + F_{k+1}G_{k+1} \\
&= (\sigma_{k+p} - \sigma_{k+p-1})G_{k+p} + \cdots \\
&\quad + (\sigma_{k+2} - \sigma_{k+1})G_{k+2} + (\sigma_{k+1} - \sigma_k)G_{k+1} \\
&= [(\sigma_{k+p} - \sigma_k) - (\sigma_{k+p-1} - \sigma_k)]G_{k+p} + \cdots \\
&\quad + [(\sigma_{k+2} - \sigma_k) - (\sigma_{k+1} - \sigma_k)]G_{k+2} + (\sigma_{k+1} - \sigma_k)G_{k+1} \\
&= \sum_{n=1}^{p} (\sigma_{k+n} - \sigma_k)G_{k+n} - \sum_{n=1}^{p-1} (\sigma_{k+n} - \sigma_k)G_{k+n+1} \\
&= \sum_{n=1}^{p-1} (\sigma_{k+n} - \sigma_k)(G_{k+n} - G_{k+n+1}) + (\sigma_{k+p} - \sigma_k)G_{k+p}.^{*}
\end{aligned}$$

---

* This result is a modified form of the "summation by parts" identity

$$a_{k+1}b_{k+1} - a_1b_1 = \sum_{n=2}^{k+1} a_n(b_n - b_{n-1}) + \sum_{n=1}^{k} b_n(a_{n+1} - a_n),$$

so called because of its similarity with the formula for integration by parts.

Thus

$$|s_{k+p} - s_k| \le \sum_{n=1}^{p-1} |\sigma_{k+n} - \sigma_k|\,|G_{k+n} - G_{k+n+1}| + |\sigma_{k+p} - \sigma_k|\,|G_{k+p}|. \tag{13-65}$$

Now let $\epsilon > 0$ be given, and let M be such that $|G_n(t)| \le M$ for all n and all t in the interval $[c, d]$. Then since $\sum_{n=1}^{\infty} F_n(x)$ is *uniformly* convergent on $[a, b]$ we can find an integer K such that

$$|\sigma_{k+n} - \sigma_k| < \frac{\epsilon}{3M}$$

for all $k > K$, $n > 0$, and *all* x in $[a, b]$. Substituting this inequality in (13–65) we obtain

$$|s_{k+p} - s_k| < \frac{\epsilon}{3M}\left[\sum_{n=1}^{p-1} |G_{k+n} - G_{k+n+1}| + |G_{k+p}|\right]. \tag{13-66}$$

But since $\{G_n(t)\}$ is monotonic nonincreasing, $G_{k+n}(t) \ge G_{k+n+1}(t)$ for all t, and

$$\begin{aligned}\sum_{n=1}^{p-1} |G_{k+n} - G_{k+n+1}| &= \sum_{n=1}^{p-1} (G_{k+n} - G_{k+n+1}) \\ &= (G_{k+1} - G_{k+2}) + (G_{k+2} - G_{k+3}) + \cdots \\ &\quad + (G_{k+p-1} - G_{k+p}) \\ &= G_{k+1} - G_{k+p}.\end{aligned}$$

Hence

$$\sum_{n=1}^{p-1} |G_{k+n} - G_{k+n+1}| + |G_{k+p}| \le 2M + M = 3M,$$

and (13–66) becomes

$$|s_{k+p} - s_p| < \frac{\epsilon}{3M}(3M) = \epsilon,$$

which is precisely what had to be shown. ▮

With this result in hand we are ready to investigate the formal solution

$$u(x, t) = \sum_{n=1}^{\infty} A_n \sin(\lambda_n x) e^{-(\lambda_n^2/a^2)t} \tag{13-67}$$

of the boundary-value problem

$$\begin{gathered}\frac{\partial^2 u}{\partial x^2} = a^2 \frac{\partial u}{\partial t}, \\ u(0, t) = 0, \quad u_x(L, t) = -hu(L, t), \quad t > 0, \\ u(x, 0) = f(x), \quad 0 < x < L,\end{gathered} \tag{13-68}$$

where the A_n are determined from the orthogonal series expansion

$$f(x) = \sum_{n=1}^{\infty} A_n \sin \lambda_n x. \tag{13-69}$$

In this case the restrictions needed to guarantee the validity of (13–67) are

(i) f and f' piecewise continuous on $[0, L]$ and

(ii) $f(x) = \frac{1}{2}[f(x^+) + f(x^-)]$ for all x in $(0, L)$,

and we now assume that they are in force. The first step in the proof consists of an argument to show that (13–67) is uniformly and absolutely convergent in the region $0 \leq x \leq L$, $t > 0$. We use the Weierstrass M-test (Appendix I), as follows.

From the general theory of Sturm-Liouville series we know that (13–69) converges (pointwise) to $f(x)$ for all x in $(0, L)$. Moreover, by Theorem 8–3 the series

$$\sum_{n=1}^{\infty} |A_n|^2$$

is convergent. Hence the sequence $\{|A_n|\}$ is bounded, and there exists a positive constant K such that

$$|A_n| \leq K$$

for all n. From this it follows that for any $t_0 > 0$

$$|A_n \sin(\lambda_n x) e^{-(\lambda_n^2/a^2)t_0}| \leq K e^{-(\lambda_n^2/a^2)t_0},$$

and the M-test will apply as soon as we have shown that

$$\sum_{n=1}^{\infty} K e^{-(\lambda_n^2/a^2)t_0} \tag{13-70}$$

converges. To this end we note that

$$\frac{e^{-(\lambda_{n+1}^2/a^2)t_0}}{e^{-(\lambda_n^2/a^2)t_0}} = e^{-(t_0/a^2)(\lambda_{n+1}-\lambda_n)(\lambda_{n+1}+\lambda_n)}.$$

But from the way in which the λ_n were determined in the preceding chapter it is clear that $\lambda_{n+1} - \lambda_n$ tends to a *positive* constant value as $n \to \infty$. Hence

$$\lim_{n\to\infty} \frac{e^{-(\lambda_{n+1}^2/a^2)t_0}}{e^{-(\lambda_n^2/a^2)t_0}} = 0,$$

and the ratio test implies that (13–70) converges.

We now differentiate (13–67) termwise with respect to t to obtain

$$\frac{\partial u}{\partial t} = -\frac{1}{a^2}\sum_{n=1}^{\infty} A_n\lambda_n^2 \sin(\lambda_n x)e^{-(\lambda_n^2/a^2)t}, \tag{13–71}$$

and repeat the above argument using the inequality

$$\left|-\frac{1}{a^2}A_n\lambda_n^2 \sin(\lambda_n x)e^{-(\lambda_n^2/a^2)t_0}\right| \le \frac{K}{a^2}\lambda_n^2 e^{-(\lambda_n^2/a^2)t_0}$$

to deduce that (13–71) is uniformly and absolutely convergent in the region $0 \le x \le L$, $t \ge t_0 > 0$. The same can also be said for the series

$$\frac{\partial^2 u}{\partial x^2} = -\sum_{n=1}^{\infty} A_n\lambda_n^2 \sin(\lambda_n x)e^{-(\lambda_n^2/a^2)t} \tag{13–72}$$

obtained by differentiating (13–67) twice with respect to x. Finally, (13–71) and (13–72) imply that

$$\frac{\partial^2 u}{\partial x^2} = a^2\frac{\partial u}{\partial t},$$

and we have therefore proved that under the hypotheses imposed above the function

$$u(x, t) = \sum_{n=1}^{\infty} A_n \sin(\lambda_n x)e^{-(\lambda_n^2/a^2)t}$$

is a solution of the one-dimensional heat equation in the region $0 < x < L$, $t > 0$.*

This brings us to the boundary conditions, the first two of which can be dispatched with ease, as follows. By the argument just given (13–67) is uniformly convergent in every (closed) region of the form $t \ge t_0 > 0$, $0 \le x \le L$, and hence represents a *continuous* function there (Theorem I–21, Appendix I). But when $x = 0$, (13–67) reduces to zero, and it follows that

$$u(0, t) = 0$$

for *all* $t > 0$. Moreover, since

$$\frac{\partial u}{\partial x} = \sum_{n=1}^{\infty} A_n\lambda_n \cos(\lambda_n x)e^{-(\lambda_n^2/a^2)t}$$

---

* At this point the reader would do well to recall the remarks made in Section 13–2 concerning solutions of boundary-value problems, and, in particular, that they need not satisfy the differential equation on the boundary of the region in question.

for $0 \leq x \leq L$, $t \geq t_0 > 0$,

$$u_x(L, t) = \sum_{n=1}^{\infty} A_n[\lambda_n \cos(\lambda_n L)]e^{-(\lambda_n^2/a^2)t}$$

and

$$-hu(L, t) = \sum_{n=1}^{\infty} A_n[-h \sin(\lambda_n L)]e^{-(\lambda_n^2/a^2)t}.$$

But

$$h \sin(\lambda_n L) + \lambda_n \cos(\lambda_n L) = 0$$

for all n (see Eq. 12–27), whence

$$u_x(L, t) = -hu(L, t)$$

for all $t > 0$, as required.

This brings us to the boundary condition

$$u(x, 0) = f(x)$$

and the only troublesome step in the proof. At first sight it might seem that we need only set $t = 0$ in (13–67) and watch the series reduce to (13–69) to complete the argument. Unfortunately this will not do. The error in such a naive argument arises from the fact that we are not in a position to assert the continuity of $u(x, t)$ for $t \geq 0$, and thus cannot guarantee that $u(x_0, t)$ approaches $f(x_0)$ as (x_0, t) approaches $(x_0, 0)$. To establish this fact we use Theorem 13–2, as follows.

The sequence

$$\{e^{-(\lambda_n^2/a^2)t}\}$$

is monotonic nonincreasing and uniformly bounded on $[0, \infty)$. Moreover, for each x_0 in $(0, L)$

$$\sum_{n=1}^{\infty} A_n \sin(\lambda_n x_0)$$

is a convergent series of constants, and hence is uniformly convergent. We now apply Theorem 13–2 to deduce that

$$u(x_0, t) = \sum_{n=1}^{\infty} A_n \sin(\lambda_n x_0)e^{-(\lambda_n^2/a^2)t}$$

is uniformly convergent for all $t \geq 0$, and thus is a *continuous* function of t. It now follows that

$$\lim_{t \to 0^+} u(x_0, t) = \sum_{n=1}^{\infty} A_n \sin(\lambda_n x_0) = f(x_0),$$

and we have proved the following theorem.

Theorem 13–3. *The series*

$$u(x, t) = \sum_{n=1}^{\infty} A_n \sin(\lambda_n x) e^{-(\lambda_n^2/a^2)t}$$

with

$$A_n = \frac{\int_0^L f(x) \sin(\lambda_n x)\, dx}{\int_0^L \sin^2(\lambda_n x)\, dx}$$

converges uniformly and absolutely in every region of the form $0 \le x \le L$, $t \ge t_0 > 0$, *and is a solution of the boundary-value problem*

$$\frac{\partial^2 u}{\partial x^2} = a^2 \frac{\partial u}{\partial t},$$

$$u(0, t) = 0, \quad u_x(L, t) = -hu(L, t), \quad t \ge 0,$$
$$u(x, 0) = f(x), \quad 0 < x < L,$$

whenever f and f' are piecewise continuous and

$$f(x) = \frac{f(x^+) + f(x^-)}{2}$$

everywhere in $[0, L]$.

Remark. The reader should appreciate that under the hypotheses imposed here this theorem is valid *only* if the last boundary condition is interpreted to mean

$$\lim_{t \to 0^+} u(x_0, t) = f(x_0)$$

for all x_0 in $(0, L)$. However, when f is continuous on $[0, L]$ and $f(0) = f(L)$, the series

$$f(x) = \sum_{n=1}^{\infty} A_n \sin \lambda_n x$$

is *uniformly* and absolutely convergent there, and

$$u(x, t) = \sum_{n=1}^{\infty} A_n \sin(\lambda_n x) e^{-(\lambda_n^2/a^2)t}$$

converges uniformly and absolutely for $0 \le x \le L$, $t \ge 0$. In this case $u(x, t)$ is continuous for $0 \le x \le L$, $t \ge 0$, and the last boundary condition is satisfied in the stronger sense

$$\lim_{\substack{t \to x_0 \\ t \to 0^+}} u(x, t) = f(x_0)$$

for all x_0 in $(0, L)$.

EXERCISES

1. Give a rigorous discussion to establish the validity of the formal solution of the boundary-value problem

$$\frac{\partial^2 u}{\partial x^2} = a^2 \frac{\partial u}{\partial t},$$
$$u(0, t) = u(L, t) = 0, \quad t > 0,$$
$$u(x, 0) = f(x), \quad 0 < x < L.$$

2. Repeat Exercise 1 for

$$\frac{\partial^2 u}{\partial x^2} = a^2 \frac{\partial u}{\partial t},$$
$$u_x(0, t) = u_x(L, t) = 0, \quad t > 0,$$
$$u(x, 0) = f(x), \quad 0 < x < L.$$

3. Prove that Theorem 13–2 remains valid if the hypothesis "nonincreasing" is replaced by "nondecreasing."

14
boundary-value problems for laplace's equation

14–1 INTRODUCTION

In this chapter we restrict our attention to boundary-value problems involving Laplace's equation

$$\frac{\partial^2 u}{\partial x^2} + \frac{\partial^2 u}{\partial y^2} + \frac{\partial^2 u}{\partial z^2} = 0 \tag{14–1}$$

in rectangular, circular, and spherical regions. The student will observe that (14–1) can be viewed as the special case of the heat equation in which $\partial u/\partial t = 0$, and with this interpretation in effect its solutions are called *steady-state* solutions since they describe temperature distributions which are independent of time. Other interpretations of (14–1) and its solutions are given in various exercises throughout this chapter.

Before Eq. (14–1) can be solved in a circular or spherical region it must be transformed to polar or spherical coordinates. The computations involved in making these changes are somewhat lengthy, and so to avoid interrupting our later work we dispose of them here and now.

The polar coordinate form of Laplace's equation is derived by introducing the change of variables

$$x = r\cos\theta, \qquad y = r\sin\theta, \tag{14–2}$$

in

$$\frac{\partial^2 u}{\partial x^2} + \frac{\partial^2 u}{\partial y^2} = 0,$$

as follows. From (14–2) and the chain rule for differentiation we obtain

$$\begin{aligned} \frac{\partial u}{\partial r} &= \cos\theta\frac{\partial u}{\partial x} + \sin\theta\frac{\partial u}{\partial y}, \\ \frac{\partial u}{\partial \theta} &= -r\sin\theta\frac{\partial u}{\partial x} + r\cos\theta\frac{\partial u}{\partial y}, \end{aligned} \tag{14–3}$$

and since the determinant of this pair of equations is nonzero everywhere in the punctured plane $r \neq 0$, (14–3) can be solved for $\partial u/\partial x$ and $\partial u/\partial y$ whenever $r \neq 0$. This gives

$$\begin{aligned}\frac{\partial u}{\partial x} &= \cos\theta \frac{\partial u}{\partial r} - \frac{1}{r}\sin\theta \frac{\partial u}{\partial \theta},\\ \frac{\partial u}{\partial y} &= \sin\theta \frac{\partial u}{\partial r} + \frac{1}{r}\cos\theta \frac{\partial u}{\partial \theta}.\end{aligned} \tag{14–4}$$

But these formulas are valid for *any* differentiable function $u = u(x, y)$, and hence can be applied to $\partial u/\partial x$ and $\partial u/\partial y$ themselves to obtain

$$\begin{aligned}\frac{\partial^2 u}{\partial x^2} &= \cos\theta \frac{\partial}{\partial r}\left(\frac{\partial u}{\partial x}\right) - \frac{1}{r}\sin\theta \frac{\partial}{\partial \theta}\left(\frac{\partial u}{\partial x}\right),\\ \frac{\partial^2 u}{\partial y^2} &= \sin\theta \frac{\partial}{\partial r}\left(\frac{\partial u}{\partial y}\right) + \frac{1}{r}\cos\theta \frac{\partial}{\partial \theta}\left(\frac{\partial u}{\partial y}\right).\end{aligned} \tag{14–5}$$

If the derivatives

$$\frac{\partial}{\partial r}\left(\frac{\partial u}{\partial x}\right), \ldots, \frac{\partial}{\partial \theta}\left(\frac{\partial u}{\partial y}\right)$$

are now computed from (14–4), substituted in (14–5), and the resulting equations added, we find that

$$\frac{\partial^2 u}{\partial x^2} + \frac{\partial^2 u}{\partial y^2} = \frac{\partial^2 u}{\partial r^2} + \frac{1}{r}\frac{\partial u}{\partial r} + \frac{1}{r^2}\frac{\partial^2 u}{\partial \theta^2}$$

(see Exercise 1). Hence *Laplace's equation in polar coordinates* is

$$\frac{\partial^2 u}{\partial r^2} + \frac{1}{r}\frac{\partial u}{\partial r} + \frac{1}{r^2}\frac{\partial^2 u}{\partial \theta^2} = 0. \tag{14–6}$$

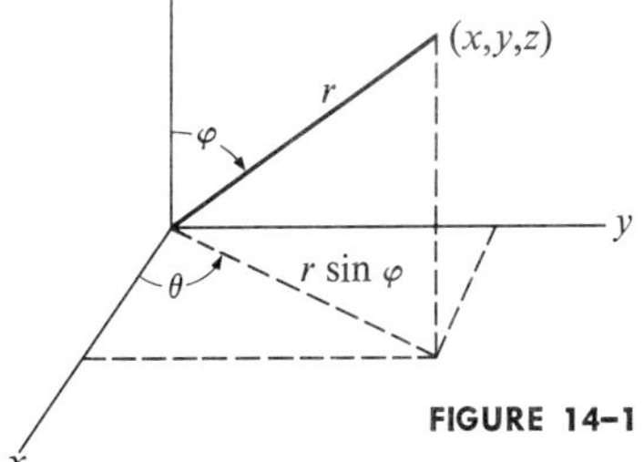

FIGURE 14–1

To convert (14–1) to spherical coordinates (r, φ, θ) we make the change of variables

$$\begin{aligned}x &= r\sin\varphi\cos\theta,\\ y &= r\sin\varphi\sin\theta,\\ z &= r\cos\varphi.\end{aligned}$$

(See Fig. 14–1.) A computation entirely similar to the one given above now yields the equality

$$\frac{\partial^2 u}{\partial x^2} + \frac{\partial^2 u}{\partial y^2} + \frac{\partial^2 u}{\partial z^2} = \frac{1}{r^2}\frac{\partial}{\partial r}\left(r^2\frac{\partial u}{\partial r}\right) + \frac{1}{r^2\sin\varphi}\frac{\partial}{\partial\varphi}\left(\sin\varphi\frac{\partial u}{\partial\varphi}\right) + \frac{1}{r^2\sin^2\varphi}\frac{\partial^2 u}{\partial\theta^2},$$

and it follows that *Laplace's equation in spherical coordinates* is

$$\frac{\partial}{\partial r}\left(r^2\frac{\partial u}{\partial r}\right) + \frac{1}{\sin\varphi}\frac{\partial}{\partial\varphi}\left(\sin\varphi\frac{\partial u}{\partial\varphi}\right) + \frac{1}{\sin^2\varphi}\frac{\partial^2 u}{\partial\theta^2} = 0. \tag{14–7}$$

Incidentally, (14–7) was the equation originally encountered by Laplace in his study of gravitational attraction in three-space, and it was only later that he found the version

$$\frac{\partial^2 u}{\partial x^2} + \frac{\partial^2 u}{\partial y^2} + \frac{\partial^2 u}{\partial z^2} = 0$$

introduced earlier.

EXERCISES

1. Complete the proof of the equality

$$\frac{\partial^2 u}{\partial x^2} + \frac{\partial^2 u}{\partial y^2} = \frac{\partial^2 u}{\partial r^2} + \frac{1}{r}\frac{\partial u}{\partial r} + \frac{1}{r^2}\frac{\partial^2 u}{\partial\theta^2}\cdot$$

2. Use the transformation formulas given in the text to derive Eq. (14–7) from (14–1).

14–2 LAPLACE'S EQUATION IN RECTANGULAR REGIONS

In this section we consider the equation

$$\frac{\partial^2 u}{\partial x^2} + \frac{\partial^2 u}{\partial y^2} = 0 \tag{14–8}$$

in a rectangular region of the form $0 < x < L$, $0 < y < M$, and begin by imposing the boundary conditions

$$\begin{aligned} u(0, y) &= 0, \qquad & u(L, y) &= 0, \\ u(x, M) &= 0, \qquad & u(x, 0) &= f(x). \end{aligned} \tag{14–9}$$

In physical terms this problem requires that we find the steady-state temperature in a thin rectangular plate three sides of which are held at 0°, and the fourth at $f(x)$, under the assumption that no heat is gained or lost across its faces.

We again find the solution by separating variables, and thus set $u(x, y) = XY$, where X and Y are, respectively, functions of x and y alone. This implies that $u_{xx} = X''Y$, $u_{yy} = XY''$, and hence (14–8) becomes

$$\frac{Y''}{Y} = -\frac{X''}{X}$$

at those points where $XY \neq 0$. This equation in turn is equivalent to the pair of ordinary differential equations

$$\begin{aligned} X'' + \lambda X &= 0, \\ Y'' - \lambda Y &= 0, \end{aligned} \tag{14–10}$$

λ a constant, and the given boundary conditions plus the requirement that $XY \neq 0$ imply

$$\begin{aligned} X(0) = X(L) &= 0, \\ Y(M) &= 0. \end{aligned} \tag{14–11}$$

Hence we can again begin with the solutions of the Sturm-Liouville system

$$\begin{aligned} X'' + \lambda X &= 0, \\ X(0) = X(L) &= 0, \end{aligned}$$

that is, with the functions

$$X_n(x) = A_n \sin \frac{n\pi x}{L}, \quad n = 1, 2, \ldots,$$

where X_n belongs to the eigenvalue $\lambda_n = n^2\pi^2/L^2$, and A_n is an arbitrary constant.

For these values of λ the second equation in (14–10) becomes

$$Y'' - \frac{n^2\pi^2}{L^2} Y = 0,$$

and has the general solution

$$Y_n(y) = B_n \sinh \frac{n\pi y}{L} + C_n \cosh \frac{n\pi y}{L}, \tag{14–12}$$

B_n and C_n arbitrary.* Since $Y(M) = 0$,

$$B_n \sinh \frac{n\pi}{L} M + C_n \cosh \frac{n\pi}{L} M = 0,$$

and we can set

$$B_n = -\cosh \frac{n\pi}{L} M, \qquad C_n = \sinh \frac{n\pi}{L} M.$$

We now substitute these values in (14–12) and use the identity

$$\sinh \alpha \cosh \beta - \cosh \alpha \sinh \beta = \sinh (\alpha - \beta),$$

valid for all α and β, to obtain

$$Y_n(y) = \sinh \frac{n\pi}{L} (M - y). \tag{14–13}$$

---

* In this problem it is more convenient to work with hyperbolic functions than with exponentials.

Thus each of the functions

$$u_n(x, y) = A_n \sin\left(\frac{n\pi x}{L}\right) \sinh \frac{n\pi}{L}(M - y), \quad n = 1, 2, \ldots, \tag{14–14}$$

is a solution of the two-dimensional Laplace equation which, in addition, satisfies the boundary conditions

$$u_n(0, y) = u_n(L, y) = u_n(x, M) = 0.$$

To find a solution which also satisfies the remaining boundary condition $u(x, 0) = f(x)$ we form the series

$$\begin{aligned} u(x, y) &= \sum_{n=1}^{\infty} u_n(x, y) \\ &= \sum_{n=1}^{\infty} A_n \sin\left(\frac{n\pi x}{L}\right) \sinh \frac{n\pi}{L}(M - y). \end{aligned}$$

When $y = 0$ this series becomes

$$u(x, 0) = \sum_{n=1}^{\infty} A_n \sinh\left(\frac{n\pi M}{L}\right) \sin \frac{n\pi x}{L},$$

and will represent the function f on the interval $(0, L)$ if the $A_n \sinh(n\pi M/L)$ are the coefficients of the Fourier series expansion of the odd extension of f to $[-L, L]$. Thus

$$A_n \sinh \frac{n\pi M}{L} = \frac{2}{L}\int_0^L f(x) \sin\left(\frac{n\pi x}{L}\right) dx$$

and

$$u(x, y) = \frac{2}{L}\sum_{n=1}^{\infty} \frac{\int_0^L f(x) \sin(n\pi x/L)\, dx}{\sinh(n\pi M/L)} \sin\left(\frac{n\pi x}{L}\right) \sinh \frac{n\pi}{L}(M - y). \tag{14–15}$$

Once again it can be shown that this solution is valid whenever f is sufficiently smooth (see Exercise 10), and that it is the only possible solution of the problem in question.

It is clear that the above argument can be applied to solve any boundary-value problem for Laplace's equation in a rectangular region given that the solution is to vanish on three sides of the rectangle. But if u_1, u_2, u_3, u_4 denote the solutions corresponding to the four possible cases with one nonzero boundary condition, the function

$$u(x, y) = \sum_{k=1}^{4} u_k(x, y)$$

will be a solution assuming nonzero values on *all* four sides of the rectangle, and it follows that we can actually solve *all* problems involving Laplace's equation and boundary conditions of the form

$$u(0, y) = f_1(y), \qquad u(L, y) = f_2(y)$$
$$u(x, 0) = f_3(x), \qquad u(x, M) = f_4(x).$$

EXERCISES

In each of the following exercises find the solution of Laplace's equation in the rectangle $0 < x < L$, $0 < y < M$ which satisfies the given boundary conditions.

1. $u(0, y) = u(L, y) = 0$, $u(x, M) = u(x, 0) = 2x(x - L)$
2. $u(L, y) = u(x, 0) = 0$,
$$u(x, M) = \begin{cases} x, & 0 < x \le L/2 \\ L - x, & L/2 \le x < L \end{cases}$$
$$u(0, y) = \begin{cases} y, & 0 < y \le M/2 \\ M - y, & M/2 \le y < M \end{cases}$$
3. $u(0, y) = u(L, y) = 0$,
$$u(x, M) = u(x, 0) = \begin{cases} 0, & 0 < x \le \dfrac{L}{2} \\ x^2 - \dfrac{3L}{2}x + \dfrac{L^2}{2}, & \dfrac{L}{2} \le x < L \end{cases}$$
4. $u(x, 0) = u(0, y) = 0$, $\quad u(x, M) = x \sin \dfrac{\pi x}{L}$, $\quad u(L, y) = y \sin \dfrac{\pi y}{M}$
5. $u(x, 0) = u(x, M) = x(x - L)$, $\quad u(0, y) = u(L, y) = \sin \dfrac{\pi y}{M}$
6. Find the series solution of the equation $u_{xx} + u_{yy} = 0$ in $0 < x < L$, $0 < y < M$ given that
$$u(0, y) = f_1(y), \qquad u(L, y) = f_2(y),$$
$$u(x, 0) = f_3(x), \qquad u(x, M) = f_4(x).$$
7. Derive the formula for the steady-state temperature in a $\pi \times \pi$ square plate whose faces are insulated so that no heat flows across them, under the assumption that
$$u_x(0, y) = u_x(\pi, y) = u(x, \pi) = 0, \qquad u(x, 0) = f(x).$$
8. Solve $u_{xx} + u_{yy} = 0$ in the region $0 < x < \pi$, $0 < y < \pi$, subject to the boundary conditions
$$u_x(\pi, y) = u_x(0, y) = u_y(x, \pi) = 0, \qquad u(x, 0) = f(x).$$

9. Find the steady-state temperature in the infinite plate shown in Fig. 14–2 if the vertical edges are kept at 0°, the edge on the x-axis at 100°, and if the temperature throughout the plate is bounded.

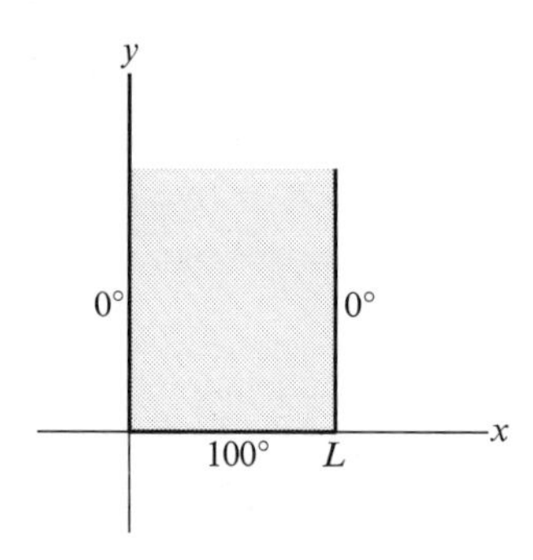

FIGURE 14–2

*10. (This Exercise assumes a knowledge of the material in Section 13–9.) When $M = L = \pi$ the boundary-value problem discussed in this section is

$$\frac{\partial^2 u}{\partial x^2} + \frac{\partial^2 u}{\partial y^2} = 0,$$

$$u(x, \pi) = u(0, y) = u(\pi, y) = 0,$$

$$u(x, 0) = f(x),$$

and the formal solution found may be written

$$u(x, y) = \sum_{n=1}^{\infty} A_n \sin(nx) \frac{\sinh n(\pi - y)}{\sinh n\pi},$$

where $\sum_{n=1}^{\infty} A_n \sin nx$ is the Fourier series expansion of the odd extension of f to the interval $[-\pi, \pi]$.

(a) Show that the sequence of functions

$$\left\{\frac{\sinh n(\pi - y)}{\sinh n\pi}\right\}, \quad n = 1, 2, \ldots,$$

is monotone nondecreasing and uniformly bounded on $0 \leq y \leq \pi$. (In fact,

$$0 \leq \frac{\sinh n(\pi - y)}{\sinh n\pi} \leq 1$$

for all n.)

(b) Assuming that f and f' are piecewise continuous on $0 \leq x \leq \pi$, show that the formal solution given above is uniformly and absolutely convergent in the rectangle $0 \leq x \leq \pi$, $y_0 \leq y \leq \pi$ for any $y_0 > 0$. [*Hint:* Use the methods of Section 13–9, including that of Exercise 3.]

(c) Show that under the assumptions given in (b) the series obtained by twice differentiating the formal solution term-by-term with respect to x and with respect to y are also uniformly and absolutely convergent on $0 \leq x \leq \pi$, $y_0 \leq y \leq \pi$.

(d) Using the results of (b) and (c), prove that the series solution $u(x, t)$ satisfies Laplace's equation in the rectangle $0 < x < \pi$, $0 < y < \pi$, and vanishes when $x = \pi$ and when $y = 0$ and π.

(e) With f and f' as above, prove that $\lim_{y \to 0^+} u(x_0, y) = f(x_0)$ for each x_0 in $(0, \pi)$, and thus show that $u(x, t)$ is a solution of the given boundary-value problem.

(f) Prove that the solution $u(x, y)$ satisfies the boundary condition $u(x, 0) = f(x)$ in the stronger sense

$$\lim_{\substack{x \to x_0 \\ y \to 0^+}} u(x, y) = f(x_0)$$

for all x_0 in $(0, \pi)$ whenever f is *continuous* and f' piecewise continuous on $[0, \pi]$.

14-3 LAPLACE'S EQUATION IN A CIRCULAR REGION; THE POISSON INTEGRAL

As our next example we solve Laplace's equation in a circular region centered at the origin. In this case we must use the polar coordinate form of the equation which, as we have seen, is

$$\frac{\partial^2 u}{\partial r^2} + \frac{1}{r}\frac{\partial u}{\partial r} + \frac{1}{r^2}\frac{\partial^2 u}{\partial \theta^2} = 0. \tag{14-16}$$

Here, in addition to the usual series solutions obtained by the method of separation of variables, there are certain exceptional solutions which depend upon only one of the variables, and we begin by examining them.

First, if u is a function of r alone, (14-16) becomes

$$\frac{d^2 u}{dr^2} + \frac{1}{r}\frac{du}{dr} = 0 \tag{14-17}$$

or

$$\frac{d}{dr}\left(r\frac{du}{dr}\right) = 0,$$

and two integrations yield

$$u = c \ln(kr), \tag{14-18}$$

where c and $k > 0$ are arbitrary constants. In this case the solutions are undefined when $r = 0$, and hence are valid only in regions of the form $r > 0$ where they provide solutions of (14-16) which are constant on circles centered at the origin.

On the other hand, if u is a function of θ alone, we have

$$\frac{d^2 u}{d\theta^2} = 0,$$

and $u = A\theta + B$, where A and B are arbitrary constants. If we now demand that u be single-valued (which is clearly the only type of solution of any physical significance), then $u(\theta + 2\pi) = u(\theta)$, $A = 0$, and it follows that (14–16) has no nonconstant solutions which depend only upon the polar angle θ.

This said, we return to Eq. (14–16), which we now solve under the assumption that u is *single-valued, continuous,* and assumes preassigned values on the boundary of the circle. For convenience we shall work in the unit circle $r \leq 1$, in which case the boundary condition can be written

$$u(1, \theta) = f(\theta), \tag{14–19}$$

where f is a continuous function such that $f(\theta + 2\pi) = f(\theta)$ for all θ.

Setting $u(r, \theta) = R(r)\Theta(\theta)$, (14–16) becomes

$$R''\Theta + \frac{1}{r}R'\Theta + \frac{1}{r^2}R\Theta'' = 0,$$

and we have

$$\frac{r^2R'' + rR'}{R} = -\frac{\Theta''}{\Theta} = \lambda, \quad \lambda \text{ a constant.}$$

This yields the pair of ordinary differential equations

$$\begin{aligned} \Theta'' + \lambda\Theta &= 0, \\ r^2R'' + rR' - \lambda R &= 0, \end{aligned} \tag{14–20}$$

the first of which is subjected to the *periodic* boundary conditions $\Theta(0) = \Theta(2\pi)$. Save for notation, this problem was solved as Example 2 in Section 12–5, where we saw that its eigenvalues and eigenfunctions are

$$\lambda_n = n^2, \quad n = 0, 1, 2, \ldots,$$

and

$$\Theta(\theta) = A_n \cos n\theta + B_n \sin n\theta. \tag{14–21}$$

Moreover, when $\lambda = \lambda_n = n^2$, the second equation in (14–20) becomes

$$r^2R'' + rR' - n^2R = 0, \tag{14–22}$$

which we recognize as an Euler equation whose solution space is spanned by the functions r^n and r^{-n} when $n \neq 0$, and 1 and $\ln r$ when $n = 0$ (see Section 4–8). We reject the solutions r^{-n} and $\ln r$ because of their discontinuity at the origin, and thus are left with

$$R_n(r) = r^n. \tag{14–23}$$

This done, we combine (14–21) and (14–23) to obtain the functions

$$u_n(r, \theta) = r^n(A_n \cos n\theta + B_n \sin n\theta),$$

each of which is continuous, satisfies Laplace's equation, and is periodic in θ with period 2π. To satisfy the boundary condition $u(1, 0) = f(\theta)$, we now form the series

$$u(r, \theta) = \frac{A_0}{2} + \sum_{n=1}^{\infty} r^n(A_n \cos n\theta + B_n \sin n\theta), \tag{14-24}$$

and set $r = 1$. This gives

$$u(1, \theta) = \frac{A_0}{2} + \sum_{n=1}^{\infty} (A_n \cos n\theta + B_n \sin n\theta),$$

and it follows that the constants A_n and B_n must be the Fourier coefficients of f; i.e.,

$$\begin{aligned} A_n &= \frac{1}{\pi} \int_{-\pi}^{\pi} f(\theta) \cos n\theta \, d\theta, \\ B_n &= \frac{1}{\pi} \int_{-\pi}^{\pi} f(\theta) \sin n\theta \, d\theta. \end{aligned} \tag{14-25}$$

Finally, we leave the reader the task of verifying that whenever f is sufficiently smooth, (14–24) and (14–25) provide a continuous, single-valued solution of Laplace's equation in the unit circle which assumes the prescribed values on the boundary of the circle. The argument needed to establish these facts is outlined in Exercise 7 below.

We have already observed that the solution of the above boundary-value problem can be interpreted as the steady-state temperature distribution in a circular region of radius one, given the temperature on the boundary of the region. There are, however, other physical interpretations of this solution (and, by implication, of the problem leading to it) which are of considerable importance in more advanced work. Most of them are derived from an integral form of the above solution, which is obtained as follows. Let s denote the variable of integration in (14–25), and substitute in (14–24). This gives

$$\begin{aligned} u(r, \theta) = \frac{1}{\pi}\Bigg\{\frac{1}{2}\int_{-\pi}^{\pi} f(s)\, ds + \sum_{n=1}^{\infty} r^n \Bigg[\left(\int_{-\pi}^{\pi} f(s) \cos ns \, ds\right) \cos n\theta \\ + \left(\int_{-\pi}^{\pi} f(s) \sin ns \, ds\right) \sin n\theta\Bigg]\Bigg\}. \end{aligned}$$

Interchanging the order of integration and summation (an operation which can be shown to be valid here), this expression may be rewritten

$$\begin{aligned} u(r, \theta) &= \frac{1}{\pi} \int_{-\pi}^{\pi} f(s) \left[\frac{1}{2} + \sum_{n=1}^{\infty} r^n (\cos ns \cos n\theta + \sin ns \sin n\theta)\right] ds \\ &= \frac{1}{\pi} \int_{-\pi}^{\pi} f(s) \left[\frac{1}{2} + \sum_{n=1}^{\infty} r^n \cos n(s - \theta)\right] ds. \end{aligned}$$

We now evaluate the sum

$$\frac{1}{2} + \sum_{n=1}^{\infty} r^n \cos n(s - \theta) \tag{14-26}$$

appearing in this integral. The easiest way to go about this is to introduce complex numbers, and set

$$z = r[\cos(s - \theta) + i \sin(s - \theta)].$$

Then

$$z^n = r^n [\cos n(s - \theta) + i \sin n(s - \theta)],$$

and it follows that (14–26) is the *real part* of

$$\frac{1}{2} + \sum_{n=1}^{\infty} z^n.$$

But when $|z| = r < 1$, as it is here, the geometric series $1 + z + z^2 + \cdots$ converges to $1/(1 - z)$. Hence

$$\begin{aligned}
\frac{1}{2} + \sum_{n=1}^{\infty} z^n &= -\frac{1}{2} + (1 + z + z^2 + \cdots) \\
&= -\frac{1}{2} + \frac{1}{1 - z} \\
&= -\frac{1}{2} + \frac{1}{1 - [r\cos(s - \theta) + ir\sin(s - \theta)]} \\
&= -\frac{1}{2} + \frac{1 - r\cos(s - \theta) + ir\sin(s - \theta)}{1 - 2r\cos(s - \theta) + r^2}.
\end{aligned}$$

Taking the real part of this expression we have

$$\begin{aligned}
\operatorname{Re}\left[-\frac{1}{2} + \frac{1}{1 - z}\right] &= -\frac{1}{2} + \frac{1 - r\cos(s - \theta)}{1 - 2r\cos(s - \theta) + r^2} \\
&= \frac{1 - r^2}{2(1 - 2r\cos(s - \theta) + r^2)},
\end{aligned}$$

and it follows that

$$u(r, \theta) = \frac{1 - r^2}{2\pi} \int_{-\pi}^{\pi} \frac{f(s)}{1 - 2r\cos(s - \theta) + r^2}\, ds. \tag{14-27}$$

This expression is known as the *Poisson Integral form* of the solution of Laplace's equation in the unit circle with boundary condition $u(1, \theta) = f(\theta)$, and, as mentioned above, appears in advanced work in this subject.

EXERCISES

1. Use the technique suggested in Example 3 of Section 4–8 to show that $u = c \ln(kr)$ is the general solution of

$$\frac{d^2u}{dr^2} + \frac{1}{r}\frac{du}{dr} = 0.$$

2. Verify that r^n and r^{-n} are solutions of

$$r^2R'' + rR' - n^2R = 0.$$

3. A thin homogeneous annular disk, having dimensions as shown in Fig. 14–3, is insulated so that no heat can flow across its faces. Find the steady-state temperature throughout the disk if the inner boundary is kept at 0°, while the outer is kept at 100°. At what points in the disk will the temperature be 50°?

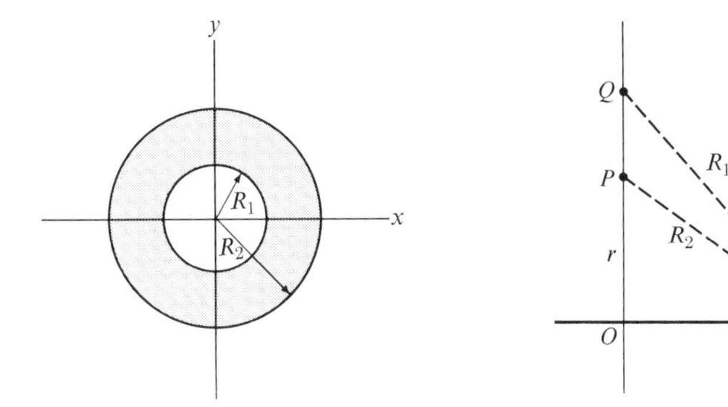

FIGURE 14–3 FIGURE 14–4

4. If a doubly infinite straight wire carries a uniform static electric charge σ per unit length ($\sigma > 0$), then the *potential* at a point P due to the charge on an element of the wire of length dx is defined to be

$$\left(\frac{1}{R_1} - \frac{1}{R_2}\right)\sigma\, dx,$$

where R_1 and R_2 are as shown in Fig. 14–4, and the distance $OQ = 1$. By experiment it can be shown that the potential $V(P)$ at P is the "sum" of the potentials due to the various elements of length, i.e., that

$$V(P) = 2\sigma \int_0^\infty \left(\frac{1}{R_1} - \frac{1}{R_2}\right) dx.$$

Set $OP = r$, and evaluate this integral. Compare the result obtained with the θ-independent solution of the two-dimensional Laplace equation.

5. Refer to Appendix I to obtain sufficient conditions on $f(\theta)$ to permit the necessary interchange of the order of integration and summation in the derivation of the Poisson integral.

6. Verify directly that the Poisson integral satisfies Laplace's equation when $r < 1$. (See Appendix I for the necessary rule for differentiating the integral.)

*7. (This exercise assumes a knowledge of the material in Section 13–9.) Assume that $f(\theta)$ and $f'(\theta)$ are piecewise continuous and periodic on $[0, 2\pi]$ with period 2π, and that

$$f(\theta) = \tfrac{1}{2}[f(\theta^+) + f(\theta^-)]$$

everywhere in this interval. Prove that

$$u(r, \theta) = \frac{A_0}{2} + \sum_{n=1}^{\infty} r^n (A_n \cos n\theta + B_n \sin n\theta),$$

where

$$f(\theta) = \frac{A_0}{2} + \sum_{n=1}^{\infty} (A_n \cos n\theta + B_n \sin n\theta)$$

is a solution of the boundary-value problem discussed in this section. Reason as follows:

(a) Show that $|A_n \cos n\theta + B_n \sin n\theta| \leq \sqrt{A_n^2 + B_n^2} \to 0$ as $n \to \infty$.

(b) Show that the solution series and all relevant series obtained from it by termwise differentiation converge uniformly and absolutely whenever $r \leq r_0 < 1$. Now conclude that the necessary termwise differentiations can be performed, and that the solution series does satisfy Laplace's equation in the region $r < 1$.

(c) Show that $\lim_{r \to 1^-} u(r, \theta_0) = f(\theta_0)$ for all θ_0 in $[0, 2\pi]$.

(d) Prove that if $f(\theta)$ is also continuous then the boundary condition is satisfied in the stronger sense

$$\lim_{\substack{\theta \to \theta_0 \\ r \to 1^-}} u(r, \theta) = f(\theta_0).$$

14–4 LAPLACE'S EQUATION IN A SPHERE; SOLUTIONS INDEPENDENT OF θ

The remaining sections of this chapter will be given over to the study of Laplace's equation in the unit sphere $r \leq 1$ where, as we have seen, the equation can be written

$$\frac{\partial}{\partial r}\left(r^2 \frac{\partial u}{\partial r}\right) + \frac{1}{\sin \varphi} \frac{\partial}{\partial \varphi}\left(\sin \varphi \frac{\partial u}{\partial \varphi}\right) + \frac{1}{\sin^2 \varphi} \frac{\partial^2 u}{\partial \theta^2} = 0. \tag{14–28}$$

As usual we shall restrict our attention to solutions which are *single-valued*, *continuous*, and *bounded* in the region under consideration, and which as a consequence are periodic in θ with 2π as a period. With these assumptions in force it is easy to show that (14–28) has no nonconstant solution involving only one of the variables. Under less restrictive hypotheses, however, solutions of the latter type do exist, but since they can be found without difficulty they have been left to the exercises.

In this section we solve (14–28) under the assumptions that the solution is independent of θ, and that u is known when $r = 1$; i.e., $u(1, \varphi) = f(\varphi)$. Then $\partial^2 u/\partial\theta^2 \equiv 0$, and (14–28) assumes the somewhat simpler form

$$r^2 \frac{\partial^2 u}{\partial r^2} + 2r \frac{\partial u}{\partial r} + \cot\varphi \frac{\partial u}{\partial \varphi} + \frac{\partial^2 u}{\partial \varphi^2} = 0. \tag{14–29}$$

We now apply the method of separation of variables, this time with $u(r, \varphi) = R(r)\Phi(\varphi)$, to obtain the pair of equations

$$r^2 R'' + 2rR' - \lambda R = 0, \quad 0 \le r < 1, \tag{14–30}$$

and

$$\Phi'' + \cot\varphi\, \Phi' + \lambda\Phi = 0, \quad 0 \le \varphi \le \pi, \tag{14–31}$$

λ a constant, which can be solved as follows.

Set $s = \cos\varphi$ in (14–31). Then

$$\frac{d\Phi}{d\varphi} = -\sin\varphi \frac{d\Phi}{ds}, \qquad \frac{d^2\Phi}{d\varphi^2} = \sin^2\varphi \frac{d^2\Phi}{ds^2} - \cos\varphi \frac{d\Phi}{ds},$$

and we have

$$\sin^2\varphi \frac{d^2\Phi}{ds^2} - 2\cos\varphi \frac{d\Phi}{ds} + \lambda\Phi = 0,$$

or

$$(1 - s^2)\frac{d^2\Phi}{ds^2} - 2s\frac{d\Phi}{ds} + \lambda\Phi = 0, \quad -1 \le s \le 1. \tag{14–32}$$

But this is none other than Legendre's equation, and hence the eigenvalues for the problem under consideration are the integers $\lambda_n = n(n + 1)$, $n = 0, 1, 2, \ldots$ (see Example 4, Section 12–6). Since the corresponding eigenfunctions for (14–32) are the Legendre polynomials $P_n(s)$, it follows that the functions

$$\Phi_n(\varphi) = P_n(\cos\varphi), \quad n = 0, 1, 2, \ldots, \tag{14–33}$$

are the eigenfunctions for (14–31) under the given boundary conditions.

Next we observe that (14–30) is an Euler equation. Moreover, when $\lambda = \lambda_n = n(n + 1)$, an easy computation shows that its solution space is spanned by the pair of functions

$$R_n(r) = r^n \quad \text{and} \quad R_n(r) = r^{-(n+1)},$$

and since the second of these solutions must be rejected because of its discontinuity at the origin, we conclude that the relevant solutions of (14–29) are

$$u_n(r, \varphi) = A_n r^n P_n(\cos\varphi), \quad n = 0, 1, 2, \ldots, \tag{14–34}$$

where A_n is an arbitrary constant and P_n is the Legendre polynomial of degree n.*

---

* By definition, $u_0(r, \varphi)$ has the value A_0 at the origin.

Finally, to satisfy the boundary condition $u(1, \varphi) = f(\varphi)$ we form the series

$$u(r, \varphi) = \sum_{n=0}^{\infty} A_n r^n P_n(\cos \varphi), \tag{14–35}$$

and determine the A_n so that

$$f(\varphi) = \sum_{n=0}^{\infty} A_n P_n(\cos \varphi) \tag{14–36}$$

for all φ in the interval $[0, \pi]$. To this end we again set $s = \cos \varphi$ and rewrite (14–36) as

$$f(\cos^{-1} s) = \sum_{n=0}^{\infty} A_n P_n(s),$$

in which form it is obvious that the A_n must be chosen as the coefficients of the Legendre series expansion of $f(\cos^{-1} s)$; i.e.,

$$\begin{aligned} A_n &= \frac{2n+1}{2} \int_{-1}^{1} f(\cos^{-1} s) P_n(s)\, ds \\ &= \frac{2n+1}{2} \int_{0}^{\pi} f(\varphi) P_n(\cos \varphi) \sin \varphi \, d\varphi. \end{aligned} \tag{14–37}$$

Using these coefficients, (14–35) will converge *in the mean* to $f(\varphi)$ when $r = 1$. Moreover, as in all of the preceding examples, it can also be shown that this series is uniformly and absolutely convergent, and twice differentiable term-by-term with respect to each variable whenever f is sufficiently smooth (see Exercise 13). This, of course, implies that (14–35) is a solution of the given boundary-value problem, and since this problem has a unique solution (see Appendix IV), we are done.

Before going on, there is one rather interesting aspect of these results which deserves closer attention. It concerns our apparent success in obtaining an orthogonal series expansion for the function $f(\varphi)$ defined on the surface of the unit sphere, which, after all, is the import of the assertion that

$$\sum_{n=0}^{\infty} A_n P_n(\cos \varphi)$$

converges *in the mean* to f whenever the A_n are given by (14–37). One can, of course, adopt the pedestrian point of view which sees this result as nothing more than a statement of standard facts concerning Legendre series. However, a more liberal and suggestive interpretation is possible by incorporating this result within the context of orthogonal series expansions, as follows.

Let $\mathcal{C}_\varphi(S)$ denote the set of all real-valued continuous functions defined on the surface of the unit sphere and dependent only on the variable φ (φ = colatitude on the sphere). Clearly $\mathcal{C}_\varphi(S)$ is a real vector space under the usual definitions of addition and scalar multiplication, and also becomes a Euclidean space if we set

$$f \cdot g = \iint_S f(\varphi)g(\varphi)\, dS, \tag{14–38}$$

where the integral in question is taken over all of S. Integrals of this sort are known as *surface integrals* and are discussed in Appendix IV. For now we need know only that they enjoy all of the standard properties of linearity (a fact which is implicit in the statement that $f \cdot g$ is an inner product), and that they can be evaluated as ordinary iterated integrals. Indeed, when S is the surface of the unit sphere, it can be shown that

$$\iint_S f(\varphi)g(\varphi)\, dS = \int_0^{2\pi}\int_0^{\pi} f(\varphi)g(\varphi) \sin\varphi\, d\varphi\, d\theta.* \tag{14–39}$$

Thus (14–38) may be rewritten

$$\begin{aligned} f \cdot g &= \int_0^{2\pi}\int_0^{\pi} f(\varphi)g(\varphi) \sin\varphi\, d\varphi\, d\theta \\ &= 2\pi\int_0^{\pi} f(\varphi)g(\varphi) \sin\varphi\, d\varphi, \end{aligned}$$

and can be evaluated in a perfectly routine fashion.

Now let $P_n(\cos\varphi)$ denote the nth Legendre polynomial, viewed as a function in $\mathcal{C}_\varphi(S)$. Then

$$P_m \cdot P_n = 2\pi\int_0^{\pi} P_m(\cos\varphi)P_n(\cos\varphi) \sin\varphi\, d\varphi,$$

and thus if we make the change of variable $s = \cos\varphi$, we obtain

$$\begin{aligned} P_m \cdot P_n &= -2\pi\int_1^{-1} P_m(s)P_n(s)\, ds \\ &= 2\pi\int_{-1}^{1} P_m(s)P_n(s)\, ds \\ &= \begin{cases} 0 & \text{if } m \neq n, \\ 2\pi\left(\dfrac{2}{2n+1}\right) & \text{if } m = n. \end{cases} \end{aligned}$$

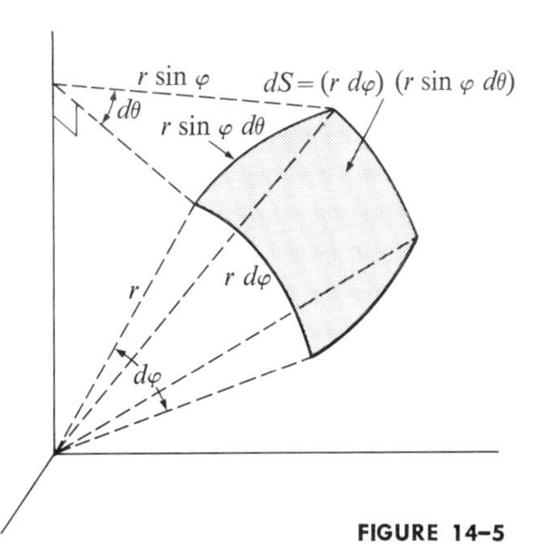

FIGURE 14–5

* This equality can be motivated by observing that the integral on the left is taken over the surface of the unit sphere, described by letting φ vary from 0 to π, θ from 0 to 2π. Since an element of surface dS on a sphere of radius r is given by the expression

$$dS = (r \sin\varphi\, d\theta)(r\, d\varphi) = r^2 \sin\varphi\, d\varphi\, d\theta,$$

(14–39) follows by setting $r = 1$. (See Fig. 14–5.)

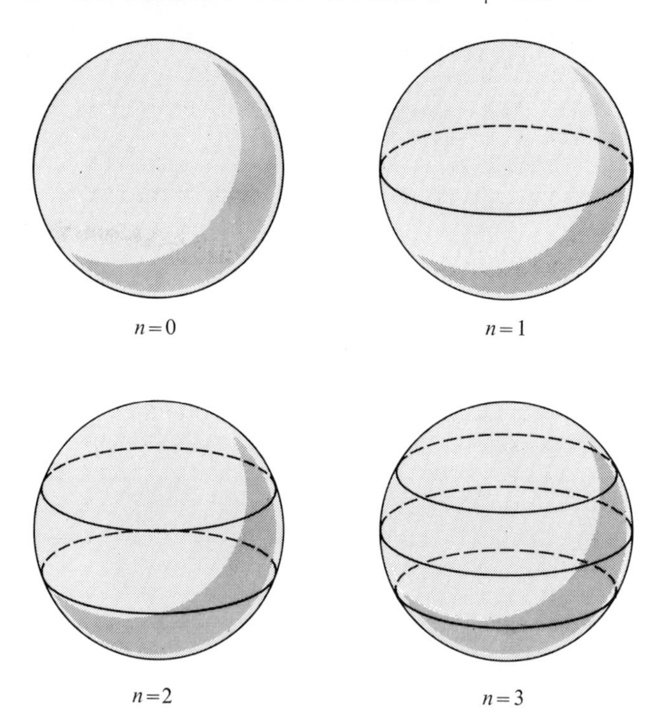

FIGURE 14–6

(See Theorem 11–6.) Thus *the functions* $P_n(\cos\varphi)$, $n = 0, 1, 2, \ldots$, *are mutually orthogonal in* $\mathcal{C}_\varphi(S)$, and the series given in (14–36), with coefficients as in (14–37), is the orthogonal series expansion of f in terms of the $P_n(\cos\varphi)$. From this point of view the statement that this series converges in the mean to f is equivalent to the assertion that the $P_n(\cos\varphi)$ form a *basis* for $\mathcal{C}_\varphi(S)$, a fact which is proved in *exactly* the same way as the corresponding statement for the Legendre polynomials in the space $\mathcal{PC}[-1, 1]$.

The functions $P_n(\cos\varphi)$ appearing in this discussion are known as *surface zonal harmonics*, or simply *zonal harmonics.** The use of the term "harmonic" is explained by the fact that *any* solution of Laplace's equation, and thus, in particular, $P_n(\cos\varphi)$, is said to be a *harmonic function.* The term "zonal" is applied because the curves on the surface S of the unit sphere along which $P_n(\cos\varphi)$ vanishes are parallel to the equator of S and thus separate S into "zones." Indeed, using the fact that the Legendre polynomial of degree n has roots $x_1, \ldots, x_n$ in the open interval $(-1, 1)$ (Theorem 11–7), it follows that the zeros of $P_n(\cos\varphi)$ consist of the n parallels of latitude on the unit sphere given by $\varphi_1 = \cos^{-1} x_1, \ldots, \varphi_n = \cos^{-1} x_n$, $0 < \varphi_i < \pi$. These curves determine the zones alluded to in the name *zonal harmonic* (see Fig. 14–6).

---

* Some authors reserve the latter term for the functions $r^n P_n(\cos\varphi)$.

EXERCISES

1. Verify that the solution space of the equation

$$r^2R'' + 2rR' - \lambda R = 0$$

is spanned by the functions r^n and $r^{-(n+1)}$ when $\lambda = n(n+1)$.

2. Find the general solution of Eq. (14–28) which depends only on r, and which vanishes at infinity.

3. Show that (14–28) has no nonconstant single-valued solution which depends only on θ.

4. Find the general solution of (14–28) which depends only on φ.

5. Use the results of the preceding exercises to show that under the conditions assumed in the text (14–28) has no nonconstant solutions which involve only one of the variables r, φ, θ.

6. A semi-infinite wire coincides with the positive z-axis, and carries a static electric charge σ per unit length ($\sigma > 0$), while a similar wire, *oppositely* charged, coincides with the negative z-axis. Given that the potential at a point P in space due to the charge element $\sigma\, dz$ is $\sigma\, dz/R$ (see Fig. 14–7), perform the appropriate integration, and find the potential at P due to the entire (doubly infinite) wire. Compare your answer with the result obtained in Exercise 4.

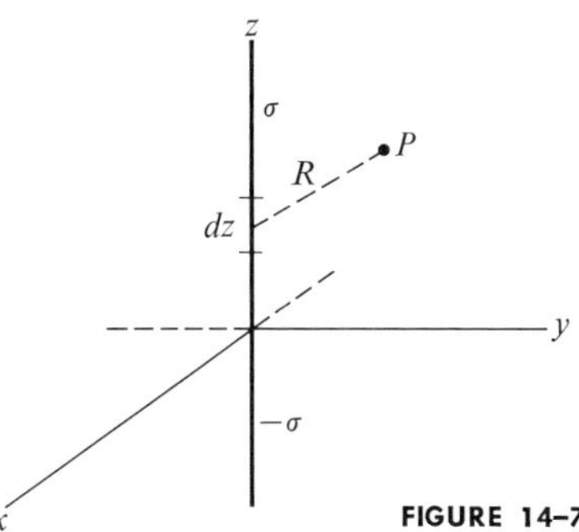

FIGURE 14–7

7. Find a solution $u(r, \varphi)$ of Laplace's equation in the region $|r| > 1$ given that $u(1, \varphi) = f(\varphi)$.

8. Let $u(r, \varphi)$ be the steady-state temperature in a spherical shell of inner radius a, outer radius b. Find $u(r, \varphi)$ if $u(a, \varphi) = f(\varphi)$, $u(b, \varphi) = 0$.

9. Find the steady-state temperature in a homogeneous sphere of unit radius when the surface is maintained at the following temperatures.

   (a) $\cos^2 \varphi$ (b) $\cos 2\varphi$ (c) $2\sin^2 \varphi - 1$ (d) $\sin^4 \varphi$

10. Suppose that

$$\sum_{k=1}^{n} \frac{\partial^2 u}{\partial x_k^2} = 0,$$

where $u = u_1(x_1) \cdots u_n(x_n)$. Prove that

$$\frac{u_k''}{u_k} = \lambda_k, \quad \lambda_k \text{ a constant}, \quad k = 1, \ldots, n,$$

and that $\lambda_1 + \cdots + \lambda_n = 0$.

11. Find the steady-state temperature in a homogeneous hemisphere of radius 1, given that the temperature is independent of longitude, that the flat surface is kept at 0°, and that the remainder of the surface has temperature $f(\varphi)$, where $f(\pi/2) = 0$.

12. Solve the problem in Exercise 11 when $f(\varphi) = \cos^3 \varphi - \cos \varphi$.

*13. Let $f(\varphi) = \frac{1}{2}[f(\varphi^+) + f(\varphi^-)]$ be piecewise continuous and periodic with period 2π, and let $f'(\varphi)$ be piecewise continuous for all φ.

(a) Prove that the series

$$u(r, \varphi) = \sum_{n=0}^{\infty} A_n r^n P_n (\cos \varphi)$$

constructed in the text is uniformly and absolutely convergent for $r \leq r_0 < 1$, $0 \leq \varphi \leq \pi$. [*Hint:* Recall that under these assumptions the Legendre series for f converges in the mean to f, and hence that there exists a real number M such that $|A_n| < M$ for all n. Then use the fact that $|P_n(x)| \leq 1$ for $-1 \leq x \leq 1$. (See Exercise 15, Section 11–3.)]

(b) Given that $|P_{n}^{(k)}(x)| \leq n^{2k}$ for $-1 \leq x \leq 1$, where $P_n^{(k)}$ denotes the kth derivative of P_n, prove that the series in (a) may be twice differentiated term-by-term with respect to either variable, and that the resulting series still converge uniformly and absolutely for $r \leq r_0 < 1$, $0 \leq \varphi \leq \pi$.

(c) Use the result in (b) to show that this series is a solution of Laplace's equation in the region $r < 1$.

(d) Prove that

$$\lim_{r \to 1^-} u(r, \varphi_0) = f(\varphi_0)$$

for fixed φ_0.

(e) Prove that

$$\lim_{\substack{r \to 1^- \\ \varphi \to \varphi_0}} u(r, \varphi) = f(\varphi_0)$$

whenever $\sum_{n=0}^{\infty} A_n P_n(\cos \varphi)$ converges uniformly to $f(\varphi)$ on the interval $[0, \pi]$.

14–5 LAPLACE'S EQUATION; SPHERICAL HARMONICS

In this section we solve Laplace's equation in the spherical region $r < 1$ subject only to the requirement that the solution be continuous, single-valued, and take on preassigned values when $r = 1$. Thus we consider the equation

$$\frac{\partial}{\partial r}\left(r^2 \frac{\partial u}{\partial r}\right) + \frac{1}{\sin \varphi} \frac{\partial}{\partial \varphi}\left(\sin \varphi \frac{\partial u}{\partial \varphi}\right) + \frac{1}{\sin^2 \varphi} \frac{\partial^2 u}{\partial \theta^2} = 0 \tag{14–40}$$

in conjunction with the boundary condition $u(1, \varphi, \theta) = f(\varphi, \theta)$, where f is continuous on the surface of the unit sphere.

Here we begin by seeking solutions of the form $u(r, \varphi, \theta) = R(r)\Phi(\varphi)\Theta(\theta)$, and substitute in (14–40) to obtain

$$\left[\frac{r^2 R'' + 2rR'}{R}\right] + \left[\frac{\Phi'' + \cot \varphi \, \Phi'}{\Phi} + \frac{1}{\sin^2 \varphi} \frac{\Theta''}{\Theta}\right] = 0. \tag{14–41}$$

Since the first term in this equation depends only on r, while the second is independent of r, we have

$$\frac{r^2R'' + 2rR'}{R} = \lambda,$$

λ a constant, or

$$r^2R'' + 2rR' - \lambda R = 0. \tag{14-42}$$

But for each value of λ the functions r^α, $\alpha = \frac{1}{2}(-1 \pm \sqrt{1 + 4\lambda})$, are a basis for the solution space of this equation (see Exercise 1). Thus if we demand that α be an integer (which we do for the sake of simplicity) it follows that $\lambda = m(m + 1)$, $m = 0, 1, 2, \ldots$, and that $\alpha = m$, or $\alpha = -(m + 1)$.* Hence, in this instance, the relevant solutions of (14–42) are

$$R_m(r) = r^m, \quad m = 0, 1, 2, \ldots, \tag{14-43}$$

the solutions $r^{-(m+1)}$ being rejected for reasons of continuity.

Next, we use the chosen values of λ to rewrite (14–41) as

$$m(m + 1) + \frac{\Phi'' + \cot\varphi\,\Phi'}{\Phi} + \frac{1}{\sin^2\varphi}\,\frac{\Theta''}{\Theta} = 0,$$

or as

$$\frac{\Theta''}{\Theta} + \left[m(m + 1)\sin^2\varphi + \frac{\Phi'' + \cot\varphi\,\Phi'}{\Phi}\sin^2\varphi\right] = 0. \tag{14-44}$$

Thus Θ''/Θ is also constant, and the requirement that Θ be periodic with period 2π implies that

$$\frac{\Theta''}{\Theta} = -n^2, \quad n = 0, 1, 2, \ldots.$$

Hence the admissible values of Θ are

$$\Theta_n(\theta) = A_n\cos n\theta + B_n\sin n\theta, \tag{14-45}$$

A_n and B_n constants, $n = 0, 1, 2, \ldots$.

Finally, when $\Theta''/\Theta = -n^2$, (14–44) becomes

$$\Phi'' + \cot\varphi\,\Phi' + \left[m(m + 1) - \frac{n^2}{\sin^2\varphi}\right]\Phi = 0, \tag{14-46}$$

an equation which we now solve by means of the following ingenious argument.

---

* The reader who feels that we have been somewhat ruthless in discarding potential values of λ should realize that we are only trying to find *a single* solution of (14–40) which satisfies the given boundary conditions, and the fact that the above choices lead to such a solution will furnish *a postiori* justification for making them. Of course, the heart of the matter lies in a uniqueness theorem which can be cited to prove that nothing is lost by this line of reasoning.

Set $s = \cos\varphi$. Then (14–46) becomes

$$(1 - s^2)\frac{d^2\Phi}{ds^2} - 2s\frac{d\Phi}{ds} + \left[m(m+1) - \frac{n^2}{1 - s^2}\right]\Phi = 0, \qquad (14\text{–}47)$$

and in this form is reminiscent of Legendre's equation of order m. To exploit this similarity we make the second change of variable

$$\Phi = (1 - s^2)^{n/2}w.$$

Then

$$\frac{d\Phi}{ds} = (1 - s^2)^{n/2}\left[\frac{dw}{ds} - \frac{ns}{1 - s^2}w\right],$$

$$\frac{d^2\Phi}{ds^2} = (1 - s^2)^{n/2}\left[\frac{d^2w}{ds^2} - \frac{2ns}{1 - s^2}\frac{dw}{ds} - n\frac{1 - s^2(n-1)}{(1 - s^2)^2}w\right],$$

and (14–47) becomes

$$(1 - s^2)\frac{d^2w}{ds^2} - 2(n+1)s\frac{dw}{ds} + [m(m+1) - n(n+1)]w = 0. \qquad (14\text{–}48)$$

On the other hand, we know that P_m, the Legendre polynomial of degree m, satisfies the equation

$$(1 - s^2)y'' - 2sy' + m(m+1)y = 0 \qquad (14\text{–}49)$$

on $[-1, 1]$. Hence $P_m^{(n)}$, the nth derivative of P_m, satisfies the equation obtained by differentiating (14–49) n-times with respect to s. Performing these differentiations we obtain, successively,

$$(1 - s^2)y^{(3)} - 2(2s)y^{(2)} + [m(m+1) - 2]y^{(1)} = 0,$$

$$(1 - s^2)y^{(4)} - 3(2s)y^{(3)} + [m(m+1) - 2(3)]y^{(2)} = 0,$$

$$\vdots$$

$$(1 - s^2)y^{(n+2)} - (n+1)(2s)y^{(n+1)} + [m(m+1) - n(n+1)]y^{(n)} = 0,$$

and since the last of these equations is just another form of (14–48), it follows that $P_m^{(n)}$ is a solution of that equation on $[-1, 1]$. When rewritten in terms of φ we find that

$$\Phi_{mn}(\varphi) = \sin^n\varphi\, P_m^{(n)}(\cos\varphi) \qquad (14\text{–}50)$$

is a nontrivial solution of (14–46) on $[0, \pi]$ for each pair of non-negative integers m and n, $n \leq m$.*

---

* Since P_m is a polynomial of degree m, $P_m^{(n)}$ will be nonzero only if $n \leq m$.

Combining the solutions found in (14–43), (14–45), and (14–50) we obtain the functions

$$r^m(A_{mn} \cos n\theta + B_{mn} \sin n\theta) \sin^n \varphi\, P_m^{(n)}(\cos \varphi),$$
$$n = 0, 1, \ldots, m, \quad m = 0, 1, 2, \ldots,$$
$$A_{mn},\ B_{mn} \text{ constants},$$

each of which is a solution of (14–40). To complete the discussion we now use these functions to construct a solution $u(r, \varphi, \theta)$ which also satisfies the boundary condition $u(1, \varphi, \theta) = f(\varphi, \theta)$. The student will recognize the method.

Let $\mathcal{C}(S)$ denote the Euclidean space composed of all continuous functions defined on the surface S of the unit sphere with inner product

$$f \cdot g = \iint_S f(\varphi, \theta) g(\varphi, \theta)\, dS$$
$$= \int_0^{2\pi} \int_0^{\pi} f(\varphi, \theta) g(\varphi, \theta) \sin \varphi\, d\varphi\, d\theta,$$

and let

$$u_{mn}(\varphi, \theta) = \cos (n\theta) \sin^n \varphi\, P_m^{(n)}(\cos \varphi),$$
$$v_{mn}(\varphi, \theta) = \sin (n\theta) \sin^n \varphi\, P_m^{(n)}(\cos \varphi), \tag{14–51}$$

where m and n are as above. These particular functions are known as *spherical harmonics*, and for each fixed value of m the $2m + 1$ functions

$$u_{m0}, u_{m1}, \ldots, u_{mm}, v_{m1}, \ldots, v_{mm}$$

are called the *spherical harmonics of order m*. In the sections which follow we will prove that these functions are a *basis* for $\mathcal{C}(S)$, and hence that $f(\varphi, \theta)$ can be written in the form

$$f(\varphi, \theta) = \sum_{m=0}^{\infty} \left[\frac{A_{m0}}{2} u_{m0} + \sum_{n=1}^{m} (A_{mn} u_{mn} + B_{mn} v_{mn}) \right], \tag{14–52}$$

where the series converges *in the mean* to f, and

$$\frac{A_{m0}}{2} = \frac{f \cdot u_{m0}}{\|u_{m0}\|^2} = \frac{\int_0^{2\pi} \int_0^{\pi} f(\varphi, \theta) \sin \varphi\, P_m(\cos \varphi)\, d\varphi\, d\theta}{u_{m0} \cdot u_{m0}},$$

$$A_{mn} = \frac{f \cdot u_{mn}}{\|u_{mn}\|^2} = \frac{\int_0^{2\pi} \int_0^{\pi} f(\varphi, \theta) \cos (n\theta) \sin^{n+1} \varphi\, P_m^{(n)}(\cos \varphi)\, d\varphi\, d\theta}{u_{mn} \cdot u_{mn}},$$

$$B_{mn} = \frac{f \cdot v_{mn}}{\|v_{mn}\|^2} = \frac{\int_0^{2\pi} \int_0^{\pi} f(\varphi, \theta) \sin (n\theta) \sin^{n+1} \varphi\, P_m^{(n)}(\cos \varphi)\, d\varphi\, d\theta}{v_{mn} \cdot v_{mn}}, \quad n \neq 0.$$

As usual, when f is sufficiently well behaved this series is twice differentiable term-by-term with respect to each variable, and the series

$$u(r, \varphi, \theta) = \sum_{m=0}^{\infty} r^m \left[\frac{A_{m0}}{2} u_{m0} + \sum_{n=1}^{m} (A_{mn}u_{mn} + B_{mn}v_{mn})\right]$$

is the solution of the given boundary-value problem.

EXERCISES

1. (a) Find the general solution of

$$r^2R'' + 2rR' - \lambda R = 0$$

on $(0, \infty)$.

(b) Prove that this equation has solutions of the form r^k, k a non-negative integer, if and only if $\lambda = m(m + 1)$, $m = 0, 1, 2, \ldots$.

2. Use mathematical induction to prove that the nth derivative of a solution of Legendre's equation of order m is a solution of

$$(1 - x^2)y^{(n+2)} - 2(n + 1)xy^{(n+1)} + [m(m + 1) - n(n + 1)]y^{(n)} = 0.$$

3. Compute the spherical harmonics of orders 0, 1, and 2.

4. Make the substitution $u(r, \varphi, \theta) = r^nF(\varphi, \theta)$ in (14–40) to obtain a partial differential equation for $F(\varphi, \theta)$, and then set $s = \cos \varphi$ to obtain

$$\frac{\partial^2 F}{\partial \theta^2} + (1 - s^2)\left\{\frac{\partial}{\partial s}\left[(1 - s^2)\frac{\partial F}{\partial s}\right] + n(n + 1)F\right\} = 0.$$

Solve this equation by the method of separation of variables, and compare the results with those in the text.

14–6 ORTHOGONALITY OF THE SPHERICAL HARMONICS; LAPLACE SERIES

To complete the discussion of the last section we must show that the spherical harmonics

$$\begin{aligned} u_{mn}(\varphi, \theta) &= \cos n\theta \sin^n \varphi\, P_m^{(n)}(\cos \varphi), \\ v_{mn}(\varphi, \theta) &= \sin n\theta \sin^n \varphi\, P_m^{(n)}(\cos \varphi), \end{aligned} \tag{14–53}$$

$n = 0, 1, \ldots, m$; $m = 0, 1, 2, \ldots$, are a basis for the Euclidean space $\mathcal{C}(S)$. (We assume throughout that $n \neq 0$ in v_{mn}.) As usual, the proof is divided into two parts; a straightforward computation to establish orthogonality, and a sequence of theorems culminating in the assertion that the u_{mn}, v_{mn} generate $\mathcal{C}(S)$. We begin with the proof of orthogonality, which in this case requires that we show

$$\begin{aligned} u_{mn} \cdot u_{rs} &= 0 \quad \text{whenever } m \neq r \text{ or } n \neq s \text{ (or both)}, \\ v_{mn} \cdot v_{rs} &= 0 \quad \text{whenever } m \neq r \text{ or } n \neq s \text{ (or both)}, \\ u_{mn} \cdot v_{rs} &= 0 \quad \text{for all } m, n, r, s. \end{aligned} \tag{14-54}$$

Recalling that inner products in $\mathcal{C}(S)$ are computed as twofold iterated integrals according to the formula

$$f \cdot g = \int_0^{2\pi} \int_0^{\pi} f(\varphi, \theta) g(\varphi, \theta) \sin \varphi \, d\varphi \, d\theta, \tag{14-55}$$

we have

$$u_{mn} \cdot u_{rs} = \int_0^{2\pi} \cos n\theta \cos s\theta \, d\theta \int_0^{\pi} \sin^{n+s+1} \varphi \, P_m^{(n)}(\cos \varphi) P_r^{(s)}(\cos \varphi) \, d\varphi.$$

Thus if $n \neq s$, $u_{mn} \cdot u_{rs} = 0$ since the first integral in this expression vanishes. On the other hand, if $n = s$, then $m \neq r$ and the equation $u_{mn} \cdot u_{rs} = 0$ will be satisfied if and only if

$$\int_0^{\pi} \sin^{2n+1} \varphi \, P_m^{(n)}(\cos \varphi) P_r^{(n)}(\cos \varphi) \, d\varphi = 0,$$

or, making the substitution $x = \cos \varphi$, if and only if

$$\int_{-1}^{1} (1 - x^2)^n P_m^{(n)}(x) P_r^{(n)}(x) \, dx = 0. \tag{14-56}$$

To establish this equality we recall that $P_m^{(n)}$ and $P_r^{(n)}$ satisfy the identities

$$\begin{aligned} (1 - x^2) P_m^{(n+2)} - 2(n+1) x P_m^{(n+1)} + [m(m+1) - n(n+1)] P_m^{(n)} &= 0, \\ (1 - x^2) P_r^{(n+2)} - 2(n+1) x P_r^{(n+1)} + [r(r+1) - n(n+1)] P_r^{(n)} &= 0, \end{aligned} \tag{14-57}$$

throughout the interval $[-1, 1]$. We now multiply the first of these expressions by $(1 - x^2)^n P_r^{(n)}$, the second by $(1 - x^2)^n P_m^{(n)}$, and subtract to obtain

$$\begin{aligned} (1 - x^2)^{n+1} [P_m^{(n+2)} P_r^{(n)} - P_r^{(n+2)} P_m^{(n)}] \\ - 2(n+1) x (1 - x^2)^n [P_m^{(n+1)} P_r^{(n)} - P_r^{(n+1)} P_m^{(n)}] \\ = [r(r+1) - m(m+1)] (1 - x^2)^n P_m^{(n)} P_r^{(n)}. \end{aligned}$$

But by setting $P_m^{(n+1)} P_r^{(n)} - P_r^{(n+1)} P_m^{(n)} = y$, the left-hand side of this identity may be rewritten

$$\frac{d}{dx} [(1 - x^2)^{n+1} y],$$

and we have

$$\frac{d}{dx} [(1 - x^2)^{n+1} y] = [r(r+1) - m(m+1)] (1 - x^2)^n P_m^{(n)} P_r^{(n)}.$$

Integrating from -1 to 1 gives

$$[r(r+1) - m(m+1)]\int_{-1}^{1} (1 - x^2)^n P_m^{(n)} P_r^{(n)}\, dx = 0,$$

and since $m \neq r$,

$$\int_{-1}^{1} (1 - x^2)^n P_m^{(n)} P_r^{(n)}\, dx = 0,$$

as desired.

To complete the proof we observe that the above argument applies equally well to the v_{mn}, and shows that $v_{mn} \cdot v_{rs} = 0$ whenever $m \neq r$ or $n \neq s$. Finally, the vanishing of $u_{mn} \cdot v_{rs}$ for all m, n, r, s is an immediate consequence of the orthogonality of $\cos n\theta$ and $\sin s\theta$ in $\mathcal{C}[0, 2\pi]$, and we have

Theorem 14–1. *The spherical harmonics are mutually orthogonal in the Euclidean space of all continuous functions on the unit sphere.*

For future use we take formal note of the following obvious consequence of this result.

Corollary 14–1. *For each fixed value of m the $(2m + 1)$ functions*

$$r^m u_{m0},\ r^m u_{m1},\ \ldots,\ r^m u_{mm},\ r^m v_{m1},\ \ldots,\ r^m v_{mm}$$

are linearly independent in the space of all continuous functions defined in the region $r \leq 1$.

Indeed, were it possible to express one of these functions as a linear combination of the others we would find by setting $r = 1$ that $u_{m0}, \ldots, u_{mm}, v_{m1}, \ldots, v_{mm}$ are linearly dependent in $\mathcal{C}(S)$. But this contradicts the known orthogonality of these functions, and the corollary follows.

On the strength of Theorem 14–1 we are justified in introducing the *formal* series expansion of a continuous function f on the unit sphere in terms of the u_{mn}, v_{mn}, even though we cannot as yet assert that this series converges in the mean to f. Using the notation of Section 8–3 we therefore have

$$f \sim \sum_{m=0}^{\infty} \left[\frac{A_{m0}}{2} u_{m0} + \sum_{n=1}^{m} (A_{mn} u_{mn} + B_{mn} v_{mn})\right], \tag{14–58}$$

where $A_{m0}/2$, A_{mn}, B_{mn} are the generalized Fourier coefficients of f and are computed according to the formulas

$$\frac{A_{m0}}{2} = \frac{f \cdot u_{m0}}{u_{m0} \cdot u_{m0}},$$
$$A_{mn} = \frac{f \cdot u_{mn}}{u_{mn} \cdot u_{mn}}, \qquad B_{mn} = \frac{f \cdot v_{mn}}{v_{mn} \cdot v_{mn}}. \tag{14–59}$$

This series is known as the *Laplace series* expansion of f, and in the next section we shall prove that it does in fact converge *in the mean* to f. But first we evaluate the various inner products $u_{mn} \cdot u_{mn}$ and $v_{mn} \cdot v_{mn}$ appearing in (14–59). The argument goes as follows.

From (14–53) and (14–55) we have

$$u_{mn} \cdot u_{mn} = \int_0^{2\pi} \cos^2 n\theta \, d\theta \int_0^{\pi} \sin^{2n+1} \varphi \, [P_m^{(n)}(\cos \varphi)]^2 \, d\varphi,$$

$$v_{mn} \cdot v_{mn} = \int_0^{2\pi} \sin^2 n\theta \, d\theta \int_0^{\pi} \sin^{2n+1} \varphi \, [P_m^{(n)}(\cos \varphi)]^2 \, d\varphi.$$

Thus if

$$I_{mn} = \int_{-1}^{1} (1 - x^2)^n [P_m^{(n)}(x)]^2 \, dx = \int_0^{\pi} \sin^{2n+1} \varphi \, [P_m^{(n)}(\cos \varphi)]^2 \, d\varphi,$$

we see that

$$\begin{aligned} \tfrac{1}{2}(u_{m0} \cdot u_{m0}) &= \pi I_{m0}, \\ u_{mn} \cdot u_{mn} = v_{mn} \cdot v_{mn} &= \pi I_{mn}, \end{aligned} \tag{14–60}$$

and it remains to find the value of I_{mn} for given integers m and n ($n \leq m$). To this end we observe that

$$I_{m0} = \int_{-1}^{1} [P_m(x)]^2 \, dx = \frac{2}{2m + 1} \tag{14–61}$$

(see Section 11–3), and then use integration by parts to express $I_{m,n+1}$ in terms of I_{mn} by setting

$$u = (1 - x^2)^{n+1} P_m^{(n+1)}, \qquad dv = P_m^{(n+1)} \, dx$$

in

$$I_{m,n+1} = \int_{-1}^{1} (1 - x^2)^{n+1} [P_m^{(n+1)}(x)]^2 \, dx.$$

This gives

$$\begin{aligned} I_{m,n+1} &= (1 - x^2)^{n+1} P_m^{(n+1)} P_m^{(n)} \Big|_{-1}^{1} \\ &\quad - \int_{-1}^{1} [(1 - x^2) P_m^{(n+2)} - 2(n + 1) x P_m^{(n+1)}](1 - x^2)^n P_m^{(n)} \, dx \\ &= - \int_{-1}^{1} [(1 - x^2) P_m^{(n+2)} - 2(n + 1) x P_m^{(n+1)}](1 - x^2)^n P_m^{(n)} \, dx. \end{aligned}$$

But since

$$(1 - x^2) P_m^{(n+2)} - 2(n + 1) x P_m^{(n+1)} + [m(m + 1) - n(n + 1)] P_m^{(n)} = 0$$

(see 14–57), it follows that

$$\begin{aligned} I_{m,n+1} &= [m(m+1) - n(n+1)] \int_{-1}^{1} (1 - x^2)^n [P_m^{(n)}(x)]^2 \, dx \\ &= [m(m+1) - n(n+1)] I_{mn}. \end{aligned}$$

Hence

$$I_{m,n+1} = (m - n)(m + n + 1) I_{mn}, \tag{14–62}$$

and an easy computation starting with the known value of I_{m0} now yields

$$I_{mn} = \frac{(m+n)!}{(m-n)!} \frac{2}{2m+1}, \quad 0 \leq n \leq m \tag{14–63}$$

(see Exercise 1).

Finally, if this result is substituted in (14–60) and (14–59) we obtain the formulas

$$A_{mn} = \frac{(2m+1)(m-n)!}{2\pi(m+n)!} \int_0^{2\pi} \int_0^{\pi} f(\varphi, \theta) \cos(n\theta) \sin^{n+1} \varphi \, P_m^{(n)}(\cos \varphi) \, d\varphi \, d\theta,$$

$$B_{mn} = \frac{(2m+1)(m-n)!}{2\pi(m+n)!} \int_0^{2\pi} \int_0^{\pi} f(\varphi, \theta) \sin(n\theta) \sin^{n+1} \varphi \, P_m^{(n)}(\cos \varphi) \, d\varphi \, d\theta,$$

for all m and n.

EXERCISES

1. Verify Formula 14–63.
2. Find the Laplace series expansion of the function

$$f(\varphi, \theta) = \sin^2 \varphi \cos^2 \varphi \sin \theta \cos \theta$$

in the following two ways:

(a) by direct application of the formulas in this section;

(b) by writing

$$f(\varphi, \theta) = \tfrac{1}{2} \cos^2 \varphi \sin^2 \varphi \sin 2\theta,$$

and expressing $\cos^2 \varphi$ as a sum of terms of the form

$$\frac{d^2}{d(\cos \varphi)^2} P_k(\cos \varphi).$$

[*Hint:* Note that

$$\cos^2 \varphi = \frac{1}{4 \cdot 3} \frac{d^2(\cos^4 \varphi)}{d(\cos \varphi)^2},$$

and then write $\cos^4 \varphi$ as a Legendre series in $\cos \varphi$.]

3. Write $\cos^2\varphi\cos^2\theta$ as a series of spherical harmonics.

4. Prove that

$$\begin{aligned}\cos^3\varphi\sin^3\varphi\sin\theta\cos^2\theta &= \left[\frac{1}{6930}P_6^{(3)}(\cos\varphi) + \frac{1}{1540}P_4^{(3)}(\cos\varphi)\right]\sin 3\theta\sin^3\varphi \\ &\quad - \left[\frac{2}{693}P_6^{(1)}(\cos\varphi) - \frac{1}{770}P_4^{(1)}(\cos\varphi) - \frac{1}{63}P_2^{(1)}(\cos\varphi)\right]\sin\theta\sin\varphi.\end{aligned}$$

*5. (a) Expand the function

$$f(\varphi, \theta) = (1 - |\cos\varphi|)(1 + \cos 2\theta)$$

in a series of spherical hamonics.

(b) Find the solution of the boundary-value problem of Section 14–5 when $f(\varphi, \theta)$ is the function in (a).

6. Solve Laplace's equation in the spherical shell with inner radius a and outer radius b ($b > a$), given that $u \equiv 0$ for $r = b$, and $u = f(\varphi, \theta)$ for $r = a$. [*Warning:* Since the point $r = 0$ is not in the shell, both of the functions r^m and $r^{-(m+1)}$ must be used.]

7. (a) Use the results of the preceding exercise to find a solution for the shell problem when $u \equiv 0$ for $r = a$, and $u = g(\varphi, \theta)$ for $r = b$.

(b) Solve the shell problem when $u = f(\varphi, \theta)$ for $r = a$, and $u = g(\varphi, \theta)$ for $r = b$.

*14–7 HARMONIC POLYNOMIALS AND THE BASIS THEOREM

To complete the discussion of the spherical harmonics we must show that these functions *generate* the Euclidean space $\mathcal{C}(S)$. This result however is an easy corollary of the fact that every function in $\mathcal{C}(S)$ can be *uniformly approximated* by a finite linear combination of the u_{mn}, v_{mn}, where by this we mean that if we are given any function f in $\mathcal{C}(S)$, and any real number $\epsilon > 0$, there exists a finite linear combination $L(\varphi, \theta)$ of the u_{mn}, v_{mn} such that

$$|f(\varphi, \theta) - L(\varphi, \theta)| < \epsilon$$

for *all* $0 \le \varphi \le \pi$, $0 \le \theta \le 2\pi$. Indeed, the passage from such an approximation to the assertion that the u_{mn}, v_{mn} generate $\mathcal{C}(S)$ is all but identical with the proof of Theorem 9–6 and will be left to the reader. With this understanding, we turn our attention to the following theorem.

Theorem 14–2. *Any continuous function on the surface of the unit sphere can be uniformly approximated by a finite linear combination of spherical harmonics.*

In broad outline, the proof consists of an argument to show that the spherical harmonics can be replaced by certain polynomials in x, y, z, which are then used to construct the desired approximation. The basic properties of the polynomials in question are established in the following lemmas.

Lemma 14–1. *For each non-negative integer m the functions $r^m u_{mn}$ and $r^m v_{mn}$ are homogeneous polynomials of degree m when expressed in rectangular coordinates.**

Proof. We begin by rewriting u_{mn} and v_{mn} as polynomials in $\cos\theta$, $\sin\theta$, $\cos\varphi$, $\sin\varphi$, as follows. From the formula

$$\cos n\theta + i\sin n\theta = (\cos\theta + i\sin\theta)^n$$

we see that $\cos n\theta$ is the real part, and $\sin n\theta$ the imaginary part of the expansion of $(\cos\theta + i\sin\theta)^n$. But, by the binomial theorem,

$$(\cos\theta + i\sin\theta)^n = \cos^n\theta + \alpha_1\cos^{n-1}\theta(i\sin\theta) + \alpha_2\cos^{n-2}\theta(i\sin\theta)^2 + \cdots + (i\sin\theta)^n,$$

where $\alpha_1, \alpha_2, \ldots$ are constants whose specific values we can ignore. Since $i^2 = -1$, the real part of this expression is a sum of terms of the form

$$c_k\cos^{n-k}\theta\sin^k\theta, \tag{14–64}$$

where c_k is a constant, and k an even integer. Similarly, the imaginary part of $(\cos\theta + i\sin\theta)^n$ is a sum of terms of the same form with k odd.

Next, we observe that the nth derivative of the Legendre polynomial $P_m(x)$ is a polynomial of degree $m - n$, composed of terms of the form $d_j x^{m-n-2j}$, d_j a constant, j a non-negative integer. (Recall that the degree of *every* term in $P_m(x)$ is either even or odd according as m is even or odd.) Hence $P_m^{(n)}(\cos\varphi)$ is a polynomial in $\cos\varphi$, each of whose terms is of the form

$$d_j\cos^{m-n-2j}\varphi. \tag{14–65}$$

Combined with (14–64), this implies that each of the spherical harmonics u_{mn}, v_{mn} can be decomposed into a sum with individual entries

$$\alpha_{jk}\cos^{n-k}\theta\sin^k\theta\sin^n\varphi\cos^{m-n-2j}\varphi$$

---

* A polynomial $p(x, y, z)$ is *homogeneous* of degree m if and only if it can be written in the form

$$p(x, y, z) = \sum_{i+j+k=m} c_{ijk}x^i y^j z^k,$$

where the c_{ijk} are real numbers. Thus $x^2 - 2y^2 + xz$ and $2xy - 3yz + z^2$ are homogeneous of degree two, while $x^2y - 4xyz - 3z^3$ is homogeneous of degree three. The reader should note that according to this definition the zero polynomial is homogeneous of every degree.

for suitable constants α_{jk} and integers j and k. Thus the functions $r^m u_{mn}$ and $r^m v_{mn}$ have a similar decomposition, with a typical term appearing as

$$\begin{aligned}\alpha_{jk} r^m \cos^{n-k} \theta \sin^k \theta \sin^n \varphi \cos^{m-n-2j} \varphi \\ = \alpha_{jk} r^{2j} [r^{n-k} \cos^{n-k} \theta \sin^{n-k} \varphi][r^k \sin^k \theta \sin^k \varphi][r^{m-n-2j} \cos^{m-n-2j} \varphi].\end{aligned} \tag{14-66}$$

To complete the proof we recall that

$$x = r \cos \theta \sin \varphi, \qquad y = r \sin \theta \sin \varphi, \qquad z = r \cos \varphi.$$

Thus when converted to rectangular coordinates, (14–66) becomes

$$\alpha_{jk}(x^2 + y^2 + z^2)^j x^{n-k} y^k z^{m-n-2j},$$

and is obviously a polynomial in x, y, z, each of whose terms is of degree m. ▌

For our purposes this lemma is important because it allows us to assert that the equation

$$\frac{\partial^2 u}{\partial x^2} + \frac{\partial^2 u}{\partial y^2} + \frac{\partial^2 u}{\partial z^2} = 0 \tag{14-67}$$

has *polynomial* solutions which are homogeneous of degree m for each integer $m \geq 0$. Any such solution of Laplace's equation is called a *homogeneous harmonic polynomial* of degree m, and the linearity of (14–67) implies that the set $\mathcal{H}_m$ of all homogeneous harmonic polynomials of degree m is a real vector space. Moreover, it follows from Corollary 14–1 that the $2m + 1$ functions

$$r^m u_{m0}, r^m u_{m1}, \ldots, r^m u_{mn}, r^m v_{m1}, \ldots, r^m v_{mn} \tag{14-68}$$

are *linearly independent* in $\mathcal{H}_m$ when expressed in rectangular coordinates. We now propose to show that they are actually a *basis* for $\mathcal{H}_m$ by proving

Lemma 14–2. *For each integer m, the vector space $\mathcal{H}_m$ is of dimension $2m + 1$.*

Proof. This is clearly true if $m < 2$. For $m > 2$ let

$$u(x, y, z) = \sum_{i+j+k=m} c_{ijk} x^i y^j z^k \tag{14-69}$$

be an arbitrary homogeneous polynomial of degree m. Then

$$\frac{\partial^2 u}{\partial x^2} + \frac{\partial^2 u}{\partial y^2} + \frac{\partial^2 u}{\partial z^2}$$

is homogeneous of degree $m - 2$, and hence is a sum of terms of the form $d_{\alpha\beta\gamma} x^\alpha y^\beta z^\gamma$, where $d_{\alpha\beta\gamma}$ is a constant and

$$\alpha + \beta + \gamma = m - 2.$$

Moreover, if for fixed values of α, β, γ we add the coefficients of $x^\alpha y^\beta z^\gamma$ appearing in the derivatives $\partial^2 u/\partial x^2$, $\partial^2 u/\partial y^2$, $\partial^2 u/\partial z^2$, we find that

$$\begin{aligned} d_{\alpha\beta\gamma} = {} & (\alpha + 2)(\alpha + 1)c_{\alpha+2,\beta,\gamma} + (\beta + 2)(\beta + 1)c_{\alpha,\beta+2,\gamma} \\ & + (\gamma + 2)(\gamma + 1)c_{\alpha,\beta,\gamma+2} \end{aligned} \tag{14–70}$$

(see Exercise 1).

Now suppose that $u(x, y, z)$ is harmonic. Then each of the $d_{\alpha\beta\gamma}$ appearing as coefficients in the polynomial

$$\frac{\partial^2 u}{\partial x^2} + \frac{\partial^2 u}{\partial y^2} + \frac{\partial^2 u}{\partial z^2}$$

must be zero, and it follows that

$$\begin{aligned} (\gamma + 2)(\gamma + 1)c_{\alpha,\beta,\gamma+2} = {} & -(\alpha + 2)(\alpha + 1)c_{\alpha+2,\beta,\gamma} \\ & - (\beta + 2)(\beta + 1)c_{\alpha,\beta+2,\gamma}. \end{aligned} \tag{14–71}$$

In other words, *a homogeneous polynomial of degree m in x, y, z is harmonic if and only if its coefficients satisfy* (14–71) *with* $\alpha + \beta + \gamma = m - 2$.

To interpret this result in terms of the vector space $\mathcal{H}_m$, we introduce the following triangular array formed from the coefficients of (14–70):

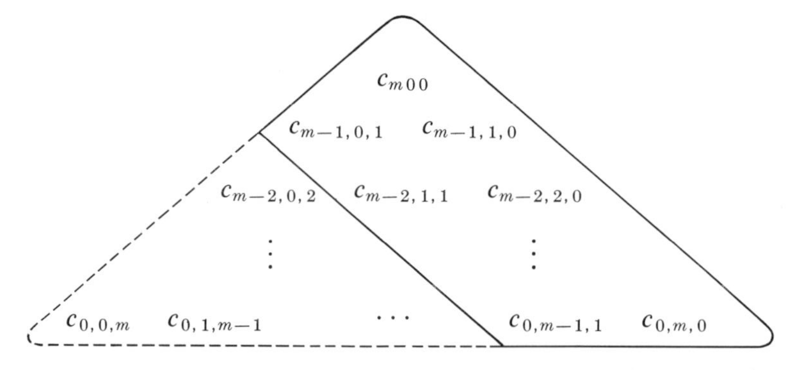

If this array is examined closely, it soon becomes apparent that the entires along the two enclosed diagonals on the upper right determine, via (14–71), **all** of the other coefficients in the triangle. Indeed, the coefficients along the uppermost diagonal determine those along the third, which in turn determine those along the fifth, etc., while the coefficients along the second diagonal determine, successively, those along the fourth, sixth, and so forth. Thus *two harmonic polynomials in* $\mathcal{H}_m$ *are identical* **if and only if** *the* $2m + 1$ *coefficients appearing along the upper diagonals of their coefficient triangles are identical.*

To conclude the proof of the lemma we now let Q_k, $0 \leq k \leq m$, denote the polynomial in $\mathcal{H}_m$ determined by setting $c_{m-k,k,0} = 1$ and all other "special" coefficients $c_{m-k,\beta,\gamma}$, $c_{m-k-1,\beta,\gamma}$ zero, and let R_k, $1 \leq k \leq m$, denote the polynomial in $\mathcal{H}_m$ obtained by setting $c_{m-k,k-1,1} = 1$ and all other special coefficients zero. (The coefficient triangles for these polynomials are those which have exactly one nonzero entry on the two uppermost diagonals, that entry being 1.) The above result implies that the $2m + 1$ polynomials $Q_0, Q_1, \ldots, Q_m, R_1, \ldots, R_m$ are linearly independent and span $\mathcal{H}_m$, and we are done. ▮

EXAMPLE 1. The most general homogeneous polynomial of degree 3 in x, y, and z is a sum of ten terms

$$p(x, y, z) = \sum_{i+j+k=3} c_{ijk} x^i y^j z^k,$$

and has the following array as coefficient triangle:

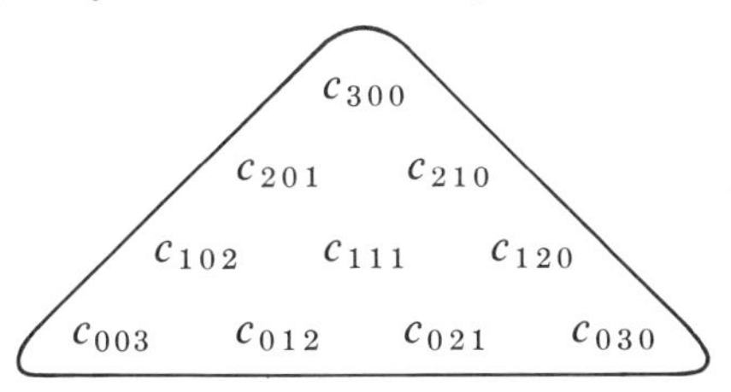

In this case an easy computation using (14–71) shows that the particular homogeneous harmonic polynomials constructed above are

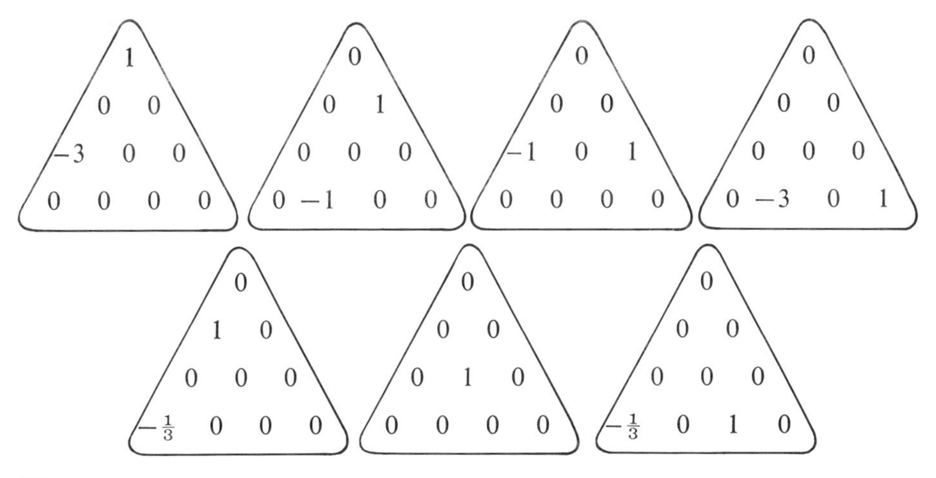

Thus

$$\begin{aligned} Q_0 &= x^3 - 3xz^2, & R_1 &= x^2 z - \tfrac{1}{3} z^3, \\ Q_1 &= x^2 y - yz^2, & R_2 &= xyz, \\ Q_2 &= -xz^2 + xy^2, & R_3 &= -\tfrac{1}{3} z^3 + y^2 z, \\ Q_3 &= -3yz^2 + y^3, & & \end{aligned}$$

and the preceding lemma implies that these polynomials are a basis for the space $\mathcal{H}_3$. Using them we can easily construct *all* homogeneous harmonic polynomials of degree 3 in x, y, z.

With these results in hand we are now in a position to prove that every continuous function $f(\varphi, \theta)$ defined on the surface of the unit sphere can be uniformly approximated by a linear combination of spherical harmonics (Theorem 14–2). Once again we use the Weierstrass approximation theorem, which in this context asserts that *every continuous function on the closed unit sphere* $r \leq 1$ *can be* **uniformly** *approximated by polynomials.** Thus, if $\epsilon > 0$ is given there exists a *polynomial* $p(x, y, z)$ such that

$$|rf(\varphi, \theta) - p(x, y, z)| < \epsilon$$

throughout the solid sphere. In particular, we note that this approximation is also valid on the surface S of the unit sphere where $r = x^2 + y^2 + z^2 \equiv 1$.

Now suppose $p(x, y, z)$ is of degree m. Then if we restrict our attention *to the surface* S we can use the identity $z^2 = 1 - (x^2 + y^2)$ to write $p(x, y, z)$ as a linear combination of the monomials

$$x^i y^j z^k, \quad k \leq 1, \quad i + j + k \leq m. \tag{14–72}$$

Thus, on S, $p(x, y, z)$ may be viewed as a member of the vector space $\mathcal{V}$ consisting of all polynomials in x, y, z which are linear combinations of the above monomials. But these monomials are certainly linearly independent in $\mathcal{V}$, and therefore are a basis for $\mathcal{V}$. Moreover, an easy counting argument shows that for each fixed value of α, $0 \leq \alpha \leq m$, there are exactly $2\alpha + 1$ monomials of type (14–72) with $i + j + k = \alpha$. Hence, since

$$\sum_{\alpha=0}^{m} (2\alpha + 1) = (m + 1)^2,$$

we have $\dim \mathcal{V} = (m + 1)^2$.

On the other hand, for each fixed value of α, the $2\alpha + 1$ functions

$$r^\alpha u_{\alpha 0}, \ldots, r^\alpha u_{\alpha\alpha}, \qquad r^\alpha v_{\alpha 1}, \ldots, r^\alpha v_{\alpha\alpha} \tag{14–73}$$

are homogeneous polynomials of degree α when expressed in rectangular coordinates (Lemma 14–1). In addition, when restricted to S these polynomials belong to the vector space $\mathcal{V}$, and are linearly independent in $\mathcal{V}$. Thus, if we let α run from 0 to m we obtain in this way $(m + 1)^2$ linearly independent polynomials in $\mathcal{V}$, and therefore have a basis for $\mathcal{V}$. It now follows that *on the surface of the sphere* $p(x, y, z)$ may be written as a linear combination of these polynomials. And with this, Theorem 14–2 is proved, since when $r = 1$ the various functions in (14–73) are none other than the spherical harmonics of order α.

---

* More generally, the three-dimensional version of the Weierstrass approximation theorem states that *every continuous function in a closed bounded region of three-space can be uniformly approximated by polynomials.*

EXERCISES

1. Verify Formula (14–70).

2. (a) Use the technique of Example 1 to find a basis for the space $\mathcal{H}_2$ of homogeneous harmonic polynomials of degree 2.

   (b) Determine which of the following polynomials belong to the space $\mathcal{H}_2$:

$$(x - z)(x + y + z), \quad x^2 + y^2, \quad 2x^2 + y^2 - xy + 2z^2,$$
$$x(y - 2z) + y(x - 2z) + z(x - y), \quad y(y - 2x) - x(x + z).$$

3. Verify that the homogeneous polynomial

$$x^3 - xz^2 - 2xy^2$$

is harmonic, and express it as a linear combination of the basis for $\mathcal{H}_3$ found in Example 1.

4. Let $R_2 = xyz$ (see Example 1). Prove that

$$R_2 = \tfrac{1}{30} r^3 v_{3,2}.$$

5. Express each of the functions

$$r^3 u_{3,n} \quad (n = 0, 1, 2, 3), \qquad r^3 v_{3,n} \quad (n = 1, 2, 3)$$

in terms of x, y, z.

6. Let $\mathcal{V}_n$ denote the vector space of all harmonic polynomials of degree n ($n \geq 2$) in x and y, and let each of the polynomials in $\mathcal{V}_n$ be written in standard form as

$$P(x, y) = \sum_{k=0}^{n} c_{n-k,k} x^{n-k} y^k.$$

   (a) Prove that $\dim \mathcal{V}_n = 2$, and that the real and imaginary parts of $(x + iy)^n$ form a basis for $\mathcal{V}_n$.

   (b) Derive a formula for the coefficients of the polynomials in $\mathcal{V}_n$ analogous to (14–70), and use it to prove that for each non-negative integer j with $k + 2j \leq n$,

$$c_{n-(k+2j),k+2j} = (-1)^j \frac{k!(n - k)!}{(k + 2j)!(n - k - 2j)!} c_{n-k,k}.$$

7. Use the result of Exercise 6(b) to find the value of each coefficient in $P(x, y)$ in terms of the coefficient $c_{n,0}$ or $c_{n-1,1}$.

8. A nonzero homogeneous harmonic polynomial in x and y is divisible by xy. Prove that the degree of the polynomial is even. (See Exercises 6 and 7.)

9. (a) Find the homogeneous harmonic polynomial of degree 6 in x and y which contains the term x^6 but no term in x^5y.

   (b) Repeat (a) for the polynomial which contains the term x^5y but no term in x^6.

   (c) Find the homogeneous harmonic polynomial of degree 6 in x and y which contains the terms $x^6 + 6x^5y$.

10. Prove that the number of terms with nonzero coefficients in a homogeneous harmonic polynomial in x and y of degree $n \geq 2$ is $n/2$, $(n/2) + 1$, or $n + 1$ if n is even, or $(n + 1)/2$ or $n + 1$ if n is odd. (See Exercises 6 and 7.)

11. Use the technique introduced in this section to find a basis for the space $\mathcal{H}_4$.

12. (a) Prove that if $f(x, y, z)$ is a harmonic function in a sphere (or in all of three-space), then so is the function $g(x, y, z) = f(-x, y, z)$. Use this result to show that for fixed values of any two variables, the even and odd parts of $f(x, y, z)$ with respect to the third variable are harmonic whenever f is harmonic.

    (b) Use the above result to prove that whenever a polynomial in $\mathcal{H}_2$ is even (or odd) with respect to a particular variable, it is a linear combination of those members of the standard basis for $\mathcal{H}_2$ which are even (or odd) with respect to that variable.

    (c) Show that a polynomial in $\mathcal{H}_2$ cannot be odd with respect to each of the three variables.

13. (a) Express each of the functions

$$r^2 u_{2,n} \quad (n = 0, 1, 2), \qquad r^2 v_{2,n} \quad (n = 1, 2)$$

    in terms of the basis for $\mathcal{H}_2$ found in Exercise 2(a). [*Hint:* Use the properties noted in Exercise 12.]

    (b) Use the result in (a) to express each of the basis polynomials found in Exercise 2(a) in terms of the basis consisting of $r^2 u_{2,n}$ $(n = 0, 1, 2)$ and $r^2 v_{2,n}$ $(n = 1, 2)$.

14. Let $P(x, y, z)$ be an arbitrary polynomial in $\mathcal{H}_3$. Prove that $\partial P/\partial x$ belongs to $\mathcal{H}_2$, and express this polynomial as a linear combination of the basis for $\mathcal{H}_2$ found in Exercise 2(a).

*15. Prove that the number of nonzero terms in $P_m(x_1, x_2, \ldots, x_k)$, the general homogeneous polynomial of degree $m \geq 1$ in $k \geq 1$ variables, is

$$\binom{m + k - 1}{k - 1}.$$

    [*Hint:* For $k > 1$ write

$$P_m(x_1, x_2, \ldots, x_k) = \sum_{j=0}^{m} x_k^j Q_{m-j}(x_1, x_2, \ldots x_{k-1}),$$

    where each Q_{m-j} is a homogeneous polynomial of degree $m - j$ in the variables $x_1, \ldots, x_{k-1}$. Then use induction on k, and the formulas

$$\binom{i}{j} + \binom{i}{j + 1} = \binom{i + 1}{j + 1}, \quad 0 \leq j \leq i - 1,$$

$$\sum_{i=0}^{s} \binom{p + i}{p} = \binom{p + s + 1}{p + 1}, \quad 0 \leq p, \quad 0 \leq s.$$

    Recall that

$$\binom{i}{j} = \binom{i}{i - j} = \frac{i!}{j!(i - j)!}.\ ]$$

*16. (a) Let m be a positive integer, and set

$$P_m(x_1, x_2, \ldots, x_k) = \sum c(m_1, m_2, \ldots, m_k) x_1^{m_1} x_2^{m_2} \ldots x_k^{m_k},$$

where the $c(m_1, m_2, \ldots, m_k)$ are constants, and the summation extends over all non-negative integral values of $m_1, m_2, \ldots, m_k$ for which $m_1 + m_2 + \cdots + m_k = m$. Prove that P_m is harmonic (i.e., is a solution of the partial differential equation $\sum_{i=1}^{k} \partial^2 u/\partial x_i^2 = 0$) if and only if

$$c(m_1, \ldots, m_{k-1}, m_k + 2) = -\sum_{i=1}^{k-1} \frac{(m_i + 2)(m_i + 1)}{(m_k + 2)(m_k + 1)} c(m_1, \ldots, m_{i-1}, m_i + 2, m_{i+1}, \ldots, m_k).$$

(b) Let $Q_m(x_1, \ldots, x_{k-1})$ be the polynomial consisting of those terms of $P_m(x_1, \ldots, x_k)$ in which the variable x_k is absent, let $x_k Q_{m-1}(x_1, \ldots, x_{k-1})$ be the polynomial in which x_k appears to the first degree, and suppose that P_m is harmonic. Prove that the coefficients of P_m are uniquely determined by the coefficients of Q_m and Q_{m-1}.

*17. Let $\mathfrak{X}$ denote the set of all homogeneous harmonic polynomials of degree $m \geq 1$ in $k \geq 2$ variables $x_1, \ldots, x_k$ with the property that each polynomial in $\mathfrak{X}$ contains exactly one term (and that with coefficient 1) for which the exponent of x_k is less than 2. Prove that $\mathfrak{X}$ is a basis for the space $\mathcal{H}_{m,k}$ of all homogeneous harmonic polynomials of degree m in k variables, and determine the dimension of this space.

*18. Can the following assertion be deduced from the results of this section?

Any harmonic function in the closed unit sphere $r \leq 1$ can be uniformly approximated by harmonic polynomials.

Why?

19. Use Theorem 14–2 to prove that the spherical harmonics are a basis for the Euclidean space $\mathcal{C}(S)$.

15 boundary-value problems involving bessel functions*

15–1 INTRODUCTION

After the succession of examples in the preceding chapters the student has probably come to the conclusion that the techniques we have developed are adequate to solve any boundary-value problem involving the wave equation, heat equation, or Laplace's equation, at least when the underlying region and boundary conditions are reasonably simple. This, however, is false, as can be seen by considering Laplace's equation in cylindrical coordinates (r, θ, z), where $x = r\cos\theta$, $y = r\sin\theta$, $z = z$. For then

$$\frac{\partial^2 u}{\partial x^2} + \frac{\partial^2 u}{\partial y^2} + \frac{\partial^2 u}{\partial z^2} = 0$$

becomes

$$\frac{\partial^2 u}{\partial r^2} + \frac{1}{r}\frac{\partial u}{\partial r} + \frac{1}{r^2}\frac{\partial^2 u}{\partial \theta^2} + \frac{\partial^2 u}{\partial z^2} = 0; \tag{15–1}$$

and if we attempt to find solutions of (15–1) which are independent of θ and have the form $u(r, z) = R(r)Z(z)$, we see that R and Z must satisfy the equations

$$Z'' - \lambda^2 Z = 0 \tag{15–2}$$

and

$$r^2R'' + rR' + \lambda^2 r^2 R = 0, \quad r > 0. \tag{15–3}$$

(The motive behind writing the constant as $-\lambda^2$ will become clear later.)

Although the first of these equations can be handled with ease, the second causes trouble because its leading coefficient vanishes at $r = 0$. Here the imaginative student might suggest using the power series method to solve (15–3) about $r = a$, $a > 0$, since its coefficients are analytic whenever $r \neq 0$. Unfortunately, this will not do, for the simple reason that the resulting series need not converge outside the interval $(0, 2a)$, whereas we seek solutions valid *for all* $r > 0$. (See Theorem 6–4.) Clearly, then, if we are to make any progress in solving this prob-

---

* The following discussion assumes a knowledge of the material in Chapter 6.

lem, or others of the same ilk, we must devise a method for studying the solutions of a linear differential equation near points where its leading coefficient vanishes. This is precisely what will be done in the sections which follow, where we introduce the celebrated *method of Frobenius* generalizing the power series technique to a large number of non-normal linear differential equations. Once this has been done, we shall use the technique in question to study the solutions of Bessel's equation and the last important class of elementary boundary-value problems.

EXERCISES

1. Use the coordinate transformations given above to show that Laplace's equation becomes

$$\frac{\partial^2 u}{\partial r^2} + \frac{1}{r}\frac{\partial u}{\partial r} + \frac{1}{r^2}\frac{\partial^2 u}{\partial \theta^2} + \frac{\partial^2 u}{\partial z^2} = 0$$

in cylindrical coordinates.

2. Assume that Laplace's equation in cylindrical coordinates has a solution of the form $u(r, z) = R(r)Z(z)$, and show that R and Z must satisfy Eqs. (15–2) and (15–3).

3. Show that the equation

$$r^2R'' + rR' + \lambda^2 r^2 R = 0$$

can be transformed into Bessel's equation of order zero by an appropriate change of variable. (See Section 6–2.)

15–2 REGULAR SINGULAR POINTS

Let

$$p(x)y'' + q(x)y' + r(x)y = 0 \tag{15–4}$$

be a second-order homogeneous linear differential equation whose coefficients are analytic in an open interval I of the x-axis, and suppose that $p(x_0) = 0$ for some x_0 in I. (Such a point is said to be a *singular point* for the equation.) Then, the assumed analyticity of the coefficients of (15–4) implies that the function p has a power series expansion

$$p(x) = \sum_{k=0}^{\infty} a_k(x - x_0)^k,$$

valid in some interval about the point x_0. Moreover, since $p(x_0) = 0$, the leading coefficient in this series must vanish, and there exists a positive integer m such that

$$\begin{aligned} p(x) &= \sum_{k=m}^{\infty} a_k(x - x_0)^k = (x - x_0)^m \sum_{k=m}^{\infty} a_k(x - x_0)^{k-m} \\ &= (x - x_0)^m p_1(x), \end{aligned}$$

where p_1 is analytic at x_0, and $p_1(x_0) \neq 0$. We now divide the coefficients of (15–4) by p_1 to obtain an equation of the form

$$(x - x_0)^m y'' + q_1(x)y' + r_1(x)y = 0, \tag{15–5}$$

whose coefficients are still analytic at x_0, and which is equivalent to (15–4) in an interval about that point. Thus, when studying the solutions of a second-order homogeneous linear differential equation about a singular point x_0 we can always write the equation in this special form.

It turns out that the behavior of the solutions of (15–5) near x_0 is strongly dependent upon the exponent m and the value of q_1 at x_0, and leads to a classification of singular points as "regular" and "irregular" according to the following definition.

Definition 15–1. A point x_0 is said to be a *regular singular point* for a second-order homogeneous linear differential equation if and only if the equation can be written in the form

$$(x - x_0)^2 y'' + (x - x_0)q(x)y' + r(x)y = 0, \tag{15–6}$$

where q and r are analytic at x_0. All other singular points are said to be *irregular.*

Thus, among the singular points for the equation

$$x^2(x + 1)^3(x - 1)y'' + xy' - 2y = 0, \tag{15–7}$$

those at 0 and 1 are regular since (15–7) may be rewritten

$$x^2y'' + \frac{x}{(x + 1)^3(x - 1)}y' - \frac{2}{(x + 1)^3(x - 1)}y = 0,$$

and

$$(x - 1)^2y'' + \frac{x - 1}{x(x + 1)^3}y' - \frac{2(x - 1)}{x^2(x + 1)^3}y = 0$$

about $x = 0$ and $x = 1$, respectively. On the other hand, -1 is an irregular singular point for (15–7) because the coefficients of y and y' in

$$(x + 1)^2y'' + \frac{x + 1}{x(x + 1)^2(x - 1)}y' - \frac{2}{x^2(x + 1)(x - 1)}y = 0$$

are undefined at $x = -1$.

As the terminology suggests, regular singular points are relatively easy to handle. In fact, with appropriate modification, the method of undetermined coefficients introduced in Chapter 6 can be used to obtain a basis for the solution space of any second-order linear differential equation about a regular singular point. Not so when x_0 is an irregular singular point. Here the situation is far more complicated, and it is not difficult to exhibit equations which fail to have series solutions of any form about such a point (see Exercise 11 below). Fortunately, all of the boundary-value problems we shall encounter involve equations whose only singularities are regular, and hence we shall have no more to say about irregular singular points.

Finally, for simplicity, we shall limit the following discussion to singularities at the origin, in which case (15–6) becomes

$$x^2y'' + xq(x)y' + r(x)y = 0.$$

The reader should appreciate that this involves no loss of generality since the change of variable $u = x - x_0$ shifts a singularity from x_0 to 0.

EXERCISES

Find and classify all of the singular points for the equations in Exercises 1 through 10.

1. $x^3(x^2 - 1)y'' - x(x + 1)y' - (x - 1)y = 0$
2. $(3x - 2)^2xy'' + xy' - y = 0$
3. $(x^4 - 1)y'' + xy' = 0$
4. $(x + 1)^4(x - 1)^2y'' - (x + 1)^3(x - 1)y' + y = 0$
5. $x^3(x - 1)y'' + (x - 1)y' + 2xy = 0$
6. $x^3(x + 1)^2y'' - y = 0$
7. $x(1 - x)y'' + (1 - 5x)y' - 4y = 0$
8. Legendre's equation:
$$(1 - x^2)y'' - 2xy' + \lambda(\lambda + 1)y = 0$$
9. Bessel's equation:
$$x^2y'' + xy' + (x^2 - \lambda^2)y = 0$$
10. Laguerre's equation:
$$xy'' + (1 - x)y' + \lambda y = 0$$
11. Prove that the equation
$$x^3y'' + y = 0$$
has no nontrivial series solution of the form
$$\sum_{k=0}^{\infty} a_k x^{k+\nu}$$
for any real number ν.

15–3 EXAMPLES OF SOLUTIONS ABOUT A REGULAR SINGULAR POINT

The simplest example of a second-order equation with a regular singular point at the origin is the Euler equation

$$x^2y'' + axy' + by = 0,$$

a and b (real) constants, studied in Chapter 4. At that time we proved that the solution space of this equation is spanned by a pair of functions y_1 and y_2 constructed from the roots ν_1, ν_2 of the equation

$$\nu(\nu - 1) + a\nu + b = 0,$$

as follows. If $\nu_1 \neq \nu_2$,

$$y_1(x) = |x|^{\nu_1}, \qquad y_2(x) = |x|^{\nu_2};$$

if $\nu_1 = \nu_2 = \nu$,

$$y_1(x) = |x|^{\nu}, \qquad y_2(x) = |x|^{\nu} \ln |x|.*$$

Save for the fact that y_1 and y_2 here appear in closed form, these expressions are typical of the solutions of second-order homogeneous linear differential equations with a regular singular point at the origin. Indeed, we shall find that about $x = 0$ the solution space of such an equation is always spanned by a pair of functions which depend upon the roots of a polynomial equation of degree two, that these functions involve powers of $|x|$, and under certain circumstances a logarithmic term as well. The following example will illustrate these remarks, and introduce the technique which is used to handle the general case.

EXAMPLE. Find the general solution of

$$x^2y'' + x(x - \tfrac{1}{2})y' + \tfrac{1}{2}y = 0 \tag{15–8}$$

on each of the intervals $(0, \infty)$ and $(-\infty, 0)$.

We begin by considering the interval $x > 0$, where we seek a solution of the form

$$y(x) = x^{\nu} \sum_{k=0}^{\infty} a_k x^k = \sum_{k=0}^{\infty} a_k x^{k+\nu}, \tag{15–9}$$

with $a_0 \neq 0$, and ν arbitrary. (This particular guess as to the form of y is motivated by the results obtained for the Euler equation, and from that point of view is not

---

* Recall that when z is complex, $|x|^z = e^{z \ln |x|}$. Hence if $\nu_1 = \alpha + \beta i$, $\nu_2 = \alpha - \beta i$, then

$$y_1(x) = |x|^{\alpha} \sin (\beta \ln |x|), \qquad y_2(x) = |x|^{\alpha} \cos (\beta \ln |x|).$$

unreasonable.) We then have

$$y'(x) = \sum_{k=0}^{\infty} (k + \nu)a_k x^{k+\nu-1},$$

$$y''(x) = \sum_{k=0}^{\infty} (k + \nu)(k + \nu - 1)a_k x^{k+\nu-2},$$

and substitution in (15–8) yields

$$\sum_{k=0}^{\infty} (k + \nu)(k + \nu - 1)a_k x^{k+\nu} + \sum_{k=0}^{\infty} (k + \nu)a_k x^{k+\nu+1} - \frac{1}{2}\sum_{k=0}^{\infty} (k + \nu)a_k x^{k+\nu} + \frac{1}{2}\sum_{k=0}^{\infty} a_k x^{k+\nu} = 0.$$

But since

$$\sum_{k=0}^{\infty} (k + \nu)a_k x^{k+\nu+1} = \sum_{k=1}^{\infty} (k + \nu - 1)a_{k-1} x^{k+\nu},$$

the above expression may be rewritten

$$[\nu(\nu - 1) - \tfrac{1}{2}\nu + \tfrac{1}{2}]a_0 x^{\nu} + \sum_{k=1}^{\infty} [(k + \nu)(k + \nu - 1) - \tfrac{1}{2}(k + \nu) + \tfrac{1}{2}]a_k x^{k+\nu} + \sum_{k=1}^{\infty} (k + \nu - 1)a_{k-1} x^{k+\nu} = 0.$$

Thus if (15–9) is to be a solution of the given equation we must have

$$\nu(\nu - 1) - \tfrac{1}{2}\nu + \tfrac{1}{2} = 0, \tag{15–10}$$

and

$$[(k + \nu)(k + \nu - 1) - \tfrac{1}{2}(k + \nu) + \tfrac{1}{2}]a_k + (k + \nu - 1)a_{k-1} = 0, \quad k \geq 1. \tag{15–11}$$

(Recall that we assumed $a_0 \neq 0$.)

Equation (15–10) determines the admissible values of ν for this problem as $\frac{1}{2}$ and 1, and is known as the *indicial equation* associated with (15–8). We now set

$$I(\nu) = \nu(\nu - 1) - \tfrac{1}{2}\nu + \tfrac{1}{2},$$

and rewrite (15–10) and (15–11) as

$$I(\nu) = 0, \tag{15–12}$$

and

$$I(k + \nu)a_k + (k + \nu - 1)a_{k-1} = 0, \quad k \geq 1. \tag{15–13}$$

In particular, when $\nu = \frac{1}{2}$, $I(k+\nu) = k(k-\frac{1}{2})$, and (15–13) becomes

$$a_k = -\frac{1}{k}a_{k-1}.$$

Thus

$$\begin{aligned} a_1 &= -a_0, \\ a_2 &= \frac{a_0}{2!}, \\ a_3 &= -\frac{a_0}{3!}, \\ &\vdots \\ a_k &= (-1)^k\frac{a_0}{k!}. \end{aligned}$$

Similarly, when $\nu = 1$,

$$a_k = -\frac{2}{2k+1}a_{k-1},$$

and

$$\begin{aligned} a_1 &= -\tfrac{2}{3}a_0, \\ a_2 &= \frac{2^2}{5\cdot 3}a_0, \\ a_3 &= -\frac{2^3}{7\cdot 5\cdot 3}a_0, \\ &\vdots \\ a_k &= (-1)^k\frac{2^k}{(2k+1)(2k-1)\cdots 5\cdot 3}a_0. \end{aligned}$$

We now set $a_0 = 1$ and substitute the above values in (15–9) to obtain the two series

$$y_1(x) = x^{1/2}\sum_{k=0}^{\infty}(-1)^k\frac{x^k}{k!}, \tag{15–14}$$

and

$$y_2(x) = x\sum_{k=0}^{\infty}(-1)^k\frac{(2x)^k}{(2k+1)(2k-1)\cdots 5\cdot 3}, \tag{15–15}$$

each of which *formally* satisfies (15–8) on the interval $(0, \infty)$.

Thus to complete the argument it remains to show that these series converge for all $x > 0$, and that the functions they define are linearly independent in $\mathcal{C}(0, \infty)$. Leaving the latter point as an exercise, we establish convergence by using the ratio test with

$$\rho_1 = \left|\frac{x^{k+3/2}}{(k+1)!}\bigg/\frac{x^{k+1/2}}{k!}\right| = \frac{|x|}{k+1}$$

for the first series, and

$$\rho_2 = \left| \frac{2^{k+1}x^{k+2}}{(2k+3)\cdots 5\cdot 3} \bigg/ \frac{2^k x^{k+1}}{(2k+1)\cdots 5\cdot 3} \right| = \frac{2|x|}{2k+3}$$

for the second. Then

$$\lim_{k\to\infty} \rho_1 = \lim_{k\to\infty} \rho_2 = 0$$

for all x, and the series do converge as desired. Thus the general solution of (15–8) on $(0, \infty)$ is

$$y(x) = c_1 y_1(x) + c_2 y_2(x),$$

where c_1 and c_2 are arbitrary constants.

Finally, to remove the restriction on the interval we observe that the above argument goes through without change if x^ν is replaced by $|x|^\nu$ in (15–9); i.e., by $(-x)^\nu$ for $x < 0$. Thus the general solution of (15–8) on any interval not containing the origin is

$$y(x) = c_1|x|^{1/2} \sum_{k=0}^{\infty} (-1)^k \frac{x^k}{k!} + c_2|x| \sum_{k=0}^{\infty} (-1)^k \frac{(2x)^k}{(2k+1)\cdots 5\cdot 3}.$$

EXERCISES

1. Prove that the functions y_1 and y_2 defined by (15–14) and (15–15) are linearly independent in $\mathcal{C}(0, \infty)$. [*Hint:* Consider the behavior of $c_1 y_1'(x) + c_2 y_2'(x)$ as $x \to 0$.]
2. The equation
$$x^2y'' + x(x - \tfrac{1}{2})y' + \tfrac{1}{2}y = 0$$
has solutions which are defined for *all* x. What are they?
3. Find the indicial equation associated with the regular singular point at $x = 0$ for each of the following equations.
   (a) $x^2y'' + xy' - y = 0$
   (b) $x^2y'' - 2x(x + 1)y' + (x - 1)y = 0$
   (c) $x^2y'' - 2xy' + y = 0$
   (d) $x^2y'' - xy' + (x^2 - \lambda^2)y = 0$, λ a constant
   (e) $xy'' + (1 - x)y' + \lambda y = 0$, λ a constant
4. Prove that the indicial equation associated with
$$x^2y'' + xq(x)y' + r(x)y = 0$$
is
$$\nu(\nu - 1) + q(0)\nu + r(0) = 0$$
whenever q and r are polynomials.

5. Find the indicial equation associated with each of the regular singular points $x = 1$ and $x = -1$ for Legendre's equation

$$(1 - x^2)y'' - 2xy' + \lambda(\lambda + 1)y = 0.$$

Use the method introduced in this section to find two linearly independent solutions of the equations in Exercises 6 through 10. In each case verify that the solutions obtained converge whenever $|x| > 0$, and that they are linearly independent in $\mathcal{C}(0, \infty)$ and $\mathcal{C}(-\infty, 0)$.

6. $2x^2y'' + xy' - y = 0$

7. $9x^2y'' + 3x(x + 3)y' - (4x + 1)y = 0$

8. $xy'' + \dfrac{x+1}{2}y' - y = 0$

9. $8x^2y'' - 2x(x - 1)y' + (x + 1)y = 0$

10. $4x^2y'' + x(2x - 7)y' + 6y = 0$

11. Use the method of this section to prove that Laguerre's equation of order λ,

$$xy'' + (1 - x)y' + \lambda y = 0,$$

has a solution which is analytic for all x, and which reduces to a polynomial when λ is a non-negative integer.

15–4 SOLUTIONS ABOUT A REGULAR SINGULAR POINT: THE GENERAL CASE

In this section we shall indicate how the technique introduced above can always be used to find *at least one* solution of

$$x^2y'' + xq(x)y' + r(x)y = 0 \tag{15–16}$$

about the origin whenever q and r are analytic at $x = 0$. We again begin by letting x be positive, in which case we seek a solution of the form

$$y(x) = x^\nu \sum_{k=0}^{\infty} a_k x^k \tag{15–17}$$

with $a_0 \neq 0$. Then

$$y'(x) = x^\nu \sum_{k=0}^{\infty} (k + \nu)a_k x^{k-1},$$

$$y''(x) = x^\nu \sum_{k=0}^{\infty} (k + \nu)(k + \nu - 1)a_k x^{k-2},$$

and substitution in (15–16) yields

$$\sum_{k=0}^{\infty} (k + \nu)(k + \nu - 1)a_k x^k + q(x)\sum_{k=0}^{\infty} (k + \nu)a_k x^k + r(x)\sum_{k=0}^{\infty} a_k x^k = 0. \tag{15–18}$$

But since q and r are analytic at $x = 0$, we can write

$$q(x) = \sum_{k=0}^{\infty} q_k x^k, \qquad r(x) = \sum_{k=0}^{\infty} r_k x^k, \tag{15–19}$$

where both series converge in an interval $|x| < R_0$, $R_0 > 0$, centered at $x = 0$. Substituting (15–19) in (15–18), we have

$$\sum_{k=0}^{\infty} (k + \nu)(k + \nu - 1)a_k x^k + \left(\sum_{k=0}^{\infty} (k + \nu)a_k x^k\right)\left(\sum_{k=0}^{\infty} q_k x^k\right) + \left(\sum_{k=0}^{\infty} a_k x^k\right)\left(\sum_{k=0}^{\infty} r_k x^k\right) = 0,$$

and if we now carry out the indicated multiplications according to the formula given in Section 6–6 we obtain

$$\sum_{k=0}^{\infty}\left[(k + \nu)(k + \nu - 1)a_k + \sum_{j=0}^{k} (j + \nu)a_j q_{k-j} + \sum_{j=0}^{k} a_j r_{k-j}\right] x^k = 0.$$

Hence (15–17) will *formally* satisfy the given differential equation in the interval $0 < x < R_0$ if and only if

$$(k + \nu)(k + \nu - 1)a_k + \sum_{j=0}^{k} [(j + \nu)q_{k-j} + r_{k-j}]a_j = 0 \tag{15–20}$$

for all $k \geq 0$.

When $k = 0$, (15–20) reduces to

$$\nu(\nu - 1) + q_0\nu + r_0 = 0 \tag{15–21}$$

(recall that $a_0 \neq 0$), and when $k \geq 1$, (15–20) reduces to

$$[(k + \nu)(k + \nu - 1) + q_0(k + \nu) + r_0]a_k + \sum_{j=0}^{k-1} [(j + \nu)q_{k-j} + r_{k-j}]a_j = 0. \tag{15–22}$$

The first of these relations is known as the *indicial equation* associated with (15–16), and its roots, which determine the admissible values of ν in (15–17), are called the *characteristic exponents* of that equation. We direct the reader's attention to the fact that since q_0 and r_0 are, respectively, the constant terms in the series expansions of q and r, (15–21) may be rewritten

$$\nu(\nu - 1) + q(0)\nu + r(0) = 0 \tag{15–23}$$

(cf. Eq. (15–19)). Thus when q and r have been explicitly given, the indicial equation can be obtained directly from (15–16) without undertaking the above computations.

To continue, we now set

$$I(\nu) = \nu(\nu - 1) + q_0\nu + r_0,$$

and let ν_1 and ν_2 denote the roots of the equation $I(\nu) = 0$. Moreover, for convenience we suppose that ν_1 and ν_2 have been labeled in such a way that $\text{Re}(\nu_1) \geq \text{Re}(\nu_2)$.* Then, when $\nu = \nu_1$, (15–22) becomes

$$I(k + \nu_1)a_k + \sum_{j=0}^{k-1} [(j + \nu_1)q_{k-j} + r_{k-j}]a_j = 0, \tag{15–24}$$

and, by the way in which ν_1 was chosen, we know that $I(k + \nu_1) \neq 0$ for $k = 1, 2, \ldots$. Hence (15–24) can be solved for a_k to yield the *recurrence relation*

$$a_k = -\frac{1}{I(k + \nu_1)} \sum_{j=0}^{k-1} [(j + \nu_1)q_{k-j} + r_{k-j}]a_j, \quad k \geq 1, \tag{15–25}$$

which serves to determine all of the a_k from $k = 1$ onward in terms of a_0. And with this we have succeeded in producing a *formal* solution of Eq. (15–16), valid in the interval $0 < x < R_0$. Moreover, if x^ν is replaced by $|x|^\nu$ throughout these computations, this result obviously holds in the interval $-R_0 < x < 0$ as well. Finally, it can be shown that the series obtained in this way always converges if $0 < |x| < R_0$.† Hence the function

$$y(x) = |x|^{\nu_1} \sum_{k=0}^{\infty} a_k x^k \tag{15–26}$$

with a_0 arbitrary and the a_k given by (15–25) is a solution of (15–16) when $0 < |x| < R_0$.

To complete the discussion we must now find a second solution of (15–16), linearly independent of the one just obtained. To this end we attempt to repeat the above argument using the second root ν_2 of the indicial equation. Of course, if $\nu_1 = \nu_2$ we get nothing new. However, if $\nu_1 \neq \nu_2$, (15–24) becomes

$$I(k + \nu_2)a_k + \sum_{j=0}^{k-1} [(j + \nu_2)q_{k-j} + r_{k-j}]a_j = 0, \tag{15–27}$$

which can again be solved for a_k *provided* $I(k + \nu_2) \neq 0$ *for all* $k \geq 1$. But when $k > 0$,

$$I(k + \nu_2) = 0$$

---

* By $\text{Re}(\nu)$ we understand the real part of the complex number ν. Thus if $\nu = \alpha + \beta i$, $\text{Re}(\nu) = \alpha$. The real part of a real number is, of course, the number itself.

† A proof of this fact can be found in the book by Coddington cited in the bibliography.

if and only if $k + \nu_2 = \nu_1$; i.e., if and only if $\nu_1 - \nu_2 = k$. [Recall that $\text{Re}(\nu_1) \geq \text{Re}(\nu_2)$.] Thus the above method will produce a second solution of (15–16) of the form

$$y(x) = |x|^{\nu_2} \sum_{k=0}^{\infty} a_k x^k, \quad a_0 \neq 0, \tag{15–28}$$

valid for $0 < |x| < R_0$ *whenever the roots of the indicial equation* $I(\nu) = 0$ *do not differ by an integer*. In this case it is easy to show that the (particular) solutions y_1 and y_2 obtained by setting $a_0 = 1$ in (15–26) and (15–28) are linearly independent, and hence that the general solution of (15–16) in a neighborhood of the origin is

$$y(x) = c_1 y_1(x) + c_2 y_2(x),$$

where c_1 and c_2 are arbitrary constants. This, for instance, is precisely what happened in the example given in the preceding section.

EXERCISES

Find two linearly independent solutions about $x = 0$ for each of the following equations, and state for what values of x these solutions are valid.

1. $x(x - 4)y'' + (x - 2)y' - 4y = 0$
2. $2x^2y'' + 5xy' - 2y = 0$
3. $8x(x + 4)y'' - 8y' + y = 0$
4. $2xy'' + 3y' - \dfrac{1}{x - 1}y = 0$
5. $3x^2y'' - \dfrac{2x}{x - 1}y' + \dfrac{2}{x - 1}y = 0$
6. $2x^2y'' + x(x - 1)y' + 2y = 0$
7. $4x^2(x + 1)y'' + x(3x - 1)y' + y = 0$
8. $x^2y'' + xy' + (x^2 - \frac{1}{9})y = 0$
9. $3x^2y'' + xy' - (1 + x)y = 0$
10. $2x^2y'' + \dfrac{x(7x + 1)}{x + 1}y' + \dfrac{1}{x + 1}y = 0$

Compute the values of the coefficients a_1, a_2, a_3 in the series solutions of each of the following equations. (Assume $a_0 = 1$.)

11. $x^2y'' + x(x + 1)y' + y = 0$
12. $16x^2y'' - 4x(x^2 - 4)y' - y = 0$
13. $x^2(x^2 - 1)y'' - xy' - 2y = 0$

14. $8x^2(x - 2)y'' + 2xy' - (\cos x)y = 0$

15. $x^2y'' + xe^x y' + y = 0$

16. Let y_1 and y_2 denote the functions obtained from (15–26) and (15–28) by setting $a_0 = 1$. Prove that y_1 and y_2 are linearly independent in $\mathcal{C}(0, R_0)$ and $\mathcal{C}(-R_0, 0)$. [*Hint:* See Exercise 1, Section 15–3.]

*17. (a) Show that $x = 1$ is a regular singular point for Legendre's equation

$$(1 - x^2)y'' - 2xy' + \lambda(\lambda + 1)y = 0,$$

and find a solution about this point of the form

$$y(x) = |x - 1|^\nu \sum_{k=0}^{\infty} a_k(x - 1)^k.$$

(b) Determine the values of x for which this solution is valid.

15–5 SOLUTIONS ABOUT A REGULAR SINGULAR POINT: THE EXCEPTIONAL CASES

To complete our study of solutions about regular singular points it remains to consider the case where the roots ν_1, ν_2 of the equation $I(\nu) = 0$ differ by an integer. Our experience with the Euler equation suggests that a solution involving a logarithmic term should arise when $\nu_1 = \nu_2$, and, as we shall see, this can also happen when $\nu_1 \neq \nu_2$. The following theorem gives a complete description of the situation, both in the general case treated above, and in each of the exceptional cases.

Theorem 15–1. *Let*

$$x^2y'' + xq(x)y' + r(x)y = 0 \tag{15–29}$$

be a second-order homogeneous linear differential equation whose coefficients are analytic in the interval $|x| < R_0$ $(R_0 > 0)$, *let* ν_1 *and* ν_2 *be the roots of the indicial equation*

$$\nu(\nu - 1) + q(0)\nu + r(0) = 0,$$

and suppose that ν_1 *and* ν_2 *have been labeled so that* $\mathrm{Re}(\nu_1) \geq \mathrm{Re}(\nu_2)$. *Then* (15–29) *has two linearly independent solutions* y_1 *and* y_2, *valid for* $0 < |x| < R_0$, *whose form depends upon* ν_1 *and* ν_2, *as follows:*

Case 1. $\nu_1 - \nu_2$ *not an integer. Then*

$$y_1(x) = |x|^{\nu_1} \sum_{k=0}^{\infty} a_k x^k, \quad a_0 = 1,$$

$$y_2(x) = |x|^{\nu_2} \sum_{k=0}^{\infty} b_k x^k, \quad b_0 = 1.$$

Case 2. $\nu_1 = \nu_2 = \nu$. *Then*

$$y_1(x) = |x|^{\nu} \sum_{k=0}^{\infty} a_k x^k, \quad a_0 = 1,$$

$$y_2(x) = |x|^{\nu} \sum_{k=1}^{\infty} b_k x^k + y_1(x) \ln |x|.$$

Case 3. $\nu_1 - \nu_2$ *a positive integer. Then*

$$y_1(x) = |x|^{\nu_1} \sum_{k=0}^{\infty} a_k x^k, \quad a_0 = 1,$$

$$y_2(x) = |x|^{\nu_2} \sum_{k=0}^{\infty} b_k x^k + c y_1(x) \ln |x|, \quad b_0 = 1, \quad c \text{ a (fixed) constant.*}$$

Finally, the values of the constants appearing in each of these solutions can be determined directly from the differential equation by the method of undetermined coefficients.

In the remainder of this section we sketch the argument leading to the solution involving a logarithmic term when $\nu_1 = \nu_2$. The reasoning in the case where $\nu_1 - \nu_2$ is a positive integer is similar, but more complicated, and will not be given. The interested reader can find the details in the text by Coddington mentioned above.

We begin by repeating a portion of our earlier argument, and attempt to determine the a_k in

$$x^{\nu} \sum_{k=0}^{\infty} a_k x^k \tag{15-30}$$

so that the resulting expression satisfies (15-29) for $0 < x < R_0$. This time, however, we also regard ν as a variable, and write

$$y(x, \nu) = x^{\nu} \sum_{k=0}^{\infty} a_k x^k. \tag{15-31}$$

Moreover, we assume from the outset that $a_0 = 1$. Then if L denotes the linear differential operator $x^2 D^2 + xq(x)D + r(x)$, we have

$$Ly(x, \nu) = I(\nu)x^{\nu} + x^{\nu} \sum_{k=1}^{\infty} \left\{ I(k + \nu)a_k + \sum_{j=0}^{k-1} [(j + \nu)q_{k-j} + r_{k-j}]a_j \right\} x^k, \tag{15-32}$$

* In certain instances it may happen that $c = 0$. Moreover, the various constants a_0 and b_0 appearing in these expressions can be assigned nonzero values different from 1 without affecting the validity of the results.

where the q_{k-j} and r_{k-j} are the coefficients in the power series expansions of q and r about the origin. [See (15–21) and (15–22).] We now use the recurrence relation

$$I(k+\nu)a_k + \sum_{j=0}^{k-1} [(j+\nu)q_{k-j} + r_{k-j}]a_j = 0$$

to determine $a_1, a_2, \ldots$ *in terms of* ν in such a way that every term but the first on the right-hand side of (15–32) vanishes. Denoting the resulting expressions by $a_k(\nu)$ and substituting in (15–31), we obtain a function

$$y_1(x, \nu) = x^\nu\left[1 + \sum_{k=1}^{\infty} a_k(\nu)x^k\right] \tag{15–33}$$

with the property that

$$Ly_1(x, \nu) = I(\nu)x^\nu. \tag{15–34}$$

(Recall that, by assumption, $a_0 = 1$.) But since ν_1 is a double root of the indicial equation $I(\nu) = 0$, $I(\nu) = (\nu - \nu_1)^2$, and (15–34) may be written

$$Ly_1(x, \nu) = (\nu - \nu_1)^2 x^\nu. \tag{15–35}$$

Thus $Ly_1(x, \nu) = 0$ when $\nu = \nu_1$, and we conclude that the function $y_1(x, \nu_1)$ is a solution of the equation $Ly = 0$. This, of course, agrees with our earlier results.

The idea for obtaining a second solution in this case originates with the observation that when (15–35) is differentiated *with respect to* ν its right-hand side still vanishes when $\nu = \nu_1$. Indeed,

$$\frac{\partial}{\partial \nu}(\nu - \nu_1)^2 x^\nu = x^\nu(\nu - \nu_1)[2 + (\nu - \nu_1)\ln x].$$

But since

$$\frac{\partial}{\partial \nu}[Ly_1(x, \nu)] = L\left[\frac{\partial}{\partial \nu} y_1(x, \nu)\right]$$

(see Exercise 13), (15–35) implies that

$$L\left[\frac{\partial}{\partial \nu} y_1(x, \nu)\right] = 0$$

when $\nu = \nu_1$. Thus if we differentiate (15–33) term-by-term with respect to ν and then set $\nu = \nu_1$, the resulting expression will formally satisfy the equation $Ly = 0$ for $0 < x < R_0$. Denoting this expression by $y_2(x, \nu_1)$ we have

$$y_2(x, \nu_1) = \frac{\partial}{\partial \nu} y_1(x, \nu)\bigg|_{\nu=\nu_1} = x^{\nu_1}\sum_{k=1}^{\infty} a_k'(\nu_1)x^k + y_1(x, \nu_1)\ln x,$$

which is precisely the form of the second solution given in the statement of Theorem 15–1 under Case 2.

EXERCISES

Find two linearly independent solutions on the positive x-axis for each of the equations in Exercises 1 through 10.

1. $x^2y'' + x(x - 1)y' + (1 - x)y = 0$
2. $xy'' + (1 - x)y' - y = 0$
3. $x^2y'' + 3xy' + (x + 1)y = 0$
4. $x^2y'' + 2x^2y' - 2y = 0$
5. $xy'' - (x + 3)y' + 2y = 0$
6. $xy'' + (2x + 3)y' + 4y = 0$
7. $xy'' + (x^3 - 1)y' + x^2y = 0$
8. $x^2y'' - 2x^2y' + 2(2x - 1)y = 0$
9. $xy'' + (1 - x)y' + 3y = 0$
10. $x^2y'' + x^2y' + (3x - 2)y = 0$
11. Use Theorem 15–1 to determine the form of two linearly independent solutions about $x = 0$ for the equation

$$x^2y'' + xy' + (x^2 - p^2)y = 0,$$

p a real number. Do not compute the solutions.

12. Prove that the solutions y_1 and y_2 given in Cases 2 and 3 of Theorem 15–1 are linearly independent in $\mathcal{C}(0, R_0)$ and $\mathcal{C}(-R_0, 0)$.
13. Verify that

$$\frac{\partial}{\partial \nu}[Ly(x, \nu)] = L\left[\frac{\partial}{\partial \nu}y(x, \nu)\right]$$

when $y(x, \nu)$ is defined by (15–31), and

$$L = x^2D^2 + xq(x)D + r(x).$$

15–6 BESSEL'S EQUATION

In this section we shall use the technique introduced above to find a basis for the solution space of the equation

$$x^2y'' + xy' + (x^2 - p^2)y = 0 \tag{15–36}$$

about the point $x = 0$ under the assumption that p is real. We recall (see Section 6–2) that (15–36) is known as *Bessel's equation of order p*, and, as we shall see, it arises in the study of boundary-value problems involving Laplace's equation and the wave equation.

Since the indicial equation associated with (15–36) is

$$\nu^2 - p^2 = 0, \tag{15–37}$$

and has roots $\pm p$, Theorem 15–1 guarantees that Bessel's equation of order p possesses a solution of the form

$$y_1(x) = x^p \sum_{k=0}^{\infty} a_k x^k, \quad p \geq 0,$$

valid for all x. To evaluate the a_k we observe that

$$(x^2 - p^2)y_1(x) = x^p \sum_{k=2}^{\infty} a_{k-2}x^k - x^p \sum_{k=0}^{\infty} p^2 a_k x^k,$$

$$xy_1'(x) = x^p \sum_{k=0}^{\infty} (k+p)a_k x^k,$$

$$x^2 y_1''(x) = x^p \sum_{k=0}^{\infty} (k+p)(k+p-1)a_k x^k.$$

Thus (15–36) implies that

$$\sum_{k=0}^{\infty} [(k+p)(k+p-1) + (k+p) - p^2]a_k x^k + \sum_{k=2}^{\infty} a_{k-2}x^k \equiv 0,$$

or

$$(2p+1)a_1 x + \sum_{k=2}^{\infty} [k(2p+k)a_k + a_{k-2}]x^k \equiv 0,$$

and we therefore have

$$a_1 = 0, \qquad a_k = -\frac{a_{k-2}}{k(2p+k)}, \quad k \geq 2.$$

From this it immediately follows that

$$a_1 = a_3 = a_5 = \cdots = 0,$$

$$a_2 = -\frac{a_0}{2(2p+2)},$$

$$a_4 = \frac{a_0}{2 \cdot 4(2p+2)(2p+4)},$$

and, in general,

$$a_{2k} = (-1)^k \frac{a_0}{2 \cdot 4 \cdot 6 \cdots (2k)(2p+2)(2p+4)\cdots(2p+2k)}$$

$$= (-1)^k \frac{a_0}{2^{2k}k!(p+1)(p+2)\cdots(p+k)}.$$

Hence

$$y_1(x) = a_0 \sum_{k=0}^{\infty} \frac{(-1)^k}{2^{2k}k!(p+1)(p+2)\cdots(p+k)} x^{2k+p}, \tag{15–38}$$

where $a_0 \neq 0$ is an arbitrary constant.

For various reasons, it turns out to be convenient to set

$$a_0 = \frac{1}{2^p \Gamma(p+1)},$$

where Γ denotes the well-known *gamma function* defined by

$$\Gamma(p) = \int_0^\infty e^{-t} t^{p-1}\, dt, \quad p > 0. \tag{15-39}$$

It can be shown that this integral converges for all $p > 0$, diverges to $+\infty$ when $p = 0$, and has the values

$$\Gamma(1) = 1,$$
$$\Gamma(p + 1) = p\Gamma(p), \quad p > 0,$$

(see Exercise 1). In particular, $\Gamma(n + 1) = n!$ whenever n is a positive integer, and for this reason the gamma function is also known as the *generalized factorial function.* Moreover, if we rewrite the identity $\Gamma(p + 1) = p\Gamma(p)$ as

$$\Gamma(p) = \frac{\Gamma(p + 1)}{p}, \tag{15-40}$$

we can use the values of $\Gamma(p + 1)$ from (15-39) to assign meaning to $\Gamma(p)$ for *nonintegral, negative p.* Indeed, as it stands the right-hand side of (15-40) can be read as the definition of $\Gamma(p)$ for $-1 < p < 0$, since $\Gamma(p + 1)$ is already defined in that interval. This done, we use (15-40) and the values just obtained to extend the definition of $\Gamma(p)$ to the interval $-2 < p < -1$. Continuing in this fashion we obtain a real-valued function defined for all values of the independent variable p, save $p = 0, -1, -2, \ldots$. The graph of the resulting function is shown in Fig. 15-1.

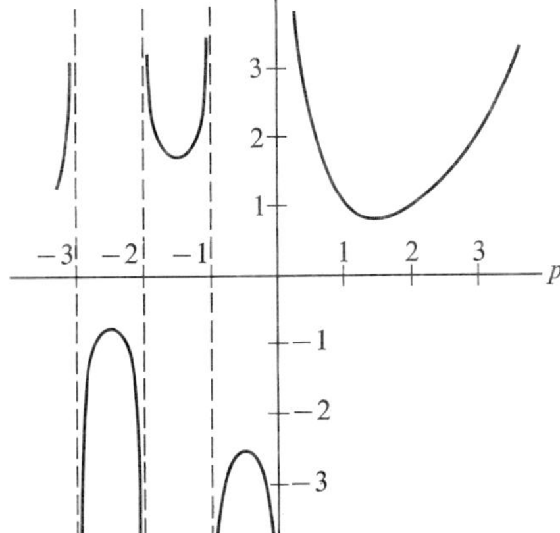

FIGURE 15-1

We now return to (15-38), and set $a_0 = 1/[2^p\Gamma(p + 1)]$ to obtain the first of the two particular solutions needed to solve Bessel's equation. This solution is known as the *Bessel function of order p of the first kind,* and is denoted by J_p. Specifically,

$$J_p(x) = \sum_{k=0}^{\infty} (-1)^k \frac{x^{2k+p}}{2^{2k+p}k!(p + 1)(p + 2)\cdots(p + k)\Gamma(p + 1)};$$

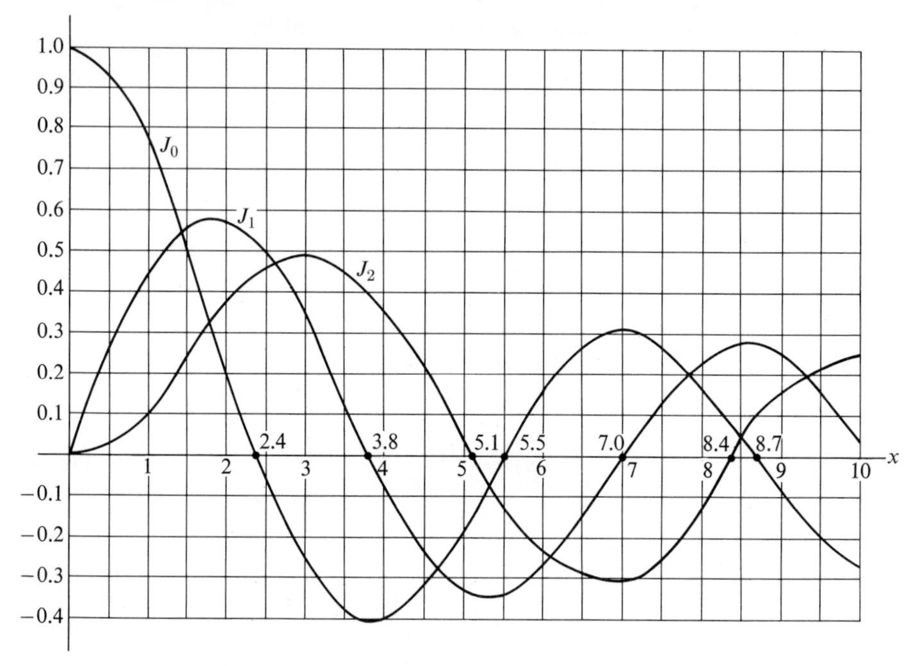

FIGURE 15–2

an expression which can be rewritten in simpler form as*

$$J_p(x) = \sum_{k=0}^{\infty} \frac{(-1)^k}{\Gamma(k+1)\Gamma(p+k+1)} \left(\frac{x}{2}\right)^{2k+p}. \tag{15–41}$$

In particular, when $p = 0$, we have

$$J_0(x) = \sum_{k=0}^{\infty} \frac{(-1)^k}{(k!)^2} \left(\frac{x}{2}\right)^{2k}, \tag{15–42}$$

and, more generally, when p is a non-negative integer n,

$$J_n(x) = \sum_{k=0}^{\infty} \frac{(-1)^k}{k!(n+k)!} \left(\frac{x}{2}\right)^{2k+n}. \tag{15–43}$$

The graphs of J_0, J_1, and J_2 are sketched in Fig. 15–2.

---

* Note that

$$\begin{aligned}\Gamma(p+1)\{(p+1)(p+2)\cdots(p+k)\} &= \Gamma(p+2)\{(p+2)\cdots(p+k)\}\\ &\;\vdots\\ &= \Gamma(p+k)\{(p+k)\}\\ &= \Gamma(p+k+1).\end{aligned}$$

To complete our discussion of Bessel's equation, it remains to find a second solution, linearly independent of J_p. Again we use Theorem 15–1, dividing the argument into cases depending upon the value of p.

Case 1. $p > 0$, *$2p$ not an integer.** Here the roots of the indicial equation associated with (15–36) do *not* differ by an integer, and a second solution can be obtained by repeating the above argument with $-p$ in place of p. Obviously this will lead to a series whose coefficients have the same form as before, and since the gamma function is defined for nonintegral negative values of its argument, the solution in question can be written

$$J_{-p}(x) = \sum_{k=0}^{\infty} \frac{(-1)^k}{\Gamma(k+1)\Gamma(k-p+1)} \left(\frac{x}{2}\right)^{2k-p} \tag{15–44}$$

for $x > 0$. (For negative values of x we must replace x^{-p} by $|x|^{-p}$. From now on, however, we shall restrict our attention to the positive x-axis.) Finally, we observe that (15–44) is defined even when p is of the form $n + \frac{1}{2}$, n an integer, and again yields a solution which is linearly independent of J_p. (See Exercises 6 and 11 below.) Hence we conclude that *the general solution of Bessel's equation of order p is*

$$y(x) = c_1 J_p(x) + c_2 J_{-p}(x),$$

whenever p is not an integer.

Case 2. $p = 0$. Here Bessel's equation becomes

$$xy'' + y' + xy = 0, \tag{15–45}$$

and its indicial equation has zero as a repeated root. Hence, by Theorem 15–1, we can find a second solution of the form

$$K_0(x) = \sum_{k=1}^{\infty} b_k x^k + J_0(x) \ln x, \tag{15–46}$$

with J_0 as above. To evaluate the b_k we note that

$$xK_0(x) = \sum_{k=3}^{\infty} b_{k-2} x^{k-1} + xJ_0(x) \ln x,$$

$$K_0'(x) = \sum_{k=1}^{\infty} k b_k x^{k-1} + J_0'(x) \ln x + \frac{J_0(x)}{x},$$

$$xK_0''(x) = \sum_{k=1}^{\infty} k(k-1) b_k x^{k-1} + xJ_0''(x) \ln x + 2J_0'(x) - \frac{J_0(x)}{x}.$$

* Note that the difference of the roots of (15–37) will be an integer if and only if p is an integer or half an integer.

Thus (15–45) yields

$$b_1 + 4b_2x + \sum_{k=3}^{\infty} [k^2b_k + b_{k-2}]x^{k-1}$$
$$+ [xJ_0''(x) + J_0'(x) + xJ_0(x)] \ln x + 2J_0'(x) \equiv 0;$$

and since $xJ_0''(x) + J_0'(x) + xJ_0(x) \equiv 0$, we have

$$b_1 + 4b_2x + \sum_{k=3}^{\infty} [k^2b_k + b_{k-2}]x^{k-1} \equiv -2J_0'(x).$$

Finally, by (15–42),

$$J_0'(x) = \sum_{k=1}^{\infty} (-1)^k \frac{2k}{2^{2k}(k!)^2} x^{2k-1},$$

whence

$$b_1 + 4b_2x + \sum_{k=3}^{\infty} [k^2b_k + b_{k-2}]x^{k-1} \equiv \sum_{k=1}^{\infty} (-1)^{k+1} \frac{4k}{2^{2k}(k!)^2} x^{2k-1}.$$

To facilitate the evaluation of the b_k we now multiply this expression by x and split the series on the left into its even and odd parts to obtain

$$b_1x + \sum_{k=1}^{\infty} [(2k+1)^2b_{2k+1} + b_{2k-1}]x^{2k+1} + 4b_2x^2$$
$$+ \sum_{k=2}^{\infty} [(2k)^2b_{2k} + b_{2k-2}]x^{2k} \equiv x^2 + \sum_{k=2}^{\infty} (-1)^{k+1} \frac{4k}{2^{2k}(k!)^2} x^{2k}.$$

Thus $b_1 = b_3 = b_5 = \cdots = 0$, while

$$4b_2 = 1, \quad \text{and} \quad (2k)^2b_{2k} + b_{2k-2} = (-1)^{k+1} \frac{4k}{2^{2k}(k!)^2}, \quad k > 1.$$

Hence

$$b_2 = \frac{1}{2^2},$$
$$b_4 = -\frac{1}{2^2 \cdot 4^2}(1 + \tfrac{1}{2}) = -\frac{1}{2^4(2!)^2}(1 + \tfrac{1}{2}),$$
$$\vdots$$
$$b_{2k} = (-1)^{k+1} \frac{1}{2^{2k}(k!)^2}\left(1 + \frac{1}{2} + \cdots + \frac{1}{k}\right) \tag{15–47}$$

(see Exercise 4), and it follows that

$$K_0(x) = \sum_{k=1}^{\infty} \frac{(-1)^{k+1}}{(k!)^2}\left(1 + \frac{1}{2} + \cdots + \frac{1}{k}\right)\left(\frac{x}{2}\right)^{2k} + J_0(x) \ln x.$$

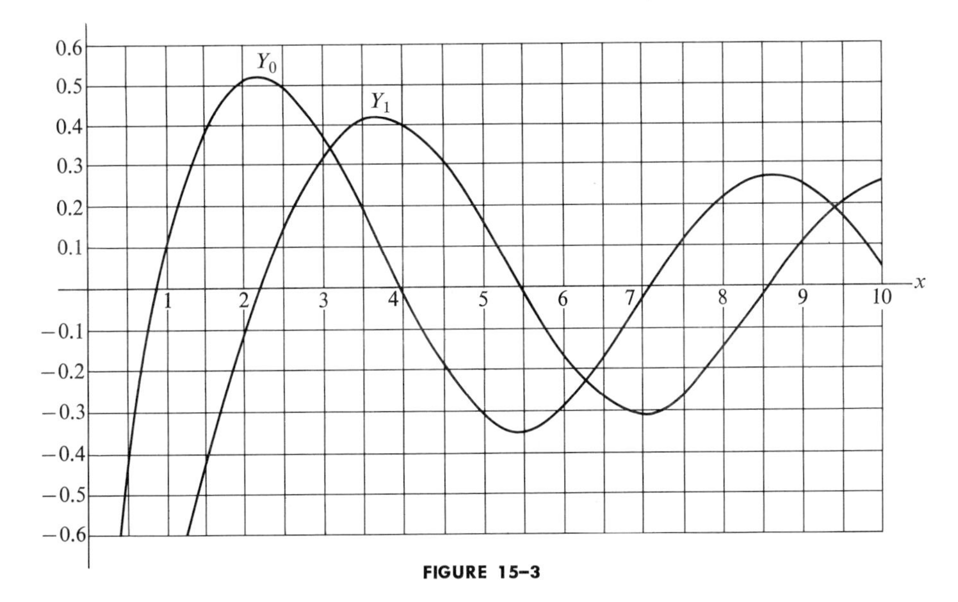

FIGURE 15–3

In theoretical work with Bessel functions it is common practice to replace K_0 by a certain linear combination of J_0 and K_0. The resulting function is known as the *Bessel function of order zero of the second kind*, and is defined by the formula

$$Y_0(x) = -\frac{2}{\pi}\sum_{k=1}^{\infty}\frac{(-1)^k}{(k!)^2}\left(1+\frac{1}{2}+\cdots+\frac{1}{k}\right)\left(\frac{x}{2}\right)^{2k} + \frac{2}{\pi}J_0(x)\left[\ln\frac{x}{2}+\gamma\right], \tag{15–48}$$

where $\gamma = 0.57721566\ldots$, and is known as *Euler's constant*.* The graph of Y_0 is shown in Fig. 15–3.

Case 3. $p = n$, *an integer.* This time the roots of (15–37) differ by $2n > 0$, and Theorem 15–1 asserts that the second solution of Bessel's equation is of the form

$$K_n(x) = \sum_{k=0}^{\infty} b_k x^{k+n} + cJ_n(x)\ln x,$$

where c is a constant. Here too the b_k and c can be evaluated by the method of undetermined coefficients, but the argument is now exceptionally long and involved. Fortunately, we shall not have to use these functions in any of our later work,

* The constant γ is defined to be the sum of the series

$$1 + \sum_{n=2}^{\infty}\left(\frac{1}{n} + \ln\frac{n-1}{n}\right).$$

and thus can omit the argument in favor of the result. It is

$$K_n(x) = -\frac{1}{2}\sum_{k=0}^{n-1} \frac{(n-k-1)!}{k!}\left(\frac{x}{2}\right)^{2k-n} - \frac{H_n}{2n!}\left(\frac{x}{2}\right)^n$$
$$-\frac{1}{2}\sum_{k=1}^{\infty} \frac{(-1)^k[H_k + H_{n+k}]}{k!(n+k)!}\left(\frac{x}{2}\right)^{2k+n} + J_n(x)\ln x,$$

where $H_n = 1 + \frac{1}{2} + \cdots + 1/n$.

Finally, we remark that it is customary to replace K_n by a linear combination of J_n and K_n, denoted Y_n, and called *the Bessel function of order n of the second kind.* It is defined by the formula

$$Y_n(x) = -\frac{1}{\pi}\sum_{k=0}^{n-1} \frac{(n-k-1)!}{k!}\left(\frac{x}{2}\right)^{2k-n} - \frac{H_n}{\pi(n!)}\left(\frac{x}{2}\right)^n$$
$$-\frac{1}{\pi}\sum_{k=1}^{\infty} \frac{(-1)^k[H_k + H_{k+n}]}{k!(n+k)!}\left(\frac{x}{2}\right)^{2k+n} + \frac{2}{\pi}J_n(x)\left[\ln\frac{x}{2} + \gamma\right].$$

EXERCISES

1. (a) Starting with the definition of $\Gamma(p)$, prove that

$$\Gamma(p+1) = p\Gamma(p).$$

(b) Prove that $\Gamma(1) = 1$, and that $\lim_{p\to 0^+}\Gamma(p) = +\infty$.

2. (a) Show that $\Gamma(\frac{1}{2}) = \sqrt{\pi}$. [*Hint:* Note that

$$[\Gamma(\tfrac{1}{2})]^2 = 4\int_0^\infty\int_0^\infty e^{-(x^2+y^2)}\,dx\,dy,$$

and evaluate the integral by changing to polar coordinates.]

(b) Find the value of $\Gamma(n/2)$, n an integer.

3. Discuss the behavior of $J_p(x)$ as $x \to 0$.

4. Prove Formula (15–47).

5. Prove that

$$\frac{d}{dx}J_0(x) = -J_1(x).$$

6. (a) Prove that the functions J_p and J_{-p} are linearly independent in $\mathcal{C}(0, \infty)$ for all nonintegral values of p.

(b) Show that the function Y_0 defined in the text is a linear combination of J_0 and K_0.

7. Prove that

$$J_{-n}(x) = (-1)^n J_n(x)$$

for all integers n.

8. (a) Prove that the general solution of

$$y'' - \frac{2\alpha - 1}{x} y' + \beta^2 y = 0, \quad x > 0,$$

is

$$y = x^\alpha Z_\alpha(\beta x),$$

where Z_α denotes the general solution of Bessel's equation of order α, and β is a (real) constant not zero.

(b) Use the result in (a) to find the general solution of

$$y'' + \frac{a}{x} y' + by = 0, \quad a \text{ and } b \text{ constants}, \ b > 0,$$

and then show that

$$J_{1/2}(x) = \sqrt{\frac{2}{\pi x}} \sin x.$$

9. (a) Prove that the general solution of the equation $u'' + x^2 u = 0$ is

$$u = \sqrt{x}\, Z_{1/4}\left(\frac{x^2}{2}\right),$$

where $Z_{1/4}$ is the general solution of Bessel's equation of order $\frac{1}{4}$.

(b) Solve the Riccati equation $y' = x^2 + y^2$. [*Hint:* Make the change of variable $y = -u'/u$ and use the result in (a).]

10. (a) Prove that the function $y = e^{\alpha x} Z_p(\beta x)$ is the general solution of the differential equation

$$y'' + \left(\frac{1}{x} - 2\alpha\right) y' + \left(\alpha^2 + \beta^2 - \frac{\alpha}{x} - \frac{p^2}{x^2}\right) y = 0$$

whenever Z_p is the general solution of Bessel's equation of order p, and α and β are (real) constants, $\beta \neq 0$.

(b) Use the result in (a) to find the general solution of

$$xy'' + (2x + 1)y' + (5x + 1)y = 0.$$

*11. (a) Prove that the Wronskian of J_p and J_{-p} satisfies the differential equation

$$\frac{d}{dx} [xW(J_p, J_{-p})] = 0.$$

[*Hint:* Use the Lagrange identity (Section 12–5).]

(b) Use the result in (a) to prove that $W[J_p, J_{-p}] = c/x$, where c is a constant, and then use the series expansions for J_p, J_{-p}, and their derivatives to show that

$$c = -\frac{2}{\Gamma(1 - p)\Gamma(p)}$$

whenever p is not an integer. (Note that this argument also provides a proof of the fact that J_p and J_{-p} are linearly independent when p is not an integer.)

15–7 PROPERTIES OF BESSEL FUNCTIONS

Now that we have the series expansions for J_p and Y_p, we are in a position to derive a number of important formulas involving Bessel functions and their derivatives. The first two are immediate consequences of (15–41), and read as follows:

$$\frac{d}{dx}[x^p J_p(x)] = x^p J_{p-1}(x), \tag{15–49}$$

$$\frac{d}{dx}[x^{-p} J_p(x)] = -x^{-p} J_{p+1}(x), \tag{15–50}$$

for all p, positive, negative, or zero.

Indeed, by (15–41) and the identity $\Gamma(p + k + 1) = (p + k)\Gamma(p + k)$, we have

$$\begin{aligned}
\frac{d}{dx}[x^p J_p(x)] &= \frac{d}{dx}\sum_{k=0}^{\infty} \frac{(-1)^k}{2^{2k+p} k!\,\Gamma(k + p + 1)} x^{2k+2p} \\
&= \sum_{k=0}^{\infty} \frac{(-1)^k 2(p + k)}{2^{2k+p} k!\,\Gamma(k + p + 1)} x^{2k+2p-1} \\
&= x^p \sum_{k=0}^{\infty} \frac{(-1)^k}{2^{2k+p-1} k!\,\Gamma(k + p)} x^{2k+(p-1)} \\
&= x^p J_{p-1}(x).
\end{aligned}$$

This proves (15–49) and, with obvious modifications, (15–50) as well. A similar pair of formulas holds for Y_p, but we omit the proof.

When the derivatives appearing in (15–49) and (15–50) are expanded, these formulas become

$$xJ_p' + pJ_p = xJ_{p-1}, \tag{15–51}$$

and

$$xJ_p' - pJ_p = -xJ_{p+1}, \tag{15–52}$$

from which, by adding and subtracting, we immediately obtain

Theorem 15–2. *The Bessel functions of the first kind satisfy the recurrence relations*

$$xJ_{p+1} - 2pJ_p + xJ_{p-1} = 0, \tag{15–53}$$

and

$$J_{p+1} + 2J_p' - J_{p-1} = 0. \tag{15–54}$$

EXAMPLE (*Bessel functions of the first kind of half-integral order*). When $p = \frac{1}{2}$, (15–41) becomes

$$J_{1/2}(x) = \frac{\sqrt{x}}{\sqrt{2}} \sum_{k=0}^{\infty} \frac{(-1)^k}{k!\,\Gamma(\frac{3}{2} + k)} \left(\frac{x}{2}\right)^{2k}.$$

But since

$$\Gamma(\tfrac{3}{2} + k) = \Gamma(\tfrac{3}{2}) \left\{ \frac{3}{2} \cdot \frac{5}{2} \cdots \frac{2k+1}{2} \right\}$$
$$= \frac{3 \cdot 5 \cdots (2k+1)}{2^k} \Gamma(\tfrac{3}{2}),$$

we find that

$$J_{1/2}(x) = \frac{\sqrt{x}}{\sqrt{2}\,\Gamma(\frac{3}{2})} \sum_{k=0}^{\infty} \frac{(-1)^k}{2^k k! 3 \cdot 5 \cdots (2k+1)} x^{2k}$$
$$= \frac{\sqrt{x}}{\sqrt{2}\,\Gamma(\frac{3}{2})} \left[1 - \frac{x^2}{2 \cdot 3} + \frac{x^4}{2 \cdot 4 \cdot 3 \cdot 5} - \frac{x^6}{2 \cdot 4 \cdot 6 \cdot 3 \cdot 5 \cdot 7} + \cdots \right]$$
$$= \frac{1}{\sqrt{2x}\,\Gamma(\frac{3}{2})} \left[x - \frac{x^3}{3!} + \frac{x^5}{5!} - \frac{x^7}{7!} + \cdots \right]$$
$$= \frac{1}{\sqrt{2x}\,\Gamma(\frac{3}{2})} \sin x.$$

Finally, referring to Exercise 2 of Section 15–6, we see that

$$\Gamma(\tfrac{3}{2}) = \sqrt{\pi}/2,$$

whence

$$J_{1/2}(x) = \sqrt{\frac{2}{\pi x}} \sin x. \tag{15–55}$$

A similar argument shows that

$$J_{-1/2}(x) = \sqrt{\frac{2}{\pi x}} \cos x, \tag{15–56}$$

and it now follows from (15–53) that *every* Bessel function of the first kind of *half-integral* order (i.e., of order $n + \frac{1}{2}$, n an integer) can be expressed in finite form in terms of elementary functions. For instance,

$$J_{3/2}(x) = -J_{-1/2}(x) + \frac{1}{x} J_{1/2}(x)$$
$$= \sqrt{\frac{2}{\pi x}} \left[\frac{\sin x}{x} - \cos x \right],$$
$$J_{5/2}(x) = -J_{1/2}(x) + \frac{3}{x} J_{3/2}(x)$$
$$= \sqrt{\frac{2}{\pi x}} \left[\frac{3 \sin x}{x^2} - \frac{3 \cos x}{x} - \sin x \right],$$

and so forth. In passing, we remark that the Bessel functions of half-integral order are the *only* Bessel functions with this special property; all others being transcendental functions which cannot be written in closed form in terms of elementary functions.*

The several formulas derived above are of crucial importance in solving boundary-value problems involving Bessel's equation. So too is the information concerning the general behavior of Bessel functions that was established in Section 6–2, and we therefore conclude this section by reviewing and sharpening some of those results.

First, we recall that *every* nontrivial solution of Bessel's equation of order p has infinitely many zeros on the positive x-axis, and that the distance between successive zeros approaches π as x approaches ∞. Thus both J_p and J_{-p} (or Y_p, as the case may be) are "oscillatory" functions, and by the Sonin-Polya theorem we can even assert that the magnitude of their oscillations decreases with increasing x. In short, the graphs of J_p and J_{-p} (or Y_p) have a *damped* oscillatory character; a fact which was borne out in the case of $J_{n+1/2}$ by the results of the preceding example. Actually, this rough description of the behavior of Bessel functions can be made much more precise. For instance, it is not too difficult to show that *every* solution of Bessel's equation of order p can be written in the form

$$y(x) = \frac{A_p}{\sqrt{x}} \sin (x + \omega_p) + \frac{r_p(x)}{\sqrt{x^3}},$$

where A_p and ω_p are constants whose values depend upon p, and r_p is a function, again dependent on p, which is *bounded* as $x \to \infty$. Thus, for large values of x, y differs very little from the damped sinusoidal function

$$\frac{A_p}{\sqrt{x}} \sin (x + \omega_p),$$

and, in particular, it follows that $y(x) \to 0$ as $x \to \infty$.†

Finally, in preparation for our study of boundary-value problems, we prove two further lemmas on the zeros of Bessel functions.

Lemma 15–1. *The zeros of J_p and J_{p+1} are distinct, and alternate on the positive x-axis.*

Proof. By (15–52),

$$xJ'_p(x) = pJ_p(x) - xJ_{p+1}(x).$$

Thus if $J_p(x_0)$ and $J_{p+1}(x_0)$ were to vanish for some $x_0 > 0$, $J'_p(x_0)$ would also

* By definition, the class of elementary functions consists of all rational functions (quotients of polynomials), trigonometric functions, exponential functions, and their inverses.

† For a proof of this fact see Chapter 8 of G. P. Tolstov, *Fourier Series*, Prentice-Hall, Englewood Cliffs, N. J., 1962.

vanish, and the uniqueness theorem for initial-value problems involving second-order linear differential equations would then imply that $J_p \equiv 0$. This is nonsense, and it therefore follows that the zeros of J_p and J_{p+1} are distinct.

To complete the proof, let $\lambda_1 < \lambda_2$ be consecutive positive zeros of J_p. Then by (15–52),

$$J_{p+1}(\lambda_1) = -J_p'(\lambda_1) \qquad \text{and} \qquad J_{p+1}(\lambda_2) = -J_p'(\lambda_2).$$

But, by assumption, $J_p'(\lambda_1)$ and $J_p'(\lambda_2)$ have opposite signs, and the above equalities then imply that J_{p+1} must vanish at least once between λ_1 and λ_2. A similar argument using (15–51) shows that J_p must vanish between consecutive zeros of J_{p+1}, and we are done. ▮

Lemma 15–2. *The function*

$$F(x) = \alpha J_p(x) + \beta x J_p'(x)$$

has infinitely many zeros on the positive x-axis for all values of p and all constants α and β.

Proof. If $\alpha = \beta = 0$ the assertion is obvious; if $\beta = 0$ it has already been proved; if $\alpha = 0$ it is an immediate consequence of our results concerning the behavior of the zeros of J_p. Thus we need only consider the case where $\alpha \neq 0$ and $\beta \neq 0$.

Here we let $\lambda_1 < \lambda_2 < \cdots$ be the positive zeros of J_p. Then since J_p is positive in the interval $(0, \lambda_1)$, negative in the interval (λ_1, λ_2), positive in (λ_2, λ_3), etc., we have

$$J_p'(\lambda_1) < 0, \quad J_p'(\lambda_2) > 0, \quad J_p'(\lambda_3) < 0, \ldots .$$

Hence $F(x)$ alternates in sign at the points $\lambda_1, \lambda_2, \ldots$, and therefore vanishes somewhere between each of them. ▮

EXERCISES

1. Prove that $d/dx[x^{-p}J_p(x)] = -x^{-p}J_{p+1}(x)$ for all p.
2. Express J_3 and J_4 in terms of J_0 and J_1.
3. Express J_5 in terms of J_0 and J_0'.
4. Prove that

$$J_{-1/2}(x) = \sqrt{\frac{2}{\pi x}} \cos x.$$

5. Show that

$$\int J_0(x) \sin x \, dx = xJ_0(x) \sin x - xJ_1(x) \cos x + c$$

and that

$$\int J_0(x) \cos x \, dx = xJ_0(x) \cos x + xJ_1(x) \sin x + c,$$

c an arbitrary constant.

6. Prove that

$$\int J_0(x)\,dx = J_1(x) + \frac{J_2(x)}{x} + \frac{1 \cdot 3}{x^2} J_3(x) + \cdots$$
$$+ \frac{(2n-2)!J_n(x)}{2^{n-1}(n-1)!x^{n-1}} + \frac{(2n)!}{2^n n!}\int \frac{J_n(x)}{x^n}\,dx.$$

7. Prove that

$$xJ_1(x) = 4\sum_{n=1}^{\infty} (-1)^{n+1} n J_{2n}(x).$$

[*Hint:* Use the formula $J_{n+1} + J_{n-1} = (2n/x)J_n$.]

8. (a) Prove that

$$\int_0^\infty J_{n+1}(x)\,dx = \int_0^\infty J_{n-1}(x)\,dx$$

for all positive integers n. [*Hint:* Integrate the recurrence relation $J_{n-1} - J_{n+1} = 2J_n'$.]

(b) Use the result in (a) to show that

$$\int_0^\infty J_n(x)\,dx = 1$$

for all integers $n > 1$. (*Remark.* This result is also valid when $n = 0$.)

9. Use the recurrence relation $J_{n-1} + J_{n+1} = (2n/x)J_n$ together with the result in Exercise 8(b) to show that

$$\int_0^\infty \frac{J_n(x)}{x}\,dx = \frac{1}{n}$$

for all integers $n > 0$.

10. (a) Prove that

$$[J_{p-1}(x)]^2 - [J_{p+1}(x)]^2 = \frac{2p}{x}\frac{d}{dx}[J_p(x)]^2.$$

(b) Use the result in (a) to show that

$$[J_p(x)]^2 = \sum_{k=0}^{\infty} \frac{2(p+2k)}{x}\frac{d}{dx}[J_{p+2k}(x)]^2.$$

11. Show that

$$J_2(x) = J_0(x) + 2J_0''(x).$$

12. Prove that

$$2^k J_n^{(k)}(x) = J_{n-k}(x) - kJ_{n-k+2}(x) + \frac{k(k-1)}{2!}J_{n-k+4}(x) + \cdots$$
$$+ (-1)^k J_{n+k}(x),$$

where

$$J_n^{(k)}(x) = \frac{d^k}{dx^k} J_n(x).$$

13. Prove that

$$\int J_p(x)\,dx = 2\sum_{k=0}^{n-1} J_{p+2k+1}(x) + \int J_{p+2n}(x)\,dx \quad \text{for all } n > 0.$$

14. (a) Prove that

$$\begin{aligned}\frac{d}{dx}[x^t J_p(x)J_q(x)] &= (t+p+q)x^{t-1}J_p(x)J_q(x)\\ &\quad - x^t[J_p(x)J_{q+1}(x) + J_q(x)J_{p+1}(x)],\\ \frac{d}{dx}[x^t J_{p+1}(x)J_{q+1}(x)] &= (t-p-q-2)x^{t-1}J_{p+1}(x)J_{q+1}(x)\\ &\quad + x^t[J_p(x)J_{q+1}(x) + J_q(x)J_{p+1}(x)].\end{aligned}$$

(b) Add the two identities in (a) and then integrate to deduce that

$$\int (t+p+q)x^{t-1}J_p(x)J_q(x)\,dx + \int (t-p-q-2)x^{t-1}J_{p+1}(x)J_{q+1}(x)\,dx$$
$$= x^t[J_p(x)J_q(x) + J_{p+1}(x)J_{q+1}(x)] + c,$$

c an arbitrary constant.

(c) Use the result in (b) to prove that

$$\int x^{2p+1}[J_p(x)]^2\,dx = \frac{x^{2(p+1)}}{2(2p+1)}\{[J_p(x)]^2 + [J_{p+1}(x)]^2\} + c.$$

15. (a) Use the formula in Exercise 14(b) to prove that

$$\begin{aligned}(p+q)\int \frac{J_p(x)J_q(x)}{x}\,dx - J_p(x)J_q(x) &= 2J_{p+1}(x)J_{q+1}(x)\\ &\quad + \left\{[(p+1)+(q+1)]\int \frac{J_{p+1}(x)J_{q+1}(x)}{x}\,dx - J_{p+1}(x)J_{q+1}(x)\right\}.\end{aligned}$$

(b) Iterate the result in (a) to obtain the formula

$$\begin{aligned}(p+q)\int \frac{J_p(x)J_q(x)}{x}\,dx &= 2\sum_{k=0}^{n} J_{p+k}(x)J_{q+k}(x) - J_p(x)J_q(x)\\ &\quad - J_{p+n}(x)J_{q+n}(x) + (p+q+2n)\int \frac{J_{p+n}(x)J_{q+n}(x)}{x}\,dx.\end{aligned}$$

(c) Use the preceding formula to deduce that

$$\int \frac{[J_n(x)]^2}{x}\,dx = -\frac{1}{2n}\left\{[J_0(x)]^2 - [J_n(x)]^2 + 2\sum_{k=1}^{n}[J_k(x)]^2\right\}.$$

16. Discuss the behavior of the positive zeros of the function J_p''.

17. Show that if $y = x^{-p}J_p$, then $xy'' + (2p+1)y' + \lambda^2 xy = 0$.

*18. (a) Prove that the Laplace transform of the function $x^pJ_p(\lambda x)$ is given by the formula

$$\mathcal{L}[x^pJ_p(\lambda x)] = \frac{\lambda^p\Gamma(2p+1)}{2^p\Gamma(p+1)(s^2+\lambda^2)^{(2p+1)/2}}$$

for all non-negative values of x and λ and all non-negative integral p. [*Hint:* Show that $y = x^p J_p(\lambda x)$ is a solution of the equation

$$xy'' + (1 - 2p)y' + \lambda^2 xy = 0,$$

and then deduce that

$$\frac{d\mathcal{L}[y]}{\mathcal{L}[y]} + (1 + 2p)\frac{s\,ds}{s^2 + \lambda^2} = 0.$$

Solve this equation for $\mathcal{L}[y]$, and complete the proof by evaluating the constant of integration.]

(b) Use the result in (a) to show that

$$\mathcal{L}[J_n(\lambda x)] = \frac{\lambda^n}{(s^2 + \lambda^2)^{1/2}[s + (s^2 + \lambda^2)^{1/2}]^n}.$$

19. (a) Use the convolution theorem and Exercise 18(a) to show that

$$\int_0^x J_0(\lambda)J_0(t - \lambda)\,d\lambda = \sin x.$$

(b) Solve the initial-value problem

$$y'' + y = J_0(x); \qquad y(0) = y'(0) = 0,$$

by Laplace transform methods, and then use the result to deduce that

$$\int_0^x \sin(t - \lambda)J_0(\lambda)\,d\lambda = xJ_1(x).$$

*15–8 THE GENERATING FUNCTION

In Section 15–6 we defined the function J_n, n a non-negative integer, to be *the* solution of Bessel's equation of order n whose series expansion about the origin converges for all x and has $1/(2^n n!)$ as its leading coefficient. Although this is certainly the most natural way of approaching the study of Bessel functions of integral order, it is also possible to define these functions by means of a *generating function* $G(x, t)$, in which case $J_n(x)$ appears as the nth coefficient in the series expansion of G developed in powers of t. (See Section 11–8.) To accomplish this we set

$$G(x, t) = e^{(x/2)[t-(1/t)]}, \quad t \neq 0, \tag{15-57}$$

and use the well-known power series expansion for the exponential function to write

$$e^{xt/2} = 1 + \frac{x}{2}t + \frac{x^2}{2^2 2!}t^2 + \cdots + \frac{x^n}{2^n n!}t^n + \cdots,$$

$$e^{-x/2t} = 1 - \frac{x}{2}t^{-1} + \frac{x^2}{2^2 2!}t^{-2} - \cdots + (-1)^n\frac{x^n}{2^n n!}t^{-n} + \cdots.$$

Hence

$$G(x, t) = \left\{ \sum_{n=0}^{\infty} \frac{x^n}{2^n n!} t^n \right\} \left\{ \sum_{n=0}^{\infty} (-1)^n \frac{x^n}{2^n n!} t^{-n} \right\},$$

and since each of these series is absolutely convergent for all x and all $t \neq 0$ we can perform the indicated multiplication and rearrange terms to obtain

$$G(x, t) = \sum_{n=-\infty}^{\infty} J_n(x) t^n,$$

where the $J_n(x)$ are functions of x alone. Moreover, an easy computation reveals that

$$J_n(x) = \left(\frac{x}{2}\right)^n \left[\frac{1}{n!} - \frac{x^2}{2^2(n+1)} + \frac{x^4}{2^4 2!(n+2)!} - \cdots + (-1)^k \frac{x^{2k}}{2^{2k} k!(k+n)!} + \cdots \right]$$

for $n \geq 0$, and that

$$J_{-n}(x) = (-1)^n J_n(x)$$

for $n < 0$, thereby proving

Theorem 15–3. *The function $e^{(x/2)[t-(1/t)]}$ generates the Bessel functions of integral order of the first kind in the sense that*

$$e^{(x/2)[t-(1/t)]} = \sum_{n=-\infty}^{\infty} J_n(x) t^n \tag{15–58}$$

for all x and all $t \neq 0$.

(Incidentally, this result motivates the choice of the coefficient a_0 in the series expansion for J_n that was made in Section 15–6.)

Having proved this theorem it is impossible to resist the temptation to do something with it, since we are now but a step away from a number of important results. To obtain them we make the change of variable $t = e^{i\theta}$ in (15–57), and use the identity $e^{i\theta} = \cos\theta + i \sin\theta$ to write

$$e^{(x/2)[t-(1/t)]} = e^{ix\sin\theta} = \cos(x\sin\theta) + i\sin(x\sin\theta).$$

Thus (15–58) becomes

$$\cos(x\sin\theta) + i\sin(x\sin\theta) = \sum_{n=-\infty}^{\infty} J_n(x)[\cos n\theta + i\sin n\theta],$$

and by equating real and imaginary parts and using the identity

$$J_{-n}(x) = (-1)^n J_n(x),$$

we find that

$$\cos(x \sin\theta) = J_0(x) + 2\sum_{n=1}^{\infty} J_{2n}(x)\cos 2n\theta,$$
$$\sin(x \sin\theta) = 2\sum_{n=0}^{\infty} J_{2n+1}(x)\sin(2n+1)\theta.* \tag{15–59}$$

Continuing, we now multiply the first of these formulas by $\cos 2k\theta$, the second by $\sin 2k\theta$, and integrate the resulting expressions term-by-term over the interval $0 \leq \theta \leq \pi$ (an operation which is legitimate here) to obtain

$$J_{2k}(x) = \frac{1}{\pi}\int_0^{\pi} \cos(x\sin\theta)\cos 2k\theta\, d\theta,$$
$$0 = \frac{1}{\pi}\int_0^{\pi} \sin(x\sin\theta)\sin 2k\theta\, d\theta. \tag{15–60}$$

(See Exercise 1.) In exactly the same way we find that

$$0 = \frac{1}{\pi}\int_0^{\pi} \cos(x\sin\theta)\cos(2k+1)\theta\, d\theta,$$
$$J_{2k+1}(x) = \frac{1}{\pi}\int_0^{\pi} \sin(x\sin\theta)\sin(2k+1)\theta\, d\theta. \tag{15–61}$$

Finally, by adding these results and using the identity

$$\cos(n\theta - x\sin\theta) = \cos n\theta\cos(x\sin\theta) + \sin n\theta\sin(x\sin\theta),$$

we deduce

Theorem 15–4. *If n is a non-negative integer*

$$J_n(x) = \frac{1}{\pi}\int_0^{\pi} \cos(n\theta - x\sin\theta)\, d\theta. \tag{15–62}$$

This formula is sometimes called *Bessel's integral form for* J_n.

Actually, the argument we have just given proves more than this. For if x is held fixed, and a_k, b_k and $\bar{a}_k, \bar{b}_k$ denote, respectively, the Fourier coefficients of the functions $\cos(x\cos\theta)$ and $\sin(x\sin\theta)$ on the interval $0 \leq \theta \leq \pi$, then (15–60) implies that

$$J_{2k}(x) = \frac{a_{2k}}{2},$$

and (15–61) that

$$J_{2k+1}(x) = \frac{\bar{b}_{2k+1}}{2}.$$

---

* The validity of these computations depends upon the fact that (15–58) is absolutely convergent for all $t \neq 0$, *real or complex*.

But by Corollary 8–1, we know that

$$\lim_{k\to\infty} a_{2k} = 0, \qquad \lim_{k\to\infty} \overline{b}_{2k+1} = 0;$$

whence the following theorem.

Theorem 15–5. *For each fixed $x \geq 0$, $J_n(x)$ approaches zero as $n \to \infty$.*

EXERCISES

1. Deduce (15–60) from (15–59) under the assumption that termwise integration is legitimate.
2. Prove that $|J_n(x)| \leq 1$ for all integers n, and all x.
3. Verify that the coefficient $J_n(x)$ appearing in the series

$$e^{(x/2)[t-(1/t)]} = \sum_{n=-\infty}^{\infty} J_n(x)t^n$$

is as asserted in the text.

4. (a) Use the identity

$$e^{(x/2)[t-(1/t)]}e^{(y/2)[t-(1/t)]} = e^{[(x+y)/2][t-(1/t)]}$$

to prove that

$$J_n(x + y) = \sum_{k=-\infty}^{\infty} J_k(x)J_{n-k}(y).$$

(b) Set $n = 0$ in the above formula and show that

$$J_0(x + y) = J_0(x)J_0(y) + 2\sum_{k=1}^{\infty} (-1)^k J_k(x)J_k(y).$$

(c) Prove that

$$[J_0(x)]^2 + 2\sum_{k=1}^{\infty} [J_k(x)]^2 = 1.$$

(d) Prove that

$$J_n(2x) = \sum_{k=0}^{\infty} J_k(x)J_{n-k}(x) + \sum_{k=1}^{\infty} (-1)^k J_k(x)J_{n+k}(x).$$

*5. (a) Show that

$$\Gamma(x) = 2\int_0^{\infty} e^{-r^2} r^{2x-1}\, dr, \quad x > 0,$$

and then use this formula to prove that

$$\frac{\Gamma(p)\Gamma(q)}{\Gamma(p + q)} = 2\int_0^{\pi/2} \sin^{2p-1}\theta \cos^{2q-1}\theta\, d\theta$$

for all $p > 0$, $q > 0$.

(b) Use the result established in (a) to rewrite the series expansion for J_p as

$$J_p(x) = \frac{(x/2)^p}{\Gamma(\frac{1}{2})\Gamma(p + \frac{1}{2})} \sum_{k=0}^{\infty} \frac{(-1)^k}{(2k)!} x^{2k} \int_0^{\pi} \sin^{2p} \theta \cos^{2k} \theta \, d\theta,$$

and then apply the Weierstrass M-test with

$$\sum_{k=0}^{\infty} \frac{|x^{2k}|}{(2k)!}$$

as the comparison series to show that

$$J_p(x) = \frac{(x/2)^p}{\Gamma(\frac{1}{2})\Gamma(p + \frac{1}{2})} \int_0^{\pi} \sin^{2p} \theta \sum_{k=0}^{\infty} \frac{(-1)^k}{(2k)!} (x \cos \theta)^{2k} \, d\theta.$$

(c) Now prove that

$$J_p(x) = \frac{(x/2)^p}{\Gamma(\frac{1}{2})\Gamma(p + \frac{1}{2})} \int_0^{\pi} \sin^{2p} \theta \cos (x \cos \theta) \, d\theta.$$

(This result is known as the *Poisson Integral Form* for J_p.)

The following exercises introduce the so-called *modified Bessel functions of integral order*, $I_n(x)$.

6. (a) Expand the function $e^{(x/2)[t+(1/t)]}$ in powers of t as

$$e^{(x/2)[t+(1/t)]} = \sum_{n=-\infty}^{\infty} I_n(x) t^n,$$

and show that

$$\begin{aligned} I_n(x) &= \frac{x^n}{2^n n!}\left[1 + \frac{x^2}{2(n+2)} + \frac{x^4}{2 \cdot 4(2n+2)(2n+4)} + \cdots\right] \\ &= \sum_{k=0}^{\infty} \frac{1}{k!(n+k)!}\left(\frac{x}{2}\right)^{n+2k} \end{aligned}$$

for all $n \geq 0$, and that

$$I_{-n}(x) = I_n(x).$$

(b) Prove that

$$I_{2n}(-x) = I_{2n}(x), \qquad I_{2n+1}(-x) = -I_{2n+1}(x),$$

and

$$I_n(x) = (-i)^n J_n(ix)$$

for all integers n.

7. (a) Prove that

$$\frac{d}{dx}[x^n I_n(x)] = x^n I_{n-1}(x)$$

$$\frac{d}{dx}[x^{-n} I_n(x)] = x^{-n} I_{n+1}(x).$$

(b) Prove that the functions I_n satisfy the following recurrence relations:

$$I_{n-1} - I_{n+1} = \frac{2n}{x} I_n,$$

$$I_{n-1} + I_{n+1} = 2I_n',$$

$$I_{n-1} = I_n' + \frac{n}{x} I_n,$$

$$I_{n+1} = I_n' - \frac{n}{x} I_n.$$

8. Show that the modified Bessel function $I_n(x)$ is a particular solution of the equation

$$x^2y'' + xy' - (x^2 + n^2)y = 0.$$

15-9 STURM-LIOUVILLE PROBLEMS FOR BESSEL'S EQUATION

In physical applications Bessel's equation usually arises in the "parametric" form

$$x^2y'' + xy' + (\lambda^2x^2 - p^2)y = 0, \tag{15-63}$$

or

$$\frac{d}{dx}\left(x\frac{dy}{dx}\right) + \left(\lambda^2 x - \frac{p^2}{x}\right)y = 0, \tag{15-64}$$

with λ a constant.* As such its solution space is spanned by the functions $J_p(\lambda x)$, and $J_{-p}(\lambda x)$ or $Y_p(\lambda x)$, depending upon the value of p. In this section we propose to establish the existence and orthogonality of eigenfunctions for a number of boundary-value problems involving Eq. (15-64) on the closed unit interval [0, 1] under the assumption that λ is a non-negative real number. Basically we shall use the method developed in Chapter 12, but must modify our arguments in certain particulars to accommodate the singularities in the equation and its solutions at the origin. For simplicity we continue to restrict our attention to Bessel functions of the first kind.

In the first place, the fact that the parameter appearing in (15-64) is multiplied by x implies that we must use the weighted inner product

$$f \cdot g = \int_0^1 f(x)g(x)x\,dx \tag{15-65}$$

in $\mathcal{PC}[0, 1]$ throughout the following discussion. Moreover, in order to ensure that the functions $J_p(\lambda x)$, $\lambda \geq 0$, belong to this space we must demand that p be non-negative. For otherwise, $J_p(\lambda x)$ is unbounded as $x \to 0$, and is not piecewise continuous. Finally, since we shall be working on the interval [0, 1], and

* The reader can verify that (15-63) is a variant of Bessel's equation by making the change of variable $s = \lambda x$ in $s^2y'' + sy' + (s^2 - p^2)y = 0$. Equation (15-64) is, of course, the self-adjoint form of (15-63).

since the leading coefficient of (15–64) vanishes when $x = 0$, we need only impose a boundary condition at the point $x = 1$.

This said, we now consider the Sturm-Liouville problem consisting of

$$\frac{d}{dx}\left(x\frac{dy}{dx}\right) + \left(\lambda^2 x - \frac{p^2}{x}\right)y = 0, \tag{15–66}$$

$p \geq 0$, $\lambda \geq 0$, and the single unmixed boundary condition

$$\beta_1 y(1) + \beta_2 y'(1) = 0, \tag{15–67}$$

$|\beta_1| + |\beta_2| \neq 0$. By the results in Section 12–8 we can assert in advance that eigenfunctions belonging to distinct eigenvalues for this problem are mutually orthogonal in $\mathcal{PC}[0, 1]$. Furthermore, the eigenfunctions belonging to a *positive* eigenvalue λ_0 will be the nonzero multiples of $J_p(\lambda_0 x)$, while the eigenfunctions belonging to the eigenvalue 0 (if 0 *is* an eigenvalue) will be the nonzero multiples of x^p, since in this case (15–66) reduces to the Euler equation

$$x^2y'' + xy' - p^2y = 0.$$

Thus, under the hypotheses imposed upon p and λ we need only examine the functions x^p and $J_p(\lambda x)$, $\lambda > 0$, as potential eigenfunctions. And here we argue as follows.

Case 1. $\beta_2 = 0$. In this case (15–67) becomes

$$y(1) = 0, \tag{15–68}$$

and it follows that $J_p(\lambda x)$ will be an eigenfunction if and only if $J_p(\lambda) = 0$. Hence the positive zeros $\lambda_1 < \lambda_2 < \cdots$ of J_p are eigenvalues and

$$J_p(\lambda_k x), \quad k = 1, 2, \ldots,$$

are eigenfunctions. Finally, since x^p does not vanish when $x = 1$, $\lambda = 0$ is *not* an eigenvalue, and the above list of eigenfunctions is complete.

Case 2. $\beta_1 = 0$. Here (15–67) becomes

$$y'(1) = 0, \tag{15–69}$$

and the positive zeros $\mu_1 < \mu_2 < \cdots$ of J'_p are now eigenvalues (see Lemma 15–2). In addition, the function x^p satisfies (15–69) when $p = 0$, and in that case $\lambda = 0$ is also an eigenvalue. Thus the eigenfunctions for this problem are

$$J_p(\mu_k x), \quad k = 1, 2, \ldots, \quad \text{when } p > 0,$$

and

$$1, \; J_0(\mu_k x), \quad k = 1, 2, \ldots, \quad \text{when } p = 0.$$

Case 3. For various reasons the case $\beta_1 \neq 0, \beta_2 \neq 0$ is not particularly interesting, and is usually replaced by the more general requirement that when $x = 1$, $J_p(\lambda x)$ satisfy the equation

$$\lambda J_p'(\lambda) - hJ_p(\lambda) = 0, \quad h \text{ a constant.} \tag{15-70}$$

(As we shall see, this type of boundary condition arises in the study of heat flow in cylindrical regions, and is not as artificial as one might think.) Since (15–70) has a variable coefficient λ, it does not fall under any of the several types of boundary conditions discussed earlier, and therefore must be treated separately.

We begin by applying Lemma 15–2 to assert that (15–70) has infinitely many positive zeros, $\nu_1 < \nu_2 < \cdots$, meaning, of course, that (15–70) is satisfied whenever $\lambda = \nu_k$. Hence the functions

$$J_p(\nu_k x), \quad k = 1, 2, \ldots,$$

satisfy (15–66) and (15–70), and as a consequence are "eigenfunctions" for this problem. Moreover, reasoning as in Section 12–8, we find that these functions are mutually orthogonal in $\mathcal{PC}[0, 1]$ with respect to the weight function x. (See Exercise 2 below.) Thus, on the face of things, the situation here would appear to be identical with that discussed in each of the preceding cases. This, however, is not quite true, for it turns out that whenever $p \leq h$, Eq. (15–70) has other roots in addition to the ν_k. Indeed, when $p = h$, Eq. (15–70) becomes

$$\lambda J_p'(\lambda) - pJ_p(\lambda) = 0,$$

which, by (15–52), can be rewritten

$$\lambda J_{p+1}(\lambda) = 0,$$

and has $\lambda = 0$ as a root. When $p < h$ the situation is much more complicated, and cannot possibly be treated with the tools we have available. Suffice it to say that the equation then admits a pair of imaginary roots $\pm i\nu_0$ —a phenomenon which does not occur with any of the boundary conditions we have considered heretofore. In the next section we will find that the existence of these additional roots introduces difficulties in the study of series expansions relative to the orthogonal set $\{J_p(\nu_k x)\}$.

In view of the now obvious fact that we will soon be computing series expansions relative to the eigenfunctions found above, we conclude this section by evaluating their norms in $\mathcal{PC}[0, 1]$. The basic formula for all of these computations is given in the following lemma.

Lemma 15–3. *For all non-negative real numbers p and λ,*

$$\int_0^1 [J_p(\lambda x)]^2 x\, dx = \tfrac{1}{2}[J_p'(\lambda)]^2 + \frac{\lambda^2 - p^2}{2\lambda^2}[J_p(\lambda)]^2. \tag{15-71}$$

*More generally,**

$$\int_0^b [J_p(\lambda x)]^2 x\,dx = \frac{b^2}{2}[J_p'(\lambda b)]^2 + \frac{\lambda^2 b^2 - p^2}{2\lambda^2}[J_p(\lambda b)]^2. \tag{15–72}$$

Proof. We begin by multiplying Bessel's equation by $2y'$ to obtain

$$2x^2 y'y'' + 2x(y')^2 + 2(x^2 - p^2)y'y = 0,$$

or

$$[x^2(y')^2]' + 2(x^2 - p^2)y'y = 0.$$

Rewriting this equation as

$$[x^2(y')^2]' + [(x^2 - p^2)y^2]' - 2xy^2 = 0,$$

we conclude that

$$2x[J_p(x)]^2 = \frac{d}{dx}\{x^2[J_p'(x)]^2 + (x^2 - p^2)[J_p(x)]^2\}.$$

Hence

$$2\int_0^\lambda [J_p(x)]^2 x\,dx = x^2[J_p'(x)]^2\Big|_0^\lambda + (x^2 - p^2)[J_p(x)]^2\Big|_0^\lambda$$
$$= \lambda^2[J_p'(\lambda)]^2 + (\lambda^2 - p^2)[J_p(\lambda)]^2;$$

the last step following from the fact that $pJ_p(0) = 0$ for all $p \geq 0$. The desired result now follows by setting $x = \lambda t$ in the above integral. ▮

When rewritten in terms of the inner product on $\mathcal{PC}[0, 1]$, Eq. (15–71) becomes

$$\|J_p(\lambda x)\|^2 = \tfrac{1}{2}[J_p'(\lambda)]^2 + \frac{\lambda^2 - p^2}{2\lambda^2}[J_p(\lambda)]^2, \tag{15–73}$$

and yields the following important formulas.

Case 1. λ_k *the kth positive zero of* J_p:

$$\|J_p(\lambda_k x)\| = \tfrac{1}{2}[J_p'(\lambda_k)]^2. \tag{15–74}$$

The reader should note that by using Formula (15–52) this result can also be written

$$\|J_p(\lambda_k x)\|^2 = \tfrac{1}{2}[J_{p+1}(\lambda_k)]^2. \tag{15–75}$$

Case 2. μ_k *the kth positive zero of* J_p':

$$\|J_p(\mu_k x)\|^2 = \frac{\mu_k^2 - p^2}{2\mu_k^2}[J_p(\mu_k)]^2. \tag{15–76}$$

---

* This formula is applied when working on the interval $[0, b]$.

In particular, when $p = 0$, we have

$$\|J_0(\mu_k x)\|^2 = \tfrac{1}{2}[J_0(\mu_k)]^2, \tag{15-77}$$

and

$$\|1\|^2 = \int_0^1 x\,dx = \tfrac{1}{2}. \tag{15-78}$$

Case 3. ν_k *the kth positive zero of* $\lambda J_p'(\lambda) - hJ_p(\lambda)$:

$$\|J_p(\nu_k)\|^2 = \frac{h^2 + \nu_k^2 - p^2}{2\nu_k^2}[J_p(\nu_k)]^2. \tag{15-79}$$

EXERCISES

1. Verify that Eq. (15-63) is Bessel's equation of order p in the variable λx.
2. Apply the argument given in Section 12-8 to prove that the functions $J_p(\nu_k x)$, $k = 1, 2, \ldots$, discussed in Case 3 above are mutually orthogonal in $\mathcal{PC}[0, 1]$ with respect to the weight function x.
3. Prove that x^p is orthogonal in $\mathcal{PC}[0, 1]$ to each of the functions $J_p(\nu_k x)$ when $h = p$ in (15-70). What does this imply about the set of functions $J_p(\nu_k x)$, $k = 1, 2, \ldots$?

15-10 BESSEL SERIES OF THE FIRST AND SECOND KINDS

Now that we have established the existence of an infinite set of mutually orthogonal eigenfunctions for each of the boundary-value problems introduced above we are faced with the task of determining whether these sets of functions are bases for $\mathcal{PC}[0, 1]$. With but one exception the answer is yes, and in general the results in this connection are analogous to those we have encountered in similar situations in the past. This time, however, we can do no more than state the relevant theorems, since their proofs are far too difficult to be given here. The first and most important of them reads as follows.

Theorem 15-6. *Let* $\lambda_1 < \lambda_2 < \cdots$ *be the positive zeros of* $J_p(x)$, *and suppose that* $p \geq 0$. *Then the functions* $J_p(\lambda_k x)$, $k = 1, 2, \ldots$, *are a basis for* $\mathcal{PC}[0, 1]$, *and every function in this space can be written uniquely in the form*

$$f(x) = \sum_{k=1}^{\infty} c_k J_p(\lambda_k x), \tag{15-80}$$

where the series in question converges in the mean to f, *and*

$$c_k = \frac{2}{[J_{p+1}(\lambda_k)]^2}\int_0^1 f(x)J_p(\lambda_k x)x\,dx. \tag{15-81}$$

Moreover, if f is piecewise smooth this series converges pointwise to

$$\tfrac{1}{2}[f(x_0^+) + f(x_0^-)]$$

for each x_0 in the interval (0, 1), *and the convergence is uniform on every closed subinterval of* (0, 1) *which does not contain a point of discontinuity of f.*

Remark. The series expansion given by (15–80) and (15–81) is also valid when $-\frac{1}{2} \leq p < 0$ even though the $J_p(\lambda_k x)$ do not belong to $\mathcal{PC}[0, 1]$ for these values of p.

The series described in this theorem is known as the *Bessel* or *Fourier-Bessel series for f of the first kind* with respect to the functions $J_p(\lambda_k x)$. The reader should note that this result actually provides us with infinitely many different series of this kind, one for each admissible value of p.

EXAMPLE 1. Find the Bessel series expansion with respect to $J_p(\lambda_k x)$ for the function x^p, $p \geq 0$.

In this case the coefficients in the series are given by the formula

$$c_k = \frac{2}{[J_{p+1}(\lambda_k)]^2} \int_0^1 x^{p+1} J_p(\lambda_k x)\, dx,$$

and can be evaluated by setting $t = \lambda_k x$ and using Formula (15–49), as follows:

$$\begin{aligned}
\int_0^1 x^{p+1} J_p(\lambda_k x)\, dx &= \frac{1}{\lambda_k^{p+2}} \int_0^{\lambda_k} t^{p+1} J_p(t)\, dt \\
&= \frac{1}{\lambda_k^{p+2}} \int_0^{\lambda_k} \frac{d}{dt}[t^{p+1} J_{p+1}(t)]\, dt \\
&= \frac{1}{\lambda_k^{p+2}} \left[t^{p+1} J_{p+1}(t)\right]_0^{\lambda_k} \\
&= \frac{1}{\lambda_k} J_{p+1}(\lambda_k).
\end{aligned}$$

Thus

$$c_k = \frac{2}{\lambda_k J_{p+1}(\lambda_k)}, \quad k = 1, 2, \ldots,$$

and

$$x^p = 2\left[\frac{J_p(\lambda_1 x)}{\lambda_1 J_{p+1}(\lambda_1)} + \frac{J_p(\lambda_2 x)}{\lambda_2 J_{p+1}(\lambda_2)} + \frac{J_p(\lambda_3 x)}{\lambda_3 J_{p+1}(\lambda_3)} + \cdots\right], \tag{15–82}$$

where the series converges in the mean in $\mathcal{PC}[0, 1]$, pointwise in (0, 1), and uniformly on any closed subinterval of (0, 1).

In particular, when $p = 0$, (15–82) yields the formula

$$\sum_{k=1}^{\infty} \frac{J_0(\lambda_k x)}{\lambda_k J_1(\lambda_k)} = \frac{1}{2}, \quad 0 < x < 1. \tag{15–83}$$

EXAMPLE 2. Expand x^2, $0 < x < 1$, as a series in $J_0(\lambda_k x)$.

Here

$$c_k = \frac{2}{[J_1(\lambda_k)]^2} \int_0^1 x^3 J_0(\lambda_k x)\, dx,$$

and, reasoning as in the preceding example, we find that

$$\begin{aligned}\int_0^1 x^3 J_0(\lambda_k x)\, dx &= \frac{1}{\lambda_k^4} \int_0^{\lambda_k} t^3 J_0(t)\, dt \\ &= \frac{1}{\lambda_k^4} \int_0^{\lambda_k} t^2 \frac{d}{dt}[t J_1(t)]\, dt \\ &= \frac{1}{\lambda_k^4} \left[t^3 J_1(t) \Big|_0^{\lambda_k} - 2 \int_0^{\lambda_k} t^2 J_1(t)\, dt \right] \\ &= \frac{J_1(\lambda_k)}{\lambda_k} - \frac{2}{\lambda_k^4} \int_0^{\lambda_k} t^2 J_1(t)\, dt.\end{aligned}$$

But

$$\begin{aligned}\int_0^{\lambda_k} t^2 J_1(t)\, dt &= -\int_0^{\lambda_k} t^2 \frac{d}{dt} J_0(t)\, dt \\ &= -t^2 J_0(t) \Big|_0^{\lambda_k} + 2 \int_0^{\lambda_k} t J_0(t)\, dt \\ &= 2 \int_0^{\lambda_k} \frac{d}{dt}[t J_1(t)]\, dt \\ &= 2\lambda_k J_1(\lambda_k),\end{aligned}$$

and it follows that

$$\int_0^1 x^3 J_0(\lambda_k x)\, dx = \frac{J_1(\lambda_k)}{\lambda_k} - \frac{4 J_1(\lambda_k)}{\lambda_k^3}.$$

Thus

$$c_k = \frac{2}{J_1(\lambda_k)} \left[\frac{1}{\lambda_k} - \frac{4}{\lambda_k^3} \right],$$

and

$$x^2 = 2\sum_{k=1}^{\infty} \frac{1}{J_1(\lambda_k)}\left(\frac{1}{\lambda_k} - \frac{4}{\lambda_k^3}\right) J_0(\lambda_k x), \quad 0 < x < 1.$$

Turning now to the functions $J_p(\mu_k x)$, we have

Theorem 15–7. *Let $\mu_1 < \mu_2 < \cdots$ be the positive zeros of $J_p'(x)$, $p \geq 0$. Then*

$$J_p(\mu_k x), \quad k = 1, 2, \ldots, \quad p > 0,$$

and

$$1,\ J_0(\mu_k x), \quad k = 1, 2, \ldots,$$

are bases for $\mathcal{PC}[0, 1]$, and series expansions relative to these bases converge in exactly the same way as the series described in the preceding theorem.

In particular, it now follows from Formulas (15–76) and (15–77) that a function f in $\mathcal{PC}[0, 1]$ can be written in the form

$$f(x) = \sum_{k=1}^{\infty} c_k J_p(\mu_k x), \quad p > 0, \tag{15–84}$$

with

$$c_k = \frac{2\mu_k^2}{[\mu_k^2 - p^2][J_p(\mu_k)]^2} \int_0^1 f(x) J_p(\mu_k x) x\, dx, \tag{15–85}$$

and

$$f(x) = 2c_0 + \sum_{k=1}^{\infty} c_k J_0(\mu_k x), \tag{15–86}$$

with

$$c_0 = \int_0^1 f(x) x\, dx, \qquad c_k = \frac{2}{[J_0(\mu_k)]^2} \int_0^1 f(x) J_0(\mu_k x) x\, dx, \quad k > 0. \tag{15–87}$$

A series of this type is usually said to be a *Bessel series expansion of f of the second kind*, or a *Dini series expansion* of f.

Finally, a similar result holds for the functions $J_p(\nu_k x)$ discussed in Case 3 of the preceding section *provided we insist that p be greater than h.* On the other hand, when $p \leq h$, an additional function must be added to the $J_p(\nu_k x)$ to obtain a basis for $\mathcal{PC}[0, 1]$. When $p = h$, the function in question is x^p, or any nonzero multiple of it (see Exercise 3, Section 15–9); but when $p < h$, it is considerably more complicated.* However, once this function is included with the $J_p(\nu_k x)$, the analog of Theorems 15–6 and 15–7 goes through unchanged.

---

* It is, in fact, $I_p(\nu_0 x)$, where ν_0 is chosen so that $i\nu_0$ is a root of the equation $\lambda J_p'(\lambda) - hJ(\lambda) = 0$, and I_p is the so-called "modified" Bessel function of order p of the first kind. (See Exercises 6 through 8 of Section 15–8.)

EXERCISES

1. Prove that for $n > 1$

$$\int_0^1 x^n J_0(\lambda x)\,dx = \frac{J_1(\lambda)}{\lambda} + \frac{n-1}{\lambda^2} J_0(\lambda) - \frac{(n-1)^2}{\lambda^2}\int_0^1 x^{n-2} J_0(\lambda x)\,dx.$$

2. Use the formula given in Exercise 1 to show that

$$x = 2\sum_{k=1}^{\infty}\left[\frac{1}{\lambda_k J_1(\lambda_k)} - \frac{1}{\lambda_k^2[J_1(\lambda_k)]^2}\int_0^1 J_0(\lambda_k t)\,dt\right] J_0(\lambda_k x),$$

for $0 < x < 1$.

3. (a) Show that

$$\int_0^1 x J_0(kx) J_0(\lambda_k x)\,dx = \frac{\lambda_k J_0'(\lambda_k) J_0(k)}{k^2 - \lambda_k^2}$$

whenever λ_k is a zero of J_0.

(b) Use the result in (a) to deduce that

$$J_0(kx) = 2J_0(k)\sum_{k=1}^{\infty}\frac{\lambda_k}{(\lambda_k^2 - k^2)J_1(\lambda_k)} J_0(\lambda_k x),$$

for $0 < x < 1$.

4. Prove that

$$1 - x^2 = 8\sum_{k=1}^{\infty}\frac{J_0(\lambda_k x)}{\lambda_k^3 J_1(\lambda_k)}$$

for $0 < x < 1$.

5. Expand the function x^p, $p \geq 0$, in $\mathcal{PC}[0, 1]$ as a series involving the functions $J_p(\mu_k x)$.

6. Prove that

$$\ln x = -2\sum_{k=1}^{\infty}\frac{J_0(\lambda_k x)}{\lambda_k^2[J_1(\lambda_k)]^2}$$

for $0 < x < 1$. [*Hint:* Use integration by parts to evaluate

$$\int_0^1 x \ln x J_0(\lambda_k x)\,dx.]$$

7. (a) Prove that

$$x^{p+1} = 4(p+1)\sum_{k=1}^{\infty}\frac{J_{p+1}(\lambda_k x)}{\lambda_k^2 J_{p+1}(\lambda_k)}$$

for all $p \geq 0$, and all x in $(0, 1)$.

(b) Use the result in (a) to show that

$$\sum_{k=1}^{\infty}\frac{J_{p+2}(\lambda_k x)}{\lambda_k J_{p+1}(\lambda_k)} = 0$$

for $0 < x < 1$. [*Hint:* Multiply the formula in (a) by $x^{-(p+1)}$ and differentiate.]

15–11 LAPLACE'S EQUATION IN CYLINDRICAL COORDINATES

At this point it should be abundantly clear that we have assembled more than enough information to solve boundary-value problems involving Bessel's equation. Indeed, given what we now know, this is largely a matter of routine computation, and involves little that could not be left to the reader's imagination—and the exercises. But, in the interest of completeness, we shall devote the following pages to a brief discussion of several problems of this type, producing a *formal* series solution for each of them. Of course, once this has been done we are still faced with the task of determining conditions under which these series actually satisfy the problems they purport to solve. Here, however, the technical difficulties are formidable, and force us to be content with the vague statement that all of our results are valid whenever the functions involved are sufficiently smooth.

This said, we turn our attention to Laplace's equation in cylindrical regions, which we first propose to solve under the assumption that the solutions are independent of the polar angle θ. In this case the relevant version of Laplace's equation is

$$\frac{\partial^2 u}{\partial r^2} + \frac{1}{r}\frac{\partial u}{\partial r} + \frac{\partial^2 u}{\partial z^2} = 0 \tag{15–88}$$

(see Section 15–1), and we remind the reader that its solutions can be interpreted as steady-state temperature distributions in the region in question.

EXAMPLE 1. Solve Eq. (15–88) in the cylindrical region $r < 1$, $0 < z < a$ (Fig. 15–4) under the assumption that

$$\begin{aligned} u(1, z) &= 0, \\ u(r, a) &= 0, \\ u(r, 0) &= f(r). \end{aligned} \tag{15–89}$$

Applying the method of separation of variables with $u(r, z) = R(r)Z(z)$ we obtain the equations

$$\begin{aligned} R'' + \frac{1}{r}R' + \lambda^2 R &= 0, \\ Z'' - \lambda^2 Z &= 0, \end{aligned} \tag{15–90}$$

λ a (positive) constant, and the endpoint conditions

$$\begin{aligned} R(1) &= 0, \\ Z(a) &= 0. \end{aligned} \tag{15–91}$$

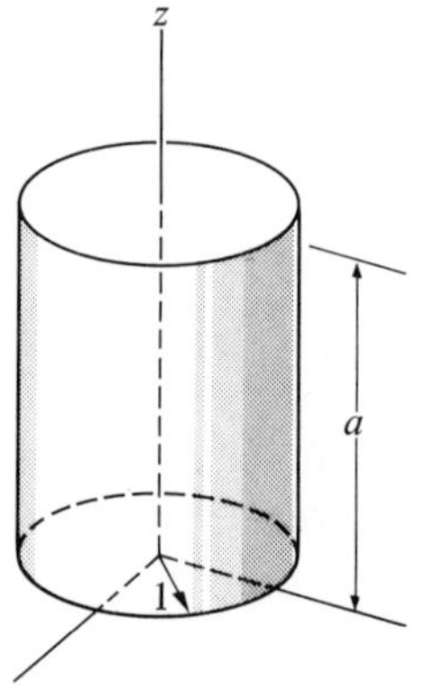

FIGURE 15–4

The first of these equations is Bessel's equation of order zero in the variable λr, and has

$$R(r) = AJ_0(\lambda r) + BY_0(\lambda r),$$

A and B constants, as its general solution. The requirement that R be continuous at the origin forces us to set $B = 0$, while (15–91) implies that $J_0(\lambda) = 0$. Thus the admissible values of λ are the positive zeros of J_0, and we have shown that up to constant multiples R must be one of the functions

$$R_k(r) = J_0(\lambda_k r), \quad k = 1, 2, \ldots.$$

Moreover, when $\lambda = \lambda_k$ the general solution of

$$Z'' - \lambda_k^2 Z = 0, \qquad Z(a) = 0,$$

is

$$Z_k(z) = A_k \sinh \lambda_k(a - z),$$

where A_k is an arbitrary constant (see p. 549). Thus the functions

$$u_k(r, z) = A_k \sinh \lambda_k(a - z) J_0(\lambda_k r), \quad k = 1, 2, \ldots,$$

are solutions of (15–88) and satisfy the boundary conditions $u(1, z) = u(r, a) = 0$.

To accommodate the remaining boundary condition we now form the series

$$u(r, z) = \sum_{k=1}^{\infty} A_k \sinh \lambda_k(a - z) J_0(\lambda_k r), \tag{15–92}$$

set $z = 0$, and replace $u(r, 0)$ by $f(r)$ to obtain

$$f(r) = \sum_{k=1}^{\infty} A_k \sinh (\lambda_k a) J_0(\lambda_k r).$$

By our earlier results we know that this equation can be satisfied for any f in $\mathcal{PC}[0, 1]$ by letting $A_k \sinh (\lambda_k a)$ be the kth coefficient in the series expansion of f relative to the functions $J_0(\lambda_k r)$. Hence

$$A_k = \frac{2}{[J_1(\lambda_k)]^2 \sinh (\lambda_k a)} \int_0^1 f(r) J_0(\lambda_k r) r \, dr,$$

and we are done.

EXAMPLE 2. Find the steady-state temperature in the cylindrical region described above, given that

$$\begin{aligned} u(r, a) &= 0, \\ u_r(1, z) &= hu(1, z), \quad h \text{ a constant}, \\ u(r, 0) &= f(r). \end{aligned} \tag{15–93}$$

Physically these conditions assert that the base of the cylinder is maintained at a known temperature $f(r)$, the top at zero, and that the lateral surface exchanges heat freely with the surrounding medium at a rate proportional to the temperature of the surface.

This time we must solve the pair of equations in (15–90) in the presence of the endpoint conditions

$$\begin{aligned} R'(1) - hR(1) &= 0, \\ Z(a) &= 0. \end{aligned} \tag{15–94}$$

Reasoning as before, we find that up to constant multiples,

$$R(r) = J_0(\lambda r),$$

and (15–94) now implies that λ must be a root of the equation

$$\lambda J_0'(\lambda) - hJ_0(\lambda) = 0.$$

Again our earlier results allow us to assert the existence of infinitely many such roots $\nu_1 < \nu_2 < \cdots$, all of which are positive, and with them, solutions

$$R_k(r) = J_0(\nu_k r), \quad k = 1, 2, \ldots.$$

Since the corresponding solutions of the equation involving Z are

$$Z_k(z) = A_k \sinh \nu_k(a - z),$$

we find that

$$u(r, z) = \sum_{k=1}^{\infty} A_k \sinh \nu_k(a - z) J_0(\nu_k r), \tag{15–95}$$

where the A_k are chosen so that

$$f(r) = \sum_{k=1}^{\infty} A_k \sinh(\nu_k a) J_0(\nu_k r).$$

(Note that the existence of this series for any f in $\mathcal{PC}[0, 1]$ is guaranteed only if $h < 0$. Otherwise a more complicated series must be used.) Thus, by Formula (15–79), we have

$$A_k = \frac{2\nu_k^2}{[h^2 + \nu_k^2][J_0(\nu_k)]^2 \sinh(\nu_k a)} \int_0^1 f(r) J_0(\nu_k r) r \, dr, \tag{15–96}$$

and the solution is complete.

EXAMPLE 3. As our final example of this type we again solve Eq. (15–88) in the region $r < 1$, $0 < z < a$, but now impose the boundary conditions

$$\begin{aligned} u(r, 0) &= 0, \\ u(r, a) &= 0, \\ u(1, z) &= f(z). \end{aligned} \tag{15–97}$$

This time the method of separation of variables leads to the pair of equations

$$\begin{aligned} Z'' + \lambda^2 Z &= 0, \\ R'' + \frac{1}{r} R' - \lambda^2 R &= 0, \end{aligned} \tag{15-98}$$

and endpoint conditions $Z(0) = Z(a) = 0$. (Here the choice of sign on λ^2 is dictated by the requirement that the solutions vanish at the endpoints of the cylinder.) It follows at once that the only possible solutions of the first equation are

$$Z_k(z) = A_k \sin \frac{k\pi z}{a}, \quad k = 1, 2, \ldots,$$

corresponding to the eigenvalues $\lambda_k = k\pi/a$, and it remains to solve

$$R'' + \frac{1}{r} R' - \lambda^2 R = 0 \tag{15-99}$$

for each of these values of λ. To this end we make the change of variable $t = i\lambda r$ $(i = \sqrt{-1})$, and rewrite (15–99) as

$$(-\lambda^2)\frac{d^2R}{dt^2} + \frac{i\lambda}{r}\frac{dR}{dt} - \lambda^2 R = 0$$

or

$$\frac{d^2R}{dt^2} + \frac{1}{t}\frac{dR}{dt} + R = 0,$$

which we recognize as Bessel's equation of order zero in the variable t. Thus, up to constant multiples, the only solutions of (15–99) which are continuous at the origin are of the form

$$R(r) = J_0(i\lambda r).$$

Using the formula for the series expansion of J_0, we have

$$\begin{aligned} J_0(i\lambda r) &= \sum_{k=0}^{\infty} (-1)^k \frac{(i\lambda r)^{2k}}{2^{2k}(k!)^2} \\ &= \sum_{k=0}^{\infty} \frac{(\lambda r)^{2k}}{2^{2k}(k!)^2}, \end{aligned}$$

and it follows that $J_0(i\lambda r)$ is a real-valued function after all. In view of this fact it is reasonable to adopt a notation which does not (misleadingly) involve i, and so we set

$$I_0(x) = \sum_{k=0}^{\infty} \frac{x^{2k}}{2^{2k}(k!)^2}. \tag{15-100}$$

The function I_0 (which the student should regard as being *defined* by this series) is called the *modified Bessel function of order zero of the first kind* (see Exercise 6, Section 15–8), and in the present case allows us to write

$$R(r) = I_0(\lambda r).$$

In particular, when $\lambda = \lambda_k = k\pi/a$,

$$R_k(r) = I_0\left(\frac{k\pi r}{a}\right),$$

and we conclude that the solution of the boundary-value problem under discussion is of the form

$$u(r, z) = \sum_{k=1}^{\infty} A_k I_0\left(\frac{k\pi r}{a}\right) \sin\left(\frac{k\pi z}{a}\right). \tag{15–101}$$

Finally, to determine the values of A_k we set $r = 1$, and $u(1, z) = f(z)$ to obtain

$$f(z) = \sum_{k=1}^{\infty} A_k I_0\left(\frac{k\pi}{a}\right) \sin\left(\frac{k\pi z}{a}\right).$$

Thus the A_k are determined by the requirement that $A_k I_0(k\pi/a)$ be the kth coefficient in the Fourier sine series expansion of f on the interval $[0, a]$, and we have

$$A_k = \frac{2}{a I_0(k\pi/a)} \int_0^a f(z) \sin\frac{k\pi z}{a}\, dz. \tag{15–102}$$

EXERCISES

1. Find the steady-state temperature in the cylindrical region $r < 1$, $0 < z < a$, given that the solution u is independent of θ, and

$$u(r, 0) = f(r), \qquad u(r, a) = g(r), \qquad u(1, z) = 0.$$

2. Solve the problem in Exercise 1 under the boundary conditions

$$u(r, 0) = f(r), \qquad u(r, a) = g(r), \qquad u(1, z) = 100.$$

3. Find the steady-state temperature distribution $u = u(r, z)$ in the cylinder $r < 1$, $0 < z < a$, given that

$$u(r, 0) = 100, \qquad u(r, a) = 0, \qquad u(1, z) = 0.$$

4. Find the steady-state temperature distribution $u = u(r, z)$ in the cylindrical region $r < 1$, $0 < z < a$, given that

$$u(r, 0) = f(r), \qquad u(r, a) = 100, \qquad u_r(1, z) = 0.$$

(The last boundary condition asserts that the lateral surface of the cylinder is insulated so that no heat flows across it.)

5. Discuss the nature of the solutions of Laplace's equation in the cylindrical region $r < 1, 0 < z < a$, which depend upon only one of the variables r, θ, z. Which of them can describe steady-state temperature distributions in the cylinder?

15-12 THE VIBRATING CIRCULAR MEMBRANE

In this section we consider the problem of describing the motion of a vibrating circular membrane such as the head of a drum under the assumption that the membrane is held fixed on the boundary of the circle and given a definite displacement and velocity at time $t = 0$. In other words, we propose to solve the two-dimensional wave equation

$$\frac{\partial^2 u}{\partial r^2} + \frac{1}{r}\frac{\partial u}{\partial r} + \frac{1}{r^2}\frac{\partial^2 u}{\partial \theta^2} = \frac{1}{a^2}\frac{\partial^2 u}{\partial t^2}, \quad a > 0, \tag{15–103}$$

in a circular region, which for convenience we take to be $r < 1$, given that

$$\begin{aligned} u(1, \theta, t) &= 0, \\ u(r, \theta, 0) &= f(r, \theta), \\ u_t(r, \theta, 0) &= g(r, \theta). \end{aligned} \tag{15–104}$$

We again begin by seeking solutions for (15–103) which are independent of θ, in which case the equation and boundary conditions become

$$\frac{\partial^2 u}{\partial r^2} + \frac{1}{r}\frac{\partial u}{\partial r} = \frac{1}{a^2}\frac{\partial^2 u}{\partial t^2}, \quad a > 0, \tag{15–105}$$

and

$$\begin{aligned} u(1, t) &= 0, \\ u(r, 0) &= f(r), \\ u_t(r, 0) &= g(r). \end{aligned} \tag{15–106}$$

Arguing in the usual fashion we set $u(r, t) = R(r)T(t)$, and find that (15–105) becomes

$$\frac{R'' + (1/r)R'}{R} = \frac{1}{a^2}\frac{T''}{T} = -\lambda^2,$$

where λ is a positive constant. (The choice of sign here is forced upon us by the requirement that T be periodic.) Thus R and T must be solutions of the equations

$$R'' + \frac{1}{r}R' + \lambda^2 R = 0, \tag{15–107}$$

$$T'' + \lambda^2 a^2 T = 0, \tag{15–108}$$

and, in addition, R must satisfy the boundary condition $R(1) = 0$. Starting with the general solution

$$R(r) = AJ_0(\lambda r) + BY_0(\lambda r)$$

of (15–107), we first set $B = 0$ in order to ensure the continuity of R at the origin, and then apply the given boundary condition to deduce that, up to constant multiples, R must be one of the functions

$$R_k(r) = J_0(\lambda_k r), \quad k = 1, 2, \ldots,$$

where λ_k is the kth positive zero of J_0. Moreover, with these as the values of λ the solutions of (15–108) are

$$T_k(t) = A_k \cos(\lambda_k at) + B_k \sin(\lambda_k at), \quad k = 1, 2, \ldots,$$

A_k, B_k constants, and it follows that the functions

$$u_k(r, t) = [A_k \cos(\lambda_k at) + B_k \sin(\lambda_k at)]J_0(\lambda_k r),$$

$k = 1, 2, \ldots$, satisfy (15–105) and the boundary condition $u(1, t) = 0$.

To find a solution $u(r, t)$ which also satisfies the given initial conditions we set

$$u(r, t) = \sum_{k=1}^{\infty} [A_k \cos(\lambda_k at) + B_k \sin(\lambda_k at)]J_0(\lambda_k r), \tag{15–109}$$

and attempt to determine the A_k and B_k so that

$$u(r, 0) = f(r),$$
$$u_t(r, 0) = g(r).$$

Thus when $t = 0$, (15–109) must reduce to

$$f(r) = \sum_{k=1}^{\infty} A_k J_0(\lambda_k r),$$

and we find that

$$A_k = \frac{2}{[J_1(\lambda_k)]^2} \int_0^1 f(r) J_0(\lambda_k r) r \, dr. \tag{15–110}$$

A similar argument applied to the series obtained by differentiating (15–109) term-by-term with respect to t yields

$$B_k = \frac{2}{\lambda_k a[J_1(\lambda_k)]^2} \int_0^1 g(r) J_0(\lambda_k r) r \, dr, \tag{15–111}$$

and the problem is solved.

We now turn to the general case with solutions dependent upon θ. Here we start by setting $u(r, \theta, t) = R(r)\Theta(\theta)T(t)$ in (15–103) to obtain

$$\frac{R''}{R} + \frac{1}{r}\frac{R'}{R} + \frac{1}{r^2}\frac{\Theta''}{\Theta} = \frac{1}{a^2}\frac{T''}{T} = -\lambda^2,$$

where, as before, λ is a positive constant. From this we obtain the pair of equations

$$T'' + \lambda^2 a^2 T = 0,$$

$$\frac{R''}{R} + \frac{1}{r}\frac{R'}{R} + \frac{1}{r^2}\frac{\Theta''}{\Theta} = -\lambda^2,$$

the second of which we rewrite as

$$-\frac{R'' + (1/r)R' + \lambda^2 R}{(1/r^2)R} = \frac{\Theta''}{\Theta} = -\mu^2,$$

where $\mu \geq 0$ is again a constant. (The choice of sign is dictated by the fact that Θ must be periodic with period 2π.) Thus we must solve the equations

$$T'' + \lambda^2 a^2 T = 0, \tag{15–112}$$

$$r^2 R'' + rR' + (\lambda^2 r^2 - \mu^2)R = 0, \tag{15–113}$$

$$\Theta'' + \mu^2 \Theta = 0, \tag{15–114}$$

subject to the boundary condition

$$R(1) = 0. \tag{15–115}$$

Starting with (15–114), we use the periodicity of Θ to deduce that the only admissible values of μ are

$$\mu_n = n, \quad n = 0, 1, 2, \ldots,$$

and hence that Θ must be a linear combination of the functions

$$\cos n\theta, \quad \sin n\theta.$$

Furthermore, when $\mu = n$, (15–113) is Bessel's equation of order n, and (15–115) implies that its solutions must be constant multiples of the function

$$R_{nk}(r) = J_n(\lambda_{nk} r), \quad k = 1, 2, \ldots,$$

where the λ_{nk} are the positive zeros of J_n. (We again reject solutions involving Y_n because of their discontinuity at the origin.) Finally, when $\lambda = \lambda_{nk}$, (15–112) yields

$$T_{nk}(t) = A_{nk} \cos (a\lambda_{nk} t) + B_{nk} \sin (a\lambda_{nk} t),$$

where A_{nk} and B_{nk} are arbitrary constants. Putting these results together, we conclude that the functions

$$u_{nk}(r, \theta, t) = [A_{nk} \cos(a\lambda_{nk}t) + B_{nk} \sin(a\lambda_{nk}t)] \cos(n\theta) J_n(\lambda_{nk}r)$$

and

$$v_{nk}(r, \theta, t) = [\overline{A}_{nk} \cos(a\lambda_{nk}t) + \overline{B}_{nk} \sin(a\lambda_{nk}t)] \sin(n\theta) J_n(\lambda_{nk}r),$$

$$n = 0, 1, 2, \ldots, \quad k = 1, 2, \ldots,$$

are solutions of (15–103) which satisfy the boundary condition $u(1, \theta, t) = 0$.

To complete the solution we now set

$$u(r, \theta, t) = \sum_{n=0}^{\infty} \sum_{k=1}^{\infty} [u_{nk}(r, \theta, t) + v_{nk}(r, \theta, t)], \tag{15–116}$$

and determine the coefficients A_{nk}, B_{nk}, $\overline{A}_{nk}$, $\overline{B}_{nk}$ so that $u(r, \theta, 0) = f(r, \theta)$ and $u_t(r, \theta, 0) = g(r, \theta)$. Thus, when $t = 0$, (15–116) must reduce to

$$f(r, \theta) = \sum_{n=0}^{\infty} \sum_{k=1}^{\infty} [u_{nk}(r, \theta, 0) + v_{nk}(r, \theta, 0)]$$

or

$$f(r, \theta) = \sum_{n=0}^{\infty} \left\{ \sum_{k=1}^{\infty} [A_{nk} J_n(\lambda_{nk}r)] \cos n\theta + \sum_{k=1}^{\infty} [\overline{A}_{nk} J_n(\lambda_{nk}r)] \sin n\theta \right\} \cdot \tag{15–117}$$

To evaluate these coefficients, we expand $f(r, \theta)$ in a Fourier series with respect to the variable θ, obtaining

$$f(r, \theta) = \sum_{n=0}^{\infty} [a_n(r) \cos n\theta + b_n(r) \sin n\theta], \tag{15–118}$$

with

$$a_0(r) = \frac{1}{2\pi} \int_{-\pi}^{\pi} f(r, \theta) \, d\theta,$$

$$a_n(r) = \frac{1}{\pi} \int_{-\pi}^{\pi} f(r, \theta) \cos n\theta \, d\theta,$$

$$b_n(r) = \frac{1}{\pi} \int_{-\pi}^{\pi} f(r, \theta) \sin n\theta \, d\theta.$$

Comparing (15–117) and (15–118), we see that

$$a_n(r) = \sum_{k=1}^{\infty} A_{nk} J_n(\lambda_{nk}r), \quad \text{and} \quad b_n(r) = \sum_{k=1}^{\infty} \overline{A}_{nk} J_n(\lambda_{nk}r),$$

for all n. Hence the A_{nk} and $\overline{A}_{nk}$ must be the coefficients in the Bessel series expansions of $a_n(r)$ and $b_n(r)$ with respect to the functions $J_n(\lambda_{nk}r)$, and we have

$$A_{nk} = \frac{2}{[J_{n+1}(\lambda_{nk})]^2} \int_0^1 a_n(r) J_n(\lambda_{nk}r) r \, dr,$$

$$\overline{A}_{nk} = \frac{2}{[J_{n+1}(\lambda_{nk})]^2} \int_0^1 b_n(r) J_n(\lambda_{nk}r) r \, dr.$$

Thus

$$A_{0k} = \frac{1}{\pi[J_1(\lambda_{0k})]^2} \int_0^1 \int_{-\pi}^{\pi} f(r, \theta) J_0(\lambda_{0k}r) r \, d\theta \, dr,$$

$$A_{nk} = \frac{2}{\pi[J_{n+1}(\lambda_{nk})]^2} \int_0^1 \int_{-\pi}^{\pi} f(r, \theta) \cos(n\theta) J_n(\lambda_{nk}r) r \, d\theta \, dr,$$

$$\overline{A}_{nk} = \frac{2}{\pi[J_{n+1}(\lambda_{nk})]^2} \int_0^1 \int_{-\pi}^{\pi} f(r, \theta) \sin(n\theta) J_n(\lambda_{nk}r) r \, d\theta \, dr.$$

A similar computation starting with the series obtained by differentiating (15-116) term-by-term with respect to t reveals that

$$B_{0k} = \frac{1}{\pi a \lambda_{0k}[J_1(\lambda_{0k})]^2} \int_0^1 \int_{-\pi}^{\pi} g(r, \theta) J_0(\lambda_{0k}r) r \, d\theta \, dr,$$

$$B_{nk} = \frac{2}{\pi a \lambda_{nk}[J_{n+1}(\lambda_{nk})]^2} \int_0^1 \int_{-\pi}^{\pi} g(r, \theta) \cos(n\theta) J_n(\lambda_{nk}r) r \, d\theta \, dr,$$

$$\overline{B}_{nk} = \frac{2}{\pi a \lambda_{nk}[J_{n+1}(\lambda_{nk})]^2} \int_0^1 \int_{-\pi}^{\pi} g(r, \theta) \sin(n\theta) J_n(\lambda_{nk}r) r \, d\theta \, dr,$$

and we are done.

EXERCISES

1. An object located at the point $x = x_0$ starts from rest and moves along the x-axis under the action of a force directed toward the origin whose magnitude is proportional to the distance of the object from the origin and to its mass, and initially is $m_0 x_0$. Determine the motion of the object if its mass m varies with time according to the formula $m = m_0(1 + t)$.
2. A uniform, flexible cable of length L is suspended vertically as shown in Fig. 15-5. At time $t = 0$ that portion of the cable between $x = 0$ and $x = \alpha L$ is given a uniform horizontal velocity $v = f(x)$. Describe the subsequent behavior of the cable, given

that its motion is governed by the equation

$$\frac{\partial^2 y}{\partial t^2} = g\frac{\partial}{\partial x}\left(x\frac{\partial y}{\partial x}\right),$$

where g is the acceleration due to gravity. [*Note.* The x-axis is directed upward with the cable suspended from the point $(L, 0)$.]

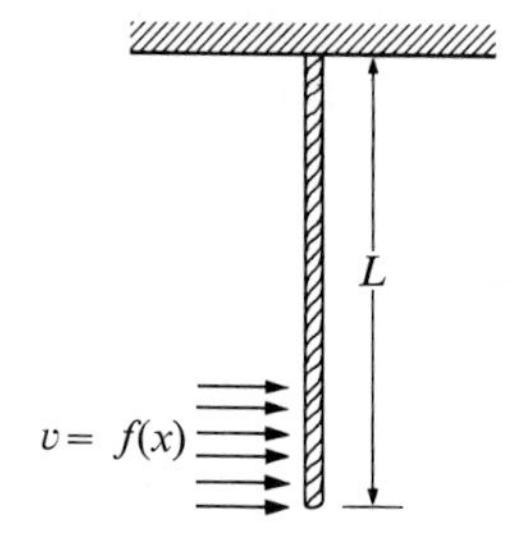

FIGURE 15–5

3. Find the steady-state temperature distribution $u(r, \theta, z)$ in the cylindrical region $r < 1, 0 < z < a$, given that

$$u(1, \theta, z) = 0, \qquad u(r, \theta, a) = 0, \qquad u(r, \theta, 0) = f(r, \theta).$$

4. Find the temperature $u(r, \theta, t)$ in the two-dimensional region shown in Fig. 15–6, given that the boundary of the region is maintained at a temperature of 0°, and that at time $t = 0$ the interior is at a uniform temperature of 100°.

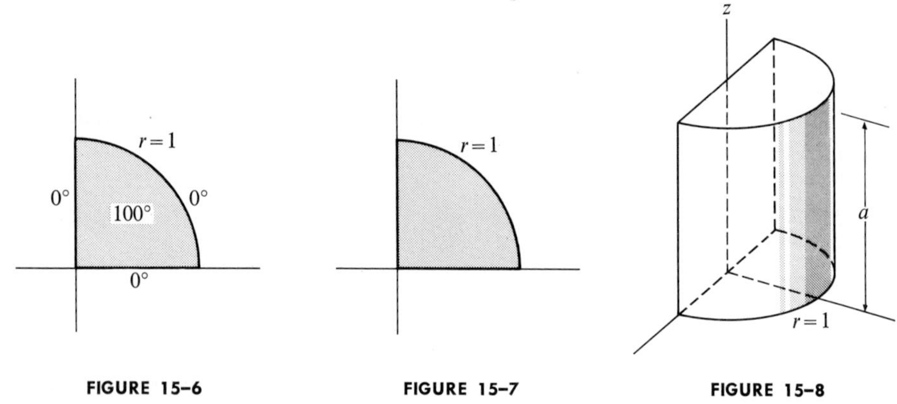

FIGURE 15–6 FIGURE 15–7 FIGURE 15–8

5. Find the equation of motion of a vibrating membrane of the shape shown in Fig. 15–7, under the assumption that the membrane is held fixed along its boundary and is released from rest at time $t = 0$ from a known position.

6. Solve Exercise 5 when the membrane is also given a known initial velocity.

7. Generalize the results of Exercises 5 and 6 to the case of an arbitrary wedge-shaped region with central angle α.

8. Find the steady-state temperature distribution in the semicylindrical region shown in Fig. 15–8, given that the temperature on the upper face is held at a known value $f(r, \theta)$, while that on all the remaining faces is zero.

APPENDIX I

infinite series

I–1 INTRODUCTION

The objective of this appendix is to provide a reasonably complete account of the material relating to the convergence of sequences and series that was used in the body of the text. From the standpoint of logical completeness this discussion ought to begin with a detailed study of the real number system, including its construction out of the rationals. This, however, is a lengthy undertaking which properly belongs in a text on advanced calculus. Hence, rather than attempt it here we shall assume that the student has a working knowledge of the real number system, at least to the extent normally taught in a first course in calculus, and we shall base our discussion upon it.

Actually it is possible to give a rigorous, self-contained account of the theory of sequential convergence if one is willing to accept, as given, a set $\mathcal{R}$ of objects called real numbers which can be added, subtracted, multiplied, and divided according to the familiar rules of arithmetic. In addition, it is necessary to assume (a) that $\mathcal{R}$ contains the ordinary integers as a subset, (b) that $\mathcal{R}$ is ordered by a relation $\leq$ having all the properties usually associated with this symbol, and (c) that $\mathcal{R}$ satisfies the so-called least upper bound principle which we now proceed to state.

> **Definition I–1.** A real number b is said to be an *upper bound* for a (nonempty) set $\mathcal{S}$ of real numbers if and only if $s \leq b$ for all s in $\mathcal{S}$. If, in addition, no number smaller than b is an upper bound for $\mathcal{S}$, then b is said to be a *least upper bound* (l.u.b.) for $\mathcal{S}$. (The terms *lower bound* and *greatest lower bound* (g.l.b.) are defined similarly.)

In these terms the least upper bound principle—which, by the way, is actually a theorem concerning the real numbers—reads as follows:

Least upper bound principle. *Every (nonempty) set of real numbers which is bounded from above has a least upper bound.* (Again there is a companion statement concerning lower bounds which we omit.)

And once this statement has been accepted we are back on solid ground where theorems can be proved and definitions given without further gaps in the reasoning.

Needless to say, we shall not attempt to give a complete treatment of the several topics mentioned above, since this would entail writing an entire text, or more on advanced calculus. Neither shall we prove every assertion that is made as we progress, since this too would result in a labored discussion. We do, however, insist upon the fact that these proofs are now within our reach, and want only time and patience to present.

I–2 SEQUENTIAL CONVERGENCE

We assume that the reader is already familiar with the notion of a *sequence* $\{a_k\}$ of real numbers, which, we recall, is simply an ordered list

$$\{a_1, a_2, \ldots, a_k, \ldots\}$$

of real numbers indexed by the positive integers, or, more formally, a real valued function F whose domain is the positive integers, and whose value $F(k)$ at k is a_k. Actually, there is no reason to insist that the indexing always begin with the subscript *one*, and when convenient we shall change it without comment.

This said, we now introduce the concept of sequential convergence, as follows.

Definition I–2. A sequence $\{a_k\}$ of real numbers is said to *converge* to the number a if, given any $\epsilon > 0$, there exists an integer K (depending in general upon ϵ) such that

$$|a_k - a| < \epsilon \tag{I–1}$$

for all $k \geq K$. When this happens we say that a is the *limit* of $\{a_k\}$, and write

$$\lim_{k \to \infty} a_k = a, \qquad \text{or} \qquad \{a_k\} \to a.$$

If, on the other hand, no such number exists, $\{a_k\}$ is said to *diverge*.

Implicit in the statement of this definition is the assertion that *the limit of a convergent sequence is unique*. To see this, suppose that $\{a_k\}$ converges to a, and let $a' \neq a$. Then if $\epsilon = |a - a'|/3$ and if K is chosen so that $|a_k - a| < \epsilon$ for all $k \geq K$, the only entries in $\{a_k\}$ which do not lie in the interval $(a - \epsilon, a + \epsilon)$ are $a_1, a_2, \ldots, a_K$, and it follows that $\{a_k\}$ does *not* converge to a'.

Having defined the notion of convergence, we now address ourselves to the problem of determining whether a given sequence converges or not. For this purpose, Definition I–2 is manifestly unsatisfactory, since it requires us to find the limit of the sequence before we can establish its convergence. Thus it is natural to seek a convergence criterion which can be applied directly to the terms of the sequence themselves. One, which is easily deduced from the least upper bound property, reads as follows.

Theorem I–1. *A monotonically nondecreasing sequence converges if and only if it is bounded from above, while a monotonically nonincreasing sequence converges if and only if it is bounded from below.*

(Recall that $\{a_k\}$ is said to be monotonically nondecreasing if $a_1 \leq a_2 \leq a_3 \leq \cdots$; monotonically nonincreasing if $a_1 \geq a_2 \geq a_3 \geq \cdots$.)

Proof. Let $\{a_k\}$ be monotonically nondecreasing. Then if $\{a_k\}$ is bounded from above it has a least upper bound a, and hence, given any $\epsilon > 0$, there exists an integer K such that $|a - a_K| < \epsilon$. Since $a_K \leq a_k \leq a$ for all $k \geq K$, it follows that $|a - a_k| < \epsilon$ for $k \geq K$, and $\{a_k\} \to a$.

Conversely, if $\{a_k\}$ is not bounded from above, then for each real number a we can find an integer K such that $a < a_K$. Setting $\epsilon = |a - a_K|$, we have $|a - a_k| > \epsilon$ for all $k > K$. Thus $\{a_k\}$ does *not* converge to a, and since a was arbitrary, we conclude that $\{a_k\}$ diverges.

This proves the first assertion in the theorem and, with obvious modifications, the second as well. ▮

Using this result, it is now relatively easy to establish a criterion which will enable us to test an arbitrary sequence for convergence merely by examining its terms. But first, a definition.

Definition I–3. A sequence $\{a_k\}$ is said to be a *Cauchy sequence* if for each $\epsilon > 0$ there exists an integer K, depending upon ϵ, such that

$$|a_m - a_n| < \epsilon \tag{I–2}$$

for all $m, n \geq K$.

The convergence criterion we now propose to establish asserts that the class of Cauchy sequences is identical with the class of convergent sequences. This is easily the most important single result on sequential convergence.

Theorem I–2. *A sequence $\{a_k\}$ of real numbers is convergent if and only if it is a Cauchy sequence.*

Proof. Suppose that $\{a_k\}$ is convergent, with a as its limit. Then, given any $\epsilon > 0$, we can find an integer K such that $|a - a_k| < \epsilon/2$ for all $k \geq K$. Thus, if m and n are both greater than K,

$$\begin{aligned} |a_m - a_n| &= |(a_m - a) + (a - a_n)| \\ &\leq |a_m - a| + |a - a_n| \\ &< \frac{\epsilon}{2} + \frac{\epsilon}{2} = \epsilon, \end{aligned}$$

and $\{a_k\}$ is a Cauchy sequence.

Conversely, suppose that $\{a_k\}$ is a Cauchy sequence. Then, in particular, $\{a_k\}$ is bounded from above and from below. Indeed, if $\epsilon > 0$ is given, and K is chosen so that $|a_m - a_n| < \epsilon$ for all $m, n \geq K$, then none of the a_k can be greater than the largest number among $a_1, a_2, \ldots, a_K, a_{K+1} + \epsilon$, and none can be smaller than the smallest number among $a_1, a_2, \ldots, a_K, a_{K+1} - \epsilon$. This said, let

$$\begin{aligned} b_1 &= \text{l.u.b. of the sequence } \{a_1, a_2, \ldots\}, \\ b_2 &= \text{l.u.b. of the sequence } \{a_2, a_3, \ldots\}, \\ &\vdots \\ b_k &= \text{l.u.b. of the sequence } \{a_k, a_{k+1}, \ldots\}. \end{aligned}$$

Then $b_1 \geq b_2 \geq b_3 \geq \cdots$, and each b_k is at least as large as the greatest lower bound of $\{a_1, a_2, \ldots\}$. Hence $\{b_k\}$ is a monotonic nonincreasing sequence bounded from below, and therefore has a limit a (Theorem I–1). We now propose to show that a is also the limit of $\{a_k\}$. To this end, let $\epsilon > 0$ be given, and let K_1 be chosen so that $|a_m - a_n| < \epsilon/3$ for all $m, n \geq K_1$. (The existence of such an integer follows from the assumption that $\{a_k\}$ is a Cauchy sequence.) Let K_2 be chosen so that $|a - b_k| < \epsilon/3$ for all $k \geq K_2$, and let K be the larger of K_1 and K_2. Then, if $k \geq K$,

$$\begin{aligned} |a - a_k| &\leq |a - b_k| + |b_k - a_k| \\ &< \frac{\epsilon}{3} + |b_k - a_k|. \end{aligned}$$

But since b_k is the least upper bound of $\{a_k, a_{k+1}, \ldots\}$, there exists an index $p \geq k$ such that $|b_k - a_p| < \epsilon/3$. Hence

$$\begin{aligned} |b_k - a_k| &\leq |b_k - a_p| + |a_p - a_k| \\ &< \frac{\epsilon}{3} + \frac{\epsilon}{3} = \frac{2\epsilon}{3}. \end{aligned}$$

Combining these results we have

$$|a - a_k| < \frac{\epsilon}{3} + \frac{2\epsilon}{3} = \epsilon,$$

and it follows that $\{a_k\}$ converges to a, as asserted. ▮

EXAMPLE 1. The sequence

$$\left\{\frac{1}{2}, \frac{1}{3}, \ldots, \frac{1}{k}, \ldots\right\}$$

is obviously a Cauchy sequence, and hence converges. Here, of course, we could just as easily have applied Definition I–2, since it is clear that the sequence in question converges to zero.

EXAMPLE 2. The sequence $\{1, -1, 1, -1, \ldots\}$ is not a Cauchy sequence, and hence does not converge.

The argument given in this section can be summarized by saying that the least upper bound principle implies that every Cauchy sequence of real numbers is convergent. It is also possible to turn things around, and deduce the least upper bound principle from Theorem I–2. In short, these two facts concerning the real number system are equivalent, and either can be taken as the starting point for the study of infinite series, and, in fact, the entire theory of real valued functions of a real variable. We omit the proof.

We conclude this section by proving an elementary computational theorem which will be needed below.

Theorem I–3. *Let $\{a_k\}$ and $\{b_k\}$ be convergent sequences with*

$$\lim_{k\to\infty} a_k = a, \qquad \lim_{k\to\infty} b_k = b.$$

Then

(i) *$\{\alpha a_k + \beta b_k\}$ is convergent for all real numbers α and β, and*

$$\lim_{k\to\infty} (\alpha a_k + \beta b_k) = \alpha a + \beta b;$$

(ii) *$\{a_k b_k\}$ is convergent, and*

$$\lim_{k\to\infty} a_k b_k = ab.$$

Proof. We leave to the reader the easy task of verifying that $\{\alpha a_k\} \to \alpha a$ whenever $\{a_k\} \to a$. This proved, (i) will follow as soon as we show that $\{a_k\} \to a$ and $\{b_k\} \to b$ imply $\{a_k + b_k\} \to a + b$. Let $\epsilon > 0$ be given. Then

$$|(a_k + b_k) - (a + b)| \le |a_k - a| + |b_k - b|,$$

and, by assumption, we can find integers K_1, K_2 such that $|a_k - a| < \epsilon/2$ for all $k \ge K_1$, $|b_k - b| < \epsilon/2$ for all $k \ge K_2$. Thus if K is the larger of K_1 and K_2, we have

$$|(a_k + b_k) - (a + b)| < \frac{\epsilon}{2} + \frac{\epsilon}{2} = \epsilon,$$

for all $k \ge K$, as required.

To prove (ii) we note that

$$\begin{aligned} |a_k b_k - ab| &= |a_k b_k - a_k b + a_k b - ab| \\ &\le |a_k|\,|b_k - b| + |b|\,|a_k - a|. \end{aligned}$$

But since $\{a_k\}$ is convergent it is a Cauchy sequence, and therefore is bounded from above and from below. (See the proof of Theorem I–2.) Thus there exists a positive constant M such that $|a_k| \leq M$ for all k, and the above inequality can be written

$$|a_k b_k - ab| \leq M|b_k - b| + |b|\,|a_k - a|. \tag{I–3}$$

Now let $\epsilon > 0$ be given, and choose integers K_1 and K_2 such that

$$|a_k - a| < \frac{\epsilon}{2|b|}$$

for all $k \geq K_1$, and

$$|b_k - b| < \frac{\epsilon}{2M}$$

for all $k \geq K_2$. (If $b = 0$, the second term in (I–3) vanishes, and we need only choose K_1.) Then with K the larger of K_1 and K_2, and $k \geq K$,

$$|a_k b_k - ab| < \frac{\epsilon M}{2M} + \frac{\epsilon|b|}{2|b|} = \epsilon,$$

and we are done. ▮

I–3 INFINITE SERIES

In this section we review the elementary facts concerning infinite series of constants, including several well-known tests for the convergence of such series.

Definition I–4. Let

$$\sum_{k=1}^{\infty} a_k = a_1 + a_2 + \cdots + a_k + \cdots \tag{I–4}$$

be an infinite series of real numbers, and let $\{s_k\}$ be the associated sequence of *partial sums*

$$\begin{aligned} s_1 &= a_1, \\ s_2 &= a_1 + a_2, \\ &\vdots \\ s_k &= a_1 + a_2 + \cdots + a_k. \end{aligned}$$

Then (I–4) is said to *converge* to the value a if and only if $\{s_k\}$ converges to a. In this case we write

$$a = \sum_{k=1}^{\infty} a_k,$$

and say that a is the *sum* of the series. Otherwise, (I–4) is said to *diverge*.

Perhaps the most familiar example of a convergent infinite series is the geometric series

$$a_0 + a_0 r + a_0 r^2 + \cdots, \tag{I–5}$$

whose *ratio* r satisfies the inequality $-1 < r < 1$. Indeed, in this case

$$\begin{aligned} s_0 &= a_0, \\ s_1 &= a_0(1 + r), \\ &\vdots \\ s_k &= a_0(1 + r + r^2 + \cdots + r^k). \end{aligned}$$

But, by the identity

$$(1 + r + \cdots + r^{k-1})(1 - r) = 1 - r^k,$$

we have

$$s_k = a_0 \frac{1 - r^{k+1}}{1 - r}, \quad r \neq 1,$$

and it follows that

$$\lim_{k\to\infty} s_k = \frac{a_0}{1 - r}$$

provided $|r| < 1$. On the other hand, if $|r| \geq 1$, the sequence $\{s_k\}$ is divergent, and hence so is (I–5).

EXAMPLE 1. The real number 0.33333 . . . is the sum of the geometric series

$$\frac{3}{10} + \frac{3}{10^2} + \frac{3}{10^3} + \cdots,$$

whose ratio is $\frac{1}{10}$. In this case the formula given above for $\lim_{k\to\infty} s_k$ yields

$$\frac{a_0}{1 - r} = \frac{3}{10} \cdot \frac{1}{1 - \frac{1}{10}} = \frac{1}{3},$$

as expected.

The question of convergence or divergence for geometric series was settled in the most satisfactory way possible, namely by obtaining a simple closed-form expression for s_k and letting k tend to infinity. The existence of such a formula for s_k and the ability to find the actual sum of the series is something of an accident, however, and for this reason we now turn our attention toward deriving convergence tests which may be applied directly to the terms of a series. We begin by restating Theorem I–2 in a form appropriate to series.

Theorem I–4. *The series $\sum_{k=1}^{\infty} a_k$ converges if and only if for each $\epsilon > 0$ there exists an integer K, depending on ϵ, such that*

$$\left| \sum_{k=m}^{n} a_k \right| < \epsilon \tag{I–6}$$

whenever $K \leq m \leq n$.

In fact we note that the expression $\sum_{k=m}^{n} a_k$ is just the difference $s_n - s_{m-1}$ of the partial sums s_n and s_{m-1}. Hence the theorem states that $\sum_{k=1}^{\infty} a_k$ converges if and only if its sequence of partial sums is a Cauchy sequence, and this is the content of Theorem I–2. ▌

Theorem I–4 provides a general convergence criterion for series, but unfortunately it is difficult to apply in practice. We devote the remainder of this section, therefore, to the derivation of several consequences of the theorem which, although lacking the generality of Theorem I–4, do provide convenient tests for convergence in a large number of cases.

Theorem I–5. *If $\sum_{k=1}^{\infty} a_k$ converges, then* $\lim_{n\to\infty} a_n = 0$.

Proof. Since

$$|a_n| = |s_n - s_{n-1}|,$$

the theorem is an immediate consequence of Theorem I–4. ▌

It is useful to restate the last result in the following form: *If a_k does* **not** *tend to zero as $k \to \infty$, then $\sum_{k=1}^{\infty} a_k$* **diverges.** Thus each of the series

$$\sum_{n=1}^{\infty} \frac{n}{1+n}, \qquad \sum_{n=1}^{\infty} \sin n, \qquad \sum_{n=1}^{\infty} \left(1 + \frac{1}{n}\right)^n$$

diverges, the first because $\lim_{n\to\infty} n/(1+n) = 1$, the second because $\lim_{n\to\infty} (\sin n)$ does not exist, and the last because

$$\lim_{n\to\infty} \left(1 + \frac{1}{n}\right)^n = e.$$

It must be emphasized, however, that the converse of Theorem I–5 does *not* hold, for, as we shall see presently, the so-called *harmonic series*

$$\sum_{k=1}^{\infty} \frac{1}{k} = 1 + \tfrac{1}{2} + \tfrac{1}{3} + \cdots$$

diverges, although $\lim_{k\to\infty} 1/k = 0$. Thus even though it is *necessary* that $\lim_{k\to\infty} a_k = 0$ in order that $\sum_{k=1}^{\infty} a_k$ converge, this condition is *not sufficient.* In the next three theorems we restrict our attention to series whose terms are *positive.*

Theorem I–6. (Comparison test.) *Let $\sum_{k=1}^{\infty} a_k$ and $\sum_{k=1}^{\infty} b_k$ be series with positive terms.*

(1) *If $\sum_{k=1}^{\infty} a_k$ converges and $b_k \leq a_k$ for every k, then $\sum_{k=1}^{\infty} b_k$ also converges.*

(2) *If $\sum_{k=1}^{\infty} a_k$ diverges and $b_k \geq a_k$ for every k, then $\sum_{k=1}^{\infty} b_k$ also diverges.*

Proof. (1) Since $b_k \geq 0$ for all k, the partial sums of $\sum_{k=1}^{\infty} b_k$ form a monotonically nondecreasing sequence. But this sequence is also bounded from above, since

$$\sum_{k=1}^{n} b_k \leq \sum_{k=1}^{n} a_k \leq \sum_{k=1}^{\infty} a_k = S,$$

where S denotes the sum of the convergent series $\sum_{k=1}^{\infty} a_k$. Hence by Theorem I–1 the sequence of partial sums of $\sum_{k=1}^{\infty} b_k$ converges.

(2) In this case $\sum_{k=1}^{n} b_k \geq \sum_{k=1}^{n} a_k$ and it follows that these partial sums are unbounded since otherwise the series $\sum_{k=1}^{\infty} a_k$ would converge. ▮

Theorem I–7. (Ratio test.) *Let $\sum_{k=1}^{\infty} a_k$ be a series of positive terms and suppose that*

$$L = \lim_{k\to\infty} \frac{a_{k+1}}{a_k}$$

exists. Then

(1) *$\sum_{k=1}^{\infty} a_k$ converges if $L < 1$,*

(2) *$\sum_{k=1}^{\infty} a_k$ diverges if $L > 1$.*

(Note that no assertion is made in case $L = 1$.)

Proof. (1) Suppose that $L < 1$ and that r is chosen to be a fixed real number with $L < r < 1$. Then for sufficiently large values of n, say $n \geq N$, we have $a_{n+1}/a_n \leq r$. Thus

$$\begin{aligned} a_{N+1} &\leq ra_N, \\ a_{N+2} &\leq ra_{N+1} \leq r^2 a_N, \\ &\vdots \\ a_{N+k} &\leq r^k a_N. \end{aligned}$$

But since $r < 1$, the geometric series $\sum_{k=0}^{\infty} a_N r^k$ converges. Thus by the *comparison test* so does

$$\sum_{k=0}^{\infty} a_{N+k} = \sum_{k=N}^{\infty} a_k,$$

and hence also the given series $\sum_{k=1}^{\infty} a_k$.

(2) The proof of this case is similar: this time $\sum_{k=1}^{\infty} a_k$ is compared with a divergent geometric series (ratio $r > 1$). We omit the details. ▮

Theorem I–8. (Integral test.) *Let $\sum_{k=1}^{\infty} a_k$ be a series of positive terms and assume that there exists a function f, continuous and monotonically nonincreasing on $1 \leq t < \infty$, such that $f(k) = a_k$ for $k = 1, 2, \ldots$. Then the series $\sum_{k=1}^{\infty} a_k$ and the improper integral $\int_1^{\infty} f(t)\, dt$ converge or diverge together.*

Proof. Assume first that $\int_1^\infty f(t)\,dt$ converges. Then a reference to Fig. I–1(a) makes it clear that

$$\sum_{k=2}^{n} a_k \le \int_1^{n+1} f(t)\,dt \le \int_1^\infty f(t)\,dt.$$

Thus the partial sums $\sum_{k=1}^n a_k$ of the given series form a monotonically nondecreasing sequence which is bounded above by the real number $a_1 + \int_1^\infty f(t)\,dt$. Convergence follows from Theorem I–1.

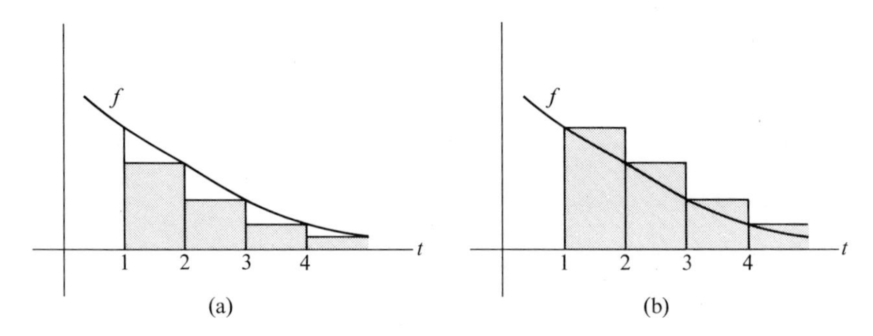

FIGURE I–1

If, on the other hand, the integral $\int_1^\infty f(t)\,dt$ diverges, then the partial sums $\sum_{k=1}^n a_k$ are unbounded, for in this case (see Fig. I–1b)

$$\sum_{k=1}^{n} a_k \ge \int_1^{n+1} f(t)\,dt,$$

and the latter integral tends to infinity as $n \to \infty$. ▮

EXAMPLE 2. The harmonic series

$$\sum_{n=1}^{\infty} \frac{1}{n} = 1 + \tfrac{1}{2} + \tfrac{1}{3} + \cdots$$

diverges, for the *integral test* may be applied in this case with $f(t) = 1/t$ to obtain

$$\int_1^\infty \frac{dt}{t} = \lim_{a\to\infty} \int_1^a \frac{dt}{t} = \lim_{a\to\infty} (\ln a) = \infty.$$

More generally given any *p-series* $\sum_{k=1}^\infty 1/k^p$, p a positive real number, we have

$$\int_1^\infty \frac{dt}{t^p} = \lim_{a\to\infty} \int_1^a t^{-p}\,dt = \lim_{a\to\infty} \frac{t^{-p+1}}{-p+1}, \quad p \ne 1,$$

and since this limit exists if and only if $p > 1$, it follows that a *p-series* $\sum_{k=1}^{\infty} 1/k^p$ *converges if* $p > 1$ *and diverges if* $p \leq 1$. In particular the series

$$\sum_{k=1}^{\infty} \frac{1}{x^2} \quad \text{and} \quad \sum_{k=1}^{\infty} \frac{1}{k^{1.01}}$$

converge, while the series

$$\sum_{k=1}^{\infty} \frac{1}{\sqrt{k}} \quad \text{and} \quad \sum_{k=1}^{\infty} \frac{1}{k^{0.99}}$$

diverge. The p-series and the geometric series constitute useful classes of series for which the question of convergence is completely settled. By using these series together with the comparison test, a large number of additional examples may be treated.

EXAMPLE 3. The series

$$\sum_{k=1}^{\infty} \frac{1}{k(k+1)} = \frac{1}{1 \cdot 2} + \frac{1}{2 \cdot 3} + \frac{1}{3 \cdot 4} + \cdots$$

converges by comparison with the series

$$\sum_{k=1}^{\infty} \frac{1}{k^2},$$

for $1/k(k+1) \leq 1/k^2$ for $k = 1, 2, \ldots$.

EXAMPLE 4. Consider the series

$$\sum_{k=0}^{\infty} = 1 + \frac{1}{1!} + \frac{1}{2!} + \frac{1}{3!} + \cdots.$$

Since

$$\frac{1}{(k+1)!} = \frac{1}{1 \cdot 2 \cdot 3 \cdots (k+1)} \leq \frac{1}{2^k},$$

the given series converges by comparison with the geometric series $\sum_{k=0}^{\infty} (\frac{1}{2})^k$. Note that the *ratio test* may also be applied in this case, since

$$L = \lim_{k \to \infty} \frac{a_{k+1}}{a_k} = \lim_{k \to \infty} \frac{1/(k+1)!}{1/k!} = \lim_{k \to \infty} \frac{1}{k+1} = 0.$$

EXAMPLE 5. Apply the ratio test to the series

$$\sum_{k=1}^{\infty} \frac{k^k}{k!} = 1 + \frac{2^2}{2!} + \frac{3^3}{3!} + \cdots.$$

Computing $\lim_{k\to\infty} a_{k+1}/a_k$, we have

$$\lim_{k\to\infty} \frac{(k+1)^{k+1}/(k+1)!}{k^k/k!} = \lim_{k\to\infty} \frac{(k+1)^{k+1}}{k^k} \cdot \frac{k!}{(k+1)!}$$

$$= \lim_{k\to\infty} \frac{(k+1)^k}{k^k}$$

$$= \lim_{k\to\infty} \left(1 + \frac{1}{k}\right)^k = e > 1.$$

Hence the series diverges.

EXAMPLE 6. For any p-series we have

$$L = \lim_{k\to\infty} \frac{a_{k+1}}{a_k} = \lim_{k\to\infty} \frac{1/(k+1)^p}{1/k^p} = \lim_{k\to\infty} \left(\frac{k}{k+1}\right)^p = 1.$$

Since such series converge if $p > 1$ and diverge if $p \leq 1$, the ratio test cannot possibly give any information in the case $L = 1$ (see Theorem I–7).

I–4 ABSOLUTE CONVERGENCE

The sequence of partial sums associated with a series whose terms are positive is monotonically nondecreasing, and hence the series converges or diverges according as the sequence is bounded above or not. For more general series, however, the question of convergence depends much more delicately on the magnitude and distribution of the positive and the negative terms. In this section we treat series which are *absolutely convergent* and series whose terms alternate in sign.

Definition I–5. The series $\sum_{k=1}^{\infty} a_k$ is said to be *absolutely convergent* if the series $\sum_{k=1}^{\infty} |a_k|$ converges.

Theorem I–9. *If $\sum_{k=1}^{\infty} |a_k|$ converges, then so does $\sum_{k=1}^{\infty} a_k$. Briefly, absolute convergence implies convergence.*

Proof. Under the assumption that $\sum_{k=1}^{\infty} |a_k|$ converges, we shall show that the partial sums of $\sum_{k=1}^{\infty} a_k$ form a Cauchy sequence. But this follows immediately from the relation

$$\left|\sum_{k=m}^{n} a_k\right| \leq \sum_{k=m}^{n} |a_k|, \quad m \leq n,$$

because the right member can be made arbitrarily small by choosing m sufficiently large. ▮

Since many important tests for convergence of series apply directly only to series whose terms are positive, Theorem I–9 provides a way, sufficient for many applications, of applying these tests to arbitrary series. The ratio test for instance, now takes the following form:

Theorem I–10. (Ratio test.) *Assume that the limit*

$$L = \lim_{n\to\infty} \left|\frac{a_{n+1}}{a_n}\right|$$

exists. Then

(1) $\sum_{k=1}^{\infty} a_k$ *converges if* $L < 1$, *and*

(2) $\sum_{k=1}^{\infty} a_k$ *diverges if* $L > 1$.

If $L = 1$, *no information is obtained.*

Proof. If $L < 1$, Theorem I–7 asserts that $\sum_{k=1}^{\infty} |a_k|$ converges; that is, $\sum_{k=1}^{\infty} a_k$ converges absolutely. Hence, by Theorem I–9, $\sum_{k=1}^{\infty} a_k$ converges.

If, on the other hand,

$$L = \lim_{n\to\infty} \left|\frac{a_{n+1}}{a_n}\right| > 1,$$

then $|a_{n+1}| > |a_n|$ for sufficiently large n. Thus a_n does not tend to zero as $n \to \infty$, and $\sum_{k=1}^{\infty} a_k$ diverges. ▮

EXAMPLE 1. Determine the set of values of x for which the series

$$\sum_{k=1}^{\infty} (-1)^k k \left(\frac{x}{2}\right)^{3k} = -\left(\frac{x}{2}\right)^3 + 2\left(\frac{x}{2}\right)^6 - 3\left(\frac{x}{2}\right)^9 + \cdots$$

converges.

Since $a_k = (-1)^k k (x/2)^{3k}$, we have

$$\left|\frac{a_{n+1}}{a_n}\right| = \left|\frac{(-1)^{n+1}(n+1)(x/2)^{3n+3}}{(-1)^n n (x/2)^{3n}}\right| = \left|\frac{n+1}{n}\left(\frac{x}{2}\right)^3\right| = \frac{n+1}{n}\left|\frac{x}{2}\right|^3.$$

Hence

$$\lim_{n\to\infty} \left|\frac{a_{n+1}}{a_n}\right| = \lim_{n\to\infty} \frac{n+1}{n}\left|\frac{x}{2}\right|^3 = \left|\frac{x}{2}\right|^3,$$

and the given series converges if $|x/2|^3 < 1$ and diverges if $|x/2|^3 > 1$. Thus the series converges for values of x lying in the interval

$$-2 < x < 2$$

and diverges if $|x| > 2$. Finally, in order to determine the behavior of the series at the points $x = \pm 2$, we note that the general term becomes $\pm(-1)^k k$ in this case, and since this quantity does not tend to zero as $k \to \infty$, the series diverges at both points.

It is an unfortunate fact that a series may converge without converging absolutely, and for such series, usually referred to as *conditionally convergent* series, the tests for convergence which we have devised so far are of no use. This situation is illustrated, for example, by the *alternating harmonic series*

$$\sum_{k=1}^{\infty} \frac{(-1)^{k+1}}{k} = 1 - \tfrac{1}{2} + \tfrac{1}{3} - \tfrac{1}{4} + \cdots,$$

whose partial sums *do* approach a limit (see Theorem I–11), despite the fact that the corresponding series of absolute values is divergent. This example is typical of *alternating series* where we have the following useful criterion for convergence.

Theorem I–11. *If the terms of the series* $\sum_{k=1}^{\infty} a_k$ *alternate in sign and satisfy*

(i) $|a_1| \geq |a_2| \geq |a_3| \geq \cdots,$

(ii) $\lim_{k\to\infty} |a_k| = 0,$

then the series converges. Moreover, if S is the sum of the series and S_n is the nth partial sum, then

$$|S - S_n| \leq |a_n|.$$

Proof. We may assume that the terms $a_1, a_3, a_5, \ldots$ are all positive and the terms $a_2, a_4, a_6, \ldots$ are all negative. Then

$$\begin{aligned} S_{2n+1} &= \sum_{k=1}^{2n+1} a_k = \sum_{k=1}^{2n-1} a_k + (a_{2n} + a_{2n+1}) \\ &= S_{2n-1} + (a_{2n} + a_{2n+1}) \\ &\leq S_{2n-1}, \end{aligned}$$

since $a_{2n} + a_{2n+1} \leq 0$. Likewise

$$\begin{aligned} S_{2n} &= \sum_{k=1}^{2n} a_k = \sum_{k=1}^{2n-2} a_k + (a_{2n-1} + a_{2n}) \\ &= S_{2n-2} + (a_{2n-1} + a_{2n}) \\ &\geq S_{2n-2}, \end{aligned}$$

since $a_{2n-1} + a_{2n} \geq 0$. It follows that the odd-numbered partial sums form a nonincreasing sequence bounded below by S_2, and that the even-numbered partial sums form a nondecreasing sequence bounded above by S_1 (see Fig. I–2). Thus each of these sequences possesses a limit, say

$$\lim_{k\to\infty} S_{2k} = S_E \qquad \text{and} \qquad \lim_{k\to\infty} S_{2k+1} = S_O,$$

where clearly $S_E \leq S_O$. But in fact since

$$\lim_{k\to\infty} |S_{k+1} - S_k| = \lim_{k\to\infty} |a_k| = 0,$$

we conclude that $S_E = S_O$, and the common limit S is the desired sum of the series. Moreover from the inequalities

$$S_{2k} \leq S \leq S_{2l+1} \quad \text{(for every } k, l\text{)},$$

we find that

$$|S - S_{2k}| \leq |S_{2k-1} - S_{2k}| = |a_{2k}|,$$

and

$$|S - S_{2k+1}| \leq |S_{2k+1} - S_{2k}| = |a_{2k+1}|,$$

completing the proof of the theorem. ▮

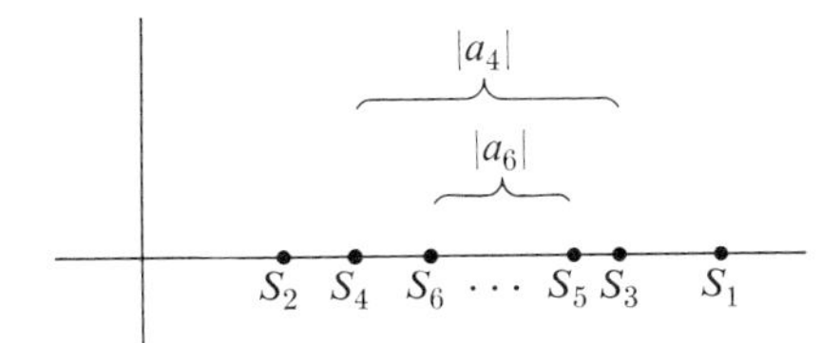

FIGURE I-2

EXAMPLE 2. The ratio test may be applied to the series

$$\sum_{k=0}^{\infty} \frac{(x+1)^k}{2k+1} \tag{I-7}$$

to show that it converges absolutely if $|x + 1| < 1$ and diverges if $|x + 1| > 1$; that is, the series converges absolutely for values of x in the interval $-2 < x < 0$. At the endpoints of this interval the ratio test gives no information. However, if we substitute $x = -2$ and $x = 0$ into the series, we obtain

$$\sum_{k=0}^{\infty} \frac{(-1)^k}{2k+1} \quad \text{and} \quad \sum_{k=0}^{\infty} \frac{1}{2k+1}$$

respectively. The first of these series converges since it is an *alternating* series whose general term tends to zero (the absolute values tending monotonically to zero), while the second series diverges. Thus the given series (I-7) converges for values of x in the interval $-2 \leq x < 0$ and diverges for x outside of this interval.

We close this section by stating without proof an important property of absolutely convergent series which is *not* shared by conditionally convergent series.

Theorem I-12. *If $\sum_{k=1}^{\infty} a_k$ is absolutely convergent and if $\sum_{k=1}^{\infty} b_k$ is any series obtained by rearranging the terms of $\sum_{k=1}^{\infty} a_k$, then $\sum_{k=1}^{\infty} b_k$ also converges absolutely and has the same sum.*

I–5 BASIC NOTIONS FROM ELEMENTARY CALCULUS

Before considering sequences and series of functions in Section I–6, we recall some of the basic facts concerning continuous and differentiable functions to which reference was made in the text. We shall give no proofs in this section although such proofs can be readily based upon the deeper properties of the real number system described in Section I–1. For a complete discussion the reader is referred to any text on advanced calculus (e.g., W. Kaplan, *Advanced Calculus*, Addison-Wesley, 1952).

> **Definition I–6.** A real valued function f defined on an interval I of the x-axis is said to be *continuous at a point* x_0 *in* I if for every $\epsilon > 0$ there exists a $\delta > 0$, depending in general on ϵ and x_0, such that $|f(x) - f(x_0)| < \epsilon$ whenever x is in I and $|x - x_0| < \delta$. If f is continuous at every point of I we say that *$f(x)$ is continuous on I.*

This is the familiar notion of continuity basic to any elementary calculus course. Not usually introduced at that level, however, is the following notion of *uniform continuity.*

> **Definition I–7.** A function f is said to be *uniformly continuous on an interval I* if for every $\epsilon > 0$ there exists a $\delta > 0$, depending in general on ϵ but *not* on x, such that $|f(x_1) - f(x_2)| < \epsilon$ whenever x_1, x_2 are in I and $|x_1 - x_2| < \delta$.

It is clear that a function which is uniformly continuous on an interval is also continuous on that interval. The following example shows, however, that the converse is false.

EXAMPLE 1. Let

$$f(x) = 1/x, \quad 0 < x \leq 1;$$

let x_0 be any point in this interval, and let $\epsilon > 0$ be given. If $x > x_0/2$, then

$$\left| \frac{1}{x} - \frac{1}{x_0} \right| = \frac{|x - x_0|}{xx_0} < \frac{2}{x_0^2} |x - x_0|.$$

If, further, $|x - x_0| < (x_0^2/2)\,\epsilon$, then

$$\left| \frac{1}{x} - \frac{1}{x_0} \right| < \epsilon.$$

Thus with

$$\delta = \min\left[\frac{x_0}{2}, \frac{x_0^2}{2}\,\epsilon\right]$$

the conditions of Definition I–6 are satisfied, and $f(x) = 1/x$ is continuous at x_0. It follows that f is continuous in the interval $0 < x \leq 1$. However, the given

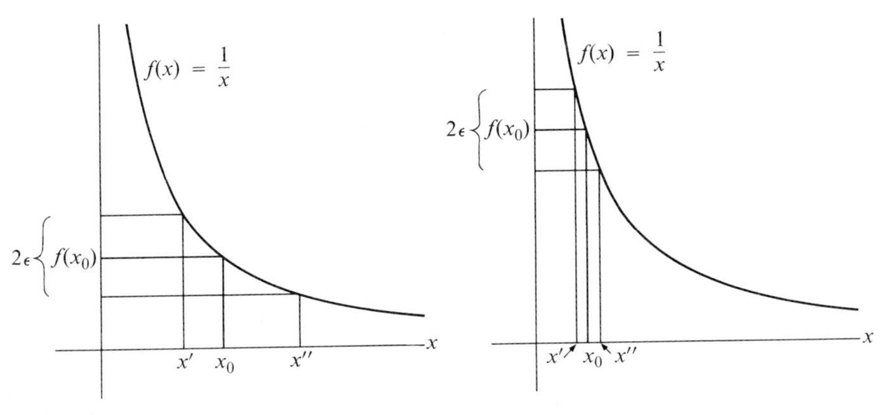

FIGURE I–3

value of δ depends on both ϵ and x_0, and a glance at Fig. I–3 convinces us that this is necessarily the case. For if any $\epsilon > 0$ is given, the points x', x'' in the figure can be made to fall as close to x_0 as desired by simply choosing x_0 sufficiently close to 0. Since δ must be chosen no larger than $|x_0 - x'|$, this shows that it cannot depend solely on ϵ. Thus f is *not* uniformly continuous on $0 < x \leq 1$. It is not hard to show, however, that the difficulty in this example is caused by the fact that the interval $0 < x \leq 1$ under consideration is not closed. This is a consequence of the following general theorem.

Theorem I–13. *If $f(x)$ is continuous on a closed interval $a \leq x \leq b$, then it is uniformly continuous on that interval.*

All of the above notions extend readily to functions of several variables. We state them here for reference.

Definition I–8. A real valued function $f(x_1, \ldots, x_n)$ defined in a region $\mathfrak{D}$ of $\mathcal{R}^n$ is *continuous at the point* $\mathbf{y} = (y_1, \ldots, y_n)$ of $\mathfrak{D}$ if for every $\epsilon > 0$ there exists a $\delta > 0$, depending in general on ϵ and $\mathbf{y}$, such that

$$|f(\mathbf{x}) - f(\mathbf{y})| = |f(x_1, \ldots, x_n) - f(y_1, \ldots, y_n)| < \epsilon$$

whenever $\mathbf{x}$, $\mathbf{y}$ are in $\mathfrak{D}$ and

$$\|\mathbf{x} - \mathbf{y}\| = \sqrt{(x_1 - y_1)^2 + \cdots + (x_n - y_n)^2} < \delta.$$

We say that f is *continuous in* $\mathfrak{D}$ if it is continuous at every point of $\mathfrak{D}$.

Definition I–9. A function $f(x_1, \ldots, x_n)$ is *uniformly continuous* in $\mathfrak{D}$ if for every $\epsilon > 0$ there exists a $\delta > 0$, depending in general on ϵ but *not* on $\mathbf{x}$, such that $|f(\mathbf{x}) - f(\mathbf{y})| < \epsilon$ whenever $\mathbf{x}$, $\mathbf{y}$ are in $\mathfrak{D}$ and $\|\mathbf{x} - \mathbf{y}\| < \delta$.

If we consider functions defined on *closed and bounded* regions $\mathfrak{D}$, we have the following theorem.*

Theorem I–14. *If $f(x_1, \ldots, x_n)$ is continuous on a closed bounded region $\mathfrak{D}$, then it is uniformly continuous on $\mathfrak{D}$.*

Two properties of continuous functions which are of the greatest interest from the viewpoint of elementary calculus are given by the next two theorems.

Theorem I–15. (The maximum-minimum property.) *If f is continuous on a closed interval $a \leq x \leq b$, then it assumes a maximum and a minimum value there. More generally, if $f(x_1, \ldots, x_n)$ is continuous on a closed, bounded region $\mathfrak{D}$ of $\mathfrak{R}^n$, then there exist real numbers m, M and points $\mathbf{y} = (y_1, \ldots, y_n)$, $\mathbf{z} = (z_1, \ldots, z_n)$ in $\mathfrak{D}$ such that*

$$m \leq f(x_1, \ldots, x_n) \leq M$$

for every $\mathbf{x} = (x_1, \ldots, x_n)$ in $\mathfrak{D}$ and such that $f(\mathbf{y}) = m$ and $f(\mathbf{z}) = M$.

Theorem I–16. (The intermediate value property.) *If f is continuous on the interval $a \leq x \leq b$ and $f(a) \neq f(b)$, then for any real number M between $f(a)$ and $f(b)$ there is a point x_0, $a \leq x_0 \leq b$ such that $f(x_0) = M$. An analogous statement holds for functions of several variables.*

While the student is certainly familiar with the maximum-minimum problem, he may fail to recognize the importance of Theorem I–16. Naively, of course, this theorem may be used to assert the existence of roots of equations; for example, if $f(a) < 0$ and $f(b) > 0$ and if f is continuous on $a \leq x \leq b$, then $f(x_0) = 0$ for some x_0 in this interval. But on a deeper level, Theorem I–16 is a consequence of the basic properties of the real number system itself and may be interpreted as connecting our intuitive notion of continuity with the abstract definition of continuity given in Definition I–6.

Turning now to definitions and theorems relating to differentiation and integration, we first state

Definition I–10. Let f be defined on an open interval containing x_0. If

$$\lim_{h \to 0} \frac{f(x_0 + h) - f(x_0)}{h}$$

exists, it is called the *derivative of f at x_0* and is denoted by $f'(x_0)$ or by

---

* A region $\mathfrak{D}$ of $\mathfrak{R}^n$ is *closed* if it contains all of its limit points. It is *bounded* if there is a real number M such that for every point $\mathbf{x} = (x_1, \ldots, x_n)$ of $\mathfrak{D}$, $\|\mathbf{x}\| = \sqrt{x_1^2 + \cdots + x_n^2} < M$.

$(d/dx)f(x_0)$. If f has a derivative at every point of an open interval $a < x < b$, then f is said to be *differentiable on* that interval.*

It is a simple exercise to show that if f is differentiable on an interval $a < x < b$, then it is also continuous in that interval. We state for reference the somewhat more important theorem.

Theorem I–17. (The mean value theorem.) *If f is continuous in the closed interval $a \leq x \leq b$ and differentiable on the open interval $a < x < b$, then there exists a point x_0, $a < x_0 < b$, such that*

$$\frac{f(b) - f(a)}{b - a} = f'(x_0).$$

The geometric content of this result is illustrated in Fig. I–4. It states that there is at least one point in the open interval $a < x < b$ where the tangent to the curve $y = f(x)$ is parallel to the secant line connecting the points $(a, f(a))$, $(b, f(b))$.

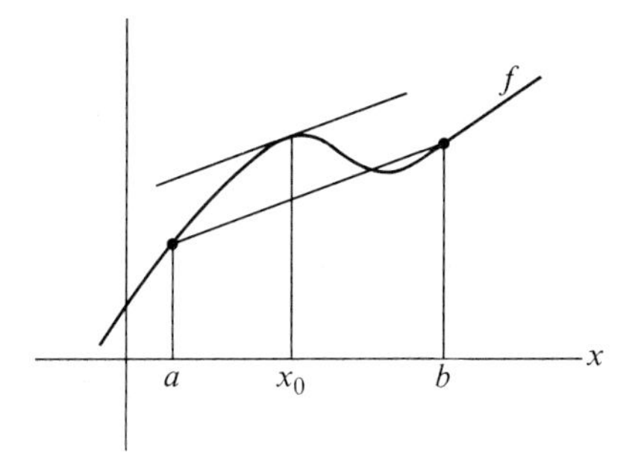

FIGURE I–4

Turning now to integration, we assume that the student already has some intuitive feeling for the definite integral $\int_a^b f(x)\,dx$ of a continuous function f. A detailed definition would be too lengthy to present here. Nevertheless, given a continuous function f, it is useful to recall the following terminology:

(1) $\int_a^b f(x)\,dx$ is called the *definite integral* of f on the interval $a \leq x \leq b$, whereas

(2) if x_0 is in the interval $a \leq x \leq b$, then the function

$$F(x) = \int_{x_0}^{x} f(t)\,dt, \quad a \leq x \leq b,$$

is called an *indefinite integral of f in $a \leq x \leq b$.*

The basic connection between differentiation and integration is provided by the *Fundamental Theorem of Calculus* (Theorem I–18).

* Later we shall need to extend this notion to include the endpoints of an interval as well.

Theorem I–18. *If f is continuous in $a \leq x \leq b$ and*

$$F(x) = \int_{x_0}^{x} f(t)\, dt$$

is an indefinite integral of f in $a \leq x \leq b$, then F is differentiable, and $F'(x) = f(x)$.

It follows almost immediately from the mean value theorem (Theorem I–17) that two indefinite integrals of f in $a \leq x \leq b$ differ by at most an additive constant. Thus we are led to the formula

$$\int_a^b f(x)\, dx = F(b) - F(a),$$

where F is any "antiderivative" of f on $a \leq x \leq b$.

Finally, we state two properties of integrals to which reference is made in the text.

Theorem I–19. *If f is continuous on $a \leq x \leq b$, then*

$$\left| \int_a^b f(x)\, dx \right| \leq \int_a^b |f(x)|\, dx.$$

Theorem I–20. (Mean value theorem for integrals.) *If f is continuous for $a \leq x \leq b$, then there is an x_0 in the open interval $a < x < b$ such that*

$$\int_a^b f(x)\, dx = (b - a) f(x_0).$$

Geometrically, $f(x_0)$ is the *average height* of f on this interval.

I–6 SEQUENCES AND SERIES OF FUNCTIONS

There are three different notions of convergence studied in connection with sequences and series of functions. Two of them, pointwise and uniform convergence are treated in this section, and the third, mean convergence, is handled in Chapter 8 of the text. We assume the elementary material of the preceding sections, especially that relating to sequences and series of constants.

Definition I–11. A sequence $\{f_k(x)\}$ of functions, each defined on an interval I, is said to *converge pointwise on I* if

$$\lim_{k \to \infty} f_k(x_0)$$

exists for each x_0 in I.

The following examples illustrate this definition and point out some of the reasons for introducing a stronger type of convergence below.

EXAMPLE 1. Let $f_k(x) = x^k, 0 \leq x \leq 1, k = 1, 2, 3, \ldots$ Then for $0 \leq x_0 < 1$ we have

$$\lim_{k\to\infty} f_k(x_0) = \lim_{k\to\infty} x_0^k = 0,$$

while, when $x = 1$, $\lim_{k\to\infty} f_k(1) = 1$. Hence the given sequence converges pointwise on the interval $0 \leq x \leq 1$ to the function

$$f(x) = \begin{cases} 0 & \text{if } 0 \leq x < 1, \\ 1 & \text{if } x = 1. \end{cases}$$

(See Fig. I–5.) Note that whereas each of the f_k is continuous on the entire interval $0 \leq x \leq 1$, the limit function f is *not* continuous in this interval (namely, it is discontinuous at $x = 1$).

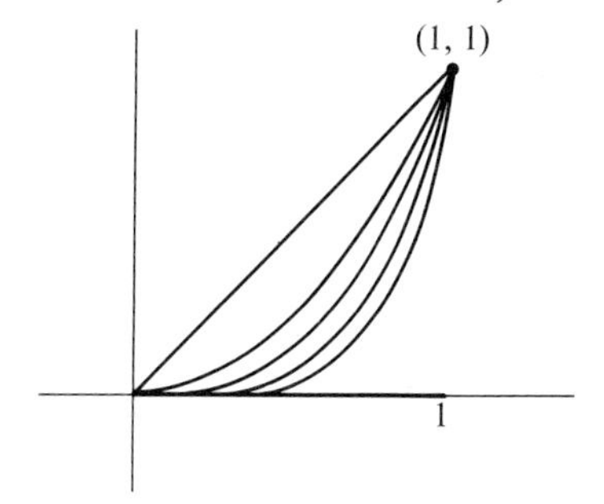

FIGURE I–5

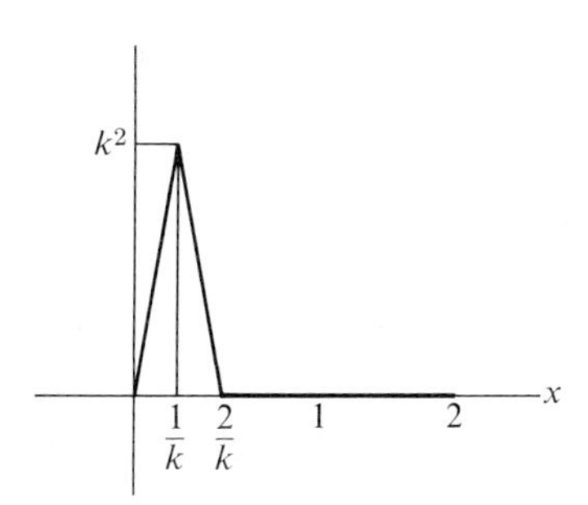

FIGURE I–6

EXAMPLE 2. Let $f_k(x)$ be the function defined on $0 \leq x \leq 2$ whose graph is indicated in Fig. I–6. Clearly $f_k(0) = 0$ for every k, so $\lim_{k\to\infty} f_k(0) = 0$. Moreover if $0 < x_0 \leq 2$, then there is some value of k, say K, such that $2/K < x_0$, and hence $f_k(x_0) = 0$ for every $k \geq K$. Thus $\lim_{k\to\infty} f_k(x_0) = 0$, and we conclude that $\{f_k(x)\}$ converges pointwise to the function which is identically zero on the interval $0 \leq x \leq 2$. This time the sequence of continuous functions converges to a *continuous limit.* Nevertheless, the individual functions f_k, regardless of how far out they lie in the sequence, may differ from the limit function $f(x) \equiv 0$ by large amounts. The members of the sequence, therefore, are not "approximations" to the limit of the sequence in the expected sense of the term. And the integrals of the members of the sequence reflect this peculiar behavior by failing to approach the integral of the limit function $f(x)$. In fact $\int_0^2 f(x)\, dx = 0$, whereas (see Fig. I–6)

$$\int_0^2 f_k(x)\, dx = \frac{1}{2} \cdot \frac{2}{k} \cdot k^2 = k,$$

and tend to infinity, not zero, as $k \to \infty$.

In order to eliminate this kind of behavior we now introduce a stronger kind of convergence, called *uniform convergence.*

Definition I–12. A sequence $\{f_k(x)\}$ is said to *converge uniformly* to the function f on the interval $a \leq x \leq b$ if for every $\epsilon > 0$ there is a positive integer K, depending on ϵ but *not* on x, such that $|f_k(x) - f(x)| < \epsilon$ whenever $k \geq K$ and x is in the given interval.

It is certainly clear that if $\{f_k(x)\}$ converges *uniformly* to f on $a \leq x \leq b$, then it also converges *pointwise* to f on this interval. In the case of uniform convergence, however, one can choose k so that

$$f(x) - \epsilon < f_k(x) < f(x) + \epsilon$$

for every x in the interval $a \leq x \leq b$. Thus $f_k(x)$ approximates f to within ϵ over the *entire* interval as shown in Fig. I–7. We noted in Example 2, that in the case of pointwise convergence, none of the members of the sequence need approximate the limit function in this way.

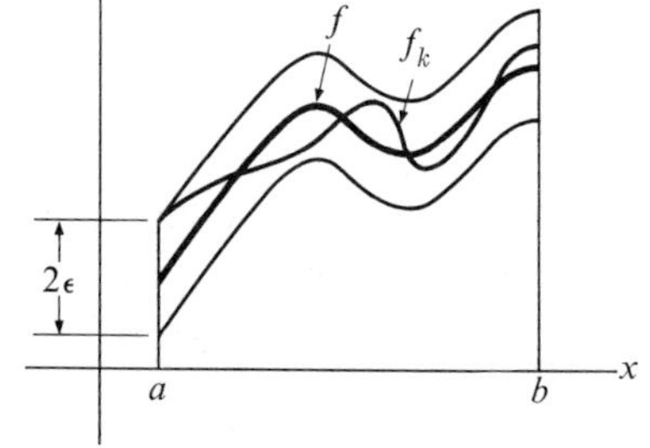

FIGURE I–7

Theorem I–21. *If a sequence $\{f_k(x)\}$ of continuous functions converges uniformly to f on $a \leq x \leq b$, then the limit function f is also continuous on this interval.*

Proof. We must show that for any x_0, $a \leq x_0 \leq b$, and any $\epsilon > 0$, there is a $\delta > 0$ such that $|f(x) - f(x_0)| < \epsilon$ whenever $a \leq x \leq b$ and $|x - x_0| < \delta$. Now

$$\begin{aligned} |f(x) - f(x_0)| &= |f(x) - f_k(x) + f_k(x) - f_k(x_0) + f_k(x_0) - f(x_0)| \\ &\leq |f(x) - f_k(x)| + |f_k(x) - f_k(x_0)| + |f_k(x_0) - f(x_0)|. \end{aligned} \tag{I–8}$$

But since the convergence is uniform, we can choose an integer k so that the first and third terms on the right-hand side of (I–8) are each less than $\epsilon/3$ for *every* x in the interval $a \leq x \leq b$. Moreover, since the function f_k thus chosen is continuous, we can also choose $\delta > 0$ such that $|f_k(x) - f_k(x_0)| < \epsilon/3$ whenever $a \leq x \leq b$ and $|x - x_0| < \delta$. The desired conclusion is now immediate. ▮

Theorem I–22. *If $\{f_k(x)\}$ is a sequence of continuous functions which converges uniformly on $a \leq x \leq b$ to the (continuous) limit function f, then for every x in this interval*

$$\lim_{k\to\infty} \int_a^x f_k(t)\,dt = \int_a^x f(t)\,dt,$$

and this convergence is uniform on $a \leq x \leq b$.

Proof. We must show that for any $\epsilon > 0$ there exists a K such that

$$\left| \int_a^x f_k(t)\,dt - \int_a^x f(t)\,dt \right| < \epsilon$$

whenever $k \geq K$ and x is in the interval $a \leq x \leq b$. For this purpose, we use the uniform convergence of $\{f_k(x)\}$ to choose K so that $|f_k(x) - f(x)| < \epsilon/(b - a)$ whenever $k \geq K$ and $a \leq x \leq b$. Then

$$\begin{aligned} \left| \int_a^x f_k(t)\,dt - \int_a^x f(t)\,dt \right| &= \left| \int_a^x [f_k(t) - f(t)]\,dt \right| \\ &\leq \int_a^x |f_k(t) - f(t)|\,dt \\ &\leq \int_a^b |f_k(t) - f(t)|\,dt \\ &< (b - a)\frac{\epsilon}{b - a} = \epsilon \end{aligned}$$

for every $k \geq K$ and every x in $a \leq x \leq b$. ▌

It would be useful to have a result similar to Theorem I–22 which would apply to differentiation instead of integration. Unfortunately this is impossible since there exist uniformly convergent sequences of differentiable functions whose limit function, although continuous, is *nowhere* differentiable.* The following theorem, however, does hold.

Theorem I–23. *If $\{f_k(x)\}$ is a sequence of continuously differentiable functions which converges pointwise to a limit f for $a \leq x \leq b$, and if the sequence $\{f_k'(x)\}$ converges uniformly on $a \leq x \leq b$, then f' exists for all x in the interval and*

$$f'(x) = \lim_{k\to\infty} f_k'(x).$$

Proof. Let g be the limit function to which $\{f_k'(x)\}$ converges uniformly. Then, applying Theorem I–22, we have

$$\begin{aligned} \int_a^x g(t)\,dt &= \lim_{k\to\infty} \int_a^x f_k'(t)\,dt \\ &= \lim_{k\to\infty} [f_k(x) - f_k(a)] \\ &= f(x) - f(a), \end{aligned}$$

and it follows from the fundamental theorem of calculus that $f'(x) = g(x) = \lim_{k\to\infty} f_k'(x)$, as desired. ▌

* See, for example, Rudin, *Principles of Mathematical Analysis*, McGraw-Hill, 1964.

All of the above results may be recast in the context of *series* of functions rather than sequences. As usual we will write

$$f(x) = \sum_{k=1}^{\infty} f_k(x), \quad a \leq x \leq b, \tag{I-9}$$

and say that the series converges (pointwise) to f, if its associated sequence of partial sums converges pointwise to f. The series is said to converge *uniformly* if the same is true of its sequence of partial sums. In this case it follows immediately from Theorems I-21 and I-22 that if each term of the series is continuous, then their sum f is also continuous, and

$$\int_a^x f(t)\,dt = \sum_{k=1}^{\infty} \int_a^x f_k(t)\,dt.$$

Finally, if each term of the series (I-9) is continuously differentiable, and if the series $\sum_{k=1}^{\infty} f_k'(x)$ obtained by differentiating each term is uniformly convergent, then

$$f'(x) = \sum_{k=1}^{\infty} f_k'(x).$$

The details of these statements are left to the reader.

In conclusion we establish the following important criterion for uniform convergence of series.

Theorem I-24. (The Weierstrass M-test.) *If $\sum_{k=1}^{\infty} M_k$ is a convergent series of positive real numbers, and if $\sum_{k=1}^{\infty} f_k(x)$ is a series of functions such that $|f_k(x)| \leq M_k$ for every k and every x in the interval $a \leq x \leq b$, then $\sum_{k=1}^{\infty} f_k(x)$ is uniformly and absolutely convergent on $a \leq x \leq b$.*

Proof. It follows from the *comparison test* that for any x_0 in the given interval the series $\sum_{k=1}^{\infty} f_k(x_0)$ converges absolutely. Thus the series converges pointwise to a limit function f on $a \leq x \leq b$. Now

$$\begin{aligned}\left| f(x) - \sum_{k=1}^{n} f_k(x) \right| &= \left| \sum_{k=n+1}^{\infty} f_k(x) \right| \\ &\leq \sum_{k=n+1}^{\infty} |f_k(x)| \\ &\leq \sum_{k=n+1}^{\infty} M_k \\ &= \sum_{k=1}^{\infty} M_k - \sum_{k=1}^{n} M_k,\end{aligned}$$

and this latter expression tends to zero as $n \to \infty$. Since it is independent of x, the convergence of $\sum_{k=1}^{n} f_k(x)$ is uniform. ▮

EXAMPLE 3. Consider the series

$$\sum_{k=1}^{\infty} \frac{\sin k^2 x}{k^2} = \sin x + \frac{\sin 4x}{4} + \frac{\sin 9x}{9} + \cdots. \tag{I-10}$$

Since

$$\left| \frac{\sin k^2 x}{k^2} \right| \leq \frac{1}{k^2}$$

for all x, and since $\sum_{k=1}^{\infty} 1/k^2$ converges, it follows from the Weierstrass M-test that the given series converges uniformly on $-\infty < x < \infty$. Let f be the limit function; i.e.,

$$f(x) = \sum_{k=1}^{\infty} \frac{\sin k^2 x}{k^2}, \quad -\infty < x < \infty.$$

Then by Theorem I–23,

$$\begin{aligned} \int_0^x f(t)\,dt &= \sum_{k=1}^{\infty} \int_0^x \frac{\sin k^2 t}{k^2}\,dt \\ &= 1 - \sum_{k=1}^{\infty} \frac{\cos k^2 x}{k^4} \\ &= 1 - \cos x - \frac{\cos 4x}{16} - \frac{\cos 9x}{81} - \cdots. \end{aligned}$$

If, on the other hand, we differentiate the terms of (I–10), we obtain the series

$$\sum_{k=1}^{\infty} \cos k^2 x = \cos x + \cos 4x + \cdots, \tag{I-11}$$

which clearly does not converge for certain values of x.

I–7 POWER SERIES

Of particular importance among series of functions are the so-called power series

$$\sum_{k=0}^{\infty} a_k x^k = a_0 + a_1 x + a_2 x^2 + \cdots,$$

where $a_0, a_1, \ldots,$ are constants. Such series enjoy special convergence properties which stem from the following theorem.

Theorem I–25. *If the power series $\sum_{k=0}^{\infty} a_k x^k$ converges for some value of x, say $x = x_0$, then it converges absolutely for every x satisfying $|x| < |x_0|$, and it converges uniformly on every interval defined by $|x| \leq |x_1| < |x_0|$.*

Proof. Since $\sum_{k=0}^{\infty} a_k x_0^k$ converges, we know that $a_k x_0^k \to 0$ as $k \to \infty$ and hence that there exists a number M such that $|a_k x_0^k| < M$, $k = 0, 1, 2, \ldots$. Now if $|x| < |x_0|$, then we have

$$\left|a_k x^k\right| < M\left|\frac{x}{x_0}\right|^k.$$

Thus since the series $\sum_{k=0}^{\infty} M|x/x_0|^k$ is a convergent geometric series it follows by the comparison test that $\sum_{k=0}^{\infty} a_k x^k$ converges absolutely on the interval $|x| < |x_0|$. In particular, for any fixed x_1, $|x_1| < |x_0|$, the series $\sum_{k=0} |a_k x_1^k|$ converges. Thus for $|x| \leq |x_1|$, we have $|a_k x^k| \leq |a_k x_1^k|$, and the Weierstrass M-test implies that the series $\sum_{k=0}^{\infty} a_k x^k$ converges uniformly on $-x_1 \leq x \leq x_1$. ▌

Now for any power series $\sum_{k=0}^{\infty} a_k x^k$, one of the following is certainly true:

(1) $\sum_{k=0}^{\infty} a_k x^k$ converges for every value of x.

(2) $\sum_{k=0}^{\infty} a_k x^k$ converges only for $x = 0$.

(3) $\sum_{k=0}^{\infty} a_k x^k$ converges for some nonzero value of x but not for all values.

In the third case the set of positive numbers x for which $\sum_{k=0}^{\infty} a_k x^k$ converges is bounded above, for otherwise, by the theorem, case (1) would apply. Letting R be the least upper bound of this set, we conclude that $\sum_{k=0}^{\infty} a_k x^k$ converges if $|x| < R$ and diverges if $|x| > R$.

Combining the above cases, we have

Theorem I–26. *For any power series $\sum_{k=0}^{\infty} a_k x^k$ there is a nonnegative number R ($R = 0$ and $R = \infty$ are included), called the radius of convergence of the series, such that the series converges (absolutely) if $|x| < R$ and diverges if $|x| > R$. Moreover, if R_1 is any number such that $0 < R_1 < R$, then $\sum_{k=0}^{\infty} a_k x^k$ converges uniformly on the interval $-R_1 \leq x \leq R_1$.*

EXAMPLE 1. Consider $\sum_{k=0}^{\infty} (1/k!)x^k$. Applying the ratio test we have

$$\left|\frac{[1/(k+1)!]x^{k+1}}{(1/k!)x^k}\right| = \frac{|x|}{k+1},$$

and for any x, this ratio tends to zero as $k \to \infty$. Thus the given series converges absolutely in $-\infty < x < \infty$ and uniformly on every finite interval $-R_1 \leq x \leq R_1$.

EXAMPLE 2. If for $\sum_{k=0}^{\infty} a_k x^k$ the limit

$$L = \lim_{k \to \infty} \left|\frac{a_{k+1}}{a_k}\right|$$

exists, then the radius of convergence of the series is $R = 1/L$. For in this case we obtain the ratio

$$\left|\frac{a_{k+1}x^{k+1}}{a_k x^k}\right| = \left|\frac{a_{k+1}}{a_k}\right| |x|,$$

and this tends to $L|x|$ as $k \to \infty$. Thus the series converges if $L|x| < 1$ and diverges if $|Lx| > 1$, i.e., converges if $|x| < R$ and diverges if $|x| > R$.

Theorems concerning differentiation and integration of power series are easily obtained from Theorem I–26.

Theorem I–27. *If the power series*

$$F(x) = \sum_{k=0}^{\infty} a_k x^k$$

has radius of convergence R, then $\int_a^b F(x)\,dx$ exists for $-R < a < b < R$ and

$$\int_a^b F(x)\,dx = \sum_{k=0}^{\infty} \int_a^b a_k x^k\,dx = \sum_{k=0}^{\infty} a_k \frac{b^{k+1} - a^{k+1}}{k+1}. \tag{I–12}$$

Proof. According to Theorem I–26, the series converges uniformly on $a \leq x \leq b$. Hence (I–12) follows from the general results on integration of uniformly convergent series. ▌

Theorem I–28. *If the power series $F(x) = \sum_{k=0}^{\infty} a_k x^k$ has radius of convergence R, then $F'(x)$ exists on $-R < x < R$ and*

$$F'(x) = \sum_{k=0}^{\infty} k a_k x^{k-1}. \tag{I–13}$$

Proof. According to the general theorem on differentiation of series, we need only show that the differentiated series (I–13) converges uniformly on every interval $-R_1 < x < R_1$ for which $R_1 < R$. For this purpose, choose x_1 such that $R_1 < x_1 < R$. Then $\sum_{k=1}^{\infty} a_k x_1^k$ converges absolutely, and hence there is a number M such that $|a_k x_1^k| \leq M$ for all k. Then for $|x| < R_1$, we have

$$\begin{aligned} |k a_k x^{k-1}| &= k|a_k|\,|x|^{k-1} \\ &\leq k|a_k| R_1^{k-1} \\ &\leq k \frac{M}{|x_1|^k} R_1^{k-1} \\ &= k \frac{M}{|x_1|} \left|\frac{R_1}{x_1}\right|^{k-1}. \end{aligned}$$

However, the series

$$\sum_{k=0}^{\infty} k \frac{M}{|x_1|} \left|\frac{R_1}{x_1}\right|^{k-1}$$

converges by the ratio test, and the uniform convergence of (I–13) now follows from the Weierstrass M-test. ▌

Stated informally, the two preceding theorems assert that a power series may be integrated or differentiated *term by term* without affecting the radius of convergence. Convergence at $|x| = R$ may be destroyed by differentiation or added by integration, however, but this can only be ascertained by examining each series individually.

EXAMPLE 3. Consider the series

$$\text{(a)} \quad \sum_{k=0}^{\infty} x^k = 1 + x + x^2 + x^3 + \cdots,$$

$$\text{(b)} \quad \sum_{k=0}^{\infty} \frac{x^{k+1}}{k+1} = x + \frac{x^2}{2} + \frac{x^3}{3} + \frac{x^4}{4} + \cdots,$$

$$\text{(c)} \quad \sum_{k=0}^{\infty} \frac{x^{k+2}}{(k+1)(k+2)} = \frac{x^2}{2} + \frac{x^3}{2 \cdot 3} + \frac{x^4}{3 \cdot 4} + \cdots.$$

Each of these series has radius of convergence $R = 1$, since (a) is a geometric series and (b) and (c) are obtained from (a) by integration. We find, however, that series (a) diverges at both endpoints $x = 1$ and $x = -1$, series (b) converges at $x = -1$ but diverges at $x = 1$, and series (c) converges at both endpoints.

There are a number of interesting implications of the theorem on differentiation of power series. If $\sum_{k=0}^{\infty} a_k x^k$ converges on $-R < x < R$, then it represents (converges to) a continuous function there:

$$F(x) = \sum_{k=0}^{\infty} a_k x^k.$$

Moreover $F'(x)$ exists on this interval and

$$F'(x) = \sum_{k=0}^{\infty} k a_k x^{k-1}, \quad -R < x < R.$$

Repeating the process of differentiation indefinitely, we obtain

$$F^{(n)}(x) = \sum_{k=0}^{\infty} k(k-1)(k-2)\cdots(k-n+1)a_k x^{k-n},$$

for $n = 1, 2, 3, \ldots$, and hence $F^{(n)}(0) = n!a_n$. With this we have proved

Theorem I–29. *If a function F can be represented by a power series $\sum_{k=0}^{\infty} a_k x^k$ on the interval $-R < x < R$, then F has derivatives of all orders on $-R < x < R$ and the coefficients of the series are uniquely determined by the relation $a_k = (1/k!)F^{(k)}(0)$, $a_0 = F(0)$.*

The theorem asserts the *uniqueness* of the power series expansion of a function F on a given interval $-R < x < R$. The *existence* of such a series is a more difficult problem, and we investigate it in the next section. We note, of course, that for a function F to have a power series representation it is necessary that F have derivatives of all orders. This condition rules out, for example, the function $\ln |x|$ which is not defined at $x = 0$ and the function $x^{8/3}$ for which the third derivative fails to exist at $x = 0$. Unfortunately the existence of infinitely many derivatives of F in $-R < x < R$ is not sufficient to assure the representability of F by a power series.*

We close this section with several arithmetic results on power series.

Theorem I–30. *If*

$$f(x) = \sum_{k=0}^{\infty} a_k x^k \qquad \text{and} \qquad g(x) = \sum_{k=0}^{\infty} b_k x^k$$

on $-R < x < R$, then for any constants α, β the series $\sum_{k=0}^{\infty} (\alpha a_k + \beta b_k)x^k$ has radius of convergence at least R and represents the function $\alpha f(x) + \beta g(x)$ on $-R < x < R$.

The proof of the theorem is an immediate consequence of Theorem I–3.

Theorem I–31. *If*

$$f(x) = \sum_{k=0}^{\infty} a_k x^k \qquad \text{and} \qquad g(x) = \sum_{k=0}^{\infty} b_k x^k$$

on $-R < x < R$, then the series $\sum_{k=0}^{\infty} c_k x^k$, where

$$c_k = \sum_{i=0}^{k} a_i b_{k-i} = a_0 b_k + a_1 b_{k-1} + \cdots + a_k b_0, \tag{I–14}$$

converges to fg in $-R < x < R$.

The reader will note that the coefficients (I–14) are exactly the ones obtained by "formal multiplication" of the given series, treating them as polynomials.

As a final comment, we note that so far we have been discussing power series $\sum_{k=0}^{\infty} a_k x^k$ whose interval of convergence was centered at the point $x = 0$. The entire discussion may be carried through equally well, however, for series of the form $\sum_{k=1}^{\infty} a_k(x - a)^k$. The interval of convergence for such series is of the form $a - R < x < a + R$ (with or without the endpoints) and the radius of con-

---

* The function defined by

$$f(x) = \begin{cases} e^{-1/(x^2)}, & x \neq 0, \\ 0, & x = 0, \end{cases}$$

has derivatives of all orders on $-\infty < x < \infty$. However, $f^{(k)}(0) = 0$ for all n, and the series $\Sigma_{k=0}^{\infty} (1/k!) f^{(k)}(0) x^k$ converges on $-\infty < x < \infty$ to the function which is identically zero, not to f.

vergence R is computed by the same techniques as before. In particular, if

$$f(x) = \sum_{k=0}^{\infty} a_k(x - a)^k$$

on $a - R < x < a + R$, then f possesses infinitely many derivatives on this interval and $a_k = (1/k!)f^{(k)}(a)$.

I-8 TAYLOR SERIES

In the preceding section we found that if a function f can be expanded in a power series as

$$f(x) = \sum_{k=0}^{\infty} a_k(x - a)^k$$

in an interval $|x - a| < R$, then f has derivatives of all orders, and

$$a_k = \frac{1}{k!} f^{(k)}(a), \quad \text{k} = 0, 1, 2, \ldots.$$

We consider now the question of *existence* of such a power series representation for a given function f.

Theorem I-32. (Taylor's formula with remainder.) *Suppose that f and its first $n + 1$ derivatives are defined and continuous on the interval I defined by $|x - a| < R$. Then for all x in I, we have*

$$f(x) = \sum_{k=0}^{n} \frac{f^{(k)}(a)}{k!}(x - a)^k + R_n(x), \tag{I-15}$$

where

$$R_n(x) = \frac{1}{n!}\int_a^x (x - t)^n f^{(n+1)}(t)\,dt. \tag{I-16}$$

Proof. We begin with the formula

$$f(x) - f(a) = \int_a^x f'(t)\,dt.$$

Transferring $f(a)$ to the right-hand side of this equation and integrating by parts [with $u = f'(t)$, $dv = dt$] we obtain*

$$f(x) = f(a) + [f'(t)(t - x)]_a^x - \int_a^x (t - x)f^{(2)}(t)\,dt,$$

or

$$f(x) = f(a) + f'(a)(x - a) + \int_a^x (x - t)f^{(2)}(t)\,dt.$$

---

* Recall that the formula for integration by parts may be written in the form $\int u\,dv = u(v + c) - \int (v + c)\,du$, where c is an arbitrary constant.

Again we integrate by parts, letting $u = f^{(2)}(t)$ and $dv = (x - t)\,dt$, to obtain

$$f(x) = f(a) + f'(a)(x - a) + \left[f^{(2)}(t) \cdot \frac{-(x - t)^2}{2!}\right]_a^x - \int_a^x \frac{-(x - t)^2}{2!} f^{(3)}(t)\,dt$$

$$= f(a) + f'(a)(x - a) + \frac{f^{(2)}(a)}{2!}(x - a)^2 + \int_a^x \frac{(x - t)^2}{2!} f^{(3)}(t)\,dt.$$

Continuing to integrate by parts (proof by mathematical induction) we arrive at the desired formulas after n integrations. (Note that the integrations can be carried out so long as the integrands are continuous.)

EXAMPLE 1. The function $f(x) = x^{8/3}$ has continuous derivatives through order two on the interval $-\infty < x < \infty$. Thus, when $a = 8$, Taylor's formula yields

$$x^{8/3} = f(8) + f'(8)(x - 8) + \int_8^x (x - t) f^{(2)}(t)\,dt$$

$$= 256 + \tfrac{256}{3}(x - 8) + \tfrac{40}{9}\int_8^x (x - t)t^{2/3}\,dt.$$

If f has derivatives of all orders at the point a, it is only natural to consider the *infinite Taylor series*

$$\sum_{k=0}^{\infty} \frac{f^{(k)}(a)}{k!}(x - a)^k, \tag{I–17}$$

and ask whether this series converges to f in an interval $|x - a| < R$. Applying Theorem I–32 we can assert that (I–17) converges to f whenever $|x - a| < R$ if and only if

$$\lim_{k \to \infty} R_k(x) = 0$$

for every x in the interval. Thus to settle this question it suffices to determine the behavior of $R_k(x)$ as $k \to \infty$. To this end the following result is particularly useful.

Theorem I–33. *If f satisfies the conditions of Theorem* I–32, *and if there exists a number M such that $|f^{(n+1)}(x)| \leq M$ for $|x - a| < R$, then*

$$|R_n(x)| \leq M \frac{|x - a|^{n+1}}{(n + 1)!}. \tag{I–18}$$

Proof. Assume first that $x \geq a$; then

$$|R_n(x)| = \left|\frac{1}{n!}\int_a^x (x - t)^n f^{(n+1)}(t)\,dt\right|$$

$$\leq \frac{M}{n!}\int_a^x (x - t)^n\,dt$$

$$= M\frac{(x - a)^{n+1}}{(n + 1)!}.$$

If, on the other hand, $x \leq a$, then

$$|R_n(x)| \leq \frac{M}{n!} \int_x^a (t - x)^n \, dt$$

$$= M \frac{(a - x)^{n+1}}{(n + 1)!}$$

Combining these results yields (I–18). ▮

EXAMPLE 2. If $f(x) = e^x$, Taylor's formula yields

$$e^x = 1 + x + \frac{x^2}{2!} + \cdots + \frac{x^n}{n!} + R_n(x),$$

with

$$R_n(x) = \frac{1}{n!} \int_0^x (x - t)^n e^t \, dt.$$

Since $|f^{n+1}(x)| = |e^x| \leq e^R$ on the interval $|x| < R$, Theorem I–33 yields

$$|R_n(x)| \leq e^R \frac{|x|^{n+1}}{(n + 1)!}, \quad -R < x < R. \tag{I–19}$$

Thus for any x in this interval

$$\lim_{n \to \infty} R_n(x) = 0.^*$$

Since R was arbitrary, it follows that

$$e^x = \sum_{k=0}^{\infty} \frac{x^k}{k!}, \quad -\infty < x < \infty, \tag{I–20}$$

i.e. that e is represented by its Taylor series on the entire real line.

The estimate (I–19) may be used for computational purposes as well. For example, let us find the number of terms of (I–20) required to compute e to an accuracy of seven decimal places. In this case we have (with $R = 1$)

$$e = e^1 = \sum_{k=0}^{n} \frac{1}{k!} + R_n(1)$$

and

$$|R_n(1)| \leq e \frac{1}{(n + 1)!} < \frac{3}{(n + 1)!}.$$

This latter expression may be made smaller than 10^{-7} by choosing $n = 10$.

---

* For fixed x, the series

$$\sum_{k=0}^{\infty} \frac{|x|^{n+1}}{(n + 1)!}$$

converges by the ratio test. Hence its general term tends to zero.

EXAMPLE 3. Since the derivatives of $\sin x$ and $\cos x$ are each bounded by $M = 1$ on $-\infty < x < \infty$, the remainder terms of their Taylor series are bounded by $|x|^{n+1}/(n+1)!$. Thus in each case the Taylor series converges to the respective function on the interval $-\infty < x < \infty$. The reader can easily show that the resulting series are

$$\sin x = x - \frac{x^3}{3!} + \frac{x^5}{5!} - \cdots,$$

$$\cos x = 1 - \frac{x^2}{2!} + \frac{x^4}{4!} - \cdots.$$

EXAMPLE 4. Combining Example 2 with Theorem I–30, we obtain

$$\sinh x = \tfrac{1}{2}(e^x - e^{-x}) = x + \frac{x^3}{3!} + \frac{x^5}{5!} + \cdots,$$

$$\cosh x = \tfrac{1}{2}(e^x + e^{-x}) = 1 + \frac{x^2}{2!} + \frac{x^4}{4!} + \cdots.$$

I–9 FUNCTIONS DEFINED BY INTEGRALS

The definition of functions by integrals is in many ways analogous to their definition by series and in this section we consider the problems of differentiating and integrating such functions.

Let the function $f(x, t)$ be defined and continuous on the rectangle R: $a \le s \le b$, $c \le t \le d$. Then the integral $\int_c^d f(s, t)\, dt$ exists for $a \le s \le b$ and defines a function $F(s)$.

Theorem I–34. *Under the conditions stated above, the function*

$$F(s) = \int_c^d f(s, t)\, dt$$

is continuous on $a \le s \le b$.

Proof. We have

$$|F(s+h) - F(s)| = \left| \int_c^d [f(s+h, t) - f(s, t)]\, dt \right|$$

$$\le \int_c^d |f(s+h, t) - f(s, t)|\, dt.$$

Since $f(s, t)$ is *uniformly* continuous on the rectangle (see Theorem I–14), we can, for any $\epsilon > 0$, find a $\delta > 0$ such that $|f(s+h, t) - f(s, t)| \le \epsilon/(d-c)$ whenever $|h| < \delta$ and both s and $s + h$ are in the given rectangle. Then $|h| < \delta$ implies that

$$|F(s+h, t) - F(s)| \le \int_c^d \frac{\epsilon}{d-c}\, dt = \epsilon. \blacksquare$$

We next ask if $F(s)$ is differentiable and if the natural formula

$$F'(s) = \int_c^d \frac{\partial}{\partial s} f(s, t)\, dt$$

holds. With mild restrictions this turns out to be true.

Theorem I–35. *If $f(s, t)$ is continuous in R* **and** *if $\partial f/\partial s$ exists and is continuous in R, then*

$$F'(s) = \frac{d}{ds}\int_c^d f(s, t)\, dt = \int_c^d \frac{\partial}{\partial s} f(s, t)\, dt.$$

Proof. We calculate

$$\frac{1}{h}[F(s + h) - F(s)] = \int_c^d \frac{1}{h}[f(s + h, t) - f(s, t)]\, dt.$$

By the mean value theorem

$$\frac{1}{h}[f(s + h, t) - f(s, t)] = \frac{\partial f}{\partial s}(s + \theta h, t),$$

where θ is some number between 0 and 1. Thus

$$F'(s) = \lim_{h\to 0} \frac{F(s + h) - F(s)}{h} = \lim_{h\to 0} \int_c^d \frac{\partial f}{\partial s}(s + \theta h, t)\, dt,$$

and since $\partial f/\partial s(s, t)$ was assumed to be continuous, Theorem I–34 yields the desired result. ▌

The formula of Theorem I–35 can be extended to allow variable limits on the defining integral, as follows.

Theorem I–36. *Let $f(s, t)$ and $\partial f/\partial s$ be continuous on the rectangle R: $a \le s \le b$, $c \le t \le d$, and let $c(s)$ and $d(s)$ be continuously differentiable functions with range in the interval $c \le t \le d$* (see Fig. I–8). *Then*

$$\frac{d}{ds} F(s) = \frac{d}{ds}\int_{c(s)}^{d(s)} f(s, t)\, dt$$

$$= \int_{c(s)}^{d(s)} \frac{\partial f}{\partial s}(s, t)\, dt$$

$$+ f(s, d(s))\, d'(s)$$

$$- f(s, c(s))c'(s).$$

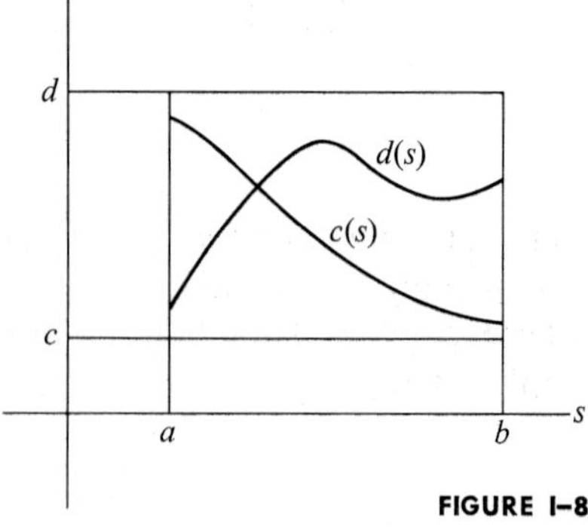

FIGURE I–8

Proof. Let $G(s, u, v)$ be the function defined by

$$G(s, u, v) = \int_u^v f(s, t)\, dt.$$

Then $F(s) = G(s, c(s), d(s))$, and the *chain rule* for functions of several variables yields

$$F'(s) = \frac{\partial G}{\partial s}(s, c(s), d(s)) + \frac{\partial G}{\partial u}(s, c(s), d(s))c'(s)$$
$$+ \frac{\partial G}{\partial v}(s, c(s), d(s))d'(s). \qquad \text{(I–21)}$$

But by the fundamental theorem of calculus

$$\frac{\partial G}{\partial v}(s, c(s), d(s)) = f(s, d(s)),$$

$$\frac{\partial G}{\partial u}(s, c(s), d(s)) = -f(s, c(s)),$$

and application of Theorem I–35 yields

$$\frac{\partial G}{\partial s}(s, c(s), d(s)) = \int_{c(s)}^{d(s)} \frac{\partial}{\partial s} f(s, t)\, dt.$$

These equations together with (I–21) give the desired formula known in the literature as *Leibnitz's formula.* ▮

We now turn our attention to *improper* integrals and let f be piecewise continuous on $0 \leq x < \infty$ (see Section 9–2).

Definition I–13. The improper integral $\int_0^\infty f(x)\, dx$ is said to *converge* if the limit

$$\lim_{B \to \infty} \int_0^B f(x)\, dx$$

exists. More precisely, the given integral is said to *converge to L,* and we write $L = \int_0^\infty f(x)\, dx$, if for every $\epsilon > 0$ there is a positive number M (depending in general on ϵ) such that

$$\left| L - \int_0^B f(x)\, dx \right| < \epsilon$$

whenever $B > M$.

In a similar fashion, $\int_{-\infty}^\infty f(x)\, dx$ is defined for a piecewise continuous function on $-\infty < x < \infty$, by the double limit

$$\lim_{\substack{A \to -\infty \\ B \to \infty}} \int_A^B f(x)\, dx.$$

Now if $\int_0^\infty f(s, t)\,dt$ exists for every value of s in an interval I, then it defines a function

$$F(s) = \int_0^\infty f(s, t)\,dt$$

on this interval. The situation here is analogous to that which arose in defining a function as the *pointwise* limit (sum) of an infinite series; for example the continuity of $f(s, t)$ on the region $0 \leq t < \infty$, s in I, does *not* imply the continuity of $F(s)$ on I. For this reason we extend the notion of uniform convergence, as follows.

Definition I–14. The integral $\int_0^\infty f(s, t)\,dt$ is said to *converge uniformly* to $F(s)$ on I if for every $\epsilon > 0$ there is a positive number M, depending in general on ϵ but *not* on s, such that

$$\left| F(s) - \int_0^B f(s, t)\,dt \right| < \epsilon$$

whenever $B > M$ and s is in I.

If $F(s) = \int_0^\infty f(s, t)\,dt$ *uniformly on* I, then the integral $\int_0^B f(s, t)\,dt$ may be viewed as approximating $F(s)$ on this interval (see the corresponding discussion concerning uniform convergence of series). Moreover we have

Theorem I–37. *If $f(s, t)$ is continuous for s in I and $0 \leq t < \infty$,* **and** *if $F(s) = \int_0^\infty f(s, t)\,dt$ converges uniformly on I, then $F(s)$ is continuous on I.*

Proof. Given $\epsilon > 0$, choose $B > 0$ so that

$$\left| F(s) - \int_0^B f(s, t)\,dt \right| < \frac{\epsilon}{3}$$

for every s in I. Then

$$\begin{aligned} |F(s + h) - F(s)| &\leq \left| F(s + h) - \int_0^B f(s + h, t)\,dt \right| \\ &\quad + \left| \int_0^B f(s + h, t)\,dt - \int_0^B f(s, t)\,dt \right| \\ &\quad + \left| \int_0^B f(s, t)\,dt - F(s) \right| \\ &\leq \frac{2\epsilon}{3} + \left| \int_0^B f(s + h, t)\,dt - \int_0^B f(s, t)\,dt \right|. \end{aligned}$$

Now choose $\delta > 0$ so that the latter term is less than $\epsilon/3$ whenever $|h| < \delta$. (Theorem I–34 provides such a δ.) ▮

The problem of integrating a function of the form

$$F(s) = \int_0^\infty f(s, t)\, dt, \quad s \text{ in } I,$$

forces us to consider the question of interchanging the order of integration in iterated integrals. We recall that for the finite rectangle R: $a \leq s \leq b, c \leq t \leq d$ we always have

$$\int_a^b \int_c^d f(s, t)\, dt\, ds = \int_c^d \int_a^b f(s, t)\, ds\, dt$$

(provided, of course, that f is continuous). However, this result is in general false if the integrations are carried out over unbounded intervals. To examine this situation more closely, we must define what is meant by *improper double integrals* and explore their relation to the improper iterated integrals.

Definition I–15. Let R be the first quadrant of the s, t plane. We say that the improper double integral

$$\iint_R f(s, t)\, ds\, dt$$

converges to L if for every $\epsilon > 0$ there is a positive number M such that

$$\left| L - \int_0^B \int_0^A f(s, t)\, ds\, dt \right| < \epsilon$$

whenever A, B are both greater than M. (Analogous definitions are given when R is a half-plane or the whole plane.)

Thus if we let

$$F(A, B) = \int_0^B \int_0^A f(s, t)\, ds\, dt,$$

then

(1) the double integral $\iint_R f(s, t)\, ds\, dt$ is the double limit

$$\lim_{\substack{A \to \infty \\ B \to \infty}} F(A, B);$$

(2) the iterated integrals are given by the iterated limits

$$\int_0^\infty \int_0^\infty f(s, t)\, ds\, dt = \lim_{B \to \infty} \lim_{A \to \infty} F(A, B),$$

$$\int_0^\infty \int_0^\infty f(s, t)\, dt\, ds = \lim_{A \to \infty} \lim_{B \to \infty} F(A, B).$$

The question of equality of the three integrals is then just a special case of the corresponding problem for limits. Thus the following result is of interest in this context.

Theorem I–38. *Let the double limit* $L = \lim_{A\to\infty, B\to\infty} F(A, B)$ *exist; i.e., assume that for every* $\epsilon > 0$ *there is an* $M > 0$ *such that* $|L - F(A, B)| < \epsilon$ *whenever* $A \geq M$ *and* $B \geq M$.

(1) *If* $\lim_{A\to\infty} F(A, B) = L(B)$ *exists for every* B, *then*

$$\lim_{B\to\infty} L(B) = L.$$

(2) *If* $\lim_{B\to\infty} F(A, B) = L(A)$ *exists for every* A, *then*

$$\lim_{A\to\infty} L(A) = L.$$

Proof. It suffices to prove (1). Now

$$|L - L(B)| \leq |L - F(A, B)| + |F(A, B) - L(B)|. \tag{I–22}$$

Thus, given $\epsilon > 0$, choose M_1 so that $|L - F(A, B)| < \epsilon/2$ whenever $A \geq M_1$ and $B \geq M_1$. Now if $B \geq M_1$ is held fixed we can find an $M \geq M_1$ such that $|F(A, B) - L(B)| < \epsilon/2$ whenever $A \geq M$. Hence $|L - L(B)| < \epsilon$, and since this can be done for any $B \geq M_1$, the proof is finished. ▮

Restated in terms of integrals, Theorem I–38 asserts that if the double integral $L = \iint_R f(s, t)\, ds\, dt$ exists, then the existence of $\int_0^\infty f(s, t)\, dt$ for every s implies that the iterated integral $\int_0^\infty \int_0^\infty f(s, t)\, dt\, ds$ exists and equals L (and similarly with the opposite order of integration). More interesting, however, is the converse problem of inferring the existence of the double integral from the existence of one of the iterated integrals. For this, the notion of uniform convergence enters once again, and we accordingly restate Definition I–14 in a form appropriate to limits.

Definition I–16. $F(A, B)$ *converges uniformly* to $L(B)$ as $A \to \infty$ if for every $\epsilon > 0$ there is a number $M > 0$ (which depends in general on ϵ but *not* on B) such that

$$|F(A, B) - L(B)| < \epsilon$$

whenever $A \geq M$.

We can now state the main result.

Theorem I–39. *Suppose that* $F(A, B)$ *converges uniformly to* $L(B)$ *as* $A \to \infty$, *and that* $L = \lim_{B\to\infty} L(B)$ *exists. Then*

(1) $\lim_{A\to\infty, B\to\infty} F(A, B) = L$, *and*

(2) *if* $\lim_{B\to\infty} F(A, B) = L(A)$ *exists for every* A, *then* $\lim_{A\to\infty} L(A) = L$.

Proof. (1) We have

$$|L - F(A, B)| \leq |L - L(B)| + |L(B) - F(A, B)|.$$

Given $\epsilon > 0$, choose M_1 so that the first term on the right-hand side is less than $\epsilon/2$ whenever $B \geq M_1$ and choose M_2 (using uniform convergence) so that the second term is less than $\epsilon/2$ whenever $A \geq M_2$. With $M = \max\{M_1, M_2\}$ we have $|L - F(A, B)| < \epsilon$ whenever A, B are both greater than M. (2) The second result now follows from Theorem I–38. ▮

For the special case of improper iterated integrals, the above theorem takes the following form:

Theorem I–40. *If the integral $\int_0^\infty f(s, t)\,dt$ converges uniformly, and if the iterated integral*

$$L = \int_0^\infty \int_0^\infty f(s, t)\,dt\,ds$$

exists, then the double integral also exists and satisfies

$$\iint_R f(s, t)\,ds\,dt = \int_0^\infty \int_0^\infty f(s, t)\,dt\,ds.$$

If, moreover, the integral $\int_0^\infty f(s, t)\,ds$ also exists, then we have

$$\int_0^\infty \int_0^\infty f(s, t)\,ds\,dt = \int_0^\infty \int_0^\infty f(s, t)\,dt\,ds.$$

Corollary. *If $f(s, t)$ is continuous in $a \leq s \leq b$, $0 \leq t < \infty$, and if*

$$F(s) = \int_0^\infty f(s, t)\,dt, \quad a \leq s \leq b,$$

the convergence to $F(s)$ being uniform on $a \leq s \leq b$, then

$$\int_a^b F(s)\,ds = \int_a^b \int_0^\infty f(s, t)\,dt\,ds = \int_0^\infty \int_a^b f(s, t)\,ds\,dt.$$

Proof. We need only apply the theorem to the function $\hat{f}(s, t)$ defined on $0 \leq s < \infty$, $0 \leq t < \infty$ by

$$\hat{f}(s, t) = \begin{cases} f(s, t), & \text{if } a \leq s \leq b, \\ 0, & \text{otherwise.} \end{cases}$$

For then $\int_0^\infty \hat{f}(s, t)\,dt$ converges uniformly on $0 \leq s < \infty$ to the function

$$\hat{F}(s) = \begin{cases} F(s), & \text{if } a \leq s \leq b, \\ 0, & \text{otherwise,} \end{cases}$$

and

$$\int_0^\infty \hat{f}(s, t)\,ds = \int_a^b \hat{f}(s, t)\,ds = \int_a^b f(s, t)\,ds$$

exists. ▮

The above theorems treat the problem of integrating functions defined by improper integrals. A useful result concerning the differentiation of such functions is the following:

Theorem I–41. *If $\partial f/\partial s(s, t)$ is piecewise continuous on $a \le s \le b$ for each t, and if*

$$F(s) = \int_0^\infty f(s, t)\, dt \qquad \textit{and} \qquad \int_0^\infty \frac{\partial f}{\partial s}(s, t)\, dt$$

both converge uniformly on $a \le s \le b$, then

$$F'(s) = \int_0^\infty \frac{\partial f}{\partial s}(s, t)\, dt.$$

Proof. Let $H(s) = \int_0^\infty \partial f/\partial s(s, t)\, dt$. Then

$$\begin{aligned}
\int_a^u H(s)\, ds &= \int_a^u \int_0^\infty \frac{\partial f}{\partial s}(s, t)\, dt\, ds = \int_0^\infty \int_a^u \frac{\partial f}{\partial s}(s, t)\, ds\, dt \\
&= \int_0^\infty [f(u, t) - f(a, t)]\, dt \\
&= F(u) - F(a),
\end{aligned}$$

whence $H(u) = F'(u)$ by differentiation. ▮

One of the main examples in the text of a function defined by an integral is the Laplace transform

$$\mathcal{L}[f](s) = \int_0^\infty e^{-st} f(t)\, dt = F(s) \tag{I–23}$$

of a piecewise continuous function. Since f need not be continuous, the uniform convergence of (I–23) (established below) is not sufficient to establish the continuity of $F(s)$. However, for functions f of exponential order, we have, for $h > 0$,

$$\begin{aligned}
|F(s + h) - F(s)| &= \left| \int_0^\infty (e^{-(s+h)t} - e^{-st}) f(t)\, dt \right| \\
&= \left| \int_0^\infty (e^{-ht} - 1) e^{-st} f(t)\, dt \right| \\
&\le \int_0^\infty (1 - e^{-ht}) e^{-st} |f(t)|\, dt.
\end{aligned}$$

Thus if α, M are any constants chosen so that $|f(t)| \le Me^{\alpha t}$, it follows for $s > \alpha$ that

$$\begin{aligned}
|F(s + h) - F(s)| &\le M \int_0^\infty (1 - e^{-ht}) e^{-(s-\alpha)t}\, dt \\
&= M \left[\frac{1}{s - \alpha} - \frac{1}{h + s - \alpha} \right],
\end{aligned}$$

and hence that $F(s + h) - F(s)$ tends to zero as $h \to 0$ through positive values. A slight modification of the argument yields the same result if $h \to 0$ through negative values, and therefore the continuity of $F(s)$ is established for $s > \alpha$. We have thus proved the following theorem.

Theorem I–42. *If f is piecewise continuous on $0 \leq t < \infty$ and is of exponential order, and if α_0 is the greatest lower bound of the set of real numbers α for which $|f(t)| \leq Me^{\alpha t}$ (for some constant M), then $\mathcal{L}[f]$ is continuous on $\alpha_0 < s < \infty$.**

Finally, we justify the formula

$$\frac{d^n}{ds^n}\mathcal{L}[f] = \frac{d^n}{ds^n}\int_0^\infty e^{-st}f(t)\,dt = (-1)^n\int_0^\infty t^n e^{-st}f(t)\,dt$$

given in Section 5–5 of the text by showing that each of the integrals

$$\int_0^\infty t^n e^{-st}f(t)\,dt, \quad n = 0, 1, 2, \ldots, \tag{I–24}$$

converges uniformly on $\alpha \leq s < \infty$ where $\alpha > \alpha_0$ (see Theorem I–42 for the definition of α_0). In fact, choosing s_1 ($\alpha_0 < s_1 < \alpha$) and M so that $|f(t)| \leq Me^{s_1 t}$, we have, for $s > \alpha$,

$$\begin{aligned}\left|\int_A^\infty t^n e^{-st}f(t)\,dt\right| &\leq M\int_A^\infty t^n e^{-(s-s_1)t}\,dt\\ &\leq M\int_A^\infty t^n e^{-(\alpha-s_1)t}\,dt.\end{aligned}$$

But the last expression tends to zero as $A \to \infty$ (t^n is of exponential order), and since it does not depend on s, the uniform convergence of (I–24) on $\alpha \leq s < \infty$ is established. In view of Theorem I–41, any number of differentiations of $\mathcal{L}[f](s)$ may be performed by differentiating under the integral sign.

---

* The number α_0, sometimes called the *order of f*, is greater than or equal to the *abscissa of convergence* s_0 of f. As was shown in the text proper, however, we may have $s_0 < \alpha_0$.

APPENDIX II

lerch's theorem

Let $f(t)$ be defined on $0 \leq t < \infty$, and be piecewise continuous on every finite interval $0 \leq t \leq A$. Assume, moreover, that $f(t)$ is of exponential order, i.e., that there exist constants α and M such that $|f(t)| \leq Me^{\alpha t}$, $0 \leq t < \infty$. It is the purpose of this appendix to demonstrate the following theorem.

> **Theorem.** (Lerch's theorem.) *If* $\mathcal{L}[f](s) = \int_0^\infty e^{-st} f(t)\, dt$ *is identically zero for all* $s > s_0$, s_0 *some constant, then* $f(t)$ *is identically zero (except possibly at its points of discontinuity).*

Proof. Let $\phi(s) = \int_0^\infty e^{-st} f(t)\, dt$, $s > s_0$. Then if

$$P(x) = \sum_{k=0}^{n} a_k x^k$$

is any polynomial with real coefficients, we have

$$\begin{aligned}\int_0^\infty e^{-st} P(e^{-t}) f(t)\, dt &= \int_0^\infty e^{-st} \sum_{k=0}^{n} a_k e^{-kt} f(t)\, dt \\ &= \sum_{k=0}^{n} a_k \int_0^\infty e^{-st} [e^{-kt} f(t)]\, dt \\ &= \sum_{k=0}^{n} a_k \phi(s + k) \equiv 0, \quad s > s_0.\end{aligned}$$

Making the change of variable $x = e^{-t}$, this last condition transforms to

$$\int_0^1 x^{s-1} P(x) f(-\ln x)\, dx \equiv 0, \quad s > s_0.$$

Now choose a fixed $s_1 > \max\{s_0, 1, \alpha + 1\}$. Then

$$x^{s_1 - 1} |f(-\ln x)| \leq M x^{s_1 - 1} e^{\alpha(-\ln x)} = M x^{s_1 - (\alpha + 1)},$$

and it follows that the function

$$G(x) = x^{s_1 - 1} f(-\ln x), \quad 0 < x \leq 1,$$

tends to zero as $x \to 0$. Let us define $G(0) = 0$, thus making G continuous at $x = 0$. Then G is a function which is bounded in the interval $0 \le x \le 1$, has only "jump" discontinuities in this interval (although there may be infinitely many such discontinuities), and satisfies

$$\int_0^1 G(x)P(x)\,dx = 0$$

for every polynomial P. We shall deduce from these conditions that $G(x) = 0$ for $0 \le x \le 1$ (except possibly at its points of discontinuity). In fact, let us choose a complete orthogonal basis for the vector space $\mathcal{PC}[0, 1]$ with inner product $\mathbf{f} \cdot \mathbf{g} = \int_0^1 f(x)g(x)\,dx$.* Then any piecewise continuous function g which satisfies

$$\int_0^1 g(x)P(x) = 0$$

for every polynomial must be identically zero (except where it is discontinuous), for it is orthogonal to every member of the chosen basis and hence must be the zero vector in $\mathcal{PC}[0, 1]$.

Except for the fact that G may have infinitely many discontinuities, the proof would be complete. Fortunately, it is not difficult to show that a complete orthonormal basis for $\mathcal{PC}[0, 1]$ is also a complete basis for the slightly larger class of functions which, like G, may have infinitely many jump discontinuities but are bounded.† We thus conclude that G is identically zero wherever it is continuous, and hence since $G(x) = x^{s_1 - 1} f(-\ln x)$, the same must be true of f. ▮

---

* We may choose, for example, the even-numbered Legendre polynomials as such a basis. For if

$$\int_0^1 g(x)P_{2k}(x)\,dx = 0, \quad k = 0, 1, 2, \ldots,$$

then for the *even extension* $\bar{g}$ of g to $-1 \le x \le 1$, we have

$$\int_{-1}^1 \bar{g}(x)P_n(x)\,dx = 0, \quad n = 0, 1, 2, \ldots$$

Thus the piecewise continuous function $\bar{g}$ must be zero (except at its points of discontinuity) and hence the same is true of g.

† The function $G(x)$ is, in fact, piecewise continuous in every interval $A \le x \le 1$ (for $0 < A < 1$) and is continuous at $x = 0$.

APPENDIX III

determinants

III–1 INTRODUCTION

We shall present here a brief introduction to determinants, with sufficient attention given to their properties to permit the usual applications. A summary of important properties, together with examples, is presented in Section III–4. The reader who wishes only a reminder concerning methods of evaluating determinants or their application to systems of linear equations may turn immediately to that section.

We wish to define a real-valued function $D(\mathbf{a}_1, \ldots, \mathbf{a}_n)$, where $\mathbf{a}_1, \ldots, \mathbf{a}_n$ are vectors in $\mathcal{R}^n$, such that

$$D(\mathbf{a}_1, \ldots, \mathbf{a}_n) = 0$$

if and only if $\mathbf{a}_1, \ldots, \mathbf{a}_n$ are linearly dependent. Such a function, when defined, will be called an $n \times n$ *determinant* and will be denoted in the more familiar form

$$\begin{vmatrix} a_{11} & a_{12} & \cdots & a_{1n} \\ a_{21} & a_{22} & \cdots & a_{2n} \\ \vdots & & & \vdots \\ a_{n1} & a_{n2} & \cdots & a_{nn} \end{vmatrix} = |a_{ij}|,$$

where the *columns* of the matrix (a_{ij}) are the components of $\mathbf{a}_1, \ldots, \mathbf{a}_n$ relative to the standard basis vectors $\mathbf{e}_1, \ldots, \mathbf{e}_n$ of $\mathcal{R}^n$.

Definition III–1. A real valued function $D(\mathbf{a}_1, \ldots, \mathbf{a}_n)$ defined for vectors $\mathbf{a}_1, \ldots, \mathbf{a}_n$ in $\mathcal{R}^n$ is called an $n \times n$ *determinant* (or a determinant of *order* n) if it satisfies the following three conditions.

I. D is linear in each of its n variables; i.e. for $i = 1, \ldots, n$, we have

$$D(\mathbf{a}_1, \ldots, \alpha\mathbf{a}_i + \beta\mathbf{a}'_i, \ldots, \mathbf{a}_n) = \alpha D(\mathbf{a}_1, \ldots, \mathbf{a}_i, \ldots, \mathbf{a}_n) + \beta D(\mathbf{a}_1, \ldots, \mathbf{a}'_i, \ldots, \mathbf{a}_n)$$

for any real numbers α, β and any vectors $\mathbf{a}'_i, \mathbf{a}_1, \ldots, \mathbf{a}_n$ in $\mathcal{R}^n$.

II. If $\mathbf{a}_i = \mathbf{a}_j$ for some i, j, $(i \neq j)$, then $D(\mathbf{a}_1, \ldots, \mathbf{a}_n) = 0$.

III. $D(\mathbf{e}_1, \ldots, \mathbf{e}_n) = 1$.

Condition II of the definition is a very special case of the desired connection between linear dependence and the vanishing of determinants. This same connection would demand, of course, that $D(\mathbf{e}_1, \ldots, \mathbf{e}_n)$ be different from zero. Condition III may therefore be regarded as normalizing the value of $D(\mathbf{a}_1, \ldots, \mathbf{a}_n)$. We shall see, in fact, that the three conditions of the definition serve to determine the function D uniquely (Theorem III–3).

EXAMPLE. If $n = 1$, we are concerned with a single vector $\mathbf{a}_1 = a_1\mathbf{e}_1$ of $\mathcal{R}^1$. Then by Conditions I and III, we have

$$D(\mathbf{a}_1) = D(a_1\mathbf{e}_1) = a_1 D(\mathbf{e}_1) = a_1.$$

Similarly, if $n = 2$ and the vectors $\mathbf{a}_1 = a_{11}\mathbf{e}_1 + a_{21}\mathbf{e}_2$ and $\mathbf{a}_2 = a_{12}\mathbf{e}_1 + a_{22}\mathbf{e}_2$ are given, then I and II yield

$$\begin{aligned} D(\mathbf{a}_1, \mathbf{a}_2) &= D(a_{11}\mathbf{e}_1 + a_{21}\mathbf{e}_2,\ a_{12}\mathbf{e}_1 + a_{22}\mathbf{e}_2) \\ &= a_{11}D(\mathbf{e}_1,\ a_{12}\mathbf{e}_1 + a_{22}\mathbf{e}_2) + a_{21}D(\mathbf{e}_2,\ a_{12}\mathbf{e}_1 + a_{22}\mathbf{e}_2) \\ &= a_{11}[a_{12}D(\mathbf{e}_1, \mathbf{e}_1) + a_{22}D(\mathbf{e}_1, \mathbf{e}_2)] \\ &\quad + a_{21}[a_{12}D(\mathbf{e}_2, \mathbf{e}_1) + a_{22}D(\mathbf{e}_2, \mathbf{e}_2)] \\ &= a_{11}a_{22}D(\mathbf{e}_1, \mathbf{e}_2) + a_{21}a_{12}D(\mathbf{e}_2, \mathbf{e}_1). \end{aligned}$$

Moreover by Condition III, $D(\mathbf{e}_1, \mathbf{e}_2) = 1$, and by Theorem III–2 below, $D(\mathbf{e}_2, \mathbf{e}_1) = -1$. Thus

$$D(\mathbf{a}_1, \mathbf{a}_2) = a_{11}a_{22} - a_{21}a_{12}.$$

Since the function D is defined for n-tuples of vectors

$$\mathbf{a}_1 = \begin{pmatrix} a_{11} \\ \vdots \\ a_{n1} \end{pmatrix}, \quad \ldots, \quad \mathbf{a}_n = \begin{pmatrix} a_{1n} \\ \vdots \\ a_{nn} \end{pmatrix},$$

we could just as easily view it as defined on the set of $n \times n$ matrices

$$M = \begin{pmatrix} a_{11} & \cdots & a_{1n} \\ \vdots & & \vdots \\ a_{n1} & \cdots & a_{nn} \end{pmatrix}.$$

In this case we follow the usual custom of denoting its value by $\det M$ or by $|a_{ij}|$. Thus

$$\begin{vmatrix} a_{11} & a_{12} & \cdots & a_{1n} \\ a_{21} & a_{22} & \cdots & a_{2n} \\ \vdots & & & \vdots \\ a_{n1} & a_{n2} & \cdots & a_{nn} \end{vmatrix} = D\left[\begin{pmatrix} a_{11} \\ a_{21} \\ \vdots \\ a_{n1} \end{pmatrix}, \ldots, \begin{pmatrix} a_{1n} \\ a_{2n} \\ \vdots \\ a_{nn} \end{pmatrix}\right].$$

The results of the example above then take the more familiar forms

$$|a_1| = a_1 \quad \text{and} \quad \begin{vmatrix} a_{11} & a_{12} \\ a_{21} & a_{22} \end{vmatrix} = a_{11}a_{22} - a_{21}a_{12}.$$

These formulas will be generalized in the next section.

III–2 BASIC PROPERTIES OF DETERMINANTS

Let D be a function satisfying Conditions I through III of Definition III–1.

Theorem III–1. *If one of the vectors* $\mathbf{a}_1, \ldots, \mathbf{a}_n$ *is the zero vector, then* $D(\mathbf{a}_1, \ldots, \mathbf{a}_n) = 0$.

Proof. Say $\mathbf{a}_i = 0$. Then by Condition I,

$$\begin{aligned} D(\mathbf{a}_1, \ldots, \mathbf{a}_i, \ldots, \mathbf{a}_n) &= D(\mathbf{a}_1, \ldots, 0 \cdot \mathbf{a}_i, \ldots, \mathbf{a}_n) \\ &= 0 \cdot D(\mathbf{a}_1, \ldots, \mathbf{a}_i, \ldots, \mathbf{a}_n) = 0. \blacksquare \end{aligned}$$

Theorem III–2. *If the sequence of vectors* $\mathbf{b}_1, \ldots, \mathbf{b}_n$ *is obtained from* $\mathbf{a}_1, \ldots, \mathbf{a}_n$ *by interchanging* $\mathbf{a}_i$ *and* $\mathbf{a}_j$ $(i < j)$, *then*

$$D(\mathbf{b}_1, \ldots, \mathbf{b}_n) = -D(\mathbf{a}_1, \ldots, \mathbf{a}_n).$$

Proof. Replacing both $\mathbf{a}_i$ and $\mathbf{a}_j$ in $D(\mathbf{a}_1, \ldots, \mathbf{a}_i, \ldots, \mathbf{a}_j, \ldots, \mathbf{a}_n)$ by $\mathbf{a}_i + \mathbf{a}_j$, and applying Conditions I and II, we have

$$\begin{aligned} 0 &= D(\mathbf{a}_1, \ldots, \mathbf{a}_i + \mathbf{a}_j, \ldots, \mathbf{a}_i + \mathbf{a}_j, \ldots, \mathbf{a}_n) \\ &= D(\mathbf{a}_1, \ldots, \mathbf{a}_i, \ldots, \mathbf{a}_i, \ldots, \mathbf{a}_n) \\ &\quad + D(\mathbf{a}_1, \ldots, \mathbf{a}_i, \ldots, \mathbf{a}_j, \ldots, \mathbf{a}_n) \\ &\quad + D(\mathbf{a}_1, \ldots, \mathbf{a}_j, \ldots, \mathbf{a}_i, \ldots, \mathbf{a}_n) \\ &\quad + D(\mathbf{a}_1, \ldots, \mathbf{a}_j, \ldots, \mathbf{a}_j, \ldots, \mathbf{a}_n). \end{aligned}$$

But the first and fourth terms of this sum vanish (by Condition I) and the sum of the second and third terms is therefore zero as desired. ▮

By applying Theorem III–2 repeatedly to adjacent vectors in the list $\mathbf{a}_1, \ldots, \mathbf{a}_n$, we obtain the following corollary.

Corollary. *If the sequence of vectors* $\mathbf{b}_1, \ldots, \mathbf{b}_n$ *is obtained from* $\mathbf{a}_1, \ldots, \mathbf{a}_n$ *by shifting one of the* $\mathbf{a}_i$ k *places to the left or right, then*

$$D(\mathbf{b}_1, \ldots, \mathbf{b}_n) = (-1)^k D(\mathbf{a}_1, \ldots, \mathbf{a}_n).$$

Now suppose that we take some permutation $\mathbf{e}_{p(1)}, \mathbf{e}_{p(2)}, \ldots, \mathbf{e}_{p(n)}$ of the standard basis vectors of $\mathcal{R}^n$.* Then, by successively interchanging pairs of vectors in this list, we can rearrange the vectors into the "natural" order $\mathbf{e}_1, \ldots, \mathbf{e}_n$, and thus by Theorem III–2 we have

$$D(\mathbf{e}_{p(1)}, \ldots, \mathbf{e}_{p(n)}) = \pm D(\mathbf{e}_1, \ldots, \mathbf{e}_n) = \pm 1, \qquad \text{(III–1)}$$

* A permutation of the set $\{1, \ldots, n\}$ is just a one-to-one function p mapping this set *onto* itself.

the plus or minus sign being chosen according as the number of interchanges required for this rearrangement is even or odd. It is of course an essential fact (which we shall not prove here) that the number of interchanges is either always even or always odd for all possible ways of carrying out the above rearrangement. The permutation p itself is accordingly said to be an *even permutation* or an *odd permutation* depending on which of these two possibilities holds.* Let us put

$$\sigma(p) = \begin{cases} +1 & \text{if } p \text{ is an even permutation,} \\ -1 & \text{if } p \text{ is an odd permutation.} \end{cases}$$

Then (III–1) becomes

$$D(\mathbf{e}_{p(1)}, \ldots, \mathbf{e}_{p(n)}) = \sigma(p). \tag{III–2}$$

We are now in a position to "compute" the value of $D(\mathbf{a}_1, \ldots, \mathbf{a}_n)$ for any n vectors of $\mathcal{R}^n$. For if

$$\begin{aligned} \mathbf{a}_1 &= a_{11}\mathbf{e}_1 + a_{21}\mathbf{e}_2 + \cdots + a_{n1}\mathbf{e}_n = \sum_{j=1}^{n} a_{j1}\mathbf{e}_j, \\ &\vdots \\ \mathbf{a}_n &= a_{1n}\mathbf{e}_1 + a_{2n}\mathbf{e}_2 + \cdots + a_{nn}\mathbf{e}_n = \sum_{j=1}^{n} a_{jn}\mathbf{e}_j, \end{aligned}$$

then repeated application of property I yields

$$\begin{aligned} D(\mathbf{a}_1, \ldots, \mathbf{a}_n) &= D\left(\sum_{j=1}^{n} a_{j1}\mathbf{e}_j, \sum_{j=1}^{n} a_{j2}\mathbf{e}_j, \ldots, \sum_{j=1}^{n} a_{jn}\mathbf{e}_j\right) \\ &= \sum_{j_1=1}^{n} a_{j_1 1} D\left(\mathbf{e}_{j_1}, \sum_{j=1}^{n} a_{j2}\mathbf{e}_j, \ldots, \sum_{j=1}^{n} a_{jn}\mathbf{e}_j\right) \\ &= \sum_{j_1=1}^{n} a_{j_1 1} \sum_{j_2=1}^{n} a_{j_2 2} D\left(\mathbf{e}_{j_1}, \mathbf{e}_{j_2}, \ldots, \sum_{j=1}^{n} a_{jn}\mathbf{e}_j\right) \\ &\;\;\vdots \\ &= \sum_{j_1=1}^{n} a_{j_1 1} \sum_{j_2=1}^{n} a_{j_2 2} \cdots \sum_{j_n=1}^{n} a_{j_n n} D(\mathbf{e}_{j_1}, \mathbf{e}_{j_2}, \ldots, \mathbf{e}_{j_n}) \\ &= \sum_{j_1=1}^{n} \sum_{j_2=1}^{n} \cdots \sum_{j_n=1}^{n} a_{j_1 1} a_{j_2 2} \cdots a_{j_n n} D(\mathbf{e}_{j_1}, \mathbf{e}_{j_2}, \ldots, \mathbf{e}_{j_n}). \end{aligned}$$

In this last sum, however, the only terms which are different from zero are those for which the sequence $j_1, \ldots, j_n$ is a permutation of the numbers $1, \ldots, n$; for only in these terms are there no repetitions among $\mathbf{e}_{j_1}, \ldots, \mathbf{e}_{j_n}$. We may rewrite

* A well-known method of determining whether p is even or odd is to count the number I of *inversions* in the list $p(1), p(2), \ldots, p(n)$, i.e., the number of pairs $p(i), p(j)$ of these integers for which $i < j$ and $p(i) > p(j)$. The permutation p is even or odd according as I is even or odd. For example the list 3, 1, 2, 5, 4 contains three inversions, hence an odd number of interchanges is required to rearrange this list into the natural order.

this sum, therefore, in the form (using III–2)

$$D(\mathbf{a}_1, \ldots, \mathbf{a}_n) = \sum_p \sigma(p) a_{p(1)1} a_{p(2)2} \cdots a_{p(n)n}, \tag{III–3}$$

where the notation indicates that the sum is extended over the $n!$ permutations of $\{1, \ldots, n\}$.

The derivation of Eq. (III–3) from Properties I through III of Definition III–1 demonstrates that *if* there is a function D which satisfies these properties, then its values must be given by Eq. (III–3). Thus there is *at most* one function (for each n) satisfying the conditions of Definition III–1. Moreover, it is not difficult to show that the function defined by Eq. (III–3) does in fact satisfy I through III. We shall omit these details and merely summarize the results in the following theorem.

Theorem III–3. *For each n, there is one and only one function D satisfying Properties* I *through* III *of Definition* III–1. *Its values are given by the formula*

$$D(\mathbf{a}_1, \ldots, \mathbf{a}_n) = \sum_p \sigma(p) a_{p(1)1} a_{p(2)2} \cdots a_{p(n)n}. \tag{III–4}$$

If we compare formula (III–4) with the matrix

$$M = \begin{pmatrix} a_{11} & a_{12} & \cdots & a_{1n} \\ a_{21} & a_{22} & \cdots & a_{2n} \\ \vdots & & & \vdots \\ a_{n1} & a_{n2} & \cdots & a_{nn} \end{pmatrix},$$

we note that the right-hand side of (III–4) consists of the sum of $n!$ terms, each one a product of n factors chosen from among the entries of M in such a way that no two of these factors occur in the same row or the same column of M.

EXAMPLE. We can also use (III–4) to obtain the result of the example in the preceding section. For the value of the 2×2 determinant is given by (III–4) as

$$\begin{aligned} D(\mathbf{a}_1, \mathbf{a}_2) &= \begin{vmatrix} a_{11} & a_{12} \\ a_{21} & a_{22} \end{vmatrix} \\ &= \sigma(p_1) a_{11} a_{22} + \sigma(p_2) a_{21} a_{12} = a_{11} a_{22} - a_{21} a_{12}, \end{aligned}$$

the signs of $\sigma(p_1)$ and $\sigma(p_2)$ being determined by counting in their respective terms the number of inversions in the arrangement of (first) subscripts.

The 3×3 determinant $D(\mathbf{a}_1, \mathbf{a}_2, \mathbf{a}_3)$ may also be evaluated by direct application of Eq. (III–4), and in this case we obtain

$$\begin{aligned} D(\mathbf{a}_1, \mathbf{a}_2, \mathbf{a}_3) &= \begin{vmatrix} a_{11} & a_{12} & a_{13} \\ a_{21} & a_{22} & a_{23} \\ a_{31} & a_{32} & a_{33} \end{vmatrix} \\ &= a_{11} a_{22} a_{33} + a_{21} a_{32} a_{13} + a_{31} a_{12} a_{23} \\ &\quad - a_{31} a_{22} a_{13} - a_{21} a_{12} a_{33} - a_{11} a_{32} a_{23}. \end{aligned}$$

For practical purposes, Eq. (III–4) is of little use for determinants of order greater than three. Indeed the expansion of a 4×4 determinant would have 24 terms, that of a 10×10 determinant would have $10! = 3{,}628{,}800$ terms, and so forth. The remainder of this section and the next are devoted, therefore, to properties of determinants which lead to simpler procedures for computing their values.

From this point on we shall use in most cases the matrix notation

$$|a_{ij}| = \begin{vmatrix} a_{11} & a_{12} & \cdots & a_{1n} \\ a_{21} & a_{22} & \cdots & a_{2n} \\ \vdots & & & \vdots \\ a_{n1} & a_{n2} & \cdots & a_{nn} \end{vmatrix}$$

for the $n \times n$ determinant $D(\mathbf{a}_1, \ldots, \mathbf{a}_n)$. In conjunction with this notation we shall refer to the vectors

$$\mathbf{a}_j = \begin{pmatrix} a_{1j} \\ a_{2j} \\ \vdots \\ a_{nj} \end{pmatrix}, \quad j = 1, 2, \ldots, n,$$

as the *columns* of the determinant and to the vectors

$$(a_{i1}, a_{i2}, \ldots, a_{in}), \quad i = 1, 2, \ldots, n,$$

as its *rows*. Multiplying a column (row) by a real number α or adding two columns (rows) is to be interpreted, therefore, as performing these operations on the corresponding vectors. We shall allow ourselves the usual misuse of language which confuses a function with its values, and in this case we shall often speak of the determinant $|a_{ij}|$ when we are really speaking about the matrix (a_{ij}) or about the value of the function D on this matrix. Context will always provide the exact meaning.

Theorem III–4. *Let $M = (a_{ij})$ be an $n \times n$ matrix and let $M' = (b_{ij})$ be the transpose of M,* i.e. *the matrix whose columns are the rows of M (thus $b_{ij} = a_{ji}$). Then*

$$\det M = \det M'.$$

Proof. From (III–4) we have

$$\det M = \sum_p \sigma(p) a_{p(1)1} a_{p(2)2} \cdots a_{p(n)n}$$

and

$$\det M' = \sum_p \sigma(p) b_{p(1)1} b_{p(2)2} \cdots b_{p(n)n}$$

$$= \sum_p \sigma(p) a_{1p(1)} a_{2p(2)} \cdots a_{np(n)}.$$

Thus the $n!$ products which enter into the expansion of det M are precisely those which occur in the expansion of det M', and we need only show that they occur with the same algebraic signs. For this purpose let

$$a_{p(1)1}a_{p(2)2}\cdots a_{p(n)n} \qquad \text{and} \qquad a_{1q(1)}a_{2q(2)}\cdots a_{nq(n)}$$

be two corresponding products, i.e. products which differ only in the order of their factors. If we apply the permutation p to the second product (or more exactly to the subscripts of its factors) we obtain

$$a_{p(1)p(q(1))}a_{p(2)p(q(2))}\cdots a_{p(n)p(q(n))}.$$

But by definition of the transpose this must agree with the first product, and hence $p(q(i)) = i$ for $i = 1, 2, \ldots, n$. This implies that either p and q are both even permutations or else both are odd. Thus $\sigma(p) = \sigma(q)$. and we are done. ▮

The last theorem may be stated informally by saying that the value of a determinant is unchanged if its rows and columns are interchanged. This result permits us to concentrate on just the columns of a determinant. Each theorem that we prove will remain true if the word "column" is replaced by "row" and vice versa.

Certain useful properties of determinants are immediate consequences of Definition III–1. For example, it follows from Property I that the value of a determinant is multiplied by the real number k if each of the entries in a *single* column or row are multiplied by k. We also easily obtain the following useful result.

Theorem III–5. *The value of a determinant is not changed by adding a multiple of the* jth *column* (*row*) *to the* ith *column* (*row*) *if* $i \neq j$.

Proof. This is an immediate consequence of Properties I and II, for

$$\begin{aligned} &D(\mathbf{a}_1, \ldots, \mathbf{a}_i + k\mathbf{a}_j, \ldots, \mathbf{a}_j, \ldots, \mathbf{a}_n) \\ &\qquad = D(\mathbf{a}_1, \ldots, \mathbf{a}_i, \ldots, \mathbf{a}_j, \ldots, \mathbf{a}_n) + kD(\mathbf{a}_1, \ldots, \mathbf{a}_j, \ldots, \mathbf{a}_j, \ldots, \mathbf{a}_n), \end{aligned}$$

and the second term of the last expression vanishes. ▮

Theorem III–6. *If the vectors* $\mathbf{a}_1, \ldots, \mathbf{a}_n$ *are linearly dependent, then*

$$D(\mathbf{a}_1, \ldots, \mathbf{a}_n) = \begin{vmatrix} a_{11} & \cdots & a_{1n} \\ \vdots & & \vdots \\ a_{n1} & \cdots & a_{nn} \end{vmatrix} = 0.$$

Proof. Suppose $\mathbf{a}_i = \sum_{j=1}^{n} c_j\mathbf{a}_j$, with $c_i = 0$. Then using Property I, we have

$$\begin{aligned} D(\mathbf{a}_1, \ldots, \mathbf{a}_i, \ldots, \mathbf{a}_n) &= D(\mathbf{a}_1, \ldots, \sum_{j=1}^{n} c_j\mathbf{a}_j, \ldots, \mathbf{a}_n) \\ &= \sum_{j=1}^{n} c_jD(\mathbf{a}_1, \ldots, \mathbf{a}_j, \ldots, \mathbf{a}_n). \end{aligned}$$

But in this sum, the ith term vanishes because $c_i = 0$, and each of the other terms also vanishes because in each case two of the vectors to which D is applied are equal. ▮

The last theorem provides half of the desired relationship between determinants and linear dependence. The converse is obtained in the next section where we introduce one further topic essential for applications.

III–3 MINORS AND COFACTORS

Let us begin this section by evaluating $D(\mathbf{e}_1, \mathbf{a}_2, \ldots, \mathbf{a}_n)$, where $\mathbf{a}_2, \ldots, \mathbf{a}_n$ are any $n - 1$ vectors of $\mathcal{R}^n$. Applying Theorem III–3, we write the value of this determinant in the form

$$|a_{ij}| = \begin{vmatrix} 1 & a_{12} & \cdots & a_{1n} \\ 0 & a_{22} & \cdots & a_{2n} \\ \vdots & & & \vdots \\ 0 & a_{n2} & \cdots & a_{nn} \end{vmatrix} = \sum_p \sigma(p) a_{p(1)1} a_{p(2)2} \cdots a_{p(n)n}. \tag{III–5}$$

But the only terms in this sum which are different from zero are those for which the permutation p satisfies $p(1) = 1$. Moreover, for every such permutation, the number of inversions in the sequence $p(1), p(2), \ldots, p(n)$ is the *same* as the number of inversions in the shorter sequence $p(2), p(3), \ldots, p(n)$. The value of (III–5), therefore, reduces to

$$\sum_q \sigma(q) a_{q(2)2} a_{q(3)3} \cdots a_{q(n)n},$$

where q ranges over all permutations of $\{2, 3, \ldots, n\}$. That is,

$$\begin{vmatrix} 1 & a_{12} & \cdots & a_{1n} \\ 0 & a_{22} & \cdots & a_{2n} \\ \vdots & & & \vdots \\ 0 & a_{n2} & \cdots & a_{nn} \end{vmatrix} = \begin{vmatrix} a_{22} & \cdots & a_{2n} \\ \vdots & & \vdots \\ a_{n2} & \cdots & a_{nn} \end{vmatrix}. \tag{III–6}$$

Now suppose we wish to evaluate the determinant

$$D(\mathbf{a}_1, \ldots, \mathbf{a}_{j-1}, \mathbf{e}_i, \mathbf{a}_{j+1}, \ldots, a_n) = \begin{vmatrix} a_{11} & \cdots & a_{1,j-1} & 0 & a_{1,j+1} & \cdots & a_{1n} \\ a_{i1} & \cdots & a_{i,j-1} & 1 & a_{i,j+1} & \cdots & a_{in} \\ \vdots & & & & & & \vdots \\ a_{n1} & \cdots & a_{n,j-1} & 0 & a_{n,j+1} & \cdots & a_{nn} \end{vmatrix}. \tag{III–7}$$

By applying the corollary to Theorem III–2, we may move the jth column $j - 1$ places to the left thus multiplying the value of the determinant by $(-1)^{j-1}$. Then applying this corollary once again (in conjunction with Theorem III–4), we move the ith *row* up $i - 1$ places, this time multiplying the value of the determinant by

$(-1)^{i-1}$. The resulting determinant now has the form (III–5) and hence can be reduced to an $(n-1) \times (n-1)$ determinant by applying Eq. (III–6). The value of $D = D(\mathbf{a}_1, \ldots, \mathbf{a}_{j-1}, \mathbf{e}_i, \mathbf{a}_{j+1}, \ldots, \mathbf{a}_n)$ is thus given by

$$D = (-1)^{i+j} M_{ij}, \tag{III–8}$$

where M_{ij} is the determinant obtained from (III–7) by deleting the ith row and the jth column. We give formal status to this new $(n-1) \times (n-1)$ determinant in the following definition.

Definition III–2. Let D be any $n \times n$ determinant. For each i, j ($1 \leq i, j \leq n$) let M_{ij} be the $(n-1) \times (n-1)$ determinant obtained from D by deleting the ith row and the jth column, and put $A_{ij} = (-1)^{i+j} M_{ij}$. Then A_{ij} is called *the cofactor of the entry* a_{ij}, and M_{ij} the *minor determinant* of this entry.

We are now in a position to prove the main theorem of this section.

Theorem III–7. *For any $n \times n$ determinant $D = |a_{ij}|$, we have*

$$D = \sum_{i=1}^{n} a_{ij} A_{ij}, \quad j = 1, 2, \ldots, n, \tag{III–9}$$

and

$$D = \sum_{j=1}^{n} a_{ij} A_{ij}, \quad i = 1, 2, \ldots, n. \tag{III–10}$$

Proof. Let j be fixed and write $\mathbf{a}_j$ in the form

$$\mathbf{a}_j = \sum_{i=1}^{n} a_{ij} \mathbf{e}_i.$$

Applying Property I of Definition III–1 followed by Eq. (III–8), we have

$$\begin{aligned} D &= D(\mathbf{a}_1, \ldots, \mathbf{a}_j, \ldots, \mathbf{a}_n) \\ &= \sum_{i=1}^{n} a_{ij} D(\mathbf{a}_1, \ldots, \mathbf{e}_i, \ldots, \mathbf{a}_n) \\ &= \sum_{i=1}^{n} a_{ij} A_{ij}. \end{aligned}$$

This proves (III–9), and (III–10) now follows from Theorem III–4, for if B_{ij} is the cofactor of b_{ij} in the transpose, then $\sum_i b_{ij} B_{ij} = \sum_i a_{ji} A_{ji} = \sum_j a_{ij} A_{ij}$. ▮

Equation (III–9) is often referred to as the *expansion of D by cofactors* of the (entries of the) jth column. Similarly (III–10) gives the expansion of D by cofactors of the ith row. Either of these formulas permits us to reduce an $n \times n$ determinant to a sum of n determinants, each of order $n - 1$. They provide, therefore, a very efficient method for evaluating arbitrary determinants, especially when used in conjunction with Theorem III–5. Before turning to examples, we obtain one important corollary of the last theorem, and apply these results to obtain the converse of Theorem III–6.

Corollary. *For any $n \times n$ determinant $D = |a_{ij}|$, we have*

$$\sum_{i=1}^{n} a_{ij}A_{ik} = 0, \quad j, k = 1, \ldots, n, \quad j \neq k, \tag{III–11}$$

and

$$\sum_{j=1}^{n} a_{ij}A_{kj} = 0, \quad i, k = 1, \ldots, n, \quad i \neq k. \tag{III–12}$$

Proof. Every term of (III–11) is unaltered if the kth column of D is changed. Hence we may set $a_{ik} = a_{ij}$, $i = 1, 2, \ldots, n$. Then

$$\sum_{i=1}^{n} a_{ij}A_{ik} = \sum_{i=1}^{n} a_{ik}A_{ik}.$$

But the latter sum vanishes since it is the expansion by cofactors of the kth column of a determinant whose jth and kth columns are equal. Similar reasoning applies to (III–12), using (III–10) in place of (III–9). ▮

Theorem III–8. *A necessary and sufficient condition that a set of n vectors $\mathbf{a}_1, \ldots, \mathbf{a}_n$ in $\mathcal{R}^n$ be linearly dependent is that $D(\mathbf{a}_1, \ldots, \mathbf{a}_n) = 0$. Thus an $n \times n$ matrix A is nonsingular if and only if* $\det(A) \neq 0$.*

Proof. The necessity of the condition has already been established in Theorem III–6. To prove the sufficiency we must show, conversely, that if $\det(A) = 0$, then the columns $\mathbf{a}_1, \ldots, \mathbf{a}_n$ of A are linearly dependent.

Consider first the case in which the cofactor A_{pq} of some entry a_{pq} of A is different from zero. Then (III–10) and (III–12) yield

$$\sum_{j=1}^{n} a_{ij}A_{pj} = 0, \quad i = 1, 2, \ldots, n, \tag{III–13}$$

and we conclude that

$$\sum_{j=1}^{n} A_{pj}\mathbf{a}_j = \mathbf{0}.$$

* A matrix A is called *nonsingular* if it has an inverse, i.e., if there is a matrix B such that $AB = BA = I$. The existence of such an inverse, usually denoted A^{-1}, is equivalent to the linear independence of the columns of A.

Since at least one of the coefficients (viz. A_{pq}) in this linear combination is *not* zero, the vectors $\mathbf{a}_1, \ldots, \mathbf{a}_n$ are linearly dependent, and we are done.

If the cofactor of every element of A is zero, then (except in the case in which $\mathbf{a}_j = \mathbf{0}$ for all j, where the theorem is obviously true) there exists an integer r, $1 \leq r \leq n - 2$, such that for some $r \times r$ submatrix M we have $\det(M) \neq 0$ but for every larger submatrix M' we have $\det(M') = 0$.* Since the order of rows and columns is immaterial, we may assume that M lies in the first r rows and columns of A, i.e. that

$$M = \begin{pmatrix} a_{11} & \cdots & a_{1r} \\ \vdots & & \vdots \\ a_{r1} & \cdots & a_{rr} \end{pmatrix}.$$

Now choose k so that $r + 1 \leq k \leq n$ and let M' be the submatrix

$$M' = \begin{pmatrix} a_{11} & \cdots & a_{1r} & a_{1,r+1} \\ \vdots & & & \vdots \\ a_{r1} & \cdots & a_{rr} & a_{r,r+1} \\ \vdots & & & \vdots \\ a_{k1} & \cdots & a_{kr} & a_{k,r+1} \end{pmatrix}.$$

We note that for $j = 1, 2, \ldots, r + 1$, the cofactor of a_{kj} in M' is independent of k. Denote it by A_j. Then (III–10) and (III–12) once again yield

$$\sum_{j=1}^{r+1} a_{ij}A_j = 0, \quad i = 1, \ldots, r, \quad \text{or} \quad i = k. \tag{III–14}$$

Since k was chosen arbitrarily from $\{r + 1, \ldots, n\}$, the relation (III–14) holds for $i = 1, 2, \ldots, n$, and hence is equivalent to

$$\sum_{j=1}^{r+1} A_j \mathbf{a}_j = \mathbf{0}.$$

Since

$$A_{r+1} = \det(M) \neq 0,$$

we thus conclude, that $\mathbf{a}_1, \ldots, \mathbf{a}_{r+1}$ are linearly dependent and hence that the same is true of $\mathbf{a}_1, \ldots, \mathbf{a}_n$. ▌

III–4 SUMMARY AND EXAMPLES

To each $n \times n$ matrix (a_{ij}) we associate a real number called the *determinant* of the matrix (or determinant of order n) and denoted by

$$|a_{ij}| = \begin{vmatrix} a_{11} & a_{12} & \cdots & a_{1n} \\ \vdots & & & \vdots \\ a_{n1} & a_{n2} & \cdots & a_{nn} \end{vmatrix}. \tag{III–15}$$

* By an $r \times s$ submatrix of a given $m \times n$ matrix A, we mean a matrix obtained from A by deleting m–r rows and n–s columns.

The value of this determinant is given by

$$|a_{ij}| = \sum_{p} \sigma(p) a_{p(1)1} a_{p(2)2} \cdots a_{p(n)n}, \tag{III–16}$$

where p ranges over all permutations of $\{1, \ldots, n\}$ and $\sigma(p)$ is $+1$ or -1 according as the number of inversions in the sequence $p(1), p(2), \ldots, p(n)$ is even or odd (see footnote on page 683).

Except for determinants of small order, however, Formula (III–16) is of little practical use. In fact even a 4×4 determinant would already have $4! = 24$ terms in its expansion, and the evaluation of a 100×100 determinant by means of this formula would be completely out of the question.*

Fortunately, quite efficient techniques exist for computing the value of $|a_{ij}|$. They are based on properties of determinants developed in the preceding sections which we review here and illustrate through examples. In so doing, we recall that when we speak of the *columns* (*rows*) of the determinant (III–15) we are really referring to the column (row) vectors of the matrix (a_{ij}). This terminology is customary, though nonsensical, and fortunately leads to no great confusion in practice. Operations performed on columns or rows (for example multiplying a column (row) by a constant or adding one column (row) to another) are understood as operations on vectors. For more formal statements of the following properties, the reader may refer to Sections III–1 through III–3.

Summary of properties of determinants

(1) If a single column (row) of $|a_{ij}|$ is multiplied by the constant k, the value of the determinant is multiplied by k.

(2) If two columns (rows) of $|a_{ij}|$ are interchanged, the value of the determinant is multiplied by -1.

(3) The value of the determinant $|a_{ij}|$ is unchanged if a multiple of one column (row) is added to another column (row).

(4) The determinant $|a_{ij}|$ vanishes if and only if its columns (rows) are linearly dependent. In particular it vanishes if two columns (rows) are equal or if one is a multiple of another.

(5) The value of the determinant $|a_{ij}|$ is unchanged if its rows and columns are interchanged.

(6) If each entry in a fixed column of $|a_{ij}|$ is multiplied by its cofactor† and the results are added, the sum is the value of the determinant; i.e., for any fixed j, $1 \leq j \leq n$,

$$|a_{ij}| = \sum_{i=1}^{n} a_{ij} A_{ij}. \tag{III–17}$$

---

* It has been estimated that the fastest modern electronic computer would require several centuries to compute and add the 100! terms of (III–16).

† See Definition III–2 for the meaning of the term cofactor.

The corresponding statement for rows is also true. Namely for any fixed i, $1 \leq i \leq n$,

$$|a_{ij}| = \sum_{j=1}^{n} a_{ij}A_{ij}. \tag{III–18}$$

Equation (III–17) is called the *expansion of* $|a_{ij}|$ *by cofactors of the* jth *column* and (III–18) *the expansion of* $|a_{ij}|$ *by cofactors of the* ith *row*. One consequence of (4) and (6) is the following property:

(7) If each entry of a fixed column (row) of $|a_{ij}|$ is multiplied by the cofactor of the corresponding entry of a second column (row), and the results are added, the sum is zero. More precisely,

$$\begin{aligned} \sum_{i=1}^{n} a_{ij}A_{ik} &= 0, \quad j, k = 1, \ldots, n, \quad j \neq k, \\ \sum_{j=1}^{n} a_{ij}A_{kj} &= 0, \quad i, k = 1, \ldots, n, \quad i \neq k. \end{aligned} \tag{III–19}$$

EXAMPLE 1. For determinants of order ≤ 3, it is not difficult to apply (III–16) directly. For the first-order determinant $|a|$ we obtain, of course, $|a| = a$, and for determinants of orders 2 and 3 we obtain, respectively,

$$\begin{vmatrix} a_{11} & a_{12} \\ a_{21} & a_{22} \end{vmatrix} = a_{11}a_{22} - a_{21}a_{12},$$

$$\begin{vmatrix} a_{11} & a_{12} & a_{13} \\ a_{21} & a_{22} & a_{23} \\ a_{31} & a_{32} & a_{33} \end{vmatrix} = a_{11}a_{22}a_{33} + a_{31}a_{12}a_{23} + a_{21}a_{32}a_{13} - a_{31}a_{22}a_{13} - a_{21}a_{12}a_{33} - a_{11}a_{32}a_{23}.$$

The last of these formulas is frequently remembered by forming the six products indicated by the arrows in Fig. III–1 and assigning a plus or minus sign according

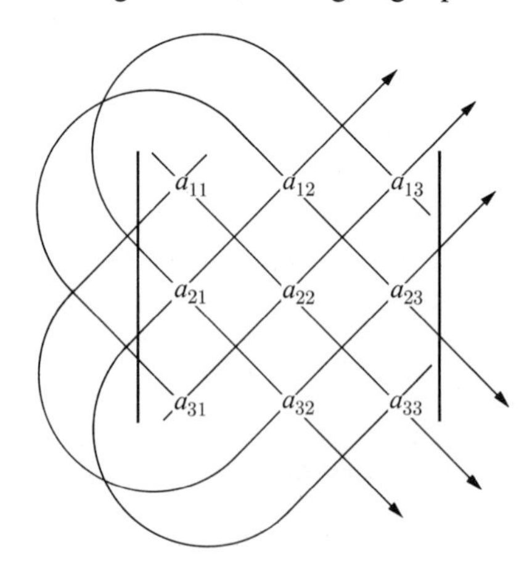

FIGURE III

as the arrow points downward or upward. Thus,

$$\begin{vmatrix} 2 & -1 & 3 \\ 0 & 1 & -6 \\ 4 & 2 & 7 \end{vmatrix} = 2 \cdot 1 \cdot 7 + 4 \cdot (-1) \cdot (-6) + 3 \cdot 0 \cdot 2 - 4 \cdot 1 \cdot 3 - (-6) \cdot 2 \cdot 2 - (-1) \cdot 0 \cdot (7)$$
$$= 14 + 24 + 0 - 12 + 24 - 0 = 50,$$

while

$$\begin{vmatrix} 3 & -7 \\ 4 & 1 \end{vmatrix} = 3 \cdot 1 - (-7) \cdot 4 = 3 + 28 = 31.$$

EXAMPLE 2. The rule for expansion by cofactors may be used to evaluate the third-order determinant of Example 1. Expanding by cofactors of the 1st row, for example, we have

$$\begin{vmatrix} 2 & -1 & 3 \\ 0 & 1 & -6 \\ 4 & 2 & 7 \end{vmatrix} = 2\begin{vmatrix} 1 & -6 \\ 2 & 7 \end{vmatrix} + (-1)\left(-\begin{vmatrix} 0 & -6 \\ 4 & 7 \end{vmatrix}\right) + 3\begin{vmatrix} 0 & 1 \\ 4 & 2 \end{vmatrix}$$
$$= 2(7 + 12) + (0 + 24) + 3(0 - 4)$$
$$= 2 \cdot 19 + 24 - 3 \cdot 4 = 50.$$

Expanding instead by cofactors of the first column, we have

$$\begin{vmatrix} 2 & -1 & 3 \\ 0 & 1 & -6 \\ 4 & 2 & 7 \end{vmatrix} = 2\begin{vmatrix} 1 & -6 \\ 2 & 7 \end{vmatrix} + 0\left(-\begin{vmatrix} -1 & 3 \\ 2 & 7 \end{vmatrix}\right) + 4\begin{vmatrix} -1 & 3 \\ 1 & -6 \end{vmatrix}$$
$$= 2(7 + 12) + 4(6 - 3) = 50.$$

EXAMPLE 3. The expansion of a determinant by cofactors is particularly simple if most of the entries in some column or row are zero, and this suggests that we first apply Property (3) above to bring about this situation. As an illustration we evaluate the fourth-order determinant

$$D = \begin{vmatrix} 3 & 7 & 9 & 11 \\ 11 & 5 & -7 & -2 \\ 6 & 2 & 4 & 9 \\ 8 & 0 & 2 & -5 \end{vmatrix}.$$

Multiplying the third column by -4 and adding it to the first column introduces a second zero into the fourth row. We then add twice the third column to the fourth column, introducing a -1 into the lower right corner. Thus

$$D = \begin{vmatrix} -33 & 7 & 9 & 29 \\ 39 & 5 & -7 & -16 \\ -10 & 2 & 4 & 17 \\ 0 & 0 & 2 & -1 \end{vmatrix} = \begin{vmatrix} -33 & 7 & 67 & 29 \\ 39 & 5 & -39 & -16 \\ -10 & 2 & 38 & 17 \\ 0 & 0 & 0 & -1 \end{vmatrix},$$

the second determinant being obtained from the first by adding twice the fourth column to the third column. If we now expand by cofactors the fourth row, we obtain

$$D = (-1)^{4+4}(-1)\begin{vmatrix} -33 & 7 & 67 \\ 39 & 5 & -39 \\ -10 & 2 & 38 \end{vmatrix} = -\begin{vmatrix} -33 & 7 & 67 \\ 39 & 5 & -39 \\ -10 & 2 & 38 \end{vmatrix},$$

and the computation could be completed as in Example 1. We continue, however, by factoring a 2 from the third row with the aid of Property (1) and then adding multiples of this row to each of the first two rows. That is,

$$D = -2\begin{vmatrix} -33 & 7 & 67 \\ 39 & 5 & -39 \\ -5 & 1 & 19 \end{vmatrix}$$

$$= -2\begin{vmatrix} 2 & 0 & -66 \\ 39 & 5 & -39 \\ -5 & 1 & 19 \end{vmatrix}$$

$$= -2\begin{vmatrix} 2 & 0 & -66 \\ 64 & 0 & -134 \\ -5 & 1 & 19 \end{vmatrix}.$$

We now expand by cofactors of the second column, obtaining

$$D = (-2)(-1)\begin{vmatrix} 2 & -66 \\ 64 & -134 \end{vmatrix} = 2(-268 + 4224) = 2 \cdot 3956 = 7912.$$

As a final example, we illustrate the application of determinants to systems of linear equations.

EXAMPLE 4. Consider the system of equations

$$\begin{aligned} a_{11}x_1 + a_{12}x_2 + \cdots + a_{1n}x_n &= b_1, \\ a_{21}x_1 + a_{22}x_2 + \cdots + a_{2n}x_n &= b_2, \\ &\ \vdots \\ a_{n1}x_1 + a_{n2}x_2 + \cdots + a_{nn}x_n &= b_n, \end{aligned} \tag{III–20}$$

and let

$$D = \begin{vmatrix} a_{11} & a_{12} & \cdots & a_{1n} \\ \vdots & & & \vdots \\ a_{n1} & a_{n2} & \cdots & a_{nn} \end{vmatrix} \tag{III–21}$$

be the determinant of coefficients of this system. If we multiply the first of these equations by A_{11}, the second by A_{21}, etc., and then add, we obtain

$$\left(\sum_{i=1}^{n} a_{i1}A_{i1}\right)x_1 + \left(\sum_{i=1}^{n} a_{i2}A_{i1}\right)x_2 + \cdots + \left(\sum_{i=1}^{n} a_{in}A_{i1}\right)x_n = \sum_{i=1}^{n} b_iA_{i1}.$$

But the coefficient of x_1 is then just the value D of $|a_{ij}|$ itself, and the coefficients of $x_2, \ldots, x_n$ are (by Eq. III–19) all zero. Thus we have

$$Dx_1 = \sum_{i=1}^{n} b_i A_{i1},$$

and if $D \neq 0$ we have determined x_1. Of course, instead of A_{11} we could have used the cofactors of the jth column of (III–21). We would then have obtained the formulas

$$Dx_j = \sum_{i=1}^{n} b_i A_{ij}, \quad j = 1, 2, \ldots, n$$

for the x_j.

If $D \neq 0$, it is easily verified that these x_j do satisfy (III–20). Moreover, the solution of (III–20) is unique if and only if the columns of (III–21) are linearly independent (see Theorem 2–12 in the text). But, by Property (4), this is equivalent to the nonvanishing of D. Thus we have the following theorem.

Theorem III–9. *The system* (III–20) *of n linear equations in n unknowns has a unique solution if and only if the determinant of its coefficients is not zero. In this case, moreover, the solution is given by*

$$x_j = \frac{\sum_{i=1}^{n} b_i A_{ij}}{D}, \quad j = 1, 2, \ldots, n. \tag{III–22}$$

It is worth noting that the expression in the numerator of (III–22) is just the expansion (by cofactors of the jth column) of the determinant obtained from the coefficient determinant $|a_{ij}|$ by replacing the jth column by the vector $(b_1, \ldots, b_n)$; that is,

$$x_j = \frac{\begin{vmatrix} a_{11} & \cdots & b_1 & \cdots & a_{1n} \\ \vdots & & & & \vdots \\ a_{n1} & \cdots & b_n & \cdots & a_{nn} \end{vmatrix}}{\begin{vmatrix} a_{11} & \cdots & a_{1j} & \cdots & a_{1n} \\ \vdots & & & & \vdots \\ a_{n1} & \cdots & a_{nj} & \cdots & a_{nn} \end{vmatrix}}, \quad j = 1, 2, \ldots, n. \tag{III–23}$$

When written in this form, (III–23) is called *Cramer's rule* for solving the given system of equations.

EXAMPLE 5. The system

$$3x - 5y = 14,$$
$$x + 2y = 3$$

has coefficient determinant

$$D = \begin{vmatrix} 3 & -5 \\ 1 & 2 \end{vmatrix} = 6 + 5 = 11,$$

and hence has the solution

$$x_1 = \frac{\begin{vmatrix} 14 & -5 \\ 3 & 2 \end{vmatrix}}{\begin{vmatrix} 3 & -5 \\ 1 & 2 \end{vmatrix}} = \frac{28 + 15}{11} = \frac{43}{11},$$

$$x_2 = \frac{\begin{vmatrix} 3 & 14 \\ 1 & 3 \end{vmatrix}}{\begin{vmatrix} 3 & -5 \\ 1 & 2 \end{vmatrix}} = \frac{9 - 14}{11} = \frac{-5}{11}.$$

It is clear, of course, that the method of Examples 4 and 5 can be applied only to systems of equations for which the coefficient matrix is a square matrix, and for this reason solutions given by (III–23) are largely of theoretical interest.

III–5 MULTIPLICATION OF DETERMINANTS

Consider the $2n \times 2n$ determinant

$$|e_{ij}| = \begin{vmatrix} a_{11} & \cdots & a_{nn} & 0 & \cdots & 0 \\ \vdots & & \vdots & \vdots & & \vdots \\ a_{n1} & \cdots & a_{nn} & 0 & \cdots & 0 \\ c_{11} & \cdots & c_{1n} & b_{11} & \cdots & b_{1n} \\ \vdots & & \vdots & \vdots & & \vdots \\ c_{n1} & \cdots & c_{nn} & b_{n1} & \cdots & b_{nn} \end{vmatrix}. \tag{III–24}$$

According to (III–16), its value is given by

$$|e_{ij}| = \sum_{p} \sigma(p) e_{p(1)1} e_{p(2)2} \cdots e_{p(2n)2n}, \tag{III–25}$$

where p ranges over the $(2n)!$ permutations of $\{1, 2, \ldots, 2n\}$. Since $e_{ij} = 0$ for $1 \leq i \leq n, n + 1 \leq j \leq 2n$, the only terms of (III–25) which are different from zero are those for which p satisfies the inequalities

$$\begin{aligned} 1 \leq p(i) \leq n \quad &\text{for} \quad 1 \leq i \leq n, \\ n + 1 \leq p(i) \leq 2n \quad &\text{for} \quad n + 1 \leq i \leq 2n. \end{aligned} \tag{III–26}$$

Moreover, for such a permutation p, the number of inversions in the sequence $p(1), p(2), \ldots, p(2n)$ is the *sum* of the number of inversions in the sequences

$$p(1), p(2), \ldots, p(n)$$

and

$$p(n + 1), p(n + 2), \ldots, p(2n).$$

Each term of (III–25), therefore, can be written as a product:

$$\sigma(r)\sigma(s)a_{r(1)1}a_{r(2)2}\cdots a_{r(n)n}b_{s(1)1}b_{s(2)2}\cdots b_{s(n)n},$$

where r and s are permutations of $\{1, \ldots, n\}$ and $\sigma(r)\sigma(s) = \sigma(p)$. Thus

$$\begin{aligned} |e_{ij}| &= \sum_p \sigma(p)e_{p(1)1}\cdots e_{p(2n)2n} \\ &= \sum_r \sum_s \sigma(r)\sigma(s)a_{r(1)1}\cdots a_{r(n)n}b_{s(1)1}\cdots b_{s(n)n} \\ &= \sum_r \sigma(r)a_{r(1)1}\cdots a_{r(n)n}\cdot \sum_s \sigma(s)b_{s(1)1}\cdots b_{s(n)n} \\ &= |a_{ij}|\cdot|b_{ij}|, \end{aligned}$$

and we have proved the following lemma.

Lemma. *Regardless of the values of the c_{ij}, the determinant* (III–24) *is the product of the two "sub-determinants" $|a_{ij}|$ and $|b_{ij}|$.*

Let us now choose $c_{ij} = 0$ if $i \neq j$ and $c_{ii} = -1$. Then we have

$$|a_{ij}|\cdot|b_{ij}| = \begin{vmatrix} a_{11} & \cdots & a_{1n} & 0 & \cdots & 0 \\ \vdots & & \vdots & \vdots & & \vdots \\ a_{n1} & \cdots & a_{nn} & 0 & \cdots & 0 \\ -1 & \cdots & 0 & b_{11} & \cdots & b_{1n} \\ & -1 & & \vdots & & \vdots \\ & & \ddots & & & \\ 0 & \cdots & -1 & b_{n1} & \cdots & b_{nn} \end{vmatrix}. \tag{III–27}$$

Denoting the first n columns of (III–27) by $\mathbf{C}_1, \ldots, \mathbf{C}_n$, and adding the sum

$$\sum_{i=1}^{n} b_{i1}\mathbf{C}_i = b_{11}\mathbf{C}_1 + b_{21}\mathbf{C}_2 + \cdots + b_{n1}\mathbf{C}_n$$

to the $(n + 1)$st column, (III–27) takes the form

$$|a_{ij}|\cdot|b_{ij}| = \begin{vmatrix} a_{11} \cdots a_{1n} & \sum a_{1i}b_{i1} & 0 \cdots 0 \\ \vdots \quad\quad \vdots & \vdots & \vdots \quad\quad \vdots \\ a_{n1} \cdots a_{nn} & \sum a_{ni}b_{i1} & 0 \cdots 0 \\ -1 \cdots 0 & 0 & b_{12} \cdots b_{1n} \\ -1 \quad\quad & \vdots & \vdots \quad\quad \vdots \\ \ddots \quad & & \\ 0 \cdots -1 & 0 & b_{n2} \cdots b_{nn} \end{vmatrix}.$$

Similarly, adding $\sum_{i=1}^{n} b_{i2}\mathbf{C}_i$ to the $(n + 2)$nd column, $\sum_{i=1}^{n} b_{i3}\mathbf{C}_i$ to the

$(n + 3)$rd column, etc., we obtain finally the determinant

$$|a_{ij}| \cdot |b_{ij}| = \begin{vmatrix} a_{11} & \cdots & a_{1n} & \sum a_{1i}b_{i1} & \cdots & \sum a_{1i}b_{in} \\ \vdots & & \vdots & \vdots & & \vdots \\ a_{n1} & \cdots & a_{nn} & \sum a_{ni}b_{i1} & \cdots & \sum a_{ni}b_{in} \\ -1 & \cdots & 0 & 0 & \cdots & 0 \\ & \ddots & & \vdots & & \vdots \\ 0 & \cdots & -1 & 0 & \cdots & 0 \end{vmatrix}, \tag{III–28}$$

which, by repeated application of Property II of Section III–4, can be written

$$|a_{ij}| \cdot |b_{ij}| = (-1)^n \begin{vmatrix} -1 & \cdots & 0 & 0 & \cdots & 0 \\ & \ddots & & \vdots & & \vdots \\ 0 & & -1 & 0 & \cdots & 0 \\ a_{11} & \cdots & a_{1n} & \sum a_{1i}b_{i1} & \cdots & \sum a_{1i}b_{in} \\ \vdots & & \vdots & \vdots & & \vdots \\ a_{n1} & \cdots & a_{nn} & \sum a_{ni}b_{i1} & \cdots & \sum a_{ni}b_{in} \end{vmatrix}.$$

We now apply the lemma once again, obtaining

$$|a_{ij}| \cdot |b_{ij}| = (-1)^n \begin{vmatrix} -1 & \cdots & 0 \\ & \ddots & \\ 0 & \cdots & -1 \end{vmatrix} \cdot \begin{vmatrix} \sum a_{1i}b_{i1} & \cdots & \sum a_{1i}b_{in} \\ \vdots & & \vdots \\ \sum a_{ni}b_{i1} & \cdots & \sum a_{ni}b_{in} \end{vmatrix}.$$

Finally noting that

$$(-1)^n \begin{vmatrix} -1 & \cdots & 0 \\ & \ddots & \\ 0 & \cdots & -1 \end{vmatrix} = (-1)^{n^2}(-1)^n = (-1)^{n(n+1)} = 1,$$

we obtain

$$|a_{ij}| \cdot |b_{ij}| = \begin{vmatrix} \sum a_{1i}b_{i1} & \cdots & \sum a_{1i}b_{in} \\ \vdots & & \vdots \\ \sum a_{ni}b_{i1} & \cdots & \sum a_{ni}b_{in} \end{vmatrix}. \tag{III–29}$$

Thus if we are given matrices

$$A = \begin{pmatrix} a_{11} & \cdots & a_{1n} \\ \vdots & & \vdots \\ a_{n1} & \cdots & a_{nn} \end{pmatrix}, \quad B = \begin{pmatrix} b_{11} & \cdots & b_{1n} \\ \vdots & & \vdots \\ b_{n1} & \cdots & b_{nn} \end{pmatrix},$$

the determinant on the right of (III–29) is seen to be the determinant of the matrix product of A and B, and we have

Theorem III–10. *If A and B are $n \times n$ matrices, then*

$$\det(AB) = \det(A) \cdot \det(B).$$

It follows from Theorem III–10 in particular, that if A is a nonsingular $n \times n$ matrix and A^{-1} is its inverse, then

$$\det(A^{-1}) = \frac{1}{\det(A)}.$$

For $\det(A^{-1}) \cdot \det(A) = \det(A^{-1}A) = \det(I) = 1$.

We conclude our discussion of determinants with one further application of Theorem III–10. Namely, let $A = (a_{ij})$ be any $n \times n$ matrix, and consider its *characteristic polynomial**

$$\det(A - \lambda I) = \begin{vmatrix} a_{11} - \lambda & a_{12} & \cdots & a_{1n} \\ a_{21} & a_{22} - \lambda & \cdots & a_{2n} \\ \vdots & & & \vdots \\ a_{n1} & a_{n2} & \cdots & a_{nn} - \lambda \end{vmatrix}.$$

Then for any nonsingular matrix P, we have

$$\begin{aligned} \det(PAP^{-1} - \lambda \mathrm{I}) &= \det(PAP^{-1} - \lambda PIP^{-1}) \\ &= \det\big(P(A - \lambda I)P^{-1}\big) \\ &= \det(P) \cdot \det(A - \lambda I) \cdot \det(P^{-1}) \\ &= \det(A - \lambda I). \end{aligned}$$

This proves the following theorem.

Theorem III–11. *If A and B are similar matrices, i.e., if there exists a nonsingular matrix P such that $B = PAP^{-1}$, then A and B have the same characteristic polynomial.*

It is not difficult to show that two $n \times n$ matrices are *similar* if and only if they represent the same linear transformation $L: \mathcal{R}^n \to \mathcal{R}^n$ (relative to possibly different bases). Since Theorem III–11 asserts that the characteristic polynomial of two such matrices is the same, it follows that the characteristic polynomial which is associated with L by choosing a matrix representation for L is independent of the particular matrix used. This allows us to conclude that the eigenvalues for L can be found by representing L in matrix form and solving the resulting characteristic equation for λ.*

* See Section 12–3.

APPENDIX IV

uniqueness theorems

IV–1 SURFACES IN $\mathcal{R}^3$, SURFACE AREA

Our objective in this appendix is to prove the uniqueness of the solutions obtained in Chapters 13 through 15 for various boundary-value problems involving partial differential equations from mathematical physics. For this purpose it will be necessary to introduce several concepts from vector field theory, among them the notion of surface integral. However these ideas will be developed only to the extent required for the stated goal.

A surface $\mathcal{S}$ in $\mathcal{R}^3$ is said to be represented *parametrically* if it is the range of a function $\mathbf{r}: \mathcal{R}^2 \to \mathcal{R}^3$. In this case $\mathcal{S}$ is the set of points (x, y, z) in $\mathcal{R}^3$ satisfying

$$\begin{aligned} x &= x(u, v), \\ y &= y(u, v), \\ z &= z(u, v), \end{aligned} \tag{IV–1}$$

where $x(u, v)$, $y(u, v)$, $z(u, v)$ are the coordinate functions of $\mathbf{r}$ and (u, v) lies in a region R_{uv} of $\mathcal{R}^2$. Equations (IV–1) are called *parametric equations* for $\mathcal{S}$. For example the sphere of radius R in $\mathcal{R}^3$ centered at the origin is represented parametrically by

$$\begin{aligned} x &= R \cos u \sin v, \\ y &= R \sin u \sin v, \\ z &= R \cos v, \end{aligned}$$

where (u, v) lies in R_{uv}: $0 \le u \le 2\pi$, $0 \le v \le \pi$ (see Fig. IV–1).

A surface $\mathcal{S}$ may also be represented *explicitly* in the form

$$z = f(x, y), \quad (x, y) \text{ in } R_{xy}, \tag{IV–2}$$

or *implicitly* as the set of points satisfying an equation

$$F(x, y, z) = 0. \tag{IV–3}$$

We are familiar with each of these forms in the case of the sphere described above. For example, the equation

$$x^2 + y^2 + z^2 = R^2$$

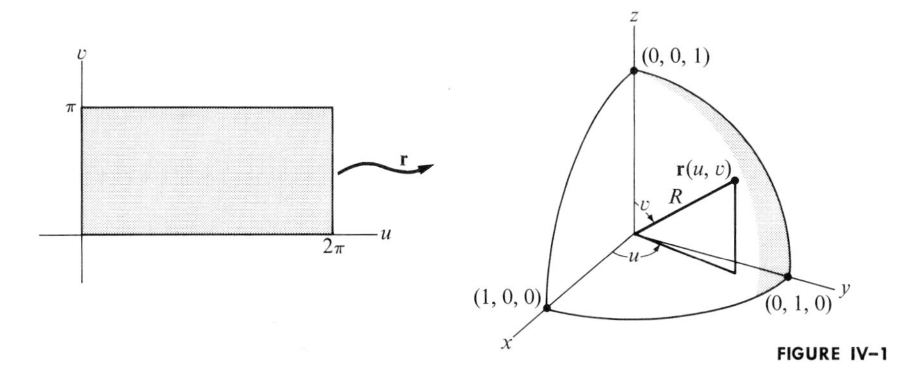

FIGURE IV-1

defines this sphere implicitly, while the upper and lower hemispheres are explicitly represented by the respective equations

$$z = \sqrt{R^2 - x^2 - y^2}, \quad x^2 + y^2 \leq R^2,$$

and

$$z = -\sqrt{R^2 - x^2 - y^2}, \quad x^2 + y^2 \leq R^2.$$

It is usually the particular application at hand that determines which of the above representations is used.

We shall say that a given surface $\mathcal{S}$ is *smooth* if it has a unique tangent plane at each of its points **P** *and* if this tangent plane varies continuously as **P** ranges over the surface. In this case we can erect at each point **P** of $\mathcal{S}$ a vector **N** which is normal to $\mathcal{S}$ at **P** (i.e., perpendicular to the tangent plane at **P**). For the explicit representation (IV–2) of $\mathcal{S}$, the property of smoothness amounts simply to the continuity of the partial derivatives $\partial f/\partial x$ and $\partial f/\partial y$, and a normal vector **N** is then given by

$$\mathbf{N} = -\frac{\partial f}{\partial x}\mathbf{i} - \frac{\partial f}{\partial y}\mathbf{j} + \mathbf{k}, \tag{IV–4}$$

where **i**, **j**, **k** are the standard basis vectors of $\mathcal{R}^3$. For implicit and explicit representations, however, we must impose assumptions in addition to continuity of the partial derivatives involved. Thus (IV–3) yields a normal vector

$$\mathbf{N} = \frac{\partial F}{\partial x}\mathbf{i} + \frac{\partial F}{\partial y}\mathbf{j} + \frac{\partial F}{\partial z}\mathbf{k} \tag{IV–5}$$

provided that this vector does not vanish, while in the case of the parametric representation (IV–1), the vectors

$$\begin{aligned} \frac{\partial \mathbf{r}}{\partial u} &= \frac{\partial x}{\partial u}\mathbf{i} + \frac{\partial y}{\partial u}\mathbf{j} + \frac{\partial z}{\partial u}\mathbf{k}, \\ \frac{\partial \mathbf{r}}{\partial v} &= \frac{\partial x}{\partial v}\mathbf{i} + \frac{\partial y}{\partial v}\mathbf{j} + \frac{\partial z}{\partial v}\mathbf{k} \end{aligned} \tag{IV–6}$$

are tangent to the surface $\mathcal{S}$ and hence determine the tangent plane only if they are *linearly independent*. In the latter case a normal $\mathbf{N}$ to $\mathcal{S}$ at the point $\mathbf{P} = \mathbf{r}(u, v)$ is orthogonal to each of the vectors (IV–6) and hence may be expressed in the form

$$\mathbf{N} = \frac{\partial \mathbf{r}}{\partial u} \times \frac{\partial \mathbf{r}}{\partial v} = \begin{vmatrix} \mathbf{i} & \mathbf{j} & \mathbf{k} \\ \frac{\partial x}{\partial u} & \frac{\partial y}{\partial u} & \frac{\partial z}{\partial u} \\ \frac{\partial x}{\partial v} & \frac{\partial y}{\partial v} & \frac{\partial z}{\partial v} \end{vmatrix}$$

$$= \begin{vmatrix} \frac{\partial y}{\partial u} & \frac{\partial z}{\partial u} \\ \frac{\partial y}{\partial v} & \frac{\partial z}{\partial v} \end{vmatrix} \mathbf{i} + \begin{vmatrix} \frac{\partial z}{\partial u} & \frac{\partial x}{\partial u} \\ \frac{\partial z}{\partial v} & \frac{\partial x}{\partial v} \end{vmatrix} \mathbf{j} + \begin{vmatrix} \frac{\partial x}{\partial u} & \frac{\partial y}{\partial u} \\ \frac{\partial x}{\partial v} & \frac{\partial y}{\partial v} \end{vmatrix} \mathbf{k} \tag{IV–7}$$

[all partial derivatives evaluated at the point (u, v)].*

EXAMPLE 1. For the sphere $\mathcal{S}$ defined parametrically by

$$x = R\cos u \sin v, \qquad y = R \sin u \sin v, \qquad z = R \cos v,$$

Eq. (IV–7) yields

$$\mathbf{N} = \begin{vmatrix} \mathbf{i} & \mathbf{j} & \mathbf{k} \\ -R\sin u \sin v & R\cos u \sin v & 0 \\ R\cos u \cos v & R \sin u \cos v & -R\sin v \end{vmatrix}$$

$$= -R^2 \cos u \sin^2 v\mathbf{i} - R^2 \sin u \sin^2 v \mathbf{j} - R^2 \sin v \cos v\mathbf{k}$$

$$= -R^2 \sin v\,(\cos u \sin v\mathbf{i} + \sin u \sin v\mathbf{j} + \cos v\mathbf{k}).$$

Here $\mathbf{N}$ is easily seen to be the inward pointing normal at $\mathbf{P}$ and $\|\mathbf{N}\| = R^2 \sin v$. Employing, instead, the explicit or implicit representations of this sphere, Eqs.

---

* For two vectors $\mathbf{a}_1 = x_1\mathbf{i} + y_1\mathbf{j} + z_1\mathbf{k}$, $\mathbf{a}_2 = x_2\mathbf{i} + y_2\mathbf{j} + z_2\mathbf{k}$ in $\mathcal{R}^3$, the *vector product* $\mathbf{a}_1 \times \mathbf{a}_2$ is defined by

$$\mathbf{a}_1 \times \mathbf{a}_2 = (y_1z_2 - z_1y_2)\mathbf{i} + (z_1x_2 - x_1z_2)\mathbf{j} + (x_1y_2 - y_1x_2)\mathbf{k}$$

$$= \begin{vmatrix} \mathbf{i} & \mathbf{j} & \mathbf{k} \\ x_1 & y_1 & z_1 \\ x_2 & y_2 & z_2 \end{vmatrix}.$$

It is easily verified that $\mathbf{a}_1 \times \mathbf{a}_2$ is orthogonal to both $\mathbf{a}_1$ and $\mathbf{a}_2$ and that $\|\mathbf{a}_1 \times \mathbf{a}_2\|$ is twice the area of the triangle determined by $\mathbf{a}_1$ and $\mathbf{a}_2$. These properties, together with the fact that the vectors $\mathbf{a}_1, \mathbf{a}_2, \mathbf{a}_1 \times \mathbf{a}_2$ (in that order) form a *right-handed triple*, serve to determine $\mathbf{a}_1 \times \mathbf{a}_2$ uniquely.

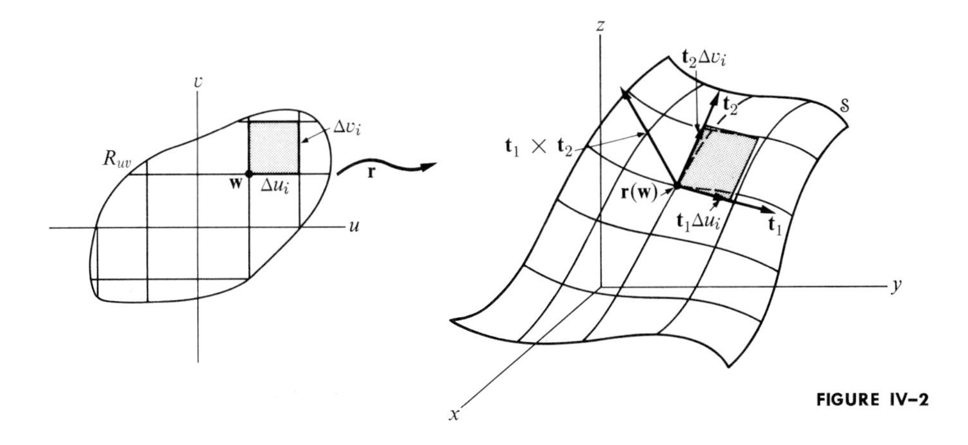

FIGURE IV-2

(IV-4) and (IV-5) yield **N** in the respective forms

$$\mathbf{N} = \frac{2x}{\sqrt{R^2 - x^2 - y^2}}\mathbf{i} + \frac{2y}{\sqrt{R^2 - x^2 - y^2}}\mathbf{j} + \mathbf{k}$$

or

$$N = 2x\mathbf{i} + 2y\mathbf{j} + 2z\mathbf{k}.$$

We now consider the notion of surface area. Here we begin with a smooth surface $\mathcal{S}$ and a parametric representation $\mathbf{r} = x(u, v)\mathbf{i} + y(u, v)\mathbf{j} + z(u, v)\mathbf{k}$ as described above. To obtain an approximation to the area of $\mathcal{S}$ we subdivide the surface into a finite number of pieces $\Delta\mathcal{S}_i$ by means of a network of intersecting curves corresponding to a grid of horizontal and vertical lines in the domain R_{uv} of **r** (see Fig. IV-2). The approximation is now found by replacing each of the $\Delta\mathcal{S}_i$ by an appropriate portion of the tangent plane to $\mathcal{S}$ at a point in $\Delta\mathcal{S}_i$ and taking area on the tangent plane as an approximation to area on the surface.

Specifically, let **w** and **r**(**w**) be as shown in Fig. IV-2 and let

$$\mathbf{t}_1 = \frac{\partial \mathbf{r}}{\partial u}(\mathbf{w}) \qquad \text{and} \qquad \mathbf{t}_2 = \frac{\partial \mathbf{r}}{\partial v}(\mathbf{w})$$

be the tangent vectors to $\mathcal{S}$ (Eq. IV-6) at the point **r**(**w**). Then if $\Delta\mathcal{S}_i$ is the image under **r** of the indicated rectangle with sides Δu_i, Δv_i, the area of the parallelogram formed by the vectors

$$\mathbf{t}_1\, \Delta u_i = \frac{\partial \mathbf{r}}{\partial u}\Delta u_i, \qquad \mathbf{t}_2\, \Delta v_i = \frac{\partial \mathbf{r}}{\partial v}\Delta v_i$$

will, intuitively at least, provide a reasonable approximation to the area of $\Delta\mathcal{S}_i$ whenever Δu_i and Δv_i are small. But, as was pointed out in the footnote on page 702, the area of this parallelogram is just the magnitude of the vector

$$\mathbf{N}\, \Delta u_i\, \Delta v_i = (\mathbf{t}_1 \times \mathbf{t}_2)\, \Delta u_i\, \Delta v_i,$$

and hence, using (IV–7), we have the required approximation in the form

$$\|\mathbf{N}\|\,\Delta u_i\,\Delta v_i = \left\{ \begin{vmatrix} \frac{\partial y}{\partial u} & \frac{\partial z}{\partial u} \\ \frac{\partial y}{\partial v} & \frac{\partial z}{\partial v} \end{vmatrix}^2 + \begin{vmatrix} \frac{\partial z}{\partial u} & \frac{\partial x}{\partial u} \\ \frac{\partial z}{\partial v} & \frac{\partial x}{\partial v} \end{vmatrix}^2 + \begin{vmatrix} \frac{\partial x}{\partial u} & \frac{\partial y}{\partial u} \\ \frac{\partial x}{\partial v} & \frac{\partial y}{\partial v} \end{vmatrix}^2 \right\}^{1/2} \Delta u_i\,\Delta v_i. \tag{IV–8}$$

We now repeat the above computation for each of the $\Delta \mathcal{S}_i$, add the results, and take the limit as the number of pieces $\Delta \mathcal{S}_i$ is allowed to increase in the usual fashion. The value of this limit (which can be shown to exist under the hypotheses in force) is, *by definition*, the area of $\mathcal{S}$, and we therefore have

$$\mathcal{A}(\mathcal{S}) = \iint_{R_{uv}} \left\{ \begin{vmatrix} \frac{\partial y}{\partial u} & \frac{\partial z}{\partial u} \\ \frac{\partial y}{\partial v} & \frac{\partial z}{\partial v} \end{vmatrix}^2 + \begin{vmatrix} \frac{\partial z}{\partial u} & \frac{\partial x}{\partial u} \\ \frac{\partial z}{\partial v} & \frac{\partial x}{\partial v} \end{vmatrix}^2 + \begin{vmatrix} \frac{\partial x}{\partial u} & \frac{\partial y}{\partial u} \\ \frac{\partial x}{\partial v} & \frac{\partial y}{\partial v} \end{vmatrix}^2 \right\}^{1/2} du\,dv. \tag{IV–9}$$

This formula is usually written in less cumbersome form as

$$\mathcal{A}(\mathcal{S}) = \iint_{R_{uv}} \left[\frac{\partial(y, z)}{\partial(u, v)}^2 + \frac{\partial(z, x)}{\partial(u, v)}^2 + \frac{\partial(x, y)}{\partial(u, v)}^2 \right]^{1/2} du\,dv, \tag{IV–10}$$

where

$$\frac{\partial(x, y)}{\partial(u, v)} = \begin{vmatrix} \frac{\partial x}{\partial u} & \frac{\partial y}{\partial u} \\ \frac{\partial x}{\partial v} & \frac{\partial y}{\partial v} \end{vmatrix},$$

etc. More generally the symbol

$$\frac{\partial(x_1, \ldots, x_n)}{\partial(u_1, \ldots, u_n)}$$

is used to denote the $n \times n$ functional determinant

$$\begin{vmatrix} \frac{\partial x_1}{\partial u_1} & \frac{\partial x_2}{\partial u_1} & \cdots & \frac{\partial x_n}{\partial u_1} \\ \frac{\partial x_1}{\partial u_2} & \frac{\partial x_2}{\partial u_2} & \cdots & \frac{\partial x_n}{\partial u_2} \\ \vdots & & & \vdots \\ \frac{\partial x_1}{\partial u_n} & \frac{\partial x_2}{\partial u_n} & \cdots & \frac{\partial x_n}{\partial u_n} \end{vmatrix}$$

and is known as the *Jacobian determinant* of the functions

$$x_1(u_1, \ldots, u_n), \quad \ldots, \quad x_n(u_1, \ldots, u_n).$$

The argument just given can easily be generalized to include integration of scalar functions defined on surfaces or, as they are called, *scalar fields*. In this case we begin with a real valued function F defined and continuous in a region of $\mathcal{R}^3$, and a smooth two-dimensional surface $\mathcal{S}$ in that region. Let $\mathcal{S}$ be subdivided into a finite number of nonoverlapping pieces $\Delta\mathcal{S}_i$, $i = 1, \ldots, n$, and choose a point $\mathbf{P}_i^*$ in each of them. Then the *surface integral* of F on $\mathcal{S}$, denoted

$$\iint_{\mathcal{S}} F\, d\mathcal{S},$$

is defined to be

$$\lim_{n\to\infty} \sum_{i=1}^{n} F(\mathbf{P}_i^*)\mathcal{A}(\Delta\mathcal{S}_i),$$

where the limit is taken in such a way that the diameter of each of the $\Delta\mathcal{S}_i$ tends to zero.* Here again it can be shown that this limit exists, and that its value is independent of the manner in which $\mathcal{S}$ is subdivided and the $\mathbf{P}_i^*$ chosen in $\Delta\mathcal{S}_i$. Finally, to obtain a formula for evaluating this integral, let

$$\begin{aligned} x &= x(u, v), \\ y &= y(u, v), \\ z &= z(u, v) \end{aligned}$$

be a parametric representation of $\mathcal{S}$ defined in a region R_{uv} of the uv-plane. Then, following the argument which led to (IV–10), we find that

$$\iint_{\mathcal{S}} F\, d\mathcal{S} = \iint_{R_{uv}} F\big(x(u, v), y(u, v), z(u, v)\big)\left[\frac{\partial(y, z)}{\partial(u, v)}^2 + \frac{\partial(z, x)}{\partial(u, v)}^2 + \frac{\partial(x, y)}{\partial(u, v)}^2\right]^{1/2} du\, dv, \tag{IV–11}$$

where the integral on the right is an ordinary double integral over R_{uv}. Integrals of this type are encountered in physical problems dealing with surface distributions of matter where they admit interpretations as mass, moments, etc.

EXAMPLE 2. Use Formula (IV–10) to calculate the surface area of a sphere $\mathcal{S}$ of radius R. Actually we have already computed the value of

$$\|\mathbf{N}\| = \left[\frac{\partial(y, z)}{\partial(u, v)}^2 + \frac{\partial(z, x)}{\partial(u, v)}^2 + \frac{\partial(x, y)}{\partial(u, v)}^2\right]^{1/2}$$

---

* By definition, the diameter of $\Delta\mathcal{S}_i$ is the least upper bound of the set of real numbers $\|\mathbf{P}_1 - \mathbf{P}_2\|$ with $\mathbf{P}_1$ and $\mathbf{P}_2$ in $\Delta\mathcal{S}_i$.

in Example 1 where we found that $\|\mathbf{N}\| = R^2 \sin v$. Hence

$$\mathcal{A}(\mathcal{S}) = R^2 \iint_{R_{uv}} \sin v \, du \, dv = R^2 \int_0^{\pi} \int_0^{2\pi} \sin v \, du \, dv = 4\pi R^2.$$

EXAMPLE 3. Find the total mass M of a hemispherical shell $\mathcal{S}$ of unit radius, given that the density of $\mathcal{S}$ is proportional to the distance from the base of the hemisphere.

Let $\mathcal{S}$ be represented (explicitly) by $z = \sqrt{1 - x^2 - y^2}$, $x^2 + y^2 \leq 1$. Then parametric equations for $\mathcal{S}$ may be given in the form

$$x = u,$$
$$y = v,$$
$$z = \sqrt{1 - u^2 - v^2},$$

with $u^2 + v^2 \leq 1$. Since the density of $\mathcal{S}$ is given by the scalar function $\rho = kz$, k a (positive) constant, Formula (IV–11) yields

$$\begin{aligned} M &= \iint_{\mathcal{S}} \rho \, d\mathcal{S} \\ &= \iint_{u^2+v^2 \leq 1} k\sqrt{1 - u^2 - v^2} \left[\frac{\partial(y, z)}{\partial(u, v)}^2 + \frac{\partial(z, x)}{\partial(u, v)}^2 + \frac{\partial(x, y)}{\partial(u, v)}^2 \right]^{1/2} du \, dv. \end{aligned}$$

But

$$\frac{\partial(y, z)}{\partial(u, v)} = \frac{u}{\sqrt{1 - u^2 - v^2}}, \quad \frac{\partial(z, x)}{\partial(u, v)} = \frac{v}{\sqrt{1 - u^2 - v^2}}, \quad \frac{\partial(x, y)}{\partial(u, v)} = 1,$$

whence

$$M = k \iint_{u^2+v^2 \leq 1} du \, dv = k\pi.$$

The last example suggests that we simplify Formula (IV–11) (and IV–10) for surfaces $\mathcal{S}$ which are represented explicitly as follows. If

$$z = f(x, y), \quad (x, y) \text{ in } R_{xy}$$

defines $\mathcal{S}$, then with

$$x = u,$$
$$y = v,$$
$$z = f(u, v), \quad (u, v) \text{ in } R_{uv},$$

we have

$$\frac{\partial(y, z)}{\partial(u, v)} = -\frac{\partial f}{\partial u}, \quad \frac{\partial(z, x)}{\partial(u, v)} = -\frac{\partial f}{\partial v}, \quad \frac{\partial(x, y)}{\partial(u, v)} = 1,$$

and (IV–11) becomes

$$\iint_{\mathcal{S}} F\, d\mathcal{S} = \iint_{R_{xy}} F(x, y, f(x, y))\sqrt{1 + \left(\frac{\partial f}{\partial x}\right)^2 + \left(\frac{\partial f}{\partial y}\right)^2}\, dx\, dy.$$

IV–2 SURFACE INTEGRALS OF VECTOR FIELDS

We come now to the problem of assigning a meaning to the integral of a continuous vector function, or *vector field*

$$\mathbf{F}(x, y, z) = P(x, y, z)\mathbf{i} + Q(x, y, z)\mathbf{j} + R(x, y, z)\mathbf{k}$$

over a smooth surface $\mathcal{S}$ in $\mathcal{R}^3$. To do so we use the integral of scalar functions introduced in the preceding section, and tentatively define the required integral to be

$$\iint_{\mathcal{S}} (\mathbf{F} \cdot \mathbf{n})\, d\mathcal{S}, \tag{IV–12}$$

where $\mathbf{n} = \mathbf{n(P)}$ is a *unit* normal vector at the point $\mathbf{P}$ of $\mathcal{S}$ chosen in such a way that $\mathbf{F} \cdot \mathbf{n}$ is a *continuous* function on $\mathcal{S}$ (Fig. IV–3). Informally this requirement means that we must designate one side of the surface as positive, and let $\mathbf{n(P)}$ be the unit normal at $\mathbf{P}$ which points in that direction. For many surfaces such a choice of positive direction is easily made. On the sphere (or any closed surface), for example, we may choose $\mathbf{n(P)}$ to be the *outward* pointing normal. And on a surface defined explicitly by

$$z = f(x, y), \quad (x, y) \text{ in } R_{xy},$$

we may choose $\mathbf{n(P)}$ to be the *upward* normal

$$\mathbf{n} = \frac{-(\partial f/\partial x)\mathbf{i} - (\partial f/\partial y)\mathbf{j} + \mathbf{k}}{\sqrt{1 + (\partial f/\partial x)^2 + (\partial f/\partial y)^2}}\,.$$

Unfortunately, there are smooth surfaces in $\mathcal{R}^3$, so-called one-sided surfaces, for which it is not possible to assign a positive direction. The most notorious example of such a surface (and aside from trivial modifications the only one) is the *Möbius strip* $\mathcal{M}$, shown in Fig. IV–4.* It is clear that $\mathcal{M}$ is genuinely one-sided in the

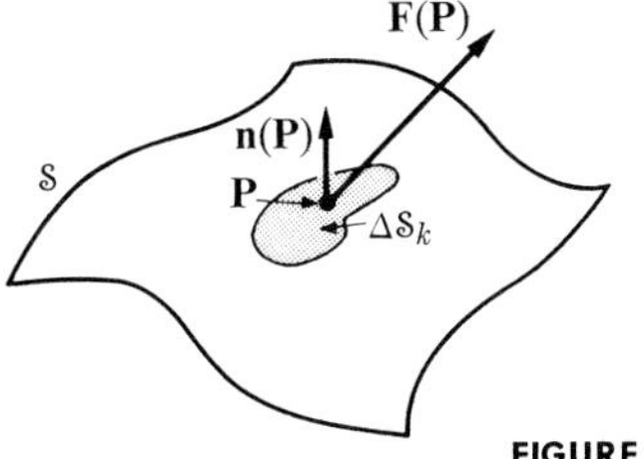

FIGURE IV–3

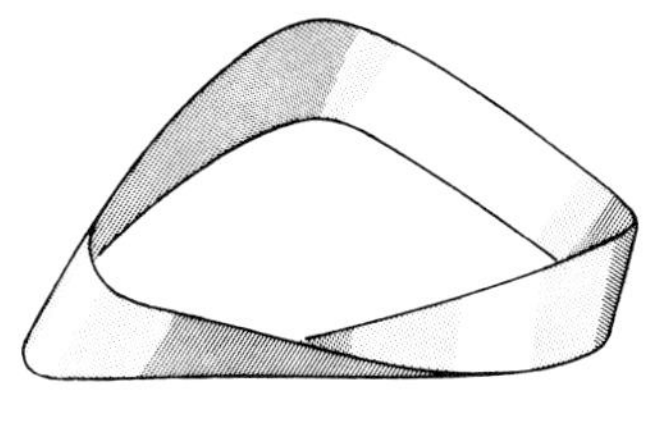

FIGURE IV–4

* August Ferdinand Möbius, German mathematician, 1790–1868.

sense that anyone who agreed to paint a single side of the strip would find that he had actually contracted to paint the entire surface. More to the point is the observation that we can pass from one of the two unit normal vectors at **P** to the other by moving the normal continuously along a smooth curve on $\mathfrak{M}$. Thus there is no natural meaning that can be attached to the integration of $\mathbf{F} \cdot \mathbf{n}$ over $\mathfrak{M}$, and we must accordingly exclude the Möbius strip and its relatives from further consideration. To this end we introduce the following definition.

Definition IV–1. A smooth surface $\mathcal{S}$ in $\mathcal{R}^3$ is said to be *orientable* if the unit normal vectors at each point **P** of $\mathcal{S}$ return to their original positions after traversing any smooth closed curve on $\mathcal{S}$. If this property fails to hold, $\mathcal{S}$ is said to be *nonorientable.* In short, a surface is orientable if it is two-sided; nonorientable otherwise.

Let us consider, now, an orientable smooth surface $\mathcal{S}$ in $\mathcal{R}^3$ with positive unit normal **n**, and let

$$\mathbf{F}(x, y, z) = P(x, y, z)\mathbf{i} + Q(x, y, z)\mathbf{j} + R(x, y, z)\mathbf{k}$$

be a continuous vector field on $\mathcal{S}$. Then the scalar function $\mathbf{F} \cdot \mathbf{n}$ is also continuous on $\mathcal{S}$, and the *surface integral of* **F** *on* $\mathcal{S}$ is defined to be

$$\iint_{\mathcal{S}} (\mathbf{F} \cdot \mathbf{n})\, d\mathcal{S}.$$

Moreover if $\mathbf{r} = \mathbf{r}(u, v)$ is a parametric representation of $\mathcal{S}$ with coordinate functions $x(u, v)$, $y(u, v)$, $z(u, v)$, and if the normal vector

$$\mathbf{N} = \mathbf{t}_1 \times \mathbf{t}_2 = \frac{\partial(y, z)}{\partial(u, v)}\mathbf{i} + \frac{\partial(z, x)}{\partial(u, v)}\mathbf{j} + \frac{\partial(x, y)}{\partial(u, v)}\mathbf{k}$$

of Eq. (IV–7) is a *positive* normal to $\mathcal{S}$ (i.e., points in the chosen positive direction from $\mathcal{S}$), then the surface integral of **F** on $\mathcal{S}$ can be evaluated as an ordinary double integral over the domain D_{uv} of **r** by the formula

$$\begin{aligned}
\iint_{\mathcal{S}} (\mathbf{F} \cdot \mathbf{n})\, d\mathcal{S} &= \iint_{R_{uv}} \left(\mathbf{F} \cdot \frac{\mathbf{N}}{\|\mathbf{N}\|}\right) \|\mathbf{N}\|\, du\, dv \\
&= \iint_{R_{uv}} (\mathbf{F} \cdot \mathbf{N})\, du\, dv \\
&= \iint_{R_{uv}} \left[P(\mathbf{r}(u, v)) \frac{\partial(y, z)}{\partial(u, v)} + Q(\mathbf{r}(u, v)) \frac{\partial(z, x)}{\partial(u, v)} \right. \\
&\qquad \left. + R(\mathbf{r}(u, v)) \frac{\partial(x, y)}{\partial(u, v)}\right] du\, dv. \qquad \text{(IV–13)}
\end{aligned}$$

Before considering examples, several remarks are in order. First, if the parametric representation of $\mathcal{S}$ is such that the vector $\mathbf{N} = \mathbf{t}_1 \times \mathbf{t}_2$ does *not* point in the positive direction from $\mathcal{S}$, the above formula must be modified to read

$$\iint_{\mathcal{S}} (\mathbf{F} \cdot \mathbf{n})\, d\mathcal{S} = -\iint_{R_{uv}} (\mathbf{F} \cdot \mathbf{N})\, du\, dv. \tag{IV–14}$$

Second, if $\mathcal{S}$ is represented explicitly by an equation of the form $z = f(x, y)$ defined in a region R_{xy} of the xy-plane, and if the upward direction from $\mathcal{S}$ is chosen as the positive direction, then

$$\mathbf{N} = \mathbf{t}_1 \times \mathbf{t}_2 = -\frac{\partial f}{\partial x}\mathbf{i} - \frac{\partial f}{\partial y}\mathbf{j} + \mathbf{k}$$

is a *positive* normal to $\mathcal{S}$, and we then have

$$\iint_{\mathcal{S}} (\mathbf{F} \cdot \mathbf{n})\, d\mathcal{S} = \iint_{R_{xy}} \left[-P(x, y, f(x, y))\frac{\partial f}{\partial x} - Q(x, y, f(x, y))\frac{\partial f}{\partial y} + R(x, y, f(x, y))\right] du\, dv. \tag{IV–15}$$

(This formula will prove quite useful later.) Finally, there is an alternative and very convenient notation for the surface integral of $\mathbf{F}$ on $\mathcal{S}$ which can be obtained by writing

$$\iint_{\mathcal{S}} (\mathbf{F} \cdot \mathbf{n})\, d\mathcal{S} = \iint_{\mathcal{S}} (P\mathbf{i} + Q\mathbf{j} + R\mathbf{k}) \cdot \mathbf{n}\, d\mathcal{S}$$

$$= \iint_{\mathcal{S}} P(\mathbf{i} \cdot \mathbf{n})\, d\mathcal{S} + Q(\mathbf{j} \cdot \mathbf{n})\, d\mathcal{S} + R(\mathbf{k} \cdot \mathbf{n})\, d\mathcal{S}.$$

For then

$$\mathbf{i} \cdot \mathbf{n} = \cos \gamma_1, \qquad \mathbf{j} \cdot \mathbf{n} = \cos \gamma_2, \qquad \mathbf{k} \cdot \mathbf{n} = \cos \gamma_3,$$

where $\gamma_1, \gamma_2, \gamma_3$ are, respectively, the angles between $\mathbf{n}$ and the vectors $\mathbf{i}, \mathbf{j}, \mathbf{k}$ (Fig. IV–5), and the quantities

$$(\mathbf{i} \cdot \mathbf{n})\, d\mathcal{S}, \quad (\mathbf{j} \cdot \mathbf{n})\, d\mathcal{S}, \quad (\mathbf{k} \cdot \mathbf{n})\, d\mathcal{S}$$

can be interpreted as the projections of an element of surface area $d\mathcal{S}$ onto the three coordinate planes. This suggests that we set

$$(\mathbf{i} \cdot \mathbf{n})\, d\mathcal{S} = dy\, dz,$$
$$(\mathbf{j} \cdot \mathbf{n})\, d\mathcal{S} = dz\, dx,$$
$$(\mathbf{k} \cdot \mathbf{n})\, d\mathcal{S} = dx\, dy,$$

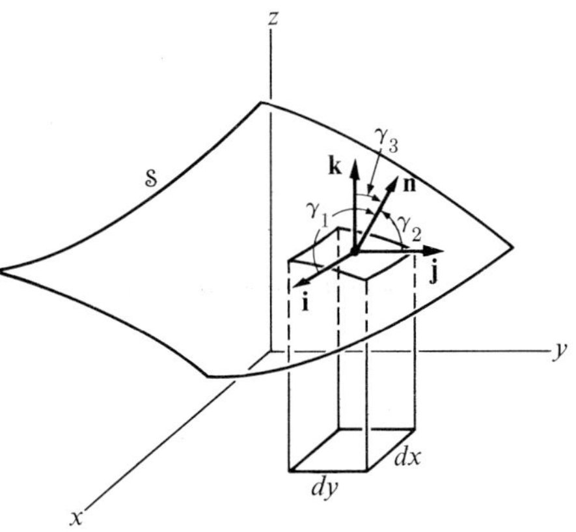

FIGURE IV–5

and write $\iint_S (\mathbf{F}\cdot\mathbf{n})\,dS$ as

$$\iint_S P\,dy\,dz + Q\,dz\,dx + R\,dx\,dy. \qquad \text{(IV–16)}$$

The value of this expression is, of course, still given by (IV–13) or (IV–14).*

EXAMPLE 1. Compute the value of

$$\iint_S (\mathbf{F}\cdot\mathbf{n})\,dS$$

when $\mathbf{F} = xy^2\mathbf{i} + \mathbf{k}$, and S is the surface of the unit sphere in $\mathbb{R}^3$ with the outward direction chosen as positive.

Since S can be represented by the equations

$$x = \cos u \sin v, \qquad y = \sin u \sin v, \qquad z = \cos v,$$

with $0 \le u \le 2\pi$, $0 \le v \le \pi$, we obtain as in Example 1 of the last section

$$\mathbf{N} = \begin{vmatrix} \mathbf{i} & \mathbf{j} & \mathbf{k} \\ \dfrac{\partial x}{\partial u} & \dfrac{\partial y}{\partial u} & \dfrac{\partial z}{\partial u} \\ \dfrac{\partial x}{\partial v} & \dfrac{\partial y}{\partial v} & \dfrac{\partial z}{\partial v} \end{vmatrix}$$

$$= -(\cos u \sin^2 v\mathbf{i} + \sin u \sin^2 v\mathbf{j} + \sin v \cos v\mathbf{k}).$$

But this vector is clearly an *inward* normal to S, and hence we must use (IV–14) to evaluate $\iint_S (\mathbf{F}\cdot\mathbf{n})\,dS$. This gives

$$\begin{aligned}\iint_S (\mathbf{F}\cdot\mathbf{n})\,dS &= \iint_{R_{uv}} (\sin^2 u \cos u \sin^3 v\mathbf{i} + \mathbf{k})\cdot(\cos u \sin^2 v\mathbf{i} \\ &\qquad + \sin u \sin^2 v\mathbf{j} + \sin v \cos v\mathbf{k})\,du\,dv \\ &= \int_0^{\pi}\int_0^{2\pi} (\sin^2 u \cos^2 u \sin^5 v + \sin v \cos v)\,du\,dv \\ &= \int_0^{\pi} \sin^5 v\,dv \int_0^{2\pi} \sin^2 u \cos^2 u\,du + \int_0^{\pi} \sin v \cos v\,dv \int_0^{2\pi} du \\ &= \frac{4\pi}{15}.\end{aligned}$$

---

* The reader should not take these remarks too seriously; they have been given only to motivate introducing (IV–16). In more advanced work, however, the expression $P\,dy\,dz + Q\,dz\,dx + R\,dx\,dy$ is given independent status, and (IV–16) is defined to be the integral of this quantity over the oriented surface S. (See Fleming, *Functions of Several Variables*, Addison-Wesley, 1965.)

EXAMPLE 2. Find the value of

$$\iint_{\mathcal{S}} xz\,dx\,dy$$

when $\mathcal{S}$ is the triangular surface in Fig. IV–6.

In this case $\mathcal{S}$ is the portion of the plane $z = 1 - x - y$ above the triangular region $0 \le x \le 1$, $0 \le y \le 1 - x$ in the xy-plane, and (IV–15) yields

$$\begin{aligned}\iint_{\mathcal{S}} xz\,dx\,dy &= \int_0^1 \int_0^{1-x} x(1 - x - y)\,dy\,dx \\ &= \int_0^1 \left[xy - x^2y - x\frac{y^2}{2}\right]_0^{1-x} dx \\ &= \frac{1}{2}\int_0^1 (x - 2x^2 + x^3)\,dx = \frac{1}{24}.\end{aligned}$$

EXAMPLE 3. Let $\mathbf{F}$ be the vector field associated with a time independent flow of fluid in a region of 3-space [i.e., $\mathbf{F}(x, y, z)$ is the velocity vector of the fluid at the point $(x\ y, z)$], and let $\mathcal{S}$ be an orientable smooth surface lying in this region. Then *if* $\mathbf{F}$ is continuous and $\mathbf{n}$ is the positive unit normal to $\mathcal{S}$ at a point $\mathbf{P}$, the quantity $[\mathbf{F}(\mathbf{P}) \cdot \mathbf{n}]\mathcal{A}(\Delta\mathcal{S})$ is an approximation to the amount of fluid flowing per unit time in the positive direction across a surface element containing $\mathbf{P}$, i.e. to the *flux* across $\Delta\mathcal{S}$ in the positive direction. The usual limiting argument now applies and allows us to assert that

$$\iint_{\mathcal{S}} (\mathbf{F} \cdot \mathbf{n})\,d\mathcal{S} \tag{IV–17}$$

is the total flux crossing $\mathcal{S}$ in the positive direction.

In an alternative computation of this same quantity, let us consider the rate at which fluid is leaving a small rectangular region B. For this purpose let

$$\mathbf{F}(x, y, z) = P(x, y, z)\mathbf{i} + Q(x, y, z)\mathbf{j} + R(x, y, z)\mathbf{k},$$

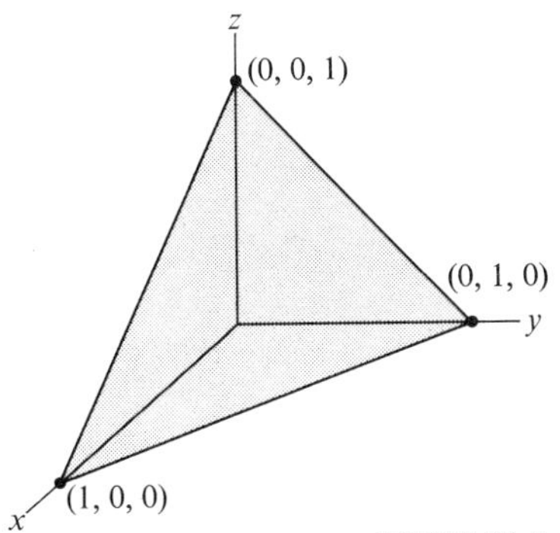

FIGURE IV–6

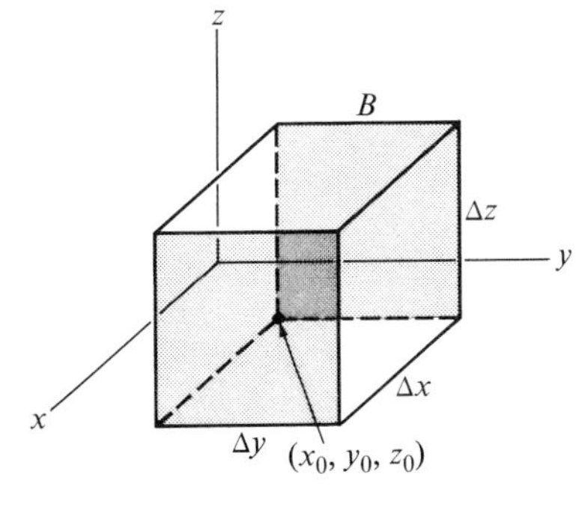

FIGURE IV–7

and suppose the box B to have edges of length Δx, Δy, Δz situated as shown in Fig. IV–7. Then the net amount of fluid leaving B through the shaded sides is (approximately)

$$P(x_0 + \Delta x, y_0, z_0)\,\Delta y\,\Delta z - P(x_0, y_0, z_0)\,\Delta y\,\Delta z,$$

which, assuming continuous differentiability of $\mathbf{F}$ and using the mean value theorem, can be rewritten

$$\left[P(x_0, y_0, z_0) + \Delta x \frac{\partial P}{\partial x}(x_0 + \theta\,\Delta x, y_0, z_0)\right]\Delta y\,\Delta z - P(x_0, y_0, z_0)\,\Delta y\,\Delta z$$
$$= \frac{\partial P}{\partial x}(x_0 + \theta\,\Delta x, y_0, z_0)\,\Delta x\,\Delta y\,\Delta z,$$

where $0 < \theta < 1$. Similar results express the rate of flow through the remaining two pairs of parallel sides of B using $\partial Q/\partial y$, $\partial R/\partial z$ instead of $\partial P/\partial x$. Hence if B is small, the total rate at which fluid is leaving B is given approximately by

$$\left(\frac{\partial P}{\partial x} + \frac{\partial Q}{\partial y} + \frac{\partial R}{\partial z}\right)\Delta x\,\Delta y\,\Delta z$$

[the partial derivatives being evaluated at $\mathbf{x} = (x_0, y_0, z_0)$]. The quantity in parentheses thus measures the rate per unit volume at which fluid is diverging from the point $\mathbf{x}$. It is accordingly called the *divergence* of the vector field $\mathbf{F}$ and is denoted by div $\mathbf{F}$; that is,

$$\operatorname{div} \mathbf{F} = \frac{\partial P}{\partial x} + \frac{\partial Q}{\partial y} + \frac{\partial R}{\partial z}. \tag{IV–18}$$

If the fluid is incompressible, then div $\mathbf{F}$ must measure the rate (per unit volume) at which fluid is being introduced at the point $\mathbf{x}$. Therefore, if $\mathcal{S}$ is a closed surface, enclosing a region R, the integral

$$\iiint_R (\operatorname{div} \mathbf{F})\,dx\,dy\,dz$$

represents the total amount of fluid introduced into the region R, and hence must also represent the total flux across the boundary surface $\mathcal{S}$. Comparison of this result with (IV–17) leads to the equation

$$\iint_{\mathcal{S}} (\mathbf{F} \cdot \mathbf{n})\,d\mathcal{S} = \iiint_R (\operatorname{div} \mathbf{F})\,dx\,dy\,dz. \tag{IV–19}$$

This result, which would almost be self-evident if one were willing to accept the physical argument just given, is one of the basic relations in vector field theory. It is known as the *divergence theorem* and is proved below.

One final remark before leaving this section. Thus far we have limited our treatment of integration to *smooth* surfaces. There is no need to be quite this restrictive, and in fact we shall want to integrate over surfaces such as the surface of a cube, or the surface of a closed cylinder (Fig. IV–8). Although not smooth, these surfaces are constructed from a finite number of nonoverlapping smooth pieces and are accordingly called *piecewise smooth* surfaces. The various integrals of these two sections are extended to a piecewise smooth surface $\mathcal{S}$ simply by adding the integrals over the smooth pieces of $\mathcal{S}$. The only question which presents real difficulty in this connection is that of orientation. However, for a piecewise smooth surface $\mathcal{S}$ which bounds a finite region of $\mathcal{R}^3$, i.e. which has an inside and an outside, the question of orientation can be settled by choosing the *outward* normal for each of the smooth pieces of $\mathcal{S}$. A general discussion of orientation of piecewise smooth surfaces is quite involved and unnecessary for our present aims. The reader is referred to a brief but excellent treatment in Protter and Morrey, *Modern Mathematical Analysis*, Addison-Wesley, 1964, pp. 602–611, where it is proved that the formulas which we have obtained in these two sections, are in fact independent of the particular parametrization used to derive them.

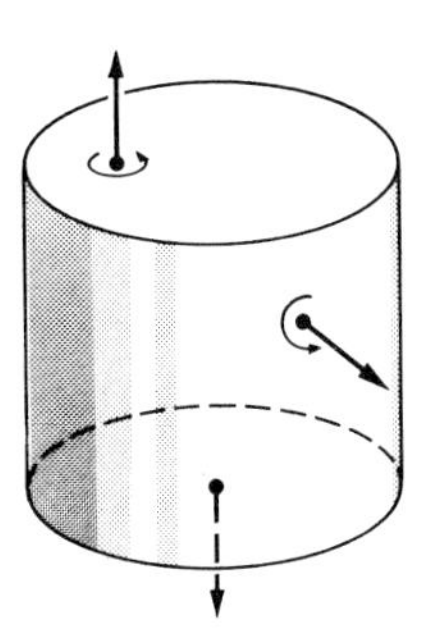

FIGURE IV–8

IV–3 THE DIVERGENCE THEOREM

In this section we shall establish the equality given in Eq. (IV–19) of the preceding section, known in the literature as the *divergence theorem*, or as *Gauss' theorem*.*

Theorem IV–1. *Let V be a region of 3-space whose boundary ∂V is a piecewise smooth two-dimensional closed surface, and let F be a continuously differentiable vector field defined in and on the boundary of V. Then if* $\mathbf{n}$ *denotes the unit outward normal to ∂V,*

$$\iint_{\partial V} (\mathbf{F} \cdot \mathbf{n})\, dS = \iiint_{V} (\operatorname{div} \mathbf{F})\, dV. \tag{IV–20}$$

Proof. We begin by establishing (IV–20) when V is an "elementary" region of the type shown in Fig. IV–9, and then pass to the general case by decomposing more general regions into elementary ones. Moreover, to simplify matters even further we rewrite (IV–20) in scalar form as

$$\iint_{\partial V} P\, dy\, dz + Q\, dz\, dx + R\, dx\, dy = \iiint_{V} \left[\frac{\partial P}{\partial x} + \frac{\partial Q}{\partial y} + \frac{\partial R}{\partial z}\right] dV,$$

* Named in honor of the famous German mathematician Karl Friedrich Gauss, 1777–1855.

and observe that the desired result will follow by additivity if we can show that

$$\iint_{\partial V} P\,dy\,dz = \iiint_V \frac{\partial P}{\partial x}\,dV,$$

$$\iint_{\partial V} Q\,dz\,dx = \iiint_V \frac{\partial Q}{\partial y}\,dV,$$

$$\iint_{\partial V} R\,dx\,dy = \iiint_V \frac{\partial R}{\partial z}\,dV.$$

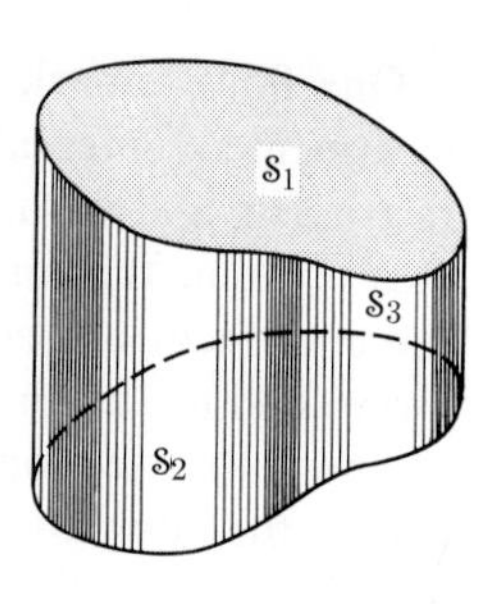

FIGURE IV–9

And finally, since these equations are all of the same type, it obviously suffices to establish just one of them.

This said, let V be an "elementary" region bounded above and below by smooth surfaces $\mathcal{S}_1$ and $\mathcal{S}_2$ described, respectively, by the functions $z = f_2(x, y)$ and $z = f_1(x, y)$ where (x, y) ranges over a region D in the xy-plane. Let $\mathcal{S}_3$ denote the lateral surface (if any) of V. Then, by Formula (IV–15),

$$\begin{aligned}\iint_{\mathcal{S}_2} R\,dx\,dy &= \iint_{\mathcal{S}_2} (R\mathbf{k})\cdot\mathbf{n}\,d\mathcal{S}\\ &= \iint_D (R\mathbf{k})\cdot\left(-\frac{\partial f_2}{\partial x}\mathbf{i} - \frac{\partial f_2}{\partial y}\mathbf{j} + \mathbf{k}\right)dA\\ &= \iint_D R(x, y, f_2(x, y))\,dA.\end{aligned}$$

Similarly

$$\iint_{\mathcal{S}_1} R\,dx\,dy = -\iint_D R(x, y, f_1(x, y))dA$$

(where the minus sign occurs because the positive normal to $\mathcal{S}_1$ points toward the interior of V), and

$$\iint_{\mathcal{S}_3} R\,dx\,dy = \iint_{\mathcal{S}_3} (R\mathbf{k})\cdot\mathbf{n}\,d\mathcal{S} = 0,$$

since $\mathbf{n}$ is orthogonal to $\mathbf{k}$ on $\mathcal{S}_3$. Hence

$$\begin{aligned}\iint_{\partial V} R\,dx\,dy &= \iint_D [R(x, y, f_2(x, y)) - R(x, y, f_1(x, y))]\,dA\\ &= \iint_V \left[\int_{f_1(x,y)}^{f_2(x,y)} \frac{\partial}{\partial z} R(x, y, z)\,dz\right] dA = \iiint_V \frac{\partial R}{\partial z}\,dV,\end{aligned}$$

as required. Thus (IV–20) holds for "elementary" regions V.

To complete the proof we now assume that V can be decomposed into a finite number of elementary regions $V_1, \ldots, V_n$ as suggested in Fig. IV–10, and apply the theorem to each of them in turn. Then, since the integrals over those portions of the bounding surfaces common to a V_i and V_j cancel in pairs, we have

$$\iint_{\partial V} (\mathbf{F} \cdot \mathbf{n})\, d\mathcal{S} = \sum_{i=1}^{n} \iint_{\partial V_i} (\mathbf{F} \cdot \mathbf{n})\, d\mathcal{S}$$

$$= \sum_{i=1}^{n} \iiint_{V_i} (\operatorname{div} \mathbf{F})\, dV$$

$$= \iiint_{V} (\operatorname{div} \mathbf{F})\, dV,$$

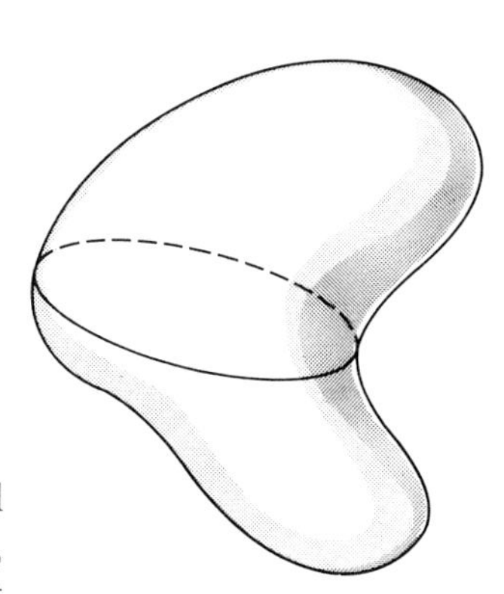

FIGURE IV–10

and (IV–20) still holds. Finally, for the most general regions considered in the statement of the theorem, we apply a limiting argument based on the case just considered.* ▌

Example 1. Use the divergence theorem to evaluate the surface integral considered in Example 1 of Section IV–2, i.e., the integral

$$\iint_{\mathcal{S}} xy^2\, dy\, dz + dx\, dy,$$

where $\mathcal{S}$ is the surface of the unit sphere $x^2 + y^2 + z^2 = 1$.

By (IV–20) we have

$$\iint_{\mathcal{S}} xy^2\, dy\, dz + dx\, dy = \iiint_{V} y^2\, dV,$$

and, changing to spherical coordinates,

$$\iint_{\mathcal{S}} xy^2\, dy\, dz + dx\, dy = \int_0^{2\pi} \int_0^{\pi} \int_0^{1} (r \sin \varphi \sin \theta)^2 r^2 \sin \varphi\, dr\, d\varphi\, d\theta$$

$$= \int_0^{2\pi} \sin^2 \theta\, d\theta \cdot \int_0^{\pi} \sin^3 \varphi\, d\varphi \cdot \int_0^{1} r^4\, dr$$

$$= \pi(\tfrac{4}{3})(\tfrac{1}{5})$$

$$= \frac{4\pi}{15}.$$

The economy of this method is too obvious to need comment.

---

* See O. D. Kellogg, *Foundations of Potential Theory*, Springer, Berlin, 1929.

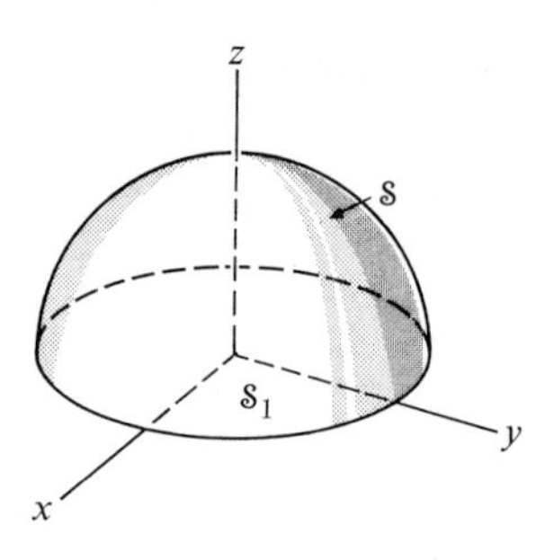

FIGURE IV–11

EXAMPLE 2. The divergence theorem can sometimes be used to evaluate surface integrals even when the surface in question is not closed. As an illustration of how this is done, consider

$$\iint_{\mathcal{S}} xy^2\,dy\,dz + x\,dz\,dx + (x + z)\,dx\,dy$$

(Fig. IV–11), where $\mathcal{S}$ is the hemispherical surface

$$z = \sqrt{1 - x^2 - y^2},\ z \geq 0.$$

To evaluate this integral let V denote the region bounded by $\mathcal{S}$ and the plane $z = 0$, and let $\mathcal{S}_1$ denote the closed disc $x^2 + y^2 \leq 1$ in the xy-plane. Then we write

$$\begin{aligned}\iint_{\mathcal{S}} xy^2\,dy\,dz &+ x\,dz\,dx + (x + z)\,dx\,dy \\ &= \iint_{\partial V} xy^2\,dy\,dz + x\,dz\,dx + (x + z)\,dx\,dy \\ &\quad + \iint_{\mathcal{S}_1} xy^2\,dy\,dz + x\,dz\,dx + (x + z)\,dx\,dy.\end{aligned}$$

(The reader should note that the plus sign appearing here is *not* a misprint. Why?) But by the divergence theorem, we have

$$\begin{aligned}\iint_{\partial V} xy^2\,dy\,dz + x\,dz\,dx + (x + z)\,dx\,dy &= \iiint_V (y^2 + 1)\,dV \\ &= \int_0^{2\pi}\int_0^{\pi/2}\int_0^1 [(r\sin\varphi\sin\theta)^2 + 1]r^2\sin\varphi\,dr\,d\varphi\,d\theta \\ &= \int_0^{2\pi}\sin^2\theta\,d\theta\cdot\int_0^{\pi/2}\sin^3\varphi\,d\varphi\cdot\int_0^1 r^4\,dr + \int_0^{2\pi}d\theta\int_0^{\pi/2}\sin\varphi\,d\varphi\int_0^1 r^2\,dr \\ &= \frac{2\pi}{15} + \frac{2\pi}{3} = \frac{4\pi}{5}.\end{aligned}$$

Finally,

$$\begin{aligned}\iint_{\mathcal{S}_1} xy^2\,dy\,dz &+ x\,dz\,dx + (x + z)\,dx\,dy \\ &= \iint_{\mathcal{S}_1} [xy^2\mathbf{i} + x\mathbf{j} + (x + z)\mathbf{k}]\cdot\mathbf{k}\,d\mathcal{S} = \iint_{\substack{x^2+y^2\leq 1\\ z=0}} (x + z)\,dA \\ &= \int_{-1}^{1}\int_{-\sqrt{1-x^2}}^{\sqrt{1-x^2}} x\,dy\,dx = 2\int_{-1}^{1} x\sqrt{1 - x^2}\,dx = 0,\end{aligned}$$

and we have

$$\iint_{\mathcal{S}} xy^2\, dy\, dz + x\, dz\, dx + (x + z)\, dx\, dy = \frac{4\pi}{5}.$$

In Eq. (IV–18) of the preceding section we defined the divergence of a continuously differentiable vector field by the equation

$$\operatorname{div} \mathbf{F} = \frac{\partial P}{\partial x} + \frac{\partial Q}{\partial y} + \frac{\partial R}{\partial z},$$

where P, Q, and R are the coordinate functions of $\mathbf{F}$ with respect to the standard basis vectors in $\mathcal{R}^3$. Although this quantity is defined in terms of a particular coordinate system, its physical interpretation suggests that its values remain unchanged under a change of basis. We are now in a position to prove this fact by giving a description of div $\mathbf{F}$ which does not involve a coordinate system in $\mathcal{R}^3$. To this end we recall that if f is real valued and continuous in a region R of 3-space, the mean-value theorem for integrals asserts the existence of a point (x^*, y^*, z^*) in R such that

$$\iiint_R f(x, y, z)\, dV = f(x^*, y^*, z^*)V,$$

where V denotes the volume of R. Thus if $\mathbf{x}$ is any point in the interior of the domain of $\mathbf{F}$, and $\mathcal{B}_\epsilon$ is the sphere of radius ϵ about $\mathbf{x}$, then

$$\iiint_{\mathcal{B}_\epsilon} (\operatorname{div} \mathbf{F})\, dV = \operatorname{div} \mathbf{F}(\mathbf{x}^*)V_\epsilon,$$

with $\mathbf{x}^*$ in $\mathcal{B}_\epsilon$, and V_ϵ the volume of $\mathcal{B}_\epsilon$. Hence, by the divergence theorem,

$$\operatorname{div} \mathbf{F}(\mathbf{x}^*) = \frac{1}{V_\epsilon} \iint_{\partial\mathcal{B}_\epsilon} (\mathbf{F} \cdot \mathbf{n})\, d\mathcal{S},$$

and passing to the limit as $\epsilon \to 0$, we have

$$\operatorname{div} \mathbf{F}(\mathbf{x}) = \lim_{\epsilon \to 0} \frac{1}{V_\epsilon} \iint_{\partial\mathcal{B}_\epsilon} (\mathbf{F} \cdot \mathbf{n})\, d\mathcal{S}. \tag{IV–21}$$

Besides providing us with a coordinate-free description of the divergence of a vector field, this expression also allows us to recapture the physical interpretation of div $\mathbf{F}(\mathbf{x})$ given in the preceding section. Indeed, if $\mathbf{F}$ is the velocity field of a time independent flow of fluid,

$$\iint_{\partial\mathcal{B}_\epsilon} (\mathbf{F} \cdot \mathbf{n})\, d\mathcal{S}$$

is, as we have seen, the flux crossing the surface of $\mathcal{B}_\epsilon$ in the positive direction. Hence, in the limit, (IV–21) measures the amount of fluid diverging from $\mathbf{x}$ per unit volume.

IV–4 BOUNDARY-VALUE PROBLEMS REVISITED: UNIQUENESS THEOREMS

In this section we shall finally prove the long postponed uniqueness theorems for boundary-value problems involving the wave equation, heat equation, and Laplace's equation. Each of these theorems is an easy consequence of the divergence theorem, or, rather, of two simple corollaries of the divergence theorem known as Green's first and second identities. However, before we can state these identities in anything approaching readable form we must simplify our notation somewhat. This can be done very effectively by introducing the symbolic vector ∇ (read "del") where

$$\nabla = \frac{\partial}{\partial x}\mathbf{i} + \frac{\partial}{\partial y}\mathbf{j} + \frac{\partial}{\partial z}\mathbf{k}. \tag{IV–22}$$

The rules for manipulating this symbol are perfectly simple and go as follows. If ψ is a differentiable scalar function, then

$$\nabla\psi = \frac{\partial\psi}{\partial x}\mathbf{i} + \frac{\partial\psi}{\partial y}\mathbf{j} + \frac{\partial\psi}{\partial z}k,$$

a quantity usually referred to as the *gradient of* ψ or *grad* ψ. Similarly, if $\mathbf{F} = P\mathbf{i} + Q\mathbf{j} + R\mathbf{k}$ is a differentiable vector field, then

$$\begin{aligned}\nabla \cdot \mathbf{F} &= \left(\frac{\partial}{\partial x}\mathbf{i} + \frac{\partial}{\partial y}\mathbf{j} + \frac{\partial}{\partial z}\mathbf{k}\right) \cdot (P\mathbf{i} + Q\mathbf{j} + R\mathbf{k}) = \frac{\partial P}{\partial x} + \frac{\partial Q}{\partial y} + \frac{\partial R}{\partial z} \\ &= \operatorname{div}\mathbf{F},\end{aligned}$$

and finally

$$\begin{aligned}\nabla \times \mathbf{F} &= \begin{vmatrix} \mathbf{i} & \mathbf{j} & \mathbf{k} \\ \dfrac{\partial}{\partial x} & \dfrac{\partial}{\partial y} & \dfrac{\partial}{\partial z} \\ P & Q & R \end{vmatrix} \\ &= \left(\frac{\partial R}{\partial y} - \frac{\partial Q}{\partial z}\right)\mathbf{i} + \left(\frac{\partial P}{\partial z} - \frac{\partial R}{\partial x}\right)\mathbf{j} + \left(\frac{\partial Q}{\partial x} - \frac{\partial P}{\partial y}\right)\mathbf{k},\end{aligned}$$

a quantity called the *curl of* $\mathbf{F}$. Finally, carrying this notation to its logical conclusion, we set

$$\nabla \cdot \nabla = \frac{\partial^2}{\partial x^2} + \frac{\partial^2}{\partial y^2} + \frac{\partial^2}{\partial z^2},$$

and write $\nabla \cdot \nabla = \nabla^2$. (The operator ∇^2 is sometimes called the *Laplacian*.)

Since $\nabla \cdot \mathbf{F}$ is just another way of writing div $\mathbf{F}$, the divergence theorem may be rewritten in terms of ∇ as

$$\iint_{\partial V} (\mathbf{F} \cdot \mathbf{n})\, d\mathcal{S} = \iiint_V (\nabla \cdot \mathbf{F})\, dV. \qquad \text{(IV–23)}$$

In particular, if u and v are twice continuously differentiable scalar functions in and on the boundary of V, and if we set

$$\mathbf{F} = u(\nabla v) = u\left(\frac{\partial v}{\partial x}\mathbf{i} + \frac{\partial v}{\partial y}\mathbf{j} + \frac{\partial v}{\partial z}\mathbf{k}\right),$$

we find that

$$\nabla \cdot \mathbf{F} = u(\nabla^2 v) + (\nabla u) \cdot (\nabla v) \qquad \text{(IV–24)}$$

and that (IV–23) becomes

$$\iint_{\partial V} [u(\nabla v) \cdot \mathbf{n}]\, d\mathcal{S} = \iiint_V [u(\nabla^2 v) + (\nabla u) \cdot (\nabla v)]\, dV. \qquad \text{(IV–25)}$$

This is *Green's first identity*. To derive the second we simply interchange u and v in (IV–25) and subtract to obtain

$$\iint_{\partial V} [u(\nabla v) - v(\nabla u)] \cdot \mathbf{n}\, d\mathcal{S} = \iiint_V [u(\nabla^2 v) - v(\nabla^2 u)]\, dV. \qquad \text{(IV–26)}$$

This done we now get on with the uniqueness theorems.

Theorem IV–2. *Let $\mathcal{S}$ be a piecewise smooth closed surface in $\mathcal{R}^3$ surrounding a region V, and let φ be a real valued continuous function on $\mathcal{S}$. Then there exists at most one solution of the boundary-value problem*

$$\nabla^2 u = 0 \quad \text{in } V,$$
$$u \equiv \varphi \quad \text{on } \mathcal{S}.$$

Proof. Let u_1 and u_2 be two solutions of the given problem and set $U = u_1 - u_2$. Then U is a solution of the boundary-value problem

$$\nabla^2 U = 0 \quad \text{in } V,$$
$$U \equiv 0 \quad \text{on } \mathcal{S},$$

and we will be done if we can show that $U \equiv 0$ in V. To this end we note that when $u = v$ Green's first identity becomes

$$\iint_{\partial V} [u(\nabla u) \cdot \mathbf{n}]\, d\mathcal{S} = \iiint_V [u(\nabla^2 u) + (\nabla u) \cdot (\nabla u)]\, dV. \qquad \text{(IV–27)}$$

Thus, by setting $u = U$, we obtain

$$\iiint_V [(\nabla U) \cdot (\nabla U)]\, dV = 0,$$

or

$$\iiint_V \left[\left(\frac{\partial U}{\partial x}\right)^2 + \left(\frac{\partial U}{\partial y}\right)^2 + \left(\frac{\partial U}{\partial z}\right)^2\right] dV = 0;$$

and since the integrand appearing here is continuous and nonnegative everywhere in V, we have

$$\left(\frac{\partial U}{\partial x}\right)^2 + \left(\frac{\partial U}{\partial y}\right)^2 + \left(\frac{\partial U}{\partial z}\right)^2 \equiv 0$$

in V. This, in turn, implies that

$$\frac{\partial U}{\partial x} = \frac{\partial U}{\partial y} = \frac{\partial U}{\partial z} = 0$$

everywhere in V, and it follows that U is constant in V. Finally, since U vanishes identically on $\partial V = \mathcal{S}$, $U \equiv 0$ in V, as required.* ▮

We omit the two-dimensional version of this result, and go on to the corresponding theorem for the heat equation.

Theorem IV–3. *If $\mathcal{S}$ and V are as above, there exists at most one solution of the three-dimensional heat equation*

$$a^2 \frac{\partial u}{\partial t} = \nabla^2 u$$

satisfying the boundary condition $u(x, y, z, t) = F(x, y, z, t)$ on $\mathcal{S}$, and the initial condition $u(x, y, z, 0) = G(x, y, z)$, where F and G are preassigned continuous functions.

Proof. As in the preceding proof let U denote the difference between any two solutions of this problem. Then U is a solution of the problem

$$a^2 \frac{\partial u}{\partial t} = \nabla^2 u,$$

$$u(x, y, z, t) \equiv 0 \quad \text{on } \mathcal{S}, \qquad u(x, y, z, 0) \equiv 0,$$

and (IV–27) implies that

$$\iiint_V [U(\nabla^2 U) + \|\nabla U\|^2]\, dV = 0.$$

---

* Recall that, by definition, the values of U in V must approach the prescribed values on $\mathcal{S} = \partial V$ as we approach $\mathcal{S}$ along any smooth curve in V. (See Section 13–2.)

But $\nabla^2 U = a^2(\partial U/\partial t)$. Thus

$$\iiint_V \left[a^2 U \frac{\partial U}{\partial t} + \|\nabla U\|^2\right] dV = 0,$$

or

$$\iiint_V \frac{\partial}{\partial t}\left(\tfrac{1}{2}a^2 U^2\right) dV = -\iiint_V \|\nabla U\|^2\, dV.$$

Since U is continuously differentiable with respect to t, we can interchange the order of integration and differentiation to obtain

$$\frac{d}{dt}\iiint_V \left(\tfrac{1}{2}a^2 U^2\right) dV = -\iiint_V \|\nabla U\|^2\, dV. \tag{IV–28}$$

Now set

$$I(t) = \iiint_V \left(\tfrac{1}{2}a^2 U^2\right) dV.$$

Then $I(0) = 0$, while $I(t) \geq 0$ and $I'(t) \leq 0$ for all $t \geq 0$. A straightforward application of the mean value theorem then implies that $I(t) \equiv 0$. Indeed, from the mean value theorem we deduce that

$$\frac{I(t) - I(a)}{t} = I'(\theta t), \quad 0 < \theta < 1,$$

or

$$I(t) = tI'(\theta t),$$

and since $I(t) \geq 0$ and $I'(t) \leq 0$, we must have $I(t) \equiv 0$. Thus the left-hand side of (IV–28) vanishes, and it follows that

$$\iiint_V \|\nabla U\|^2\, dV = 0.$$

As before, this implies that $U \equiv 0$ in V, and the proof is complete. ▮

(Note that this argument can be used equally well for any of the other boundary conditions considered earlier.)

Finally we consider the uniqueness problem for the wave equation.

Theorem IV–4. *With $\mathcal{S}$ and V as above, there exists at most one solution of the three-dimensional wave equation*

$$a^2 \frac{\partial^2 u}{\partial t^2} = \nabla^2 u$$

satisfying the boundary condition $u(x, y, z, t) = F(x, y, z, t)$ on $\mathcal{S}$, and the *initial conditions*

$$u(x, y, z, 0) = G(x, y, z), \qquad \frac{\partial u}{\partial t}(x, y, z, 0) = H(x, y, z)$$

in V, where F, G, and H are preassigned continuous functions.

Proof. Once again, let U denote the difference between any two solutions of the given problem. Then U is a solution of the problem

$$a^2 \frac{\partial^2 u}{\partial t^2} = \nabla^2 u,$$

$$u(x, y, z, t) = 0 \quad \text{on } \mathcal{S},$$

$$u(x, y, z, 0) = \frac{\partial u}{\partial t}(x, y, z, 0) = 0 \quad \text{in } V,$$

and hence (IV–25), with $u = \partial U/\partial t$ and $v = U$, yields

$$\iiint_V \left[\frac{\partial U}{\partial t} \nabla^2 U + \nabla\left(\frac{\partial U}{\partial t}\right) \cdot \nabla U\right] dV = \iint_{\mathcal{S}} \frac{\partial U}{\partial t} \nabla U \cdot \mathbf{n}\, d\mathcal{S} = 0. \tag{IV–29}$$

But

$$\frac{\partial U}{\partial t} \nabla^2 U = \frac{\partial U}{\partial t}\left(a^2 \frac{\partial^2 U}{\partial t^2}\right) = \frac{1}{2} a^2 \frac{\partial}{\partial t}\left(\frac{\partial U}{\partial t}\right)^2,$$

and

$$\nabla\left(\frac{\partial U}{\partial t}\right) \cdot \nabla U = \left[\frac{\partial}{\partial t}(\nabla U)\right] \cdot \nabla U = \frac{1}{2} \frac{\partial}{\partial t} \|\nabla U\|^2;$$

so that (IV–29) becomes

$$\frac{1}{2} \frac{\partial}{\partial t} \iiint_V \left[a^2 \left(\frac{\partial U}{\partial t}\right)^2 + \|\nabla U\|^2\right] dV = 0. \tag{IV–30}$$

Thus the integral in (IV–30) does not depend on t, and since the initial conditions require it to be zero when $t = 0$, we conclude that it is identically zero. This implies that

$$a^2 \left(\frac{\partial U}{\partial t}\right)^2 + \|\nabla U\|^2 \equiv 0,$$

and hence U must be a constant. Again using the initial conditions, this constant must be zero, and we are done. ▮

As in the case of Theorem IV–2, we omit the lower dimensional versions of Theorems IV–3 and IV–4.

recommendations for further reading

LINEAR ALGEBRA

FINKBEINER, D., *Introduction to Matrices and Linear Transformations;* Freeman, San Francisco, 1960.

GEL'FAND, I., *Lectures on Linear Algebra;* Interscience, New York, 1961.

HALMOS, P., *Finite Dimensional Vector Spaces*, 2nd Ed.; Van Nostrand, Princeton, 1958.

HOFFMAN, K., and R. KUNZE, *Linear Algebra;* Prentice-Hall, Englewood Cliffs, N.J., 1961.

NEARING, E., *Linear Algebra and Matrix Theory;* Wiley, New York, 1963.

SHILOV, G., *Introduction to the Theory of Linear Spaces;* Prentice-Hall, Englewood Cliffs, N.J., 1961.

DIFFERENTIAL EQUATIONS

BIRKHOFF, G., and G. C. ROTA, *Ordinary Differential Equations;* Ginn, Boston, 1962.

CODDINGTON, E., *Introduction to Ordinary Differential Equations;* Prentice-Hall, Englewood Cliffs, N.J., 1961.

PONTRYAGIN, L., *Ordinary Differential Equations;* Addison-Wesley, Reading, Mass., 1962.

TRICOMI, F., *Differential Equations;* Hafner, New York, 1961.

WEINBERGER, H., *A First Course in Partial Differential Equations;* Blaisdell, New York, 1965.

YOSHIDA, K., *Lectures on Differential and Integral Equations;* Interscience, New York, 1960.

ORTHOGONAL FUNCTIONS AND SERIES EXPANSIONS

CHURCHILL, R., *Fourier Series and Boundary Value Problems*, 2nd Ed.; McGraw-Hill, New York, 1963.

DAVIS, H., *Fourier Series and Orthogonal Polynomials;* Allyn and Bacon, Boston, 1963.

JACKSON, D., *Fourier Series and Orthogonal Polynomials;* Mathematical Association of America (Carus Monograph), 1941.

LEBEDEV, N., *Special Functions and Their Applications;* Prentice-Hall, Englewood Cliffs, N.J., 1965.

RAINVILLE, E., *Special Functions;* Macmillan, New York, 1960.

SANSONE, G., *Orthogonal Functions*, Revised English Ed.; Interscience, New York, 1959.

TOLSTOV, G., *Fourier Series*, Prentice-Hall, Englewood Cliffs, N.J., 1962.

APPLIED MATHEMATICS

COURANT, R., and D. HILBERT, *Methods of Mathematical Physics*, Vol. I; Interscience, New York, 1953.

FRIEDMAN, B., *Principles and Techniques of Applied Mathematics;* Wiley, New York, 1956.

SOBOLEV, S., *Partial Differential Equations of Mathematical Physics;* Addison-Wesley, Reading, Mass., 1964.

ADVANCED CALCULUS AND VECTOR FIELD THEORY

BARTLE, R., *Elements of Real Analysis;* Wiley, New York, 1964.

BUCK, R., *Advanced Calculus*, 2nd Ed.; McGraw-Hill, New York, 1965.

CROWELL, R., and R. WILLIAMSON, *Calculus of Vector Functions;* Prentice-Hall, Englewood Cliffs, N.J., 1962.

FLEMING, W., *Functions of Several Variables;* Addison-Wesley, Reading, Mass., 1965.

answers to odd-numbered exercises

Chapter 1

Section 1–1

1. $(-1, 3), (-3, 9)$
3. $(-\frac{5}{2}, -\frac{2}{3}), (\frac{5}{2}, \frac{2}{3})$
5. $(-6, -3), (18, 9)$
7. $\tan^2 x + 1, -\tan^2 x - 1$
9. $\dfrac{10x + 5}{x^2 + x - 6}, \dfrac{2x + 1}{x^2 + x - 6}$
11. No; undefined at $x = 1$
13. Yes
15. No; undefined at $x = 0$
17. No; discontinuous at $x = 0$
19. No; discontinuous at $x = 0$
21. $(-3, -\frac{3}{2})$
23. $(-\frac{9}{5}, -\frac{106}{35})$
25. $\frac{1}{2}(1 + \sec^2 x)$
27. $\ln (x + 2)$

Section 1–2

3. (a) $(-1, 5)$ (b) $(0, \frac{13}{2})$ (c) $(0, -2)$
5. (a) $-x^3 + 2x^2 - 8x + 3$ (b) 7 (c) $\frac{3}{2}x^3 - 4x^2 - \frac{1}{2}x + \frac{17}{20}$
9. (a), (b), (d), (e), (g)
11. No; we cannot assert that $\mathbf{x} + \mathbf{0} = \mathbf{x}$.

Section 1–4

1. (a) No; not closed under addition (b) Yes; satisfies subspace criterion
   (c) Yes; satisfies subspace criterion (d) Yes; satisfies subspace criterion
   (e) No; not closed under addition
3. (a) Yes; satisfies subspace criterion
   (b) No; not closed under scalar multiplication
   (c) Yes; satisfies subspace criterion (d) No; not closed under addition
   (e) Yes; satisfies subspace criterion
7. (a), (b), (f)
9. (a), (f)
11. All constant functions
13. (a) $x(\alpha x + \beta)$ (b) $(x + 1)(\alpha x + \beta)$ (c) $\alpha x^2 + \beta x + y$
17. If $\mathcal{X}$ consists of the two vectors $(0, 1, 1)$, $(0, -1, 1)$ and $\mathcal{Y}$ the two vectors $(1, 0, 1)$, $(-1, 0, 1)$, then $\mathcal{S}(\mathcal{X} \cap \mathcal{Y})$ is the trivial subspace of $\mathcal{V}$, but $\mathcal{S}\mathcal{X} \cap \mathcal{S}\mathcal{Y}$ comprises all vectors of the form $(0, 0, x_3)$. If $\mathcal{X}$ consists of the two vectors $(0, 0, 1)$, $(0, 1, 1)$ and $\mathcal{Y}$ the two vectors $(0, 0, 1)$, $(1, 0, 1)$, then $\mathcal{S}(\mathcal{X} \cap \mathcal{Y})$ and $\mathcal{S}\mathcal{X} \cap \mathcal{S}\mathcal{Y}$ each comprise all vectors of the form $(0, 0, x_3)$.
21. (c) $\mathcal{W}_1 \cap \mathcal{W}_2$ contains only $\mathbf{f}(x) = \mathbf{0}$.

Section 1–5

1. (a) All subsets which contain 1, 2, or 3 vectors
   (b) All subsets which contain 1, 2, or 3 vectors
   (c) All subsets which contain 1, 2, or 3 vectors and which do not include both (1, 1, 1) and (2, 2, 2)
   (d) All subsets which contain 1, 2, or 3 vectors and which do not include (0, 0, 0)

3. No; $(0, 2, -1)$ and $(0, \frac{1}{2}, -\frac{1}{2})$, or $(0, \frac{1}{2}, -\frac{1}{2})$ and $(0, \frac{2}{3}, -\frac{1}{3})$

5. $$\mathbf{i} = 1(1, 0) + 0(0, -1) = \cos\theta(\cos\theta, \sin\theta) - \sin\theta(-\sin\theta, \cos\theta)$$
$$= \frac{1}{\alpha}(\alpha, 0) + 0(0, \beta) = 1(1, 1) - 1(0, 1),$$
$$\mathbf{j} = 0(1, 0) - 1(0, -1) = \sin\theta(\cos\theta, \sin\theta) + \cos\theta(-\sin\theta, \cos\theta)$$
$$= 0(\alpha, 0) + \frac{1}{\beta}(0, \beta) = 0(1, 1) + 1(0, 1),$$
$$\mathbf{i} + \mathbf{j} = 1(1, 0) - 1(0, -1) = (\cos\theta + \sin\theta)(\cos\theta, \sin\theta) + (-\sin\theta + \cos\theta)(-\sin\theta, \cos\theta)$$
$$= \frac{1}{\alpha}(\alpha, 0) + \frac{1}{\beta}(0, \beta) = 1(1, 1) + 0(0, 1),$$
$$\alpha'\mathbf{i} + \beta'\mathbf{j} = \alpha'(1, 0) - \beta'(0, -1) = (\alpha'\cos\theta + \beta'\sin\theta)(\cos\theta, \sin\theta) + (-\alpha'\sin\theta + \beta'\cos\theta)(-\sin\theta, \cos\theta)$$
$$= \frac{\alpha'}{\alpha}(\alpha, 0) + \frac{\beta'}{\beta}(0, \beta)$$
$$= \alpha'(1, 1) + (-\alpha' + \beta')(0, 1)$$

7. $$(2, -2, 1, 3) = -1(1, 0, 0, 0) - 5(0, 1, 0, 0) - 2(0, 0, 1, 0) + 3(1, 1, 1, 1)$$
$$= 0(1, 1, 0, 0) + 3(0, 0, 1, 1) - 2(-1, 0, 1, 1) + 2(0, -1, 0, 1)$$
$$= 2(2, -1, 0, 1) + 0(1, 3, 2, 0) + 1(0, -1, -1, 0) + 1(-2, 1, 2, 1)$$
$$= -1(1, -1, 2, 0) + 0(1, 1, 2, 0) + 3(3, 0, 0, 1) - 3(2, 1, -1, 0)$$

15. $$x^2 = \tfrac{2}{3}(\tfrac{3}{2}x^2 - \tfrac{1}{2}) + \tfrac{1}{3}(1),$$
$$x^3 = \tfrac{3}{5}(\tfrac{5}{3}x^3 - \tfrac{3}{2}x) + \tfrac{9}{10}(x)$$

21. $$x^3 = x(x - 1)(x - 2) + 3x(x - 1) + x,$$
$$x^3 + 3x - 1 = x(x - 1)(x - 2) + 3x(x - 1) + 4x - 1$$

23. (a) $x^3 + 2x + 5 = 1(x^3 + 2x + 5)$
    (b) $x^2 + 1 = \frac{1}{3}(3x^2 + 2) + \frac{1}{18}(6)$
    (c) $2x^3 - x^2 + 10x + 2 = 2(x^3 + 2x + 5) - \frac{1}{3}(3x^2 + 2) + 1(6x) - \frac{11}{9}(6)$

Section 1–6

1. (a) $-1, 1, 0$
   (b) $-3, 0, 1$
   (c) $0, -1, 1$
   (d) $-\frac{3}{2}, -\frac{1}{2}, 1$
   (e) $6, -4, 2$
   (f) $-2, 1, 2$

3. (a) $(2, 1, 0)$
   (b) $(4, 3, 3)$
   (c) $(-2, -1, -5)$
   (d) $(7, \frac{9}{2}, 3)$
   (e) $(6, 2, 0)$
   (f) $(10, 6, 3)$

5. One such basis is $(-3, 0, 0, 0)$, $(0, 1, 0, 0)$, $(0, 0, 2, 0)$, $(0, 0, 0, -1)$.

7. Yes; one such basis is $(\frac{1}{2}, 0)$, $(0, \frac{1}{3})$.

Section 1–7

1. (a) 2 (b) 3
3. (a) One such basis is 1, x.
   (b) One such basis is 1, $x, x^2, \ldots, x^{n-1}$.
5. (a) 3 (b) 2
7. 2
9. One such pair is $\mathbf{x}_3 = (1, 0, 0, 0)$, $\mathbf{x}_4 = (0, 1, 0, 0)$.

Section 1–8

1. $\overline{A'B'}$ and $\overline{A''B''}$, where, for example,
   $A' = (2, 1)$, $B' = (4, 1)$, $A'' = (1, 2)$, $B'' = (3, 2)$
3. $\overline{A'B'}$ and $\overline{A''B''}$, where, for example,
   $A' = (2, -1)$, $B' = (4, 1)$, $A'' = (1, 0)$, $B'' = (3, 2)$
5. $\overline{A'B'}$ and $\overline{A''B''}$, where, for example,
   $A' = (1, 2)$, $B' = (-2, -1)$, $A'' = (0, 3)$, $B'' = (-3, 0)$
7. $\mathbf{0}$
9. $\mathbf{v}(\overrightarrow{OE})$, where $O = (0, 0)$, $E = (-\frac{1}{2}, -3)$
11. $(3, 0)$
13. $(-3, -4)$
15. $(-13, -6)$
17. $\mathbf{v}(\overrightarrow{OC})$, where $O = (0, 0)$, $C = (2, 2)$
19. $\mathbf{v}(\overrightarrow{OC})$, where $O = (0, 0)$, $C = (-11, 5)$
21. $(x_3 + x_2 - x_1, y_3 + y_2 - y_1)$
23. $\mathbf{v}(\overrightarrow{OE})$, where $O = (0, 0)$, $E = (x_2 - x_1 + x_4 - x_3, y_2 - y_1 + y_4 - y_3)$

Section 1–9

3.
$$\begin{aligned}
m = 2:\ [0] &= \{2k\} &&= \{0, \pm 2, \pm 4, \pm 6, \ldots\}\\
[1] &= \{2k + 1\} &&= \{\pm 1, \pm 3, \pm 5, \ldots\}\\
m = 3:\ [0] &= \{3k\} &&= \{0, \pm 3, \pm 6, \pm 9, \ldots\}\\
[1] &= \{3k + 1\} &&= \{\ldots, -8, -5, -2, 1, 4, 7, \ldots\}\\
[2] &= \{3k + 2\} &&= \{\ldots, -7, -4, -1, 2, 5, 8, \ldots\}\\
m = 4:\ [0] &= \{4k\} &&= \{0, \pm 4, \pm 8, \pm 12, \ldots\}\\
[1] &= \{4k + 1\} &&= \{\ldots, -11, -7, -3, 1, 5, 9, \ldots\}\\
[2] &= \{4k + 2\} &&= \{0, \pm 2, \pm 6, \pm 10, \ldots\}\\
[3] &= \{4k + 3\} &&= \{\ldots, -9, -5, -1, 3, 7, 11, \ldots\}
\end{aligned}$$

Chapter 2

Section 2–1

1. Central reflection in the origin
3. Compression (toward the x_1 axis) by a factor $\frac{1}{3}$, followed by a magnification by a factor 2
5. Rotation through 45° followed by magnification by a factor 2
7. Reflection in the line $x_2 = -x_1$
9. The whole plane is mapped on (0, 0)

11. Not linear 13. Linear 15. Linear

21. $A(x) = cx$, c an arbitrary real number

Section 2–2

5. (b) If $\mathcal{X} = \{\mathbf{0}\}$, then $\mathcal{A}(\mathcal{X})$ is the space of all linear transformations from $\mathcal{V}_1$ to $\mathcal{V}_2$; if $\mathcal{X} = \mathcal{V}_1$, then $\mathcal{A}(\mathcal{X})$ contains only the zero transformation.

Section 2–3

3. (a) $LDy(x) = y(x) - y(a)$, $DLy(x) = y(x)$

(b) $$L^n D^n y(x) = y(x) - \left[y(a) + y'(a)(x - a) + \frac{y''(a)}{2!}(x - a)^2 + \cdots + \frac{y^{(n-1)}(a)}{(n-1)!}(x - a)^{n-1} \right]$$

$D^n L^n y(x) = y(x)$

5. (a) $x^2D^2 + 3xD + 1$ (b) $2xD^2 + (1 - 2x)D - 1$
(c) $2xD^2 + 2(1 - x)D - 1$ (d) $x^2D^2 + 2x(1 - x^2)D - 6x^2$
(e) $x^2D^2 + 2x(2 - x^2)D + 2 - 4x^2$

7. (a) $0, 0, 0$
(b) $0, -6 \sin 2x, 2e^x$
(c) $-3 \sin x - \cos x, -2e^x, 2 - 2x - 2x^2$
(d) $0, 0, 2e^{-2x}$
(e) $4x^2, 4x^2 - 2/x, e^x(x^2 + 2x - 2)$

11. If $B = O$, then the subspace is $\mathcal{V}$; if $B = I$, then the subspace contains only O.

Section 2–4

1. (a) $\mathfrak{N}(A) = \mathcal{O}$; $A^{-1}(y_1, y_2) = (\frac{1}{2}y_1, -\frac{1}{2}y_2)$
(b) $\mathfrak{N}(A)$ is the line $x_2 = 0$; there is no inverse.
(c) $\mathfrak{N}(A)$ is the line $x_1 + x_2 = 0$; there is no inverse.
(d) $\mathfrak{N}(A) = \mathcal{O}$; $A^{-1}(y_1, y_2) = \left(\frac{y_1 + y_2}{2}, \frac{y_1 - y_2}{2}\right)$

3. (a) $\mathfrak{N}(A)$ comprises all constant polynomials; there is no inverse.
(b) $\mathfrak{N}(A) = \mathcal{O}$; there is no inverse. ($A$ is not onto.)
(c) $\mathfrak{N}(A)$ comprises all constant polynomials; there is no inverse.
(d) $\mathfrak{N}(A) = \mathcal{O}$ unless $q(x) \equiv 0$, in which case $\mathfrak{N}(A) = \mathcal{P}$; there is no inverse. ($A$ is not onto.)

5. $\alpha_1\beta_2 - \alpha_2\beta_1 \neq 0$

Section 2–5

3. Representation of αA is $\begin{pmatrix} \alpha\alpha_1 & \alpha\beta_1 \\ \alpha\alpha_2 & \alpha\beta_2 \end{pmatrix}$.

Representation of $A + B$ is $\begin{pmatrix} \alpha_1 + \alpha_1' & \beta_1 + \beta_1' \\ \alpha_2 + \alpha_2' & \beta_2 + \beta_2' \end{pmatrix}$.

Representation of AB is $\begin{pmatrix} \alpha_1\alpha_1 + \beta_1\alpha_2' & \alpha_1\beta_1' + \beta_1\beta_2' \\ \alpha_2\alpha_1' + \beta_2\alpha_2' & \alpha_2\beta_1' + \beta_2\beta_2' \end{pmatrix}$.

5. $\dim \mathcal{L}(\mathcal{R}^2, \mathcal{R}^2) = 4$

Section 2–6

1. (a) $\begin{bmatrix} 1 & 1 & 0 \\ 1 & 0 & 1 \\ 0 & 0 & 0 \end{bmatrix}$ (b) $\begin{bmatrix} 1 & 0 & 1 \\ 0 & 1 & -1 \\ 0 & -1 & 1 \end{bmatrix}$ (c) $\begin{bmatrix} 1 & -1 & 0 \\ 1 & 2 & 1 \\ 0 & 0 & 0 \end{bmatrix}$

3. (a) $(1, \frac{1}{2}, \frac{1}{3})$ (b) $(1, -\frac{1}{2}, -\frac{1}{6})$ (c) $(\frac{1}{2}, -\frac{1}{4}, -\frac{1}{12})$

5. (a) $\begin{bmatrix} 0 & -1 & 0 \\ 0 & 0 & -2 \\ 0 & 1 & 0 \\ 0 & 0 & 2 \end{bmatrix}$ (b) $\begin{bmatrix} 0 & 0 & 0 \\ 0 & 2 & 4 \\ 0 & 1 & 6 \\ 0 & 0 & 2 \end{bmatrix}$ (c) $\begin{bmatrix} 0 & 3 & 6 \\ 0 & 4 & 14 \\ 0 & 1 & 8 \\ 0 & 0 & 2 \end{bmatrix}$

7. If $\mathcal{B}_1 = \{\mathbf{e}_1, \mathbf{e}_2, \ldots, \mathbf{e}_n\}$ and $\mathcal{B}_2 = \{\mathbf{e}'_1, \mathbf{e}'_2, \ldots, \mathbf{e}'_m\}$, then

$$A(\mathbf{x}) = A\left(\sum_{j=1}^{n} x_j \mathbf{e}_j\right) = \sum_{i=1}^{m}\left(\sum_{j=1}^{n} \alpha_{ij} x_j\right)\mathbf{e}'_i.$$

Section 2–7

1. $\begin{bmatrix} -2 & 1 & -6 \\ 3 & 7 & -11 \\ 4 & -8 & 8 \end{bmatrix}$ 3. $\begin{bmatrix} 20 & \frac{8}{3} & -4 \\ \frac{4}{3} & 0 & \frac{16}{3} \end{bmatrix}$

5. $\begin{bmatrix} -\frac{2}{9} & \frac{1}{4} & \frac{1}{3} & \frac{8}{3} \\ \frac{1}{3} & -\frac{5}{2} & \frac{1}{2} & -\frac{1}{6} \\ -\frac{35}{18} & \frac{7}{18} & \frac{1}{4} & 0 \end{bmatrix}$

7. For standard bases, $\begin{bmatrix} -1 & -2 & -2 \\ 1 & 1 & 0 \\ 0 & 1 & 1 \\ 0 & 0 & 1 \end{bmatrix}$,

for bases $\mathcal{B}_1, \mathcal{B}_2$, $\begin{bmatrix} 0 & 0 & 0 \\ 1 & 2 & 0 \\ 0 & 1 & 2 \\ 0 & 0 & 1 \end{bmatrix}$

11. If $\mathcal{B}_1 = \{(e_{11}), (e_{12}), (e_{21}), (e_{22})\}$, $\mathcal{B}_2 = \{(e'_{11}), (e'_{12}), (e'_{13}), (e'_{21}), (e'_{22}), (e'_{23})\}$,

then $[A:\mathcal{B}_1, \mathcal{B}_2] = \begin{bmatrix} 1 & 0 & 0 & 0 \\ 0 & 1 & 0 & 0 \\ 0 & 0 & 1 & 0 \\ 0 & 0 & 0 & 1 \\ 0 & 0 & 0 & 0 \\ 0 & 0 & 0 & 0 \end{bmatrix}$.

13. (b) $\begin{bmatrix} 0 & 2 & 0 & 0 & 0 \\ -2 & 0 & 0 & 0 & 0 \\ 0 & 0 & -3 & -4 & 4 \\ 0 & 0 & 2 & -1 & 2 \\ 0 & 0 & -2 & 2 & -1 \end{bmatrix}$ 15. $m_D(x) = x^3$

Section 2–8

1. (a) $\begin{bmatrix} 18 & -16 \\ 19 & 11 \\ 31 & -25 \end{bmatrix}$ (b) Product undefined (c) $\begin{bmatrix} 1 & 0 & 0 \\ 0 & 1 & 0 \\ 0 & 0 & 1 \end{bmatrix}$

   (d) (3) (e) $\begin{bmatrix} 8 & 2 & -4 & 12 \\ 0 & 0 & 0 & 0 \\ 4 & 1 & -2 & 6 \\ -2 & -\frac{1}{2} & 1 & -3 \end{bmatrix}$

3. $\alpha_{12} = \alpha_{21}$, or else $\alpha_{11} = \alpha_{22}$ and $\alpha_{12} = -\alpha_{21}$

5. (a) $\begin{bmatrix} \frac{1}{5} & \frac{1}{5} \\ -\frac{4}{5} & \frac{1}{5} \end{bmatrix}$ (b) $\begin{bmatrix} \frac{1}{2} & 0 & \frac{1}{2} \\ -1 & 1 & -1 \\ -\frac{1}{2} & 0 & \frac{1}{2} \end{bmatrix}$

   (c) $\begin{bmatrix} \frac{1}{13} & \frac{5}{13} & \frac{3}{13} \\ -\frac{2}{13} & \frac{3}{13} & -\frac{6}{13} \\ -\frac{5}{13} & \frac{1}{13} & -\frac{2}{13} \end{bmatrix}$

7. (a) Invertible if and only if $\alpha\beta \neq 0$;

   inverse $\begin{bmatrix} -\dfrac{2}{\alpha\beta} & \dfrac{1}{\beta} \\ \dfrac{1}{\alpha} & 0 \end{bmatrix}$

   (b) Invertible if and only if $\alpha\beta(\beta - \alpha) \neq 0$;

   inverse $\begin{bmatrix} \dfrac{\beta}{\alpha(\beta - \alpha)} & -\dfrac{1}{\alpha(\beta - \alpha)} \\ -\dfrac{\alpha}{\beta(\beta - \alpha)} & \dfrac{1}{\beta(\beta - \alpha)} \end{bmatrix}$

   (c) Invertible if and only if $\alpha \neq 0$;

   inverse $\begin{bmatrix} 0 & \dfrac{\beta}{\alpha} & \dfrac{1}{\alpha} \\ 0 & -1 & 0 \\ \dfrac{1}{\alpha} & -\dfrac{\beta}{\alpha^2} & -\dfrac{1}{\alpha^2} \end{bmatrix}$

9. $\begin{bmatrix} \alpha & \beta \\ \dfrac{1 - \alpha^2}{\beta} & -\alpha \end{bmatrix}$, α, β arbitrary except that $\beta \neq 0$; also $\begin{bmatrix} 1 & 0 \\ \gamma & -1 \end{bmatrix}$ and $\begin{bmatrix} -1 & 0 \\ \gamma & 1 \end{bmatrix}$,

   γ arbitrary; $\begin{bmatrix} 1 & 0 \\ 0 & 1 \end{bmatrix}$ and $\begin{bmatrix} -1 & 0 \\ 0 & -1 \end{bmatrix}$

11. $\begin{bmatrix} \alpha & \beta \\ -\dfrac{\alpha^2}{\beta} & -\alpha \end{bmatrix}$, α, β arbitrary except that $\beta \neq 0$; also $\begin{bmatrix} 0 & 0 \\ \gamma & 0 \end{bmatrix}$,

    $\gamma \neq 0$ but otherwise arbitrary

13. $\begin{bmatrix} 2 & 0 & 1 & 0 \\ 0 & 2 & 0 & 1 \\ 0 & 0 & -1 & 0 \\ 0 & 0 & 0 & -1 \end{bmatrix}$

15. (a) $\begin{bmatrix} 0 & -1 \\ \frac{5}{2} & -1 \end{bmatrix}$ (b) $\begin{bmatrix} -6 & 0 \\ 0 & 3 \end{bmatrix}$ (c) $\pm\begin{bmatrix} 1 & 0 \\ 9 & 0 \end{bmatrix}$

17. One such pair is $A = \begin{bmatrix} 1 & 1 \\ 1 & 1 \end{bmatrix}$, $B = \begin{bmatrix} -1 & -1 \\ 1 & 1 \end{bmatrix}$.

19. (b) An $n \times n$ diagonal matrix (α_{ij}) is invertible if and only if $\alpha_{ii} \neq 0$ for $i = 1, 2, \ldots, n$. The inverse is an $n \times n$ diagonal matrix (α'_{ij}) in which

$$\alpha'_{ii} = 1/\alpha_{ii}, \quad i = 1, 2, \ldots, n.$$

(d) The matrix is an $n^2 \times 1$ matrix with entries corresponding to the n^2 matrices (e_{ij}) of the image basis. Those entries which correspond to (e_{11}), (e_{22}), (e_{33}), $\ldots$, (e_{nn}) are each 1 and all other entries are 0.

Chapter 3

Section 3–1

1. (a) $6e^{2x}$ (b) $2 \cos x - \sin x$
   (c) $2 - 2x \ln x$ (d) $2e^x(x \sin x - 3 \sin x - 2x \cos x)$

3. $a = \frac{3}{2}$, $b = 0$, $c = -\frac{1}{2}$

5. $D(xD) = xD^2 + D$, $(xD)D = xD^2$

7. (a) $(a_1D + 1)(b_1D + 1) = (a_1 + b_1)D + 1$, for given a_1 and b_1
   (b) D and $\sqrt{x}\, D$ are each linear on $[0, \infty)$ but $D[\sqrt{x}\, D]$ is undefined at $x = 0$.

9. (a) $3xD + 2$ (b) $(e^x + e^{-x})D^2$
   (c) $xD + Dx + 1$

13. (a) $(D - 1)(D - 2)$ (b) $(2D + 1)(D + 2)$
    (c) $(2D + 1)^2$ (d) $(D + 1)(D - 2)^2$
    (e) $(D - 1)(D + 2)(4D^2 + 1)$ (f) $(D - 1)(D + 1)(D^2 + 1)$
    (g) $(D^2 + \sqrt{2}\, D + 1)(D^2 - \sqrt{2}\, D + 1)$
    (h) $[D - 1][D^2 - \frac{1}{2}(1 + \sqrt{5})\, D + 1][D^2 - \frac{1}{2}(1 - \sqrt{5})\, D + 1]$

19. (a) $\sum_{i=0}^{n} c_i a_i x^k$ where $c_0 = 1$, $c_i = k(k - 1) \cdots (k - i + 1)$, $i = 1, 2, 3, \ldots, n$

Section 3–2

1. (a) 2 (b) 3 (c) 3, 3 (d) 2 (e) 3, 3, 3

3. (b) $y = c_1e^{ax} \cos bx + c_2e^{ax} \sin bx$
   (c) $y = be^{ax} \cos bx - \dfrac{a(b + 1)}{b} e^{ax} \sin bx$

5. (a) The given solutions are not linearly independent.
   (b) $y = c_1 \sin^3 x + c_2$

7. (a) $y = c_1 \sinh x + c_2 \cosh x$
   (b) $y = c_1 x^3 + c_2 x^3 \ln x$
   (c) $y = c_1 \sin 2x + c_2 \cos 2x$
   (d) $y = c_1 x + c_2 x \ln \dfrac{1 + x}{1 - x}$

Section 3–3

1. $y = \dfrac{c}{x^2}$, on $(-\infty, 0)$ or $(0, \infty)$

3. $y = \dfrac{c}{\sin x}$, on $k\pi < x < (k + 1)\pi$, $k = 0, \pm 1, \pm 2, \ldots$

5. $y = e^{-x} + ce^{-(3/2)x}$, on $(-\infty, \infty)$

7. $i = \dfrac{E}{R} + ce^{-(R/L)t}$, on $(-\infty, \infty)$

9. $y = \dfrac{ce^x - (2x + 1)e^{-x}}{4(x^2 + 1)}$, on $(-\infty, \infty)$

11. $y = \dfrac{e^x(x - 1) + c}{x \sin x}$, on $k\pi < x < (k + 1)\pi$, $k = 0, \pm 1, \pm 2, \ldots$

13. $y = \left(\dfrac{1 + \sqrt{1 - x^2}}{x}\right)(e^x + c)$, on $(-1, 0)$ or $(0, 1)$

15. $y = -\dfrac{\sin x + \ln|1 - \sin x| + c}{(1 + \sin x)^2}$

   on $\dfrac{2k - 1}{2}\pi < x < \dfrac{2k + 1}{2}\pi$, $k = 0, \pm 1, \pm 2, \ldots$

17. $y = (\sin x \ln|\csc 2x + \cot 2x| + c \sin x)^{-1}$
   For a given c the solution is valid on any interval which does not include any point $k\pi/2$, $k = 0, \pm 1, \pm 2, \ldots$, or at which the quantity in the parentheses vanishes; $y = 0$ is also a solution on the intervals $k\pi/2 < x < (k + 1)\pi/2$.

19. $y = e^x[\frac{1}{2} + c(x^2 + 1)^{-3/2}]^{2/3}$, on $(-\infty, \infty)$

21. $y = \left[3x + \dfrac{3}{2}\left(\dfrac{x^2 - 6x}{\ln x}\right) - \dfrac{3}{2}\left(\dfrac{x^2 - 12x}{(\ln x)^2}\right) + \dfrac{1}{4}\left(\dfrac{3x^2 - 72x + c}{(\ln x)^3}\right)\right]^{1/3}$
   on $(0, 1)$ and $(1, \infty)$

23. $y = \{(1 - x)[c + \frac{1}{2}\ln(-x + \sqrt{x^2 - 1})] - \sqrt{x^2 - 1}\}^2$, on $(-\infty, -1)$;
   $y = -\{(1 - x)[c + \frac{1}{2}\cos^{-1} x] + \sqrt{1 - x^2}\}^2$, on $(-1, 1)$;
   $y = \{(x - 1)[c + \frac{1}{2}\ln(x + \sqrt{x^2 - 1})] - \sqrt{x^2 - 1}\}^2$, on $(1, \infty)$;
   also, $y = 0$ is a solution on $-\infty < x < \infty$.

25. $y = -\dfrac{5}{x^3 + 5x + c\sqrt{|x|}}$

   For a given c the solution is valid on any interval on which the denominator nowhere vanishes; $y = 0$ is also a solution, on $(-\infty, \infty)$.

27. (a) $y = -\frac{6}{x} + \frac{1}{x}\int_2^x e^{-x^2/2}\,dx$; on $(-\infty, 0)$ and $(0, \infty)$

(b) $y(1) = -6.3407$, $y'(1) = 6.9472$

29. $y = 1 + \dfrac{1}{1 - x + ce^{-x}}$

31. $y = x + \dfrac{2x}{c - \ln|x|}$

33. $y = \dfrac{\cos x}{\sin x}[1 + (ce^{-\sin^2 x} - \frac{1}{2})^{-1}]$

35. (b) (i) $y = \dfrac{ce^x - 2}{1 - ce^x}$; also $y = -1$

(ii) $y = \dfrac{3}{2}\left(\dfrac{ce^{12x} + 1}{ce^{12x} - 1}\right)$; also $y = \frac{3}{2}$

(iii) $y = 1 + \dfrac{1}{x + c}$; also $y = 1$

(iv) $y = \dfrac{1}{6}\left(\dfrac{2ce^{(5/6)x} + 3}{ce^{(5/6)x} - 1}\right)$; also $y = \frac{1}{3}$

37. $v' + v^2 - 1 = 0$, $v = 1 + \dfrac{2}{ce^{2x} - 1}$, also $v = 1$; $y = c_1e^x + c_2e^{-x}$

39. (a) $y = c_1e^{2x} + c_2e^{3x}$
(b) $y = c_1e^{-(3/2)x} + c_2e^x$
(c) $y = c_1e^{-x} + c_2e^{2x}$
(d) $y = c_1e^{-(5/4)x} + c_2e^{(4/3)x}$
(e) $y = c_1e^{-(\sqrt{6}/2)x} + c_2e^{(\sqrt{6}/2)x}$

41. (a) $c_1e^{-x} + c_2xe^{-x}$
(b) $c_1e^{(3/2)x} + c_2xe^{(3/2)x}$
(c) $c_1e^{(3/2)x} + c_2xe^{(3/2)x}$
(d) $c_1e^{(1/6)x} + c_2xe^{(1/6)x}$
(e) $c_1e^{(\sqrt{2}/2)x} + c_2xe^{(\sqrt{2}/2)x}$

Section 3–4

1. $y = -\dfrac{1}{x^2}$, on $x > 0$

3. $y = e^{-x} - (e^{-3/2} + 3e^{-9/2})e^{-(3/2)x}$, on $(-\infty, \infty)$

5. $y = \dfrac{1}{x \sin x}\displaystyle\int_{-1/2}^x \frac{e^x}{x}\,dx$, on $(-\pi, 0)$

7. $y = \dfrac{1}{\sin x}\left[1 + (1 + \sin x)\ln\dfrac{1 + \sin x}{2}\right]$, on $(0, \pi)$

9. $y = \frac{3}{2}\ln\dfrac{x - 1}{3(x + 1)}$ on $(-\infty, -1)$

11. $y = \sqrt{\dfrac{2}{\pi x}}\left[-\dfrac{9 + \sqrt{3}\,\pi}{12}\sin x + \dfrac{\sqrt{3} - \pi}{4}\cos x\right]$, on $(0, \infty)$

13. $y = (1 + 2\ln 2) - 2\ln|x|$, on $(-\infty, 0)$

15. The equation $xy' + y = 0$ is of the first order on $(-\infty, \infty)$ but is not normal on any interval which contains $x = 0$. If $y(x)$ is any solution then the equation implies that $y(0) = 0$, and so if $y_0 \neq 0$ there is no solution which satisfies the initial condition $y(0) = y_0$.

23. G is the linear operator which transforms $h(x)$ into $e^{k(x)}\int_{x_0}^x h(x)e^{-kx}\,dx + y_0e^{k(x-x_0)}$.

Section 3–5

1. (a) Interval is $(-\infty, \infty)$.
   (b) One representation of y is
   $y = c_1 \cdot 1 + c_2 e^{-x} + c_3 e^{2x}$.
   (c) $y_1 = 1$, $y_2 = \frac{1}{2} - \frac{2}{3}e^{-x} + \frac{1}{6}e^{2x}$,
   $y_3 = -\frac{1}{2} + \frac{1}{3}e^{-x} + \frac{1}{6}e^{2x}$
3. (a) Interval is $(-\infty, \infty)$.
   (b) One representation of y is
   $y = c_1 e^x + c_2 x e^x + c_3 e^x \cos x + c_4 e^x \sin x$.
   (c) $y_1 = 2e^x - 2xe^x - e^x \cos x + e^x \sin x$,
   $y_2 = -2e^x + 4xe^x + 2e^x \cos x - 3e^x \sin x$,
   $y_3 = e^x - 3xe^x - e^x \cos x + 3e^x \sin x$,
   $y_4 = xe^x - e^x \sin x$
5. (a) Interval is $(0, \infty)$.
   (b) One representation of y is
   $$y = c_1\left(x^2 + \frac{1}{x}\right) + c_2(3x^2).$$
   (c) $y_1 = \dfrac{x^2}{3} + \dfrac{2}{3x}$, $y_2 = \dfrac{x^2}{3} - \dfrac{1}{3x}$

Section 3–6

1. $W[1, e^{-x}, 2e^{2x}] = -12e^x$
3. $W[1, x, x^2, \ldots, x^n] = 1! \cdot 2! \cdot 3! \cdots n!$
5. $W[x^{1/2}, x^{1/3}] = -\frac{1}{6}x^{-1/6}$
7. $W[e^x, e^x \sin x] = e^x \cos x$
9. $W[1, \sin^2 x, 1 - \cos x] = 2 \sin^3 x$
11. $W[\sqrt{1 - x^2}, x] = 1/\sqrt{1 - x^2}$
23. $$a_1(x) = -\frac{u_1(x)u_2''(x) - u_2(x)u_1''(x)}{u_1(x)u_2'(x) - u_2(x)u_1'(x)},$$
    $$a_2(x) = \frac{u_1'(x)u_2''(x) - u_2'(x)u_1''(x)}{u_1(x)u_2'(x) - u_2(x)u_1'(x)}$$
25. (a) $x^2y'' - x(x + 2)y' + (x + 2)y = 0$
    (b) $x^2y'' - 2xy' + 2y = 0$
    (c) $y'' + y = 0$
    (d) $(x \cos x - \sin x)y'' + (x \sin x)y' - (\sin x)y = 0$
    (e) $x^2(1 - \ln x)y'' + xy' - y = 0$

Section 3–7

1. (a) $-1/x$ (b) $-4/(1 - x^2)$
   (c) $2x^3$ (d) -1
   (e) $e^{1-\cos x}$ (f) $e^{(2/3)[(1+x^3)^{1/2}-2^{1/2}]}$
3. xe^{ax}
5. $\dfrac{1}{\sin^2 x}$
7. $-1 + \dfrac{x}{2} \ln\left|\dfrac{1 + x}{1 - x}\right|$

Chapter 4

Section 4–1

3. $x^2 - 2ax + (a^2 + b^2) = 0$
5. (a) $(D + 1)^2(D + 2)$
(b) $(D - 1)(D^2 + 1)$
(c) $(D + \sqrt{5})(D - \sqrt{5})(D^2 + 2D + 5)$
(d) $(D + 1)(D - 1)(D + 2)(D - 2)$
(e) $(D^2 + \sqrt{2\sqrt{10} - 2}\, D + \sqrt{10})(D^2 - \sqrt{2\sqrt{10} - 2}\, D + \sqrt{10})$

Section 4–2

1. $y = c_1e^{-2x} + c_2e^x$
3. $y = c_1e^{3x/4} + c_2e^{-5x/2}$
5. $y = c_1 \cos 2x + c_2 \sin 2x$
7. $y = c_1e^{-2x} \cos 2x + c_2e^{-2x} \sin 2x$
9. $y = c_1e^x \cos x + c_2e^x \sin x$
11. $y = c_1e^{-x} \cos \sqrt{3}\, x + c_2e^{-x} \sin \sqrt{3}\, x$
13. $y = c_1e^{2\sqrt{3}x} + c_2e^{\sqrt{3}x/2}$
15. $y = c_1e^{(3/8)x} \cos \dfrac{\sqrt{2}}{4} x + c_2e^{(3/8)x} \sin \dfrac{\sqrt{2}}{4} x$
17. $y = (\frac{1}{2} - \frac{7}{3}x)e^{4x}$
19. $y = 2 \cos \sqrt{2}\, x + 2 \sin \sqrt{2}\, x$
21. $y = -\frac{2}{3}e^{-2x} \sin 3x$
23. $y = 3xe^{\sqrt{5}x}$
25. $y = \sqrt{2}\, e^{(\sqrt{2}/2)x} \left(\cos \dfrac{\sqrt{2}}{2} x + \sin \dfrac{\sqrt{2}}{2} x\right)$
29. (a) $y'' + 6y' + 9y = 0$ (b) $y'' - 2y' + 5y = 0$
(c) $y'' + 4y' + 4y = 4$ (d) $y'' + 4y' + 3y = 3x + 16$
(e) $y'' + 9y = 3x$
31. (b) $y = c_1 \cos (2x + c_2)$
33. (a) $y = e^x \cos 5x + \frac{1}{5}e^x \sin 5x$
(b) $y = (\frac{1}{2} - \frac{1}{10}i)e^{(1+5i)x} + (\frac{1}{2} + \frac{1}{10}i)e^{(1-5i)x}$

Section 4–3

1. $y = c_1e^x + c_2e^{-x} + c_3e^{-3x}$
3. $y = c_1 + (c_2 + c_3x)e^{-(3/2)x}$
5. $y = c_1e^{-2x} + c_2e^{2x} + c_3e^{-(1/2)x}$
7. $y = c_1 + c_2x + c_3e^{-x} + c_4e^x$
9. $y = (c_1 + c_2x) \cos 3x + (c_3 + c_4x) \sin 3x$
11. $y = c_1 + c_2x + c_3e^{-(1/2)x} \cos \dfrac{\sqrt{3}}{2} x + c_4e^{-(1/2)x} \sin \dfrac{\sqrt{3}}{2} x$
13. $y = (c_1 + c_2x + c_3x^2 + c_4x^3)e^x$
15. $y = (c_1 + c_2x + c_3x^2)e^{-2x} + c_4 \cos \sqrt{3}\, x + c_5 \sin \sqrt{3}\, x$

17. (a) $(D - 1)^3$ (b) $D^2 - 4D + 8$ (c) $(D^2 + 1)^3$
(d) $D^2(D + 2)$ (e) $(D^2 + 4)^3$ (f) $(D^2 - 2D + 5)^3$
(g) $(D^2 + 1)^2$
(h) $(D - 3)^3(D^2 - 6D + 10)^3$
(i) $D(D - 1)^2(D - 2)^3(D - 3)^4$
(j) $D^{n+1}(D - 1)^n(D - 2)^{n-1} \cdots (D - k)^{n-k+1} \cdots (D - n)$

21. (d) The results of (a) and (b) show that the solutions are linearly independent in $\mathcal{C}(-\infty, \infty)$.

Section 4–4

(*Note.* Sometimes a simplified particular solution may be obtained by deleting from the solution given by variation of parameters, or the use of Green's functions, any terms which satisfy the homogeneous equation.)

1. $y = (c_1 + \ln|\cos x|)\cos x + (c_2 + x)\sin x$

3. $y = \dfrac{2x - 1}{32} e^{2x} + (c_1 + c_2 x)e^{-2x}$

5. $y = \frac{1}{4}e^{-(1/2)x}(-x \sin x - 2\cos x + c_1 + c_2 x)$

7. $y = \frac{1}{84}(7e^{2x} - 14 - 3e^{-2x}) + c_1 e^{(-5+\sqrt{37})x} + c_2 e^{(-5-\sqrt{37})x}$

9. $y = \frac{1}{5}e^{2x}(\sin x - 2\cos x) + c_1 e^x \cos x + c_2 e^x \sin x$

11. $y_p = x^3(\frac{1}{2}\ln x - \frac{3}{4})$

13. $y_p = \dfrac{1}{16}\left[\ln|\tan x| + (\cos 2x)\ln\dfrac{\cos^2 2x}{(1 + \cos 2x)|\sin 2x|}\right]$

15. $y_p = -x$

17. $K(x, t) = \frac{1}{3}[e^{2(x-t)} - e^{-(x-t)}]$

19. $K(x, t) = 2e^{x-t}\sin\frac{1}{2}(x - t)$

21. $K(x, t) = \dfrac{x(x - t)}{t}$

23. $K(x, t) = \dfrac{1 - t^2}{2}\ln\left|\dfrac{(1 + x)(1 - t)}{(1 - x)(1 + t)}\right|$

25. If $a \neq 0$ or if $\omega^2 \neq b^2 - a^2$, then

$$y = \frac{1}{2c}\left[\frac{\sin(\omega t + \varphi_1)}{D_1} - \frac{\sin(\omega t + \varphi_2)}{D_2}\right] + \frac{1}{2c}\left[e^{-at}\frac{\sin(ct - \varphi_1)}{D_1} + \frac{\sin(ct + \varphi_2)}{D_2}\right],$$

where

$$c = (b^2 - a^2)^{1/2}, \quad D_1 = [a^2 + (\omega + c)^2]^{1/2}, \quad D_2 = [a^2 + (\omega - c)^2]^{1/2},$$

and φ_1, φ_2 are determined by the relations

$\cos\varphi_1 = (\omega + c)/D_1$, $\sin\varphi_1 = a/D_1$, $\cos\varphi_2 = (\omega - c)/D_2$, $\sin\varphi_2 = a/D_2$, $0 \leq \varphi_1, \varphi_2 < 2\pi$.

If $a = 0$ and $\omega^2 = b^2$, then

$$y = \frac{1}{2\omega^2}(\sin\omega t - \omega t\cos\omega t).$$

Section 4–5

1. $y = (\frac{1}{3}x^3 - \frac{1}{2}x^2 + \frac{1}{2}x - \frac{1}{4})e^x + c_1e^{-x} + (c_2 + c_3x)e^x$
3. $y = -(\frac{1}{3}x^3 + \frac{3}{2}x^2 + \frac{3}{2}x + \frac{3}{4}) + c_1 + c_2x + c_3e^{2x}$
5. $y = \frac{1}{25}(4\cos x + 3\sin x) + c_1e^x + c_2e^{2x} + c_3e^{-3x}$
7. $y = \frac{1}{2}\cos x + c_1 + c_2e^{-x} + c_3e^x$
9. $y = \left(\frac{x^2}{2} - \frac{5}{2}x + c_1\right)e^x + c_2 + c_3x + c_4e^{-x}$
11. $K(x, t) = e^{x-t} - (x - t) - 1$
13. $K(x, t) = \frac{1}{2}[e^{x-t} - 2e^{2(x-t)} + e^{3(x-t)}]$
15. $K(x, t) = \sinh(x - t) - (x - t)$

Section 4–6

1. $y = 3x + k_1\int e^{-(x^2/2)}\,dx + k_2$
3. $y = k_1x^2 + k_2$
5. $y = x^{3/2}(k_1 \ln x + k_2) \quad x > 0$
7. $y = e^{-\int[a_0(x)/a_1(x)]dx}\left[\int \frac{h(x)}{a_1(x)}e^{\int[a_0(x)/a_1(x)]dx} + k\right]$
9. Let $y_1 = J_0(x)\int \frac{dx}{x[J_0(x)]^2}$; $\lim_{x\to 0} y_1 = \infty$, but $\lim_{x\to 0} \frac{y_1}{\ln x} = 1$

Section 4–7

1. $y_p = \frac{3}{2}e^x + x^2 - 2x$
3. $y_p = \frac{1}{2}\cos x - \frac{1}{2}\sin x$
5. $y_p = -\frac{1}{2}xe^{-2x}$
7. $y_p = (x^2 - 3x + \frac{30}{7})e^x$
9. $y_p = \frac{1}{625}(899 + 1305x + 675x^2 + 125x^3)$
11. $y_p = \frac{1}{6250}(2374 - 2940x + 3200x^2 - 1500x^3 + 625x^4)$
13. $y_p = \frac{1}{3}(x - 1)e^x$
15. $y_p = 2 + 2x + xe^{-x}$
17. $y_p = -(\sin x + \cos x)$
19. $y_p = \frac{1}{12}(39x + 9x^2 + 2x^3 + 12xe^x)$
21. $y_p = (c_1 + c_2x)x^2e^{2x} + (c_3 + c_4x + c_5x^2)\cos x + (c_6 + c_7x + c_8x^2)\sin x$
23. $y_p = c_1e^{-x} + x^3[(c_2 + c_3x + c_4x^2)\cos x + (c_5 + c_6x + c_7x^2)\sin x]$
25. $y_p = x^2[(c_1 + c_2x)e^x + (c_3 + c_4x)e^{-x} + (c_5 + c_6x + c_7x^2 + c_8x^3)\cos x + (c_9 + c_{10}x + c_{11}x^2 + c_{12}x^3)\sin x]$
27. $y_p = x^3(c_1 + c_2x + c_3x^2)e^x + (c_4 + c_5x)\cos x + (c_6 + c_7x)\sin x$
29. $y_p = c_1 + c_2x + x^2e^{x/2}[(c_3 + c_4x)\cos x + (c_5 + c_6x)\sin x + (c_7 + c_8x + c_9x^2)\cos x + (c_{10} + c_{11}x + c_{12}x^2)\sin x$

Section 4–8

1. $y = c_1|x| + c_2|x|^{-2}$
3. $y = c_1 \sin(3\ln|x|) + c_2\cos(3\ln|x|)$
5. $y = c_1|x|^p + c_2|x|^{-p}$, if $p \neq 0$; $y = c_1 + c_2\ln|x|$, if $p = 0$
7. $y = |x|^{-1}(c_1 + c_2\ln|x|) + c_3|x|^7$

9. $y = c_1|x|^{-1} + c_2|x|^{\sqrt{2}} + c_3|x|^{-\sqrt{2}}$

11. $y = c_1|x| + c_2|x|^{-1} + c_3 \cos(\ln|x|) + c_4 \sin(\ln|x|)$

13. Let $d = (a_1 - 1)^2 - 4a_0$.

(i) If $d > 0$, set $\alpha_1 = \frac{1}{2}[1 - a_1 - \sqrt{d}]$,

$\alpha_2 = \frac{1}{2}[1 - a_1 + \sqrt{d}]$; then $K(x, t) = \dfrac{x^{\alpha_1}[x^{\sqrt{d}} - t^{\sqrt{d}}]}{\sqrt{d}\, t^{\alpha_2 - 1}}$.

(ii) If $d = 0$, set $\alpha = (1 - a)/2$; then $K(x, t) = \dfrac{x^{\alpha}[\ln x - \ln t]}{t^{\alpha - 1}}$.

(iii) If $d < 0$, set $a = \dfrac{1 - a_1}{2}$, $b = \dfrac{\sqrt{-d}}{2}$; then

$$K(x, t) = \frac{x^a}{bt^a} \cos[b(\ln x - \ln t)].$$

17. (a) $(xD + 2)(xD - 1)$ (b) $(xD + 3)(xD - 3)$
(c) $(xD + 1)^2(xD - 7)$ (d) $(xD - 1)(xD - 2)(xD - 3)$
(e) $(x^2D^2 + xD + 1)(xD + 1)(xD - 1)$

19. $y = -\frac{1}{6}(\ln x)\cos(\ln x^3)$ 21. $y = \frac{1}{2}x$

Section 4–9

(Numerical values are approximate)

1. (a) $x = x_0 e^{-[(\ln 2)/5600]t} = x_0 e^{-0.00012t}$
(b) $t = 18{,}600$

3. (a) $x = (x_0 + 72.13k)e^{-0.01386t} - 72.13k$
(b) $t = \dfrac{\ln(x_0 + 72.13k) - \ln(72.13k)}{0.01386}$

5. 132.9 hours

7. (a) 9.00 years
(b) 221,000,000; 856,000,000

9. $r = r_0 - kt$; evaporation complete when $t = r_0/k$.

11. $h = h_0 \cos\sqrt{2g/L}\, t$

Section 4–10

1. $i = i_0 + \dfrac{bE_0L}{Y^2} e^{-(R/L)t} + \dfrac{E_0}{Y} e^{-at} \sin(bt - \beta)$,

where $Y = \sqrt{(R - aL)^2 + b^2L^2}$, $R - aL = Y\cos\beta$,

$bL = Y\sin\beta$, $0 < \beta < \pi$.

3. (a) $q = 0.03e^{-(1/RC)t}$
(c) 3.5676 volts

5. If $R > 2(L/C)^{1/2}$, set $\Delta = [R^2 - (4L/C)]^{1/2}$; then

$$q = 0.03e^{-[R/(2L)]t}\left[\cosh\frac{\Delta}{2L}t + \frac{R}{\Delta}\sinh\frac{\Delta}{2L}t\right].$$

If $R = 2(L/C)^{1/2}$, then $q = 0.03[1 + (R/2L)t]e^{-[R/(2L)]t}$.

If $R < 2\left(\frac{L}{C}\right)^2$, set $\Delta' = [(4L/C) - R^2]^{1/2}$; then

$$q = 0.03e^{-[R/(2L)]t}\left[\cos\frac{\Delta'}{2L}t + \frac{R}{\Delta'}\sin\frac{\Delta'}{2L}t\right].$$

7. (i) $q = q(0)\cos\frac{t}{\sqrt{LC}} + \sqrt{LC}\left[i(0) - \frac{\omega EC}{1 - \omega^2 LC}\right]\sin\frac{t}{\sqrt{LC}} + \frac{EC}{1 - \omega^2 LC}\sin\omega t$

   (ii) $q = q(0)\cos\omega t + \frac{1}{\omega}\left[i(0) + \frac{E}{2\omega L}\right]\sin\omega t - \frac{E}{2\omega L}t\cos\omega t$

   (Resonance in this case)

Chapter 5

Section 5–1

1. (a) Piecewise continuous on $[0, \infty)$, from definition
   (b) Piecewise continuous on $[0, \infty)$, from definition
   (c) Not piecewise continuous on $[0, \infty)$, since function becomes infinite at $t = 1$
   (d) Piecewise continuous on $[0, \infty)$; note that

$$\lim_{t \to 2^+} \frac{t - 2}{t^2 - t - 2} = \lim_{t \to 2^-} \frac{t - 2}{t^2 - t - 2} = \frac{1}{3}.$$

   Note that the function, as given, is not continuous on $[0, \infty)$, since it is undefined at $t = 2$.

   (e) Not piecewise continuous on $[0, \infty)$, since it becomes infinite as $t \to 0$

7. $\frac{4}{\pi}$

9. $-\frac{2}{3}$

11. $\lim_{t \to 0^+} (t \sin 1/t) = 0$; $t \sin 1/t$ is piecewise continuous on $[0, \infty)$.

Section 5–2

1. $\mathcal{L}[t] = \frac{1}{s^2}$; $s_0 = 0$; abscissa of convergence is 0.

3. $\mathcal{L}[f](s) = \frac{1}{s}(1 - e^{-s})$, if $s \neq 0$; $\mathcal{L}[f](0) = 1$; $s_0 = -\infty$

5. $\mathcal{L}[\sin at] = \frac{a}{s^2 + a^2}$; $s_0 = 0$

Section 5–4

1. $\frac{\cos a + s \sin a}{s^2 + 1}$

3. $n!\left[\frac{a^n}{n!s} + \frac{a^{n-1}}{(n-1)!s^2} + \frac{a^{n-2}}{(n-2)!s^3} + \cdots + \frac{a}{1!s^n} + \frac{1}{s^{n+1}}\right]$

5. $\frac{s}{s^2 - a^2}$

7. $\frac{2a^2}{s(s^2 + 4a^2)}$

9. $\dfrac{4s}{(s^2+4)^2}$

13. $\dfrac{1}{a}e^{-as}$

17. If $|f(t)| \leq Ce^{\alpha t}$, then $\lim_{t\to\infty} e^{-st}f(t) = 0$ if $s > \alpha$.

21. $y = e^{2t} + 2e^t$

23. $y = -\frac{1}{2} + \frac{1}{5}e^{-t} + \frac{3}{10}\cos 2t + \frac{3}{5}\sin 2t$

25. $y = -\frac{1}{2}t^2 + \frac{4}{5}e^{3t} - \frac{9}{5}e^{-2t}$

27. $y = -2 + 2\cos\frac{1}{2}t + \sin\frac{1}{2}t$

Section 5–5

1. $\dfrac{3}{(s-2)^2+9}$

3. $\dfrac{72s(s^2-9)}{(s^2+9)^4}$

5. $\dfrac{(s+3)\cos 4 - 2\sin 4}{(s+3)^2+4}$

7. If $\mathfrak{L}[f] = \varphi(s)$, then $\mathfrak{L}[te^{2t}f'(t)] = -\left[(s-2)\dfrac{d}{ds}\varphi(s-2) + \varphi(s-2)\right]$.

9. $\dfrac{(3s+2)e^{-s/2}}{2s^2}$

11. $\dfrac{1-e^{-2\pi s}}{s^2+1}$

13. $\dfrac{1}{s^2}[1 - 4e^{-2s} + 4e^{-3s} - e^{-4s}]$

15. $\dfrac{2}{(s-2)^2+4}$

17. $\dfrac{1}{s^2+1}\left(\dfrac{1+e^{-\pi s}}{1-e^{-\pi s}}\right)$

19. $\dfrac{2(s-2)}{s[(s-2)^2+1]^2}$

21. $\dfrac{8(5s^2-1)}{(s^2+1)^4} + \dfrac{2(\cos 1 - \sin 1)}{s^3}$

23. $$\frac{3s^4 - 16s^3 + 96s - 108}{(s-1)^2[(s-3)^2+1]^3} + \frac{1}{5(s-1)^2} + \frac{e^{2a}(2\sin a - 5a\sin a - \cos a)}{5(s-1)^2}$$
$$= \frac{s^4 - 16s^3 + 120s^2 - 400s + 460}{5[(s-3)^2+1]^3} + \frac{e^{2a}(2\sin a - 5a\sin a - \cos a)}{5(s-1)^2}$$

25. $1 - e^{-t}$

27. $\frac{1}{4} - \frac{1}{4}e^{-2t} - \frac{1}{2}te^{-2t}$

29. $\dfrac{1}{(n-1)!}t^{n-1}e^{at}$

31. $\frac{1}{5}e^{-2t}\sin 5t$

33. $t\sin t$

35. $\frac{3}{2}\sin t + \frac{3}{2}t\cos t$

37. $$\sqrt{3}\,u_2(t)\sin\frac{1}{\sqrt{3}}(t-2) = \begin{cases} 0 & \text{if } t \leqq 2,\\ \sqrt{3}\sin\dfrac{1}{\sqrt{3}}(t-2) & \text{if } t > 2 \end{cases}$$

39. $$\frac{1}{\sqrt{2}}\left[\cosh\frac{t}{\sqrt{2}}\sin\frac{t}{\sqrt{2}} - \sinh\frac{t}{\sqrt{2}}\cos\frac{t}{\sqrt{2}}\right]$$

41. $$1 + u_1(t) = \begin{cases} 1 & \text{if } t \leqq 1,\\ 2 & \text{if } t > 1 \end{cases}$$

43. $\frac{1}{3a^2}e^{-at} - \frac{1}{3a^2}e^{at/2}\cos\frac{\sqrt{3}}{2}at + \frac{4\sqrt{3}}{9a^2}e^{at/2}\sin\frac{\sqrt{3}}{2}at$

45. $\frac{1}{t}(e^{-2t} - e^{-3t})$

49. (b) $\mathcal{L}^{-1}\left[\frac{s}{(s^2+a^2)^{n+1}}\right] = \frac{1}{2^n a n!}\underbrace{t\int_0^t t\int_0^t t\cdots\int_0^t}_{n-1 \text{ times}} t\sin at\,dt\,dt\ldots dt, \quad n \geq 2$

Section 5–6

1. $y = \frac{1}{4}e^t - \frac{1}{4}e^{-t} - \frac{1}{2}te^{-t}$

3. $y = \frac{2}{3}e^{-t}\cos\sqrt{2}\,t + \frac{\sqrt{2}}{3}e^{-t}\sin\sqrt{2}\,t + t - \frac{2}{3}$

5. $y = -\frac{1}{k}[1 - e^{-kt}] + \frac{2}{k}\sum_{j=0}^{\infty}(-1)^j u_j(t)[1 - e^{-k(t-j)}]$

7. $y = -\frac{27}{50}\cos 2t + \frac{57}{25}\sin 2t - \frac{3}{5}te^t + \frac{1}{25}e^t + \frac{1}{2}e^{2t}$

9. $y = te^{-2t} + e^{-2t} + u_2(t)[e^{-(t-2)} - (t-1)e^{-2(t-2)}]$
$= \begin{cases} te^{-2t} + e^{-2t} & \text{if } t \leq 2 \\ te^{-2t} + e^{-2t} + e^{-(t-2)} - (t-1)e^{-2(t-2)} & \text{if } t > 2 \end{cases}$

11. $y = 2\sin 2t + \frac{3}{2}e^{-(t-\pi)} - \frac{5}{2}e^{t-\pi}$

13. $y = 1 - 2te^{-t} - 2u_1(t)[1 - te^{-(t-1)}] + u_2(t)[1 - (t-1)e^{-(t-2)}]$
$= \begin{cases} 1 - 2te^{-t} & \text{if } t < 1, \\ -1 - 2te^{-t} + 2te^{-(t-1)} & \text{if } 1 \leq t \leq 2, \\ -2te^{-t} + 2te^{-(t-1)} - (t-1)e^{-(t-2)} & \text{if } 2 < t \end{cases}$

17. $J_0(t) = \sum_{k=0}^{\infty}\frac{(-1)^k}{2^{2k}(k!)^2}t^{2k}$

Section 5–7

1. $\int_0^t f(t-\xi)\sin\xi\,d\xi$

3. $e^{-t} + t - 1$

5. $\frac{3}{2}[t\cos t + \sin t]$

7. $\frac{e^{bt} - e^{at}}{b - a}$ if $a \neq b$; te^{at} if $a = b$

9. $\frac{a}{a^2 - b^2}(\cos bt - \cos at)$ if $a^2 \neq b^2$; $\frac{1}{2}t\sin at$ if $b = \pm a$

11. $e\int_1^{t+1} f(t-\xi)e^{-\xi}g(\xi)\,d\xi$

17. $\frac{t^{n-1}}{(n-1)!}$

21. $\frac{\pi}{2} - \tan^{-1}\frac{s}{a}$

23. $\frac{3\pi}{2} - \frac{1}{2}s\ln\left(1 + \frac{9}{s^2}\right) - 3\tan^{-1}\frac{s}{3}$

27. The minimum is between $x = 1$ and $x = 2$.

31. (a) $\sqrt{\pi}$ (b) 2

Section 5–8

1. $K(t, \xi) = (t - \xi)e^{2(t-\xi)}$

3. $K(t, \xi) = \frac{1}{2}e^{-3(t-\xi)} \sin 2(t - \xi)$

5. $K(t, \xi) = (t - \xi)e^{(t-\xi)/6}$

7. $K(t, \xi) = \dfrac{1}{3}\left[e^{-(t-\xi)} - e^{(t-\xi)/2} \cos \dfrac{\sqrt{3}}{2}(t - \xi) + \sqrt{3}\, e^{(t-\xi)/2} \sin \dfrac{\sqrt{3}}{2}(t - \xi)\right]$

9. $$K(t, \xi) = \frac{\sqrt{2}}{4}\left\{e^{-\sqrt{2}(t-\xi)/2}\left[\sin \frac{\sqrt{2}}{2}(t - \xi) + \cos \frac{\sqrt{2}}{2}(t - \xi)\right] + e^{\sqrt{2}(t-\xi)/2}\left[\sin \frac{\sqrt{2}}{2}(t - \xi) - \cos \frac{\sqrt{2}}{2}(t - \xi)\right]\right\}$$

11. $$y = u_1(t)\{1 - t + \tfrac{1}{2}[e^{t-1} - e^{-(t-1)}]\} = \begin{cases} 0 & \text{if } 0 \le t \le 1, \\ 1 - t + \tfrac{1}{2}[e^{t-1} - e^{-(t-1)}] & \text{if } t > 1 \end{cases}$$

17. $y = \frac{1}{6}e^{-(t-\pi)} - \frac{4}{15}e^{(t-\pi)/2} - \frac{3}{10} \sin t - \frac{1}{10} \cos t$

19. $y = \frac{3}{5}e^{4(t-2)} + \frac{12}{5}e^{-(t-2)} + \frac{9}{25}e^{-t} + \frac{1}{25}e^{4t-10} - \frac{1}{5}te^{-t}$

21. $$y = e^{t/2}[\tfrac{1}{24}(t - a) \sin 3t - \tfrac{1}{72} \sin 3a \sin 3(t - a)] + e^{(t-a)/2}[7 \cos 3(t - a) - \tfrac{11}{6} \sin 3(t - a)]$$

23. $y = t - a - \frac{3}{2} \sin (t - a) + \frac{1}{2}(t - a) \cos (t - a)$

Section 5–9

3. $$y = \tfrac{1}{2}e^t + \tfrac{1}{2}e^{-t} + \frac{2\sqrt{3}}{3}e^{t/2} \sin \frac{\sqrt{3}}{2}t + u_1(t)\left[\tfrac{1}{3}e^{-(t-1)} - \tfrac{1}{3}e^{(t-1)/2} \cos \frac{\sqrt{3}}{2}(t - 1) + \frac{\sqrt{3}}{3}e^{(t-1)/2} \sin \frac{\sqrt{3}}{2}(t - 1)\right]$$
$$= \begin{cases} \tfrac{1}{2}e^t + \tfrac{1}{2}e^{-t} + \dfrac{2\sqrt{3}}{3}e^{t/2} \sin \dfrac{\sqrt{3}}{2}t & \text{if } 0 \leqq t \leqq 1, \\ \tfrac{1}{2}e^t + \tfrac{1}{2}e^{-t} + \dfrac{2\sqrt{3}}{3}e^{t/2} \sin \dfrac{\sqrt{3}}{2}t + \tfrac{1}{3}e^{-(t-1)} - \tfrac{1}{3}e^{(t-1)/2} \cos \dfrac{\sqrt{3}}{2}(t - 1) \\ \qquad + \dfrac{\sqrt{3}}{3}e^{(t-1)/2} \sin \dfrac{\sqrt{3}}{2}(t - 1) & \text{if } t > 1 \end{cases}$$

5. $$y = \tfrac{1}{2} \sin 2t + \tfrac{1}{6}u_{\pi/2}(t)[2 \cos t - \sin 2t] = \begin{cases} \tfrac{1}{2} \sin 2t & \text{if } 0 \leqq t \leqq \pi/2, \\ \tfrac{1}{2} \sin 2t + \tfrac{1}{3} \cos t - \tfrac{1}{6} \sin 2t & \text{if } t > \pi/2 \end{cases}$$

7. $y = 2t - \dfrac{\sqrt{3}}{3} \sin \sqrt{3}\, t$

9. $y = a \cos \sqrt{\dfrac{k}{m}}\, t + b\sqrt{\dfrac{m}{k}} \sin \sqrt{\dfrac{k}{m}}\, t$

13. (a) $y = e^{-t}(-\frac{1}{8} + \frac{9}{8}e^{-8t})$
 (b) $t = \frac{1}{4} \ln 3 = 0.27465$
 (c) $\sqrt{3}/27 = 0.064150$ above position of equilibrium

(d) Mass starts at time $t = 0$ at the point 1 unit below the position of equilibrium with velocity 10, and moves upward. It passes through the position of equilibrium at $t = \frac{1}{4}\ln 3$, and at $t = \frac{1}{2}\ln 3$ it reaches its maximum height, $\sqrt{3}/27$, above the position of equilibrium, after which it descends asymptotically toward the position of equilibrium.

15. $y = e^{-\sigma t}[a + (b + \sigma a)t]$

17. $y = e^{-\lambda t/(2m)}\left[a \cos \frac{\sqrt{4km - \lambda^2}}{2m} t + \frac{2bm + \lambda a}{\sqrt{4km - \lambda^2}} \sin \frac{\sqrt{4km - \lambda^2}}{2m} t\right] \to 0.$

Chapter 6

Section 6–1

5. (b) No; if $f(x) \equiv 0$ on $(0, \infty)$ except possibly on a finite closed interval, then the equation cannot have infinitely many solutions on $(0, \infty)$.

Section 6–2

7. (a) $u'' + \left(1 + \frac{1 - 4p^2}{4x^2}\right) u = 0$, on $(-\infty, 0)$ and $(0, \infty)$

(b) $u'' + \frac{1 + p(p + 1)(1 - x^2)}{(1 - x^2)^2} u = 0$, on $(-\infty, -1)$, $(-1, 1)$ or $(1, \infty)$

(c) $u'' + \frac{x^2 + 2 + 4p^2(1 - x^2)}{4(1 - x^2)^2} u = 0$, on $(-\infty, -1)$, $(-1, 1)$ or $(1, \infty)$

(d) $u'' + (2p + 1 - x^2)u = 0$, on $(-\infty, \infty)$

Section 6–3

1. (a) $\frac{d}{dx}\left[(1 - x^2)\frac{dy}{dx}\right] + 6y = 0$, on $(-\infty, -1)$, $(-1, 1)$ or $(1, \infty)$

(b) $\frac{d}{dx}\left(e^{-x^2}\frac{dy}{dx}\right) + \left(1 - \frac{4}{x^2}\right)e^{-x^2}y = 0$, on $(-\infty, \infty)$

(c) $\frac{d}{dx}\left[(x^3 - 2)^{-1/3}\frac{dy}{dx}\right] - 3(x^3 - 2)^{-4/3}y = 0$, on $(-\infty, -2^{1/3})$, $(-2^{1/3}, 2^{1/3})$ or $(2^{1/3}, \infty)$

(d) $\frac{d}{dx}\left[(-\ln x)^{3/2}\frac{dy}{dx}\right] + \frac{(-\ln x)^{1/2}\sin x}{2x} y$, on $(0, 1)$,

$\frac{d}{dx}\left[(\ln x)^{3/2}\frac{dy}{dx}\right] - \frac{(\ln x)^{1/2}\sin x}{2x} y$, on $(1, \infty)$

(e) $\frac{d}{dx}\left[\frac{1}{x + 1}\frac{dy}{dx}\right] + \frac{2x}{(x + 1)^2} y = 0$, on $(-\infty, -1)$ or $(-1, \infty)$

3. $\frac{d^2y}{dt^2} + (e^{2t} - p^2)y = 0$, on $(-\infty, 0)$ or $(0, \infty)$

5. (a) Every solution has infinitely many zeros on every interval of the form $(-\infty, -a)$ or (a, ∞), where $a > 1$, and does not oscillate on $(-1, 0)$ or $(0, 1)$.

(b) Every solution has infinitely many zeros on every interval of the form $(-\infty, a)$ and does not oscillate on $(1, \infty)$.

(c) Every solution has infinitely many zeros on every interval of the form (a, ∞), where $a \geq 1$, but does not oscillate on $(-\infty, 0)$ or $(0, 1)$.

7. (a) Every solution has infinitely many zeros on every interval of the form (a, ∞), where $a \geq 0$, but does not oscillate on $(-\infty, 0)$.

Section 6–5

1. $y = a_0 y_0 + a_1 y_1$; $y_0 = \sum_{k=0}^{\infty} \frac{(-1)^k}{(2k)!} x^{2k}$,

$$y_1 = \sum_{k=0}^{\infty} \frac{(-1)^k}{(2k+1)!} x^{2k+1}, \text{ on } (-\infty, \infty)$$

3. $y = a_0 y_0 + a_1 y_1$; $y_0 = 1 + \sum_{k=1}^{\infty} \frac{1}{k![2 \cdot 5 \cdot 8 \cdots (3k-1)]} x^{3k}$,

$$y_1 = \sum_{k=0}^{\infty} \frac{1}{k![1 \cdot 4 \cdot 7 \cdots (3k+1)]} x^{3k+1}, \text{ on } (-\infty, \infty)$$

5. $y = a_1(x + x^3) + 3a_0 \sum_{k=0}^{\infty} \frac{(-1)^k}{(2k-3)(2k-1)} x^{2k}$, on $(-1, 1)$

7. $y = a_1(x + \frac{8}{3}x^3)$

$$+ a_0\left[1 + 9x^2 + \tfrac{15}{2}x^4 + 3\sum_{k=3}^{\infty} (-1)^k \frac{(2k-5)!(2k+1)}{2^{k-3}[k!(k-3)!]} x^{2k}\right],$$

on $\left(-\frac{\sqrt{2}}{2}, \frac{\sqrt{2}}{2}\right)$

9. $y = a_0 y_0 + a_1 y_1$, where $y_0 = \sum_{k=0}^{\infty} \frac{(-1)^k}{k!3^k} x^{3k}$,

$$y_1 = \sum_{k=0}^{\infty} \frac{(-1)^k}{1 \cdot (1 + 3 \cdot 1)(1 + 3 \cdot 2) \cdots (1 + 3k)} x^{3k+1}, \text{ on } (-\infty, \infty)$$

11. $y = \frac{1}{6}x^2 + \frac{7}{6}\sum_{k=2}^{\infty} \frac{1}{k!(2k-1)} (\frac{1}{6})^{k-1} x^{2k} + \sum_{k=1}^{\infty} \frac{(k-1)!}{(2k+1)!} (\frac{2}{3})^k x^{2k+1}$

$$+ a_0\left[1 - \frac{1}{6}\sum_{k=1}^{\infty} \frac{1}{k!(2k-1)} (\tfrac{1}{6})^{k-1} x^{2k}\right] + a_1 x, \text{ on } (-\infty, \infty)$$

13. $y = a_0 y_0 + a_1 y_1 + a_2 y_2$, where

$$y_0 = 1 + \sum_{k=1}^{\infty} \frac{1 \cdot 10 \cdot 19 \cdots (9k-8)}{(3k)!} x^{3k},$$

$$y_1 = x + \sum_{k=1}^{\infty} \frac{4 \cdot 13 \cdot 22 \cdots (9k-5)}{(3k+1)!} x^{3k+1},$$

$$y_2 = x^2 + 2\sum_{k=1}^{\infty} \frac{7 \cdot 16 \cdot 25 \cdots (9k-2)}{(3k+2)!} x^{3k+2}, \text{ on } (-\infty, \infty)$$

15. $y = a_0y_0 + a_1y_1$, where

$$y_0 = 1 + \sum_{k=1}^{\infty} \frac{2^{2k}[(0)(-1) - 1][(1)(0) - 1][(2)(1) - 1]\cdots[(k-1)(k-2) - 1]}{(2k)!} x^{2k},$$

$$y_1 = x + \sum_{k=1}^{\infty} \frac{[(1)(-1) - 4][(3)(1) - 4][(5)(3) - 4]\cdots[(2k-1)(2k-3) - 4]}{(2k+1)!} x^{2k+1},$$

on $(-1, 1)$

17. $y = y_p + a_0y_0 + a_1y_1$, where

$$y_p = -\tfrac{1}{8}x^2 - \tfrac{1}{12}x^3 + \tfrac{7}{128}x^4 + \tfrac{1}{60}x^5 - \tfrac{21}{5120}x^6 - \tfrac{1}{5}\sum_{k=4}^{\infty} \frac{(2k-7)!(2k+1)(2k+3)}{2^{4(k-1)}k!(k-4)!} x^{2k},$$

$$y_0 = 1 - \frac{25}{8}x^2 + \frac{175}{128}x^4 - \frac{105}{1024}x^6 - 5\sum_{k=4}^{\infty} \frac{(2k-7)!(2k+1)(2k+3)}{2^{4(k-1)}k!(k-4)!} x^{2k},$$

$y_1 = x - x^3 + \frac{1}{5}x^5$, on $(-2, 2)$

19. $y = a_0 \sum_{k=0}^{\infty} (-1)^{k+1} \frac{k+1}{2^k(3k-1)} x^{3k} + a_1x$, on $(-2^{1/3}, 2^{1/3})$

21. $y = \frac{1}{2}x^2 + \frac{1}{6}x^3 + \frac{1}{24}x^4 + \sum_{k=3}^{\infty} a_{2k}x^{2k} + \sum_{k=2}^{\infty} a_{2k+1}x^{2k+1}$, where

$$a_{2k} = \frac{1}{(2k)!}[1 - 2(k-2) + 2^2(k-2)(k-3) - 2^3(k-2)(k-3)(k-4) + \cdots + (-2)^{k-2}(k-2)!],$$

$$a_{2k+1} = \frac{1}{(2k+1)!}[1 - (2k-3) + (2k-3)(2k-5) - (2k-3)(2k-5) \times (2k-7) + \cdots + (-1)^{k-1}(2k-3)(2k-5)(2k-7)\cdots(1)],$$

on $(-\infty, \infty)$

23. There exists a solution which is a polynomial of degree zero (a nonzero constant) if and only if $\gamma = 0$; there exists a polynomial of degree 1 which is a solution if and only if $\beta + \gamma = 0$; there exists a solution which is a polynomial of degree $n \geq 2$ if and only if $n(n-1) + \beta n + \gamma = 0$ and either $\alpha = 0$ or $k(k-1) + \beta k + \gamma \neq 0$ for all nonnegative integers k which are less than n and differ from n by an even integer.

25. (b) A basis is $\{y_0, y_1\}$ where

$$y_0 = 1 + \sum_{k=1}^{\infty} (-1)^k \frac{2^{k-1}\lambda(\lambda-2)(\lambda-4)\cdots[\lambda - 2(k-1)]}{k[(2k-1)!]} x^{2k},$$

$$y_1 = x + \sum_{k=1}^{\infty} (-1)^k \frac{2^k(\lambda-1)(\lambda-3)(\lambda-5)\cdots[\lambda-(2k-1)]}{(2k+1)!} x^{2k+1},$$

on $(-\infty, \infty)$

Section 6–6

1. $y = 1 - \frac{1}{6}x^3 + \frac{1}{120}x^5 + \frac{1}{180}x^6 + \cdots$, on $(-\infty, \infty)$
3. $y = -1 - x + \frac{1}{2}x^2 - \frac{1}{6}x^3 + \cdots$, on $(-1, 1)$
5. $y = 1 + x + x^2 + \frac{1}{3}x^3 + \cdots$, on $(-\frac{1}{2}, \frac{1}{2})$
7. $y = 6x + \frac{1}{3}x^3 + \frac{2}{9}x^4 + \frac{1}{20}x^5 + \cdots$, on $(-3, 3)$
9. $y = \frac{9}{2}(x - 3)^2 + \frac{1}{40}(x - 3)^5 - \frac{1}{240}(x - 3)^6 + \frac{1}{1260}(x - 3)^7 + \cdots$, on $(0, 6)$
11. $y = \frac{1}{8}x^2 + \frac{1}{18}x^3 + \frac{1}{72}x^4 + \frac{1}{240}x^5 + \cdots$, on $(-\infty, \infty)$
13. $x - \frac{1}{3}x^3 + \frac{1}{10}x^5 - \frac{1}{42}x^7 + \cdots$, on $(-\infty, \infty)$

Chapter 7

Section 7–1

1. (a) -5 (b) $\frac{11}{3}$ (c) 0 (d) -10 (e) 2
3. (a) 0 (b) $\frac{1}{4}(e^2 - 1)$ (c) $3 - e$ (d) 0 (e) $\dfrac{2\pi e - 4}{\pi^2 + 4}$
5. (b)
9. (a) No; $\mathbf{f} \cdot \mathbf{f} = 0$ does not imply $f(x) \equiv 0$ on $[a, b]$.
   (b) Yes; Definition 7–1 applies. (*Note.* If $[P(x)]^2 \equiv 0$ on $[a_1, b_1]$, then $P(x) \equiv 0$ on $[a, b]$.)
11. (b) $\alpha > 0$

Section 7–2

1. (a) 3 (b) $\frac{5}{2}$ (c) $\sqrt{35}$ (d) $\dfrac{5\sqrt{2}}{6}$ (e) $\sqrt{39}$
3. (a) $\dfrac{\sqrt{3}}{3}$ (b) $\sqrt{e - 1}$ (c) $\dfrac{2\sqrt{30}}{15}$ (d) $\dfrac{\sqrt{2}}{2}$ (e) $\sqrt{2}(1 - \ln 2)$
5. (a) (i) 0
   (ii) 0
   (iii) $-\frac{1}{2}$
   (b) (i) $\sqrt{3}/2$
   (ii) $\sqrt{15}/4$
   (iii) $\frac{1}{2}$
7. (a) (i) π, if $m = n \neq 0$; 0 otherwise
   (ii) 0
   (iii) 2π if $m = n = 0$; π if $m = n \neq 0$; 0 if $m \neq n$.
   (b) Orthogonal, that is, the cosine of the angle between any two distinct vectors (functions) is zero.

Section 7–3

5. $\dfrac{\sqrt{2}}{2}, \dfrac{\sqrt{6}}{2}x, \dfrac{3\sqrt{10}}{4}(x^2 - \frac{1}{3})$
7. $A(x^3 - \frac{3}{5}x)$, A any nonzero constant
9. $\pm\dfrac{\sqrt{6}}{3}(1, 0, -1, 2)$
11. $ae^x - \dfrac{a}{2}(e^2 - 1)e^{-x}$, a arbitrary

Section 7–4

1. (a) $(1, 1, 0)$, $(-1, 1, 0)$, $(0, 0, 1)$
   (b) $(\frac{1}{2}, 0, 2)$, $\frac{1}{17}(-20, 51, 5)$, $\frac{55}{356}(12, 5, -3)$
   (c) $(2, -1, 1)$, $\frac{1}{18}(4, -11, -19)$, $\frac{2}{83}(5, 7, -3)$
   (d) $(1, 2, -1)$, $\frac{1}{6}(13, 2, 17)$, $\frac{24}{77}(-6, 5, 4)$
   (e) $(1, 0, 0)$, $(0, 1, 0)$, $(0, 0, 1)$

3. (a) $(1, 0, 0, 1)$, $(-1, 0, 2, 1)$, $\frac{1}{3}(2, 3, 2, -2)$, $\frac{2}{7}(-1, 2, -1, 1)$
   (b) $\frac{1}{2}(4, 0, 1, -2)$, $\frac{1}{21}(-4, 0, 62, 23)$, $\frac{23}{209}(7, 0, -4, 12)$, $(0, 1, 0, 0)$
   (c) $(1, 0, 0, 0)$, $(0, 1, 0, 0)$, $(0, 0, 1, 0)$, $(0, 0, 0, 1)$

5. (a) e^x, $e^{-x} - \dfrac{2}{e^2 - 1}\, e^x$ (b) e^x, $e^{2x} - \dfrac{2(e^2 + e + 1)}{3(e + 1)}\, e^x$
   (c) 1, $2x - 1$, $e^x + 6(e - 3)x + 10 - 4e$

7. $1, x, x^2, x^3, \ldots$

11. (a) $\alpha_i = \dfrac{\mathbf{x} \cdot \mathbf{e}_i}{\mathbf{e}_i \cdot \mathbf{e}_i}$, $i = 1, 2, 3, \ldots, n$ (b) $\mathbf{x} \cdot \mathbf{y} = \displaystyle\sum_{i=1}^{n} \frac{(\mathbf{x} \cdot \mathbf{e}_i)(\mathbf{y} \cdot \mathbf{e}_i)}{(\mathbf{e}_i \cdot \mathbf{e}_i)}$

13. $\pm \dfrac{1}{\sqrt{21}}(4, -1, 2)$

15. (b) $(1, 1, 1)$

17. (a) $P_0(x) = 1$, $P_1(x) = x$, $P_2(x) = \frac{3}{2}(x^2 - \frac{1}{3})$
    $P_3(x) = \frac{5}{2}(x^3 - \frac{3}{5}x)$, $P_4(x) = \frac{35}{8}(x^4 - \frac{6}{7}x^2 + \frac{3}{35})$
    (b) (i) $2P_2(x) - 2P_1(x) + 2P_0(x)$,
    (ii) $-2P_3(x) + 6P_2(x) - 6P_1(x) + P_0(x)$,
    (iii) $3P_3(x) + 3P_2(x) + \frac{7}{2}P_0(x)$

Section 7–5

1. (a) $\dfrac{\sqrt{2}}{2}$ (b) 2 (c) $\frac{51}{40}$ (d) $\dfrac{30\sqrt{29}}{29}$

3. (a) 1 (b) $\dfrac{\sqrt{43}}{2}$ (c) $\frac{2}{17}\sqrt{1785}$ (d) 3

5. The line spanned by $(4, 3)$ and that spanned by $(3, -4)$

7. (a) $\frac{11}{9}\sqrt{3}$ (b) $\frac{6}{497}\sqrt{497}$ (c) 1

9. $(\frac{17}{9}, \frac{4}{9}, \frac{19}{9})$

11. (a) Projection, $2 \sin x$; $\|\mathbf{d}\| = \pi\sqrt{6\pi}/3$
    (b) Projection, $\frac{1}{2} + \frac{1}{2}\cos 2x$; $\|\mathbf{d}\| = 0$
    (c) Projection, $\frac{1}{3}\pi^2 - 4\cos x + \cos 2x$; $\|\mathbf{d}\| = \frac{1}{15}\sqrt{5\pi(8\pi^4 + 2295)}$

13. $a_n = 0$; $b_n = (-1)^{n+1} \cdot \dfrac{2}{n}$

15. $a_0 = 1$, $a_2 = -\frac{1}{2}$, $a_n = 0$ for $n \neq 0, 2$; $b_n = 0$, all n

17. $a_0 = \pi$, $a_n = \dfrac{2}{n^2\pi}[(-1)^n - 1]$, when $n \neq 0$; $b_n = 0$

19. $a_n = \dfrac{(-1)^n(e^\pi - e^{-\pi})}{(n^2 + 1)\pi}$, $b_n = \dfrac{(-1)^{n+1}n(e^\pi - e^{-\pi})}{(n^2 + 1)\pi}$

21. (a) Projection, $\frac{1 + (-1)^n}{2(n+1)} \cdot 1 + \frac{3[1 - (-1)^n]}{2(n+2)} x + \frac{15n[1 + (-1)^n]}{4(n+1)(n+3)} (x^2 - \frac{1}{3})$;

$$\|\mathbf{d}\| = \left\{\frac{2}{2n+1} - \frac{3(2n^2 + 2n + 3)}{(n+1)^2(n+3)^2}[1 + (-1)^n] - \frac{3}{(n+2)^2}[1 - (-1)^n]\right\}^{1/2}$$

(b) Projection, $3(\sin 1 - \cos 1)x$; $\|\mathbf{d}\| = \frac{1}{2}\sqrt{22 \sin 2 - 20}$

(c) Projection, $\frac{1}{2} \cdot 1 + \frac{15}{16}(x^2 - \frac{1}{3})$; $\|\mathbf{d}\| = \frac{\sqrt{6}}{24}$

(d) Projection, $1 + 3(x^2 - \frac{1}{3})$; $\|\mathbf{d}\| = 0$

23. If $\mathcal{X} = \mathcal{V}$, then $\mathcal{X}^\perp$ contains only $\mathbf{0}$; if $\mathcal{X}$ contains only $\mathbf{0}$, then $\mathcal{X}^\perp = \mathcal{V}$.

27. (a) $\alpha\mathbf{y} + \beta\mathbf{z}$, where $\mathbf{y}$ and $\mathbf{z}$ are linearly independent and orthogonal to $(1, 1, 0)$ and α, β are arbitrary reals; for example:

$$\alpha(1, -1, 0) + \beta(0, 0, 1)$$

(b) $\alpha(1, 1, 0)$, α an arbitrary real

(c) $\alpha\mathbf{y} + \beta\mathbf{z}$, where $\mathbf{y}$ and $\mathbf{z}$ are linearly independent and orthogonal to $(1, 2, -1)$ and α, β are arbitrary reals; for example:

$$\alpha(1, 0, 1) + \beta(-2, 1, 0)$$

(d) $\alpha(1, -1, 0)$, α an arbitrary real

Section 7–6

1. (a) A: 1.4082; B: 1.4136
   (b) B's data (Variances 9.056×10^{-5} and 4.744×10^{-5} for A and B, respectively)

3. (a) A: 16.06, variance 0.00236; B: 16.12, variance 0.00092; B's results are the more consistent.
   (b) 16.09, variance 0.00254, greater than for either A's or B's data alone

5. (2.25, 2)

7. (a) $(-0.8, 0.4, 1.7)$ (b) $(3.4, -1.3, 3.6)$ (c) $(1.9, 0.4, 0.8)$

9. (b) $m_1 = 1$, $m_2 = 5$, $m_3 = 3$, $m_4 = 1$

13. (a) $54x - 35y = 0$ (b) $417x - 142y = 0$ (c) $26x + 17y = 0$

15. $y = \frac{4}{3}x_1 + \frac{5}{3}x_2$

17. $y = x_1 - x_2$

19. $y = \frac{1}{2}x_1 - \frac{1}{4}x_2 - \frac{7}{4}x_3$

21. (a) $y = -\frac{5}{22}x^2 + \frac{69}{22}x$
    (b) $y = \frac{65}{34}x^2 - \frac{1}{10}$
    (c) $y = -\frac{9}{37}x^2 - \frac{36}{37}x$

23. $y = -\frac{3}{8}x^4 + \frac{15}{8}x^2 - \frac{1}{10}$

Chapter 8

Section 8–1

1. $\lim_{k\to\infty} \{a_k\} = 0$

3. $\{a_k\}$ does not converge.

5. $\lim_{k\to\infty} \{a_k\} = 1$

7. $\lim_{k\to\infty} \{a_k\} = 2$

9. $\lim_{k\to\infty} \{x_k\} = \mathbf{e}_1 + \mathbf{e}_2 + \mathbf{e}_3$

11. $\{x_k\}$ does not converge.

Section 8–3

3. With respect to the first set

$$c_0 = \frac{e^{\pi} - e^{-\pi}}{\sqrt{2\pi}};$$

$$c_k = \frac{(-1)^k}{(1 + k^2)\sqrt{\pi}}(e^{\pi} - e^{-\pi}), k = 1, 2, 3, \ldots$$

With respect to the second set

$$c_k = (-1)^{k+1}\frac{k}{(1 + k^2)\sqrt{\pi}}(e^{\pi} - e^{-\pi}),\ k = 1, 2, 3, \ldots$$

Section 8–4

3. $f(x) = \frac{a_0}{2} + \sum_{k=1}^{\infty}(a_k \cos kx + b_k \sin kx)$, where

$$a_k = \frac{1}{\pi}\int_{-\pi}^{\pi} f(x) \cos kx\, dx, \qquad k = 0, 1, 2, 3, \ldots,$$

$$b_k = \frac{1}{\pi}\int_{-\pi}^{\pi} f(x) \sin kx\, dx, \qquad k = 1, 2, 3, \ldots$$

Section 8–5

3. (b) ℓ_2

13. (b) The vector $\mathbf{y} = (y_1, y_2, y_3, \ldots)$ is in $S(\mathfrak{X})$ if and only if for some integer $k \geq 2$, $y_{k+1} = y_{k+2} = y_{k+3} = \cdots = 0$ and $y_1 = y_2 + y_3 + \cdots + y_k$.

Chapter 9

Section 9–2

1. (a) $f(1^-) = 1$, $f(1^+) = -1$; in $\mathcal{PC}[0, 2]$
   (b) $f(0^+)$ does not exist; not in $\mathcal{PC}[0, 2]$
   (c) $f(1^-) = 0$, $f(1^+) = 1$; in $\mathcal{PC}[0, 2]$
   (d) $f(1^-) = f(1^+) = 0$; in $\mathcal{C}[0, 2]$ and so in $\mathcal{PC}[0, 2]$
   (e) $f(\frac{1}{2}^-) = 1$, $f(\frac{1}{2}^+) = f(\frac{3}{2}^-) = 0$, $f(\frac{3}{2}^+) = \frac{1}{2}$; in $\mathcal{PC}[0, 2]$
   (f) $f(1^-)$ does not exist; $f(1^+) = 2$; not in $\mathcal{PC}[0, 2]$

3. (a) In $\mathcal{C}[0, 1]$ and so in $\mathcal{PC}[0, 1]$
   (b) Not in $\mathcal{PC}[0, 1]$ (Infinitely many discontinuities)
   (c) Not in $\mathcal{PC}[0, 1]$ (Infinitely many discontinuities)

7. Yes; a linear combination would vanish identically only if all the coefficients were zero.

Section 9–3

1. (a) Odd (b) Even (c) Neither (d) Even (e) Odd
   (f) Odd (g) Neither (h) Even (i) Neither (j) Even

7. (a) $f_E(x) = \dfrac{x^2+1}{x^2-1}$, $f_O(x) = \dfrac{2x}{x^2-1}$

(b) $f_E(x) = \dfrac{1}{1-x^2}$, $f_O(x) = -\dfrac{x}{1-x^2}$

(c) $f_E(x) = -\cos 2x$, $f_O(x) = x\cos x$

(d) $f_E(x) = \dfrac{x^2+1}{x^2-1}$, $f_O(x) = \dfrac{x^3+x}{x^2-1}$

(e) $f_E(x) = \dfrac{ax^2+c}{a^2x^4+(2ac-b^2)x^2+c^2}$,

$f_O(x) = -\dfrac{bx}{a^2x^4+(2ac-b^2)x^2+c^2}$

9. $f_E(x) = \dfrac{a_0}{2} + \sum_{n=1}^{\infty} a_n \cos nx$, $f_O(x) = \sum_{n=1}^{\infty} b_n \sin nx$

Section 9–4

1. $2\sum_{k=1}^{\infty} \dfrac{(-1)^{k+1}}{k} \sin kx$

3. $\dfrac{e^{\pi}-e^{-\pi}}{\pi}\left[\dfrac{1}{2} + \sum_{k=1}^{\infty} \dfrac{(-1)^k}{1+k^2}(\cos kx - k\sin kx)\right]$

5. $\dfrac{1}{4\pi} - \dfrac{1}{2} + \sum_{n=1}^{\infty}\left\{\dfrac{1}{n\pi}\sin\dfrac{n}{2}\cos nx + \left[(-1)^{n+1}\dfrac{2}{n} + \dfrac{1}{n\pi}\cos\dfrac{n}{2} - \dfrac{2}{n^2\pi}\sin\dfrac{n}{2}\right]\sin nx\right\}$

7. $\frac{2}{3}\pi^2 + 4\sum_{k=1}^{\infty} \dfrac{(-1)^{k+1}}{k^2}\cos kx$

9. $-2\sum_{k=1}^{\infty} \dfrac{1}{k}\sin kx$

11. (a) On $\left[2\pi, \dfrac{5\pi}{2}\right)$ and $\left(\dfrac{5\pi}{2}, 3\pi\right)$ the series converges to -1 and 1, respectively; at $x = 5\pi/2$ and $x = 3\pi$ the series converges to zero. On $[-2\pi, -3\pi/2)$ the series converges to 1, on $(-3\pi/2, -\pi/2)$ to -1, and on $(-\pi/2, 0]$ to 1. At $x = -3\pi/2$ and $x = -\pi/2$ the series converges to zero.
(b) When $x = k\pi$, the series converges to ± 1, according as k is even or odd; when $x = (2k+1)\pi/2$, the series converges to zero.

13. (b) When $x = k\pi$, the series converges to $\pi/3$; when $x = (4k+1)\pi/3$ the series converges to $\pi/6$ or $\pi/3$, according as k is or is not an integral multiple of 3.

15. (a) $\dfrac{\pi}{4} + \sum_{k=1}^{\infty}\left(\dfrac{(-1)^k - 1}{k^2\pi}\cos kx + \dfrac{(-1)^{k+1}}{k}\sin kx\right)$.

17. (a) $\cos \alpha x = \dfrac{\sin \alpha\pi}{\pi}\left[\dfrac{1}{\alpha} + 2\alpha\sum_{k=1}^{\infty}\dfrac{(-1)^{k+1}}{k^2-\alpha^2}\cos kx\right]$.

19. (a) Periodic; fundamental period 8
(b) Periodic; fundamental period 2π
(c) Not periodic
(d) Periodic; fundamental period 2π
(e) Not periodic
(f) Not periodic
(g) Periodic, every rational number a period; no fundamental period

23. (a) $-\dfrac{2}{\pi}\displaystyle\sum_{k=1}^{\infty}\frac{\sin(2k-1)x}{2k-1}$
(b) $\dfrac{\pi}{2}-\displaystyle\sum_{k=1}^{\infty}\frac{\sin 2kx}{k}$

(c) $\pi + 2\displaystyle\sum_{k=1}^{\infty}\frac{\sin kx}{k}$

25. (a) $\frac{1}{2}-\frac{1}{2}\cos 2x$
(b) $\frac{1}{2}\sin 2x$
(c) $\frac{3}{4}\sin x - \frac{1}{4}\sin 3x$
(d) $\frac{1}{2}\sin x + \frac{1}{4}\sin 2x$
(e) $\frac{1}{2}\cos x + \frac{1}{2}\cos 5x$
(f) $\frac{3}{8}+\frac{1}{2}\cos 2x + \frac{1}{8}\cos 4x$

27. Let $f(x) \sim \dfrac{a_0}{2} + \displaystyle\sum_{k=1}^{\infty}(a_k\cos kx + b_k \sin kx)$; then

(a) $f(-x) \sim \dfrac{a_0}{2} + \displaystyle\sum_{k=1}^{\infty}(a_k\cos kx - b_k\sin kx)$;

(b) $f_E(x) \sim \dfrac{a_0}{2} + \displaystyle\sum_{k=1}^{\infty} a_k\cos kx,\quad f_O(x) \sim \sum_{k=1}^{\infty} b_k \sin kx.$

(c) 1, $\cos x$, $\cos 2x$, ... and $\sin x$, $\sin 2x$, ... are bases for $\mathcal{E}[-\pi,\pi]$ and $\mathcal{O}[-\pi,\pi]$, respectively.

Section 9–5

1. (i) $E_f = 1,\ -\pi \le x \le \pi,$

$$O_f = \begin{cases} -1, & -\pi \le x < 0, \\ 1, & 0 < x \le \pi \end{cases}$$

(ii) $E_f = x^2,\ -\pi \le x \le \pi,$

$$O_f = \begin{cases} -x^2, & -\pi \le x < 0 \\ x^2, & 0 < x \le \pi \end{cases}$$

(iii) $$E_f = \begin{cases} \pi + x, & -\pi \le x \le 0, \\ \pi - x, & 0 \le x \le \pi \end{cases}$$

$$O_f = \begin{cases} -\pi - x, & -\pi \le x < 0, \\ \pi - x, & 0 < x \le \pi \end{cases}$$

(iv) $$E_f = \begin{cases} e^{-x}, & -\pi \le x \le 0, \\ e^{x}, & 0 \le x \le \pi \end{cases}$$

$$O_f = \begin{cases} -e^{-x}, & -\pi \le x < 0, \\ e^{x}, & 0 < x \le \pi \end{cases}$$

3. $\dfrac{8}{\pi}\displaystyle\sum_{k=1}^{\infty}\frac{k}{4k^2-1}\sin 2kx$

5. $\dfrac{2}{\pi}\displaystyle\sum_{k=1}^{\infty}\frac{k}{1+k^2}[1+(-1)^{k+1}e^{\pi}]\sin kx$

7. $2\displaystyle\sum_{k=1}^{\infty}\frac{1}{k}\sin kx$

9. $4\displaystyle\sum_{k=1}^{\infty}\frac{1}{(2k-1)^2}\left[(-1)^{k+1}-\frac{2}{\pi(2k-1)}\right]\sin(2k-1)x$

11. $$A_k = \begin{cases} \dfrac{4}{\pi}\displaystyle\int_0^{\pi/2} f(x)\cos kx\,dx & \text{if } k = 1, 3, 5, \ldots, \\ 0 & \text{if } k = 0, 2, 4, 6, \ldots \end{cases}$$

13. $-\cos x$

Section 9–6

3. $f(x) = \frac{1}{4} + \sum_{k=1}^{\infty} A_k \cos \frac{k\pi x}{2}$, where

$$A_k = \begin{cases} \frac{2}{k\pi} - \frac{4}{k^2\pi^2} & \text{if } k = 1, 5, 9, \ldots, \\ -\frac{8}{k^2\pi^2} & \text{if } k = 2, 6, 10, \ldots, \\ -\frac{2}{k\pi} - \frac{4}{k^2\pi^2} & \text{if } k = 3, 7, 11, \ldots, \\ 0 & \text{if } k = 4, 8, 12, 16, \ldots \end{cases}$$

5. $\frac{2}{\pi} - \frac{4}{\pi} \sum_{k=1}^{\infty} \frac{1}{16k^2 - 1} (\cos kx + 4k \sin kx),\ 0 < x < \frac{\pi}{2}$

7. $\frac{3}{4} + \sum_{k=1}^{\infty} \left[\frac{(-1)^k - 1}{k^2\pi^2} \cos k\pi x + \frac{1}{k\pi} \sin k\pi x \right],\ 8 < x < 10$

9. $-\frac{2}{\pi} \sum_{k=1}^{\infty} \frac{1}{k} \sin k\pi x,\ 1 < x < 2$

11. $\frac{16}{\pi^3} \sum_{k=1}^{\infty} \left(\frac{(-1)^{k+1} 3\pi}{(2k-1)^2} - \frac{2}{(2k-1)^3} \right) \sin \frac{(2k-1)\pi x}{2},\ 2 < x < 3$

13. $-2 \sum_{k=1}^{\infty} \frac{1}{k} \sin kx,\ -\pi < x < \pi$

Section 9–8

1. (b) $\|f_i(x) g_j(y)\| = \|f_i(x)\| \, \|g_j(y)\|$

3. If $F(-x, y) = F(x, y)$ and $F(x, -y) = F(x, y)$, then

$$F(x, y) = \sum_{p,q=0}^{\infty} a_{pq} \cos px \cos qy.$$

If $F(-x, y) = -F(x, y)$ and $F(x, -y) = -F(x, y)$, then

$$F(x, y) = \sum_{m,n=1}^{\infty} a_{mn} \sin mx \sin ny.$$

5. $2 \sum_{m=1}^{\infty} \frac{(-1)^{m+1}}{m} \sin mx$

7. $\frac{1}{2} + \frac{1}{2} \cos 2x \cos 2y - \frac{1}{2} \sin 2x \sin 2y$

9. $\frac{2\pi^2}{3} \sum_{m=1}^{\infty} \frac{(-1)^{m+1}}{m} \sin mx + 8 \sum_{m,q=1}^{\infty} \frac{(-1)^{m+q+1}}{mq^2} \sin mx \cos qy$

Chapter 10

Section 10–3

1. $A_0 = 2\sum_{k=1}^{\infty} b_k/k; \qquad A_k = -b_k/k, \quad k \neq 0; \qquad B_k = a_k/k$

5. (a) $|f(x_1) - f(x_2)| < M|x_1 - x_2|^{\alpha}$, where
   (i) $M > 0, \alpha > 0$; (ii) $M > 1, \alpha = 1$; (iii) $M > 2, \alpha = 1$;
   (iv) $M > 2\pi, \alpha = 1$; (v) $M > 3\pi^2, \alpha = 1$.
   (c) Only constant functions satisfy such a condition.

Section 10–6

3. No; the formula $x/2 = \sum_{k=1}^{\infty} (-1)^{k+1} (\sin kx/k)$ is invalid outside $(-\pi, \pi)$.

Section 10–7

1. With obvious conventions it is a vector space.
3. (a) 0 (b) 0

Chapter 11

Section 11–2

1. $P_1(x) = x$; $P_2(x) = \frac{3}{2}x^2 - \frac{1}{2}$; $P_3(x) = \frac{5}{2}x^3 - \frac{3}{2}x$;
   $P_4(x) = \frac{35}{8}x^4 - \frac{15}{4}x^2 + \frac{3}{8}$; $P_5(x) = \frac{63}{8}x^5 - \frac{35}{4}x^3 + \frac{15}{8}x$

Section 11–3

1. $P_3(x) = \frac{5}{2}x^3 - \frac{3}{2}x$; $P_4(x) = \frac{35}{8}x^4 - \frac{15}{4}x^2 + \frac{3}{8}$
3. $P_6(x) = \frac{231}{16}x^6 - \frac{315}{16}x^4 + \frac{105}{16}x^2 - \frac{5}{16}$

Section 11–4

1. If $f(x)$ is an even function in $\mathcal{PC}[-1, 1]$ then

$$f(x) = \sum_{k=0}^{\infty} a_{2k}P_{2k}(x) \quad \text{(mean)},$$

   where $a_{2k} = (4k + 1)\int_0^1 f(x)P_{2k}(x)\,dx$.
   If $f(x)$ is an odd function in $\mathcal{PC}[-1, 1]$, then

$$f(x) = \sum_{k=0}^{\infty} a_{2k+1}P_{2k+1}(x) \quad \text{(mean)},$$

   where $a_{2k+1} = (4k + 3)\int_0^1 f(x)P_{2k+1}(x)\,dx$.

3. $$|x| = \frac{1}{2} + \sum_{k=1}^{\infty} (-1)^{k+1} \frac{(2k - 2)!(4k + 1)}{2^{2k}(k - 1)!(k + 1)!} P_{2k}(x), \quad -1 < x < 1$$

5. (a) $\frac{3}{5}P_1(x) + \frac{2}{5}P_3(x)$
   (b) $2P_0(x) - \frac{6}{35}P_1(x) + \frac{2}{45}P_3(x) + \frac{8}{63}P_5(x)$
   (c) $\frac{22}{15}P_0(x) - P_1(x) + \frac{76}{21}P_2(x) + \frac{32}{35}P_4(x)$

Section 11–5

1. $n^{1/2}\left[\int_{-1}^{1} g(x)P_n(x)\,dx\right] = \left(n\cdot\frac{2}{2n+1}\right)^{1/2}\left(\frac{2n+1}{2}\right)^{1/2}\left|\int_{-1}^{1} g(x)P_n(x)\,dx\right|$

Section 11–6

1. All the functions belong to $\mathcal{G}_2^w$, except those in (a) and (g).
7. (b) $H_6(x) = x^6 - 15x^4 + 45x^2 - 15,$
   $H_7(x) = x^7 - 21x^5 + 105x^3 - 105x$

Section 11–7

9. (b) $L_0^{(2)} = 1,$ $\quad L_2^{(2)} = x^2 - 8x + 12,$
   $L_1^{(2)} = x - 3,$ $\quad L_3^{(2)} = x^3 - 15x^2 + 60x - 60$

Chapter 12

Section 12–1

1. $y = \cos x + c \sin x$
3. No solutions
5. $y = c\cos 3x$
7. No solutions
9. $\lambda_n = n^2, n = 0, 1, 2, \ldots;\ y_n = c_1 \cos nx + c_2 \sin nx$

Section 12–2

1. $\mathcal{V}$ and the trivial subspace

Section 12–3

1. Using standard basis $\mathbf{e}_1, \mathbf{e}_2$:
   (a) $\lambda = 1$, with eigenvectors of form $x_1\mathbf{e}_1$;
   $\lambda = 2$, with eigenvectors of form $x_1(\mathbf{e}_1 + \mathbf{e}_2)$
   (b) $\lambda = 0$, with eigenvectors of form $x_2\mathbf{e}_2$;
   $\lambda = 1$, with eigenvectors of form $x_1\mathbf{e}_1$
   (c) $\lambda = 0$, with eigenvectors of form $x_1(\mathbf{e}_1 - \mathbf{e}_2)$;
   $\lambda = 2$, with eigenvectors of form $x_1(\mathbf{e}_1 + \mathbf{e}_2)$
   (d) $\lambda = 3$, with eigenvectors of form $x_2(2\mathbf{e}_1 + \mathbf{e}_2)$;
   $\lambda = -1$, with eigenvectors of form $x_2(-2\mathbf{e}_1 + \mathbf{e}_2)$

3. Using standard basis $\mathbf{e}_1, \mathbf{e}_2, \mathbf{e}_3$:
   (a) $\lambda = 0$, with eigenvectors of form $x_1(\mathbf{e}_1 - \mathbf{e}_3)$;
   $\lambda = 1$, with eigenvectors of form $x_2\mathbf{e}_2$;
   $\lambda = 2$, with eigenvectors of form $x_1(\mathbf{e}_1 + \mathbf{e}_3)$
   (b) $\lambda = 2$, with eigenvectors of form $x_2\mathbf{e}_2$;
   $\lambda = \sqrt{2}$, with eigenvectors of form $x_1(\mathbf{e}_1 + [\sqrt{2} - 1]\mathbf{e}_3)$;
   $\lambda = -\sqrt{2}$, with eigenvectors of form $x_1(\mathbf{e}_1 - [1 + \sqrt{2}]\mathbf{e}_3)$
   (c) $\lambda = 1$, with eigenvectors of the form $x_2(-3\mathbf{e}_1 + \mathbf{e}_2 - 3\mathbf{e}_3)$;
   $\lambda = 2$, with eigenvectors of the form $x_2(2\mathbf{e}_1 + \mathbf{e}_2) + x_3(2\mathbf{e}_1 + \mathbf{e}_3)$

(d) $\lambda = 0$, with eigenvectors of the form $x_3(3\mathbf{e}_1 - 2\mathbf{e}_2 + \mathbf{e}_3)$;

$\lambda = \dfrac{5 + \sqrt{5}}{2}$, with eigenvectors of the form $x_2\left(\mathbf{e}_1 + \mathbf{e}_2 + \dfrac{\sqrt{5} - 1}{2}\mathbf{e}_3\right)$;

$\lambda = \dfrac{5 - \sqrt{5}}{2}$, with eigenvectors of the form $x_2\left(\mathbf{e}_1 + \mathbf{e}_2 - \dfrac{1 + \sqrt{5}}{2}\mathbf{e}_3\right)$

5. (a) $\mathbf{x} = -\frac{2}{3}\mathbf{e}_1 + \mathbf{e}_2 + \frac{4}{3}\mathbf{e}_3$ (b) $\mathbf{x} = \frac{1}{3}\mathbf{e}_1 + \frac{1}{3}\mathbf{e}_3$
   (c) $\mathbf{x} = -\frac{8}{3}\mathbf{e}_1 - 2\mathbf{e}_2 + \frac{10}{3}\mathbf{e}_3$
9. (b) $\lambda = 0$ is an eigenvalue of multiplicity $n + 1$.
   $\mathcal{S}_0$ is the subspace of constant functions; $\dim \mathcal{S}_0 = 1$.

Section 12–4

1. (a) $\lambda_1 = \frac{1}{2}[c + a + \sqrt{(c - a)^2 + 4b^2}\,]$,
   $\lambda_2 = \frac{1}{2}[c + a - \sqrt{(c - a)^2 + 4b^2}\,]$
   (b) $[c - a + \sqrt{(c - a)^2 + 4b^2}\,]\mathbf{e}_1 - 2b\mathbf{e}_2$,
   $-2b\mathbf{e}_1 + [c - a - \sqrt{(c - a)^2 + 4b^2}\,]\mathbf{e}_2$
   [*Note.* If $b = 0$, one of these vectors is the zero vector; in this case the required basis is $\{\mathbf{e}_1, \mathbf{e}_2\}$.]
3. $\lambda = n^2$, $n = 0, 1, 2, \ldots$
5. (b) The null space is the set of all functions $F(k)$ in $\mathcal{F}$ of the form
   $$F(k) = c_1k + c_2,\quad k = 0, \pm 1, \pm 2, \ldots,$$
   where c_1 and c_2 are arbitrary real constants.

Section 12–5

3. $\lambda = n^2$, $n = 0, 1, 2, 3, \ldots$, with eigenfunctions of the form
   $y = c \cos nx \quad (c \neq 0)$,
   $\lambda = \dfrac{(2n + 1)^2}{4}$, $n = 0, 1, 2, 3, \ldots$ with eigenfunctions of the form
   $y = c \sin \dfrac{(2n + 1)x}{2} \quad (c \neq 0)$

Section 12–7

1. $y = \dfrac{128}{\pi}\displaystyle\sum_{n=0}^{\infty} \frac{1}{(2n + 1)^5} \sin \frac{2n + 1}{2} x$
3. No solution; $\lambda_0 = 0$, $c_0 = -\dfrac{2}{\pi} \neq 0$
5. $y = \displaystyle\sum_{k=1}^{\infty}\left[\frac{(-1)^{k+1}4}{(2k - 1)^4\pi} - \frac{1}{(2k - 1)^3}\right]\sin(2k - 1)x + \left[\frac{1 + (-1)^{k+1}}{8k^3}\right]\sin 2kx$
7. If $\int_0^{2\pi} h(x)\,dx \neq 0$, there is no solution; if $\int_0^{2\pi} h(x)\,dx = 0$, then
   $$y = c + \sum_{n=1}^{\infty} a_n \cos nx + b_n \sin nx,$$
   where
   $$a_n = \frac{1}{\pi}\int_0^{2\pi} h(x) \cos nx\,dx, \qquad b_n = \frac{1}{\pi}\int_0^{2\pi} h(x) \sin nx\,dx,$$
   and c is arbitrary.

Section 12–8

1. $\lambda_n = n^2 - 1$, $\varphi_n(x) = \sin nx$, $n = 1, 2, 3, \ldots$; orthogonal in $\mathcal{C}[0, \pi]$
3. $\lambda_n = -n^2\pi^2$, $\varphi_n(x) = e^{-x} \sin n\pi x$, $n = 1, 2, 3, \ldots$; orthogonal in $\mathcal{C}[0, 1]$ with respect to the weight function e^{2x}
5. $$\lambda_n = n^2\pi^2,\quad \varphi_n(x) = \begin{cases} e^{x/2} \sin \dfrac{n\pi x}{2}, & n = 2, 4, 6, \ldots, \\ e^{x/2} \cos \dfrac{n\pi x}{2}, & n = 1, 3, 5, \ldots; \end{cases}$$
   orthogonal in $\mathcal{C}[-1, 1]$ with respect to the weight function e^{-x}
7. $\lambda_n = -n^2, \varphi_n(x) = e^{-x}(n \cos nx + \sin nx), n = 1, 2, 3, \ldots,$ and $\lambda = 1, \varphi(x) = 1$; orthogonal in $\mathcal{C}[0, \pi]$ with respect to the weight function e^{2x}
9. Here the set S of eigenvalues is not a sequence but a region of the complex plane, comprising those points $\lambda = \xi + i\eta$ such that $-1 < \xi$ and $|\eta| < 2(\xi + 1)^{1/2}$, with the addition of one boundary point, $\lambda = -1$. To obtain the eigenfunction $\varphi(\lambda; x)$ corresponding to a given $\lambda \neq 0$ of S, set $\lambda = \rho e^{i\theta}$ (where $\rho = [\xi^2 + \eta^2]^{1/2}$, $\xi = \rho \cos \theta$, $\eta = \rho \sin \theta$, $0 \leqq \theta < 2\pi$) and let $\alpha + i\beta = \rho^{1/2}e^{i(\theta-\pi)/2}$, so that $\alpha + i\beta = (-\lambda)^{1/2}$; then on $0 < x$,
   $\varphi(\lambda; x) = x^{1+\alpha}[\cos(\beta \ln x) + i \sin(\beta \ln x)] - x^{1-\alpha}[\cos(\beta \ln x) - i \sin(\beta \ln x)]$;
   also, $\varphi(0, x) = x \ln x$, $0 < x$.

Section 12–9

1. (a) One-to-one (b) Not one-to-one (c) One-to-one
   (d) One-to-one (e) Not one-to-one
7. $$K(x, t) = \begin{cases} \frac{1}{3}(x + 1)(t - 2), & x \leq t, \\ \frac{1}{3}(x - 2)(t + 1), & x \geq t. \end{cases}$$

Section 12–10

3. (a) $$K(x, t) = \begin{cases} \dfrac{1}{k}(\cot k \sin kt - \cos kt) \sin kx, & x \leq t, \\ \dfrac{1}{k}(\cot k \sin kx - \cos kx) \sin kt, & x \geq t \end{cases}$$
   [*Note.* The constant k may be chosen as any real number except that k must not be an integral multiple of π, since then $L = D^2 + k^2$ is not one-to-one when restricted to $\mathcal{S}$.]

   (b) $$y(x) = \frac{1}{k}\left\{(\cot k \sin kx - \cos kx)\int_0^x h(t) \sin kt\, dt + (\sin kx)\int_x^1 h(t)(\cot k \sin kt - \cos kt)\, dt\right\}$$

Chapter 13

Section 13–2

1. (a) $a(x, y, z)D_x + b(x, y, z)D_y + c(x, y, z)D_z + d(x, y, z)$
   (b) $a(x, y)D_{xx} + b(x, y)D_{xy} + c(x, y)D_{yy} + d(x, y)D_x + e(x, y)D_y + f(x, y)$

Section 13–4

1. $u(x, t) = \cos \dfrac{n\pi at}{L} \sin \dfrac{n\pi x}{L}$

3. $u(x, t) = \frac{4L^2}{\pi^3} \sum_{n=1}^{\infty} \frac{1}{n^3} [1 + (-1)^{n+1}] \cos \frac{n\pi at}{L} \sin \frac{n\pi x}{L}$

5. $u(x, t) = \sum_{n=1}^{N} \left(A_n \cos \frac{n\pi at}{L} + \frac{L}{n\pi a} B_n \sin \frac{n\pi at}{L} \right) \sin \frac{n\pi x}{L}$

7. $u(x, t) = \frac{8d}{\pi^2} \sum_{n=1}^{\infty} \frac{1}{n^2} \sin \frac{n\pi}{2} \cos \frac{n\pi at}{L} \sin \frac{n\pi x}{L}$

$$= \frac{4d}{\pi^2} \sum_{n=1}^{\infty} \frac{1}{n^2} \left[\sin n\pi \left(\frac{1}{2} + \frac{at}{L} \right) + \sin n\pi \left(\frac{1}{2} - \frac{at}{L} \right) \right] \sin \frac{n\pi x}{L}$$

When $t = L/(2a)$ or $t = 3L/(2a)$, the string is in the position of equilibrium ($y = 0$); when $t = L/a$, it is in a position which is the reflection in the x-axis of its initial position; when $t = 2L/a$, its position is identical with its initial position.

9. (a) $u(x, y) = \frac{2L}{\pi^2 a\rho\delta} \sum_{n=1}^{\infty} \frac{1}{n^2} \sin \frac{n\pi}{2} \sin \frac{n\pi\delta}{L} \sin \frac{n\pi at}{L} \sin \frac{n\pi x}{L}$

(b) $\lim_{\delta \to 0} u(x, y) = \frac{2}{\pi a\rho} \sum_{n=1}^{\infty} \frac{1}{n} \sin \frac{n\pi}{2} \sin \frac{n\pi at}{L} \sin \frac{n\pi x}{L}$

The vibration represented in (b) may be interpreted as that approximated by a string when it is given a sharp blow at the point $x = L/2$.

11. (a) $\varphi(x, t) = \sum_{n=1}^{\infty} \left(C_n \cos \frac{n\pi at}{L} + D_n \sin \frac{n\pi at}{L} \right) \sin \frac{n\pi x}{L}$,

where

$$C_n = \frac{2}{L} \int_0^L f(x) \sin \frac{n\pi x}{L}\, dx, \quad D_n = \frac{2}{n\pi a} \int_0^L g(x) \sin \frac{n\pi x}{L}\, dx$$

(b) $\varphi(x, t) = \frac{1}{2}(A_0 + B_0 t) + \sum_{n=1}^{\infty} \left(A_n \cos \frac{n\pi at}{L} + B_n \sin \frac{n\pi at}{L} \right) \cos \frac{n\pi x}{L}$,

where

$$A_n = \frac{2}{L} \int_0^L f(x) \cos \frac{n\pi x}{L}\, dx, \quad n = 0, 1, 2, \ldots$$

$$B_0 = \frac{2}{L} \int_0^L g(x)\, dx, \quad B_n = \frac{2}{n\pi a} \int_0^L g(x) \cos \frac{n\pi x}{L}\, dx, \quad n = 1, 2, \ldots$$

(c) $\varphi(x, t) = \sum_{n=1}^{\infty} \left[A_{2n-1} \cos \frac{(2n-1)\pi at}{2L} + B_{2n-1} \sin \frac{(2n-1)\pi at}{2L} \right] \sin \frac{(2n-1)\pi x}{2L}$,

where

$$A_{2n-1} = \frac{2}{L} \int_0^L f(x) \sin \frac{(2n-1)\pi x}{2L}\, dx,$$

$$B_{2n-1} = \frac{4}{(2n-1)\pi a} \int_0^L g(x) \sin \frac{(2n-1)\pi x}{2L}\, dx$$

Section 13–6

1. $u(x, t) = \frac{4L}{\pi^2} \sum_{m=0}^{\infty} \frac{1}{(2m+1)^2} e^{-[(2m+1)^2\pi^2 t/L^2a^2]} \sin \frac{(2m+1)\pi x}{L}$

3. $u(x, t) = \frac{8L^2}{\pi^3} \sum_{m=0}^{\infty} \frac{1}{(2m+1)^3} e^{-[(2m+1)^2\pi^2 t/L^2a^2]} \sin \frac{(2m+1)\pi x}{L}$

5. $u(x, t) = \frac{2k}{\pi} \left\{ 1 - 2 \sum_{m=1}^{\infty} \frac{1}{4m^2 - 1} e^{-(4m^2\pi^2 t/L^2a^2)} \cos \frac{2m\pi x}{L} \right\}$

7. $u(x, t) = \frac{A_0}{2} + \sum_{n=1}^{\infty} A_n e^{-(n^2\pi^2 t/L^2a^2)} \cos \frac{n\pi x}{L},$

where

$$A_n = \frac{2}{L} \int_0^L f(x) \cos \frac{n\pi x}{L}\, dx, \quad n = 0, 1, 2, 3, \ldots$$

9. $u = 100 \left(1 - \frac{x}{L} \right) + \frac{200}{\pi} \sum_{n=1}^{\infty} \frac{1}{n(4n^2 - 1)} e^{-(n^2\pi^2 t/L^2a^2)} \sin \frac{n\pi x}{L}$

11. (a) $u(x, t) = \frac{x}{L} + \frac{2}{\pi} \sum_{n=1}^{\infty} \frac{(-1)^n}{n} e^{-(n^2\pi^2 t/a^2L^2)} \sin \frac{n\pi x}{L}$ (b) $u(x, t) = \frac{x}{L}$

13. $u(x, t) = \frac{100x}{L} + \frac{2}{\pi} \sum_{n=1}^{\infty} \frac{k_0 + (-1)^{n+1}(k_1 - 100)}{n} e^{-(n^2\pi^2 t/L^2a^2)} \sin \frac{n\pi x}{L}$

Section 13–7

3. $u(x, y, t) = \left(\frac{8M^2}{\pi^3} e^{-(\pi^2 t)/(a^2L^2)} \cos \frac{\pi x}{L} \right) \sum_{k=0}^{\infty} \frac{e^{-[(2k+1)^2\pi^2 t]/(a^2M^2)}}{(2k+1)^3} \sin \frac{(2k+1)\pi y}{M}$

5. (a) $u(x, y, z) = \sum_{\substack{m=1 \\ n=1}}^{\infty} \alpha_{mn} (\tanh \pi\sqrt{m^2 + n^2} \cosh \sqrt{m^2 + n^2}\, z - \sinh \sqrt{m^2 + n^2}\, z) \sin mx \sin ny$

where

$$\alpha_{mn} = \frac{4}{\pi^2 \tanh \pi\sqrt{m^2 + n^2}} \int_0^\pi \int_0^\pi f(x, y) \sin mx \sin ny\, dx\, dy$$

(b) Let $u_i(x, y, z), v_i(x, y, z), i = 1, 2, 3$, be solutions corresponding respectively to the boundary value 0 on all but a single face, on which they have the values indicated.
$u_1(x, y, 0) = f_1(x, y), u_2(0, y, z) = f_2(y, z), u_3(x, 0, z) = f_3(x, z);$
$v_1(x, y, \pi) = g_1(x, y), v_2(\pi, y, z) = g_2(y, z), v_3(x, \pi, z) = g_3(x, z).$
Then u_1 is the u of part (a), with $f(x, y)$ replaced by $f_1(x, y)$ and

$$v_1 = \sum_{\substack{m=1 \\ n=1}}^{\infty} F_{mn} \sinh \sqrt{m^2 + n^2}\, z \sin mx \sin ny,$$

where

$$F_{mn} = \frac{4}{\pi^2 \sinh \pi\sqrt{m^2 + n^2}} \int_0^\pi \int_0^\pi g_1(x, y) \sin mx \sin ny\, dx\, dy.$$

The formulas for u_2, u_3 are obtained from that for u_1 by permuting cyclically the subscripts 1, 2, 3 and the variables x, y, z. Similarly the formulas for v_2, v_3 are obtained from that for v_1. The solution of the given boundary-value problem is $u = u_1 + u_2 + u_3 + v_1 + v_2 + v_3$.

7. $$u(x, y, t) = \sum_{\substack{m=1 \\ n=1}}^{\infty} (E_{mn} \cos a\sqrt{m^2 + n^2}\, t + F_{mn} \sin a\sqrt{m^2 + n^2}\, t) \sin mx \sin ny,$$

where

$$E_{mn} = \frac{4}{\pi^2} \int_0^{\pi} \int_0^{\pi} f(x, y) \sin mx \sin ny \, dx \, dy,$$

$$F_{mn} = \frac{4}{\pi^2 a\sqrt{m^2 + n^2}} \int_0^{\pi} \int_0^{\pi} g(x, y) \sin mx \sin ny \, dx \, dy$$

Section 13–9

1. The formal solution is

$$u(x, t) = \sum_{n=1}^{\infty} B_n e^{-(n^2\pi^2 t)/(L^2 a^2)} \sin \frac{n\pi x}{L}, \text{ where } B_n = \frac{2}{L} \int_0^L f(x) \sin \frac{n\pi x}{L} \, dx.$$

Chapter 14

Section 14–2

1. $$u(x, y) = \sum_{m=0}^{\infty} C_m \left[\sinh \frac{(2m + 1)\pi(M - y)}{L} + \sinh \frac{(2m + 1)\pi y}{L} \right] \sin \frac{(2m + 1)\pi x}{L},$$

where

$$C_m = -\frac{16L^2}{(2m + 1)^3\pi^3 \sinh [(2m + 1)\pi M/L]}$$

3. $$u(x, y) = \sum_{n=1}^{\infty} C_n \left[\sinh \frac{n\pi(M - y)}{L} + \sinh \frac{n\pi y}{L} \right] \sin \frac{n\pi x}{L},$$

where

$$C_n = \frac{L^2}{n^3\pi^3 \sinh (n\pi M/L)} \left\{ n\pi \sin \frac{n\pi}{2} + 4\left[(-1)^n - \cos \frac{n\pi}{2} \right] \right\}$$

5. $$u(x, y) = \frac{\{\sinh (\pi x/M) + \sinh [\pi(L - x)/M]\} \sin (\pi y/M)}{\sinh (\pi L/M)}$$
$$+ \sum_{m=0}^{\infty} C_m \left[\sinh \frac{(2M + 1)\pi(M - y)}{L} + \sinh \frac{(2m + 1)\pi y}{L} \right]$$
$$\times \sin \frac{(2m + 1)\pi x}{L},$$

where

$$C_m = -\frac{8L^2}{(2m + 1)^3\pi^3 \sinh [(2m + 1)\pi M/L]}$$

7. $$u(x, y) = \frac{a_0}{2\pi} (\pi - y) + \sum_{n=1}^{\infty} a_n \frac{\sinh n(\pi - y) \cos nx}{\sinh n\pi},$$

where

$$a_n = \frac{2}{\pi} \int_0^{\pi} f(x) \cos nx \, dx$$

9. $u(x, y) = \dfrac{400}{\pi} \displaystyle\sum_{m=0}^{\infty} \frac{1}{2m+1} \left[\cosh \frac{(2m+1)\pi y}{L} - \sinh \frac{(2m+1)\pi y}{L} \right]$
$\times \sin \dfrac{(2m+1)\pi x}{L}$

Section 14–3

3. $u(r) = 100 \dfrac{\ln r - \ln R_1}{\ln R_2 - \ln R_1}$; $u(\sqrt{R_1 R_2}) = 50$

5. It is sufficient that $f(\theta)$ belong to $\mathcal{PC}[-\pi, \pi]$. On each subinterval of continuity apply Theorem I–15 to the functions $f(\theta) \cos n(s - \theta)$, Theorem I–24 to the series

$$\sum_{n=1}^{\infty} r^n f(\theta) \cos n(s - \theta),$$

Theorem I–22 to the sequence of partial sums, and combine the results for the subintervals.

Section 14–4

7. $u(r, \varphi) = \displaystyle\sum_{n=0}^{\infty} A_n r^{-(n+1)} P_n(\cos \varphi)$,

where

$$A_n = \frac{2n+1}{2} \int_0^{\pi} f(\varphi) P_n(\cos \varphi) \sin \varphi \, d\varphi$$

9. (a) $u(r, \varphi) = \frac{1}{3} + r^2(\cos^2 \varphi - \frac{1}{3})$
(b) $u(r, \varphi) = -\frac{1}{3} - \frac{2}{3}r^2 + 2r^2 \cos^2 \varphi$
(c) $u(r, \varphi) = \frac{1}{3} + \frac{2}{3}r^2 - 2r^2 \cos^2 \varphi$
(d) $u(r, \varphi) = \frac{8}{15} + \frac{8}{21}r^2(1 - 3\cos^2 \varphi) + \frac{1}{35}(3 - 30\cos^2 \varphi + 35\cos^4 \varphi)$

11. $u(r, \varphi) = \displaystyle\sum_{k=0}^{\infty} B_k r^{2k+1} P_{2k+1}(\cos \varphi)$,

where

$$B_k = (4k+3) \int_0^{\pi/2} f(\varphi) P_{2k+1}(\cos \varphi) \sin \varphi \, d\varphi$$

Section 14–5

1. (a) $R = \begin{cases} c_1 r^{(-1+\sqrt{1+4\lambda})/2} + c_2 r^{(-1-\sqrt{1+4\lambda})/2} & \text{if } \lambda > -\frac{1}{4}; \\ r^{-1/2}(c_1 + c_2 \ln r) \quad \text{if } \lambda = -\frac{1}{4}; \\ r^{-1/2}(c_1 \cos [\frac{1}{2}\sqrt{-(1+4\lambda)}] \ln r + c_2 \sin [\frac{1}{2}\sqrt{-(1+4\lambda)}] \ln r) & \text{if } \lambda < -\frac{1}{4} \end{cases}$

3. $u_{00} = 1$ $\quad u_{10} = \cos \varphi$ $\quad u_{11} = \cos \theta \sin \varphi$
$v_{11} = \sin \theta \sin \varphi$ $\quad u_{20} = \frac{3}{2}\cos^2 \varphi - \frac{1}{2}$ $\quad u_{21} = 3 \cos \theta \sin \varphi \cos \varphi$
$u_{22} = 3 \cos 2\theta \sin^2 \varphi$ $\quad v_{21} = 3 \sin \theta \sin \varphi \cos \varphi$ $\quad v_{22} = 3 \sin 2\theta \sin^2 \varphi$

Section 14–6

3. $\cos^2 \varphi \cos^2 \theta \sim \dfrac{1}{6} u_{00} + \dfrac{1}{3} u_{20} + \dfrac{1}{24} u_{22} + \dfrac{1}{4} \displaystyle\sum_{k=2}^{\infty} \frac{4k+1}{k(k+1)(4k^2-1)} u_{2k,2}$

5. (a) $$f(\varphi, \theta) \sim \frac{1}{2} u_{00} + \sum_{k=1}^{\infty} \frac{(-1)^k (4k+1)}{2^{2k+1}(4k^2-1)[(k+1)!]^2} \times \left[\frac{(2k+2)!}{2} u_{2k,0} - (2k-2)!(2k^2+k+2)u_{2k,2}\right]$$

(b) $$u(r, \varphi, \theta) = \frac{1}{2} u_{00} + \sum_{k=1}^{\infty} \frac{(-1)^k (4k+1) r^{2k}}{2^{2k+1}(4k^2-1)[(k+1)!]^2} \times \left[\frac{(2k+2)!}{2} u_{2k,0} - (2k-2)!(2k^2+k+2)u_{2k,2}\right]$$

7. (a) $$u(r, \varphi, \theta) = \sum_{m=0}^{\infty} [r^m - a^{2m+1} r^{-(m+1)}] \left[\tfrac{1}{2}\bar{A}_{m0} u_{m0} + \sum_{n=1}^{\infty} \bar{A}_{mn} u_{mn} + \bar{B}_{mn} v_{mn}\right],$$

where

$$\bar{A}_{mn} = \bar{A}_{mn}(a, b; g) = \frac{b^{m+1}}{b^{2m+1} - a^{2m+1}} A_{mn}(g),$$

$$\bar{B}_{mn} = \bar{B}_{mn}(a, b; g) = \frac{b^{m+1}}{b^{2m+1} - a^{2m+1}} B_{mn}(g),$$

with

$$A_{mn}(g) = \frac{2m+1}{2\pi} \cdot \frac{(m-n)!}{(m+n)!} \times \int_0^{2\pi} \int_0^{\pi} g(\varphi, \theta) \cos n\theta \sin^{n+1} \varphi P_m^{(n)} (\cos \varphi)\, d\varphi\, d\theta,$$

$$B_{mn}(g) = \frac{2m+1}{2\pi} \cdot \frac{(m-n)!}{(m+n)!} \times \int_0^{2\pi} \int_0^{\pi} g(\varphi, \theta) \sin n\theta \sin^{n+1} \varphi P_m^{(n)} (\cos \varphi)\, d\varphi\, d\theta$$

(b) $$u(r, \varphi, \theta) = \sum_{m=0}^{\infty} \left\{\tfrac{1}{2}\alpha_{m0}(r) u_{m0} + \sum_{n=1}^{\infty} [\alpha_{mn}(r) u_{mn} + \beta_{mn}(r) v_{mn}]\right\},$$

where

$$\alpha_{mn}(r) = \bar{A}_{mn}(b, a; f)[r^m - b^{2m+1} r^{-(m+1)}] + \bar{A}_{mn}(a, b; g) \times [r^m - a^{2m+1} r^{-(m+1)}],$$

$$\beta_{mn}(r) = \bar{B}_{mn}(b, a; f)[r^m - b^{2m+1} r^{-(m+1)}] + \bar{B}_{mn}(a, b; g) \times [r^m - a^{2m+1} r^{-(m+1)}],$$

with notation as in part (a)

Section 14-7

3. $Q_0 - 2Q_2$

5. $r^3 u_{30} = -\tfrac{3}{2}x^2 z - \tfrac{3}{2}y^2 z + z^3$
$r^3 u_{31} = -\tfrac{3}{2}x^3 - \tfrac{3}{2}xy^2 + 6xz^2$
$r^3 u_{32} = 15(x^2 z - y^2 z)$
$r^3 u_{33} = 15(x^3 - 3xy^2)$
$r^3 v_{31} = -\tfrac{3}{2}x^2 y - \tfrac{3}{2}y^3 + 6yz^2$
$r^3 v_{32} = 30xyz$
$r^3 v_{33} = 15(3x^2 y - y^3)$

7. $$C_{n-2j,2j} = (-1)^j \frac{n!}{(2j)!(n-2j)!} C_{n,0},$$

$$C_{n-2j-1,2j+1} = (-1)^j \frac{(n-1)!}{(2j+1)!(n-2j-1)!} C_{n-1,1}$$

9. (a) $x^6 - 15x^4y^2 + 15x^2y^4 - y^6$
   (b) $x^5y - \frac{10}{3}x^3y^3 + xy^5$
   (c) $x^6 + 6x^5y - 15x^4y^2 - 20x^3y^3 + 15x^2y^4 + 6xy^5 - y^6$

11. $x^4 - 6x^2z^2 + z^4$
   $x^2y^2 - x^2z^2 - y^2z^2 + \frac{1}{3}z^4$
   $y^4 - 6y^2z^2 + z^4$
   $x^2yz - \frac{1}{3}yz^3$
   $y^3z - yz^3$
   $x^3y - 3xyz^2$
   $xy^3 - 3xyz^2$
   $x^3z - xz^3$
   $xy^2z - \frac{1}{3}xz^3$

13. (a) $r^2u_{20} = -\frac{1}{2}(x^2 - z^2) - \frac{1}{2}(y^2 - z^2)$
   $r^2u_{21} = 3(xz)$
   $r^2u_{22} = 3(x^2 - z^2) - 3(y^2 - z^2)$
   $r^2v_{21} = 3(yz)$
   $r^2v_{22} = 6(xy)$

   (b) $x^2 - z^2 = -r^2u_{20} + \frac{1}{6}r^2u_{22}$
   $xy = \frac{1}{6}r^2v_{22}$
   $y^2 - z^2 = -r^2u_{20} - \frac{1}{6}r^2u_{22}$
   $xz = \frac{1}{3}r^2u_{21}$
   $yz = \frac{1}{3}r^2v_{21}$

17. $\dim \mathcal{H}_{m,k} = \binom{m+k-2}{k-2} + \binom{m+k-3}{k-2}$

Chapter 15

Section 15–1

3. Set $x = \lambda r$

Section 15–2

1. $x = \pm 1$ regular singular points; $x = 0$ irregular singular point
3. $x = \pm 1$ regular singular points
5. $x = 1$ regular singular point; $x = 0$ irregular singular point
7. $x = 0$ and $x = 1$ regular singular points
9. $x = 0$ regular singular point

Section 15–3

3. (a) $\nu^2 - 1 = 0$
   (b) $\nu^2 - 3\nu - 1 = 0$
   (c) $\nu^2 - 3\nu + 1 = 0$
   (d) $\nu^2 - 2\nu - \lambda^2 = 0$
   (e) $\nu^2 = 0$

5. $\nu^2 = 0$, in each case.

7. $y_1 = |x|^{1/3}(1 + \frac{1}{5}x)$, $y_2 = |x|^{-1/3} \sum_{k=0}^{\infty} \frac{(-1)^k}{k!3^k(3k-5)(3k-2)} x^k$

9. $y_1 = |x|^{1/2}$, $y_2 = |x|^{1/4} \sum_{k=0}^{\infty} \frac{1}{k!2^{2k}(4k-1)} x^k$

11. $y = a_0\left[1 + \sum_{k=1}^{\infty} (-1)^k \frac{\lambda(\lambda-1)\cdots(\lambda-k+1)}{(k!)^2} x^k\right]$.

   If $\lambda = 0$, $y = a_0$; if λ is a positive integer, then

   $$y = a_0 \sum_{k=0}^{\lambda} (-1)^k \frac{\lambda!}{(\lambda-k)!(k!)^2} x^k.$$

Section 15–4

1. $y_1 = |x|^{1/2} \sum_{k=0}^{\infty} a_k x^k$, where $a_0 = 1$, $a_k = \dfrac{2}{k(2k+1)} \sum_{j=0}^{k-1} \dfrac{2j-15}{4^{k-j-1}} a_j$, $k \geqq 1$; $0 < |x| < 4$;

$y_2 = \sum_{k=0}^{\infty} a_k x^k$, where $a_0 = 1$, $a_k = \dfrac{1}{k(2k-1)} \sum_{j=0}^{k-1} \dfrac{j-8}{4^{k-j}} a_j$, $k \geqq 1$; $|x| < 4$

3. $y_1 = |x|^{5/4} \sum_{k=0}^{\infty} a_k x^k$, where $a_0 = 1$, $a_k = \dfrac{(-1)^k}{4^{k+1}k(4k+5)} \sum_{j=0}^{k-1} (-1)^j 4^j (4j+7) a_j$, $k \geqq 1$; $0 < |x| < 4$;

$y_2 = \sum_{k=0}^{\infty} a_k x^k$, where $a_0 = 1$, $a_k = \dfrac{(-1)^k}{2^{2k+1}k(4k-5)} \sum_{j=0}^{k-1} (-1)^j 4^j (2j+1) a_j$, $k \geqq 1$; $|x| < 4$

5. $y_1 = x \sum_{k=0}^{\infty} a_k x^k$, where $a_0 = 1$, $a_k = -\dfrac{2}{3k(k+1)} \sum_{j=0}^{k-1} j a_j$, $k \geqq 1$; $|x| < 1$;

$y_2 = |x|^{-2/3} \sum_{k=0}^{\infty} a_k x^k$, where $a_0 = 1$, $a_k = -\dfrac{2}{3k(3k-5)} \sum_{j=0}^{k-1} (3j-5) a_j$, $k \geqq 1$; $0 < |x| < 1$

7. $y_1 = x \sum_{k=0}^{\infty} a_k x^k$, where $a_0 = 1$, $a_k = \dfrac{(-1)^k}{k(4k+3)} \sum_{j=0}^{k-1} (-1)^j (4j+3) a_j$, $k \geqq 1$; $|x| < 1$;

$y_2 = |x|^{1/4} \sum_{k=0}^{\infty} a_k x^k$, where $a_0 = 1$, $a_k = \dfrac{(-1)^k 4}{k(4k-3)} \sum_{j=0}^{k-1} (-1)^j j a_j$, $k \geqq 1$; $0 < |x| < 1$

9. $y_1 = x\left[1 + \sum_{k=1}^{\infty} \dfrac{1}{k![7 \cdot 10 \cdot 13 \cdots (3k+4)]} x^k\right]$, $|x| < \infty$

$y_2 = |x|^{-1/3}\left[1 + \sum_{k=1}^{\infty} \dfrac{1}{k![(-1) \cdot 2 \cdot 5 \cdots (3k-4)]} x^k\right]$, $0 < |x| < \infty$

11. $\nu_1 = i$, $a_0 = 1$ give $a_1 = -\frac{2}{5} - \frac{1}{5}i$, $a_2 = \frac{1}{10} + \frac{1}{20}i$, $a_3 = -\frac{17}{780} - \frac{1}{130}i$;
$\nu_2 = -i$, $a_0 = 1$ give $a_1 = -\frac{2}{5} + \frac{1}{5}i$, $a_2 = \frac{1}{10} - \frac{1}{20}i$, $a_3 = -\frac{17}{780} + \frac{1}{130}i$.

This yields $y = c_1 y_1 + c_2 y_2$, where

$$y_1 = (1 - \tfrac{2}{5}x + \tfrac{1}{10}x^2 - \tfrac{17}{780}x^3 + \cdots)\cos(\ln|x|) + (\tfrac{1}{5}x - \tfrac{1}{20}x^2 + \tfrac{1}{130}x^3 + \cdots)\sin(\ln|x|),$$

$$y_2 = (-\tfrac{1}{5}x + \tfrac{1}{20}x^2 - \tfrac{1}{130}x^3 + \cdots)\cos(\ln|x|) + (1 - \tfrac{2}{5}x + \tfrac{1}{10}x^2 - \tfrac{17}{780}x^3 + \cdots)\sin(\ln|x|).$$

This solution is valid for $0 < |x| < \infty$.

13. $\nu_1 = \sqrt{2}\,i$, $a_0 = 1$ give $a_1 = 0$, $a_2 = \dfrac{-4 + \sqrt{2}\,i}{12}$, $a_3 = 0$;

$\nu_2 = -\sqrt{2}\,i$, $a_0 = 1$ give $a_1 = 0$, $a_2 = \dfrac{-4 - \sqrt{2}\,i}{12}$, $a_3 = 0$.

This yields $y = c_1y_1 + c_2y_2$, where

$$y_1 = (1 - \tfrac{1}{3}x^2 + \cdots)\cos(\sqrt{2}\ln|x|) + \left(-\frac{\sqrt{2}}{12}x^2 + \cdots\right)\sin(\sqrt{2}\ln|x|),$$

$$y_2 = \left(\frac{\sqrt{2}}{12}x^2 + \cdots\right)\cos(\sqrt{2}\ln|x|) + (1 - \tfrac{1}{3}x^2 + \cdots)\sin(\sqrt{2}\ln|x|).$$

This solution is valid for $0 < |x| < 1$.

15. $\nu_1 = i$, $a_0 = 1$ give $a_1 = \frac{1}{5}(-2 - i)$, $a_2 = \frac{1}{80}(3 - i)$, $a_3 = \frac{1}{9360}(67 + 81i)$, $\nu_2 = -1$, $a_0 = 1$ give $a_1 = \frac{1}{5}(-2 + i)$, $a_2 = \frac{1}{80}(3 + i)$, $a_3 = \frac{1}{9360}(67 - 81i)$. This yields $y = c_1y_1 + c_2y_2$ where

$$y_1 = (1 - \tfrac{2}{5}x + \tfrac{3}{80}x^2 + \tfrac{67}{9360}x^3 + \cdots)\cos(\ln|x|) + (\tfrac{1}{5}x + \tfrac{1}{80}x^2 - \tfrac{9}{1040}x^3 + \cdots)\sin(\ln|x|),$$

$$y_2 = (-\tfrac{1}{5}x - \tfrac{1}{80}x^2 + \tfrac{9}{1040}x^3 + \cdots)\cos(\ln|x|) + (1 - \tfrac{2}{5}x + \tfrac{3}{80}x^2 + \tfrac{67}{9360}x^3 + \cdots)\sin(\ln|x|).$$

This solution is valid for $0 < |x| < \infty$.

17. (a) $\nu = 0$ gives $y = \sum_{k=0}^{\infty} a_k(x - 1)^k$,

where

$$a_k = \frac{(-1)^{k+1}}{2^k k^2}\sum_{j=0}^{k-1}(-1)^j 2^j[\lambda(\lambda + 1) - j]a_j, \quad k \geqq 1.$$

(*Note.* It can be shown that, with $a_0 = 1$, the formulas above imply that

$$y = 1 + \sum_{k=1}^{\infty}\frac{[\lambda(\lambda+1) - (k-1)k][\lambda(\lambda+1) - (k-2)(k-1)]\cdots[\lambda(\lambda+1) - 0\cdot 1]}{2^k(k!)^2} \times (x - 1)^k$$

so that if λ is a non-negative integer n, then y is a polynomial of degree n.)

(b) $-1 < x < 3$. (If λ is a non-negative integer, then the solution is valid for all x.)

Section 15–5

1. $y_1 = x$, $y_2 = \sum_{k=1}^{\infty}\frac{(-1)^k}{k!\cdot k}x^{k+1} + x\ln x$

3. $y_1 = x^{-1}\left[1 + \sum_{k=1}^{\infty}(-1)^k\frac{1}{(k!)^2}x^k\right]$,

$$y_2 = 2x^{-1}\sum_{k=1}^{\infty}(-1)^{k+1}\frac{\sum_{j=1}^{k}1/j}{(k!)^2}x^k + x^{-1}\left[1 + \sum_{k=1}^{\infty}(-1)^k\frac{1}{(k!)^2}x^k\right]\ln x$$

5. $y_1 = 4!\sum_{k=0}^{\infty}\frac{k + 1}{(k + 4)!}x^{k+4}$, $y_2 = 1 + \frac{2}{3}x + \frac{1}{6}x^2$

7. $y_1 = x^2 + \sum_{k=1}^{\infty}(-1)^k\frac{1}{5\cdot 8\cdots(3k + 2)}x^{3k+2}$, $y_2 = \sum_{k=0}^{\infty}(-1)^k\frac{1}{3^k k!}x^{3k}$

9. $y_1 = 1 - 3x + \frac{3}{2}x^2 - \frac{1}{6}x^3$,

$$y_2 = 7x - \tfrac{23}{4}x^2 + \tfrac{11}{12}x^3 - 6\sum_{k=4}^{\infty}\frac{(k - 4)!}{(k!)^2}x^k + (1 - 3x + \tfrac{3}{2}x^2 - \tfrac{1}{6}x^3)\ln x$$

11. If $2p$ is not an integer, Case I; if $p = 0$, Case II; if $2p$ is an integer other than zero, Case III.

Section 15–6

3. If $p = 0$, $J_p(x) \to 1$ as $x \to 0$; if $p > 0$, $J_p(x) \to 0$ as $x \to 0$; if $p < 0$ and p is not an integer, then $J_p(x) \to \pm\infty$ as $x \to 0$.

9. (b) $y = -\dfrac{Z_{1/4}(x^2/2) + 2x^2 Z'_{1/4}(x^2/2)}{2xZ_{1/4}(x^2/2)}$.

[*Note.* Every y of this form is a solution of $y' = x^2 + y^2$ on every interval on which $x \neq 0$ and $Z_{1/4}(x^2/2)$ is defined and nonvanishing. Any given solution y may be so represented by choosing the pair of arbitrary constants in $Z_{1/4}(x^2/2) = x^{-1/2}u(x)$ in infinitely many different ways, corresponding to the solutions u of the differential equation $y(x) = -u'(x)/u(x)$.]

Section 15–7

3. $J_5(x) = \left(\dfrac{12}{x} - \dfrac{192}{x^3}\right) J_0(x) - \left(1 - \dfrac{72}{x^2} + \dfrac{384}{x^4}\right) J_0'(x)$

19. (b) $y = \mathcal{L}^{-1}[(1 + s^2)^{-3/2}]$

Section 15–9

3. If $h = p$ in Case III, the functions $J_p(\nu_k x)$, $k = 1, 2, 3, \ldots$, do not form a basis for $\mathcal{PC}[0, 1]$.

Section 15–10

5. If $p > 0$, $x^p = 2\displaystyle\sum_{k=1}^{\infty} \frac{\mu_k J_{p+1}(\mu_k)}{[\mu_k^2 - p^2][J_p(\mu_k)]^2} J_p(\mu_k x)$, $0 < x < 1$,

where μ_k is the kth positive zero of $J_p'(x)$. If $p = 0$ then $x^p = 1$ and the Dini series contains a single nonvanishing term, $2c_0 = 1$.

Section 15–11

1. $u = \displaystyle\sum_{k=1}^{\infty} [A_k \sinh \lambda_k(a - z) + B_k \sinh \lambda_k z] J_0(\lambda_k r)$,

where λ_k is the kth positive zero of $J_0(x)$, and

$$A_k = \frac{2}{[J_1(\lambda_k)]^2 \sinh \lambda_k a} \int_0^1 f(r) J_0(\lambda_k r) r\, dr,$$

$$B_k = \frac{2}{[J_1(\lambda_k)]^2 \sinh \lambda_k a} \int_0^1 g(r) J_0(\lambda_k r) r\, dr$$

3. $u = \displaystyle\sum_{k=1}^{\infty} A_k \sinh \lambda_k(a - z)\, J_0(\lambda_k r)$,

where λ_k is the kth positive zero of $J_0(\lambda) = 0$ and

$$A_k = \frac{200}{\lambda_k J_1(\lambda_k) \sinh \lambda_k a}$$

5. $u = c_1 \ln r + c_2$; $u = c_1\theta + c_2$; $u = c_1 z + c_2$. In the third situation u is a possible temperature distribution in the cylinder; in the first two situations this is true only in the special cases in which $c_1 = 0$.

Section 15–12

1. $$x = x_0\left[\frac{Y_0'(1)J_0(1+t) - J_0'(1)Y_0(1+t)}{J_0(1)Y_0'(1) - J_0'(1)Y_0(1)}\right]$$

3. $$u(r,\theta,z) = \sum_{n=0}^{\infty}\sum_{k=1}^{\infty}[A_{nk}\cos n\theta + \overline{A}_{nk}\sin n\theta]J_n(\lambda_{nk}r)\sinh\lambda_{nk}(a-z),$$

where λ_{nk} is the kth positive zero of $J_n(\lambda)$, and

$$A_{0k} = \frac{1}{\pi[J_1(\lambda_{0k})]^2\sinh\lambda_{0k}a}\int_{-\pi}^{\pi}\int_0^1 f(r,\theta)J_0(\lambda_{0k}r)r\,dr\,d\theta,$$

$$A_{nk} = \frac{2}{\pi[J_{n+1}(\lambda_{nk})]^2\sinh\lambda_{nk}a}\int_{-\pi}^{\pi}\int_0^1 f(r,\theta)J_n(\lambda_{nk}r)r\cos n\theta\,dr\,d\theta,\quad n > 0,$$

$$\overline{A}_{nk} = \frac{2}{\pi[J_{n+1}(\lambda_{nk})]^2\sinh\lambda_{nk}a}\int_{-\pi}^{\pi}\int_0^1 f(r,\theta)J_n(\lambda_{nk}r)r\sin n\theta\,dr\,d\theta$$

5. $$u(r,\theta,t) = \sum_{n=1}^{\infty}\sum_{k=1}^{\infty}A_{2n,k}J_{2n}(\lambda_{2n,k}r)\cos(\lambda_{2n,k}\,at)\sin 2n\theta,$$

where $\lambda_{2n,k}$ is the kth positive zero of $J_{2n}(\lambda)$ and

$$A_{2n,k} = \frac{8}{\pi[J_{2n+1}(\lambda_{2n,k})]^2}\int_0^{\pi/2}\int_0^1 f(r,\theta)J_{2n}(\lambda_{2n,k}r)r\sin 2n\theta\,dr\,d\theta,$$

with $u(r,\theta,0) = f(r,\theta)$

7. $$u(r,\theta,t) = \sum_{n=1}^{\infty}\sum_{k=1}^{\infty}[A_{nk}\cos(\lambda_{n\pi/\alpha,k}at) + B_{nk}\sin(\lambda_{n\pi/\alpha,k}at)]$$
$$\times J_{n\pi/\alpha}(\lambda_{n\pi/\alpha,k}r)\sin(n\pi\theta/\alpha)$$

where $\lambda_{n\pi/\alpha,k}$ is the kth positive zero of $J_{n\pi/\alpha}(\lambda)$ and

$$A_{nk} = \frac{4}{\alpha[J_{(\alpha+n\pi)/\alpha}(\lambda_{n\pi/\alpha,k})]^2}\int_0^{\alpha}\int_0^1 f(r,\theta)J_{n\pi/\alpha}(\lambda_{n\pi/\alpha,k}r)r\sin(n\pi\theta/\alpha)\,dr\,d\theta,$$

$$B_{nk} = \frac{4}{a\alpha\lambda_{n\pi/\alpha,k}[J_{(\alpha+n\pi)/\alpha}(\lambda_{n\pi/\alpha,k})]^2}\int_0^{\alpha}\int_0^1 g(r,\theta)J_{n\pi/\alpha}(\lambda_{n\pi/\alpha,k}r)r\sin(n\pi\theta/\alpha)\,dr\,d\theta,$$

with $u(r,\theta,0) = f(r,\theta)$ and $u_t(r,\theta,0) = g(r,\theta)$

index

ABCDE69876